ɔk i

Comprehensive Brachytherapy

Physical and Clinical Aspects

D0303171

SWIMS

C20344572

IMAGING IN MEDICAL DIAGNOSIS AND THERAPY

William R. Hendee, Series Editor

Forthcoming titles in the series

CX0344572
WN360   CGH

IMAGING IN MEDICAL DIAGNOSIS AND THERAPY

William R. Hendee, Series Editor

Comprehensive Brachytherapy

Physical and Clinical Aspects

Edited by

Jack L. M. Venselaar
Dimos Baltas
Ali S. Meigooni
Peter J. Hoskin

CRC Press
Taylor & Francis Group
Boca Raton London New York

CRC Press is an imprint of the
Taylor & Francis Group, an **informa** business

A TAYLOR & FRANCIS BOOK

CRC Press
Taylor & Francis Group
6000 Broken Sound Parkway NW, Suite 300
Boca Raton, FL 33487-2742

First issued in paperback 2016

© 2013 by Taylor & Francis Group, LLC
CRC Press is an imprint of Taylor & Francis Group, an Informa business

No claim to original U.S. Government works

Version Date: 20120801

ISBN 13: 978-1-138-19855-5 (pbk)
ISBN 13: 978-1-4398-4498-4 (hbk)

This book contains information obtained from authentic and highly regarded sources. Reasonable efforts have been made to publish reliable data and information, but the author and publisher cannot assume responsibility for the validity of all materials or the consequences of their use. The authors and publishers have attempted to trace the copyright holders of all material reproduced in this publication and apologize to copyright holders if permission to publish in this form has not been obtained. If any copyright material has not been acknowledged please write and let us know so we may rectify in any future reprint.

Except as permitted under U.S. Copyright Law, no part of this book may be reprinted, reproduced, transmitted, or utilized in any form by any electronic, mechanical, or other means, now known or hereafter invented, including photocopying, microfilming, and recording, or in any information storage or retrieval system, without written permission from the publishers.

For permission to photocopy or use material electronically from this work, please access www.copyright.com (http://www.copyright.com/) or contact the Copyright Clearance Center, Inc. (CCC), 222 Rosewood Drive, Danvers, MA 01923, 978-750-8400. CCC is a not-for-profit organization that provides licenses and registration for a variety of users. For organizations that have been granted a photocopy license by the CCC, a separate system of payment has been arranged.

Trademark Notice: Product or corporate names may be trademarks or registered trademarks, and are used only for identification and explanation without intent to infringe.

Library of Congress Cataloging-in-Publication Data

Comprehensive brachytherapy : physical and clinical aspects / editors, Jack Venselaar ... [et al.].
    p. ; cm. -- (Imaging in medical diagnosis and therapy)
  Includes bibliographical references and index.
  ISBN 978-1-4398-4498-4 (hardcover : alk. paper)
  I. Venselaar, Jack. II. Series: Imaging in medical diagnosis and therapy.
  [DNLM: 1. Brachytherapy--methods. 2. Neoplasms--radiotherapy. WN 250.5.B7]

616.99'40642--dc23   2012030326

Visit the Taylor & Francis Web site at
http://www.taylorandfrancis.com

and the CRC Press Web site at
http://www.crcpress.com

"The simple is the seal of the true.

And beauty is the splendor of truth."

Subrahmanyan Chandrasekhar

(1910–1995)

Contents

SECTION I   Introduction

SECTION II   Brachytherapy Dosimetry

SECTION III   Brachytherapy Treatment Planning and Imaging

SECTION IV Uncertainties in Clinical Brachytherapy

SECTION V Clinical Brachytherapy

SECTION VI Developments in Clinical Brachytherapy

SECTION VII Radiation Protection and Quality Management in Brachytherapy

List of Abbreviations

1D (2D, 3D, 4D)	1 dimensional (2 dimensional,…)
AAPM	American Association of Physicists in Medicine
ABS	American Brachytherapy Society
AC	Alternating current
ACR	American College of Radiology
ACRO	American College of Radiation Oncology
ADCL	Accredited Dosimetry Calibration Laboratory (USA)
ADT	Androgen deprivation therapy
AEC	Atomic Energy Commission (USA)
AHRQ	Agency for Healthcare Research and Quality (USA)
AIMDD	Active implantable medical device directive (EEC)
AKR	Air kerma rate
ALARA	As low as reasonably achievable
AO	Abnormal occurrence
AP	Anterior-posterior
APBI	Accelerated partial breast irradiation
ASBS	American Society of Breast Surgeons
ASTRO	American Society of Radiation Oncology
AT	Applicator tube
BCF	Biochemical failure
BD	Basal dose (points)
BED	Biologically effective dose
BEV	Beam's eye view
BFFS	Biochemical failure-free survival
BFP	Boltmann-Fokker-Plank equation
BIPM	Bureau International des Poids et Mesures
bNED	Biochemical non-evidence of disease
BRAPHYQS	Brachytherapy Physics Quality Assurance System (GEC-ESTRO)
bRSF	Biochemical relapse-free survival
BSR	Brachytherapy Source Registry Working Group (AAPM)
bSSFP	Balanced steady state free precession
BT	Brachytherapy
BTSC	Brachytherapy Subcommittee (AAPM)
BTV	Biological target volume (= MTV)
CBCT	Cone beam computerized tomography
CC	Collapsed cone
CCTV	Closed circuit television system
CDRH	Center for Devices and Radiological Health (FDA)
CE	Conformité Européenne (EEC)
CEMR	contrast enhanced MR
CERN	Conseil Européen pour la Recherche Nucléaire (European Organization for Nuclear Research)
CHF	Congestive heart failure
CIPM	Comité International des Poids et Mesures (International Committee for Weights and Measures)
CMC	Calibration and Measurement Capabilities (BIPM)
CMS	Center for Medicare and Medicaid Services
CN	Conformation number
CNS	Central nervous system
COIN	Conformal index
COMS	Collaborative Ocular Melanoma Study (USA)
COPD	Chronic obstructive pulmonary disease
CPE	Charged particle equilibrium
CPU	Central processing unit
CRT	Conformal (external-beam) radiation therapy
CSDA	Continuous slowing down approximation
CSS	Cause-specific survival
CT	Computerized tomography
CTV	Clinical target volume
DANTSYS	Diffusion Accelerated Neutral Particle Transport Code System
DC	Direct current
DCEMR	Dynamic contrast-enhanced MR
DCIS	Ductal carcinoma in situ
DFEM	Discontinuous finite element method
DGMP	Deutsche Gesellschaft für Medizinische Physik (Germany)
DICOM	Digital Imaging and Communications in Medicine
DIL	Dominant intra-prostatic lesion
DIN	Deutsches Institut für Normung (German Institute for Standardization)

DNR	Dose non-uniformity ratio		GYN	Gynecology
DRR	Digitally reconstructed radiograph		HAM	Harrison-Anderson-Mick applicator
DSA	Diffusion synthetic acceleration		HDE	Humanitarian Device Exemption
DSB	Double-strand break		HDR	High dose rate
DTA	Distance-to-agreement		HEBD	High-Energy Brachytherapy Dosimetry
DVH	Dose volume histogram		HIFU	High intensity focused ultrasound
DWI	Diffusion weighted imaging		HPV	Human papilloma virus
EAU	European Association of Urology		HR	High risk
EBRT	External beam Radiotherapy		HU	Hounsfield unit
EBS	Electronic brachytherapy source		HVL	Half value layer
EC	Electron capture		IAEA	International Atomic Energy Agency
ECOG	Eastern Cooperative Oncology Group		IBBT	Intracavitary balloon brachytherapy
ED	Erectile dysfunction		IC	Internal conversion
E(E)C	European (Economic) Community		ICBT	Intracavitary brachytherapy
EFOMP	European Federation of Organisations for Medical Physics		ICRP	International Commission on Radiological Protection
EI	External index		ICRU	International Commission on Radiation Units and Measurements
EIC	Extensive intraductal component			
EMBRACE	International study on MRI-guided brachytherapy in locally advanced cervical cancer		IEC	International Electrotechnical Commission
			IED	Isoeffective dose
			IFU	Instructions for use
EORTC	European Organisation for Research and Treatment of Cancer		IGBT	Image guided brachytherapy
			IMBT	Interstitial multicatheter brachytherapy
EPE	Extraprostatic extension		IMRT	Intensity modulated radiation therapy
EPID	Electronic portal imaging device		INMRI	Istituto Nazionale di Metrologia delle Radizioni Ionizzanti (National Institute of Metrology of Ionizing Radiation, Italy)
EPR	Electron paramagnetic resonance			
EQD2	Equivalent total dose delivered in 2 Gy fractions of photon irradiation, producing the same biological effect			
			IOERT	Intraoperative electron beam radiotherapy
			IOHDR	Introperative HDR
ESR	Electron spin resonance		IORT	Intraoperative radiotherapy
ESRO	International School of Radiotherapy and Oncology (ESTRO)		IPEM	Institute of Physics and Engineering in Medicine (UK)
ESTRO	European Society for Radiotherapy and Oncology		IPSA	Inverse planning with simulated annealing
			IPSS	International prostate symptom score
ETC	Education and Training Committee (ESTRO)		IR	Intermediate risk
FDA	Food and Drug Administration (USA)		ISA	Inter-seed attenuation
FDAMA	FDA Modernization Act		ISO	International Organization for Standardization
FDG	Fluoro deoxy glucose		IVB(T)	Intravascular brachytherapy
FIT	Flexible intraoperative template		IVDD	In Vitro Diagnostic Device Directive (EEC)
FMEA	Failure mode effect analysis		IVUS	Intravascular US
fMRI	Functional magnetic resonance imaging		KPS	Karnofsky performance status
FSD	Fletcher-Suit-Delclos		LAN	Local area network
FSE	Fast spin-echo (MR)		LBTE	Linear Boltzmann transport equation
GBBS	Grid-based Boltzmann equation solver		LDR	Low dose rate
GEC	Groupe Européen de Curiethérapie (ESTRO)		LET	Linear energy transfer
GOG	Gynecologic Oncology Group		LNHB	Laboratoire National Henri Becquerel (France)
GRE	Gradient-echo (MR)		LNT	Linear no threshold (model)
GROVEX	Large volume parallel plate extrapolation chamber (PTB)		LPL	Lethal and potentially lethal (model)
			LQ	Linear-quadratic (model)
GTV	Gross tumor volume		LVSI	Lymphovascular invasion
GU	Genito-urinary		MAA	Macroaggregated albumin

MBDCA	Model based dose calculation algorithm	PBSI	Permanent breast seed brachytherapy implant
MC	Monte Carlo	PDR	Pulsed dose rate
MCD	Mean central dose	PET	Positron emission tomography
MCPT	Monte Carlo photon transport	PIB	Permanent interstitial brachytherapy
MDD	Medical Devices Directive (EEC)	PMA	Premarket Approval application (FDA)
MDR	Medium dose rate	PMMA	Poly(methyl methacrylate), with Trade Marks: Plexiglas, Lucite, Perspex
MFP	Mean free path		
MIRD	Medical Internal Radiation Dose	PMT	Photomultiplier tube
MO	Multi objective	POI	Point of interest
MOSFET	Metal-Oxide-Semiconductor-Field-Effect-Transistors	PORTEC	Post operative radiation therapy in endometrial carcinoma
MPD	Minimum (or matched) peripheral dose	PPI	Permanent prostate implantation
MR	Magnetic resonance	PROBATE	GEC-Brachytherapy prostate working group (Gec-ESTRO)
MRI	Magnetic resonance imaging		
MRS(I)	Magnetic resonance spectroscopic (imaging)	PSA	Prostate specific antigen
MTD	Minimum target dose	PSD	Plastic scintillation detectors
MTV	Metabolic tumor volume (= BTV)	PSDL	Primary Standard Dosimetry Laboratory
MUPIT	Martinez Universal perineal interstitial template	PSS	Primary-scatter separation
NDR	Natural dose ratio	PTB	Physikalisch-Technische Bundesanstalt (Germany)
NEA	Nuclear Energy Agency (USA)		
NED	Non-evidence of disease	PTL	Planning target length
NCI	National Cancer Institute (USA)	PTV	Planning target volume
NCRP	National Council on Radiation Protection and Measurements (USA)	PVT	Polyvinyl toluene
		QA	Quality Assurance
NCS	Netherlands Commission on Radiation Dosimetry	QC	Quality control
		QI	Quality index
NDVH	Natural dose volume histogram	QM	Quality management
NIBB	Non-invasive image guided breast brachytherapy	QRRO	Quality Research in Radiation Oncology (ACR)
NIST	National Institute of Standards and Technology (USA)	QUATRO	Quality Assurance Team for Radiation Oncology (IAEA)
NMi	National Metrology institute	RAKR	Reference air kerma rate
NNDC (NUDAT)	National Nuclear Data Center (NUDAT)	RAL	Remote afterloader
		RBE	Relative biological effectiveness
NPL	National Physical Laboratory (UK)	RCA	Root cause analysis
NRC	National Research Council (Canada)	RF	Radiofrequency
NRC	Nuclear Regulatory Commission (USA)	RI	Radiation incident
NSABP	National Surgical Adjuvant Breast and Bowel Project group (NCI)	RIL	Reference isodose length
		RL	Radioluminescence
NSD	Nominal standard dose	RMF	Repair-misrepair-fixation (model)
NT	Normal tissue	RMR	Repair–misrepair (model)
NTCP	Normal tissue complication probability	RMS	Root mean square
NVB	Neurovascular bundle	ROI	Region of interest
OAR	Organ at risk	ROSIS	Radiation Oncology Safety Information System (ESTRO)
OR	Operating room		
ORNL	Oak Ridge National Laboratory (USA)	RPC	Radiological Physics Center (USA)
OS	Overall survival	RPN	Risk priority number
OSD	Occurrence, Severity, Detectability	RTOG	Radiation Therapy Oncology Group (NCI)
OSL	Optically stimulated luminescence	RTP	Radiotherapy (treatment) planning
PA	Posterior-anterior	RTT	Radiotherapy technologist
PBI	Partial breast irradiation	SCD	Source chamber distance
PBQAG	Prostate brachytherapy quality assurance group	SDP	Source dwell position

SE	Spin-echo (MR)		TPS	Treatment planning system
SI	Système international d'unités, International System of units		TR	Repetition time (MR)
			TRAK	Total reference air kerma
SIB	Simultaneous integrated boost		TRUS	Transrectal ultrasound
SIOP	International Society of Paedriatric Oncology		TSD	Time-dose factor
SLFB	Biochemical failure-free survival rates		TURP	Transurethral resection of the prostate
SNR	Signal to noise ratio		TVL	Tenth value layer
SO	Single objective		UI	Uniformity index
SPECT	Single-photon emission computerized tomography		UNSCEAR	UN Scientific Committee on the Effects of Atomic Radiation
SSDL	Secondary Standard Dosimetry Laboratory			
SSDS	Stepping source dosimetry system		US	Ultrasound
SUV	Standardized uptake value (PET)		USNRC	US Nuclear Regulatory Commission
TCP	Tumor control probability		VBT	Vaginal brachytherapy
TCU	Treatment console unit		VCB	Vaginal cuff brachytherapy
TDF	Time-dose factor		VOI	Volume of interest
TE	Echo-time (MR)		VSL	Van Swinden Laboratory (Netherlands)
TG	Task group		WAFAC	Wide-Angle-Free-Air Chamber
TLD	Thermoluminescence dosimetry		WBI	Whole breast irradiation

Series Preface

Advances in the science and technology of medical imaging and radiation therapy are more profound and rapid than ever before since their inception over a century ago. Further, the disciplines are increasingly cross-linked as imaging methods become more widely used to plan, guide, monitor, and assess treatments in radiation therapy. Today the technologies of medical imaging and radiation therapy are so complex and so computer-driven that it is difficult for the persons (physicians and technologists) responsible for their clinical use to know exactly what is happening at the point of care, when a patient is being examined or treated. The persons best equipped to understand the technologies and their applications are medical physicists, and these individuals are assuming greater responsibilities in the clinical arena to ensure that what is intended for the patient is actually delivered in a safe and effective manner.

The growing responsibilities of medical physicists in the clinical arenas of medical imaging and radiation therapy are not without their challenges, however. Most medical physicists are knowledgeable in either radiation therapy or medical imaging, and expert in one or a small number of areas within their discipline. They sustain their expertise in these areas by reading scientific articles and attending scientific talks at meetings. In contrast, their responsibilities increasingly extend beyond their specific areas of expertise. To meet these responsibilities, medical physicists periodically must refresh their knowledge of advances in medical imaging or radiation therapy, and they must be prepared to function at the intersection of these two fields. How to accomplish these objectives is a challenge.

At the 2007 annual meeting of the American Association of Physicists in Medicine in Minneapolis, this challenge was the topic of conversation during a lunch hosted by Taylor & Francis Publishers and involving a group of senior medical physicists (Arthur L. Boyer, Joseph O. Deasy, C.-M. Charlie Ma, Todd A. Pawlicki, Ervin B. Podgorsak, Elke Reitzel, Anthony B. Wolbarst, and Ellen D. Yorke). The conclusion of this discussion was that a book series should be launched under the Taylor & Francis banner, with each volume in the series addressing a rapidly advancing area of medical imaging or radiation therapy of importance to medical physicists. The aim would be for each volume to provide medical physicists with the information needed to understand technologies driving a rapid advance and their applications to safe and effective delivery of patient care.

Each volume in the series is edited by one or more individuals with recognized expertise in the technological area encompassed by the book. The editors are responsible for selecting the authors of individual chapters and ensuring that the chapters are comprehensive and intelligible to someone without such expertise. The enthusiasm of volume editors and chapter authors has been gratifying and reinforces the conclusion of the Minneapolis luncheon that this series of books addresses a major need of medical physicists.

Imaging in Medical Diagnosis and Therapy would not have been possible without the encouragement and support of the series manager, Ms. Luna Han of Taylor & Francis Group. The editors and authors, and most of all I, are indebted to her steady guidance of the entire project.

William Hendee
Series Editor
Rochester, MN

Foreword

The discovery of radium at the end of the nineteenth century was a revolution that is hard to appreciate today. Cancer was a relatively rare disease, but with one painful exception: cervix cancer. This was, and still is, often a disease of women occurring at an age when children are still at home. It is a deeply disrupting event in family life and, in some countries, is still a stigma; families are broken, the children left alone, and the patient abandoned by her relatives. The only treatment, one century ago, was surgery with a high mortality risk (20%) from the primitive anesthesia or postoperative infection.

Suddenly, thanks to the discovery of the Curie–Sklodowska couple, a curative treatment appeared that could save lives without any of these dangers. It appeared as a miracle, the miracle of brachytherapy. The present book explains how these initial successes have been followed by many more. Radium is not used anymore and has been replaced by artificial radionuclides with more favorable characteristics: lower energy, shorter half-life, miniaturization, etc. The recent computer revolution has allowed another step toward better dosimetry, better source localization, and, above all, better control of the implant positioning relative to the target.

In a world of IMRT, IGRT, respiratory-gated irradiation, and charged particles, it was imperative to reconsider brachytherapy techniques and their future. What this book offers is a wide range of timely reflections on the achievements and future prospects for brachytherapy. It helps to understand the current state of the technique and its indications, as well as future areas for development and/or improvement. Clearly brachytherapy is here to stay.

As a bridge between surgery and radiotherapy, brachytherapy has always been practiced by enthusiastic radiation oncologists (and gynecologists), many of the authors of the present book among them. But there is no brachytherapy without skilled physicists, dosimetrists, and radiation therapy technologists. As in any of the radiotherapy domains, teamwork is central for producing a safe environment and complete clinical success.

This book, therefore, addresses the topic of brachytherapy from many vantage points, oncological, physical, and biological, and will satisfy readers from the entire radiotherapy community.

Pierre Scalliet

editors have tried to bring together as much as possible different views and approaches, aiming at a complete overview of best brachytherapy practices, in the hopes that this book will gain value for practitioners worldwide.

Recommendations may change in the future with the availability of more clinical, physical, dosimetric, and technological

data. However, we believe that the material presented in the book defines the basis of modern brachytherapy, and we are confident that it will remain useful for our readers in the future.

Jack L. M. Venselaar, Dimos Baltas, Ali S. Meigooni,
Peter Hoskin

Acknowledgments

The editors of this book express their gratitude to all the contributors. The authors have made a tremendous effort to provide us with a fully up-to-date overview of all aspects of brachytherapy. Many of them have helped the editors with a review of one or more of other chapters. Also, other colleagues have contributed, either in the form of reviews or by providing us with their figures and data. Special thanks are directed to Chris Melhus, Csaba Polgar, Rob van der Laarse, and Mark Rivard for their invaluable work.

About the Editors

Dr. Jack L. M. Venselaar received his PhD degree in 2000 at the Leiden University in The Netherlands on the topic of accuracy requirements of external beam therapy treatment planning. He is a senior specialist medical physicist with experience in the field of radiotherapy and brachytherapy physics, radiation protection, and hyperthermia. He has pursued research in the field of dosimetry, quality assurance, and technology of brachytherapy since the 1980s. His contributions to brachytherapy physics and technology include dosimetry of high-activity sources, development and establishment of dosimetry protocols, advanced quality control procedures and systems, development of quality audit systems, and brachytherapy radiation protection data. He was a board member of The Netherlands Commission of Radiation Dosimetry (NCS) and participated in several of the NCS committees including one chair position. From 2001 to 2008, he was the chair of the Brachytherapy Physics Quality Systems (BRAPHYQS) group of GEC-ESTRO. Dr. Venselaar acts as a liaison between the GEC-ESTRO Committee of ESTRO and the AAPM Brachytherapy Subcommittee and the ABS Physics Committee. He has been the national coordinator of the Patterns of Care in Brachytherapy study project of GEC-ESTRO, and has been a consultant in several assignments for the International Atomic Energy Agency (IAEA) for both teaching and advisory activities. From 2001 to 2009, he was a teacher for brachytherapy physics lectures in the international ESTRO Teaching Course on Modern Brachytherapy Techniques.

Prof. Dr. Dimos Baltas is the director of the Department of Medical Physics and Engineering of the State Hospital in Offenbach, Germany. Since 1996, he has been an associated adjunct research professor for Medical Physics and Engineering of the Institute of Communication and Computer Systems (ICCS), Department of Electrical and Computer Engineering, National Technical University of Athens, and since 2005 he has been an honorary Scientific Associate of the Nuclear Physics and Elementary Particles Section, Physics Department, University of Athens, Greece. He received his PhD degree in medical physics in 1989 at the University of Heidelberg, Germany, and did research in applied radiation biology, modeling, and radiobiology-based planning. He pursued research in the field of dosimetry, quality assurance, and technology in brachytherapy since the late 1980s. His contribution in brachytherapy physics and technology includes experimental dosimetry of high activity sources, development and establishment of dosimetry protocols, advanced quality assurance procedures and systems, experimental and Monte Carlo–based dosimetry of new high and low activity sources, development of new algorithms for dose calculation, inverse planning and optimization, and imaging-based treatment planning in brachytherapy as well as development of new radionuclides for brachytherapy. Since 2000, he has been the chairman of the Task Group on Afterloading Dosimetry of the German Association of Medical Physics (DGMP). Since 2004, he has been a member of the medical physics experts group of the Brachytherapy Physics Quality Systems (BRAPHYQS) group of GEC-ESTRO. He has been a member of the scientific advisory board of the *Journal Strahlentherapie und Onkologie* and *Journal of Radiation Oncology Biology Physics* since 2005, a physics editor of the *European Journal for Brachytherapy* and *Journal of Contemporary Brachytherapy* since 2008, and a member of the scientific advisory board of the *Journal of Computational and Mathematical Methods* in medicine since 2010. Since 2007, he has been a member of the GEC-ESTRO Committee, and since 2009, he has been a faculty member for the GEC-ESTRO Teaching Course on Modern Brachytherapy Techniques. In 2011, he became an honorary member of the Austrian Society for Radiation Oncology, Radiation Biology and Medical Radiation Physics (ÖGRO).

Dr. Ali S. Meigooni received his PhD degree in physics from the Department of Physics and Astronomy in Ohio University in 1984. Immediately after that, he started his career in the field of medical physics. He gained his experiences by working in major institutions such as the University of Minnesota, Yale University, Washington University, and the University of Kentucky. Dr. Meigooni contributed in all aspects of radiation therapy and fulfilled the position of postdoctoral fellow at the University of Minnesota in 1984 followed by a full professor position at the University of Kentucky in 2002. He is certified by the American Board of Radiology (ABR) for the therapeutic radiology on 1997. In addition, he was certified by The University of the New York State for Medical Physicist-Therapeutic Radiological Physics in the state of New York and by the Department of Health and Human Services in the state of Nevada as an authorized medical physicist. He has been and still is actively involved in several different AAPM Task Groups such as TG-43, TG-143, TG-129, TG-137, TG-138, TG-55, HEBD, and LEBD. One of his main contributions to medical radiation physics was in the field of brachytherapy, particularly for development of the TG-43 recommendation. In addition, Dr. Meigooni had a major contribution on this field for his teaching contribution in the medical physics graduate program at Washington University and the University of Kentucky. Presently, Dr. Meigooni is active as the chief medical physicist in The Comprehensive Cancer Center of Nevada, as well as the adjunct professor at the University of Nevada Las Vegas. He has several different MS or PhD students around the globe who are working on their way for this field.

Prof. Peter Hoskin MD trained in clinical oncology at the Royal Marsden Hospital London and has been a consultant in clinical oncology at Mount Vernon Cancer Centre, Northwood, U.K., since 1992. He is also a professor in clinical oncology at the University College London. Research interests have ranged from palliative radiotherapy to the use of radiosensitizers and brachytherapy. Current research program in brachytherapy focus on the role of HDR brachytherapy in prostate cancer and the EMBRACE project for image-guided gynecological brachytherapy. He has been the national coordinator of the Patterns of Care in Brachytherapy study project of GEC-ESTRO for the United Kingdom and a consultant for the International Atomic Energy Agency (IAEA) for brachytherapy and palliative radiotherapy. He is the chair of the prostate brachytherapy group PROBATE in ESTRO, the course director for the ESTRO prostate brachytherapy course, and a teacher on the international ESTRO Teaching Course on Modern Brachytherapy Techniques. Other appointments include Editor of Clinical Oncology, Gynecology Section Editor for the journal *Brachytherapy*, and member of the editorial boards of *Radiotherapy and Oncology* and the *Journal of Oncology*.

Contributors

Facundo Ballester, Ph.D.
Departamento de Física Atómica,
  Molecular y Nuclear
Universitat de València
Burjassot, Spain

Dimos Baltas, Ph.D.
Department of Medical Physics and
  Engineering
Klinikum Offenbach GmbH
Offenbach, Germany

Luc Beaulieu, Ph.D.
Département de Radio-Oncologie et
  Centre de Recherche en Cancérologie
Centre Hospitalier Universitaire de
  Québec
Québec, Canada
and
Département de Physique, de Génie
  Physique et d'Optique
Université Laval
Québec, Canada

Sam Beddar, Ph.D., FCCPM, FAAPM
Department of Radiation Physics
M.D. Anderson Cancer Center
Houston, Texas

Margaret Bidmead, Ph.D.
Department of Physics
Royal Marsden NHS Foundation Trust
  and Institute of Cancer Research
London

Hans Bjerke, M.Sc.
Norwegian Radiation Protection
  Authority
Section on Dosimetry and Medical
  Applications
Østerås, Norway

David J. Carlson, Ph.D.
Department of Therapeutic Radiology
Yale University School of Medicine
New Haven, Connecticut

Åsa Carlsson Tedgren, Ph.D.
Faculty of Health Sciences
Linköping University
Linköping, Sweden
and
Swedish Radiation Safety Authority
Stockholm, Sweden

Zhe J. Chen, Ph.D.
Department of Therapeutic Radiology
Yale University School of Medicine
New Haven, Connecticut

Joanna E. Cygler, Ph.D., FCCPM, FAAPM
Department of Medical Physics
The Ottawa Hospital Cancer Center
Ontario, Canada
and
Department of Radiology
University of Ottawa
Ontario, Canada

Rupak K. Das, Ph.D.
Department of Human Oncology and
  Medical Physics
University of Wisconsin
Madison, Wisconsin

Carlos E. de Almeida, Ph.D., FAAPM, DABMP
Radiological Sciences Laboratory
Rio de Janeiro University
Rio de Janeiro, Brazil

Larry A. DeWerd, Ph.D., FAAPM
Department of Medical Physics
University of Wisconsin
Madison, Wisconsin

Beth Erickson, M.D., FACR
Medical College of Wisconsin
Department of Radiation Oncology
Milwaukee, Wisconsin

David S. Followill, Ph.D., FAAPM
Department of Radiation Physics
M.D. Anderson Cancer Center
Houston, Texas

Michel Ghilezan, M.D., Ph.D.
Michigan Health Professionals/21st
  Century Oncology
Pontiac, Michigan

Peter Grimm, D.O.
Prostate Cancer Center of Seattle
Seattle, Washington

Christine Haie-Meder, M.D.
Brachytherapy Service
Institut Gustave-Roussy
Villejuif, France

Peter J. Hoskin, M.D.
Cancer Centre
Mount Vernon Hospital
Middlesex, United Kingdom

Christian Kirisits, Ph.D.
Department of Radiotherapy
Comprehensive Cancer Center Vienna
Medical University of Vienna
Vienna, Austria

Eric E. Klein, Ph.D., FAAPM
Radiation Oncology Department
Washington University
Saint Louis, Missouri

I

Introduction

1

Introduction and Innovations in Brachytherapy

Jack L. M. Venselaar
Institute Verbeeten

Dimos Baltas
Klinikum Offenbach GmbH

Peter Hoskin
Mount Vernon Hospital

Ali S. Meigooni
Comprehensive Cancer Centers of Nevada

1.1 A Short History of Nearly Everything in Brachytherapy

Paraphrasing the title of Bill Bryson's book, this section makes a quick tour through an amazing century of brachytherapy developments. Rivard et al. (2009) have recently described these marked highlights in a Vision 20/20 paper in *Medical Physics Journal* (largely followed here) providing the reader with the references recommended for a pleasant further journey of reading.

Often, technological discoveries are distinctly marked in history. Independent from each other, two of these discoveries launched the start of the radiation oncology era. Wilhelm Röntgen discovered x-rays in November 1895, and shortly afterward, Henri Becquerel accidentally exposed a photographic plate to uranium in 1896, identifying the phenomenon of emitted radiation (Dutreix et al. 1998). Immediately following these almost coincidental discoveries, Pierre and Marie Curie extracted radium from pitchblende ore in 1898 and determined the origin of this penetrating radiation. This was a physics Nobel Prize–winning achievement, which was awarded to the couple and Henri Becquerel in 1903.

Shortly after, new pathways were explored to apply these radiations in the first treatments of patients. Becquerel himself experienced the effects of radiation exposure by carrying a tube containing decigrams of radium chloride in his vest pocket. He recorded the evolution of his skin reaction (Becquerel and Curie 1901). The first experiences of clinical applications belonged to Danlos and Bloch (1901) in Paris, and Abbé (1904) in New York. Soon after, laboratories and institutes were created such as the Radium Biological Laboratory in Paris in 1906, and then in 1909 in London, Finze started treatments with radium. A book on radium therapy, which is now known as brachytherapy, was published in 1909 by Wickham and Degrais (1909).

These early twentieth century achievements established the medical application of radiation within one decade of its discovery. The basic principles of systematic use of radiation were established somewhat later after World War I in the Radium Hemmet in Stockholm, the Memorial Hospital in New York, and the Radium Institute in Paris. For brachytherapy, ^{226}Ra was the only radionuclide used for these sources (Mould 1993). Arrangement of the radioactive sources in certain geometric patterns, with definitions of the strength, distance, and treatment time, developed into a set of rules for patient treatments. For intracavitary treatments, the Stockholm and Paris methods were described in 1914 and 1919, respectively. Paterson and Parker (1938) set the basis of the 1930 rules for the Manchester system, as described in a book by Meredith (1967).

Two other moments are important for further development of brachytherapy. The first one was the discovery of artificial radioactivity (also known as manmade radionuclide) in 1934, allowing the use of artificially produced radioactive materials in radiotherapy. The second was the development of remote afterloading devices, in the 1950s and 1960s, which provided improved personnel radiation protection and gave more flexibility to the applications. The new artificial radionuclides ^{60}Co, ^{137}Cs, ^{182}Ta, and ^{198}Au were initially designed to have the penetration abilities, as far as possible, similar to the ^{226}Ra sources. Ulrich Henschke was the first to clinically explore the use of ^{192}Ir (Henschke et al. 1963). This radionuclide is currently the most widely applied radioactive source in the field of brachytherapy.

New rules for interstitial therapy using low-dose-rate (LDR) ^{192}Ir wire sources, which could be applied in various lengths and strengths, were developed in a very precise and consistent manner in Paris (Chassagne et al. 1963; Pierquin et al. 1978). Hollow tubes, in the form of either needles or plastic catheters, were surgically implanted, positioned equidistant and parallel, to allow placement of radioactive wires or ribbons with an afterloading procedure. Using conventional x-ray and dummy sources, a form of 2D imaging for implant localization became possible, which almost completely eliminated operating room staff exposure.

The first remote afterloading systems were developed simply to minimize the radiation exposure from the radioactive sources. Using a cable attachment, these machines just pushed and pulled the sources mechanically in and out of the preinserted applicators (Walstam 1962; Henschke et al. 1966). Only later, its function changed toward multiple-programmable source trains and eventually to a miniature stepping source technology. High-activity sources for high-dose-rate (HDR) and pulsed-dose-rate (PDR) applications, with the latter type mostly used in Europe, have largely replaced the use of LDR afterloading—the exception is permanent implant technology for LDR prostate brachytherapy. The tiny ^{192}Ir sources with a typical outer diameter of 1 mm replaced the ^{137}Cs tube and pellet applications. Optimization, modulating the dose distribution by varying the dwell times, has become the standard feature in brachytherapy systems, providing the users of the afterloader with much greater flexibility for patient treatment.

The use of the low-energy photon-emitting sources loaded with ^{125}I and later ^{103}Pd and ^{131}Cs for permanent implant techniques, mainly for treatment of prostate cancer, was introduced widely in the 1980s. Many thousands of patients have been treated with the small seeds since the transperineal approach (Holm et al. 1983) was adopted, first in the United States and nowadays with rapidly increasing numbers in Europe and other parts of the world (Guedea et al. 2010). The low energy results in radiation levels to the companions of the patients at a negligible level, while the one-stop procedure is well appreciated by the patients. Clinical descriptions of the indications for modern brachytherapy and the most commonly applied current techniques are provided in the clinical chapters of this book.

Further development and history of brachytherapy technology and medical physics practice during the last 50 years, covering the period in which the remotely controlled afterloaders were introduced and refined, have been described in a review article by Williamson (2006) in *Physics in Medicine and Biology*. Ibbott et al. (2008) have described 50 years of the involvement of the American Association of Physicists in Medicine (AAPM) in radiation dosimetry, including dosimetric issues in brachytherapy, in *Medical Physics*. Issues and trends in brachytherapy physics were further discussed in the *Medical Physics* Anniversary Paper by Thomadsen et al. (2008). The medical physics practice of brachytherapy treatment planning and the research activities for advancement were discussed in the above-mentioned Vision 20/20 article (Rivard et al. 2009), focusing further on current trends and challenges in brachytherapy treatment planning.

1.2 Competitive Therapy Strategies

Although brachytherapy has a definite place in radiation oncology, especially for the treatment of gynecological, breast, and prostate tumors, it is also faced with strong competition from other emerging technologies. This book is written by all contributors having a strong belief in the value of brachytherapy as the treatment modality suited for a wide range of treatment indications. Notwithstanding this, for many of those indications, there may be other treatment strategies with or without the use of ionizing radiation. Often, choices are made in which the advantages of one strategy are to be weighed against the associated disadvantages or the advantages of the other treatment option or options.

1.2.1 Therapies Based on the Use of Ionizing Radiation

External beam therapy is a rapidly evolving treatment modality for various cancer types. The manufacturers of linear accelerators have refined their products to make them more versatile, flexible, faster, and reliable. Built-in tools such as multileaf collimators and dynamic jaws eliminate the need for heavy-metal shielding blocks and allow for treatment of patients with any arbitrary field shape. Electronic portal imaging devices (EPIDs) and cone beam CT (CBCT) provide the user with images to verify and correct the patient positioning on the treatment couch and to compare with pretreatment patient planning, using off-line or online protocols. Intensity modulated radiation therapy (IMRT) can create dose distributions by using multiple fields and field segments in order to be more conformal to the treatment target plus the margin around it. Internal checks and verification steps, integrated in network solutions for data recording and storage, are built in. These steps all aim at safe dose administrations to the patient and to a reduction of the margins in order to deposit the prescribed dose to where it is needed while reducing the dose to the surrounding normal tissues. As a result, it is possible to select dose escalation to the target volume and/or a reduction of the complications of the treatment (Whelan et al. 2010; Ritter 2008; Junius et al. 2007; Shen et al. 2010; Hoskin et al. 2012).

A completely different approach is the use of a different form of ionizing radiation other than the photon and electron beams as used in the above-mentioned systems: heavy particle therapy. Proton and heavy ion beams have specific properties that make them suitable for the creation of highly targeted treatment plans. By modifying the beam size and energies, and also by adapting the direction of the beam incident on the patient, a highly conformal dose distribution is achieved with good sparing of normal tissue. The technology is well beyond the investigational stage (Suit et al. 2010). Although the application of particles heavier than electrons for medical treatments is now several decades old, and in spite of the relatively expensive infrastructure for a proton facility, there is an increased interest in its application from many sides: institutions, health insurance companies, banks and investors, and governments. Simple economics tells us that to be efficient, these high-cost investments must be applied with a large throughput of patients, thus reducing the numbers for the conventional radiation oncology, including brachytherapy.

Furthermore, there is a tendency to shift one of the paradigms of radiation oncology itself. There are, for example, an increasing number of studies evaluating hypofractionated EBRT (Akimoto et al. 2005; Martinez et al. 2005). This paradigm shift is in keeping with many brachytherapy schedules as well. The patient group with prostate cancer is one of the target groups (Demanes et al. 2011; Barkati et al. 2012).

Finally, there is, among many radiation oncologists themselves, a strong belief that the above-mentioned IMRT or dynamic arc therapy techniques will, at the end, solve all problems, so why bother about an intervention-based technique?

Observed from the point of view of politics or health care management, there is a great disparity in the reimbursement system in many countries. Apparently, in many countries, governments or insurance companies are reluctant to start a project in which patients are treated with radioactive seeds left in their body at the cost of several thousands of dollars or euros, while they are used to having the relatively cheap applications using afterloading technology. Issues like these are clearly demonstrated in the reports from patterns of care surveys for brachytherapy, for example, in Canada and Europe (Tai et al. 2004; Guedea et al. 2010).

1.2.2 Competing Therapies Not Based on Ionizing Radiation

There is other competition from nonionizing radiation offering radical focal treatment. Each modality is seeking to harness modern technology to optimize its ability to ablate a local tumor, minimizing effects on surrounding normal tissue. Thus, there are challenges from surgical robotic tools such as the Da Vinci robotic system, laser ablation, cryotherapy, and high intensity focused ultrasound (HIFU) techniques. The choice faced by the patient is often bewildering, and the evidence of superiority of one over the other is lacking. In this setting, the influence of individual clinicians (choosing either to refer the patient for radiotherapy/brachytherapy or to use other surgical or physical therapies) and the media often has greatest impact.

1.3 Strengths of Brachytherapy

By definition, brachytherapy brings the desired radiation dose directly to the target: using sealed radioactive sources, which are positioned within or at the vicinity of the tumor. It should be noted that this definition has been slightly revised by including the miniaturized electronic brachytherapy sources as an alternative to the sealed radioactive sources.

Generally speaking, brachytherapy takes advantage of the fact that the sources are connected directly to the target volume and they move with the target when the target itself is moving: there is a minimal in-patient variation during treatment.

The other advantage, compared to external beam techniques, is that the target receives a sufficiently high dose while the inverse square law ensures that even in the near proximity, the dose to the surrounding normal tissue (i.e., organ at risk) is reduced considerably.

The overall treatment time for a temporary brachytherapy procedure is a factor that largely depends on the selection of the treatment modality (i.e., HDR versus LDR) or the type of the source (low-energy versus high-energy sources) that could provide the best clinical results and patient's comfort. Most brachytherapy courses are short or very short compared to the schedules employed in external beam therapy.

1.4 Innovations in Brachytherapy

The traditional brachytherapy treatments were limited to a few body sites with well-defined geometrical locations, especially to GYN, breast, and skin. However, as diagnostic imaging techniques advance with our knowledge of radiation dosimetry and radiation delivery, more complex tumor sites are being treated with this modality. Development of HDR and its application in gynecology and breast balloon therapy are among these innovations in the field of brachytherapy. Introducing more advanced technology always brings new questions and challenges. The following is a brief overview of such issues, which the reader will find described in greater detail in the respective following chapters.

- There are new afterloading devices, developed from the existing experience with greater flexibility of source size, source strength, radionuclide, number of sources, catheter length, and capacity for a greater number of catheters.
- Several investigations have shown the limitations of existing treatment planning approaches in brachytherapy. Groups worldwide are working to safely introduce the new model-based dose calculation algorithms to achieve more realistic dose calculation software.
- In the last 15 years, brachytherapy has moved from a 2D dose distribution using conventional orthogonal x-ray film-based dosimetry toward a full 3D volumetric treatment modality. This change enables us to provide a more conformal dose delivery for very complex target volumes, with a better sparing of healthy tissues. Published results are impressive and have demonstrated their clinical potentials.
- Strategies are explored to use present-day knowledge of radiobiology in new models to determine the biological equivalent dose (BED) in order to optimize treatment schedules.
- Robotic brachytherapy has been developed for prostate seed implant, and it can be considered promising for future brachytherapy.
- Electronic brachytherapy sources have been introduced as an alternative to sealed radioactive sources, with potential enhancement of brachytherapy treatments by custom selection of peak kilovoltage and milliampere setting to optimize the treatment delivery. Further development in

terms of the size of the source and the flexibility of the source transfer system is needed in order to compete with the present miniaturized HDR sources.

- New applicator devices have been designed, with greater flexibility to combine intracavitary and interstitial procedures in the treatment of patients with asymmetric patterns of tumor around the cervical canal. Together with positron emission tomography-computerized tomography (PET-CT) imaging, this applicator will add a new dimension to the field of brachytherapy.
- Surface and eye plaque applicators, in combination with low-energy brachytherapy sources, will provide an excellent treatment modality for ocular melanoma.
- Although not a sealed brachytherapy source, ^{90}Y microspheres are developed for use in specific organs like the liver.
- The new technological developments of in vivo dosimetry in brachytherapy allow the measurement of dose to the organ at risk (i.e., bladder and rectum in HDR gynecology therapy) while it can be used as a QA device for a complex treatment modality, which simple hand calculation cannot achieve in verification of the dose delivery.

Within the existing technologies, there are definitive needs for further clarification of the potentials and also identifications of the areas that need improvements.

- What are the uncertainties in the various steps of a brachytherapy procedure regarding the physics and the clinical aspects of dose delivery?
- What are the weakest points in these analyses, that is, where can we achieve the best possible gain and obtain the highest benefits from our efforts for improvement?
- Can we reduce the tumor margins in order to reduce the toxicity and increase the dose delivered to the tumor to reduce the recurrence rate and hence obtain a better outcome for the patients?
- Is it possible to demonstrate the potential of brachytherapy in comparison to other treatment modalities?
- What is the integral dose delivered to a patient when different radiation treatment modalities are compared: brachytherapy versus IMRT, versus proton treatment for the same target?

Most of these issues are discussed extensively in the different chapters of this book by authors who are themselves active as the experts at the frontline of the innovative research.

1.5 Cooperations in Brachytherapy

Several different active task groups, supported by their parent societies [e.g., AAPM, European Society for Radiotherapy and Oncology (ESTRO), American Brachytherapy Society (ABS), American Society for Radiation Oncology (ASTRO), etc.], are enthusiastically working to enhance the efficacy of brachytherapy modalities for treatment of cancer patients. In the United States, the AAPM has a structure within the Therapy Committee,

the Brachytherapy Subcommittee (BTSC), with the task to guide and initiate new developments in the field of brachytherapy. In Europe, the European Society for Radiotherapy and Oncology (ESTRO) has a subcommittee for brachytherapy, the GEC-ESTRO (Groupe Européen de Curiethérapie), supervising the physics and several clinical expert groups in different areas of the field.

One of the major advances in brachytherapy during the first decade of the twenty-first century has been the strengthening of the cooperation between these groups around the globe. This has increased the speed of communications, mutual understanding, and acceptance of new recommendations with mutual support. It is expected that groups from other continents will further join into these cooperations, thus strengthening worldwide networks of clinical and physics brachytherapy.

Coordinated research, clinical practice, clinical trials, training, and education of the practitioners are within the reach of the societies. Of these, the execution of coordinated clinical trials to demonstrate the efficacy of modern image-guided brachytherapy to deliver the robust scientific clinical evidence for brachytherapy is among the current key task group projects.

1.6 Summary

One hundred years later, brachytherapy is still a very dynamic field. Recent focus has been on the technology of radiation delivery, on dose calculation, and on recommendations for quality assurance. Key advancements have been made by introduction of modern concepts of dose delivery making optimal use of the available imaging modalities. Technology has improved, allowing a wider range of dose rates, energies of the emitted photons, and applicator forms and sizes, while providing maximum radiation safety and protection for the personnel. New paradigms in radiobiology models open new pathways for treatment schedules. Proper training of the staff for any new procedure and equipment along with clinical trials are required to validate the efficacy of these new approaches.

References

Abbé, R. 1904. Notes on the physiologic and therapeutic action of radium. *Wash Med Ann* 2:363–77.

Akimoto, T., Ito, K., Saitoh, J. et al. 2005. Acute genitourinary toxicity after high-dose-rate (HDR) brachytherapy combined with hypofractionated external-beam radiation therapy for localized prostate cancer: Correlation between the urethral dose in HDR brachytherapy and the severity of acute genitourinary toxicity. *Int J Rad Oncol Biol Phys* 63:463–71.

Barkati, M., Williams, S.G., Foroudi, F. et al. 2012. High-dose-rate brachytherapy as a monotherapy for favorable-risk prostate cancer: A phase II trial. *Int J Radiat Oncol Biol Phys* 82:1889–96.

Becquerel, H. and Curie, P. 1901. Action physiologique des rayons du radium. *Acad Sci Paris CR* 132: 1289–91.

Chassagne, D., Raynal, M., and Pierquin, B. 1963. Technique of endocurietherapy by iridium 192 with plastic tubes in the breast tumors. *J Radiol Electrol Med Nucl* 44:269–71.

Danlos, H. and Bloch, P. 1901. Note sur le traitement du lupus érythèmateux. par des applications de radium. *Ann Dermatol Syphil Paris* 2:986–8.

Demanes, D.J., Martinez, A.A., Ghilezan, M. et al. 2011. High-dose-ratemonotherapy: safe and effective brachytherapy for patients with localized prostate cancer. *Int J Radiat Oncol Biol Phys* 81:1286–92.

Dutreix, J., Tubiana, M., and Pierquin, B. 1998. The hazy dawn of brachytherapy. *Radiother Oncol* 49:223–32.

Guedea, F., Venselaar, J., Hoskin, P. et al. 2010. Patterns of care for brachytherapy in Europe: updated results. *Radiother Oncol* 97:514–20.

Henschke, U., Hilaris, B.S., and Mahan, G.D. 1963. Afterloading in interstitial and intracavitary radiation therapy. *Am J Roentgenol, Radium Ther Nucl Med* 90:386–95.

Henschke, U.K., Hilaris, B.S., and Mahan, G.D. 1966. Intracavitary radiation therapy of cancer of the uterine cervix by remote afterloading with cycling sources. *Am J Roentgenol Radium Ther Nucl Med* 96:45–51.

Holm, H.H., Juul, N., Pedersen, J.F., Hansen, H., and Stroyer, I. 1983. Transperineal 125iodine seed implantation in prostatic cancer guided by transrectal ultrasonography. *J Urol* 130:283–6.

Hoskin, P., Rojas, A., Lowe, G. et al. 2012. High-dose-rate brachytherapy alone for localized prostate cancer in patients at moderate or high risk of biochemical recurrence. *Int J Radiat Oncol Biol Phys* 82:1376–84.

Ibbott, G.S., Ma, C.M., Rogers, D.W.O., Seltzer, S.M., and Williamson, J.F. 2008. Anniversary paper: Fifty years of AAPM involvement in radiation dosimetry. *Med Phys* 35:1418–27.

Junius, S., Haustermans, K., Bussels, B. et al. 2007. Hypofractionated intensity modulated irradiation for localized prostate cancer, results from a phase I/II feasibility study. *Radiother Oncol* 2:29.

Martinez, A.A., Demanes, D.J., Galalae, R. et al. 2005 Lack of benefit from a short course of androgen deprivation for unfavorable prostate cancer patients treated with an accelerated hypofractionated regime. *Int J Rad Oncol Biol Phys* 62:1322–31.

Meredith, W.J. 1967. *Radium Dosage: The Manchester System*. Livingston, Edinburgh.

Mould, R.F. 1993. *A Century of X-rays and Radioactivity in Medicine*. Institute of Physics, Bristol.

Paterson, R. and Parker, H.M. 1938. A dosage system for interstitial radium therapy. *Br J Radiol* 1:252–340.

Pierquin, B., Chassagne, D.J., Chahbazian, C.M., and Wilson, J.F. 1978. *Brachytherapy*. W.H. Green, St Louis, MO.

Ritter, M. 2008. Rationale, conduct, and outcome using hypofractionated radiotherapy in prostate cancer. *Semin Radiat Oncol* 18:249–56.

Rivard, M.J., Venselaar, J.L.M., and Beaulieu, L. 2009. The evolution of brachytherapy treatment planning. *Med Phys* 36:2136–53.

Shen, Y., Zhang, H., Wang, J. et al. 2010. Hypofractionated radiotherapy for lung tumors with online cone beam CT guidance and active breathing control. *Radiother Oncol* 5:19.

Suit, H., DeLaney, T., Goldberg, S. et al. 2010. Proton vs. carbon ion beams in the definitive radiation treatment of cancer patients. *Radiother Oncol* 95:3–22.

Tai, P., Yu, E., Battista, J., and Van Dyk, J. 2004. Radiation treatment of lung cancer—Patterns of practice in Canada. *Radiother Oncol* 71:167–74.

Thomadsen, B.R., Williamson, J.F., Rivard, M.J., and Meigooni, A.S. 2008. Anniversary paper: Past and current issues, and trends in brachytherapy physics. *Med Phys* 35:4708–23.

Walstam, R. 1962. Remotely-controlled afterloading radiotherapy apparatus. *Phys Med Biol* 7:225–8.

Whelan, T.J., Pignol, J.P., Levine. 2010. Long-term results of hypofractionated radiation therapy for breast cancer. *N Engl J Med* 362:513–20.

Wickham, L. and Degrais, P. 1909. *Radiumthérapie*. Bailliére, Paris.

Williamson, J.F. 2006. Brachytherapy technology and physics practice since 1950: A half-century of progress. *Phys Med Biol* 51:R303–25.

Standard Technology in Brachytherapy

Rupak K. Das
University of Wisconsin

Jack L. M. Venselaar
Instituut Verbeeten

2.1 Introduction

Brachytherapy was first introduced at the beginning of the twentieth century, where the radioactive sources were manually implanted into the tumor, thereby subjecting the physician and other medical personnel to unwanted radiation exposure. In the middle of last century, most brachytherapy procedures were performed using manual afterloading techniques, where hollow needles or applicators were placed in the tumor volume and then radioactive sources placed in the needles or applicators, thereby reducing the radiation exposure. Remote afterloaders (RALs), where radioactive sources are placed remotely in the needles or applicators placed in the operating room, were introduced at the end of the last century, thereby minimizing the exposure to nurses and caregivers. Historically, first low–dose-rate (LDR) and then ^{60}Co-based high-dose-rate (HDR) technology was introduced. In the 1980s, programmable ^{137}Cs Selectron-LDR

and later miniaturized ^{192}Ir stepping sources for HDR and pulsed-dose-rate (PDR) applications became popular. Currently, new interest in use of other source types focusing both on high energy source (^{60}Co in the afterloader of Eckert & Ziegler BEBIG GmbH) and low energy source (^{169}Yb and ^{170}Tm) is being introduced or is in the process of being tested clinically.

2.1.1 HDR RAL

HDR brachytherapy devices delivering dose rates up to 700 cGy/min at 1 cm from the source, administering discrete fractions in a temporary implant, have become common in the treatment of gynecological, breast, and prostate cancer. Treatment is delivered by a RAL, which is a computer-driven system that transports the radioactive source (typically 370 GBq of ^{192}Ir or 74 GBq of ^{60}Co) from a shielded safe into the applicator placed in the patient and back to its safe. The currently available HDR RALs use stepping-source technology, consisting of a single source at the end of a cable that moves the source in steps through the applicators placed in the treated volume. The treatment unit can treat implants consisting of many needles or catheters. Each catheter or needle is connected to the RAL through a channel. The computer drives the cable with the source from the safe through a given channel to the programmed position in the applicator (dwell position) for a specific amount of time (dwell time). In any applicator, there may be many dwell positions. After treating all the positions in a given catheter (channel), the source is retracted to its safe and then driven to the next channel. The dwell positions and the dwell time in each channel are independently programmable, thereby giving a high level of flexibility of dose delivery.

2.1.2 LDR RAL

The LDR RAL (Selectron-LDR) uses ^{137}Cs pellets as its source, each with a nominal strength of 5 mgRaeq (36.135 U, 470 MBq) and 2.5 mm in diameter [see Nath et al. (1995) for the definition of U]. ^{137}Cs was chosen as a replacement for ^{226}Ra because an "equivalent" dose rate and dose distribution could be achieved, allowing the brachytherapy community to build on several decades of prior clinical experience. The scattering and attenuation characteristics of ^{137}Cs gamma rays in tissue are similar to those of ^{226}Ra gamma rays, providing very nearly the same relative dose distribution. By proper choice of activity of the ^{137}Cs pellets, the ^{226}Ra dose rate could be approximated. The inactive pellets are ferromagnetic, and the active pellets are

nonmagnetic. The RAL is controlled by a microprocessor. The positions of the active pellets in each channel are programmed into the microprocessor. The pellets are transferred from the main safe to an intermediate safe and are sorted by a magnetic sorting technique. They are stored in their programmed order in the intermediate safe. When treatment is initiated, the pellets are transferred from the intermediate safe in the afterloader to the brachytherapy applicators. When treatment is interrupted for nursing care or visitors, the pellets are transferred back to the intermediate safe. At the end of treatment, the pellets are transferred to the main safe. All pellet transfers are performed with compressed air. The Selectron-LDR was successfully marketed for more than 20 years by Nucletron (Veenendaal, The Netherlands), but the systems are now taken out-of-service. Several systems may still be in clinical practice, but many users changed their treatment protocols to be used either with a PDR or an HDR modality.

2.1.3 PDR RAL

The PDR RAL platforms are identical or at least very similar for PDR and HDR designs. Also, the physical construction of the PDR source is almost identical to that of the HDR source. The active length of the source pellets can be shorter, and there is the lower activity of the PDR source (between 37 and 74 GBq at installation, depending on the institution's needs and room shielding) compared with the HDR source. Some details are given in Table 2.1. Figure 2.1a through c shows the construction details of three RAL ^{192}Ir HDR sources.

TABLE 2.1 Features of Modern HDR and PDR Afterloading Equipment

1. Vendor and Product Specification					
Name of Vendor	Nucletron B.V.	Nucletron B.V.	Eckert & Ziegler BEBIG GmbH	Varian Medical Systems Inc.	Varian Medical Systems Inc.
Website	www.nucletron.com	www.nucletron.com	www.ibt-bebig.eu	www.varian.com	www.varian.com
Name of product(s)	microSelectron Digital 6CH - 18CH - 30CH	Flexitron 40CH	MultiSource: 20 channels (extended to 40 channels in 2011) GyneSource: with 5 channels but same specs as MultiSource	Varisource iX	GammaMedplus iX; GammaMedplus iX 3/24 with 3 channels but same specs
Specify capability for use as HDR and/or PDR	HDR and PDR	HDR and PDR	HDR only	HDR only	HDR for iX models, PDR for non iX models
2. Specifications of Source or Sources					
Single or dual source capability	2 drives	3 drives	Single source	Single source	Single source
Possible types of source (radionuclide), available now and/or under development	^{192}Ir	^{192}Ir	Co-60: source type Co0.A86 Ir-192: source type Ir2.A85-2	^{192}Ir	^{192}Ir

(*continued*)

TABLE 2.1 (Continued)   Features of Modern HDR and PDR Afterloading Equipment

2. Specifications of Source or Sources					
Maximum source strength for each possible radionuclide, with approval for marketing by authorities	HDR: 57 mGy.h−1 @ 1m (518 GBq (14Ci) of Ir-192) PDR: 10.2 mGy.h−1 @ 1m (92.5 GBq (2.5 Ci) of Ir-192) Maximum storage source capacity: 518 GBq (14 Ci)	HDR: 49 mGy.h−1 @ 1m (444 GBq (12 Ci) of Ir-192) PDR: 10.2 mGy.h−1 @ 1m (92.5 GBq (2.5 Ci) of Ir-192) Maximum storage capacity: 814 GBq (22 Ci) of Ir192	Co0.A86: 22.6 mGy.h−1 @ 1m (2 Ci, 74 GBq) values + 10 % Ir2.A85-2: 40 mGy.h−1 @ 1m (10 Ci, 370 GBq) values +30 % − 10 %	44.8 mGy.h−1 @ 1m (11 Ci, 407 GBq)	61.1 mGy.h−1 @ 1m (15 Ci, 555 GBq) Local regulations may prohibit to install more than 10 Ci
Source outer dimensions for each of the source types (L = length, OD = outer diameter, in mm)	3.5 mm active length × 0.6 mm diameter Source capsule outer diameter: 0.9mm	3.5 mm active length × 0.6 mm diameter Source capsule outer diameter: 0.86 mm	Co0.A86 1.0(OD) × 5(L) mm Ir2.A85-2 0.9(OD) × 5(L) mm	0.59(OD) × 5(L) mm	HDR: 0.9(OD) × 4.52(L) mm; PDR: 0.9(OD) × 2.97(L) mm
Guarantee for the maximum number of source transfers, or source cycles	25,000 cycles	30,000 cycles	Co0.A86 100,000 (>400.000 tested) Ir2.A85-2 25,000	1,000	5,000
3. Specification of Applicators					
Outside diameter of the applicators (needles and or tubes, in mm)	Needles 1.3 mm Flexibles 4F	Needles 1.3 mm Flexibles 4F	needles 1.5 mm (17G) catheters 1.65mm for both sources	18G (1.27 mm) needles constructed from robust think wall tubing. 4.7F robust thick walled catheters	17G (1.5 mm) needles. 5F catheters.
Minimum curvature of an applicator (plastic loop radius, c.g. for an 180 degree curve, in mm)	13 mm for all applicators with a fixed geometry and an inner diameter of 2.5 mm or larger; 15 mm for flexible applicators with an inner diameter of 1.5 mm or higher; 20 mm for flexible applicators with an inner diameter of 1.3 mm up to 1.5 mm	13 mm for applicators with a fixed geometry; 15 mm for flexible applicators with an inner diameter of 1.2 mm and outer diameter of 1.5mm or higher;	10 mm (loop of 90°) 15 mm (loop of 180°) for both sources	17 mm	13 mm
4. Source to Cable Attachment					
Method of source attachment to cable	Laser welded to ultra flexible drive wire	Laser welded to ultra flexible drive wire, including weld protection	Laser welded	Embedded in the Nitinol (nickel-titanium) source drive wire	Laser welded to ultraflexible drive cable
5. Source Extension and Movement					
Maximum source extension (in mm from the indexer)	1500 mm	1400 mm	1500 mm	1500 mm	1300 mm
Speed of source movement, in seconds over maximum source extension	500 mm/s, typ. outdrive time 4 s for 1500 mm	1400mm in 3.7 sec	300 mm/s 5 s for 1500 mm	600 mm/s	630 mm/s
6. Number of Channels					
Number of hardware applicator channels	30	40	MultiSource: 20 GyneSource: 5	20	iX model: 24; iX 3/24 model: 3
Maximum number of channels that can be used in one plan/treatment	90	40	MultiSource (2011): 40 GyneSource model: 5	unlimited	unlimited

(*continued*)

TABLE 2.1 (Continued)   Features of Modern HDR and PDR Afterloading Equipment

7. Method of Source Movement					
Method of source movement	Forward stepping, 48 dwell positions with 2.5/5 or 10 mm stepsize	Forward stepping, 400 dwell positions with any stepsize of multiples of 1 mm	source pulling (step backward)	source pulling (step backward) 60 steps 2–99 mm step size in 1 mm increments	source pulling (step backward) 60 steps 1-10 mm step size in 1 mm increments
Treatment window for defining possible steps and dwell positions (in mm)	775 mm	400 mm	max 600 mm; step size selectable 1–10 mm in 1 mm increments, max 100 dwell positions	700 – 1500 mm	710 – 1300 mm
Method of counting dwell positions	From catheter-tip to indexer	From applicator entry to catheter-tip	User defined; default: from catheter-tip to indexer	From catheter-tip to indexer	From catheter-tip to indexer
8. Source Arrangement and Dose Calculation					
Source arrangements and dose calculations	Dwell times in 0.1 s increments, 0.1 – 999.9 s per dwell position	Dwell times in 0.1 s increments, 0.1 – 999.9 s per dwell position	1 s to 3,600 s 0.001 s increments	Dwell times in 0.1 s increments, 0.1 – 999.9 s at 60 positions	Dwell times in 0.1 s increments, 0.1 – 999.9 s at 60 positions
9. Safety Features					
Method of source retraction event of failure	Backup DC motor with friction clutch, backup batteries Manual handcrank as second backup	Backup DC motor with friction clutch, backup batteries Manual handcrank as second backup	Independent backup-retraction motor powered by backup battery system, and additional hand crank	Backup motor and independent backup drive assembly; backup battery and additional backup hand crank	Backup battery and additional backup hand crank
Independent measurement system to detect safe source retraction	Triple safety check by optodetector, mechanical end-switch and radiation detector	Integrated radiation detectors and proximity switches	Radiation measurement and several photoelectric barriers	On board radiation detectors and mechanical switches	On board radiation detectors and mechanical switches
Other safety features	Independent (hardware) secondary timer per dwell position, independent shaftencoders for motion detection and position verification, independent (software) verification of treatment record	Independent (hardware) secondary timer per dwell position, independent shaftencoders for motion detection and position verification, independent torque measurement per drive, independent (software) verification of treatment record	Database for applicators and applicator-geometry; automated calibrationand guide tube verification; position check by camera; secondary timer, independent room monitoring	Wire length check, independent radiation detector, secondary timmer and software watchdog checks, mechanical source home postional check, physical wire overtravel check, wire force feedback system, wire slipage check, plan data checksum verification	Independent radiation detector, secondary timmer and software watchdog checks, mechanical source home postional check, applicator / guide tube length check, plan data checksum verification
What is the maximum level of leakage dose, in µGy.h−1 @ 0.1 m from any point of the surface of the machine, at maximum source strength loading	< 1.1 µGy.h−1 @ 0.1 m	< 1 µGy.h−1 @ 0.1 m	According to IEC60601-2-17 Co0. A86: < 10 µGy/h @ 1 m Ir2.A85-2: < 1 µGy/h @ 1 m	6 µGy.h−1 at 0.1 m for 11Ci source	4.5 µGy.h−1 at 0.1 m for 15Ci source

(*continued*)

TABLE 2.1 (Continued)   Features of Modern HDR and PDR Afterloading Equipment

10. Control Unit and Treatment Planning System					
Digital communication system	Separate control system with network connection. Extended graphical user interface with patient and library/QA plan database	Separate control system with network connection. Extended graphical user interface with patient and library/QA plan database	PC based: OS Windows XP, Windows 7 or Vista, control PC networked to planning PC	Application user interface running on Windows XP platform. Patient database, QA features, user defined access rights, network interface	Application user interface running on Windows XP platform. Patient database, QA features, user defined access rights, network interface
Remote access to System Control Unit	Yes	Yes	Yes, if internet connection is available	No	No
Dicom RT Plan import interface for treatment plan input data	Yes	Yes	At TPS side DICOM RT Plan export available, no import	At TPS side DICOM RT Plan export available	At TPS side DICOM RT Plan export available
Dicom RT Report for R&V	Yes	Yes	Export DICOM RT Plan Dose and Structure at TPS	Export DICOM RT Plan Dose and Structure at TPS	Export DICOM RT Plan Dose and Structure at TPS
TPS dedicated to vendor's afterloader or able to create plans for other makes of afterloading equipment	TPS dedicated to Nucletron afterloader	TPS dedicated to Nucletron afterloader	Export DICOM RT or Bebig proprietary data format	Export in DICOM RT or Varian proprietary format (for any radioactive source)	Export in DICOM RT or Varian proprietary format (for any radioactive source)
Dedicated TPS supporting all afterloader features and brachytherapy applications	Yes	Yes	Dedicated 3D TPS, all kinds of image modalities available. 2D x-ray images import either by x-ray scanner or DICOM. Fusion. Real-time online US prostate planning	Yes, supports 2D and 3D planning without limitations	Yes, supports 2D and 3D planning without limitations
11. Special Features					
User ID and password protection	Yes, configurable permissions per user	Yes, protection by user ID and password, file protection by checksum	Yes, protection by user ID and password, file protection by checksum	Yes, configurable permissions per user	Yes, configurable permissions per user
Details of available applicators:					
MR/CT compatibility with applicators available	Yes, full range of carbon/plastic GYN applicators, interstitial plastic and titanium needles	Yes, full range of carbon/plastic GYN applicators, interstitial plastic and titanium needles	Yes, full range of titanium/plastic Gyn applicators (MR/CT compatible), plastic needles and tubes for interstitial applications	Yes, full range	Yes, full range
Capability to add interstitial needles to intracavitary gyn-applicators	Yes, for ovoids and for ring	Yes, for ovoids and for ring	Yes, for ring applicator	Yes, for ring applicator	Yes, for ring applicator
Sterilization	All applicators (incl CT/MR) can be steam-sterilized at 134° C or are delivered sterile (e.g. catheters)	All applicators (incl CT/MR) can be steam-sterilized at 134° C or are delivered sterile (e.g. catheters)	Steam sterilization 134°C 3 bar for 5 min, or delivered sterile (e.g.)catheters	All Gyn applicators and majority of others can be steam sterilized	All Gyn applicators and majority of others can be steam sterilized
Other features			Integrated in vivo dosimetry system Integrated calibration of source positioning		

(a)

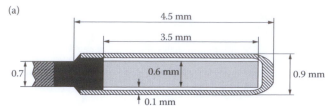

Nucletron HDR ^{192}Ir model 192-Ir-mHDR-v2-revised source

(b)

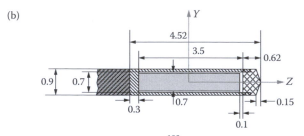

GammaMed HDR ^{192}Ir model plus source

(c)

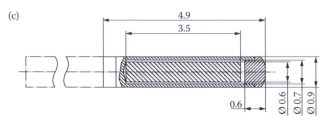

FIGURE 2.1 Construction details: (a) Nucletron MicroSelectron HDR v2 source; (b) Varian GammaMed HDR Plus; (c) BEBIG model Ir2. A85-2 source. (Courtesy of Nucletron, Veenendaal, The Netherlands; Varian, Palo Alto, California, USA Eckert & Ziegler BEBIG GmbH, Berlin, Germany.)

The PDR RAL simulates continuous LDR treatments by delivering the same total dose at the same total time. With the higher activity source, this is accomplished by exposing the source from the afterloader for only a fraction of the time for each hour. Consider this example: the desired dose rate is 50 cGy/h at the point of interest for a total treatment time of 48 h. The PDR source delivers an instantaneous dose rate of 500 cGy/h at that point. In this instance, the source should be exposed for a tenth of an hour, or 6 min, each hour for the same total treatment time of 48 h. This scheme delivers the same total dose at the same total time yielding the same average dose rate. More detailed descriptions of the features of modern RALs, such as the ones shown in Figure 2.2, the essential components, and an overview of the available equipment are given later in this chapter. Technical details are summarized in Table 2.1.

2.2 Regulatory Bodies

2.2.1 Current Regulatory Environment for Brachytherapy Equipment

Development and safe use of medical equipment such as afterloading systems are subject to approval and certification by regulatory bodies. Overseeing practices may be different in different countries. The next paragraphs describe procedures followed by the Nuclear Regulatory Commission (NRC) and the Food and Drug Administration (FDA) in the United States and the role of CE Marking in the area of the European Union. The information below is mainly based on the specific sections on this subject in the recent report of AAPM task group 167 "American Association of Physicists in Medicine" and "European Society for Radiotherapy and Oncology" Recommendations on Dosimetry Requirements for New or Innovative Brachytherapy Sources, Devices, and Applications (Nath et al. submitted).

2.2.2 Nuclear Regulatory Commission

The NRC is a U.S. government agency that was established by the Energy Reorganization Act in 1974 to take over the role of oversight of nuclear energy matters and nuclear safety from the Atomic Energy Commission (AEC). Like its predecessor AEC, the NRC oversees reactor safety, reactor licensing and renewal, radioactive material safety and licensing, and waste management (storage and disposal). The NRC's mission is, among other things, to regulate the U.S. civilian use of by-product, source, and special nuclear materials to ensure adequate protection of public health and safety. By-product material is, for example, used in radiation oncology in brachytherapy sources and devices, gamma stereotactic radiosurgery devices, and teletherapy units and in some calibration sources. The NRC oversees medical uses of nuclear material through licensing, inspection, and enforcement programs. It issues medical use licenses to medical facilities and authorized physician users, develops guidance and regulations for use by licensees, and maintains a committee of medical experts to obtain advice about the use of by-product materials in medicine. The Advisory Committee on the Medical Uses of Isotopes, composed of physicians, medical scientists, and other health care professionals, provides advice to the NRC staff on current initiatives in the medical uses of radioactive materials. The NRC (or the responsible Agreement State) also regulates the manufacture and distribution of these products.

2.2.3 Food and Drug Administration

The FDA oversees the good practices in the manufacturing of radiopharmaceuticals, medical devices, and radiation-producing products (e.g., the ones used in radiation oncology). The states regulate the practices of medicine and pharmacy and administer programs associated with radiation-producing x-ray machines and accelerators. The Center for Devices and Radiological Health (CDRH) is the branch of the FDA responsible for the approval of all medical devices before they can be legally marketed in the United States, as well as overseeing the manufacturing, performance, and safety of these devices. CDRH regulatory powers include the authority to require technical reports from the manufacturers or importers of regulated products, to require that radiation-emitting products meet mandatory safety performance standards, to declare regulated products defective, and

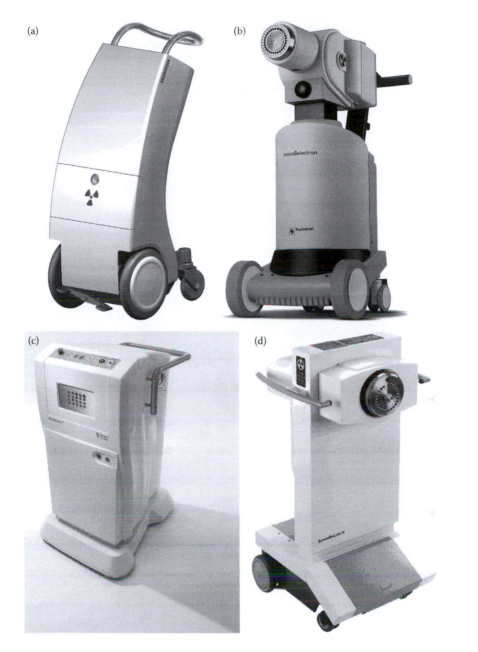

FIGURE 2.2 Remote afterloading systems: (a) Flexitron; (b) MicroSelectron; (c) BEBIG MultiSource; (d) GammaMed Plus iX. (Courtesy of Nucletron, Veenendaal, The Netherlands; Eckert & Ziegler BEBIG GmbH, Berlin, Germany; and Varian, Palo Alto, California, USA.)

to order the recall of defective or noncompliant products. The CDRH also conducts limited amounts of direct product testing.

FDA approval for medical devices started in 1976. Therefore, any device that was designed and fabricated prior to 1976 has been *grandfathered* FDA approval for its use at that time. Further, if a device is made by an investigator for one case with special situations and is not marketed, FDA needs only to be informed but no approval is required. The FDA's role is to assure the public that medical devices are safe and effective when used as described by the "label," that is, the label on the product, the package insert, or other documents that accompany the device. The FDA issues approval or clearance for the device when it is determined that the device is safe and effective if used as described in the Instructions for Use (IFU). The FDA clearance allows the manufacturer to promote and sell the device for the indications described in the labeling. If used as described in the label, payors following the guidelines of the Centers for Medicare and Medicaid Services (CMS) usually reimburse medical providers for that specific use in patients. However, licensed practitioners of medicine have the right to use approved devices for an individual patient in any way they deem appropriate, either according to the label or "off-label." Many novel drugs and devices have resulted from creative off-label use by experienced medical practitioners.

Any new device, including a new or innovative brachytherapy source, is generally evaluated for its approval as a medical device by the FDA based on relation to a legally marketed predicate device, and includes its risk factor. If a medical device is being introduced, it can be categorized into two groups: (1) similar to an existing device, and (2) one for which no comparable device exists. If the device is not substantially equivalent as the second group, then sufficient information including clinical and nonclinical data must be provided showing that it is safe and effective for its intended use. It will also include clinical data such as the specific indication for use, trial protocol, test hypothesis, statistical analysis including sample size needed to test the hypothesis, specific primary safety and effectiveness end points for trial, secondary end points for trial, patient inclusion criteria, patient exclusion criteria, informed consent form, and case report forms. As part of the information submitted by the manufacturer for a Premarket Approval Application (PMA), the vendor has to provide the FDA with sufficient clinical and nonclinical data to show that the device provides reasonable assurance of safety and effectiveness for its intended use. The intended use mirrors the conditions in the clinical trial for that device.

Off-label use is allowed under the practice of medicine only when a drug or device is used by a qualified licensed healing arts practitioner. The FDA Modernization Act of 1997 (FDAMA) includes a section on the practice of medicine that explicitly allows a clinician to make a medical decision (practice of medicine) to use a medical device that has been approved for one indication in an off-label manner for another indication. However, it is then the clinician's (or clinical team's) responsibility to evaluate the safety and effectiveness of this off-label use of the device for the individual patient being treated.

2.2.4 European Union Regulatory Structure

The CE Mark, or CE Marking, is an important step before a medical device can be sold in Europe. CE stands for *Conformité Européenne*, which means as much as *in agreement with the European regulations*. Since June 1998, the CE Marking is a legally binding statement by the manufacturer. It is not a mark of quality, rather a sign that the product has reached the requirements of special European device directives [Medical Devices Directive (MDD 93/42/EEC, or MDD for short), in vitro Diagnostic Device Directive (IVDD 98/79/EC), or the Active Implantable Medical Device Directive (AIMDD 90/385/EEC)]. Most medical devices, like brachytherapy devices, are in the scope of the MDD 93/42/EEC. All manufacturers that can demonstrate compliance with the MDD are allowed to apply the CE Marking on their products. At the time of writing, in 27 countries of the European Union, the CE Marking is required. Additionally, Norway, Iceland, and Liechtenstein are signatories of the European Economic Area (EEA) and require CE Marking. Switzerland has transposed the Medical Devices Directives into their national law, and thus the CE Marking requirements. In some European countries, the national legislation is stricter, and the requirement is extended in the sense that CE Marking must be present for all medical devices, even if developed in-house and for in-house usage only.

The medical devices are classified into different groups. Class I devices are low-risk devices. Medium-risk devices are subdivided into two groups: Class IIa and Class IIb devices. Class IIa devices are, for example, diagnostic ultrasound devices or hearing aids. Class IIb devices have a greater risk than Class IIa devices. Surgical lasers, heart defibrillators, and radiation treatment devices (including brachytherapy devices), for example, belong to this group, where product failure can result in a serious health risk for the patient or the personnel. The highest-risk devices are in Class III. Included in this category are, for instance, stents, intracardiac catheters, and breast implants. Classification rules are specified in Annex IX of the MDD. Furthermore, for devices of classes I (sterile or measuring devices), II, IIa, IIb, and III, a Quality Management System must be implemented that complies with Annex II or V of the MDD. Many manufacturers use ISO 13485 to meet those requirements. In this ISO standard, a comprehensive management system for medical devices is described. After the Quality Management System is implemented, a Technical File must be prepared for medical devices of all classes. It includes, among other components, a device description, product specification, technical drawings, performance testing, and risk analysis. If the company has no place of business in Europe, the appointment of an authorized qualified representative in Europe is necessary. This representative acts as a liaison between the company and the competent authorities in Europe. After the successful review of the Technical File and the Quality Management System by a notified body, the CE Marking certificate is issued. To maintain the certification, the auditing must be repeated every year. Finally, a declaration of conformity must be prepared. In this short document, the companies state that they are in full compliance with the MDD. Moreover, it is required to translate documentation and labeling of the medical device for those European countries where the device should be sold. Even if this point is not addressed in the MDD, it is necessary to be in concordance to the national laws.

2.3 Features of Currently Available Afterloaders

This section provides the reader with a table of features and components of modern RALs available worldwide. Such tables have been published previously, for example, in the book *Achieving Quality* (Thomadsen 2000), in Chapter 2 of ESTRO Booklet #8 (Venselaar and Pérez-Calatayud 2004), and the Chapter 63 of the book "*Quality Assurance and Safety in Radiotherapy*" edited by T. Pawlicki et al. (Venselaar 2011). Table 2.1 is an update of the latter publication accounting for the merging of the companies Nucletron and Isodose Control. Features and components of RALs are discussed below.

2.3.1 Number of Channels

Afterloaders have a sufficient number of channels to cover large implants. Except for the specific designs of gynecological afterloaders, the numbers vary from 20 in the Varisource iX up to 40 in the Flexitron machines. In general, only those institutions with experience in performing very large implants will consider this as an important condition in the purchase of a system. It is noted that both the BEBIG MultiSource and the Varian GammaMedplus iX offer an alternative for those clinics that only perform gynecological treatments. Then, a total of three channels are required to connect a tandem-ovoid or tandem-ring type of applicator. Eckert & Ziegler BEBIG GmbH offers the GyneSource with five channels, and Varian offers the GammaMedplus iX 3/24 with three channels.

2.3.2 Single/Dual Source Capability/ Other Radionuclides

The Flexitron is designed with two source drive mechanisms. The intention has been to create a machine that offers more flexibility than other makes. In principle, two identical sources, two sources of different strength, or even two sources with different radionuclides can be used at the same time: the system can keep track of the dwell positions and dwell times of each source separately. Treatment times can become shorter, and a decayed source can be used longer. To the knowledge of the authors, the approval for clinical use with two sources has not yet been granted.

2.3.3 Guarantee for Running Cycles

There is a large variation in the maximum number of source transfer cycles guaranteed by the vendor for their devices. For the average clinic, this number is not at all a problem. But a very busy clinic should make the calculation for the planned workload. In one specific case, this number can be a serious problem: the potential use of an HDR ^{60}Co in an afterloader would tremendously increase the duration of the interval between two source exchanges, maybe up to once every 5 years. When many patients are to be treated with this source, the guaranteed number might become critical. A user in these circumstances should be aware of this and is advised to make the calculation of the expected number of cycles.

2.3.4 Size of Source in Relation to Catheter/Needle Diameter

The companies have effectively reduced the size of their source designs over the years in order to be able to use smaller needle and flexible tube diameters, while maintaining the high activity contents of nominal 370 GBq (10 Ci) for their ^{192}Ir sources. The Varisource iX has a source diameter of only 0.59 mm; other machines operate with 0.85- or 0.9-mm outer diameter for their ^{192}Ir sources. The BEBIG ^{60}Co source has 1.0-mm diameter. The smallest outer diameter recorded for needles and 4F for flexible tubes is now 1.27 mm for the Varian machine, going up to 1.3 and 1.5 mm for the other makes. The use of smaller needle

diameters claims to be less traumatic and may thus form a benefit for the patient. It is not expected that the source size can be reduced much further.

2.3.5 Minimum Curvature in Flexible Tubes

The smaller sources in the newer designs are welded to more flexible drive cables as compared with those in the classic designs. This is reflected in the minimum curvature of a (tube) applicator that can be used without too much friction between the source drive cable and the tube. Typical values of 15- and 17-mm radius for a 180° loop in a flexible tube are recorded, with 13-mm radius for the GammaMedplus source. A smaller value might be important, for example, in cases where loop techniques are used for a floor-of-mouth implant.

2.3.6 Speed of Source Movement

The source driving mechanism determines the speed with which the source is moved, for example, from the source safe to the first dwell position. A slow movement with a high active source leads to the so-called source transfer dose to healthy tissue outside the implant. Depending on a multitude of factors, this source transfer dose might become significant. Travel speeds of 300 up to 630 mm/s are recorded for the different makes.

2.3.7 Method of Source Movement

The present-day RALs have generally a choice for source dwell positions every millimeter in the treatment window (every 2.5 mm in the MicroSelectron system). A drive cable and dummy source are used before the real source enters a source channel to make sure that there are no obstructions in the catheter or catheter connection. In the Nucletron systems, the real source then moves forward stepping through the programmed trajectory; the other makes send out the real source to the most distal position and then retract the source.

2.3.8 Safety Features

Safety features are important parts to any system. The machines therefore have many interlocks, a source retraction mechanism, built-in radiation detector, and other safety features. These are discussed in Section 2.6.

2.3.9 Leakage Dose

The shielded safe to house the high active source is generally made of tungsten of sufficient thickness to provide shielding for personnel involved in the treatments. Typical levels of exposure are recorded to be of less than 6 µGy/h at 0.1 m or close to 1 µGy/h at 1 m. The higher penetration of the ^{60}Co radiation is reflected in a level close to 10 µGy/h at 1 m. IEC60601-2-17 standard states that the radiation level due to leakage radiation of storage container(s) in general must be such that the dose equivalent rate at any position at 1 m from the surface of the container shall not exceed 1 μSv/h. In case of storage container(s) specified for use in a treatment room with restricted access, the dose equivalent rate at any position at 1 m from the surface of the container shall not exceed 10 μSv/h. A license from the national authorities may be more restrictive, and acceptance test measurements should demonstrate that the machine complies with their upper limits. From a general viewpoint of safety, a source-containing machine should always be stored in a vault when not in clinical use.

2.3.10 Network Connectivity

All modern machine types can be connected to a hospital network. In this way, the treatment plan, created by the brachytherapy treatment planning system (TPS), can be transferred directly and without manual interference into the afterloader. The algorithms in both TPS and in the software of the afterloader ensure that the source decay is properly accounted for. Manually entering the data for a complex plan or the cumbersome transfer with memory devices is nowadays considered obsolete.

2.4 Components of an HDR RAL

In general, all available HDR RALs consist of the same general components, although they differ somewhat in their technological details. A general overview of the systems with the major parts is described below.

2.4.1 Shielded Safe

A stepping-source RAL uses either a ^{192}Ir source (370 GBq) or a ^{60}Co (74 GBq) to provide a dose rate up to 700 or 400 cGy/min, respectively, at 1 cm from the source. To house this highly radioactive source, a shielded safe made of tungsten or depleted uranium of sufficient thickness to provide enough radiation shielding is an integral part of the treatment unit. The safe is visible in Figure 2.3. When the source is not in treatment mode, the shielding reduces the air kerma rate to 1–6 μGy/h on the surface of the afterloader. Once in treatment mode, the source is driven out of the safe while it follows the program through the dwell positions. In the event of an interruption or termination of the treatment, the source is driven back to the shielded safe.

2.4.2 Radioactive Source

While delivering an HDR requires an intense source, passing the source through needles placed through a tumor requires a source of a small size. The radionuclides used for all HDR RALs now marketed are ^{192}Ir and ^{60}Co. Since ^{192}Ir has a high specific activity of about 341 GBq/mg and effective gamma-ray energy of 0.398 MeV (Baltas et al. 2007), 370 GBq (10 Ci) sources made out of this radioactive material can be smaller and easier to shield compared to ^{60}Co or ^{137}Cs. The radioactive source in an HDR RAL is usually 3 to 10 mm in length and less than 1 mm in diameter, fixed at the end of a steel cable (see Figure 2.1). The Nucletron source is placed

FIGURE 2.3 Top cover of the afterloader removed to show the position of the shielded safe. (Courtesy of Nucletron, Veenendaal, The Netherlands.)

in a stainless-steel capsule and welded to the cable, whereas the Varian source is placed in a hole drilled into the cable and closed by welding. Currently, Varian markets a source with a length of 5 mm. Since ^{192}Ir has a half-life of only 74 days, these sources need to be changed every 3–4 months to keep the treatment in the HDR range. Recently, Eckert & Ziegler BEBIG GmbH has introduced an HDR RAL with a 74-GBq (2 Ci) ^{60}Co as its source. It has an active core 0.5 mm in diameter and 3.5 mm in length covered by cylindrical stainless-steel capsule 0.15 mm thick, with an external diameter of 1 mm. With a half-life of 5.2 years, the ^{60}Co HDR source needs to be changed once in every 5–10 years. Recent interest on ^{169}Yb as an HDR source due to its intermediate half-life of 32 days, relatively high specific activity, and mean photon energy of 93 keV is drawing the attention of clinicians and vendors. ^{169}Yb decays via electron capture to several excited states of ^{169}Tm. The de-excitation to the ground state occurs via gamma-ray emission, which results in emission of several characteristic x-rays. The electrons emitted are of low energy so that they are easily filtered out either by the ytterbium itself or by thin layers of encapsulation materials. Since the mean energy is lower than ^{192}Ir and ^{60}Co, less shielding is needed for the construction of a treatment vault. Another source with a relatively low photon energy is ^{170}Tm. This thulium isotope has a half-life of 128 days, a relative high specific activity of 221 GBq mg^{-1}, and a mean energy of the emitted photons of 66 keV (Baltas et al. 2007). Radiation protection data for both ^{169}Yb and ^{170}Tm were published, indicating that ^{170}Tm could be of interest for sources for temporary implants (Granero et al. 2006). In the decay scheme, several beta components are present, with relatively high maximum energy of 0.968 MeV. Thulium is rather expensive due to high costs of production. The specific activity of ^{170}Tm is lower than for ^{192}Ir making it more suitable for application as a PDR source. Influence of possible designs of encapsulated ^{170}Tm sources has recently been investigated (Ballester et al. 2010).

As required by national or state regulations, physicists need to calibrate the source after each installation using a well-type chamber and an electrometer, both of which have been calibrated by Accredited Dosimetry Calibration Laboratories (ADCLs). The resulting source calibration should be verified against the vendor's calibration, and the two values should be within 3% of each other (Kutcher et al. 1994). For clinical implementation, the user can then utilize either the manufacturer's or the institution's source strength.

2.4.3 Source Drive Mechanism

When a treatment is started, the check cable (an exact duplicate of the radioactive source along with its cable, except not radioactive) stepper motor drives the check cable to the programmed length plus a couple of millimeters to verify the integrity of the system. This consists of checking the proper attachment of the transfer tube to the indexer ring and also the attachment of the transfer tube to the applicator or catheters used for the treatment. A noneventful run of the check cable initiates the source cable stepper motor connected to the reel containing the source drive cable to turn. This causes the source cable to advance from the shielded safe along a path constrained by transfer tubes to the first treated dwell position in the applicator attached to the first channel. The source dwells at that position for a predetermined duration (dwell time) as calculated by the treatment-planning system. After completing that dwell it goes on to the subsequent dwell positions. Some units step as the source drives out (microSelectron and Isodose Control), stopping first at the dwell position most proximal to the afterloader, while in the others (VariSource and GammaMed), the source travels first to the most distal dwell (toward the tip of the applicator), and a

bit farther, and then steps as the source returns toward the safe. Stepping on the outward drive obviates any concern about the effect of slack in the drive mechanism affecting the accuracy of the source position. The unit that lets the source step backward from a distal to a proximal dwell position and then back into the unit includes correction for slack in the calibration of the source location. Upon completion of the treatment for the first channel, the source is retracted into the safe and redirected to travel to the second channel after a safety check run by the check cable. The process is repeated for all the subsequent treatment channels. The programmed movement of the source is verified by means of an optical encoder or other devices that compare the angular rotation of the stepper motor or cable length ejected or retracted with the number of pulses sent to the drive motor. This system is capable of detecting catheter obstruction or constriction as increased friction in the cable movement. Under certain fault conditions, such as if the stepper motor fails to retract the source, a high torque, direct-current (DC) emergency motor will retract the source. The confirmation of the source's exit from and return to the safe is carried out by an "opto-pair," consisting of a light-sensitive detector and an infrared light source, which detects the cable when its tip obstructs the light path.

2.4.4 Indexer

The indexer (Figure 2.4) is the part of the RAL that directs the source cable from the exit of the safe to one of the exit ports from the unit (channels). This mechanism gives the option of connecting a number of catheters (N = 2, 3, 18, 20, 24, up to 40) to the RAL.

The various catheters or applicator parts connect to these channels, usually through connecting guides called "transfer tubes." Different units have between 2 and 24 channels available for connection. If a patient's treatment requires more than the number of channels on a given treatment unit, the treatment must be broken

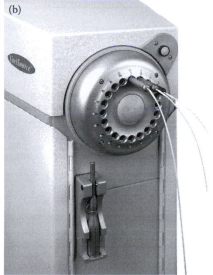

FIGURE 2.4 Frontal view of two afterloaders and their indexers: (a) BEBIG MultiSource; (b) Varian Varisource. (Courtesy of Eckert & Ziegler BEBIG GmbH, Berlin, Germany and Varian, Palo Alto, California, USA.)

into sessions, where the catheters are connected up to the number of channels available and treated. Then the transfer tubes are disconnected from the catheters just treated and reconnected to the next set of catheters for continuation of the treatment. If a treatment plan contains more catheters than the maximum channels that the machine can handle, the computer treatment-planning system breaks the plan into two or three sessions as needed.

2.4.5 Transfer Tubes

As the name suggests, transfer tubes are long tubes that act as a conduit to transfer the source from the RAL to the applicators or catheters for treatment. One end of the transfer tube is attached to the indexer of the RAL (Figure 2.5), whereas the other end is attached to the interstitial, intracavitary, or transluminal applicators (Figure 2.6).

Often, the applicator end of the transfer tube contains spring-loaded ball bearings that block the path through the tube if no applicator is attached. When an applicator is inserted, it pushes aside the ball bearings, opening the path for the source cable. When the check cable makes its test run, if no applicator is

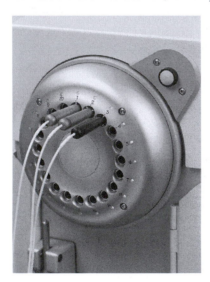

FIGURE 2.5 Transfer tubes attached to RAL. (Courtesy of Varian, Palo Alto, California, USA.)

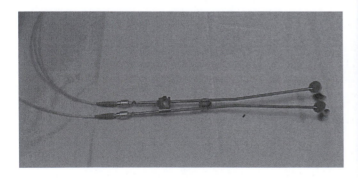

FIGURE 2.6 Transfer tubes attached to applicator. (Courtesy of Rupak Das.)

attached to the transfer tube, the check cable hits the obstacle of the ball bearings and prevents ejection of the source. Each type of applicator has its own type of transfer tube.

2.4.6 Treatment Control Station

The treatment control station (Figure 2.7) allows the user to select the dwell positions and dwell times to be used in each channel. There are three ways of entering this information: (1) manually at the control station by the keyboard/mouse; (2) recalling a standard plan from the computer, which has been saved before, and then editing the data without affecting the standard plan from which it originated; or (3) importing the data from a treatment planning system via a transfer medium [floppy disk, compact disk (CD), or a memory stick] or a local area network (LAN) connection to the treatment control station.

2.4.7 Nurse Station (PDR Equipment)

All PDR RALs should be located close to the nurses' station, so that the patient and the room can be under constant surveillance. In order to communicate and observe the patient during the treatment, this station should also be fitted with an intercom and a closed-circuit television system.

2.4.8 Treatment Control Panel

The treatment control station transfers the data to the treatment control panel. A hard or soft Start button initiates the execution of the treatment according to the program. In addition, there is an Interrupt button, which when pressed retracts the source and stops the timer, allowing the user to enter the treatment room without receiving radiation exposure. A Resume or Start button resumes the treatment from the time and the dwell position where it was interrupted. A master Emergency Off button initiates the high-torque DC emergency motor to retract the source.

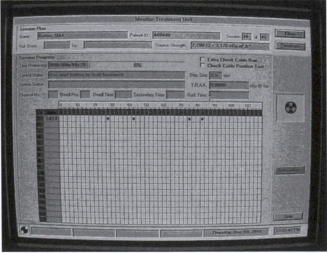

FIGURE 2.7   TCS, treatment control station. (Courtesy of Rupak Das.)

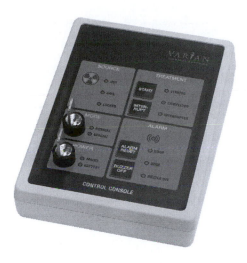

FIGURE 2.8 Treatment control console. (Courtesy of Varian, Palo Alto, California, USA.)

In the normal course of a successful termination of the treatment, the timer runs to zero and the machine automatically retracts the source. Figure 2.8 shows an example of the treatment control console.

2.5 Facility Considerations

2.5.1 Facility Considerations for LDR

Planning for the facility to house an LDR RAL involves shielding, location, ancillary services, and audiovisual communication issues. To design appropriate shielding, one must decide what treatments will occur in the room and what will be the maximum air kerma strength per hour and maximum total reference air kerma (TRAK) per week for these treatments. The room shielding must limit the dose equivalent in uncontrolled areas to less than 1.0 mSv per year, with a maximum dose equivalent of 0.02 mSv in any 1 h (NCRP 1993). Radiation workers may receive up to 0.05 Sv per year, but based on ALARA (as low as reasonably achievable), rooms should generally be designed to limit exposure of radiation workers to 0.1 mSv per week in controlled areas (NCRP 1993). In some other countries, lower limits are defined; see Chapter 29 on radiation protection. Some states also limit the instantaneous dose equivalent rate received for both the general public and radiation workers. The instantaneous dose equivalent rate limit is usually not an issue for LDR RAL. As always, a facility and an individual must comply with the applicable regulations in their state. Although a bedside shield may be used to reduce the dose equivalent levels to comply with the regulations, adequately shielded walls are the preferable alternative. This situation does not rely on someone to place the shield in the appropriate position for each case and provides more usable floor space in the room. LDR RAL rooms are typically hospital rooms that should be located close to the nurses' station where the nurse can keep the patient and the room under surveillance. If there are multiple shielded rooms, they should be adjacent to each other and remote from unshielded hospital rooms. The nurses need an intercom and a closed-circuit television system to observe and to communicate with the patient during treatment. The LDR RAL treatment control console is placed immediately outside the patient room, with a remote at the nurses' station. The power for the LDR RAL should be on the emergency power circuit. Compressed air is also required for the Selectron LDR RAL. The room door should be interlocked. An area radiation monitor should be in the room with a remote outside the room to indicate when the sources are exposed. As with any brachytherapy room, there should be emergency equipment and an emergency container (bail out pig) in the room. The door should have a plaque for mounting radiation signs and instructions when patients are being treated.

2.5.2 Facility Considerations for PDR

The facility requirements for a PDR RAL will be the same as for the Selectron LDR RAL, except for the compressed air. If the room has been shielded for LDR treatments, the shielding should be adequate for the equivalent PDR treatments. The instantaneous dose equivalent rate will be higher with the PDR RAL than for LDR treatments, but the dose equivalent rate averaged over an hour should be the same as for the LDR treatment. However, if local regulations include an instantaneous dose rate limit, it must be verified that these limits will not be violated. From a shielding standpoint, one advantage that PDR has over LDR is that the half-value layer (HVL) in concrete for ^{192}Ir is less than the concrete HVL for ^{137}Cs (NCRP 1972) because of the lower average energy of the ^{192}Ir. One caveat though is that if the new PDR treatments are planned that were not performed previously with LDR, these treatments may result in a higher average dose equivalent rate or higher total dose equivalent. For instance, large interstitial implants may result in higher average dose equivalent rate and higher total dose equivalent than the gynecological treatments. In this case, the adequacy of the shields must be verified.

2.5.3 Facility Considerations for HDR

The radioactive source in the HDR machine starts at about 370 GBq (10 Ci) with an exposure rate at a distance of 1 m from the source of about 46 mSv/h. According to the rules and regulations of the U.S. NRC (USNRC 1994), the annual limit for radiation exposure to the public is 1 mSv, and the annual occupational limit is 50 mSv. Following the principle of maintaining exposures as low as reasonably achievable, the NRC usually expects licensees to create warning alert levels effectively to limit occupational personnel exposures to 10% of the NRC limit, which becomes 5 mSv for whole body exposure. In addition to the annual limit, the NRC requires that in an unrestricted area, the dose equivalent rate may not exceed 0.02 mSv in any hour. Thus, the HDR machine needs to be housed in an adequately shielded room. To meet these requirements in an HDR suite, where the walls and the ceiling are at least 1.5 m from the machine head, concrete walls of about 43 to 50 cm (or 4 to 5 cm of lead) are needed.

For larger rooms, the concrete wall thickness will be lower since the exposure rate is inversely proportional to the square of the distance from the radioactive source. Details on the procedures for calculating the thickness of barriers for a particular facility are elaborated in the works of Cember (1996) and McGinley (2002). Several floor plans for remote afterloading suites were shown by Glenn P. Glasgow in Chapter 10 of the AAPM/ABS 2005 Summer School Book (Glasgow 2005).

2.6 Safety Features

HDR RALs are complicated devices containing very high activity radioactive sources. Serious accidents can happen quickly. All such units have many safety features and operational interlocks to prevent errant source movement or to facilitate rapid operator response in the event of a system failure.

2.6.1 Emergency Switches

Numerous Emergency Off switches are located at convenient places and easily accessible, in case a situation arises. One Emergency Off switch is located on the control panel. Another Emergency Off button is located on the top of the RAL treatment head. Vendors usually install one or two emergency switches in the walls of the treatment room. In the event a treatment is initiated with someone other than the patient in the treatment room, that person can stop the treatment and retract the source by pressing the Emergency Off button. Figure 2.9 shows the Emergency Off switch on the treatment control panel.

2.6.2 Emergency Crank

All HDR RALs have emergency cranks to retract the source cable manually if the source fails to retract normally and the emergency motor also fails to reel in the source. Figure 2.10 shows such a crank for the microSelectron RAL. Using the crank requires the operator to enter the room with the source

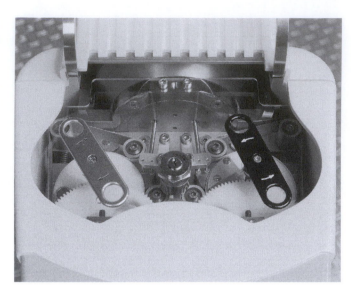

FIGURE 2.10 Emergency crank. (Courtesy of Nucletron, Veenendaal, The Netherlands.)

unshielded. Exposure rates for this situation are considered below.

2.6.3 Door Interlock

Interlock switches prevent initiation of a treatment with the door open. When a treatment is in progress, opening the door interrupts the treatment. This safety feature protects the medical personnel from radiation exposure in the event somebody enters the treatment room without the knowledge of the operator. If a door is inadvertently opened during the treatment, the treatment is interrupted and the source returns to the safe. The treatment can be resumed at the same point where it was interrupted by closing the door and pressing the Start or the Resume button at the control panel.

2.6.4 Audio/Visual System

All HDR suites are equipped with a closed-circuit television (CCTV) system or shielded windows and/or mirrors for observing the patient, and a two-way audio system to communicate with the patient during treatment.

2.6.5 Radiation Monitor and "Treatment On" Indicator

Three separate independent systems alert personnel when the source is not shielded. One radiation detector is part of the treatment unit and indicates on the control panel when it detects radiation. An independent unit, usually mounted on the treatment room wall with displays both inside and outside the room, also alerts the operator and other personnel when the radioactive source is out of the safe. A Treatment On indicator outside the

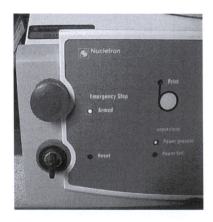

FIGURE 2.9 Emergency button TCP. (Courtesy of Rupak Das.)

room, activated when the source passes the reference optical pair discussed above, also indicates that a treatment is in progress.

2.6.6 Emergency Service Instruments

In the event the radioactive source fails to retract after termination, interruption, pushing the Emergency Off switch, or cranking the stepper motor manually, the immediate priority is to remove the source from the patient. Table 2.2 gives the exposure rates at various distances from an unshielded 370 GBq ^{192}Ir source. The table shows that the dose to the patient, with the source in contact, can cause injury in a very short time. On the other hand, the operator, working at a greater distance, is unlikely to receive a dose exceeding regulatory limits for a year, let alone one that would cause health problems. Once the source is removed from the patient and moved to a distance of even a meter, the exposure rate is quite low, and whatever actions that need to be taken to remove the patient from the room can be performed safely. Most institutions set the effective annual limit to the body at 10 times less than the U.S. NRC limit of 0.05 Sv in keeping with the principle to keep exposures as low as reasonably achievable. Ideally, such a dose should not be received in one, short exposure. The allowed exposure to the hands is 10 times that to the body. The preferred approach to a source that cannot be retracted by any method is to remove the applicator from the patient as quickly as possible and to place the applicator containing the source in a shielded container (Figure 2.11).

If it is clear that the cable is caught in the transfer tube and not in the applicator itself, the applicator or catheter may be disconnected from the transfer tube and the source pulled from the applicator. In some cases, this will be faster than removing the applicator. The reason for avoiding disconnecting the applicator from the transfer tube is that a source may stay in the applicator if the source capsule shatters. In that case, removing the

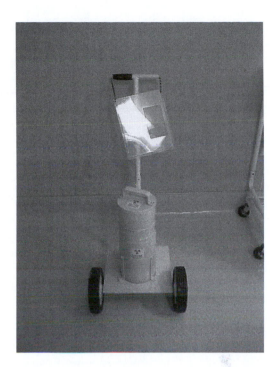

FIGURE 2.11 Shielded container for emergency. (Courtesy of Rupak Das.)

applicator attached to the transfer tube keeps the system closed, while disconnecting the two opens a path for parts of a broken source to fall from the applicator into body cavities or crevices, or roll onto the floor. A situation may arise when the source needs to be detached manually from the treatment unit. One (still unlikely) scenario would be if the source were stuck out of the treatment unit, the sources or the closed applicator had been removed from the patient, a person was pinned very close to the source so neither that person nor the treatment unit could be moved, and the source on the cable could not reach the shielded container. In this special situation, the source cable should be cut from the unit and the source placed in the shielded container that is always present in the room. In cutting the source cable, it must be clear that the cut is not through the source capsule. For units with the capsule welded on the cable, the cut must be through the braided cable as opposed to the smooth steel capsule. For sources imbedded in the cable, a sufficient length of the cable must be seen to assure the cut occurs behind the source (typical source size is indicated in Figure 2.1). Thus, emergency tools that must be present in the treatment room and always readily accessible include a wire cutter, a pair of forceps, and a shielded service container.

2.6.7 Backup Battery

In case of a power failure during the treatment, the machine is equipped with a backup battery to provide retraction of the source to its safe. The batteries should be tested with each source exchange.

TABLE 2.2 Exposure Rates from Unshielded 370-GBq ^{192}Ir Source

Typical Situation	Distance (m)	Dose Equivalent Rates (Sv/h)	Time to Receive	
			10 Sv (Likely Injury)	0.05 Sv (Annual Body Limit)
In patient	0.01	444 Sv/h	1.35 min	0.007 min = 0.4 s
Handling with Kelly Clamps	0.1	4.44	2.3 h	0.68 min (6.8 min for hand limit)
Handling with Kelly Clamps	0.3	0.5	20 h	0.10 h (6 min)
Standing near	1	0.044	9.5 days	1.1 h
Standing far	2	0.011	37.5 days	4.5 h

2.7 Applicators and Appliances

In Section 2.2.1 "Current Regulatory Environment for Brachytherapy Equipment," the role of the authorities is described in overseeing the development and clinical use of medical products. It is noted that the same regulatory system is valid for the use of applicators and appliances to brachytherapy units. These parts need approval and marking as well. This section describes a number of applicator types used for brachytherapy and some specific instructions on how to deal with these components.

Generally speaking, there is a strong interaction between manufacturers and customers for the design of applicators with specific design features. The comments from experts in the field lead to further improvement and enhanced functionality. There is a rather great similarity in design of applicators from different vendors, although material use and size may be different. Mixing applicators from one vendor to afterloaders from other vendors is virtually impossible as connectors are specific for each system. Only if the technical details of the afterloading systems are the same, for example, for systems that can have both HDR and PDR modalities, the use of just one type of applicator sets may be possible, always according to the instructions from the manufacturer.

2.7.1 Gynecology

Endocervical treatments are performed either with a tandem-ovoid combination [based on the original design from the Manchester experience: Fletcher-Suit or Fletcher-Suit-Delclos (FSD) types] or with a tandem-ring combination (the Stockholm experience). The ovoid dimension is typically 20 mm as a standard diameter, while often, a 16 mm diameter is sold as a "mini" ovoid. Caps over the ovoids are sometimes available to enlarge the diameter to either 25 or 30 mm. Manufacturers usually offer a choice, for example, alternatives in the form of full solid plastic ovoids versus the shielded versions. In the latter case, a reduction of dose toward bladder and rectum can be achieved due to the absorption of the direct radiation in the tungsten shield segments. Intrauterine tandems are delivered in lengths up to 80 mm, with angulations of 15°, 30°, and 45°. No shielding is available in tandem-ring applicators. Variation in dose delivery to bladder and rectum is then found in the choice of the user to adapt the loading pattern by shifting the dwell positions in the ring from dorsal to ventral or vice versa.

With the upcoming modern 3D image-based brachytherapy, all vendors offer a series of CT- and MR-compatible applicators with plastic or titanium source tubes in order to avoid distortion in the images that might lead to problems with the reconstruction. The latest modification to the design of both ovoid- and ring-type applicators is the drilling of a number of holes as an option to place extra needles in the parametria (Figure 2.12). In that way, lateral tumor extension can be covered with sufficient dose by choosing extra dwell positions in these needles (see Chapter 21 on gynecological brachytherapy). Some gynecological applicators are equipped with specific retractors to allow for sparing of the rectum during an HDR treatment.

Cylindrical applicators, dome-shaped in the tip, are used to load the source in the vagina. Cylinder diameter varies typically from 15 up to 40 mm. Usually, the cylindrical applicator is constructed from a number of segments with a length of 20 or 25 mm, which can be combined to the desired total length. Other cylinder types are constructed as one-piece applicators. Dose specification is done at a certain depth from the surface of the cylinder in the vaginal wall, which means that the user must carefully know the curvature of the first, dome-shaped segment. Variations on this design are made by some vendors by inserting tungsten shields in the cylinder to protect the rectum from overdose. Other designs have more than one central tube inside that can be connected to the afterloader with a location at peripheral positions in the cylinder, thus allowing one to create asymmetric transversal dose distributions. Templates are used to help the fixation of the cylindrical applicator to the perineum to avoid movements of the applicator during the treatment session.

2.7.2 Bronchus

Bronchial applicators are relatively simple in their design. Usually, the applicator is a narrow plastic tube as long as 1 m and with a diameter of 1.7 mm (5F). A coupling device is attached

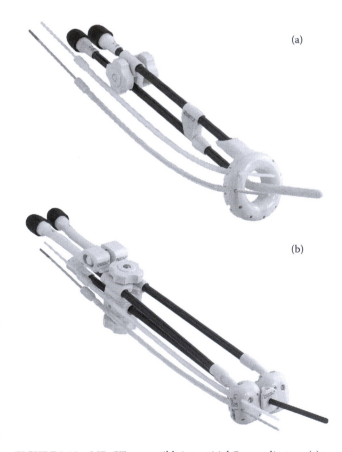

(a)

(b)

FIGURE 2.12 MR-CT compatible interstitial Gyn applicators: (a) tandem ring; (b) tandem ovoids. (Courtesy of Nucletron, Veenendaal, The Netherlands.)

to the tube for connection with the afterloader, directly to the machine's indexer. The tube can be introduced into a broncho-scope through the open channel. Most institutions use closed end tubes. Open-ended tubes must be provided with a sealing plug before the source can be inserted. Bronchus applicators are not reusable. Dose specification is usually done at a given distance from the heart of the catheter.

2.7.3 Esophagus

Esophagus applicators have much larger diameters than bron-chus catheters, with a choice from 5 up to 20 mm. Diameters of 10–15 mm are most common. Many institutions will therefore limit their stored set of applicators, thus aiming for standardiza-tion and cost reduction. The applicators consist of an outer tube and an inner source guiding tube. The outer tube has a tapered end for insertion in the (stenosis of the) esophagus. Specific guide wires and x-ray markers are used to enable the insertion and to determine the target volume. A fixation mask is used to fix the tube to the mouth of the patient. Dose specification is rec-ommended at a given distance from the surface of the applica-tor over a treatment length that is to be defined by the radiation oncologist based on the clinical findings.

2.7.4 Nasopharynx and Oropharynx

Specific types of applicators have been designed to position the source as near as possible to the target for tumors in the naso-pharynx and oropharynx.

2.7.5 Rectum

Some vendors provide cylindrical applicators that can be inserted in the rectum of the patient to treat the inner rectum wall. A more recent version makes use of an inflatable balloon design with a number of catheters placed in the periphery of the applicator. In this way, the catheters can be differentially loaded leading to asymmetric dose distributions (Figure 2.13).

FIGURE 2.13 Intracavitary rectal applicator. (Courtesy of Nucletron, Veenendaal, The Netherlands.)

2.7.6 Breast

The conventional technique of treating breast tumors with a brachytherapy technique relies on the insertion of either rigid needles or flexible tubes. Several appliances are available to help the user to create and maintain a regular implant, for example, with source trajectories that are parallel and equidistant such as recommended in the rules of the Paris dosimetry system. Templates with predrilled holes, template holders, "bridges," and compression mechanisms have been designed for this pur-pose. Implants are then made in either a single plane or using double or triple planes according to the clinical findings of the radiation oncologist. Another type of breast implant device was marketed as the MammoSite Balloon. This type of applicator is left in the breast by the surgeon performing the tumorectomy at the site of the surgical cavity. The intention is to irradiate the wall of the cavity shortly after the operation. The original design of the applicator has a centralized catheter and a dwell position in the middle of the cavity. Dose specification is at a given depth from the balloon surface in the wall of the cav-ity. Modifications have been made to the design by varying the diameter of balloons and by using balloons with several cath-eters inside. In the same way as with other multicatheter appli-cators, the peripheral loading patterns that can be obtained allow relative sparing, in this case to avoid too high doses to the skin of the patient.

A completely different and very recent development is the AccuBoost system. Together with the microSelectron after-loader Nucletron, the AccuBoost system enables clinicians to provide a boost treatment option, as part of the whole breast irradiation procedure. The AccuBoost procedure is a real-time image-guided radiation therapy technique using radiographic (mammographic) equipment to image and pinpoint the tissue that needs to be irradiated. The applicator is constructed as a tungsten cylinder, in which the source travels through a catheter in the upper lumen. Direct and scattered radiation defines a dose pattern with a given relative depth dose, more or less compa-rable with an orthovoltage x-ray beam. Several of these beams can be combined, for example, AP-PA to a compressed breast, or even with a four-field beam technique. The AccuBoost system is capable of recording and archiving the therapeutic dose.

2.7.7 Needles and Tubes

Steel interstitial needles are available in lengths varying from 80 to 200 mm. Blunt, beveled, and trocar points can be chosen for different purposes. Needles are used individually in interstitial implants or in a combination with an applicator, for example, a perineal implant for rectum or anal canal applicators. The length of a needle is a critical feature in stepping source afterloaders where the source travels a fixed distance. Then, the interstitial needle must be used with matching source guide tubes so that the tube plus needle is a fixed length.

Tubes have similar outer dimensions as needles of either 1.7 mm (5F) or 1.9 mm (6F). Again, the length of the tube may be

FIGURE 2.14   Skin applicator, Valencia type. (Courtesy of Nucletron, Veenendaal, The Netherlands.)

a critical parameter in having the source dwell at the intended position. The user must be aware and experienced in the IFU by the vendor.

2.7.8 Skin Applicators

For treatment of small skin lesions using an HDR afterloading machine, specific applicators have been designed. These have names like the "Leipzig" or "Valencia" applicator (Figure 2.14). The HDR source has a dwell position defined in the tip of a tungsten cone. The aperture creates a circular dose distribution with a unique depth dose pattern. Data can be found in the literature. Treatment consists of a number of fractions to be determined by the radiation oncologist. The application is further discussed in Chapter 23.

2.7.9 Templates and other Appliances

Several templates have been designed in order to obtain a proper geometry of the implant, as with the breast implant technique mentioned above, and with the perineum and rectal implants. The diameter of the needles is a critical feature as the applicator is meant to rigidly fixate the needles. Differences of 0.1–0.2 mm in diameter may prevent this fixation.

The vendors offer several solutions for positioning and fixation of the applicators, in the form of clamps, base plates, and other accessories. With multicatheter breast implants, couplings have been designed to be able to interrupt the treatment (or to disconnect the patient between pulses in a PDR scheme), with the set of transfer tubes connected to one part of the connector and the implanted tubes to the other part. Sets of x-ray catheters are available to determine the source trajectories in the catheters for accurate source positioning and treatment planning. These sets are often typical for a certain type of applicator.

2.7.10 Applicator Libraries

Compared to older designs of applicator sets, several of the modern sets are nowadays rigidly attached to avoid movements between the parts. As an example, there was no connection between the intrauterine tandem and the ovoid tubes in the original versions. Then, a connection was created, but with some degrees of freedom in the fixation. Now, some gynecological applicators can be inserted in parts, but are then interconnected in a rigid and standard way. This latter solution allows predefining the applicator geometry in the treatment planning system as a standard that can be taken from an applicator library. Such an approach simplifies the reconstruction process and may help prevent errors.

2.7.11 Cleaning and Sterilization

Cleaning and sterilization of reusable applicators and transfer tubes should be performed only with methods given by the manufacturer of the product. More and more, the sterilization units of a hospital require explicitly written statements by the vendors detailing how their products should be sterilized. A user is therefore responsible in requiring these written instructions at purchase of any new applicator. Applicators specified for single use by the manufacturer may not be reused.

References

Ballester, F., Granero, D., Pérez-Calatayud, J., Venselaar, J.L.M., and Rivard, M.J. 2010. Study of encapsulated 170Tm sources for their potential use in brachytherapy. *Med Phys* 37:1629–37.

Baltas, D., Sakelliou, L., and Zamboglou, N. 2007. *The Physics of Modern Brachytherapy for Oncology*. Taylor & Francis Group, Boca Raton.

Cember, H. 1996. *Introduction to Health Physics*, 3rd ed., McGraw-Hill, New York.

Glasgow, G.P. 2005. Brachytherapy facility design. In: *Brachytherapy Physics*, 2nd ed. Proc. of the Joint AAPM/ABS Summer School. Medical Physics Publishing, Madison, Wisconsin 127–51.

Granero, D., Pérez-Calatayud, J., Ballester, F., Bos, A.J., and Venselaar, J.L.M. 2006. Broad-beam transmission data for new brachytherapy sources Tm-170 and Yb-169. *Radiat Prot Dosim* 118(1):11–5.

Kutcher, G.J., Coia, L., Gillin, M. et al. 1994. Comprehensive QA for radiation oncology: Report of AAPM Radiation Therapy Committee Task Group 40. *Med Phys* 21:581–618.

McGinley, P. 2002. *Shielding Techniques for Radiation Oncology Facilities*, 2nd ed. Medical Physics Publishing, Madison, Wisconsin.

Nath, R., Anderson, L.L., Luxton, G., Weaver, K.A., Williamson, J.F., and Meigooni, A.S. 1995. Dosimetry of interstitial brachytherapy sources: Recommendations of the AAPM Radiation Therapy Committee Task Group No. 43. *Med Phys* 22:220–34.

Nath, R., DeWerd, L.A., Dezarn, W.A. et al. AAPM and ESTRO recommendations on dosimetry requirements for new or innovative brachytherapy sources, devices, and applications. AAPM Brachytherapy Subcommittee Task Group No. 167. Submitted to *Med Phys*.

NCRP. 1972. Protection Against Radiation From Brachytherapy Sources. National Council on Radiation Protection and Measurements. Report No. 40. Bethesda, MD.

NCRP. 1993. Recommendations on Limits for Exposure to Ionizing Radiation. National Council on Radiation Protection and Measurements. Report No. 116 Bethesda, MD.

Thomadsen, B.R. 2000. Achieving Quality in Brachytherapy. Institute of Physics Press, Bristol, U.K.

USNRC 1994 U.S. Nuclear Regulatory Commission. 1994. Title 10, Chapter 1, Code of Federal Regulations—Energy, Part 35, Medical Use of By-product Material. Government Printing Office, Washington, DC.

Venselaar, J.L.M. 2011. Quality Assurance of HDR Brachytherapy. In: *Quality assurance and safety for radiotherapy*, Eds. T. Pawlicki, P.B. Dunscombe, A.J. Mundt, and P. Scalliet. Taylor & Francis Group, Boca Raton: Taylor & Francis/CRC Press. pp. 381–96.

Venselaar, J.L.M., and Pérez-Calatayud, J. 2004. A practical guide to quality control of brachytherapy equipment. ESTRO Booklet No. 8. European Society for Therapeutic Radiology and Oncology, Brussles, Belgium. pp. 13–29.

Radionuclides in Brachytherapy: Current and Potential New Sources

Facundo Ballester
Universitat de València

Ravinder Nath
Yale University School of Medicine

Dimos Baltas
Klinikum Offenbach GmbH

3.1 Introduction

Only a year after the discovery of radioactivity in 1895 by Henri Becquerel, radium was discovered as a new radioactive element by Madame Curie. Soon it was realized that radium had a tremendous therapeutic potential in treating many proliferative diseases including cancer. First successful clinical results were reported in the first decade of the twentieth century for treating gynecological cancers. One of the advantages of radium in treating aggressive cancers was that it emitted a spectrum of radiations that included some high-energy photons. This allowed a high-dose irradiation of the central area of a target volume while still providing a significant dose to distant points several centimeters away such as the pelvic walls in treating cervical cancers. Radium ushered in a new era of medicine, which started to offer life-saving treatments to patients with untreatable diseases. For many decades, radium and encapsulated radon seeds (encapsulated sources filled with radon gas collected from radium salts) offered this new treatment modality called brachytherapy to hundreds of thousands of cancer patients. As late as 1980s, radium was still a treatment of choice for selected cancers as noted in many standard textbooks on radiation therapy.

With the development of techniques for the artificial production of new radionuclides in nuclear reactors and particle accelerators in the 1950s, alternatives to radium and radon began to appear on the clinical scene. ^{226}Ra had the advantage of a very long half-life of 1620 years, but the great disadvantage of producing as decay product of ^{222}Rn, gaseous, soluble, and highly radiotoxic in tissue. The possibility of breakage of the capsule and release of ^{222}Rn while the source was in the treatment created an undesirable condition for use of ^{226}Ra in brachytherapy. Moreover, its γ-radiations contain some very high energy components that require large thicknesses of lead for radiation protection of personnel. Also, the protection of medical staff performing brachytherapy with hot-loading techniques using radium presents serious challenges. These drawbacks and the possibility of artificial radionuclide production in nuclear reactors or particle accelerators (usually cyclotrons) since the 1950s led to the search for substitutes for ^{226}Ra in brachytherapy. Today, radium is not used for brachytherapy because of the challenges described above, and acceptable substitutes with much lower risk profile are commonly used in clinics worldwide.

In the 1960s, ^{137}Cs became an acceptable alternative to radium for gynecological brachytherapy. However, there was considerable debate in the clinical community that it may not be an adequate substitute for radium because of differences in dose distributions. Therefore, efforts were made to make the dose distributions from ^{137}Cs sources mimic those produced by radium.

Usually, γ-radiation (with photon energies greater than 20 keV) is used for therapy in brachytherapy using encapsulated sources. In these sources, soft x-rays, α- and β-particles, and Auger and internal conversion (IC) electrons emitted by the radionuclide are absorbed in capsules made of biologically inert materials such as platinum, titanium, etc. In contrast, the soft radiations offer great potential when used as radiopharmaceuticals for internal uptake by tumors especially using targeted drug-delivery techniques such as monoclonal antibodies and nanoparticles. These radionuclides can be used for targeted therapies with the goal of tracking and killing individual cancer cells and would open the next new era of microscopic brachytherapy.

Our main focus in this chapter is on radionuclides suitable for brachytherapy using photons emitted from encapsulated sources. The essential features that these γ-emitting sources must meet for use in brachytherapy can be summarized as follows:

(1) Charged particles or soft particle radiations emitted from the radionuclide must be either absent or easily absorbable by thin layers of materials like titanium.

(2) No toxic gases should be produced in the radionuclide decay.

(3) The radionuclide must have a high specific activity so that reasonably small sources can be produced with an adequate dose rate.

(4) Other physical properties such as the emitted energies and half-life should fit the purpose of their application.

(5) The radionuclide must be capable of fabrication in a radioactive material that is not highly soluble and toxic in biological tissue in order to minimize the risk presented in case of unintentional leakage of radioactivity in the patient or in handling by medical personnel.

(6) The radionuclide should be capable of fabrication as a radioactive material that maintains physical integrity in the event of damage to the source, for example, it should not become a fine powder or dust and should not sublimate or vaporize.

(7) The radionuclide should allow fabrication of encapsulated sources of different sizes and shapes.

(8) The radionuclide in the fabricated radioactive source should not be susceptible to damage during at least one of the common methods of sterilization used in surgical suites.

Today, many radionuclides with the above characteristics are produced for clinical distribution. These are generated (1) in a nuclear reactor as a fission product within a spent uranium fuel rod, or as an activation product of a neutron–capture reaction (n, γ) through β^- of a stable nucleus [an alternative to neutron capture reaction can also take place in some cases through the proton knockout (n, p) reaction]; and (2) in a cyclotron using particle beams for activation of stable nuclides.

In this chapter, we describe the characteristics of the radionuclides currently used in brachytherapy as well as some of the other ones that could be of interest in the near future (Lymperopoulou et al. 2006; Van Damme et al. 2008; Ballester et al. 2010).

3.2 Brachytherapy Radionuclide Properties

Among the many physical and chemical properties of the radionuclides used in brachytherapy, we have focused our attention on the following physical properties:

(1) Decay scheme. It describes the type of emitted radiation from the decay of the radionuclide (α, β, Auger, and IC

electrons; x-ray and γ-rays), energy E_i, and intensity Y_i of the radiations emitted.

(2) Half-life, $T_{1/2}$. It is the time in which half of the activity decays. It determines whether the source could be used in permanent or temporary implants.

(3) Specific activity, A_{spe}. It is the amount of activity per unit mass of the radionuclide. It determines how small a source could be manufactured and the maximum dose rate it could deliver with that mass. It is calculated as

$$A_{spe} = \frac{\ln 2}{T_{1/2}} \frac{N_A}{M} \quad (3.1)$$

where M is the molar mass, and N_A is the Avogadro number.

(4) Weighted mean energy, E_{ave}. It determines the amount of energy deposited per decay and the depth of penetration in tissue. In this chapter, it is provided for the electron and photon spectra emitted for each radionuclide using the equation:

$$E_{ave} = \frac{\sum_i Y_i E_i}{\sum_i Y_i}. \quad (3.2)$$

(5) Air kerma-rate constant is the air kerma rate at a distance of 1 cm from a point source containing the bare radionuclide in air. It provides a measure of the radiation field in the vicinity of a point source of the radionuclide. It is calculated using the equation

$$\Gamma_\delta = \frac{1}{4\pi} \sum_i \left(\frac{\mu_k}{\rho} \right)_i Y_i E_i \quad (3.3)$$

where $\left(\dfrac{\mu_k}{\rho} \right)_i$ is the mass energy-transfer coefficient in air for photons of energy E_i. The value of δ, the cutoff energy for inclusion in the calculations, is taken as 10 keV in this chapter. The air kerma-rate constant is the SI equivalent of the quantity referred to as the specific gamma constant.

The physical properties of the radionuclides described in this chapter have been taken from ICRP 107 (Eckerman and Endo 2008). Updated data can be found from the National Nuclear Data Center (NNDC) (NUDAT2.6 2010). Baltas et al. (2007) also provided a detailed description of these radionuclides. The half-value layers (HVLs) of lead for photons emitted from bare point sources have been taken from the study of Papagiannis et al. (2008).

3.3 High-Energy Photon Sources

Physical properties of the high-energy photon emitting brachytherapy radionuclides that are most commonly used are shown in Table 3.1. ^{226}Ra is not included in the table as it is considered obsolete in present-day use, but for completeness, it is discussed in the following paragraphs.

3.3.1 Radium-226

3.3.1.1 Physical Characteristics

^{226}Ra (half-life is about 1620 years) is the sixth member of the ^{238}U decay series that ends with stable ^{206}Pb, and it is an example of a long-lived parent followed by radionuclides of short half-lives reaching secular equilibrium within a few weeks. As a result of the decay processes from ^{226}Ra to ^{206}Pb, starting with an activity A of ^{226}Ra in a sealed source, when secular equilibrium is reached, the same activity A of ^{222}Rn, ^{218}Po, ^{214}Pb, ^{214}Bi, ^{214}Po, and ^{210}Ti is obtained. As a consequence, a large number of photons, monoenergetic electrons, β-particles, α-particles, and recoil nucleus are emitted. Of all these radiations, γ-ray of 1.44 MeV (^{214}Pb) and 2.42 MeV (^{214}Po) are those useful for clinical treatments with sealed ^{226}Ra sources. The seventh radionuclide of this series is ^{210}Po with a half-life of 22 years that does not reach secular equilibrium before hundred years after encapsulation.

3.3.1.2 Production and Construction Methods

In the early ^{226}Ra sources, the radioactive material was supplied mostly in the chemical form of radium sulfate or radium chloride (powder) mixed with an inert filler and loaded into cells about 1 cm long and 1 mm in diameter. A typical radium source might contain 1 to 3 cells, depending on the source length. The specific activity was about 36.6 TBq kg^{-1}. Since ^{222}Rn is a heavy, inert, and toxic gas, which results from the decay of ^{226}Ra, these sources were manufactured and sealed perfect to keep the radioactive material inside. The encapsulation material was platinum with a minimum thickness of 0.5 mm. The high-energy β- and

α-particles emitted are absorbed by the encapsulation material. These sources were available as needles or tubes in a variety of lengths and activities.

3.3.1.3 Radiation Protection

The average energy of the γ-rays from ^{226}Ra in equilibrium with its daughter products and filtered by 0.5 mm of platinum is about 0.83 MeV. The high energy of these photons makes it difficult to shield health professionals and others from unwanted radiation exposure. For these reasons, several other radionuclides (described below), which emit lower-energy photons, have been introduced in the past replacing ^{226}Ra from clinical practice. The HVL of the γ-rays from an encapsulated source containing this radionuclide is about 14 mm of lead.

3.3.1.4 Clinical Applications

^{226}Ra and ^{222}Rn are virtually unused today primarily because of the hazards of chemical and radioactive toxicity of radium and its byproducts and because of the radiation protection issues arising from the high energy of photons emitted by these sources. The long half-life is nowadays considered as a disadvantage from the perspective of radioactive waste disposal.

3.3.2 Gold-198

3.3.2.1 Physical Characteristics

^{198}Au has a half-life of 2.695 days that disintegrates following three β⁻ transitions that combine in a spectrum with mean and maximum energies of 312.2 and 1372.2 keV, respectively. The following radiations are emitted in addition to β-electrons: 3 γ-rays [mean energy = 415.1 keV; yield = 0.97/(Bq s)]; 62 x-rays [3.7 keV, 0.58/(Bq s)]; 18 IC electrons [353.0 keV, 4.3×10^{-2}/(Bq s)]; and 15 Auger electrons [804 eV, 0.50/(Bq s)]. The weighted mean energy spectra of electrons and photons are 212.4 and 259.9 keV, respectively. Removing electrons and photons with energies below 5 keV, the weighted mean energy spectra become 307.4 and 400.6 keV (see Figure 3.1).

TABLE 3.1 Physical Properties of High-Energy Photon Emitting Brachytherapy Radionuclides

	^{198}Au	^{192}Ir	^{137}Cs	^{60}Co
Half-Life	2.695 days	73.83 days	30.07 year	5.27 year
Type of Disintegration	β⁻	β⁻ (95.1%) EC (4.9%)	β⁻	β⁻
Mean γ Energy (keV)	415.1	372.2	661.7	1252.0
Mean X-ray Energy (keV)	3.7	3.6	1.8	0.51
Mean β⁻ray Energy (keV)	312.2	180.7	188.4	96.5
Mean IC Electron Energy (keV)	353.0	266.9	630.3	1233.0
Mean Auger Electron Energy (eV)	804	844	538	1150
Air Kerma-Rate Constant, $\Gamma_{\delta=10\,\text{keV}}$ ($\times 10^{-18}$ Gy m^2 (Bq s)$^{-1}$)	15	32	6.1×10^{-5}	85
Nominal Specific Activity, A_{spe} ($\times 10^5$ TBq kg^{-1})	90	3.4	3.2×10^{-2}	0.41

Note: Air kerma-rate constants have been calculated for δ > 10 keV.

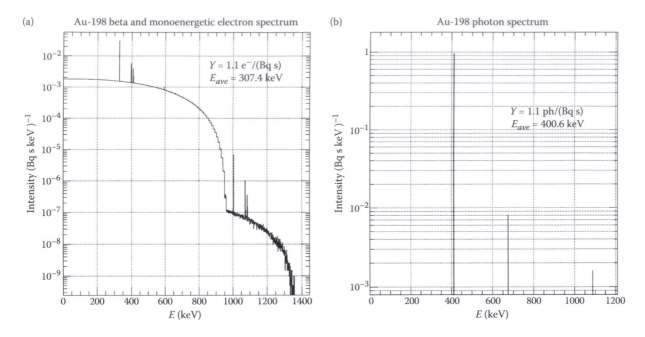

FIGURE 3.1   Radiation emissions of ^{198}Au. (a) Electron spectrum. (b) Photon spectrum. Energy mean values and intensities have been calculated for $\delta > 5$ keV.

3.3.2.2 Production and Construction Methods

^{198}Au is produced in a reactor by bombarding a ^{197}Au target with neutrons ($\sigma = 98.8$ b). A typical gold seed, also known as a gold "grain," is encapsulated in 0.1 mm of platinum, which is sufficient to absorb the electrons emitted. The outside dimensions of a ^{198}Au source are 2.5 mm long and 0.8 mm in diameter. Although the specific activity of ^{198}Au is 9.0×10^6 TBq kg^{-1}, these sources in clinical applications are generally of low strength.

3.3.2.3 Radiation Protection

The HVL of ^{198}Au is 3.3 mm of lead.

3.3.2.4 Clinical Applications

Because of their short half-life, ^{198}Au seeds are used in permanent implants only. ^{198}Au "grains" have been extensively used in the past for treatment of various tumors including gynecological, breast, prostate, head and neck, and other soft tissue cancers. The use of this radionuclide has decreased in recent years because of the availability of competing radionuclides ^{125}I, ^{103}Pd, and ^{131}Cs, which have longer half-lives (making shipment and scheduling of treatments more convenient) and lower photon energies (improving radiation safety compared to ^{198}Au).

3.3.3 Iridium-192

3.3.3.1 Physical Characteristics

^{192}Ir has a half-life of 73.83 days. It decays by β^- emission (95.1%, six β^- rays with a maximal energy of 675.1 keV and mean energy of 180.7 keV) to excited levels of ^{192}Pt and by EC (4.9%) to excited levels of ^{192}Os. ^{192}Ir emits, on average, 29 γ-rays [mean

energy = 372.2 keV, yield = 2.2/(Bq s)]; 122 x-rays [3.6 keV, 2.7/(Bq s)]; 174 IC electrons [266.9 keV, 0.16/(Bq s)]; and 29 Auger electrons [844 eV, 2.3/(Bq s)]. The weighted mean energy spectra of electrons and photons are 64.5 and 168.3 keV, respectively. Removing electrons and photons with energies below 5 keV, the weighted mean energy spectra become 176.2 and 346.1 keV (see Figure 3.2).

3.3.3.2 Production and Construction Methods

^{192}Ir is produced from enriched ^{191}Ir targets (37% natural abundance) in a nuclear reactor via ^{191}Ir$(n,\gamma)^{192}$Ir reaction. Its high-specific activity (3.4×10^5 TBq kg^{-1}) makes it practical to make small radioactive sources with activities exceeding 4.4 TBq, which can be used in high-dose-rate (HDR) brachytherapy afterloaders. HDR ^{192}Ir sources are encapsulated in a thin titanium or stainless-steel capsule (0.5–1 mm diameter × 0.5–3.5 mm length cylinders), and these sources are laser-welded to the end of a flexible cable, which can be controlled remotely by microprocessors.

^{192}Ir sources are also available in the form of small seeds (3 mm long with a diameter of 0.5 mm) placed in nylon ribbons at 1-cm intervals for low-dose-rate (LDR) brachytherapy. Two different seed designs are commercially available in the United States. One of them has an inner core alloy composed of 30% iridium and 70% platinum, encapsulated in stainless steel. The other has an inner core alloy composed of 10% iridium and 90% platinum surrounded by a 0.1-mm-thick cladding of platinum.

^{192}Ir sources are also available as wires composed of an alloy of 25% iridium and 75% platinum enveloped in pure platinum. The wires are available in two sizes with outer diameters of 0.3 and 0.5 mm. These wires are more popular in Europe.

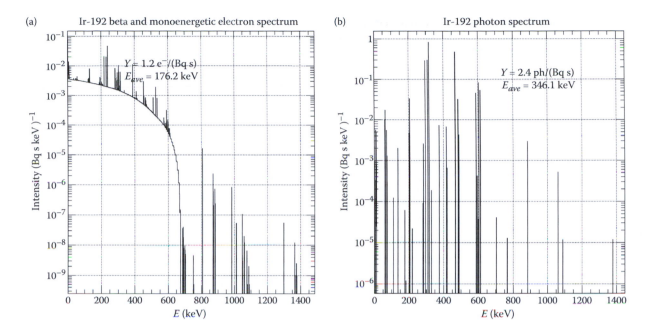

FIGURE 3.2 Radiation emissions of ^{192}Ir. (a) Electron spectrum. (b) Photon spectrum. Energy mean values and intensities have been calculated for $\delta > 5$ keV.

3.3.3.3 Radiation Protection

The HVL thickness for this radionuclide is 2.5 mm of lead.

3.3.3.4 Clinical Applications

^{192}Ir wires and sources in ribbons are particularly suitable for afterloading techniques. These sources are used for temporary brachytherapy implants only because their high average photon energy makes them unsuitable for permanent implants due to the possibility of excessive radiation exposure of family members and public from the release of implanted patients from the hospital. The application is generally in the LDR modality. In the form of a miniaturized HDR source, the high specific activity of the radionuclide has made it the most successful source applied in afterloading systems for (fractionated) HDR treatment schedules. ^{192}Ir is the only γ-ray isotope that has been clinically used for intravascular brachytherapy (IVB) thus far.

3.3.4 Cobalt-60

3.3.4.1 Physical Characteristics

The half-life of ^{60}Co is 5.27 years. ^{60}Co undergoes β^- decay (99.9%, mean energy 95.6 keV, maximum energy 317.3 keV) to the excited states of ^{60}Ni. De-excitation to the ground state occurs mainly via emission of γ-rays of 1173.2 and 1332.5 keV each with an absolute intensity of nearly 100%. The following radiations are emitted following β^- decay: 6 γ-rays [mean energy = 1253.0 keV, yield = 2.0/(Bq s)]; 25 x-rays [509 eV, 1.6 × 10^{-3}/(Bq s)]; 24 IC electrons [1233.0 keV, 2.9 × 10^{-4}/(Bq s)]; and 7 Auger electrons [1150 eV, 1.2 × 10^{-3}/(Bq s)]. The weighted mean energy spectra of electrons and photons are 96.8 and 1252.8 keV, respectively (see Figure 3.3).

3.3.4.2 Production and Construction Methods

^{60}Co is produced through neutron capture by ^{59}Co$(n,\gamma)^{60}$Co in a nuclear reactor, but its long half-life requires long irradiation times (the duration depends on the neutron flux available, but it is of the order of 5 years) for sufficient source strength. The maximum specific activity reached is about 4.1 × 10^4 TBq kg^{-1}. HDR source having dimensions similar to those of ^{192}Ir can be fabricated. Such sources have a cylindrical core of 3.5 mm in length and 0.5 mm in diameter; the capsule is 5 mm in length and 1 mm in diameter. The sources are laser-welded to the end of a flexible steel wire.

3.3.4.3 Radiation Protection

Due to the high energy of the γ-ray emitted, the use of ^{60}Co sources in clinical applications requires the use of remotely controlled afterloaders in heavily shielded rooms such as linear accelerator bunkers. The HVL is 12 mm of lead.

3.3.4.4 Clinical Applications

The long half-life of ^{60}Co and its high specific activity make it practical for HDR brachytherapy implants and offer the advantage of less frequent replacements of HDR sources compared with those of ^{192}Ir.

3.3.5 Cesium-137

3.3.5.1 Physical Characteristics

^{137}Cs has a half-life of 30.17 years and decays through β^- emissions (93.5%, mean energy 174.3 keV, maximum energy of 514.0 keV) to the metastable state of ^{137m}Ba (half-life 2.55 min) or to the ^{137}Ba ground state (5.6%, mean energy 416.3 keV, maximum energy

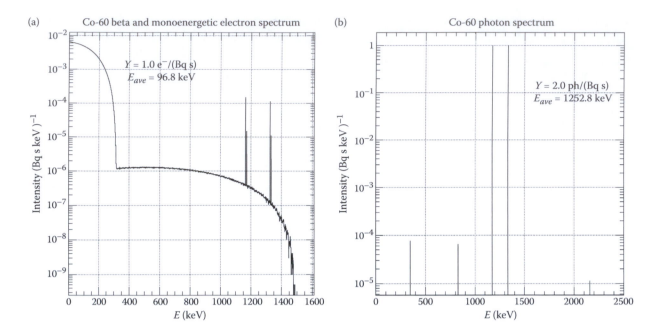

FIGURE 3.3   Radiation emissions of ^{60}Co. (a) Electron spectrum. (b) Photon spectrum. Energy mean values and intensities have been calculated for δ > 5 keV.

1175.6 keV). The metastable state de-excites emitting a γ-ray of 661.7 keV in competition with IC electrons. In addition to β$^-$ emissions, the following radiations are emitted: 1 γ-ray [mean energy = 661.7 keV, yield = 0.90/(Bq s)]; 50 x-rays [1.8 keV, 1.5/(Bq s)]; 6 IC electrons [630.3 keV, 0.10/(Bq s)]; and 15 Auger electrons [528 eV, 1.3/(Bq s)]. The weighted mean energy spectra of electrons and photons are 109.5 and 253.5 keV, respectively. Removing electrons and photons with energies below 5 keV, the weighted mean energy spectra become 225.8 and 612.1 keV (see Figure 3.4).

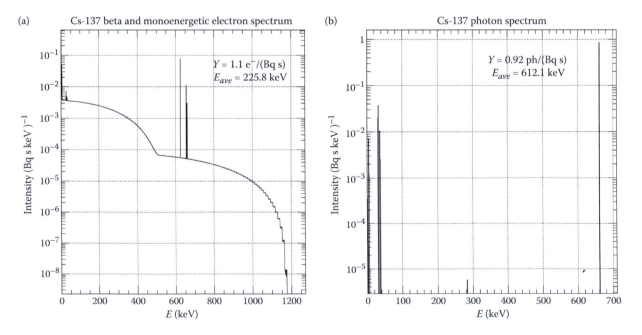

FIGURE 3.4   Radiation emissions of ^{137}Cs. (a) Electron spectrum. (b) Photon spectrum. Energy mean values and intensities have been calculated for δ > 5 keV.

3.3.5.2 Production and Construction Methods

^{137}Cs is a byproduct of ^{235}U nuclear fission in nuclear reactors. For brachytherapy sources, the ^{137}Cs is trapped in an inert matrix material such as gold, ceramic, or borosilicate glass and is supplied in the form of insoluble powders or ceramic microspheres.

Needles, tubes, and spheres of ^{137}Cs are available for interstitial and intracavitary brachytherapy. The encapsulation can be stainless steel or platinum being less than a millimeter thick. The β particles and low-energy characteristic x-rays are absorbed by the encapsulation, making the clinical source a pure γ-emitter. The needles can be up to 60 mm long. The tubes are usually 2 cm in length, although there are smaller ones for use with dome applicators in the irradiation of the vaginal vault. The spheres are used as source trains for use in remote-afterloading intracavitary brachytherapy catheters.

3.3.5.3 Radiation Protection

The HVL is 6.2 mm of lead.

3.3.5.4 Clinical Applications

The ^{137}Cs half-life of 30.17 years enables use over a long period of time. Its low-specific activity makes it practical for LDR implants. ^{137}Cs sources have been used in both interstitial and intracavitary brachytherapy.

3.4 Low-Energy Photon Sources

The physical properties of three low-energy photon emitting brachytherapy radionuclides that are widely used are shown in Table 3.2.

3.4.1 Iodine-125

3.4.1.1 Physical Characteristics

^{125}I has a half-life of 59.4 days decaying exclusively by EC process to an excited state of ^{125}Te followed by spontaneous decay to the ground state with the emission of 1 γ-ray [mean energy = 35.5 keV, yield = 6.7×10^{-2}/(Bq s)]; 43 x-rays [26.0 keV, 1.6/(Bq s)];

TABLE 3.2 Physical Properties of Low-Energy Photon Emitting Brachytherapy Radionuclides

	^{125}I	^{103}Pd	^{131}Cs
Half-Life (days)	59.4	17.0	9.7
Type of Disintegration	EC	EC	EC
Mean γ Energy (keV)	35.5	137.1	–
Mean X-ray Energy (keV)	1.5	1.0	1.8
Mean IC Electron Energy (keV)	7.7	35.3	–
Mean Auger Electron Energy (eV)	520	641	585
Air Kerma-Rate Constant, $\Gamma_{\delta=10\,keV}$ ($\times 10^{-18}$ Gy m^2 (Bq s)$^{-1}$)	9.9	9.0	4.3
Nominal Specific Activity, A_{spe} ($\times 10^5$ TBq kg^{-1})	6.5	27	38

Note: Air kerma-rate constants have been calculated for δ > 10 keV.

6 IC electrons [7.7 eV, 0.95/(Bq s)]; and 13 Auger electrons [520 eV, 23/(Bq s)]. The weighted mean energy spectra of electrons and photons are 0.8 and 1.1 keV, respectively. Removing electrons and photons with energies below 5 keV, the weighted mean energy spectra become 27.1 and 28.4 keV (see Figure 3.5).

3.4.1.2 Production and Construction Methods

^{125}I is obtained as a decay product of ^{125}Xe, which is produced in a nuclear reactor via the reaction ^{124}Xe$(n,\gamma)^{125}$Xe. Since the introduction in 1960s of two seed models of the ^{125}I seeds, a large number of other seed models have become commercially available. In one of the original models, ^{125}I was adsorbed onto a silver rod that is the central core of the source. The rod was encapsulated in 0.5 mm of titanium. In the other original model, the seeds consisted of ion exchange resin spheres containing ^{125}I inside a titanium tube including gold or tungsten markers. Titanium encapsulation served to absorb low-energy electrons and x-rays. The newer models are variations on the original designs using different substrates and radiographic markers. Seeds are available in strengths up to 50 U (see Chapter 10 for use of "source strength" in the TG-43 formalism). Basically, all seed models are cylinders 5 mm long and 0.8 mm in diameter.

3.4.1.3 Radiation Protection

The HVL thickness for the photons emitted by encapsulated sources containing this radionuclide is 0.025 mm of lead. Thin layers of lead are sufficient for radiation protection.

3.4.1.4 Clinical Applications

Except for their use in ^{125}I sources in eye plaques and brain tumors for temporary brachytherapy, ^{125}I seeds are used principally for permanent implants.

3.4.2 Palladium-103

3.4.2.1 Physical Characteristics

With a half-life of 17 days, ^{103}Pd decays via EC process to the excited state of ^{103m}Rh (branching ratio 0.9988). In this process, ^{103}Pd emits 8 γ-rays [mean energy = 359.6 keV, yield = 3.0×10^{-4}/(Bq s)]; 40 x-rays [1.6 keV, 9.2/(Bq s)]; 48 IC electrons [108.6 keV, 1.9×10^{-5}/(Bq s)]; and 11 Auger electrons [782 eV, 7.4/(Bq s)]. In the de-excitation to the ground state, ^{103m}Rh emits 1 γ-ray [mean energy = 39.8 keV, yield = 6.8×10^{-4}/(Bq s)]; 40 x-rays [243 eV, 7.0/(Bq s)]; 6 IC electrons [35.3 keV, 0.99/(Bq s)]; and 11 Auger electrons [462 eV, 5.9/(Bq s)]. All these radiations combine in weighted mean energy spectra of electrons and photons of 3.0 and 1.0 keV, respectively. Removing electrons and photons with energies below 5 keV, the weighted mean energy spectra become 32.6 and 20.7 keV (see Figure 3.6).

3.4.2.2 Production and Construction Methods

^{103}Pd can be produced in a nuclear reactor when stable ^{102}Pd captures a thermal neutron. It can also be produced in a

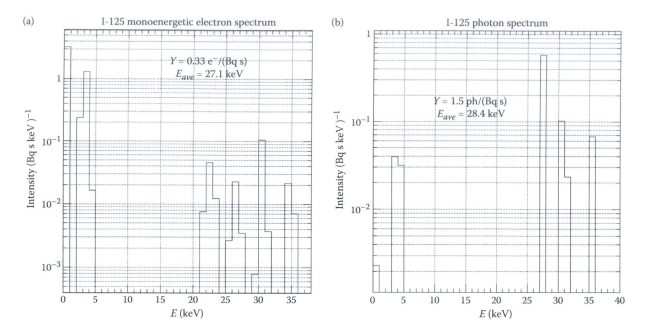

FIGURE 3.5 Radiation emissions of ^{125}I. (a) Electron spectrum. (b) Photon spectrum. Energy mean values and intensities have been calculated for δ > 5 keV.

cyclotron by bombarding protons into a rhodium target. Original ^{103}Pd sources consisted of a cylindrical titanium tube sealed at both ends with laser-welded titanium end cups. Enclosed in the tube were two ^{103}Pd graphite cylinders and a lead rod x-ray marker for radiographic identification. The cylindrical tube was about 5 mm long and 0.8 mm in diameter. In the last decade, numerous other models of ^{103}Pd sources

that are variations of the original design have become commercially available.

3.4.2.3 Radiation Protection

The HVL of photons emitted from encapsulated sources of ^{103}Pd is 0.0085 mm of lead. Thin layers of lead are sufficient for radiation protection.

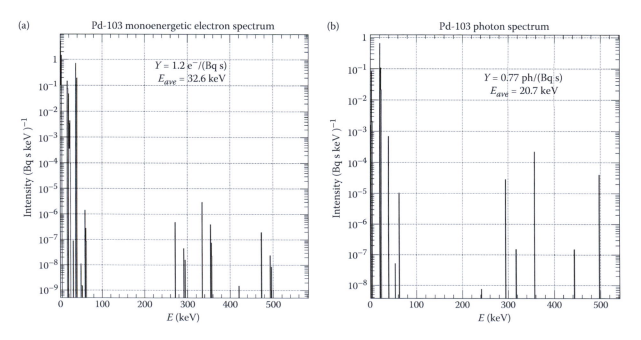

FIGURE 3.6 Radiation emissions of ^{103}Pd. (a) Electron spectrum. (b) Photon spectrum. Energy mean values and intensities have been calculated for δ > 5 keV.

3.4.2.4 Clinical Applications

[103]Pd seeds are primarily used for permanent implants. However, high-activity [103]Pd sources have been fabricated, which makes it possible to use [103]Pd for temporary brachytherapy applications.

3.4.3 Cesium-131

3.4.3.1 Physical Characteristics

With a half-life of 9.7 days, [131]Cs decays via EC process with the emission of 49 x-rays [mean energy = 1.8 keV, yield = 12.7/(Bq s)] and 13 Auger electrons [585 keV, 10.9/(Bq s)]. The weighted mean energy spectra of electrons and photons are 0.6 and 1.8 keV, respectively. Removing electrons and photons with energies below 5 keV, the weighted mean energy spectra become 26.3 and 30.3 keV (see Figure 3.7).

3.4.3.2 Production and Construction Methods

[131]Cs is produced by neutron capture in nuclear reactors via the indirect [130]Ba(n,γ)[131]Ba–[131]Cs process, which requires high flux reactors with a capability of changing targets, which are available only in a few centers in the world.

3.4.3.3 Radiation Protection

The HVL of photons emitted from encapsulated sources of [131]Cs is 0.022 mm of lead. Thin layers of lead are sufficient for radiation protection.

3.4.3.4 Clinical Applications

[131]Cs seeds are primarily used for permanent implants. Theoretically, [131]Cs may be more effective than ^{125}I against faster-growing tumors such as those found in the brain, lung, and other malignancies because of shorter half-life, higher average x-ray energy, and higher effective biological dose (Murphy et al. 2004; Rivard 2007). Their use in ocular brachytherapy has also been investigated (Melhus and Rivard 2008; Rivard et al. 2008).

3.5 Beta Sources

3.5.1 Phosphor-32

3.5.1.1 Physical Characteristics

[32]P is a pure β⁻ emitter with 14.3-day half-life and with average and maximum energy of 694.8 and 1710 keV, respectively (see Figure 3.8).

3.5.1.2 Production and Construction Methods

This radionuclide is produced in a nuclear reactor via the [32]S(n,p)[32]P reaction.

3.5.1.3 Clinical Applications

[32]P applicators as flexible plastic disc-shaped devices are used for the treatment of skin lesions. Dose rates on the surface of the applicators range from 12 to 25 Gy/h. In the form of a ^{32}P wire source, it has also been used in an IVB system.

3.5.2 Strontium-90, Yttrium-90

3.5.2.1 Physical Characteristics

[90]Sr is a pure β⁻ emitter with a half-life of 28.9 years emitting electrons of average and maximum energies of 196 and 546 keV,

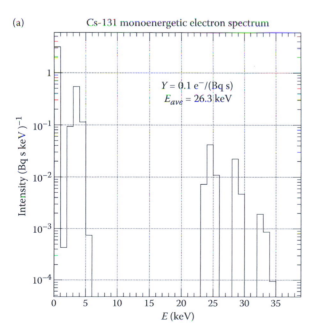
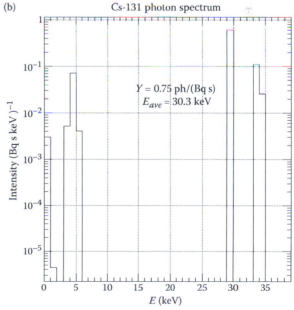

FIGURE 3.7 Radiation emissions of [131]Cs. (a) Electron spectrum. (b) Photon spectrum. Energy mean values and intensities have been calculated for δ > 5 keV.

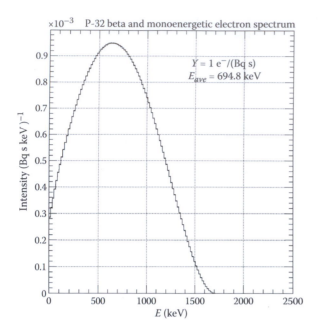

FIGURE 3.8 β^- spectrum of ^{32}P.

respectively. Its daughter ^{90}Y decays with a 64.1-h half-life emitting only beta rays of average and maximum energies of 932.9 and 2280.0 keV, respectively. A tiny fraction of ^{90}Y decays results in one of two excited levels of the daughter isotope ^{90}Zr, which relaxes by emission of low-intensity gamma rays. Both radionuclides are in secular equilibrium resulting in a combined electron spectrum with an average energy of 564.0 keV. It is primarily the ^{90}Y betas that are useful for therapy (see Figure 3.9).

3.5.2.2 Production and Construction Methods

Highly purified ^{90}Sr is obtained from stored nuclear fission waste materials containing also ^{137}Cs.

3.5.2.3 Clinical Applications

This radionuclide is suitable for treatment of lesions in the eye where the depth of penetration needed is a few millimeters. The sources are made so that the lowest energy β radiation is slowed, using only the high-energy betas for therapy. Concave and planar applicators have been built for ophthalmic applications and nasopharyngeal treatments. The second most common application for ^{90}Sr is IVB for prevention of restenosis in coronary or peripheral arteries.

3.5.3 Ruthenium-106/Rhenium-106

3.5.3.1 Physical Characteristics

^{106}Ru disintegrates to the stable nuclide ^{106}Pd via ^{106}Rh. The half-life of ^{106}Ru is 373.6 days, and it is a pure β-particle emitter with a maximum energy of 39.4 keV and a mean energy of 10.0 keV. In the later disintegration of ^{106}Rh (half-life is 29.8 s), the mean energy of the continuous β-particle spectra is 1410.0 keV with a maximum energy of 3500.0 keV. Other emissions are 86 γ-rays [mean energy = 602.2 keV, yield = 0.34/(Bq s)]; 40 x-rays [1.8 keV, 1.4×10^{-2}/(Bq s)]; 524 IC electrons [548.4 keV, 1.6×10^{-3}/(Bq s)]; and 10 Auger electrons [847 eV, 1.2×10^{-2}/(Bq s)] (see Figure 3.10).

3.5.3.2 Production and Construction Methods

^{106}Ru/^{106}Rh is produced by nuclear fission of ^{235}U and is obtained from spent nuclear fuel.

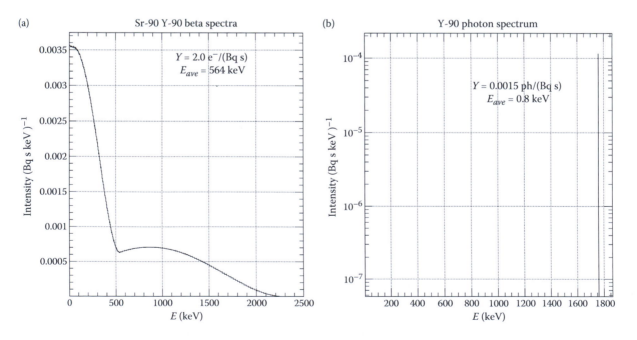

FIGURE 3.9 (a) Combined β^- spectrum of ^{90}Sr–^{90}Y in secular equilibrium. (b) Photon spectrum.

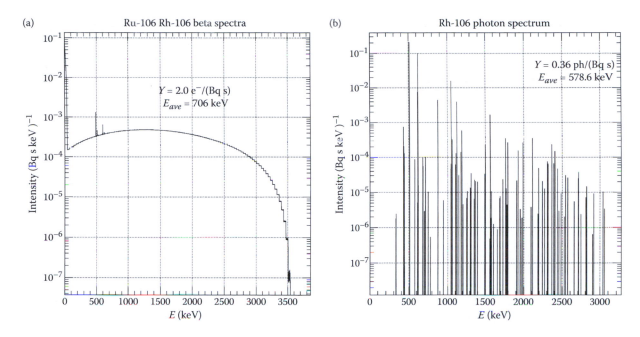

FIGURE 3.10 Radiation emissions of ^{106}Ru–^{106}Rh in secular equilibrium. (a) Electron spectrum. (b) Photon spectrum.

3.5.3.3 Clinical Applications

^{106}Ru has been used in removable ophthalmic plaques as a convenient and effective alternative to enucleation for the treatment of malignant melanoma, uvea, and other tumors of the eye. The choice of the most adequate plaque for each specific treatment depends mainly on the height of the tumor apex. For small-sized tumors, applicators of ^{106}Ru (or ^{90}Sr) may be preferred over ^{125}I plaques.

3.6 Other Radionuclides of Potential Use in Brachytherapy

The choice of radionuclides for use in brachytherapy is a complex decision. The aspects to be weighed are of many types: technical, economical, clinical, and public health issues related to radiation protection. The exploration of the periodic table provides an initial screening of radionuclides, which are, prima facie, suitable for brachytherapy. For permanent implants, we find among those with a lifetime short enough and photon energy low enough that the patients can be released from the hospital safely after the procedure. For temporary implants, we seek radionuclides with half-life long enough that the cost of source replacement is not prohibitive and also the photon energy is high enough for sufficient depth of penetration in specific applications. In all cases, we seek radionuclides with high specific activity so that even smaller radioactive sources can be fabricated for newer applications. Besides the technical aspects, it must be considered that the production costs are reasonably low and the fabrication and clinical use of these radioactive sources do not generate unacceptable or uncontrollable pollutants, or that these sources are easily separable or leak out, etc. The search of new radionuclides for brachytherapy remains an active research field.

In Table 3.3, the basic physical characteristics of two potential radionuclides for use in brachytherapy are presented.

3.6.1 Ytterbium-169

3.6.1.1 Physical Characteristics

^{169}Yb has a half-life of 32.0 days. It decays via EC to several excited states of ^{169}Tm, which results in emission of 74 γ-rays [mean energy = 142.4 keV, yield = 1.5/(Bq s)], 53 x-rays [3.4 keV, 33.1/(Bq s)], 434 IC electrons [38.3 keV, 3.1/(Bq s)], and 15 Auger electrons [1034 eV, 26/(Bq s)]. All these radiations combine in weighted mean energy spectra of electrons and photons of 5.1

TABLE 3.3 Physical Properties of Two Radionuclides of Potential Use in Brachytherapy

	^{169}Yb	^{170}Tm
Half-Life (days)	32.0	128.6
Type of Disintegration	EC	β$^-$, EC
Mean γ Energy (keV)	142.4	84.3
Mean X-ray Energy (keV)	3.4	1.4
Mean IC Electron Energy (keV)	38.3	64.9
Mean Auger Electron Energy (eV)	1034	878
Air Kerma-Rate Constant, $\Gamma_{\delta=10\,keV}$ ($\times 10^{-18}$ Gy m^2 (Bq s)$^{-1}$)	13	15
Nominal Specific Activity, A_{spe} ($\times 10^5$ TBq kg^{-1})	8.8	2.2

Note: Air kerma-rate constants have been calculated for δ > 10 keV.

and 9.5 keV, respectively. Removing electrons and photons with energies below 5 keV, the weighted mean energy spectra become 29.1 and 78.6 keV (see Figure 3.11).

3.6.1.2 Production

^{169}Yb is produced from enriched ^{168}Yb targets (0.13% natural abundance) in a nuclear reactor by the (n, γ) reaction reaching maximum activities of up to 620 GBq/mm^3.

3.6.1.3 Radiation Protection

The HVL of photons emitted from bare point sources of ^{169}Yb is 1.7 mm of lead.

3.6.1.4 Clinical Applications

^{169}Yb has potential for permanent or temporary implants due to its intermediate half-life and its relatively high specific activity (Piermattei et al. 1995). The average energy of ^{169}Yb is energetic enough to allow uniform radiation exposure to a clinical target while its shielding requirements are lower compared with those of ^{192}Ir. Movable shielding may be developed and applied as a substitute to permanent room shielding. In addition, the low energy of ^{169}Yb allows for potentially interesting clinical applications such as the design of shielded applicator to modulate the dose distribution to the specific requirements of a patient, thereby adding to conformal brachytherapy treatment planning. Several studies have been published about the potential use of ^{169}Yb, and even some source designs have been proposed and manufactured (Perera et al. 1994; Das et al. 1995; MacPherson and Battista 1995; Piermattei et al. 1995; Medich

and Munro 2010; Cazeca et al. 2010a,b; Medich 2010; Leonard et al. 2011). ^{169}Yb sources are not commercially available at present.

3.6.2 Thulium-170

3.6.2.1 Physical Characteristics

^{170}Tm is another potential radionuclide for use in brachytherapy (Enger et al. 2011). It decays via β^- (99.9%) and EC with a half half-life of 128.6 days. In these processes, the following radiations are emitted: 2 γ-rays [mean energy = 84.3 keV, yield = 2.5×10^{-2}/(Bq s)], 106 x-rays [1.4 keV, 1.4/(Bq s)], 12 IC electrons [64.9 keV, 0.16/(Bq s)], and 30 Auger electrons [818 eV, 1.2/(Bq s)]. All these radiations combine in weighted mean energy spectra of electrons and photons of 138.5 and 2.8 keV, respectively. Removing electrons and photons with energies below 5 keV, the weighted mean energy spectra become 260.5 and 46.5 keV (see Figure 3.12).

3.6.2.2 Production

^{170}Tm has a high specific activity of 2.2×10^5 TBq kg^{-1}. It is produced in a nuclear reactor using as a target the naturally occurring ^{169}Tm isotope, which has a 100% natural abundance, and therefore does not require expensive isotopic enrichment. With a relatively high thermal neutron capture cross section of 105 barns, production could be efficient and cost effective.

3.6.2.3 Radiation Protection

The low mean photon energy will require less shielding. The HVL of photons emitted from bare point sources of ^{170}Tm is 0.34 mm of

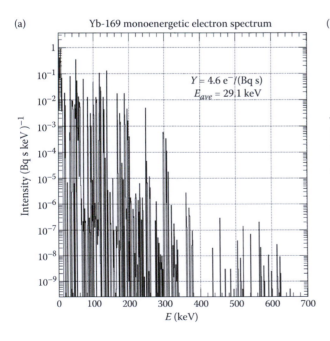

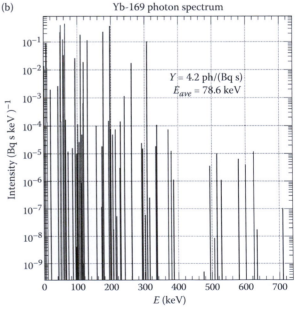

FIGURE 3.11 Radiation emissions of ^{169}Yb. (a) Electron spectrum. (b) Photon spectrum. Energy mean values and intensities have been calculated for $\delta > 5$ keV.

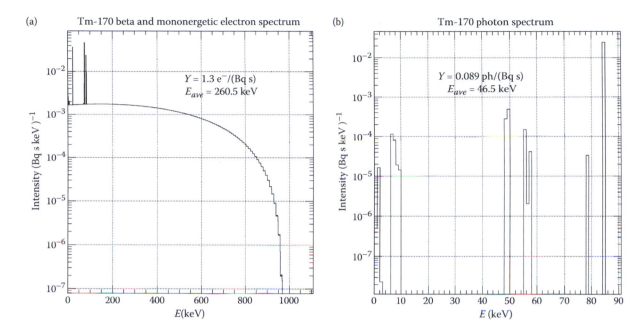

FIGURE 3.12 Radiation emissions of ^{170}Tm. (a) Electron spectrum. (b) Photon spectrum. Energy mean values and intensities have been calculated for $\delta > 5$ keV.

lead. The major drawback of ^{170}Tm for its use in HDR applications is the low yield of photons emitted compared to the electron yield (about 9 photons with energy > 5 keV per 100 electrons emitted).

References

Ballester, F., Granero, D., Perez-Calatayud, J. et al. 2010. Study of encapsulated ^{170}Tm sources for their potential use in brachytherapy. *Med Phys* 37(4):1629–37.

Baltas, D., Sakelliou, S., and Zamboglou, N. 2007. *The Physics of Modern Brachytherapy for Oncology*. Taylor & Francis Group, New York.

Cazeca, M.J., Medich, D.C., and Munro, J.J. 3rd. 2010a. Effects of breast-air and breast-lung interfaces on the dose rate at the planning target volume of a MammoSite catheter for Yb-169 and Ir-192 HDR sources. *Med Phys* 37(8):4038–45.

Cazeca, M.J., Medich, and Munro, J.J. 3rd 2010b. Monte Carlo characterization of a new Yb-169 high dose rate source for brachytherapy application. *Med Phys* 37(3):1129–36.

Das, R.K., Mishra, V., Perera, H. et al. 1995. A secondary air kerma strength standard for Yb-169 interstitial brachytherapy sources. *Phys Med Biol* 40(5):741–56.

Eckerman, K.F. and Endo A. 2008. ICRP 38 user guide to the ICRP CD and the DECDATA software. *Ann ICRP* 38:e1–e25.

Enger, S.A., Michel D'Amours, M., and Beaulieu, L. 2011. Modeling a hypothetical 170Tm source for brachytherapy applications. *Med Phys* 38(10), 5307–10

Leonard, K.L., DiPetrillo, T.A., Munro, J.J. et al. 2011. A novel ytterbium-169 brachytherapy source and delivery system for use in conjunction with minimally invasive wedge resection of early-stage lung cancer. *Brachytherapy* 10(2):163–9.

Lymperopoulou, G., Papagiannis, P., Angelopoulos, A. et al. 2006. A dosimetric comparison of 169Yb and 192Ir for HDR brachytherapy of the breast, accounting for the effect of finite patient dimensions and tissue inhomogeneities. *Med Phys* 33(12):4583–9.

MacPherson, M.S. and Battista, J.J. 1995. Dose distributions and dose rate constants for new ytterbium-169 brachytherapy seeds. *Med Phys* 22(1):89–96.

Medich, D.C. and Munro, J.J. 3rd 2010. Dependence of Yb-169 absorbed dose energy correction factors on self-attenuation in source material and photon buildup in water. *Med Phys* 37(5):2135–44.

Medich, D.C., Tries, M.A., and Munro, J.J. 2006. Monte Carlo characterization of an ytterbium-169 high dose rate brachytherapy source with analysis of statistical uncertainty. *Med Phys* 33(1):163–172.

Melhus, C.S. and Rivard, M.J. 2008. COMS eye plaque brachytherapy dosimetry simulations for 103Pd, 125I, and 131Cs. *Med Phys* 35(7):3364–71.

Murphy, M.K., Piper, R.K., Greenwood, L.R. et al. 2004. Evaluation of the new cesium-131 seed for use in low-energy x-ray brachytherapy. *Med Phys* 31(6):1529–38.

NUDAT 2.6. 2010. National Nuclear Data Center, Brookhaven National Laboratory, www.nndc.bnl.gov/nudat2/.

Papagiannis, P., Baltas, D., Granero, D. et al. 2008. Radiation transmission data for radionuclides and materials relevant to brachytherapy facility shielding. *Med Phys* 35(11):4898–906.

Perera, H., Williamson, J.F., Li, Z. et al. 1994. Dosimetric characteristics, air-kerma strength calibration and verification of Monte Carlo simulation for a new Ytterbium-169 brachytherapy source. *Int J Radiat Oncol Biol Phys* 28(4):953–70.

Piermattei, A., Azario, L., and Montemaggi, P. 1995. Implantation guidelines for [169]Yb seed interstitial treatments. *Phys Med Biol* 40(8):1331–8.

Piermattei, A., Azario, L., Rossi, G. et al. 1995. Dosimetry of [169]Yb seed model X1267. *Phys Med Biol* 40(8):1317–30.

Rivard, M.J. 2007. Brachytherapy dosimetry parameters calculated for a [131]Cs source. *Med Phys* 34(2):754–62.

Rivard, M.J., Melhus, C.S. Sioshansi, S. et al. 2008. The impact of prescription depth, dose rate, plaque size, and source loading on the central axis using [103]Pd, ^{125}I, and [131]Cs. *Brachytherapy* 7(4):327–35.

Van Damme, J.J., Culberson, W.S., DeWerd, L.A. et al. 2008. Air-kerma strength determination of a [169]Yb high dose rate brachytherapy source. *Med Phys* 35(9):3935–42.

Quality Assurance of Equipment

Jack L. M. Venselaar
Instituut Verbeeten

Firas Mourtada
Christiana Care Health System
Helen F. Graham Cancer Center

4.1 Introduction

Several national and international medical physics societies established recommendations on the frequencies and tolerances of brachytherapy quality control (QC) procedures. The published reports present an extensive description of the required procedures; hence, the simple repetition of their contents is out of the scope of this chapter. Comprehensive overviews can, for example, be found in the books and book chapters of Thomadsen (2000), Williamson et al. (1994), and the European Society of Radiotherapy and Oncology (ESTRO) Booklet No. 8 (Venselaar and Pérez-Calatayud 2004). In this chapter, concise descriptions of these procedures are given in combination with frequency tables that provide the medical physicists with useful information to assist with creating and maintaining quality assurance (QA) tasks for their clinical brachytherapy program. The frequencies listed here are essentially taken from consensus American Association of Physicists in Medicine (AAPM) and ESTRO recommendations, with a short discussion on the differences in the approaches expressed in their reports. In clinical practice, an increase in test frequency is required when the stability of a system is suspect or when a specific patient treatment method demands special attention. It is also important to note that any national set of legal requirements should be followed.

Since QC procedures differ between high dose rate (HDR), pulsed dose rate (PDR), medium dose rate (MDR), and low dose rate (LDR), often these modalities are addressed separately. This chapter includes HDR and LDR. According to the definition in the International Commission on Radiation Units and Measurements Report 38 (ICRU 1985), MDR systems will operate with a dose rate in the range between 2 and 12 Gy/h. Essentially, MDR systems apply higher active sources than LDR systems using similar hardware. Differences between HDR and PDR systems are minor, except for the specific application issues and software to control the timer setting of the pulses. In general, PDR treatments are delivered on the same hardware and applicators as the HDR modality. Hence, PDR QA requires some additional tests to check the treatment console unit (TCU) software operations that are different from HDR.

The main reports from the AAPM that provide guidance for QA and the procedures necessary to maintain safety and efficacy in brachytherapy are as follows: (1) TG-40 report published in 1994 on comprehensive QA for radiation oncology (Kutcher et al. 1994), (2) TG-56 report published in 2004 that narrowed the focus to a code of practice for brachytherapy physics (Nath et al. 1997), and (3) TG-64 report published in 1999 that dealt only with permanent prostate brachytherapy (Yu et al. 1999, not discussed here any further). The dose calculation formalism for brachytherapy treatments that was developed in (4) the AAPM TG-43 report (Nath et al. 1995), its update TG-43U1 (Rivard et al. 2004), and its supplement TG-43U1S1 (Rivard et al. 2007) has proven to be a great success, allowing rapid and widespread adoption and providing a robust methodology for reproducible dose distributions among different treatment planning systems (TPSs). Kubo et al. (1998), on behalf of TG-59, examined the HDR treatment delivery practices and provided a document for the users to assure safe delivery of HDR treatments. The document consists of detailed HDR procedures for design of an HDR brachytherapy program, staffing and training, treatment-specific QA, and emergency procedures. It reviews all aspects of HDR treatment delivery safety, including prescription, treatment plan, treatment delivery, and radiation safety. These reports provide a consistent set of recommendations regarding QA. Recommendations for calibration of sources for brachytherapy implants are included in these reports. A more recent paper by Butler and Merrick (2008) dealt with the clinical practice and QA challenges in modern brachytherapy sources and dosimetry. It focused primarily on LDR low-energy photon-emitting sources, but the rationale of the procedures described in their paper is similarly of importance for high activity and high-energy photon-emitting sources used in HDR brachytherapy.

As of this writing, other task groups under the AAPM Brachytherapy Subcommittee (AAPM-BTSC) are working on the development of recommendations to meet the needs of modern brachytherapy practices. TG-138 has recently developed recommendations regarding brachytherapy dose evaluation uncertainties (DeWerd et al. 2011). The High-Energy Brachytherapy Source Dosimetry Working Group (HEBD) (Pérez-Catalayud et al. 2012) is developing guidelines for the application of TG-43 to higher-energy (>50 keV) photon-emitting sources. For that purpose, a close cooperation with the Brachytherapy Source Registry Working Group (BSR) is essential. TG-143 is tasked to address dosimetry for elongated brachytherapy sources. TG-167 is developing recommendations on investigational brachytherapy sources. TG-182 is working on electronic brachytherapy quality management. These latter groups are close to publication. There are other groups working on eye plaque dosimetry, microspheres, and image-guided robotic brachytherapy. A new task group, TG-186, to investigate the future of brachytherapy dose calculations was tasked to develop recommendations on commissioning model-based dose calculation algorithms for heterogeneity-based brachytherapy dosimetry.

Further interesting reading is found in a number of recent publications, of which the following can be mentioned in the present context:

- *Anniversary Paper: Past and Current Issues, and Trends in Brachytherapy Physics*, by Thomadsen et al. (2008)
- *Anniversary Paper: Fifty Years of AAPM Involvement in Radiation Dosimetry*, by Ibbott et al. (2008)
- The overview on *Brachytherapy Technology and Physics Practice Since 1950: a Half-Century of Progress*, by Williamson (2006)
- *Current Brachytherapy Quality Assurance Guidance: Does It Meet the Challenges of Emerging Image-Guided Technologies?* by Williamson (2008)
- *Quality Assurance Issues for Computed Tomography, Ultrasound, and Magnetic Resonance Imaging-Guided Brachytherapy*, by Cormack (2008)
- A Vision 20/20 paper on *The Evolution of Brachytherapy Treatment Planning*, by Rivard et al. (2009)
- A Vision 20/20 paper on model-based dose calculation algorithms, by Rivard et al. (2010)

The terms "quality assurance" and "quality control" are often used for the same purpose. According to the book *Achieving Quality in Brachytherapy* (Thomadsen 2000), much of what medical physicists call QA falls more to the realm of QC by definition, but obviously, these "two concepts share many features and it often becomes unclear whether a particular action serves to control or to demonstrate quality." Internationally agreed definitions for the terminology used to sort out the various facets of this topic are discussed in this book, and from this most of the following definitions are cited, with reference to ISO 9000 and several other publications (see, e.g., Thomadsen 2000, ISO 1994).

Quality management. "All activities of the overall management function that determine the quality policy, objectives and responsibilities, and implement them by means such as quality planning, quality control, quality assurance, and quality improvement…." The goal of quality management is to achieve a desired level of quality, in this case, for brachytherapy applications.

Quality policy. The overall intentions and direction of an organization related to quality as formally expressed by top management. The purpose of the quality policy is to give the intentions and direction concerning quality in brachytherapy in the departments.

Quality assurance. "The part of the quality management focused on providing confidence that quality requirements will be fulfilled." The goal of QA is to demonstrate quality. Similar definitions of QA were given before as "…The activity of providing the evidence needed to establish confidence… that the quality function is being effectively performed," and "quality assurance is: all planned and systematic activities implemented within the quality system that can be demonstrated to provide confidence that a product or service will fulfil requirements for quality."

Quality control. According to ISO 9000, QC is "the part of quality management focused on fulfilling quality requirements." QC follows the general process of

(1) Evaluating actual operating performance
(2) Comparing actual performance to goals
(3) Acting on the difference.

Quality terms were defined as follows:

- *Quality* is "the degree to which a set of inherent characteristics fulfils requirements." In brachytherapy, parameters controlled in daily or weekly tests are such inherent characteristics.
- *Requirement* is "the need or expectation that is stated, generally implied or obligatory." The requirements are stated in standards by authorities or in local procedures taken from recommendations.

These definitions may be helpful in setting up a brachytherapy QA program within the context of a general quality management program in the radiotherapy institution. However, it is up to the user of this material to define the goals and requirements for such a purpose along the guidelines presented here. The separate steps and QA procedures are described in more detail to help the user develop and implement a comprehensive program.

Note that the main focus of this chapter is on the QA procedures of the equipment itself, the QA of treatment planning and the imaging procedures used in brachytherapy are dealt with in separate chapters in this book. See Chapters 13 and 14, respectively.

4.2 Safety and Physics Aspects of QC of Afterloading Equipment

A description of QC procedures for afterloading equipment is separated in this chapter in the procedures of (1) safety systems and (2) physical parameters. Methods for source calibration and source strength monitoring are discussed in a separate section and in full detail in Chapter 5. This discussion assumes that an HDR or PDR afterloader (see Figure 4.1) makes use of an ^{192}Ir source, whereas an LDR system is based on the use of ^{137}Cs sources. Application of other radionuclides has been suggested in some publications as an alternative to iridium in HDR systems due to the specific physical properties such as a longer half-life (^{60}Co) or a lower mean energy of the emitted photons (^{169}Yb, ^{170}Tm). A quality management system is not affected by the choice for a specific radionuclide. Only the frequencies of the checks may be influenced to adapt to the source exchange periodicity. The use of these radionuclides is not further discussed here.

No attempt is made by the AAPM or ESTRO to recommend a separate set of procedures for commissioning of afterloading equipment. In most of the published materials, the commissioning phase is indicated with a comment "at acceptance." According to local regulations, a report should be available of the tests performed to demonstrate the proper functions and safety aspects of the equipment.

Tests on the safety systems of the afterloader are performed for a number of reasons. They are meant to demonstrate a proper function of the safety systems, to protect the patient from treatment delivery errors, to minimize unnecessary radiation exposure to the patient and staff, and to ensure treatment delivery as intended.

Most of the included items listed in the next paragraphs can be tested by different methods. Since individual treatment equipment and installation details in the treatment room as well as the tools available at the department may differ, the exact method to be used for safety checks has to be adapted to a specific situation. The following is a list of functions and/or items to be tested and contains a short description of the methods that can be employed. In the reports of the professional organizations, some alternative methods are described.

4.2.1 Tests of Safety Systems

- *Visual and voice monitoring equipment.* Observe that the visual monitoring camera and intercom systems are working properly.
- *Applicator attachment.* Program the HDR afterloading unit to send the source into each of the channels without attaching catheters or transfer tubes to the unit. Try to initiate a source run. The HDR unit should produce an error message.

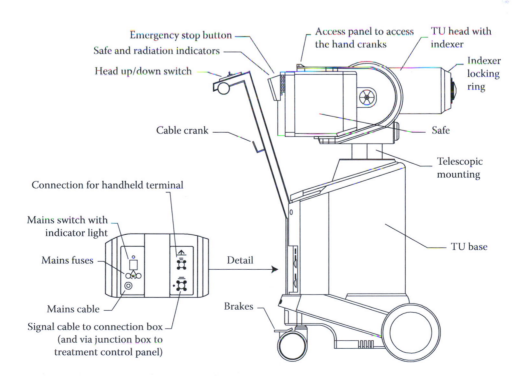

FIGURE 4.1 Components that are typical for a modern afterloading device. (Courtesy of Nucletron, Veenendaal, The Netherlands.)

- *Indexer locking ring.* Attach catheters to each of the channels but do not lock them in place, and try to initiate a source movement. The HDR unit should produce an error message. Use of one catheter is sufficient for this test.
- *Door interlock.* Attach and lock a catheter to each channel. Program the source to dwell at the tip of the catheter. Leave the door open and try to initiate the source run. Close the door. Initiate the source run. Open the door while the source is out. Check to see that the run is aborted. Inspect the fault indication on the console and the printout from the unit to ensure that a correct record of the fault has been made. Use of one catheter is sufficient for this test.
- *Warning lights.* Observe warning lights during a source run.
- *Room monitor.* Listen through the intercom for an audio signal during a source run. Use the room cameras as visual monitors of the radiation area monitor.
- *Handheld radiation survey meter.* Immediately on opening the door during a source run, hold the handheld monitor in the doorway and see whether it indicates the presence of radiation. The reading of the monitor itself should be checked regularly against a known source (level) of radiation. Keep the exposure to the observer as low as possible. The radiation exposure is expected to be background, indicating source retraction into the afterloader safe.
- *Treatment interrupt.* During a source exposure, press the interrupt button (see Figure 4.2) to abort the run and ensure that the source is retracted. Inspect the fault indication on the console and the printout from the unit to ensure that a correct record of the fault has been made. Resume and complete the treatment and inspect the final printout.
- *Emergency stop.* During a source exposure, press the emergency stop button (see Figure 4.2) to abort the run

and check that the source is retracted. Inspect the fault indication on the console and the printout from the unit to ensure that a correct record of the fault has been made.
- *Timer termination.* Test that a source exposure continues until the time elapsed equals the time set on the timer.
- *Obstructed catheter.* With HDR and PDR equipment, attach and lock an obstructed catheter or a catheter that has been curled into a loop with a radius of curvature too small for the source to negotiate near the end. Check that the obstruction or the restriction due to curvature being too tight is detected. Note that before starting the procedure, verify with the documentation to the system and—eventually—with the manufacturer to avoid possible damage to the source wire or afterloader.
- *Power loss.* Check that an interrupt of the AC power during treatment (open the circuit or unplug the unit; see manufacturer's instruction manual) results in immediate source retraction. Check that upon restoring the power, the treatment parameters and the remaining dwell time are correctly recalled. If the machine has a backup power supply, so that the treatment continues normally despite an AC power loss, a check on the equipment should be performed that would demonstrate that the treatment is not interrupted by the power failure. Inspect the indication on the console and the printout from the unit to ensure that a correct record of the treatment and the fault has been made.
- *Air pressure loss.* If the system works with a pneumatic source drive system (e.g., in the Selectron-LDR design; see Figure 4.3), the system is programmed with a train of dummy sources for one catheter connected to an applicator

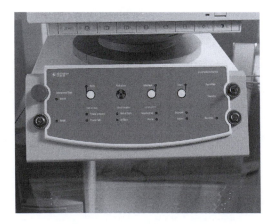

FIGURE 4.2 Console of the afterloading device with the interrupt and emergency buttons. (Courtesy of Nucletron, Veenendaal, The Netherlands.)

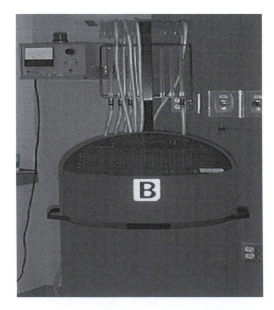

FIGURE 4.3 Selectron-LDR (Nucletron) designed to operate with a pneumatic source drive system available at the MD Anderson brachytherapy department. (Courtesy of M.D. Anderson, Houston, Texas.)

tube. Start a "treatment." Detach the wall mounted compressed air connector and check that an immediate retraction is activated of the source train. Inspect the indication on the console and the printout from the unit to ensure that a correct record of the treatment and the fault has been made.

- *Integrity of transfer tubes and applicator.* Perform a visual inspection of the transfer tubes and applicators.
- *Radiation leakage.* Check that the radiation exposure level at 10 and 100 cm from the afterloader safe with the source retracted is lower than the level specified in the legal requirements. Repeat the check in all relevant directions (front, back, sideways, top). If the afterloader itself is stored in a storage room/area, the test is (also) performed around the storage room, at the shielded door, hallways, nurse stations, or adjacent rooms.
- *Contamination test on applicators and transfer tubes.* Wipe tests can be performed by using filter paper, cotton-tipped swabs, or wipe test "swipes" (absorbent cotton pads). The radiation from the swab can then be analyzed with a NaI-crystal counter or scintillator-based nuclear pulse counter in order to detect any gamma-emitting nuclides. Damping the wipe pad with a suitable solvent such as water or alcohol can be useful when trying to collect the maximum amount of radioactive material present. As the radiation level to be detected is generally low, a good counting geometry and a counting time long enough to collect a suitable signal, which is high enough to differentiate from background, are essential. Another possible test is to measure the radiation level close to an applicator with a sensitive contamination detector. As sources are not directly accessible for wipe testing, one should perform the test to a set of (used, preferably plastic) applicators and transfer tubes.
- *Contamination test on the check cable.* Attach an applicator of the HDR or PDR system, position the check cable in the applicator (with the manual controller), and disconnect the applicator. Perform a wipe test of the check cable. If this procedure is not possible, see the manufacturer's instruction manual for methods to perform a similar test to the check cable.
- *Emergency equipment [forceps, emergency safe, (transport) container, wire cutter].* The presence of emergency equipment close to the afterloading unit should be checked. An HDR or PDR source should never be touched by hand, and long forceps and/or tweezers are therefore needed. A wire cutter and a transport container should be available. Surgical supplies, emergency instructions, and the operator's manual must be available at the machine. It is noted that for LDR sources, the exposure levels are significantly lower; however, any direct contact with the LDR source(s) must be minimized according to the As Low As Reasonably Achievable (ALARA) principle. If applicable, a list of error codes and their meaning must be available near the equipment.

- *Hand crank functioning.* The function of the manual source retraction crank of HDR and PDR systems must be checked. Detailed instructions are provided by the manufacturer of the system.
- *Emergency procedures.* Emergency situations must be practiced by all personnel involved in brachytherapy treatments. The goal of such procedures should be to provide maximum safety to the patient and to keep exposure to the patient and to the personnel as low as possible.

4.2.2 Tests on Physical Parameters

- *Source positioning accuracy.* Accurate brachytherapy treatment delivery requires that the source is delivered to the correct location inside the patient for the programmed dwell position. In a stepping source device (i.e., the afterloader), the motion encoder usually requires the distance along the catheter corresponding to a specific dwell position, often the first, to be able to send the source to the correct location. The distance may refer to the length from some part of the unit or from an arbitrary point (e.g. the point at the indexer locking ring). The dwell positions will then be relative to this absolute position, and therefore, the precise localization of this position is important. To locate this position, the unit is provided with a set or sets of radio-opaque markers, which can be inserted into the catheter in well-defined positions. The markers may consist of a long wire with nubs attached, which correspond to one or more specific source dwell point(s) in an applicator.

  One of the methods to verify the correct source position uses a specially designed ruler that connects directly to the afterloader using a similar transfer tube. Attaching the ruler and focusing a camera on the scale allows the position of the tip of the source to be verified during a source run. Modern HDR and PDR afterloaders are provided with high-resolution cameras that can be directed to the ruler. The source position can thus be safely observed up to a precision of 0.25 mm. In order to have consistency in the check, one must ensure a well-defined (e.g., straight) positioning of the transfer tube because any curvature would influence the accuracy of the measurement.

  Another method to determine source positioning, suitable for LDR systems as well, utilizes autoradiographs. Tape a small-diameter plastic catheter to a film (e.g., radiochromic) and indicate by pinholes on both sides of the catheter the position of the marker corresponding to the reference position (usually the first dwell position). Program the source to stop at the marked position and execute the run. On the film, the dark "blot" indicates the effective center of the source, and this should fall on the line between the two pinholes. The irradiation time will have to be chosen according to the activity of the source and type of film used, among other secondary characteristics. Figure 4.4 shows a useful commercial tool that incorporates a strip of scaled radiochromic film for auto-radiograph localization into a position check jig with visual scale.

Another jig that can be used for source positioning QC and can accommodate different types of applicators is equipped with a permanently fixed array of diode detectors. If this device is calibrated with a radiograph where x-ray markers are inserted in the catheter, it is possible to evaluate the source position relative to the markers. Commercially available tools based on this principle may come to the market.

Utilization of a well-type chamber to perform a number of regular QC tests of an HDR system was described by DeWerd et al. (1995) who developed a special insert replacing the standard calibration insert of the chamber. The insert is a lead cylinder with a channel for a central catheter and one or two spacers of acrylic at precisely known distances from the bottom of the well chamber. The reading of the chamber is taken as a function of the position of the source in the catheter. Maximum readings are observed when the source is at the level of the acrylic spacers. In a relatively short time period, information can be derived from the reading versus position curve such as position verification, timer consistency, and consistency of the source strength measurement.

- *Length of transfer tubes.* Transfer tubes (or source guide tubes) that are used to connect the afterloader to the catheters are a critical element in the overall source positioning accuracy for cable-driven HDR and PDR systems. Often the user has a number of transfer tubes at his or her disposal to accommodate different catheters and/or needles. In some systems, these tubes are rigid, and after an initial check of the proper lengths, any visual inspection of their integrity is sufficient. Only when this visual inspection raises any doubts or when the user suspects any problems

(e.g., after extreme force exerted on the transfer tube) a verification of the length should be considered. In other systems, the length of the transfer tubes can be adjusted to correct for small individual deviations, for example, using a bolt and nut adjustment. In such systems, it is recommended to perform a systematic length test on each of the tubes. The manufacturer provides tools for this purpose, for example, a dedicated ruler system that allows measuring the overall length of the catheter and tube combination. The methods described above for checking the source position accuracy can be used as well. It is recommended to perform a few checks for a number of possible combinations of the available transfer tubes and source channels to demonstrate that the batch of transfer tubes is sufficiently uniform in length. As noted before, the curvature of the transfer tubes may influence the results of the check.

- *Transit time effect.* The effect of the transit time increases healthy tissue exposure beyond that due to treatment dwell time(s) alone. The transit dose contribution is directly related to the speed of the source transfer from the safe to the irradiation position(s). In general, the further the point of interest from the dwell position(s), the greater the fractional dose contribution due to source transit. It is noted that the transit dose is not accounted for in any dose calculation algorithm as it is dependent on several absolute and relative parameters. Detailed comments are found in the chapter on "Treatment Planning and Optimization" by Thomadsen et al. (1994), showing that the effect can be in the order of several percent. A fixed geometry is essential in order to check the consistency of the transit time effect. The transit time correction factor can be derived from

$$f_{tr}(t) = 1 - \frac{Mt_0}{Mt_{(t)}}$$

where t is the dwell time, Mt_0 is the electrometer reading at $t = 0$ (zero dwell time, only dose contribution during source transport), and Mt is the electrometer reading for dwell time t. The value for $t = 0$, Mt_0, is determined for the specific geometry by programming dwell times in the range of 5 to 120 s and then extrapolating to $t = 0$. The source transfer speed QC check can demonstrate deviations over time.

- *Timer consistency.* Comparing the programmed treatment time with a reference timer, for example, a stopwatch, will perform the timer consistency check. Any timer registration independent from the machine timer will do the job. Check over a useful or clinically reasonable range of times, for example, 0–200 s for HDR and PDR systems, and use a longer time span for a timer check in LDR systems.

- *Timer linearity.* Program a range of dwell times in a fixed geometry close to an ionization chamber or directly in

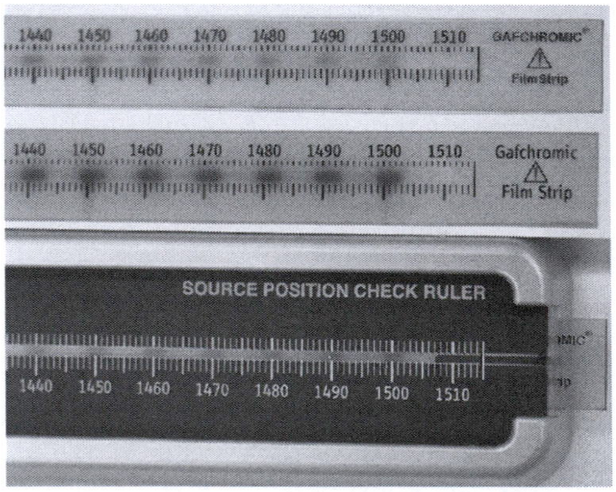

FIGURE 4.4 Radiochromic film test in a source position check ruler (Nucletron), showing test results with very good agreement (top) and a 1-mm systematic shift (bottom). (Courtesy of M.D. Anderson, Houston, Texas.)

a well-type ionization chamber, and then check that the chamber readings increase linearly. For the most accurate result, the readings have to be adjusted for the transit time effect using the transit time correction factor (see above). Another method is to set the electrometer to a charge integration mode and program a dwell position for a nominal duration, for example, 30 s, and then set the electrometer for a shorter duration for integration, for example, 20 s. One can push the start button on the electrometer after the source arrives inside the well chamber, and the electrometer will stop measuring before source retraction. This can be done with several electrometer types.

- *Source calibration.* The medical physicist is responsible for performing in-house calibration of the afterloading source. Methods for source calibration and monitoring the source strength are described in the next section. Before clinical use, the measured value must be compared with vendor's source strength certificate.
- *Radioactive decay.* Correction for radioactive decay in the afterloading system should be made at suitably frequent intervals. An interval of 1 day for ^{192}Ir is generally considered sufficient as it corresponds with approximately 1% difference in the source strength. For a ^{137}Cs-based system such as the Nucletron Selectron-LDR afterloader, the delivery system does not have the software to correct for source decay. This is then taken into account in the treatment planning phase, and typically, a midyear decay point is acceptable. For HDR/PDR systems, most TPSs and afterloader treatment consoles make use of more accurate time measurement functions and therefore have a decay correction factor with resolution better than 1%. The radioactive decay constant used in the afterloader and in the accompanying TPS should match.
- *Unintended radionuclide contamination.* A check can be performed on the radionuclide composition of the delivered source. Any short-lived contamination in a newly produced source would exhibit a deviating decay factor. Some users therefore repeat the initial calibration check on the ^{192}Ir source after 1 or 2 weeks and compare the result with the expected value. Some authorities such as the German Ministry for Environment, Nature Conservation, and Reactor Security even officially require that the source strength is in addition after a period of 2 weeks from the first calibration, measured to check the purity of the radionuclide. This is mainly focused on the use of ^{192}Ir-based afterloading systems where there exists the problem of possible contamination with ^{194}Ir, but the procedural rule is applicable to other radionuclides as well.

4.2.3 PDR-Specific QC

In general, PDR treatments are delivered with the same hardware components (afterloader, transfer tubes, applicators, etc.) as the HDR modality. Therefore, a PDR QA program is largely the same as an HDR program, although it requires some additional tests to check the TCU software operation, which is different compared to that in the HDR program. In the (inter)national recommendations and protocols, usually tables of PDR QC are not provided separately.

PDR treatments are given in series of pulses with pulse widths on the order of 10–30 min per hour and with programmable intervals. Pulsed doses are given (on the order of 40–100 cGy/h) at the prescription isodose level and with an overall treatment duration of 48–72 h. Typical user-defined pulse schemes should be tested at acceptance of a machine before a first patient is treated. As the treatment reports are different for PDR and HDR, the user must become familiar with the pretreatment and posttreatment records. For long overall durations, the user must check how a system deals with the source decay (correction) during treatments.

Safety checks must be performed on the additional equipment such as the operating console and treatment panel, as well as on the nurse station panel, which is usually located at a distance from the patient room. The recommended frequency is similar to the frequency of checks for warning lights and door interlocks (see Table 4.2).

The user should be well aware that the number of source and dummy source transfers to the catheters is significantly higher than with an HDR afterloader. Wear and tear can lead to error messages on a heavily used system. It is therefore recommended that a user replace the dummy source wire with the same interval as the source exchange.

4.3 Test of Ancillary Equipment and Procedures

4.3.1 Calibration of Brachytherapy Sources

Calibration of brachytherapy sources at the hospital is an essential component of a well-designed QA program. The aim of the calibration is twofold: (1) to ensure that the value entered into the TPS agrees with the source calibration certificate to within a predetermined limit, and (2) to ensure traceability to (inter) national standards. The traceability is important as it simplifies national and international comparison of treatment results and improves consistency in clinical outcome.

The recommended quantities to specify the source strength are the reference air-kerma rate K_R and air-kerma strength S_K, which are mutually related through the inverse-square law to the reference distance r_0 as $K_R = S_K \times 1/r_0^2$. Note that the specification in terms of air-kerma strength S_K is adopted in the AAPM TG-43U1 formalism (Rivard et al. 2004), whereas the reference air-kerma rate K_R is recommended by ICRU 38 (1985) and ICRU 58 (1997). This latter quantity is also used in the report from the International Atomic Energy Agency TecDoc 1274 (IAEA 2002). Other quantities for specifying source strength such as the apparent activity (in becquerels or curies) are now considered obsolete.

The calibration can be done with a number of methods that should all have traceability to a primary calibration standard. The methods are either based on a so-called "in-air" measurement technique, on the use of a well-type ionization chamber, or on a solid phantom dedicated for calibration purposes. In principle, any source can be calibrated with these methods, but there are some practical limitations. With in-air and solid phantom calibrations, the signals typically obtained when using low strength sources (i.e., LDR sources) are small, and the final uncertainty in the air-kerma strength or reference air-kerma rate may be unnecessarily high. For HDR sources, however, all calibration methods discussed above may be considered.

Most of the present recommendations rely on the use of a dedicated well-type ionization chamber (Figure 4.5). Although such chambers provide an easy, fast, and reliable method for source calibration, it must be kept in mind that in-air calibration is a more fundamental method. Still, the well-type chambers offer the best of practice: reliability, reproducibility, and ease of use. Several primary or secondary standards laboratory or (in the United States) Accredited Dosimetry Calibration Laboratories (ADCLs) provide the users all over the world with a calibration factor for their instrument, which is to be used with its accompanying electrometer and inserts. The procedure for in-house calibration is then simple and can be repeated at a few intervals in the lifetime of the local HDR afterloader source.

Chapter 5 is dedicated to the calibration of radioactive sources used in brachytherapy. Therefore, no further details are given in this chapter.

A general requirement is to use the available dosimetry equipment at each source exchange before any clinical treatments take place. In many institutions, the procedure is repeated at the end of the lifetime of the source. In this way, the old and the new sources are checked in a short period of time, the stability of the measurement system is demonstrated, and a confidence in the use of the correct decay factor is established. The instrument (the ion chamber, electrometer, and cables) itself should be checked for linearity, leakage, and consistency of the readings with a measurement at regular intervals with a long-lived source, for example, a ^{137}Cs source, once a year. Any deviation for the source under consideration larger than 3% must be inspected by repetition of the measurement and/or with independent means.

In this context, it is worthwhile mentioning that there is noticeably a renewed interest in the calibration issues regarding brachytherapy sources at the level of the European primary standard laboratories and more specifically in the development of a dose-to-water calibration standard. At the end of 2008, this goal was accepted as an explicit part of a new research activity in the European Association of National Metrology Institutes (EURAMET) network of national measurement institutes. The iMERA joint research project on brachytherapy aims at developing a primary standard of absorbed dose to water for ^{192}Ir at three national measurements institutes (NMIs) and also for ^{125}I at four NMIs within a time frame of 3 years, 2009–2011. Regardless if this research is successful and how this advance is further implemented, there is still much work to be done before direct chamber calibrations for liquid water dose rate are available to clinical users. It is the responsibility of the international radiation metrology community to coordinate efforts to uniformly transition to dose-to-water standards worldwide and to disseminate these standards jointly with the professional societies to the clinical end user of the sources. Coordination for this transition from reference air kerma to dose to water across the entire field of brachytherapy is key. A distinct requirement from the clinical user's point of view is that such a new approach should at least give the same, but preferably, lower source calibration uncertainties.

4.3.2 QA of Imaging Techniques

As described in a *Medical Physics* Vision 20/20 paper (Rivard et al. 2009) and elsewhere, the use of x-ray, CT, and other 3D imaging modalities such as MR, US, and, more recently, PET marks a distinct improvement in brachytherapy planning, to be considered as a departure from the surgical practice of brachytherapy.

The step forward is made primarily due to direct availability of the radiological equipment for interventional procedures. Expensive imaging modalities are nowadays readily available in every modest-sized hospital, whereas in many radiotherapy departments, 3D imaging has replaced the conventional radiotherapy simulator for patient setup. Transrectal ultrasound (TRUS)-guided permanent and HDR temporary implants have become standard practice for prostate brachytherapy in North America and Europe. For other body sites, especially gynecology, 3D imaging is now or will be soon included in the newest recommendations as a standard guidance for volume definition and dose prescription. A new generation of applicators,

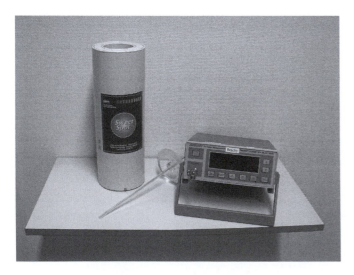

FIGURE 4.5 Example of a well-type chamber with electrometer. (Courtesy of Institute Verbeeten, Tilburg, The Netherlands.)

in which a combination of intracavitary and interstitial techniques is pursued to increase the flexibility for source position optimization in the implant geometry, is coming into the field, based on the use of carbon fiber or titanium as construction material. A new generation of the Fletcher-Suite-Delclos colpostat with movable shields has been shown to reduce CT artifact without compromising the rectal and bladder shields. Although MRI is only slowly entering the field, it is widely recognized that this technique readily adds to the quality of the treatments for its ability to discriminate between different tissue types. Contouring will increasingly be based on this imaging technology. Functional MRI depends on its ability to demonstrate active parts of the tissue under consideration, allowing suborgan identification. If tissue contouring can be performed in this way, this would allow a spatially modulated dose painting procedure, in which the full organ, for example, a prostate, is treated to the required dose, but subparts are administered a much higher dose. There is a similar expectation for the use of PET and PET-CT imaging for identifying active tumor areas. One of the challenges in brachytherapy imaging is the further development of registration for different modalities and image sets made at different points in time. A key to this challenge is the need to address deformable image registration. In brachytherapy, not only organ movements or the increase or decrease in tumor volume but also the required transportation of a patient from one location to another, for example, operating area to radiology department, may influence the relative position of the applicators and the organs (see Figure 4.6).

The imaging facilities available at the departments are considered to be an integral part of the dosimetry chain for brachytherapy. Chapter 14 deals with the QA aspects of imaging procedures, so no details are given in this chapter.

4.3.3 QA of Treatment Planning in Brachytherapy

Treatment planning relies (1) on the proper reconstruction of the implant geometry, as discussed in the previous paragraph, and (2) on accurate algorithms for calculation of dose deposition from sources in the medium. The evolution of treatment planning in brachytherapy was extensively discussed in the Vision 20/20 paper in *Medical Physics* (Rivard et al. 2009), describing the history, the conventional dose calculation approach, the present TG-43 and TG-43U1 status, and the ongoing developments in Monte Carlo and deterministic-based methodology. The latter evolution is aiming at an improved accuracy of dose calculation in inhomogeneous media (shields, nonwater conditions, lack of scatter situations). It is not the intention to go into any of the details here. Until new types of algorithms come into clinical practice, it has to be assumed that the standard of the algorithms in the clinically applied TPSs for brachytherapy is the TG-43 and TG-43U1 approach (Nath et al. 1995; Rivard et al. 2004). If this is not the case, at least the dose and/or dose rate in "along and away" tables found for the specific sources under the TG-43 formalism should be used as benchmark data under assumption of infinite homogeneous water media.

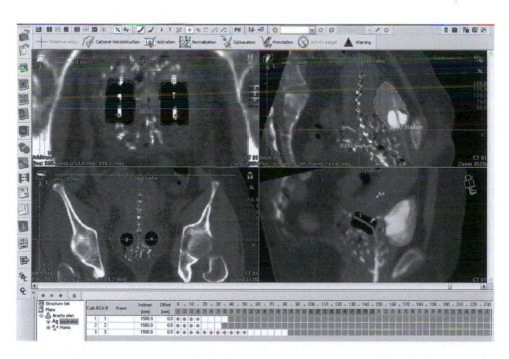

FIGURE 4.6 Set of orthogonal images reconstructed from a CT scan of a patient with a uterine tube and two ovoids. (Courtesy of M.D. Anderson, Houston, Texas.)

QA of brachytherapy treatment planning should cover at least the following issues.

(1) What exactly is the source model used in the HDR afterloader: avoiding misinterpretation of terminology such as "New" or "Classical," or otherwise due to technological developments?

(2) Where do we find the (benchmark) data sets that belong to the individual source: dose rate constant, radial dose function, geometry function, anisotropy factor or function, in the TG-43 formalism?

(3) How does one validate the calculation by the TPS of the dose distribution around the individual source?

(4) What else is there to check or validate: source decay, optimization, dose-volume histogram calculation, etc.?

These aspects will be dealt with in detail in Chapter 13 on QA of brachytherapy treatment planning procedures.

4.3.4 Dosimetry Equipment

The calibration system, any handheld radiation monitor, survey detectors, or any other radiation detection device used in the brachytherapy department should be checked with an independent method, for example, to detect any dysfunction or drift of the equipment. In case of HDR afterloaders, one generally makes use of sources with relative short half-lives. A long-lived source such as a ^{137}Cs source is then suitable for the purpose. Simple procedures with well-described measurement setup are sufficient to demonstrate stability of the reading of the devices against such sources over the years.

4.3.5 Protective Means

Lead aprons and lead shields may be available in the department. Although the aprons do not provide any significant protection against the relatively high energy of the photons emitted by an ^{192}Ir source, they are useful in case of lower energy sources (below 100 keV) and with the x-ray procedures that may be applied for imaging. The aprons suffer from wear and tear over the years and should be inspected on an annual basis. A standardized procedure for lead aprons can be to inspect them on a therapy simulator table or a similar device. Lead shields are sometimes used as additional shields in treatment rooms in case the walls are not sufficiently thick. These shields are unlikely to change shielding properties, but visual inspection may show damage, for example, from falling over. The same holds for lead slabs if mounted onto the walls. If mounted invisibly in a wall, a radiation survey with suitable equipment may be necessary to show that it is still in place and not broken off from its fastening bolts.

4.3.6 Preventive Maintenance

A service contract with the vendor of the afterloading equipment should cover proper functioning of the machine itself.

Service engineers execute a complete set of tests of critical parts and will replace them if needed, often according to an expected life cycle. An example of this is the dummy check as the critical part in a heavily occupied PDR system cable because of the many runs in highly fractionated treatment schedules. As indicated above, some departments therefore choose to replace the dummy cable at each source exchange instead of the standard annual replacement.

4.4 Summary of Recommendations for QA

Table 4.1 was adapted from the AAPM TG-40 report (Kutcher et al. 1994) to summarize the verification steps for brachytherapy, which is here adapted to include both the original recommendations for LDR (manual and afterloading) applications and HDR or PDR afterloading applications (see also Tables 4.2 and 4.3).

4.5 Applicators

It has to be verified after receiving brachytherapy applicators that the delivered product is in accordance with what has been ordered. Whenever applicable, the instructions for use must be checked for availability and studied for their contents. One should check that all individual parts have been delivered (see Figure 4.7 as example). The mechanical properties of the applicators should be checked, and verification should be made that the applicators and transfer tubes are of the correct length. Checks should be made to ensure that the product is functioning as described in the instructions for use. Any mechanical code, which is intended to force the connection of a specific applicator with a specific transfer tube, should be verified. Applicators for a specific irradiation device are not to be used with other irradiation devices.

For applicators with shields, checks should be performed to ensure that the built-in shields are in the correct position, for example, by using an x-ray image of the applicator. For applicators that are more complex to handle (e.g., Fletcher applicators), each user should become familiar with the use of this device by instruction and training. Before use in a patient, it is advised to practice the correct assembly of the applicator.

Instructions for sterilization must be provided with the products and shared with the sterilization facility manager. When selecting products, care must be taken to have the proper sterilization procedures available. Not all hospitals have facility for gas sterilization. In some cases, one might be obliged to find a nearby hospital that does gas sterilization and arrange a contract with them for the sterilization of the applicators. Many users therefore prefer to order presterilized products, such as plastic tubes for interstitial and endoluminal cases.

A detailed discussion on the QA procedures of applicators and appliances can be found in the ESTRO Booklet No. 8 (Venselaar and Pérez-Calatayud 2004) and the book of Thomadsen (2000). Steps and procedures that can be used for verification of the

TABLE 4.1 Procedure-Specific Verification

Endpoint	Procedure	When? (LDR)	When? (HDR/PDR)
Accuracy of OR implant description	Direct observation	During procedure	Before start of treatment
Prescription accuracy and consistency	Consistency of loading and prescription with disease stage, therapy chart treatment plan, department treatment policies	First half of treatment	Before start of treatment
Verify correct source used	Inspection of documentation in plan and pretreatment report of afterloader for agreement with information of last source exchange	Source preparation and source loading	Before start of treatment
	Verify performance of last spot check according to QA protocol	Before start of treatment	Before first treatment/day
	Therapist or physicist (or individual knowledgeable in loading or connecting) always assists physician	At source loading	At connecting catheter(s)
Treatment plan	Calculation of plan and check for accuracy/consistency	First half of treatment	Before start of treatment
Implant removal	Disconnection of catheter(s) by therapist or physicist jointly with nursing staff or physician for implant removal	At expected removal time	At end of treatment/ last fraction
Sources all removed	Patient survey count	At removal	
	Source inventory	Next working day	
Review treatment	Verify treatment time and report	After completion of procedure	After completion of procedure
Record, QA audit	All QA, treatment, and radiation safety records complete	After completion of procedure	After completion of procedure

Source: Adapted from the original Table XII in Kutcher, G.J., Coia, L., Gillin, M. et al., *Med Phys* 21, 581–618, 1994, to include a column for HDR/PDR applications.

TABLE 4.2 Frequencies and Tolerances of QC Tests for HDR/PDR Afterloading Equipment

Description	ESTRO Booklet #8 Minimum Requirements		AAPM TG-56 1997 Minimum Requirements	
Safety Systems	Test Frequency	Action Level	Test Frequency	Action Level
Warning lights	Daily/3M*	–	Daily	–
Room monitor	Daily/3M*	–	Daily	–
Communication equipment	Daily/3M*	–	Daily	–
Emergency stop	3M	–	3M	–
Treatment interrupt	3M	–	3M	–
Door interlock	3M	–	Daily	–
Power loss	3M	–	3M	–
Applicator and catheter attachment	6M	–	Daily	–
Obstructed catheter	3M	–	3M	–
Integrity of transfer tubes and applicators	3M	–	3M	–
Timer termination	Daily	–	Daily	–
Contamination test on r.a. leak	A	–	6M (NRC)	–
Leakage radiation, survey around machine	A	–	3M	–
Emergency equipment (forceps, emergency safe, survey meter)	Daily/3M*	–	Daily	–
Practicing/training (emergency) procedures	A	–	A	–
Hand crank functioning	A	–	A	–
Handheld monitor	3M/A**	–	Daily	–
Physical Parameters				
Source calibration	SE	>5%	Daily/3M	5% (d) / 3% (3M)
Source positioning	Daily/3M*	>2 mm	Daily/3M	>1 mm (d) / >0.5 mm (3M)
Length of treatment tubes	A	>1 mm	3M	>1 mm
Irradiation timer	A	>1%	Daily	>2%
Date, time, and source strength in treatment unit	Daily	–	Daily	–
Transit time effect	A	–	A	–

Notes: 3M—quarterly; A—annual; SE—source exchange.

*Daily checks are assumed to be an implicit part of normal operation. The department's policy determines whether a separate logbook of these daily checks should be kept. A "formal" check by the medical physicist should be performed at least at the lower indicated frequency, for example, quarterly.

**The lower frequency determines the interval to verify the proper function of the handheld monitor, for example, with a known source of radiation.

TABLE 4.3 Frequencies and Tolerances of QC Tests for LDR Afterloading Equipment

Description	ESTRO Booklet #8 Minimum Requirements		AAPM TG-56 1997 Minimum Requirements	
Safety Systems	Test Frequency	Action Level	Test Frequency	Action Level
Warning lights	Daily/3M*	–	Daily	–
Room monitor	Daily/3M*	–	Daily	–
Communication equipment	Daily/3M*	–	Daily	–
Emergency stop	6M	–	3M	–
Treatment interrupt	6M	–	3M	–
Door interlock	6M	–	3M	–
Power loss	6M	–	3M	–
Air pressure loss	6M	–	3M	–
Applicator and catheter attachment	6M	–	3M	–
Obstructed catheter	6M	–	A	–
Integrity of transfer tubes and applicators	6M	–	A	–
Timer termination	Daily	–	Daily	–
Leakage radiation, survey around machine	A	–	6M (NRC)	–
Contamination test on r.a. leak	A	–	A	–
Emergency equipment (forceps, emergency safe, survey meter)	Daily/3M*	–	Daily	–
Practicing/training (emergency) procedures	A	–	A	–
Hand crank functioning	A	–	A	–
Handheld monitor	3M/A**	–	Daily	–
Physical Parameters				
Source calibration, mean of batch	SE	>3%	A	>3%
Source calibration, individual source; decay	SE	>5%	A	>5%
Linear uniformity	SE	>5%	SE	–
Source position, source length	6M	>2 mm	3M	>1 mm
Irradiation timer	A	>2 s	3M	–
Date, time, and source strength in treatment unit	Daily	–	Daily	–

Notes: 3M—quarterly; A—annual; SE—source exchange.

*Daily checks are assumed to be an implicit part of normal operation. The department's policy determines whether a separate logbook of these daily checks should be kept. A "formal" check by the medical physicist should be performed at least at the lower indicated frequency, for example, quarterly.

**The lower frequency determines the interval to verify the proper function of the handheld monitor, for example, with a known source of radiation.

correct functioning, as mentioned in this section, are summarized in Table 4.4.

4.5.1 Regular Tests of Applicators and Transfer Tubes

Due to clinical usage, radiation damage (applicators with plastic parts), and sterilization process, the mechanical properties of applicators may change. The transfer tubes can also age due to residue accumulation from source cable friction and possible kinks during treatment (notably for PDR cases with large patients). The mechanical integrity of all applicators and transfer tubes, including the connectors, should be regularly checked. It is especially necessary to check the length of reusable plastic applicators and (nonmetallic reinforced) transfer tubes because of the possibility of expansion or shrinkage over time. Depending on the type of the afterloading machine, applicators and transfer tubes that are too short may lead to an overlap of

several dwell positions if the advancing source hits the end of the applicator before it reaches the first programmed position if there is not a properly functioning interlock system. The length of the applicator–transfer tube combination can simply be measured with a wire of appropriate length, a check ruler, or a high-resolution camera system. Batches of transfer tubes should be labeled and tracked to differentiate transfer tubes and their time of service length (Tables 4.4 and 4.5).

4.5.2 Contamination, Cleaning, and Sterilization

High activity sources such as ^{192}Ir HDR sources should never be touched by hand during wipe tests performed for verification that the surface of the source is not contaminated with radioactivity. Therefore, a procedure must be used as described in Section 4.2. If any signal is detected above background radiation, the contaminated parts must be checked in more detail,

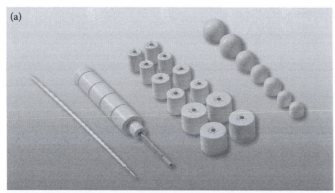

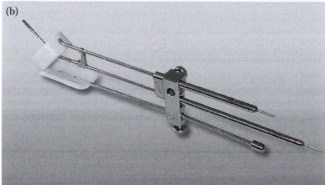

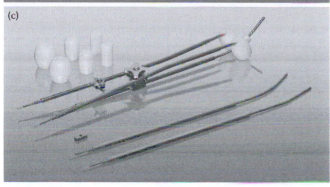

FIGURE 4.7 CT and MR compatible applicator sets. (a) Segmented vaginal set. (b) Ring applicator set. (c) Fletcher-Suit-Delclos ovoids and intrauterine tandem set. (Courtesy of Varian Medical Systems, Palo Alto, CA, USA.)

TABLE 4.4 Steps and Procedures for Verification of Correct Functioning of Applicators and Appliances

Article	Procedure
Integrity of applicator materials	Visual inspection of the applicators, depending on their use: before or after each treatment (such as in the case of reusable materials)
Fixation mechanisms	Check each fixation screw and mechanism for proper functioning
Shielding in applicators	Check for the presence and position of shields included in the applicator, at acceptance (radiography)
Source positioning	Autoradiography whenever applicable for verification of source position, at acceptance or when there is suspicion of (length) changes in the applicators
Identification of connecting mechanisms	Check the integrity of the applicator in relation to its use or to its connection to the afterloader, at acceptance
Sterilization procedures	Check the presence of the instructions for sterilization and follow these instructions meticulously to avoid unintended damaging
Validity of dose distributions in relation to specific applicators	Carefully check the applicability of any dosimetrical "atlas" for precalculated and tabulated treatment times, at acceptance
Radioactive contamination	Careful handling with the applicators at detachment and check the tubes in a NaI crystal to detect possible leakage or contamination

Source: Venselaar, J.L.M. and Pérez-Calatayud, J. (eds.). *A Practical Guide to Quality Control of Brachytherapy Equipment*, 1st ed., Chapter 6. ISBN 90-804532-8, ©2004 by ESTRO.

4.6 Frequencies and Tolerances

Frequencies and tolerances for the individual QC tests and items for HDR (and PDR) systems are listed in Table 4.2; for LDR systems, see Table 4.3. These tables contain two sets of recommendation, illustrating some differences in interpretation. The first set is taken from the ESTRO Booklet No. 8 (Venselaar and Pérez-Calatayud 2004), and the second set from AAPM TG-56 (Nath et al. 1997). The ESTRO table is rather concise indicating the main aspects of QC and the associated frequencies, whereas the AAPM report is very valuable in that it is much more detailed and also contains more information on the methods for the QC tests and their alternatives. Therefore, the reader is referred to the original documents.

In this ESTRO booklet, the recommendations for a given test frequency must be considered as a *minimal* and not as an *optimal test frequency*. An increase in the frequency of the test is required whenever the stability of the system is suspect, or when the specific treatment method demands a special accuracy. The medical physicist should carefully consider which recommended test frequency is applicable in his or her clinical situation, considering

and there should be no clinical use of the source and/or afterloader until the cause of the contamination is clarified. The radiation safety officer must be informed immediately. Any further spread of radioactive contamination must be avoided.

Cleaning and sterilization of reusable applicators and transfer tubes should be performed only with methods given by the manufacturer of the product. If there are no cleaning and sterilization methods described in the instructions for use, the manufacturer should be contacted unless it is possible to use a normal procedure, for example, for the cleaning and sterilization of metallic applicators. Applicators specified for single use by the manufacturer may not be reused.

TABLE 4.5 QC Checks for Various Treatment Appliances for HDR and LDR Brachytherapy

Article	Feature to be Tested	Frequency
	Gynecological appliances	
Tandems	Flange screws function	Each use
	Curvature	Each use
	Closure caps function	Each use
	Plastic sleeve and rod fit and slide	Each use
Fletcher-type ovoids	Source carrier function	Each use
	Integrity of welds	Each use
	Position of shields	Semiannual/after repair
	Identification markers	Each use
	Attenuation of ovoids	Acceptance
	Bridge integrity/thumb screws	Each use
Henschke applicators	Placement of caps on stem	Each use
	Bridge integrity, screws in bridge	Each use
	Closure caps function	Each use
Tandem-based cylinders and tandem checks	Flanges function	Each use
	Identification markers	Each use
	Cylinders fit snugly	Each use
Solid cylinders	Source carrier function	Each use
	Closure caps function	Each use
Intraluminal catheters	Integrity	Each use (after steril.)
	Strength of tip	Each use (after steril.)
	Interstitial equipment	
Source holding needles	Straightness	Each use
	Patency	Each use
	Integrity	Each use
	Sharpness	Each use
	Bevel	Each use
	Diameter	Each use
	Length	Each use
	Connection	Each use
	Collar (template)	Each use
	Funnel integrity	Each use
Catheter-inserting needles	Straightness	Each use
	Sharpness	Each use
	Bevel	Each use
	Diameter	Each use
Catheters	Integrity	Each use
	Diameter	Each use
	Length	Each use
Plastic buttons	Fit snugly yet slide	Each use
	Do not narrow catheter	Each use
Metal buttons	Slide onto catheters	Each use
Catheters with buttons attached	Buttons firmly attached	Just before use

(continued)

TABLE 4.5 (Continued) QC Checks for Various Treatment Appliances for HDR and LDR Brachytherapy

Article	Feature to be Tested	Frequency
	Interstitial equipment	
Templates	Hole placement	Acceptance
	Hole angulation	Acceptance
	Needle guidance	Acceptance
	Needle fixation	After each use
	Obturator fit	Each use
	Obturator screw function	Each use
Ultrasound templates	Rotational alignment	Semiannually
	Scaling	Semiannually
	Surface applicators	
Skin applicators	Thickness	Acceptance
	Source position	Acceptance
	Applicator fixation	Each use

Source: Slightly adapted from Thomadsen, B.R. *Achieving Quality in Brachytherapy.* Chapter 8. Institute of Physics Publishing. Medical Science Series. Bristol and Philadelphia, 2000.

- The likelihood of the occurrence of a malfunction
- The seriousness of the possible consequences of an unnoticed malfunction to patients and/or to the personnel
- The chances that, if a malfunction occurs, this will not be identified during normal treatment applications

For example, it is recommended to *formally* check the performance of the warning lights in the treatment room, the proper functioning of the room radiation detector, and the audio and/or video communication system for the patient only once every 3 to 4 months and then to record the results of the check in a logbook. The reason is that it can be assumed that a malfunction of any of these systems will be quickly detected by the radiation technologists during their routine work. Starting the treatment and signing the documents for that treatment may implicitly assume that the daily tests were performed and that the results were satisfactory, according to a department's written policy. Other departments may wish to develop special daily check forms to record and sign for the execution of these tests on satisfactory completion.

The ESTRO report states that the tables of their recommendations should not be interpreted as a rigid prescription of tests and frequencies, but rather as a starting point for developing a written QA program, individualized to the needs of each institution. Nevertheless, due to differences in the liability issues between countries, the practical interpretation of the reports from the professional organization leaves less room to the medical physicists in the United States. The requirements must be strictly followed, and any deviation from these recommendations must be well documented in the institution. In addition, local authorities like the NRC and agreement states in the United States may have developed their own sets of guidelines, which then must be followed.

Note that daily tests should be executed on a routine basis before treating the first patient of the day (1 h at least to allow for time to fix any issues, and, if unable to fix, then to call the service engineer), and only on days when patients are treated. For most of the tests in the tables, a quarterly interval is suggested because this is usually the frequency with which HDR sources are replaced. Some departments may apply a 4-month interval instead, if source replacement takes place only three times annually. The QC checks, which are performed quarterly or with a lower frequency, must be explicitly logged in a logbook, which is kept by the physicist.

Note further that the data, provided in the tables as "action level," reflect the upper limit in clinical conditions. For an acceptance test, the design specifications must be compared. Often the design of the system is such that a much better performance can be obtained under reference conditions, such as positional checks on straight catheters with autoradiography or with a high-resolution video camera system on a check ruler. Larger deviations will be observed with curvatures in the catheters or in the applicators.

It is the task of the medical physicist to inspect the performance history of the system thoroughly using the data in the logbook noted during the clinical lifetime of the equipment.

As indicated above, a quick detection of any malfunction means that the risks for mistreatment for the patient are minimal. A thorough analysis of risks in a department may reveal the weaker points of a clinical practice. This is the approach favored by many, thus advocating the introduction of a *Failure Mode and Effects Analysis* (FMEA) system. Thomadsen, for example, discussed this in "Critique of Traditional Quality Assurance Paradigm" (Thomadsen 2008). Referring to his own book on QA procedures for brachytherapy alone, he counted a total of 239 pages on the routine aspects of QA (Thomadsen 2000). Thus, "the conglomeration of all these reports and recommendations has left the clinical physicist with a rapidly expanded job just addressing QA. Even though most of the reports and texts suggest that not every test is appropriate for every institution, the standard was set such that failure to follow all recommendations appeared negligent, certainly to many state radiation regulators." Just performing all the recommended QA might not provide protection against all possible events. Instead, each facility should consider its own needs for addressing potentially hazardous situations and, particularly, target measures for control.

An FMEA would also include assessing the efficacy of any QA procedure in their setting. Such an assessment should demonstrate potentially hazardous situations, the frequencies with which they can occur, and the associated risks for patient, personnel, or environment. Risk analysis is in itself a time-consuming exercise. But the situations to describe and analyze certain clinical practices are often very similar for different users. The AAPM and other national or international organization have recognized this. As an example, AAPM TG-100 *Method for Evaluating QA Needs in Radiation Therapy* (M. Saiful Huq) was, in part, charged with the task (see www.aapm.org) of identifying a structured and systematic QA program that balances patient safety and quality versus commonly available resources and strikes a good balance between prescriptiveness and flexibility.

It is explicitly stated that the task group members are also working on FMEA analysis for HDR brachytherapy. Still, until this promising approach has been established and received full support from stakeholders, the existing reports that form the basis of this chapter should be followed.

References

Butler, W.M. and Merrick, G.S. 2008. Clinical practice and quality assurance challenges in modern brachytherapy sources and dosimetry. *Int J Radiat Oncol Biol Phys* 71(Supplement):S142–6.

Cormack, R.A. 2008. Quality assurance for computed tomography, ultrasound, and magnetic resonance imaging-guided brachytherapy. *Int J Radiat Oncol Biol Phys* 71(Supplement): S136–41.

DeWerd, L.A., Jursinic, P., Kitchen, R., and Thomadsen, B. 1995. Quality assurance tool for high dose rate brachytherapy. *Med Phys* 22:435–40.

DeWerd, L.A., Ibbott, G.S., Meigooni, A.S. et al. 2011. A dosimetric uncertainty analysis for photon-emitting brachytherapy sources: Report of AAPM Task Group No. 138 and GEC-ESTRO. *Med Phys* 38:782–801.

Ibbott, G.S., Ma, C.M., Rogers, D.W.O., Seltzer, S.M., and Williamson J.F. 2008. Anniversary paper: Fifty years of AAPM involvement in radiation dosimetry. *Med Phys* 35:1418–27.

International Atomic Energy Agency (IAEA). 2002. Calibration of photon and beta ray sources used in brachytherapy. Guidelines on Standardized Procedures at Secondary Standards Dosimetry Laboratories (SSDLs) and Hospitals. IAEA-TECDOC-1274, March 2002. IAEA, Vienna.

International Commission on Radiation Units and Measurements (ICRU). 1985. *Dose and Volume Specification for Reporting and Recording Intracavitary Therapy in Gynecology. Report 38 of ICRU*. ICRU Publications, Bethesda, MD.

International Commission on Radiation Units and Measurements (ICRU). 1997. *Dose and Volume Specification for Reporting Interstitial Therapy. Report 58 of ICRU*. ICRU Publications, Washington, DC.

ISO 8402. 1994. Quoted by Peach, R.W. 1992 in *The ISO 9000 Handbook*. McGraw-Hill, New York.

Kubo, H.D., Glasgow, G.P., Pethel, T. et al. 1998. High dose-rate brachytherapy treatment delivery: Report of the AAPM Radiation Therapy Committee Task Group No. 59. *Med Phys* 25:375–403.

Kutcher, G.J., Coia, L., Gillin, M. et al. 1994. Comprehensive QA for radiation oncology: Report of AAPM Radiation Therapy Committee Task Group 40. *Med Phys* 21:581–618.

Nath, R., Anderson, L.L., Luxton, G. et al. 1995. Dosimetry of interstitial brachytherapy sources: Recommendations of the AAPM Radiation Therapy Committee Task Group No. 43. *Med Phys* 22:209–34.

Nath, R., Anderson, L.L., Meli, J.A. et al. 1997. Code of practice for brachytherapy physics: Report of the AAPM Radiation Therapy Committee Task Group No. 56. *Med Phys* 24:1557–98.

Pérez-Calatayud, J., Ballester, F., Das, R.K. et al. 2012. Dose calculation for photon-emitting brachytherapy sources with average energy higher than 50keV: report of the AAPM and ESTRO. *Med Phys* 39(5):2904–29.

Rivard, M.J., Coursey, B.M., DeWerd, L.A. et al. 2004. Update of AAPM Task Group No. 43 Report: A revised AAPM protocol for brachytherapy dose calculations. *Med Phys* 31:633–74.

Rivard, M.J., Butler, W.M., DeWerd, L.A. et al. 2007. Supplement to the 2004 update of the AAPM Task Group No. 43 Report. *Med Phys* 34:2187–205.

Rivard, M.J., Venselaar, J.L.M., and Beaulieu, L. 2009. The evolution of brachytherapy treatment planning. *Med Phys* 36:2136–53.

Rivard, M.J., Beaulieu, L., and Mourtada, F. 2010. Enhancements to commissioning techniques and quality assurance of brachytheraphy treatment planning systems that use model-based dose calculation algorithms. *Med Phys* 37:2645–58.

Thomadsen, B.R., Houdek, P.V., Van Der Laarse, R. et al. 1994. Treatment planning and optimization. In: *High Dose Rate Brachytherapy: A Textbook*. Library of Congress Cataloging-in-Publication Data, S. Nag (ed.). Futura Publishing Company Inc., Armonk, New York.

Thomadsen, B.R. 2000. *Achieving Quality in Brachytherapy*. Institute of Physics Publishing. Medical Science Series. Bristol and Philadelphia.

Thomadsen, B.R. 2008. Critique of traditional quality assurance paradigm. *Int J Radiat Oncol Biol Phys* 71(Supplement):S166–9.

Thomadsen, B.R., Williamson, J.F., Rivard M.J., and Meigooni, A.S. 2008. Anniversary paper: Past and current issues, and trends in brachytherapy physics. *Med Phys* 35:4708–23.

Venselaar, J.L.M. and Pérez-Calatayud, J. (eds.). 2004. *A Practical Guide to Quality Control of Brachytherapy Equipment*, 1st ed., ISBN 90-804532-8, ©2004 by ESTRO, Brussels. www .estro-education.org/publications/Documents/booklet8_ Physics.pdf.

Williamson, J.F., Ezzell, G.A., Olch, A., and Thomadsen, B.R. 1994. Quality assurance for high dose rate brachytherapy. In: *High Dose Rate Brachytherapy: A Textbook*. Library of Congress Cataloging-in-Publication Data, S. Nag, (ed.). Futura Publishing Company Inc., Armonk, New York.

Williamson, J.F. 2006. Brachytherapy technology and physics practice since 1950: A half-century of progress. *Phys Med Biol* 51:R303–25.

Williamson, J.F. 2008. Current brachytherapy quality assurance guidance: Does it meet the challenge of emerging image-guided technologies? *Int J Radiat Oncol Biol Phys* 71(Supplement):S18–22.

Yu, Y., Anderson, L.L., Li, Z. et al. 1999. Permanent prostate seed implant brachytherapy: Report of the American Association of Physicists in Medicine Task Group No. 64. *Med Phys* 26:2054–76.

Brachytherapy Dosimetry

Source Calibration

Hans Bjerke
*Norwegian Radiation
Protection Authority*

Larry A. DeWerd
University of Wisconsin

Jan P. Seuntjens
Montreal General Hospital

Margaret Bidmead
*Royal Marsden NHS
Foundation Trust and Institute
of Cancer Research*

5.1 Introduction: Clinical Background

Clinical dosimetry relies on a foundation of source calibration. This chapter discusses the framework in which the calibration of brachytherapy sources takes place, with the aim of achieving a high level of accuracy in clinical dose delivery when applying this treatment modality. Precise knowledge of the strength of the source is one of the steps in the dosimetry chain. The dosimetry of brachytherapy sources depends on measurements of the radiation from the source defined in the unit of gray. The quantity for dosimetry is the air kerma, either as in the reference air-kerma rate (RAKR, here the notation $\dot{K}_R$ is used) or the air-kerma strength, S_K. The quantity of interest in all radiation therapy is the absorbed dose to water, D_W, which is calculated from the quantity air kerma.

The value of the source strength is used by medical physicists in the algorithms that are clinically applied for dose calculations, sometimes for simple manual calculations but usually in the treatment planning system(s). At present, there is one de facto standard of a dose calculation formalism that is used by virtually all treatment planning systems worldwide. This formalism was developed in the United States and published as a task group report of AAPM Brachytherapy Subcommittee in the report TG-43 (Nath et al. 1995). In the revised TG-43 2D

formalism (Rivard et al. 2004, 2007; see Chapter 10 for further details), the dose rate is given by

$$\dot{D}(r,\theta) = S_K \Lambda \frac{G_L(r,\theta)}{G_L(r_0,\theta_0)} g_L(r) F(r,\theta) \tag{5.1}$$

but for the source calibration, we determine the dose rate at the reference point (in the polar coordinate system for the source orientation; see Figure 10.1) and the equation reduces to

$$\dot{D}(r_0,\theta_0) = S_K \Lambda \tag{5.2}$$

for the quantity air-kerma strength or to

$$\dot{D}(r_0,\theta_0) = \dot{K}_R \Lambda \tag{5.3}$$

for the quantity RAKR. The dose rate constant, Λ, converts the air-kerma strength S_K or RAKR into $\dot{D}_W$. The reference point for the source calibration is chosen to be at a distance of 1 cm from the source center ($r_0 = 1$ cm, at $\theta_0 = \pi/2$). When using SI units for the quantities in Equation 5.1, the values of Λ for actual sources are close to 1×10^4, with a range of 0.7–1.3 ($\times 10^4$), maintaining the same unit of time (h^{-1}) and with the $\dot{K}_R$ defined at 1 m in vacuo. If the S_K is used for the source calibration, it usually refers to air-kerma rate at 1 cm, and the distance correction (10^4) is not needed. When using air-kerma strength, there is often confusion about different units (μ or c), distance (cm and m), and time units (h, min, or s), and care must be taken to use the correct, consistent units.

The aim of clinical source calibration is twofold: (1) to ensure traceability to international standards, and (2) to determine a value to enter into the treatment planning system and the treatment machine. The traceability of the dose, either through absorbed dose or air kerma, is important as it simplifies national and international comparison of treatment results. In the following sections, the calibration procedure both at the level of the standards laboratories and the clinics is described.

5.2 Framework and Formalism for Calibration

5.2.1 Framework

The system for radiation metrology at Primary Standard Dosimetry Laboratories (PSDLs) and Secondary Standard Dosimetry Laboratories [SSDLs; in the United States, these secondary laboratories are called Accredited Dosimetry Calibration Laboratories (ADCLs) with direct traceability to the National Institute of Standards and Technology (NIST)] provides the framework for consistency in radiation dosimetry by providing a service for calibration of users' radiation instruments. The Bureau International des Poids et Mesures (BIPM) (http://www.BIPM.org) was set up by the Metre Convention in 1875 as the international center for metrology, with the laboratory and offices in Sèvres (Paris, France), in order to ensure worldwide uniformity on matters relating to metrology, including radiation metrology. The PSDLs have developed several primary standards for brachytherapy dosimetry. Those national laboratories that maintain primary standards calibrate the secondary standards of the ADCLs or SSDLs, which, in turn, calibrate the reference instruments of the users. Some PSDLs also calibrate users' reference instruments.

As an example, the BIPM governs an ongoing comparison of the $\dot{K}_R$ for high-dose-rate (HDR) ^{192}Ir [BIPM.RI(I)-K8 2011] in which 11 laboratories are participating. The recommended units used at present in the calibration certificates are Gy s^{-1} or μGy h^{-1}.

The transfer of a "calibration" from a standards laboratory can be done either by an instrument calibrated for the same source (source type) or by using a source that is calibrated at the lab. According to the recommendations, the medical physicist is responsible for the in-house calibration of the source or sources used clinically (Butler et al. 2006). It is a common practice to have the instruments calibrated regularly, for example, every 2 years, at one of the laboratories.

Methods used in the past for clinical source calibration were with a so-called "in-air" measurement technique or by using a well-type ionization chamber. A third method is to use a solid phantom specifically designed for calibration purposes. These methods are extensively described in the literature, and sometimes, these appear explicitly in national regulations as the recommended method of calibration. In the United States, a calibrated, traceable well chamber is the only recommended method for clinical calibrations.

For the in-air method, a chamber is positioned at one or several well-determined distances from the source, and the reading is recalculated to the reference distance with the inverse square law. A scatter-free rigid jig is used to keep the chamber at the distance of—usually—5 to 20 cm from the source. The method cannot be used for ^{125}I or ^{103}Pd due to the low energy and low intensity of the photons emitted from these sources. Some reasons for the unsuitability are as follows:

- The uncertainty in the air-kerma calibration coefficient for an air cavity chamber at these low photon energies is unacceptably high.
- In general, a low-energy photon source does not have a sufficiently high RAKR for in-air measurements. This, in combination with a possibly high leakage current, means that such measurements are subject to a large uncertainty.

Air humidity may affect the attenuation of the low-energy photons, thus affecting the measured current more than is the case, for example, in measurements with ^{192}Ir brachytherapy sources. In this chapter on in-air calibration techniques, correction factors are given for the calibration of ^{192}Ir HDR sources only. However, the correction factors have only minor energy dependence, and they can therefore be used, without loss of accuracy, in calibration of ^{60}Co and ^{137}Cs brachytherapy sources.

A solid phantom to keep an ion chamber at a fixed distance of, for example, 5 cm from the source allows for more precision in the distance between the source and the detector but requires standardized dimension and known composition. The uncertainties due to the influence of the material composition on the measurement results are especially relevant for the low-energy photon-emitting sources, and therefore, this method is considered by many as only being suitable for performing relative measurements. The in-phantom measurement technique is not discussed in greater detail in this chapter.

For both phantom and "in-air" methods, the measured charge or current is strongly dependent on the source-to-detector distance, and errors in the distance may yield large uncertainties in the source calibration. The signal-to-noise ratio can be very low, requiring long measurement times. For the phantom and "in-air" methods, a Farmer-type chamber may be used as a reference instrument.

Present-day recommendations such as from AAPM for clinical brachytherapy calibration procedures essentially all focus on the use of dedicated well-type chambers. For well chambers, measurements may be dependent on the source position inside the chamber, and the user must identify the optimal position of the source to be measured at the central axis of the well chamber. Thus, specialized inserts are made for source types to precisely locate the source within the chamber.

Air-kerma calibration coefficients for ionization chambers are available from PSDLs, SSDLs, and ADCLs for several radiation qualities, which can be weighted to the spectrum of the given brachytherapy source. The same laboratories may also directly calibrate well chambers in RAKR or S_K. Absorbed dose-to-water calibration coefficients are also offered at some laboratories. At PSDLs and some ADCLs, the calibration will usually be done in house, whereas most SSDLs need to go out to the clinic to perform a cross calibration.

5.2.2 Formalisms for Measuring RAKR

Generally, the PSDL or ADCL measures the RAKR or S_K using some standard reference method. The clinic then measures the quantity using a calibrated ionization chamber (a well chamber).

5.2.2.1 In-Air Measurement Technique

The RAKR is a quantity specified at the distance of 1 m compared to the air-kerma strength, which is specified at 1 cm. $\dot{K}_R$ and S_K are numerically the same at 1 m. For low-energy sources (energy below 50 keV), the quantity is specified in vacuo. $\dot{K}_R$ is usually determined with an ionization chamber as the measuring device. A direct measurement at 1 m in air, however, is not always practical due to low signals and the possibly high leakage currents of the ionization chambers used. Often shorter distances are used, and the reading is recalculated to the reference distance.

The RAKR can be determined from measurements made in-air using the equation

$$\dot{K}_R = N_K \left(M_u / t \right) k_{\text{air}} k_{\text{scatt}} k_n \left(d / d_{\text{ref}} \right)^2 \qquad (5.4)$$

where N_K is the air-kerma calibration coefficient of the ionization chamber at the actual photon energy; M_u is the measured charge collected during the time t and corrected for ambient temperature and pressure, recombination losses, and transit effects during source transfer in the case of afterloading systems; k_{air} is the correction for attenuation of the primary photons by the air between the source and the chamber; k_{scatt} is the correction for scattered radiation from the walls, floor, measurement setup, air, etc.; k_n is the nonuniformity correction factor, accounting for the nonuniform electron fluence within the air cavity; d is the measurement distance, that is, the distance between the center of the source and the center of the ionization chamber; and d_{ref} is the reference distance of 1 m.

Some details of the correction factors in Equation 5.4 will be discussed later in this chapter.

5.2.2.2 Well-Type Chamber Measurements

In a well-type chamber, the RAKR $\dot{K}_R$ in terms of Gy s^{-1} at 1 m can be determined from

$$\dot{K}_R = M k_{\text{ion}} k_{\text{sg}} N_{\dot{K}_R} \qquad (5.5)$$

where M is the instrument response (in amperes), at the sweet spot, corrected to standard atmospheric conditions; k_{ion} is the ion recombination correction factor determined with the hospital source at the time of measurement; k_{sg} is the source geometry factor (in case of a source different from the source at primary standards laboratory); and $N_{\dot{K}_R}$ is the calibration coefficient for the radiation quality to convert the corrected instrument reading to RAKR.

The instrument response, M, is given by

$$M = I_{\text{raw}} f_{\text{elec}} k_{\text{dec}} k_{\text{Tp}} \qquad (5.6)$$

where I_{raw} is the ionization current displayed on the electrometer; f_{elec} is the electrometer correction factor to convert the displayed current (in amperes) to the actual current; k_{dec} is the decay correction factor to correct the measured current to a reference time; and k_{Tp} is the air density correction factor to normalize the measured current to standard atmospheric conditions if this is required for the type of instrument:

$$k_{\text{Tp}} = \frac{(273.15 + T)}{293.15} \times \frac{1013.25}{p} \qquad (5.7)$$

where T is the air temperature (in degree Celsius), and p is the ambient air pressure (in hectopascals).

5.2.3 Formalism for Dose Rate in Water

When the calibration coefficient in absorbed dose rate to water at 1 cm is available for the well chamber, the $\dot{K}_R \Lambda$ in Equation 5.3 will be exchanged with $\dot{D}_{\text{W,1}}$:

$$\dot{D}\left(r_0, \theta_0 \right) = \dot{D}_{\text{W,1}} \qquad (5.8)$$

and the equation for the source calibration is

$$\dot{D}_{\text{W,1}} = N_{\text{D,W}} \times \left(M_u / t \right) \qquad (5.9)$$

where $N_{\text{D,W}}$ is the absorbed dose-to-water calibration coefficient for the actual source, and M_u and t are the same as in Equation 5.4.

The change in the formalism is intuitive and simple, but the change to absorbed dose-to-water formalism requires control of both the new value for the calibration coefficient for the well chamber and a treatment planning system that is prepared for new input values. It is noted that the well-chamber method is recommended, but alternative dose-to-water methods based on their respective dose-to-water calibration factors could be used as well. Then, the source calibration method itself would be, in principle, the same as for the air-kerma measurement.

5.3 Primary Standards for Brachytherapy Source Calibrations

Measurements of RAKR and absorbed dose rate to water from brachytherapy sources at PSDLs establish the reference conditions for source calibration. For clinical work, valuable information on methods, instrumentation, and correction factors may be found here. Under controlled laboratory condition, not only are the standards developed, but also source characterization like the source anisotropy function values are determined. The air-kerma standards for HDR sources have many things in common with the in-air measurement technique, whereas air-kerma standards for the low-energy sources are large free air chambers or extrapolation chambers with variable volume. Absorbed dose-to-water standards for low energy are also based on free air chambers, but modified with water-equivalent materials, special design, and small phantom. Absorbed dose-to-water standards for HDR sources are calorimeters (water or graphite) determining the dose rate at 1-cm distance from the source. All techniques use Monte Carlo simulations for determination of correction factors.

5.3.1 Air-Kerma Standards

In 2004, the National Physical Laboratory (NPL) set up a new calibration service for dosemeters used to determine the RAKR of HDR ^{192}Ir brachytherapy sources (Sander and Nutbrown 2006). The NPL primary standard is a spherical cavity chamber of 102.52 cm^3. Other HDR primary standards use the interpolation method; the RAKR of the HDR source is measured with chambers (Farmer type or spherical) having a flat energy response function. Reports from comparisons reveal that these standards are in good agreement well below 1% (Douysset et al. 2005, 2008). Also, all afterloaders with their associated ^{192}Ir sources agree with each other to within 1% (Rasmussen et al. 2011).

The primary air-kerma-strength standard for low-energy sources is maintained at the NIST using a Wide-Angle-Free-Air Chamber (WAFAC) (Mitch and Soares 2009; Seltzer et al. 2003). The chamber design is based on the free air-kerma standard for x-ray dosimetry. At Physikalisch-Technische Bundesanstalt (PTB), Braunschweig, Germany, a large volume parallel plate extrapolation chamber (GROVEX) serves as the primary standard (Selbach et al. 2008). The measuring principle is based on the fact that the air-kerma rate is proportional to the difference between the ionization currents divided by the difference between the respective electrode distances, as long as it is guaranteed that the electrode distance is larger than the range of the secondary electrons emerging in front of the rear electrode. The GROVEX is thus a quasi-wall-less air-filled ionization chamber. The low-energy brachytherapy source is placed with the source axis 30 cm from the chamber and rotated during measurements. In both instruments, an aluminum filter is used to eliminate the low-energy emissions (<5 keV) that have no clinical significance.

(a)

(b)

FIGURE 5.1 (a) WAFAC maintained at the NIST. (Courtesy of M. Mitch, 2009.) (b) GROVEX chamber developed at PTB, Braunschweig. (Courtesy of H.J. Selbach.)

Figure 5.1 shows the WAFAC and the GROVEX chambers in their laboratory environment.

5.3.2 Absorbed Dose-to-Water Standards

A new and direct absorbed dose-rate-to-water calibration of a brachytherapy source allows for a simpler, more consistent, procedure for the dosimetry formalism. If the standards laboratories provided the community with a $\dot{D}_{W,1}$ calibration factor, some uncertainties may be reduced since the dose rate function quantity, Λ, would be eliminated.

For that reason, in 2008, PSDLs in Europe collaborated in the EURAMET iMERA project (http://www.EMRP.org) and developed several standards for directly measuring $\dot{D}_{W,1}$, the dose rate at the reference point with an uncertainty level below 2% ($k = 1$) for a number of photon sources. The traceability may, in the future, therefore be directly applied to absorbed dose-to-water and not to air-kerma standards. However, do note that the air-kerma standards have an uncertainty of 1.5% ($k = 1$). The quantities air-kerma strength (or RAKR) and the dose rate constant will then be replaced by D_w.

The principles of the standards have been described in the literature. The absorbed dose-to-water calorimeter for external beams was proposed by Domen (1980), and the realizations of these present standards for HDR ^{192}Ir brachytherapy sources started after 2005 (Seuntjens and Palmans 1999; Sarfehnia and Seuntjens 2010). The absorbed dose rate is determined at 1-cm distance from the source, applying the same distance as clinical reference and input to treatment planning systems. The PTB developed a water calorimeter for the calibration service of HDR ^{192}Ir and ^{60}Co sources (Selbach and Meier 2010). The Van Swinden Laboratory (VSL) standard for HDR ^{192}Ir source calibration is also based on water calorimeter (De Pooter and De Prez 2010). At the McGill University, the standard is, as the PTB standard, a parallel-plate calorimeter (Sarfehnia and Seuntjens 2010). The NPL standard for HDR ^{192}Ir source calibration is based on a ring-shaped graphite calorimeter (Palmans 2007), and the Istituto Nazionale di Metrologia delle Radizioni Ionizzanti Agenzia Nazionale per le Nuove Tecnologie, l'Energia e lo Sviluppo Economico sostenible Italy (ENEA-INMRI) standard is also based on a ring-shaped graphite calorimeter (Toni 2010).

Reference is given to three absorbed dose-rate-to-water primary standards for ^{125}I low-dose-rate (LDR) brachytherapy sources. The Laboratoire National Henri Becquerel (LNE-LNHB) ^{125}I LDR primary standard is based on measurements using a torus free-air chamber and a spherical water-equivalent phantom, with radius of 1 cm; the source inside is perpendicular to the chamber. Monte Carlo simulations are performed for the correction factor calculations (Aubineau-Lanice et al. 2010). The PTB ^{125}I LDR primary standard is a large air-filled, parallel-plate extrapolation chamber with water-equivalent walls (Schneider and Selbach 2010). The measurement depth within the water phantom is defined by the entrance plate. The ENEA-INMRI ^{125}I LDR standard is based on the same principle as the PTB standard and is a large-angle, variable volume chamber. Monte Carlo simulation converts the measurements to absorbed dose rate to water at 1 cm from the source (Toni 2010).

5.4 Equipment

5.4.1 Ionization Chambers to Be Used

For HDR sources and in-air or in-phantom measurement techniques, ionization chambers with volumes greater than 0.5 cm^3 can be used (Farmer-type 0.6-cm^3 chamber). For LDR sources, ionization chambers of larger volumes, up to about 1000 cm^3, may be needed to obtain a sufficient signal especially at larger distances up to 1 m. For very large chambers, there is some uncertainty in their suitability for brachytherapy source calibrations (Verhaegen et al. 1992). Small chamber volumes that are too small may give problems due to a lack of signal and thus a low signal-to-noise ratio. The large-volume ionization chambers

are usually not part of the measuring equipment of the medical physicist.

Well chambers are the preferred instruments for use in the clinic. Those that are specially developed for brachytherapy dosimetry have many advantages for source calibration. The requirements for these are given in the standard (IEC 2009). The combination of the well-type chamber and the electrometer is unique, and together, it forms the "instrument" to be calibrated by the standards laboratory, although in the United States, the electrometer and the well chamber are calibrated separately.

Older types of well-type chambers, often found in departments of nuclear medicine, are not recommended for use for brachytherapy. The clinical user must be aware of the fact that if it is an open type of chamber prone to atmospheric changes, pressure and temperature corrections are required. In addition, if the chamber has been subjected to other environmental conditions, adequate stabilization times must be allowed; generally, letting it stabilize overnight is very adequate. The heat production by an HDR source may also influence the temperature in the chamber and may require specific attention. Some vendors have made inserts that eliminate this problem. Stability checks of such equipment should be performed on a semiannual or yearly basis. A redundant check with another brachytherapy well chamber is adequate for this check.

5.4.2 Measuring Assembly

A measuring assembly includes an electrometer for the measurements of current or charge and a power supply for the ionization chamber's polarizing voltage. The electrometer should have a good resolution (0.1% on the reading), and the long-term stability should not exceed ±0.5% over 1 year. The polarizing voltage should be easy to vary and reverse in order to determine the recombination and polarity effects. The electrometer must measure currents up to 200 nA and must be traceably calibrated in the current range –1 to –150 nA.

5.4.3 Jigs for In-Air Measurements and Solid Phantoms

Commercial jigs for the purpose of in-air measurements have been sold by Nucletron in a construction shown in Figure 5.2. Typical distances of 10 cm were used between the source and the ion chamber (two measurements could be taken, one with the source placed on the left and one on the right side of the ion chamber; results were averaged). Others have used the same principle, sometimes with unilateral position of the chamber, and sometimes with other distances between 5 and 20 cm or using multiple distances. The use of such a jig has decreased gradually since the 1990s, as most departments choose to use a well-type chamber instead. The uncertainty of such jigs is greater than the well chamber, and the measurement is more cumbersome.

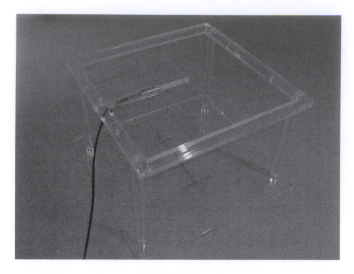

FIGURE 5.2 Calibration jig suitable to calibrate an HDR or pulsed-dose-rate (PDR) [192]Ir source to be positioned 10 cm left and right from a centrally placed Farmer-type ionization chamber. A small plastic tube is used in this jig to keep the catheters carefully at 20 cm apart. Readings with the source in the left and right catheters are averaged to correct for positional inaccuracies. (Courtesy of J. Venselaar.)

5.4.4 Solid Phantoms

Meertens (1990) described the use of a solid phantom for calibration of the Selectron-LDR [137]Cesium pellet sources. Three standard straight intrauterine metal tubes were placed at 5-cm distance around a Farmer ionization chamber centrally positioned in the phantom, and the sources were programmed to remain equally distributed in the treatment tubes around the chamber for a given amount of time. The chamber was then in a region of low-dose gradient. The solid phantom was made of a 15-cm-diameter Perspex (poly(methyl methacrylate) [PMMA]) cylinder of 15-cm height. The only known solid phantom ever to be formally part of a national requirements report on HDR dosimetry was the Krieger phantom, used in Germany (Krieger 1991; Krieger and Baltas 1999). As indicated above, the solid phantom methods are not further discussed in this report. Effects of phantom size were discussed by several investigators, for example, Pérez-Calatayud et al. (2004) and Granero et al. (2008).

5.4.5 Air-Kerma Calibration of Ionization Chambers

If an in-air calibration is to be done, the air-kerma calibration factor, N_K, is obtained from an SSDL or directly from a PSDL. Again, note that if a well chamber is used, it is calibrated directly for the brachytherapy source. In 2004, the NPL commissioned a primary standard for the realization of RAKR of HDR ^{192}Ir brachytherapy sources. This has meant that it is now possible to calibrate ionization chambers so that they are directly traceable to an air-kerma standard using an ^{192}Ir source (Sander and Nutbrown 2006). The chamber needs to be calibrated using

existing standards at other qualities (Goetsch et al. 1991; Borg and Rogers 1999; Borg et al. 2000; Van Dijk 2002; Van Dijk et al. 2004; Mainegra-Hing and Rogers 2006). The air-kerma calibration factor is then obtained by either an interpolation procedure or by polynomial fitting. The differences may be up to 1%, dependent on the method of fit (Van Dijk 2002). It is, however, not the aim of this chapter to address the different methods used for determining N_K for HDR [192]Ir. A recent paper compares the various methods of calibration and the various afterloaders for HDR ^{192}Ir and shows that all fall within ±1% (Rasmussen et al. 2011).

It is worth mentioning that primary standards exist for some types of ^{192}Ir LDR sources; for example, the NIST acting as the PSDL of the United States maintains a standard for calibrating ^{192}Ir seeds, and thus, the ADCLs can provide a calibration of a well chamber, which is the only instrument officially recommended for use by AAPM.

5.4.6 Correction Factors for In-Air Measurements

Equation 5.4 converts the reading of the ionization chamber to a $\dot{K}_R$ value. To obtain the RAKR value with the least possible uncertainty necessitates the use of up-to-date correction factors and careful performance of the in-air measurements. In this section, the various correction factors that are needed in the equation are discussed in detail.

5.4.6.1 Measurement Distances

Increasing the distance decreases the uncertainty in the measurement of the calibration distance and the effect of the finite size of the ionization chamber. However, this improvement results in a reduced signal and an increased relative importance of room and equipment scatter. There are four effects that contribute to the uncertainties in calibration of brachytherapy sources using an ionization chamber. These effects, expressed as a function of the source-to-chamber distance (SCD), are as follows:

- Uncertainty due to the effects of the chamber size. The uncertainty in the nonuniformity correction factor decreases with increasing SCD.
- Scatter, which as a percentage of the total signal increases with increasing SCD.
- Positional uncertainty, which follows the inverse square law and thus decreases with increasing SCD.
- Leakage current relative to the ionization reading, the effect of which increases with increasing SCD.

The measurement distance should be selected so that the combined uncertainty due to the above effects will be minimized. This would generally be the distance where the various correction factors, when combined in quadrature, have a minimum value. For a combination of ^{192}Ir HDR source and a Farmer-type chamber, the optimum distance has been shown to be 16 cm (DeWerd et al. 1994). With the possible exception of scattered radiation, it can be noted that the different contributions listed above have only minor energy dependence. Thus, the optimum distance for ^{60}Co source calibrations should be approximately the same as that for an ^{192}Ir HDR source.

It must be pointed out that the nonuniformity correction factors used in this chapter are calculated assuming point source geometry. Thus, in all in-air measurements, in HDR as well as LDR, the distances used must be large enough so that the source can be considered as a point source. Furthermore, the inclusion of the inverse square relation in Equation 5.2 implies that sufficiently large distances must be used. A practical criterion is that the distance between the chamber center and the center of the source must be at least 10 times the length of the source in order to ensure that the error introduced due to the point source approximation is less than 0.1%.

5.4.6.2 Scatter Correction Factor

To maintain the contribution of scattered radiation at a minimum, the source and the chamber should be placed in the center of the room and well above the floor (at least 1 m from any wall or floor). All measurements should preferably be carried out using the same jig (example shown in Figure 5.2) position within the room. Two methods have been used to determine the scatter correction: the multiple distance method (Goetsch et al. 1991) and the shadow shield method (Verhaegen et al. 1992; Drugge 1995; Piermattei and Azario 1997). In the former method, the air-kerma rate due to scattered radiation is assumed to be constant over the measurement distances. Monte Carlo analysis was performed to confirm constant room scatter at all distances. In addition, the Monte Carlo analysis confirmed that the scatter from the source holder, the seven-distance apparatus, and the HDR afterloader did not contribute to the air-kerma measurement results as long as the distance from the wall or floor is about 1.5 m (Rasmussen et al. 2011).

Both methods have their advantages and disadvantages. In the multiple-distance method, readings are made at a series of distances with carefully measured separations. These readings reflect the inverse square law differences between them and an assumed constant amount of scatter. It is essential in this method that the changes in distance be precise and accurate in order to derive the correction that yields the "true" center-to-center SCDs. Deviations from the inverse square law are then due to scattered radiation. On the other hand, if the measurement distances are kept within reasonable limits, for example, between 10 and 50 cm, then the scattered radiation is unlikely to vary too much, and the assumption of constant scatter is valid within certain limits. The advantage of the method is that it is simple to use and seems to agree quite well with measured scatter corrections.

5.4.6.3 Nonuniformity Correction Factor

In the measurements of brachytherapy sources in-air, the non-collimated geometry, with high divergence of the incident photons, differs from the geometry of collimated photon beams such as those external beams used for calibrating the chamber. Consequently, the chambers are not calibrated under the same conditions that are present during brachytherapy source calibrations, that is, the calibration does not include the effect of divergence of the photons.

The secondary electrons entering the air cavity are mainly generated when the photons interact with the inner wall of the ionization chamber. Due to the nonuniform photon fluence over the wall, the generation of secondary electrons from the wall varies significantly from place to place in the wall. The net result of this is nonuniform electron fluence in the air cavity of the chamber. In order to take account of this nonuniformity, it is necessary to apply a nonuniformity correction factor, k_n. This factor depends on the following:

- Shape and dimensions of the ionization chamber (spherical, cylindrical, internal radius, and length)
- Measurement distance and the source geometry ("point source," line source, etc.)
- Material in the inner wall of the chamber
- Energy of the photons emitted from the source

The most widely used nonuniformity correction factors are those given by Kondo and Randolph (1960). In their theory, the electron fluence in the air cavity of the ionization chamber is assumed to be isotropic. The theory was later extended by Bielajew (1990a,b) who included a more realistic angular distribution of electron fluence in the air cavity of the chamber. In contrast to the isotropic theory, this anisotropic theory predicts the wall material and photon energy dependence in the nonuniformity correction factor. The relationship between the two theories is given by

$$A_{pn}(d) = A_{pn}^{KR}(d) + \omega A'_{pn}(d) \tag{5.10}$$

where $1/A_{pn}^{KR}(d)$ is the nonuniformity correction factor obtained from the isotropic theory of Kondo and Randolph, and $1/A_{pn}(d)$ is the nonuniformity correction factor according to the anisotropic theory of Bielajew. $A'_{pn}(d)$ takes into account the anisotropic electron fluence within the air cavity, and the degree of anisotropy is given by the energy- and material-dependent factor ω. Thus, the theory by Bielajew predicts an energy and inner wall material dependence in the nonuniformity correction factor. In contrast, the theory by Kondo and Randolph is independent of both these factors. For Farmer-type chambers, it has been shown that the theory by Bielajew agrees better with experiments than that of Kondo and Randolph (Tölli et al. 1997). It is therefore recommended in this chapter that the factor $1/A_{pn}(d)$ according to the theory by Bielajew be used for determination of k_n. Thus,

$$k_n(d) = 1/A_{pn}(d). \tag{5.11}$$

For cylindrical ionization chambers, it has been shown that the nonuniformity correction factor obtained with the anisotropic theory is, for commonly used chamber wall materials, quite insensitive to the ω-values (Tölli et al. 1997). Values of ω for some commonly used inner wall materials, calculated for

[192]Ir photon energy, were presented in the ESTRO Booklet No. 8 (Venselaar and Pérez-Calatayud 2004). For materials that are not included in the table, a good approximation is to use the value for that material with similar dosimetric properties as that listed in Table 5.1. For example, the ω value for C552 plastic can be taken to be the same as that for graphite, that is, 0.992. It should be noted that the wall material referred to is the material of the inner wall of the ionization chamber, and not the material of the buildup cap.

The values in Table 5.1 were calculated for an unfiltered ^{192}Ir source. As shown for graphite (the inner wall material of an NE2571 chamber) in Figure 5.3, the nonuniformity correction factor has only minor energy dependence. Other wall materials listed in Table 5.1 show similar behavior. Without loss of accuracy, these values can therefore be used in ^{137}Cs and ^{60}Co calibrations.

The parameters, $A_{pn}^{KR}(d)$ and $A'_{pn}(d)$, for the calculation of the nonuniformity correction factor for cylindrical chambers are given in the tables in ESTRO Booklet No. 8 as a function of the cylindrical chamber's shape factor, $\sigma = R_c/L_c$, and the distance factor, $\alpha = R_c/d$. In these formulas, R_c is the chamber's internal radius, L_c is the internal half-length of the chamber, and d is the measurement distance. Interested readers are referred to the work of Venselaar and Pérez-Calatayud (2004), which is directly accessible on the Internet.

TABLE 5.1   Material- and Photon-Energy-Dependent Factors, ω, for [192]Ir

Inner Wall Material	ω
A-150	1.066
PMMA	1.014
Graphite	0.992

Source: Venselaar, J.L.M. and Pérez-Calatayud, J. (eds.). *A Practical Guide to Quality Control of Brachytherapy Equipment*, 1st ed., ©2004 by ESTRO, Brussels.

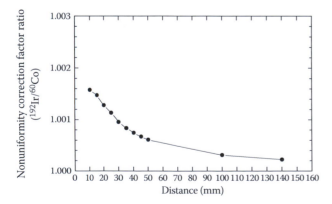

FIGURE 5.3 Ratio of nonuniformity correction factor for an NE2571 ionization chamber at ^{192}Ir and ^{60}Co qualities. (From Venselaar, J.L.M. and Pérez-Calatayud, J. (eds.). *A Practical Guide to Quality Control of Brachytherapy Equipment*, 1st ed., ©2004 by ESTRO, Brussels. With permission. Courtesy of ESTRO, Brussels.)

TABLE 5.2   Correction Factors for Air Attenuation of Primary Photons from ^{192}Ir, ^{137}Cs, and [60]Co Brachytherapy Sources

Distance (cm)	[192]Ir	[137]Cs	[60]Co
10	1.001	1.001	1.001
20	1.002	1.002	1.001
30	1.004	1.003	1.002
40	1.005	1.004	1.003
50	1.006	1.005	1.003
60	1.007	1.006	1.004
70	1.009	1.007	1.005
80	1.010	1.008	1.005
90	1.011	1.009	1.006
100	1.012	1.009	1.007

Source: Venselaar, J.L.M. and Pérez-Calatayud, J. (eds.). *A Practical Guide to Quality Control of Brachytherapy Equipment*, 1st ed., ©2004 by ESTRO, Brussels.

5.4.6.4 Attenuation Correction Factor of Primary Photons in Air

For determination of the RAKR from the measured air kerma at the distance d, it is necessary to correct for the attenuation of the primary photons between the source and the ionization chamber. Table 5.2 gives the k_{air} correction factors at different distances between the source and the ionization chamber (Verhaegen et al. 1992; Drugge 1995; Palani Selvam et al. 2002).

5.4.7 Correction Factors for Well-Type Chamber Measurements

Equation 5.5 converts readings of the well-type chamber into the $\dot{K}_R$ value. Two correction factors are discussed.

5.4.7.1 Correction for Ion Recombination

k_{ion} is determined with the hospital source at the time of measurement. The ion recombination factor should be determined. A suggested method is to use the Attix (1984) two-voltage approach. For a new ^{192}Ir source with an initial activity of 370 GBq, for example, with the PTW 33004/Nucletron SDS well chamber, the recombination factor is typically around 1.002, whereas for the Standard Imaging, well chambers' typical values are around 1.001 or less. The factor will also depend on the RAKR.

5.4.7.2 Correction for Source Position/ Geometry of Measurement Setup

k_{sg} is included in the equation to correct for in case of a source different from the source at the primary standards laboratory. Usually, the position of maximum response of the well chamber (sweet spot) is used. This can be found by stepping the ^{192}Ir source through the chamber at a maximum of 2.5-mm intervals between measurement positions. Minimum and maximum measurement positions should be at least ±10 mm from the expected "sweet spot" position to ensure a good regression on the curve. It is useful to have a previously defined maximum range ±2.5 mm as an aid to ensure that the catheter is in the same position.

The current should be measured once the source is at the sweet spot and the reading is stable. Current should be corrected to standard temperature, pressure, and humidity for a reference time and date, and background. At least three separate source transfers should be performed for current measurements. The mean chamber response should then be calculated, and the maximum spread of readings between transfers should be <0.1% (IAEA 2002).

5.5 Implementation and Clinical Dosimetry

5.5.1 General Topics

Once a new source is installed in a clinical department, the clinical user is required to perform a definitive calibration of their source in their own treatment room using the chamber–electrometer combination that has been calibrated by a National Standards' laboratory. Efforts are made to maintain low scatter conditions, stable temperature and humidity, reproducible and known source positioning, and a standard transfer tube and well-chamber insert. Issues discussed below aim at maintaining the high standards for in-house calibration procedures of the source or sources. General quality control aspects of the performance of the afterloader are not part of this section (see Chapter 4).

5.5.1.1 Quality Assurance

Quality control of the timer, source position, and room safety checks should follow a planned service whether or not the source is exchanged. These checks should also be performed after any major repair or system modification. Checks may be adapted depending on the work performed.

5.5.1.2 Source Calibration in Clinic

The definitive calibration is carried out in two parts independently and using different measuring equipment. The first part consists of measuring the RAKR of the new source assuming the use of the calibrated well chamber and electrometer.

In order to reduce the scatter component to less than 0.1% of the reading, the well chamber should be at least 1 m from any wall and 1 m above floor level on a low scatter surface (Chang et al. 2008). If the water-equivalent thickness of the tabletop (at least 1 m above floor level) is between 2.5 and 15 mm, the scatter contribution to the measured ionization current should be within tolerance.

A suitable transfer tube and catheter are attached to the brachytherapy treatment unit and the source holder of the well chamber. It is important to ensure that the catheter is pushed to the bottom of the source holder. It is also useful to have an alignment mark on the source holder to ensure reproducibility.

The chamber should be allowed to reach thermal equilibrium with the surrounding air. As an example, the ESTRO Booklet No. 8 reports that it takes 400 min to eliminate a 4°C temperature difference between the room and an HDR 1000 Plus well chamber.

The correct polarizing voltage should be set to agree with that used at the primary standards laboratory. Owing to the differences between the electrometers that are available, care should be taken by the user to ensure that the correct polarity is selected when the chamber is operated in the hospital, and sufficient time for the electrometer to stabilize after change in polarizing voltage should be allowed. The electrometer should be zeroed and a background reading taken.

5.5.1.3 Well Chamber Stability Check

As a stability check for the well chambers, the above process can be repeated using a tertiary well chamber. The ratio of the two measured currents can be calculated and compared with the reference value. The ratio should agree with the reference to within 2%. If a larger discrepancy exists, then this must be investigated before continuing. The problem could lie with either chamber. Because of the suspicion of drift in the response of a well-type chamber over prolonged periods of time, it is quite common to use the reading from a long-lived source, for example, a ^{137}Cs source, as a reference. By using a uniquely defined insert for reproducible positioning of such a source, one can verify the chamber signal.

Generally, a simple method for checking well chambers is to use three calibrated chambers to measure the same source. Two chambers can be used, but then it is difficult to always determine which chamber is problematic. If one of the three chambers disagrees with the other two, it is easy to determine the problem chamber. When the same source is measured with three calibrated chambers, the agreement after applying calibration coefficients should be less than 1% and generally is less than 0.5%. Use of the same electrometer makes the variation in this agreement smaller.

5.5.1.4 Transit Time

The transit time for the well chamber setup is measured first. The transit time T_r is not the time taken for the source to travel to and from the dwell position, but rather the effective time required to correct for the air kerma recorded by the chamber during source transit. The transit time is measured by plotting a graph of charge recorded by the well chamber against the programmed dwell time and determining the intercept for $t = 0$. As a note, if current is measured, wait until stability of the current is achieved in the well chamber, and then there is no need for correction of transit time. The dose may still be desired to know what the patient receives during transit time.

A tertiary chamber with a recently derived calibration coefficient is used and set up as for the first part of the source calibration. The electrometer is set to charge mode, the electrometer nulled, and background charge recorded for 60 s.

The HDR source is programmed to dwell at the position of maximum response for 10, 20, 30, 40, 60, 120, and 240 s where at least two exposures are carried out for each set time. The results are averaged and corrected for background. A graph of recorded charge against the programmed treatment time is produced, a straight line is fitted to the points, and the gradient G, intercept I, and correlation coefficient R are calculated. The timer linearity will be related to the correlation coefficient R. The transit time for the setup, $T_r = I/G$, is calculated. The transit time measurement is performed to ensure that source transfer is consistent within the limitations imposed by the variability of transfer tube positioning. The gradient, G, is used in place of a measured current to calculate the value of the source RAKR. This value is compared with that from the first definitive calibration of the source. Results should agree within 2%. This check compares the unit timer with the secondary standard electrometer.

Depending on the equipment used, several other techniques might be used to eliminate the transit component of the signal. One simple solution would be to use an externally triggered electrometer to collect charge during an interval after the source has stopped moving.

5.5.1.5 Types of Well-Type Ionization Chambers

The well-type chamber for brachytherapy source calibrations should be of the type designed especially for radiotherapy applications and preferably capable of measuring the RAKR of both LDR and HDR sources. One should note that if the chamber is sealed and the pressure of the gas is at a higher level than the ambient atmospheric pressure, it might develop a problem of slow leakage of the gas. In this case, a change in the calibration coefficient would result. Chambers open to the atmosphere need correction for temperature and pressure since the calibration coefficient is based upon a density of air corresponding to standard ambient conditions, usually 20°C or 22°C and 101.3 kPa. Also, note that depending on the altitude, the density of air correction for LDR sources may be an overcorrection. For further details on correcting for this phenomenon, reference is made to Bohm et al. (2005) and Griffin et al. (2005).

It should be noted that pressurized well-type ionization chambers used in nuclear medicine departments are not recommended for brachytherapy measurements for the following reasons:

- The chambers measure only in units of activity.
- The chambers have settings for given radionuclides but not for brachytherapy sources.
- Without close control, the general use of the chamber may result in contamination from nuclear medicine procedures.
- Since the gas may leak from the pressurized volume, the response may change over time.
- The thick walls required for the pressurization may absorb a significant part of the radiation to be measured. Since this results in high-energy dependence, small variations in the relative peak intensities are unduly emphasized.

Only correction for source decay and—if applicable—temperature and air pressure is needed to obtain confidence in the stability of the instrument's reading. Any sudden deviation of

more than 0.5% in the check reading might indicate a problem. If the check source corrected reading remains within 2% of that at the time of the initial calibration, the assays may proceed, but any possible reason for the deviation should be investigated.

5.5.1.6 Precautions for Use

Well-type chambers (e.g., Figure 5.4) provide an easy and reliable method for calibrating brachytherapy sources. The calibration point of a well-type chamber is defined as the point at which the center of the source is positioned during the calibration procedure; this point may differ from one source to another depending on the source length. Some chambers have a fixed, nonremovable, spacer in the well, and the source is then conveniently placed on the top of the spacer. Other models, on the other hand, have a mechanism to move and fix the source holder to different heights, and the source is then placed at the bottom of the movable holder during the calibration procedures. The location of the calibration point must be stated on the chamber's calibration certificate. Possible spacers and the outer dimensions of the source used to calibrate the chamber must also be stated in the calibration certificate. Spacers should be easily identified to avoid use of wrong spacer lengths. In most commercial well-type chambers, a guide tube is provided to hold the source catheter along the axis of the cylindrical well. The sensitivity of the chamber versus the source position along the guide tube must be checked (determination of the "sweet spot"). This can be done by varying the position of a small source along the length of the guide tube. Usually, the signal is within some 1% over a trajectory of several millimeters of the guide tube, indicating that it is very easy to have highly reproducible readings.

It should be noted that well-type chambers with thick internal walls may show energy dependence, which is particularly emphasized when calibrating low-energy photon sources, such as ^{125}I and [103]Pd. For instance, the filtration of low-energy photons depends on the thickness of the wall of the source holder. It

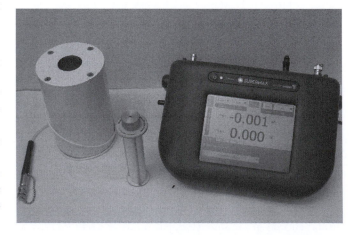

FIGURE 5.4 Standard Imaging well-type ionization chamber. Dedicated inserts can be used for calibration of specified source types at well-defined source positions. (Courtesy of H. Bjerke.)

is important to understand that a well-type chamber, in general, exhibits larger dependence on the source design compared with Farmer-type chambers. The well-type chamber's calibration coefficient is valid only for the type of source for which it has been calibrated. This is specifically true for low-energy photon emitters. In ^{192}Ir HDR calibrations, the differences of calibrations between the source types are all within 1% (Rasmussen et al. 2011). In some cases, this dependence is stated on the well-type chamber's calibration certificate, but not always. In such cases, when there is no such statement, care should be taken if using the chamber under conditions different from the calibration conditions.

In the calibration procedure at the hospital, the source is positioned within the chamber in a manner that reproduces the position that was used during the calibration. However, a separate measurement of sweet spot should be carried out to ensure consistency. The source position during the calibration is stated on the chamber's calibration certificate.

5.5.1.7 Validation Against Vendors' Certificate

The measured AKR should be compared with the value on the supplier's certificate, correcting the latter value for source decay. The values should agree within 5%. The calculation should be checked by a second physicist. The RAKR should then be entered into the brachytherapy planning system. Values with discrepancies greater than 5% should be checked for reason of discrepancy with the vendor, and one should attempt to reconcile any differences.

Vendors have various ways of establishing apparent activity and its conversion to AKR. There can be a large range in which the vendors would operate, and they can definitely fall at the edges of their range. For HDR sources, discrepancies of up to 10% have been reported (DeWerd and Thomadsen 1994). More recent anecdotal experience has not found a great discrepancy. LDR manufacturers bin their sources since they are dealing with 100 sources per order. The bins are generally 3.5% ($k =$ 1), and sometimes, if there are not enough sources to fill an order, sources from above and below the bin are used, expecting that they average. Thus, a general principle to be followed is to trust the clinics' measured value, but still try to resolve the difference.

5.6 Uncertainty in the Determination of Absorbed Dose to Water

In this chapter, three methods are explained for the source calibration or reference dosimetry for brachytherapy. In addition, two quantities are given: the RAKR or air-kerma strength and the absorbed dose to water. An estimation of relative uncertainty of $\dot{D}_{W,1}$ for the alternative methods is given in Chapter 15. The user is encouraged to set up an uncertainty budget for his or her own equipment and measuring practice. An uncertainty budget will give information about weak and strong points in the source calibration routine.

5.7 Vendor Requirements

Brachytherapy sources have multiplied, including new vendors and new sources. There are requirements set down for these new (and old) sources so that the dose to the patients can be controlled. The sources are generally divided by energy: those less than 50 keV and those greater than 50 keV. The majority of sources less than 50 keV are LDR ^{125}I or [103]Pd, and those greater than 50 keV are [192]Ir. In all cases, the treatment planning system vendors need to provide the appropriate dosimetric parameters in their software.

5.7.1 LDR ^{125}I or [103]Pd Sources

The ADCLs in the United States have been able to demonstrate that vendors generally have been able to maintain their calibrations to within ±2% of the established standard. However, some vendors have been noted to have variation with time. Thus, a procedure for the establishing and maintenance of consistent traceability of the air-kerma strength standard was set forth (DeWerd et al. 2004). There is a procedure for a new low-energy source passing through the PSDL through the ADCL and then for dosimetric measurements. After this and for prior low-energy sources, there is a maintenance procedure that involves a check on the calibration on a 6- or 12-month schedule. The source is sent through the PSDL and then via each ADCL. This procedure is elaborated upon in the work of DeWerd et al. (2004).

5.7.2 HDR Sources of Energy Greater than 50 keV

Again, there is a procedure for prior sources and any new sources introduced by the vendors Li et al. (2007). The vendor is responsible for determining by measurement and Monte Carlo analysis that the dosimetric parameters are measured and produced by an independent researcher. The calibrations of air-kerma strength or RAKR must be traceable to a primary laboratory and available to the clinic via calibrated instruments (well chambers).

5.8 Publications by Professional Organizations

Further reading on calibration and procedures can be found in several publications and recommendations by national and international organizations and expert groups, for example, in ICRU Report 33 (1980), Attix 1986, AAPM Report 21 (1987), Aird et al. (1993), AAPM Report 46 (1994), DeWerd and Thomadsen (1995), AAPM Report 59 (Nath et al. 1997), DeWerd (1997, 2002), AAPM Report 66 (Nath et al. 1999), IAEA TecDoc 1079 (1999) and its update IAEA TecDoc 1274 (2002), DeWerd et al. (2004), ESTRO Booklet No. 8 (Venselaar and Pérez-Calatayud 2004), NCS (2004), and NPL Report DQL-RD 004

(Sander and Nutbrown 2006). Of these publications, DeWerd (2002) and Nath et al. (1999) discuss details of dosimetry of intravascular brachytherapy sources, whereas IAEA (2002) and NCS (2004) discuss more details of the dosimetry of beta sources.

5.9 Conclusions

The calibration of the sources used in patients is the essential starting point for treating cancer. The measurement of the source RAKR or the air-kerma strength described in this chapter is the key information for the determination of treatment time. The agreement for all source calibrations between all countries is very good, but it is a requirement for all brachytherapy sources to be calibrated before use in the hospital, even if the manufacturer gives the source information. Three source calibration methods are described; the well chamber method is recommended. Traceability to primary absorbed dose-to-water standards *is* available, and source calibration in absorbed dose rate to water at 1-cm distance is *an* expected practice in the future.

References

AAPM. 1987. Specification of brachytherapy source strength. American Association of Physicists in Medicine Report No. 21. American Institute of Physics, New York.

AAPM. 1994. Comprehensive QA for radiation oncology. American Association of Physicists in Medicine Report No. 46 of AAPM Radiation Therapy Committee Task Group 40. *Med Phys* 21:581–618.

Aird, E.G.A., Jones, C.H., Joslin, C.A.F. et al. 1993. Recommendations for Brachytherapy dosimetry. Report of a joint working party of the BIR and the IPSM. The British Institute of Radiology, London.

Attix, F.H. 1984. Determination of A_{ion} and P_{ion} in the new AAPM radiotherapy dosimetry protocol. *Med Phys* 11:714–6.

Attix, F.H. 1986. *Introduction to Radiological Physics and Radiation Dosimetry*. John Wiley and Sons, New York.

Aubineau-Lanice, I., Avles-Lucas, P., Bordy, J.-M. et al. 2010. Development of a primary standard in terms of absorbed dose to water for ^{125}I brachytherapy seeds. In: *Book of Extended Synopsys of International Symposium on Standards, Applications and Quality Assurance in Medical Radiation Dosimetry*–IAEA–CN–182-129, Vienna.

Bielajew, A.F. 1990a. Correction factors for thick-walled ionization chambers in point source photon beams. *Phys Med Biol* 35:501–16.

Bielajew, A.F. 1990b. An analytic theory of the point-source non-uniformity correction factor for thick-walled ionization chambers in photon beams. *Phys Med Biol* 35:517–38.

BIPM.RI(I)-K8. 2011. Available at http://kcdb.bipm.org/appendixB/appbresults/BIPM.RI(I)-K8/BIPM.RI(I)-K8_Technical_Protocol%20.pdf. Accessed August 30, 2011.

Bohm, T.D., Griffin, S.L., DeLuca, P.M. Jr,, and DeWerd, L.A. 2005. The effect of ambient pressure on well chamber response: Monte Carlo calculated results for the HDR 1000 plus. *Med Phys* 32:1103–14.

Borg, J. and Rogers, D.W.O. 1999. Spectra and air kerma strength for encapsulated ^{192}Ir sources. *Med Phys* 26:2441–4.

Borg, J., Kawrakow, I., Rogers, D.W., and Seuntjens, J.P. 2000. Monte Carlo study of correction factors for Spencer-Attix cavity theory at photon energies at or above 100 keV. *Med Phys* 27:1804–13.

Butler, W.M., Huq, M.S., Li, Z. et al. 2006. Third party brachytherapy seed calibrations and physicist responsibilities. *Med Phys* 33:247–8.

Chang, L., Ho, S.Y., Chui, C.S., Lee, J.H., Du, Y.C., and Chen, T. 2008. A statistical approach to infer the minimum setup distance of a well chamber to the wall or to the floor for 192Ir HDR calibration. *Med Phys* 35:2214–7.

De Pooter J.A. and De Prez L.A. 2010. Development of a water calorimeter as a primary standard for absorbed dose to water measurements for HDR brachytherapy sources. In: *Book of Extended Synopsys of International Symposium on Standards, Applications and Quality Assurance in Medical Radiation Dosimetry*–IAEA–CN–182-267, Vienna.

DeWerd, L.A. and Thomadsen, B.R. 1995. Source strength standards and calibration of HDR/PDR sources. In: *Brachytherapy Physics*. J. Williamson, B.R. Thomadsen, and R. Nath (eds.). AAPM 1994 Summer School, Med Phys Publishing, Madison, WI. pp. 541–55.

DeWerd, L.A. 1997. Brachytherapy dosimetric assessment: source calibration. In: *Radiological Society of North America, Categorical Course in Brachytherapy Physics*. B. Thomadsen (ed.). RSNA, Oak Brook, IL.

DeWerd, L.A. 2002. Source standardization and calibration for intravascular brachytherapy. In: *Intravascular Brachytherapy/Fluoroscopically Guided Interventions*. S. Balter, R.C. Chan, and T.B. Shope Jr. (eds.). AAPM 2002 Summer School Proceedings. Medical Physics Monograph 28. Madison, WI. pp. 423–43.

DeWerd, L.A., Huq, M.S., Das, I.J. et al. 2004. Procedures for establishing and maintaining consistent air-kerma strength standards for low-energy, photon-emitting brachytherapy sources: recommendations of the Calibration Laboratory Accreditation Subcommittee of the American Association of Physicists in Medicine. *Med Phys* 31:675–81.

Domen, S.R. 1980. Absorbed dose water calorimeter. *Med Phys* 7:157–9.

Douysset, G., Gouriou, J., Delaunay, F., DeWerd, L., Stump, K., and Micka, J. 2005. Comparison of dosimetric standards of USA and France for HDR brachytherapy. *Phys Med Biol* 50:1961–78.

Douysset, G., Sander, T., Gouriou, J., and Nutbrown, R. 2008. Comparison of air kerma standards of LNE-LNHB and NPL for ^{192}Ir HDR brachytherapy sources: EURAMET project no 814. *Phys Med Biol* 53:N85–97.

Drugge, N. 1995. Determination of the Reference Air Kerma Rate for clinical ^{192}Ir sources. Thesis, Internal Report, University of Göteborg.

Goetsch, S.J., Attix, F.H., Pearson, D.W., and Thomadsen, B.R. 1991. Calibration of 192Ir high-dose-rate afterloading systems. *Med Phys* 18:462–7.

Granero, D., Pérez-Calatayud, J., Pujades-Claumarchirant, M.C. et al. 2008. Equivalent phantom sizes and shapes for brachytherapy dosimetric studies of ^{192}Ir and [137]Cs. *Med Phys* 35:4872–77.

Griffin, S.L., DeWerd, L.A., Micka, J.A., and Bohm, T.D. 2005. The effect of ambient pressure on well chamber response: Experimental results with empirical correction factors. *Med Phys* 32:700–9.

IAEA. 1999. Calibration of brachytherapy sources. Guidelines on standardized procedures for the calibration of brachytherapy sources at secondary standard dosimetry laboratories (SSDLs) and hospital. International Atomic Energy Agency. IAEA-TECDOC-1079. IAEA, Vienna.

IAEA. 2002. Calibration of photon and beta ray sources used in brachytherapy. Guidelines on standardized procedures at secondary standards dosimetry laboratories (SSDLs) and hospitals. International Atomic Energy Agency. IAEA-TECDOC-1274. IAEA, Vienna.

ICRU. 1980. Radiation Quantities and Units. Report 33 of International Commission on Radiation Units and Measurements. Bethesda, MD.

IEC 62467-1. 2009. Medical electrical equipment—Dosimetric instruments as used in brachytherapy—Part 1: Instruments based on well-type ionization chambers.

Kondo, S. and Randolph, M.L. 1960. Effect of finite size of ionization chambers on measurement of small photon sources. *Rad Res* 13:37–60.

Krieger, H. 1991. Messung der Kenndosisleistung punkt- und linienförmiger HDR-^{192}Ir-Afterloadingstrahler mit einem PMMA-Zylinderphantom. *Z. f. Med Phys* 1, 38–41.

Krieger, H. and Baltas, D. (eds.). 1999. Praktische Dosimetrie in der HDR -Brachytherapie. Deutsche Gesellschaft für Medizinische Physik. DGMP Report Nr. 13.

Li, Z., Das, R.K., DeWerd, L.A. et al. 2007. Dosimetric prerequisites for routine clinical use of photon emitting brachytherapy sources with average energy higher than 50 kev. *Med Phys* 34:37–40.

Mainegra-Hing, E. and Rogers, D.W. 2006. On the accuracy of techniques for obtaining the calibration coefficient N(K) of 192Ir HDR brachytherapy sources. *Med Phys* 2006 33:3340–7.

Meertens, H. 1990. In-phantom calibration of Selectron-LDR sources. *Radiother Oncol* 17:369–78.

Mitch, M.G. and Soares, C.G. 2009. Primary Standards for Brachytherapy Sources. In: AAPM Summer School. http://www.aapm.org/meetings/09SS/documents/16Mitch-BrachyPrimaryStandards.pdf.

Nath, R., Anderson, L.L., Luxton, G. et al. 1995. Dosimetry of interstitial brachytherapy sources: Recommendations of the AAPM Radiation Therapy Committee Task Group No 43 AAPM Report No. 51. *Med Phys* 22:209–34.

Nath, R., Anderson, L.L., Meli, J.A. et al. 1997. Code of practice for brachytherapy physics. Report No. 59 of AAPM Radiation Therapy Committee Task Group 56. *Med Phys* 24:1557–98.

Nath, R., Amols, H., Coffey, C. et al. 1999. Intravascular brachytherapy physics. Report No. 66 of AAPM Radiation Therapy Committee Task Group 60. *Med Phys* 26:119–52.

NCS. 2004. Quality control of sealed beta sources in brachytherapy. Recommendations on detectors, measurement procedures and quality control of beta sources. NCS Report No. 14 of the Netherlands Commission on Radiation Dosimetry. NCS, Delft. Available at http://www.stralingsdosimetrie.nl/ncs-report.php.

Palani Selvam, T., Govinda Rajan, K.N., Nagarajan, P.S., Bhatt, B.C., and Sethulakshmi, P. 2002. Room scatter studies in the air kerma strength standardization of the Amersham CDCS-J type 137Cs source: A Monte Carlo study. *Phys Med Biol* 47:N113–9.

Palmans, H. 2007. An overview of radiotherapy dosimetry at the NPL. Teddington. June 2007. Available at http://www.npl.co.uk/upload/pdf/20050605_rsug_palmans_1.pdf.

Pérez-Calatayud, J., Granero, D., and Ballester, F. 2004. Phantom size in brachytherapy source dosimetric studies. *Med Phys* 31:2075–81.

Piermattei, A. and Azario, L. 1997. Applications of the Italian protocol for the calibration of brachytherapy sources. *Phys Med Biol* 42:1661–9.

Rasmussen, B.E., Davis, S.D., Schmidt, C.R., Micka, J.A., and DeWerd, L.A. 2011. Comparison of air-kerma strength determinations for HDR ^{192}Ir sources. *Med Phys* 38:6721–9.

Rivard, M., Coursey, B., Hanson, W. et al. 2004. Update of AAPM Task Group No 43 Report: A revised AAPM protocol for brachytherapy dose calculations. *Med Phys* 31:633–74.

Rivard, M.J., Butler, W.M., DeWerd, L.A. et al. 2007. Supplement to the 2004 update of the AAPM Task Group No. 43 Report. American Association of Physicists in Medicine. *Med Phys* 34:2187–205.

Sander, T. and Nutbrown, R.F. 2006. The NPL air kerma primary standard TH100C for high dose rate ^{192}Ir brachytherapy sources. National Physics Laboratory. NPL Report DQL-RD 004 ISSN 1744-0637. Teddington, UK. http://publications.npl.co.uk.

Sarfehnia, A. and Seuntjens, J. 2010. Development of a water calorimetry-based standard for absorbed dose to water in HDR 192Ir brachytherapy. *Med Phys* 37:1914–23.

Schneider, T. and Selbach, H.J. 2010. Determination of the absorbed dose to water for ^{125}I interstitial brachytherapy sources. In: *Book of Extended Synopsys of International Symposium on Standards, Applications and Quality Assurance in Medical Radiation Dosimetry*–IAEA–CN–182-101, Vienna.

Selbach, H.-J., Kramer, H.-M., and Culberson, W.S. 2008. Realization of reference air-kerma rate for low-energy photon sources. *Metrologia* 45:422–8.

Selbach H.J. and Meier M. 2010. Calibrations of high-dose rate and low-dose rate brachytherapy sources. In: *Book of Extended Synopsys of International Symposium on Standards, Applications and Quality Assurance in Medical Radiation Dosimetry–IAEA–CN–182-081*, Vienna.

Seltzer, S.M., Lamperti, P.J., Loevinger, R., Mitch, M.G., Weaver, J.T., and Coursey, B.M. 2003. New national air-kerma-strength standards for ^{125}I and ^{103}Pd brachytherapy seeds. *J Res Natl Inst Stand Technol* 108:337–58.

Seuntjens, J. and Palmans, H. 1999. Correction factors and performance of a 4 degrees C sealed water calorimeter. *Phys Med Biol* 44:627–46.

Tölli, H., Bielajew, A.F., Mattsson, O. et al. 1997. Fluence non-uniformity effects in air kerma determination around brachytherapy sources. *Phys Med Biol* 42:1301–18.

Toni, M.P. 2010. New brachytherapy standards paradigm shift. In: *Book of Extended Synopsys of International Symposium on Standards, Applications and Quality Assurance in Medical Radiation Dosimetry–IAEA–CN–182-INV015*, Vienna.

Van Dijk, E. 2002. Comparison of two different methods to determine the air kerma rate calibration factor, Nk, for 192Ir. In: *Proceedings of the International Symposium on Standards and Codes of Practice in Medical Radiation Physics*, IAEA-CN-9675, Vienna.

Van Dijk, E., Kolkman-Deurloo, I.-K.K., and Damen, P.M.G. 2004. Determination of the reference air kerma rate for ^{192}Ir brachytherapy sources and the related uncertainty. *Med Phys* 31:2826–33.

Venselaar, J.L.M. and Pérez-Calatayud, J. (eds.). 2004. *A Practical Guide to Quality Control of Brachytherapy Equipment*, 1st ed., ESTRO, Brussels. http://www.estro-education.org/publications/Documents/booklet8_Physics.pdf.

Verhaegen, F., van Dijk, E., Thierens, H., Aalberts, A., and Seuntjens, J. 1992. Calibration of low activity ^{192}Ir brachytherapy sources in terms of reference air-kerma rate with large volume spherical ionization chambers. *Phys Med Biol* 37:2071–82.

Experimental Dosimetry Methods

Dimos Baltas
Klinikum Offenbach GmbH

Brigitte Reniers
Maastro Clinic

Ali S. Meigooni
*Comprehensive Cancer
Centers of Nevada*

6.1 Introduction

6.1.1 The Place of Experimental Dosimetry in Brachytherapy

The methodology for quantifying the strength of a brachytherapy source in terms of air-kerma strength, S_K, measurement as well as the established dosimetry protocols for dose rate calculations around such a source based on its strength are extensively discussed in Chapters 5 and 10.

As discussed in Chapter 10, for determination of different dosimetric parameters implemented in the internationally established TG-43 (Nath et al. 1995) and TG-43U1 (Rivard et al. 2004) dosimetry protocol, experimentally validated MC simulation results are required. This recommendation has been extended in the AAPM and ESTRO report on dose calculation for photon-emitting brachytherapy sources with average energy higher than 50 keV (Li et al. 2007; Pérez-Calatayud et al. 2012). Furthermore, the need for investigating and quantifying the dose perturbation arising from simple or complex applicator geometries and applicator materials (plastic or metal), shielding materials, and tissue inhomogeneities necessitates the availability of appropriate experimental dosimetry systems in modern brachytherapy.

6.1.2 General Requirements for Detectors in Brachytherapy

The radiation field around brachytherapy sources is characterized by (1) high dose gradients, (2) an extended dose rate range, and (3) photon energies typically lower than those in the standard dosimetry of external beam fields. Brachytherapy field spreads from those of high (^{60}Co and ^{137}Cs) and intermediate (^{198}Au, ^{192}Ir, and ^{169}Yb), to those of low-energy radionuclides (^{125}I, ^{103}Pd, and

^{131}Cs). Given these three characteristics of the radiation field, experimental brachytherapy dosimetry does place severe demands on candidate detectors (Williamson and Meigooni 1995): (1) wide dynamic range, (2) flat energy response, (3) small active volume, (4) high sensitivity, and (5) isotropic angular response.

The size of a detector defines both the maximum spatial resolution that can be achieved as well as the minimum meaningful distance of measurement. Especially in intermediate- and high-energy radionuclides, a softening of the photon spectrum is observed with increased distance from the source due to the gradual buildup of scattered radiation (Baltas et al. 2007). The same effect can be seen in the new electronic brachytherapy sources (EBSs). Due to the energy softening effect, the energy response of a detector becomes crucial regarding its suitability for experimental studies. Finally, pronounced angular anisotropic response of detectors is expected to lead to underestimation of the dose with increased distance from the source since the fraction of backscattered photons increases with distance.

6.1.3 Detectors for Use in Brachytherapy

A variety of experimental dosimeters have been used for dosimetry around brachytherapy sources, such as ionization chambers, thermoluminescence dosimetry (TLD), diodes, plastic scintillators, diamond detectors, radiographic and radiochromic film, polymer gels, and chemical dosimeters. Table 6.1 summarizes the sensitivity of several detectors based on the data published by Perera et al. [see Perera et al. (1992) and Williamson (1995), taken from Baltas et al. (2007)].

Based on these values, ionization chambers are not appropriate for measurements with low-energy radionuclide low-dose-rate (LDR) sources such as ^{125}I, ^{103}Pd, and ^{131}Cs. Diodes (silicon) and

TABLE 6.1 Absolute Response Characteristics of 1.0-mm³ Detector to 0.001-Gy Absorbed Dose

Detector	Energy Dissipated per Created Quantum (eV)	No. Quanta Emitted	Typical Quantum Efficiency × Geometric Collection Efficiency	Practical Response Relative to Ionization Chamber
LiF-TLD-100	8400	1.8×10^6	0.20×0.08	0.12
Silicon diode	3.6	4.0×10^9	1.00×1.00	1.7×10^4
Ionization chamber (air)	33.8	2.4×10^5	1.00×1.00	1.00
Plastic scintillator	100	6.2×10^7	0.20×0.05	2.60

Note: Data are according to Perera et al. (1992).

plastic scintillators demonstrate higher sensitivity and very small dimensions, thus allowing for an adequate spatial resolution. On the other hand, due to the potential energy dependence of their response, they have to be carefully used (Williamson et al. 1999). New types of polyvinyl toluene (PVT) mixtures with medium atomic-number atoms very closely approximate the radiological properties of water (within 10%) in the energy range 0.020–0.662 MeV and yield improved energy response. Unfortunately, there is no extended experience with such types of plastic detectors.

Ionization chambers, diodes, plastic scintillators, and diamond detectors present the advantage of allowing direct measurement in water phantoms, thus obviating the need for solid-state phantoms and the influence of corresponding corrections and their associated uncertainties. Currently, LiF thermoluminescence is the most commonly used detector for experimental brachytherapy dosimetry, for both absolute and relative dose rate measurement, in the whole energy range of brachytherapy sources (Rivard et al. 2004; DGMP 1999; DIN 1993; Meigooni et al. 1988, 1994; Meli et al. 1988; Rashid et al. 1993; Luxton 1994; Wallace 2002; Hill et al. 2005). This chapter will further focus on two experimental dosimetry methods: the standardized (1) ionization dosimetry, and (2) TLD. However, because of the use of film dosimetry methods by several groups, and because of the favorable characteristics of the diamond dosimetry technique, the next two paragraphs provide a short introduction for further reading.

6.1.4 Film Dosimetry in Brachytherapy

Radiographic films such as Kodak X-V demonstrate a supralinearity of dose response for doses above some tens of centigrays and a pronounced photon energy dependence for energies below 0.127 MeV (reaching up to a factor of 10 for photon energies of 0.020–0.040 MeV; Muench et al. 1991). Different kinds of radiochromic films are now available. Extended studies have been performed on the MD-55 (ISP Corp., Wayne, NJ) regarding the dose response curve over an extended dose range (up to hundreds of grays), and a small photon energy dependence (i.e., gaining sensitivity as photon energy rises) in the low-energy range below 0.127 MeV (Muench et al. 1991) has been demonstrated. This gain of sensitivity (i.e. about 40%–45% higher response ^{137}Cs or ^{60}Co than for 0.020–0.040 MeV photons; Muench et al. 1991; Chiu-Tsao et al. 1994; Bohm et al. 2001) is much lower than that of radiographic films. Despite these advantages, dosimetry using earlier models of radiochromic films required exposures at several grays.

For example, a dose above 100 Gy is required for low-energy photons such as those emitted by ^{125}I to achieve an optical density of 1.0, for the low-sensitivity-type GafChromic (HD-800 or MD-55) films. This low sensitivity limited the use of radiochromic films for dosimetry studies for low-energy radionuclides. However, for the new models of GafChromic films such as the EBT film (ISP Corp., Wayne, NJ), the film sensitivity has been increased, and thus, much lower doses (0.01–8 Gy) are needed to obtain the same response. The first version of the EBT film seemed to show very favorable characteristics for brachytherapy dosimetry with very low energy dependence (Butson et al. 2006; Chiu-Tsao et al. 2005, 2008). The second version of the film (i.e., EBT2 and EBT3) had lost this property (Aldelaijan et al. 2010; Lindsay et al. 2010). In the present state, this film type has not been regularly utilized for dosimetry of low-energy brachytherapy sources.

6.1.5 Diamond Detectors in Brachytherapy

The diamond detectors of small sensitive volume (1.8–1.9 mm³) present promising properties for high-energy photon beams and for the intermediate energy of ^{192}Ir. Their response has been shown to be nearly independent of incident photon energy for energies above 0.1 MeV since the ratio of carbon to water mass energy absorption coefficient is nearly constant in this energy region (Baltas et al. 2007). This implies that for intermediate- and high-energy radionuclides (^{169}Yb, ^{192}Ir, ^{198}Au, ^{137}Cs, and ^{60}Co), diamond detectors offer a very low energy dependence of their response. For low energies, (i.e., 0.020–0.030 MeV), however, a significant change (approximately 50%) in their energy response is expected (Figure 6.1). The angular dependence of diamond detector response is very low in both high-energy photon beams (maximum 2%, Rustgi 1995) and the intermediate photon energies of ^{192}Ir (maximum 1.5%; Nakano et al. 2003). Furthermore, the temperature dependence of their response was found to be less than 2.5% in the range 14°C–40°C (Nakano et al. 2003). In addition, diamond detectors exhibit a good spatial resolution due to their small sensitive volume and a high sensitivity due to their high density. The combination of the properties in comparison with the detector types mentioned in Table 6.1 constitutes diamond detectors appropriate for both relative and absolute dosimetry for the intermediate- and high-energy brachytherapy sources. Unfortunately, there is only very limited experience with diamond detectors, which is mainly due to their small availability and their high price. Finally, their validity for obtaining brachytherapy dosimetry parameters has not

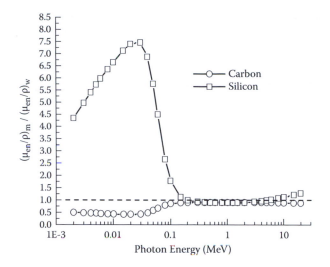

FIGURE 6.1 Ratio of mass energy absorption coefficients of carbon and silicon to that of water, $(\mu_{en}/\rho)_m/(\mu_{en}/\rho)_w$, in dependence on photon energy. (From Baltas, D. et al. *The Physics of Modern Brachytherapy for Oncology.* CRC Press, Taylor & Francis Group, LLCC, Boca Raton, FL, 2007. With permission.)

been convincingly demonstrated in the vicinity of low-energy photon-emitting brachytherapy sources.

6.2 Phantom Material

The reference medium for the dosimetry of brachytherapy sources is water (Rivard et al. 2004; DGMP 1999; DIN 1993). In the TG-43U1 report (Rivard et al. 2004), pure, degassed water is defined to be composed of two parts hydrogen atoms and one part oxygen atoms, with a mass density of $\rho = 0.998$ g cm^{-3} at 22°C as the recommended water composition for reference dosimetry. However, for experimental procedures of radiation dosimetry around the brachytherapy source, a water phantom

has some limitations. For example, most of the detectors that are useful in the brachytherapy field, such as TLD or radiochromic films, are not water-proof detectors. Moreover, accurate source–detector positioning is crucial in the dosimetry of brachytherapy sources due to the high dose gradients in their vicinity. It is nearly impossible to suspend the source and detectors with the required accuracy in the water phantom without introduction of a solid medium.

Several solid phantom materials have been investigated as water substitutes for brachytherapy experimental dosimetry using TLD detectors (Rivard et al. 2004; DGMP 1999; DIN 1993; Meigooni et al. 1988, 1994; Meli et al. 1988; Rashid et al. 1993; Luxton 1994; Wallace 2002; Hill et al. 2005). These solid phantoms can be precisely machined in order to accommodate the source and detectors with high precision for intermediate- and high-energy radionuclides. Appropriate solid phantom machining can also facilitate measurements using the ionization chamber and film dosimeters, which normally cannot be used in liquid water medium.

Dosimetric characteristics of these phantom materials depend on the energy spectrum of the emitted photons and thus on the specific source under consideration. Furthermore, due to the energy softening effect (i.e., Compton scattering will introduce lower energy relative to the incident beam energy) with increasing distance from a source, the dosimetric characteristics of a specific phantom material are also expected to be distance dependent. The physical properties and the chemical composition of all common dosimetry materials as well as a detailed evaluation of the distance dependence of the water equivalence of the phantom material are discussed in more detail in Chapter 7 (Baltas et al. 2007; Meigooni et al. 2006).

A summary of the evaluation of the water equivalence of the different phantom materials for the several radionuclides based on Monte Carlo simulation results presented by Baltas et al. (2007) is shown in Table 6.2. Due to the fact that the water equivalence

TABLE 6.2 Summary of Water Equivalence of Different Phantom Materials for Seven Brachytherapy Radionuclides of Interest Subdivided According to Energy of Emitted Photons

Material	Low-Energy Radionuclides ^{103}Pd, ^{125}I, and ^{131}Cs	Intermediate-Energy Radionuclide ^{169}Yb and ^{170}Tm	Intermediate-Energy Radionuclides ^{192}Ir and ^{198}Au	High-Energy Radionuclides ^{137}Cs and ^{60}Co
Plastic water[a] (PW or PW2030)	++++ Distance dependent $k_{m,w}$	+++++ Distance dependent $k_{m,w}$	+++++ Slightly distance dependent $k_{m,w}$	+++++ Distance independent $k_{m,w}$
Plastic water[a] (PW-LR)	++++ Distance dependent $k_{m,w}$	+++++ Distance dependent $k_{m,w}$	+++++ Slightly distance dependent $k_{m,w}$	+++++ Distance independent $k_{m,w}$
Solid water	+++ Distance dependent $k_{m,w}$	++++ Distance dependent $k_{m,w}$	+++++ Slightly distance dependent $k_{m,w}$	+++++ Distance independent $k_{m,w}$
Polymethyl methacrylate (PMMA)	— —	+++ Distance dependent $k_{m,w}$	+++++ Slightly distance dependent $k_{m,w}$	++++ Distance dependent $k_{m,w}$
Polystyrene	— —	— —	+++ Distance dependent $k_{m,w}$	++++ Distance dependent $k_{m,w}$
RW3 (Polystyrene plus 2% TiO$_2$)	— —	++++ Distance dependent $k_{m,w}$	+++++ Slightly distance dependent $k_{m,w}$	+++++ Distance independent $k_{m,w}$

Source: Baltas, D. et al. *The Physics of Modern Brachytherapy for Oncology.* CRC Press, Taylor & Francis Group, LLCC, Boca Raton, FL, 2007.

Note: $k_{m,w}$ is the correction factor required for absorption and scattering of the radiation at a specific radial distance r from a source in a phantom when compared to liquid water as defined in Equation 6.1

[a] Computerized Imaging Reference Systems (CIRS) Inc., Norfolk, VA.

correction is more energy-dependent in the lower photon energy region, exact distance correction factors have to be calculated for the specific design of a source (see, e.g., Meigooni et al. 2006).

According to the scope of the experimental measurements, adequate sizes for the phantoms have to be selected in order to warrant full scatter conditions (Pérez-Calatayud et al. 2004; Melhus and Rivard 2006). This phantom size depends on both the energy of the radionuclide and the phantom material and is correlated with the mean free path (mfp) of photons in this material. In case that this condition is not met, additional corrections for the missing scatter have to be considered for the actual experimental setup.

In an ideal condition, for the same radial distance from a brachytherapy source, the absorbed dose to a water-equivalent phantom material is identical to the dose to water. However, due to the contribution of absorbed dose in a medium via photoelectric absorption, and its dependence on the atomic number of the chemical elements of the material, this may not be true for all the brachytherapy sources. Therefore, in order to be able to compare the absorbed dose to a specific phantom material to the dose to water, a correction is required for a specific radial distance r from a source, $k_{m,w}$, that can be calculated using the following equation:

$$k_{m,w}(r) = \frac{\dot{D}_m(r)}{\dot{D}_w(r)} \quad (6.1)$$

where $\dot{D}_m(r)$ and $\dot{D}_w(r)$ are the dose rate to water in a given medium and water, respectively, per unit source strength at the same radial distance.

6.3 Ionization Dosimetry

Ionization chambers can be used either in water phantoms or in solid material phantoms for the measurement of relative dose distributions as well as of absolute dose rate and dose values. In general, ionization chambers with small dimensions and thus small collecting volumes, usually in the range 0.01–0.6 cm³, should be considered in order to achieve appropriate spatial resolution and to keep dose-gradient-related effects within an acceptable level. Thus, an optimum chamber volume has to be selected to achieve, on the one hand, adequate spatial resolution (the lower the better), and, on the other hand, proper chamber signal (the higher the better). Depending on the source strength and due to their small collecting volume (lower chamber response), certain restrictions will be placed to the maximum possible measurement distance, which will result in proper chamber readings. In addition, the chamber should demonstrate a flat angular response over the angle range of its orientation relative to the source position.

Ionization chambers with the lowest possible energy dependence in their response at the energy region of interest should be considered, especially for intermediate energy radionuclides such as ^{169}Yb, ^{170}Tm, and ^{192}Ir where a clear energy softening effect is expected.

It is noted that Chapter 5 described the calibration issues to determine the reference source strength in terms of reference air-kerma rate or air-kerma strength (S_k) of the brachytherapy source with traceability to the standard laboratories. The formalism for in-air measurements was presented, but for the purpose of a calibration measurement in the clinic, a well-type chamber method was recommended. The following subsection describes the relevant aspects of ionization dosimetry in the surrounding of brachytherapy source for the purpose of the experimental validation of MC simulation estimated source characteristics (such as for the TG-43 parameters) or of calculated (MC-based or using treatment planning systems) dose distributions around sources and applicators.

6.3.1 Measurement of Dose or Dose Rate

The estimation of the required dose to water in water, D_w, from the ionization chamber reading depends on the available calibration factor N of the chamber under consideration.

The most commonly used calibration protocol for ionization dosimetry in radiotherapy is that for dose to water in water (DIN 1993, 1996, 1997; DGMP 1999; ICRU 2001; IPEMB, 1996, 2005; IAEA 2000; Ma et al. 2001). Thus, for most of the ionization chambers available in radiotherapy departments, calibration factors in terms of dose to water in water, N_w, for a reference beam quality are available. Usually, ^{60}Co beams are considered as a reference beam quality.

In general, the dose to water in water medium, D_w, expressed in grays can be determined from the detector measurement M in a medium m (phantom material) using the following equations:

- Chamber calibrated in absorbed dose to water

$$D_w = N_w \cdot M \cdot k_P \cdot k_T \cdot k_{pol} \cdot k_{ion} \cdot k_V \cdot k_{appl} \cdot k_{m,w} \cdot k_{wp} \quad (6.2)$$

- Chamber calibrated in exposure

$$D_w = N_X \cdot \left(\frac{W}{e}\right) \cdot \left(\frac{\mu_{en}}{\rho}\right)_\alpha^w \cdot k_{RC} \cdot M \cdot k_P \cdot k_T \cdot k_{pol} \cdot k_{ion} \cdot k_V \cdot k_{appl} \cdot k_{m,w} \cdot k_{ap} \quad (6.3)$$

- Chamber calibrated in air kerma

$$D_w = N_K \cdot (1 - g_\alpha) \cdot \left(\frac{\mu_{en}}{\rho}\right)_\alpha^w \cdot M \cdot k_P \cdot k_T \cdot k_{pol} \cdot k_{ion} \cdot k_V \cdot k_{appl} \cdot k_{m,w} \cdot k_{ap} \quad (6.4)$$

Variables in the above equations are defined as follows:

N_w the absorbed dose in-water calibration factor of the chamber for the γ-energy of the radionuclide considered, usually expressed in Gy C^{-1}.

N_X the exposure calibration factor of the chamber for the γ-energy of the radionuclide considered, usually expressed in R C^{-1}.

N_K the air-kerma calibration factor of the chamber for the γ-energy of the radionuclide considered, usually expressed in Gy C^{-1}.

M the measured charge (C).

k_P the correction factor for the current air pressure conditions P, expressed in Pa, other than the reference value in the calibration certificate of the chamber, that is, the reference air pressure of 1013.25 hPa:

$$k_P = \left(\frac{1013.25 \ \text{hPa}}{P} \right).$$

If the pressure was expressed in mm Hg, the conversion factor would be $P_{\text{mmHg}} = 0.75 \times Pa$.

k_T the correction factor for the current air temperature conditions T in degrees Celsius other than the reference value in the calibration certificate of the chamber, that is, the reference temperature T_0 in degrees Celsius.

$$k_T = \left(\frac{273.15 + T(°\text{C})}{273.15 + T_0(°\text{C})} \right).$$

This reference temperature is usually $T_0 = 22°\text{C}$. The product of the air pressure and temperature correction coefficients is commonly defined as $k_{TP} = k_T.k_P$ (see Chapter 5).

Alternatively to air pressure and temperature correction factors, a low activity check radioactive source (^{90}Sr) can be used to determine $k_T.k_P$ according to the DIN recommendations (DIN 1993, 1996, 1997).

k_{pol} the correction factor for the polarity effect of the bias voltage for the photon energy of the radionuclide.

k_{ion} the correction factor that accounts for the unsaturated ion collection efficiency and thus for the charge lost to recombination for the specific radionuclide photon energy and the applied nominal voltage V (see also Chapter 5). k_{ion} is the reciprocal of the ion collection efficiency A_{ion} (Attix 1984).

k_V a correction factor to account for the effect of the chamber's finite size (i.e., volume) when the center of the chamber air cavity volume is considered as the point of reference for the positioning of the chamber and thus as the reference point for the measurements.

k_{appl} the correction factor to account for the attenuation in the applicator/catheter wall used for positioning or fixing the source in the in-phantom setup when the measured dose has to exclude this.

$k_{\text{m,w}}$ the correction required for absorption and scattering of the radiation at the specific point of measurement in the phantom material when compared to liquid water (see also Equation 6.1). For the specific radionuclide and source design, especially for the low-energy sources, this factor is distance and polar angle dependent. $k_{\text{m,w}}$ has to be estimated experimentally or using MC simulation calculations for the specific measurement setup.

$k_{\alpha p}$ the perturbation correction factor accounting for differences when changing from a medium of air surrounding the ion chamber to a phantom material. This is chamber type dependent. For ^{192}Ir and PMMA material, a good approximation is $k_{\text{ap}} = 1.0$ (DGMP 1999; DIN 1993).

k_{wp} the perturbation correction factor accounting for differences when changing from a medium of water surrounding the ion chamber to a phantom material. This is chamber type dependent. For phantom materials with density and effective atomic number that closely approximate those of water, $k_{\text{wp}} = 1.0$ (DGMP 1999; DIN 1993, 1997).

g_α the energy fraction of the electrons that is liberated by photons in air that are lost to radiative processes (Bremsstrahlung). g_α is for the photon energies relevant to brachytherapy practically zero (Baltas et al. 2007).

$\dfrac{W}{e}$ the ionization constant; $W/e = 33.97$ J C^{-1} and is energy independent for energies above 1 keV (Boutillon and Perroche-Roux 1987), with e being the elementary charge and W the mean energy expended in air per ion pair formed.

k_{RC} the roentgen to C kg^{-1} conversion factor, with $k_{RC} = 2.58 \times 10^{-4}$ C kg^{-1} R^{-1}.

$\left(\dfrac{\mu_{\text{en}}}{\rho} \right)_\alpha^w$ the ratio of mass energy absorption coefficient of water to that of air.

Table 6.3 summarizes the values of this ratio, which are given for the different brachytherapy radionuclides based on the corresponding effective energies. The photon energy spectrum at the place of measurement has to be considered for an accurate estimation of $\left(\dfrac{\mu_{\text{en}}}{\rho} \right)_\alpha^w$.

TABLE 6.3 Values of the Ratio of Mass Energy Absorption Coefficient of Air to That of Water $\left(\dfrac{\mu_{en}}{\rho} \right)_\alpha^w$ for Different Brachytherapy Radionuclides Using Their Effective Energy and Mass Energy Absorption Coefficients for Dry Air and Liquid Water (Taken by Baltas et al. 2007)

	Radionuclide								
Parameter	^{137}Cs	^{131}Cs	^{60}Co	^{198}Au	^{125}I	^{192}Ir	^{103}Pd	^{169}Yb	^{170}Tm
Effective energy E_{eff} (MeV)	0.652	0.030	1.257	0.417	0.028	0.398	0.021	0.131	0.067
$\left(\dfrac{\mu_{en}}{\rho} \right)_\alpha^w$	1.113	1.013	1.114	1.112	1.014	1.112	1.020	1.105	1.061

Note: Data for ^{131}Cs and ^{170}Tm have been calculated by the authors according to Hubbell and Seltzer (1995), where a log–log interpolation has been applied to these tables.

6.3.1.1 Calibration Factors N_K, N_W, and N_X

For the calibration factors, the discussion and methods described in Chapter 5 for the source calibration are also valid here. These are provided for well-defined measurement geometries for the purpose of absolute calibration of the source. It is, however, emphasized that, in general, additional calibration points (i.e., calibration factors at specific photon energies) to those usually given for ^{60}Co or ^{137}Cs are required, depending on the energy of the radionuclide being considered. Currently and internationally N_w calibration factors for ^{60}Co reference photon beam are provided for clinically used ionization chambers. Consequently, Equation 6.2 is the one that is most applicable for dose rate calculations in current clinical practice.

6.3.1.2 Chamber Finite Size Effect Correction Factor k_V

The correction factor k_V accounts for the radiation field inhomogeneity within the chamber volume (Kondo and Randolph 1960; Bielajew 1990a,b). Detailed description of this correction and values for k_V for different ionization chambers and geometries can be found in the work of Baltas et al. (2007).

6.3.1.3 Phantom Dimensions

Depending on the radionuclide, the dimensions of the phantom used, and the measurement geometry, additional correction for missing scattering at chamber positions lying at the periphery of the phantom (few backscatter material available) could become necessary for the individual experimental setup (Pérez-Calatayud et al. 2004; Melhus and Rivard 2006).

6.3.1.4 Room Scatter Effects

In an experimental procedure with an ionization chamber, attention must be paid to encounter any potential influence of scattered radiation from the room wall or floor on the measurements. For small distances from, for example, room wall, an increase in the chamber readings of some percent can be the case with in-air measurements and a smaller increase in the case of an in-phantom setup. This can be investigated by repeating measurements with variation of distance to the room wall. Generally, the effects can largely be avoided by keeping a minimum distance of 1.0 m from all possible scattering materials; see, for example, ESTRO Booklet No. 8 (*Venselaar and Pérez-Calatayud* 2004).

6.3.1.5 Other Effects

It is important that the phantom temperature is in equilibrium with the ambient/air temperature. Furthermore, if a radioactive check source is used to obtain the $k\rho \cdot k_T$ correction, then it must be ensured that this source has also achieved room air temperature. In any other case, correction factors must be implemented to account for these temperature differences.

6.3.1.6 Calculating Dose Rate from Dose

When the dose rate values, $\dot{D}_w$, are of interest, these can be calculated based on the estimated dose to water in water D_W (Equations 6.2 to 6.4) using the following equation:

$$\dot{D}_w = \frac{D_w}{\tau} \cdot k_u \tag{6.5}$$

where τ is the time interval during which the electrometer reading (charge) was taken. For high–dose-rate (HDR) sources such as ^{169}Yb, ^{192}Ir, and ^{60}Co and distances up to 10.0 cm, τ is normally in the order of some minutes. k_u is a unit conversion factor, to convert the result of Equation 6.5 to the required dose rate unit, usually cGy h^{-1}, cGy min^{-1}, or Gy min^{-1}, which depends on the strength of the source, that is, on its use with LDR, pulsed-dose-rate (PDR), or HDR sources.

6.4 TLD

A thermoluminescent dosimeter, or TLD, is a type of radiation detector that measures ionizing radiation by measuring the amount of visible light emitted from a crystal in the detector when it is heated (Cameron et al. 1968). These crystals contain one or more impurities to produce trap states for energetic electrons when irradiated with ionizing radiation. The trapped electrons will be released to their ground state and release visible light, when they are heated to a specific temperature. The amount of light emitted is directly related to the amount of radiation dose absorbed by the crystal. A phototube detector is able to measure the amount of the visible light released. Materials exhibiting thermoluminescence properties in response to ionizing radiation include calcium fluoride, lithium fluoride, calcium sulfate, lithium borate, calcium borate, and potassium bromide. The TLDs most commonly used in medical applications are composed of LiF with some impurities such as LiF:Mg, Ti, LiF:Mg, Cu, P, and Li2B4O7:Mn, because they are tissue equivalent. Other TLDs, such as CaSO4:Dy, Al2O3:C, and CaF2:Mn are used because of their high sensitivity.

TLDs are available in various forms (e.g., powder, chips, and rods) and sizes; see, for example, Figure 25.1. The TLD chips and rods can be reused after each irradiation by resetting all of the excited or trapped electrons following a procedure known as annealing. Well-established and reproducible annealing cycles, including the heating and cooling rates, should be used. As described in detail by Meigooni et al. (1995), there are two commonly used annealing techniques: pre-irradiation and pre-readout techniques. Table 6.4 shows the comparison of these two techniques.

Meigooni et al. (1995) concluded that pre-readout annealing increases the TLD sensitivity by about 40% by eliminating the low-temperature peaks in the glow curve of the TLD. However, the linearity reduces by about 10% at 600 cGy. Moreover, one could save about 48 h waiting time between the irradiation and readout that is needed in the pre-irradiation annealing technique.

A basic TLD reader system consists of a silver plate known as a planchet for placing and heating the TLD and a photomultiplier tube (PMT) to detect the thermoluminescence light emission and converting that to electrical signal. A plot of the intensity of thermoluminescence as a function of time is called the TLD glow

TABLE 6.4 Comparison between the Pre-irradiation and Pre-readout Annealing Techniques

Parameter	Pre-irradiation Annealing	Pre-readout Annealing
400°C	1 h	1 h
Rapid cooling to the room temperature	30–45 min	N/A
80°C	24 h	N/A
Post-irradiation annealing at 100°C	N/A	10 min
Post-irradiation waiting period before readout	24 h	N/A

curve (Cameron et al. 1968). The magnitude of the electric signal is proportional to the detected photon fluence. An electrometer will record the PMT signal as a charge or current. A thermocouple is located directly beneath the planchet to provide the heat. For reproducibility of TLD results, the heating and cooling pattern from the thermocouple is fixed. Normally this pattern is set to be at an initial temperature of 50°C followed by heating at the rate of 7°C/s to a final temperature of 250°C, and then remaining at 250°C for 10 s before it starts cooling down with the same rate using a fan installed on the system. The planchet and thermocouple are placed on a sliding drawer that is located in a light-tight (i.e., dark) chamber within the TLD reader system to minimize any readout of the visible light from outside. The constancy of the reader is examined by a fixed-output light source that would be exposed to the PMT when the planchet is completely pulled out from the chamber. It is recommended to examine this constancy at the beginning of the TLD reading and possibly several times during the reading process. These values should be recorded for approval of the functionality of the reader. In addition, the background contribution from the light leak to the chamber is examined by several readings with no TLD in the planchet and also several with some unexposed TLD chips in the chamber. It should be noted that the oxygen of the air may also create some visible light due to the heating of the planchet. Meigooni et al. (1995) have demonstrated that for accurate measurement of the low dose values (i.e., less than 10 cGy) with TLDs, one may require some nitrogen gas in order to flush the oxygen out of the chamber and minimize its contribution to the background reading.

After reading the irradiated TLD chips or rods around a brachytherapy source, the responses are converted to absorbed dose using the following equations:

$$\dot{D}(r) = \frac{TL_{net}}{T \cdot \varepsilon_{calib} \cdot d(T) \cdot E(r)} \quad (6.6)$$

where TL_{net} is the net TLD response found in the irradiation of n chips or rods and corrected for background (TL_{BKG}), differences between the chips using a relative correction factor (C_{ij}) known as the chip factor, and the linearity of the TLD response (F_{lin}) (Meigooni et al. 1995)

$$TL_{net} = \frac{\frac{1}{n} \sum_{\substack{i=1 \\ j=1}}^{n} \left[\left(TL'_{ij} - TL'_{BKG} \right) / C_{ij} \right]}{F_{lin}}. \quad (6.7)$$

ε_{calib} is the TLD sensitivity calibration factor (nC/cGy) using a calibrated radiation source. This sensitivity calibration factor varies with the TLD size and mass. T is the irradiation time period (in hours), $d(T)$ is the correction for the decay of the source strength during the irradiation time period

$$d(T) = \frac{1 - e^{-\lambda T}}{\lambda T} \text{ with } \lambda = \text{decay constant (in the unit of h}^{-1})$$

and $E(r)$ is the energy-dependent correction of the TLD response, between the calibration source and the measuring source

$$E(r) = \left(\frac{TL_{net}}{D_{water}(cGy)} \right)_{calib.}^{Brachy}.$$

Analysis of the uncertainty of the TLD is very critical for the final measured data. Chapter 15 presents details of this topic. It should be noted that, depending on the location of the point of interest relative to the source, the number of the chips used for dosimetry may affect the statistical fluctuation of the final data. Also, smaller detectors are recommended at shorter distances, whereas the larger chips are better for the larger distances. Various investigators have shown comparisons between the Monte Carlo simulated data and TLD measured data, indicating that these two techniques are complementing each other.

References

Aldelaijan, S., Devic, H., Mohammed, N. et al. 2010. Evaluation of EBT-2 model GAFCHROMIC film performance in water. *Med Phys* 37:3687–93.

Attix, H.F. 1984. Determination of A_{ion} and P_{ion} in the new AAPM radiotherapy dosimetry protocol. *Med Phys* 11:714–6.

Baltas, D., Sakelliou, L., and Zamboglou, N. 2007. *The Physics of Modern Brachytherapy for Oncology*. CRC Press, Taylor & Francis Group, LLCC, Boca Raton, FL.

Bielajew, A.F. 1990a. Correction factors for thick-walled ionisation chambers in point-source photon beams. *Phys Med Biol* 35:501–16.

Bielajew, A.F. 1990b. An analytic theory of the point-source non-uniformity correction factor for thick-walled ionisation chambers in photon beams. *Phys Med Biol* 35:517–38.

Bohm, D., Pearson, D.W., and Das, R.K. 2001. Measurements and Monte Carlo calculations to determine the absolute detector response of radiochromic film for brachytherapy dosimetry. *Med Phys* 28:142–6.

Boutillon, M. and Perroche-Roux, A.M. 1987. Re-evaluation of the W value for electrons in dry air. *Phys Med Biol* 32:213–9.

Butson, M.J., Cheung, T., and Yu, P.K. 2006. Weak energy dependence of EBT gafchromic film dose response in the 50 kVp–10 MVp X-ray range. *Appl Radiat Isot* 64:60–2.

Cameron, J.R., Suntharalingam, N., and Kenny, G.N. 1968. *Thermoluminescent Dosimetry*. The University of Wisconsin Press, London.

Chiu-Tsao, S.T., de la Zerda, A., Lin, J., and Kim, J.H. 1994. High-sensitivity GafChromic film dosimetry for ^{125}I seed. *Med Phys* 21:651–7.

Chiu-Tsao, S.T., Ho, Y., Shankar, R., Wang, L., and Harrison, L.B. 2005. Energy dependence of response of new high sensitivity radiochromic films for megavoltage and kilovoltage radiation energies. *Med Phys* 32:3350–4.

Chiu-Tsao, S.T., Medich, D., and Munro, J. 2008. The use of new GAFCHROMIC EBT film for ^{125}I seed dosimetry in Solid Water phantom. *Med Phys* 35:3787–99.

DGMP. 1999. *Praktische Dosimetrie in der HDR-Brachytherapie*. DGMP-Bericht Nr. 13 of Deutsche Gesellschaft für Medizinische Physik. H. Krieger and D. Baltas (eds.). DGMP, Berlin.

DIN. 1993. Clinical Dosimetry. Part 2. Brachytherapy with Sealed Gamma Sources; Klinische Dosimetrie. Teil 2. Brachytherapie mit umschlossenen gammastrahlenden radioaktiven Stoffen. Deutsches Institut für Normung e. V. DIN 6809. DIN, Berlin.

DIN. 1996. Clinical dosimetry—Part 5: Application of X-rays with peak voltages between 100 and 400 kV in radiotherapy; linische Dosimetrie. Teil 5: Anwendung von Roentgenstrahlen mit Roehrenspannungen von 100 bis 400 kV in der Strahlentherapie. Deutsches Institut für Normung e. V. DIN 6809. DIN, Berlin.

DIN. 1997. Procedures of dosimetry with probe-type detectors for photon and electron radiation—Part 2 Ionisation dosimetry; Dosismessverfahren nach der Sondenmethode fuer Photonen- und Elektronenstrahlung. Teil 2: ionisationsdosimetrie. Deutsches Institut für Normung e. V. DIN 6800. DIN, Berlin.

Hill, R., Holloway, L., and Baldock, C. 2005. A dosimetric evaluation of water equivalent phantoms for kilovoltage x-ray beams. *Phys Med Biol* 50:N331–44.

Hubbell, J.H. and Seltzer, S.M. 1995. Tables of X-ray mass attenuation coefficients and mass energy-absorption coefficients Version 1.4 on http://physics.nist.gov/xaamdi. National Institute of Standards and Technology, Gaithersburg. Originally published as NISTIR 5632.

IAEA. 2000. Absorbed dose determination in external beam radiotherapy. International Atomic Energy Agency, IAEA. Technical Reports Series, No 398. IAEA, Vienna.

ICRU. 2001. Dosimetry of High-Energy Photon Beams based on Standards of Absorbed Dose to Water. International Commission on Radiation Units and Measurements. ICRU Report 64. ICRU, Bethesda.

IPEMB. 1996. The IPEMB code of practice for the determination of absorbed dose for x-rays bellow 300 kV generating potential (0.035 mm Al-4 mm Cu HVL; 10-300 kV generating potential). Institute of Physicists and Engineers in Medicine. *Phys Med Biol* 41:2605–25.

IPEMB. 2005. Addendum to the IPEMB code of practice for the determination of absorbed dose for x-rays below 300 kV generating potential (0.035 mm Al-4 mm Cu HVL). Institute of Physicists and Engineers in Medicine. *Phys Med Biol* 50:2739–48.

Kondo, S. and Randolph, M.L. 1960. Effect of finite size of ionization chambers on measurement of small photon sources. *Rad Res* 13:37–60.

Lindsay, P., Rink, A., Ruschin, M., and Jaffray, D. 2010. Investigation of energy dependence of EBT and EBT-2 gafchromic film. *Med Phys* 37:571–6.

Li, Z., Das, R.K., DeWerd, L. et al. 2007. Dosimetric prerequisites for routine clinical use of photon emitting brachytherapy sources with average energy higher than 50 kev. *Med Phys* 34:37–40.

Luxton, G. 1994. Comparison of radiation dosimetry in water and in solid phantom materials for I-125 and Pd-103 brachytherapy sources: EGS4 Monte Carlo study. *Med Phys* 21:631–41.

Ma, C.M., Coffey, C.W., DeWerd, L.A. et al. 2001. AAPM protocol for 40-300 kV x-ray beam dosimetry in radiotherapy and radiobiology. *Med Phys* 28:868–93.

Meigooni, A.S., Meli, J.A., and Nath, R. 1988. A comparison of solid phantoms with water for dosimetry of ^{125}I brachytherapy sources. *Med Phys* 15:695–701.

Meigooni, A.S., Li, Z., and Williamson, J.F. 1994. A comparative study of dosimetric properties of Plastic water and Solid Water in brachytherapy applications. *Med Phys* 21:1983–7.

Meigooni, A.S., Panth, H., Mishra, V., and Williamson, J.F. 1995. Instrumentation and dosimeter-size artifacts in quantitative thermoluminescence dosimetry of low-dose fields. *Med Phys* 22:555–61.

Meigooni, A.S., Awan, S.B., Thompson, N.S., and Dini, S.A. 2006. Updated Solid Water™ to water conversion factors for ^{125}I and ^{103}Pd brachytherapy sources. *Med Phys* 33:3988–92.

Meli, J.A., Meigooni, A.S., and Nath, R. 1988. On the choice of phantom material for the dosimetry of ^{192}Ir sources. *Int J Rad Oncol Biol Phys* 14:587–94.

Melhus, C.S. and Rivard, M.J. 2006. Approaches to calculating AAPM TG-43 brachytherapy dosimetry parameters for ^{137}Cs, ^{125}I, ^{192}Ir, ^{103}Pd and ^{169}Yb sources. *Med Phys* 33:1729–37.

Muench, P.J., Meigooni, A.S., Nath, R., and McLaughlin, W.L. 1991. Photon energy dependence of the sensitivity of radiochromic film and comparison with silver halide film and LiF TLDs used for brachytherapy dosimetry. *Med Phys* 18:769–75.

Nakano, T., Suchowerska, N., Bilek, M.M., McKenzie, D.R., Nag, N., and Kron, T. 2003. High dose-rate brachytherapy source localisation: Positional resolution using a diamond detector. *Phys Med Biol* 48:2133–46.

Nath, R., Anderson, L., Luxton, G., Weaver, K., Williamson, J.F., and Meigooni, A.S. 1995. Dosimetry of interstitial brachytherapy sources: Recommendations of the AAPM Radiation Therapy Committee Task Group 43. *Med Phys* 22:209–34.

Perera, H., Williamson, J.F., Monthofer, S.P. et al. 1992. Rapid two-dimensional dose measurement in brachytherapy using plastic scintillator sheet: Linearity, signal-to-noise ratio and energy response characteristics. *Int J Radiat Oncol Biol Phys* 23:1059–69.

Pérez-Calatayud, J., Granero, D., and Ballester, F. 2004. Phantom size in brachytherapy source dosimetric studies. *Med Phys* 31:2075–81.

Pérez-Calatayud, J., Ballester, F., Das, R.K. et al. 2012. Dose calculation for photon-emitting brachytherapy sources with average energy higher than 50 keV: Report of the AAPM and ESTRO. *Med Phys* 39(5):2904–29.

Rashid, H., Bjarngard, B.E., Chin, L.M. and Rice, R.K. 1993. Dosimetry of ^{125}I sources in a low-density material using scaling. *Med Phys* 20:765–8.

Rivard, M., Coursey, B., Hanson, W. et al. 2004. Update of AAPM Task Group No. 43 Report: A revised AAPM protocol for brachytherapy dose calculations. *Med Phys* 31: 633–74.

Rustgi, S.N. 1995. Evaluation of the dosimetric characteristics of a diamond detector for photon beam measurements. *Med Phys* 22:567–70.

Venselaar, J.L.M. and Pérez-Calatayud, J. (eds.). 2004. *A Practical Guide to Quality Control of Brachytherapy Equipment*. 1st ed., ESTRO, Brussels. http://www.estro-education.org/publications/Documents/booklet8_Physics.pdf.

Wallace, R.E. 2002. Evaluated phantom material for ^{125}I and ^{103}Pd dosimetry. Poster SU-DD-EXH-12, AAPM Annual Meeting, Montreal, Canada.

Williamson, J.F. 1995. Recent developments in basic brachytherapy physics in radiation therapy physics. A.R. Smith (ed.). In: *Medical Radiology Diagnostic Imaging and Radiation Oncology*. A.L. Baert, L.W. Brady, H.P. Heilmann, M. Molls, K. Sartor (series eds.). Springer Verlag, Berlin, pp. 247–302.

Williamson, J.F. and Meigooni, A. S. 1995. Quantitative dosimetry methods in brachytherapy, In: *Brachytherapy Physics AAPM Summer School 1994*. J.F. Williamson, B.R. Thomadsen, and R. Nath, (eds.). Medical Physics Publishing Corporation, Madison, WI, pp. 87–133.

Williamson, J.F., Dempsey, J.F., Kirov, S.V., Monroe, J.I., Binns, W.R., and Hedtjärn, H. 1999. Plastic scintillator response to low-energy photons. *Phys Med Biol* 44:857–72.

Computational Methods for Dosimetric Characterization of Brachytherapy Sources

Panagiotis Papagiannis
University of Athens
Medical School

Luc Beaulieu
Centre Hospitalier Universitaire
de Québec and Université Laval

Firas Mourtada
Helen F. Graham Cancer Center

7.1 Preface

Computational dosimetry is primarily used for obtaining the prerequisite dosimetric parameters for every brachytherapy source intended for routine clinical use in the widespread formalism of the AAPM TG-43 (see Chapter 10). Hence, it forms the basis of conventional brachytherapy treatment planning practice.

The purpose of this chapter is to provide an insight to the fundamental principles of computational dosimetry methods. Besides enhancing physical intuition, this will establish awareness in that data used as treatment planning system input, or for commissioning purposes, are physical quantities. As such, they are subject to the limitations of the assumptions used for their calculation as well as uncertainty, and their accuracy should not be taken for granted.

This chapter is not intended to provide an in-depth presentation of computational methods used for brachytherapy dosimetry. Some suggestions for additional reading are provided from the wealth of literature available on the subject. The novice following methods presented in this chapter will also benefit from the discussion on potential pitfalls in the context of brachytherapy and associated cited literature.

Finally, computational dosimetry methods are currently entering everyday clinical practice in the form of advanced dose calculation algorithms used in contemporary treatment planning systems. Although this chapter is confined to dosimetry of sources in homogeneous water for illustrative purposes, it will assist readers in their familiarization with the main points of computational methods, their strengths, and their limitations.

7.2 Introduction: Effects to Be Considered by a Computational Dosimetry Method

Use of computational methods to determine the dose distribution around a brachytherapy source will be briefly described. Given the source physical characteristics (radionuclide, design, materials, etc.), this dose distribution is evaluated in a bounded homogeneous water geometry. The vast majority of brachytherapy sources can be considered pure photon emitters. Taking into account that their photon energies are relatively low such that photon attenuation is negligible within the range of the most energetic secondary electrons liberated in water, charged particle equilibrium (CPE) exists at all points of the geometry separated from its boundaries or the source by a distance at least equal to the aforementioned range. Since dose is equal to kerma under conditions of CPE, and radiative energy loss of secondary electrons is negligible in water in the energy range of brachytherapy (i.e., energy transfer equals energy absorption or $\mu_{tr} = \mu_{en}$), dose rate can be calculated as

$$\dot{D}(\vec{r}) \equiv \dot{K}(\vec{r}) = \int \Phi(E,\vec{r},t) E\left[\frac{\mu_{en}(E)}{\rho}\right] dE \qquad (7.1)$$

where $\Phi(E,\vec{r},t)$ represents the energy distribution of fluence rate in units of $MeV^{-1} \, cm^{-2} \, s^{-1}$. In other words, to know the photon fluence rate spectrum, $\Phi(E,\vec{r},t)$, at all points in the geometry is to know the dose rate distribution.

Since the photon energy spectra emitted by radionuclides used for brachytherapy are known with sufficient accuracy (see, e.g., NUDAT 2.5, National Nuclear Data Center, Brookhaven National Laboratory, http://www.nndc.bnl.gov/nudat2/), one might make the hasty assumption that the brachytherapy source dosimetry

problem is solved. This assumption disregards the effect of several factors such as the spatial distribution of radioactivity within the source, photon attenuation within the source, scatter radiation originating from the source, photon attenuation in the water medium surrounding the source, and scatter generation within it. No simple solution exists for computational dosimetry since one of the above factors cannot be accounted for analytically.

To illustrate this, let us begin with the simplest case of a bare (unencapsulated) point source in void space emitting one photon per disintegration with an energy of 355 keV, which is the emission weighted mean photon energy of ^{192}Ir. As expected, the photon fluence from this source (Figure 7.1a) is isotropic, and its

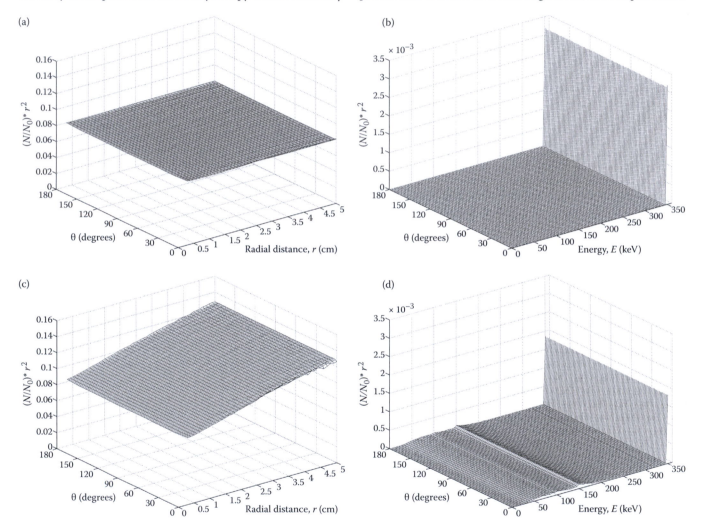

FIGURE 7.1 (a) Depiction of the photon fluence in vacuo for a bare point source emitting one 355-keV photon per disintegration. The number of photons per square centimeter, N, divided by the number of initially emitted photons, N_0, and multiplied by r^2 to remove the influence of the geometry factor is plotted versus radial distance, r, and polar angle, θ. As expected, the fluence is isotropic and equals $\frac{1}{4\pi}$ in the axis used in this plot. (b) Depiction of the photon fluence spectrum in vacuo for the same source, in the form of the number of photons per square centimeter, N, reaching $r = 5$ cm divided by the number of initially emitted photons, N_0, plotted versus energy, E, and polar angle, θ. As expected, the energy fluence is zero except for the 345–355 keV energy bin where it is equal to $\frac{1}{4\pi(5cm)^2}$. (c) Same as (a) but with the source centered in a homogeneous water sphere of 15-cm radius. In contrast to (a), since the influence of the geometry factor is removed, the fluence appears to increase with distance due to the contribution of photons scattered in water. (d) Same as (b) but with the source centered in a homogeneous water sphere of 15-cm radius. An energy fluence of photons that have undergone scattering in water is discerned, and the fluence of primary photons in the 345–355 keV energy bin is reduced relative to that in (b) by $\exp(-0.111*5) = 0.57$, where 0.111 $cm^2 \, g^{-1}$ is the linear attenuation coefficient of water for the 355-keV energy. (Courtesy of P. Papagiannis.)

reduction with distance can be easily calculated using a geometry factor, which assumes the form of the inverse square law for a point source. Since this source is situated in void space, the photon fluence spectrum is known (see Figure 7.1b). According to Equation 7.1, for source activity A, the dose rate to a small volume of water at position r relative to the source can be calculated as

$$\dot{D}(r) = \frac{1}{4\pi r^2} AE \left[\mu_{en}(E) \bigg/ \rho \right]_{water} \tag{7.2}$$

Activity, however, is not an appropriate quantity for source strength characterization. Air-kerma strength, S_K, is the correct quantity. According to its definition (see Chapter 5), S_K can be expressed similar to Equation 7.2 as $S_K = \frac{1}{4\pi} AE \left[\mu_{en}(E) \bigg/ \rho \right]_{air}$, so that we arrive at a simple expression of dose rate per unit air-kerma strength at any point in our geometry:

$$\dot{D}(r) \bigg/ S_K = \frac{1}{r^2} \left[\mu_{en}(E) \bigg/ \rho \right]_{air}^{water} . \tag{7.3}$$

If our source was centered in a homogeneous water phantom, the photon fluence would be different (Figure 7.1c and d). The attenuation of the emitted photon fluence spectrum by water can be taken into account using the law of exponential attenuation. Therefore, the primary dose rate (the component of dose deposited by photons that have not undergone any interaction before they reach r) can be calculated as

$$\dot{D}(r) \bigg/ S_K = \frac{1}{r^2} \left[\mu_{en}(E) \bigg/ \rho \right]_{air}^{water} e^{-\mu(E)r}. \tag{7.4}$$

The scattered photon fluence spectrum that builds up as photons interact in water (see, e.g., the spectrum of energies lower than 355 keV in Figure 7.1d) cannot be analytically calculated. At each interaction site of a primary photon, there is a finite probability that the photon will undergo one of a number of possible interaction types; according to the type of interaction occurring, there might be a probability distribution determining the new direction of a photon and its energy degradation, and this process might be repeated several times for each of the primary photons due to multiple scattering. It is not the stochastic nature of the process that precludes an analytical solution, but its complexity due to the dispersion of photon tracks, the progressive energy degradation, and multiple scattering. After all, the primary dose rate is also a stochastic quantity attributable to the stochastic nature of photon interactions that reduce the primary photon fluence with distance traversed within a material. Its expectation value, however, can be easily calculated by Equation 7.4, in contrast to that of the scattered dose rate.

The situation is the same with actual brachytherapy sources. The effect of the radioactivity distribution within the source volume on the photon fluence spectrum at each point (compare Figures 7.1a and 7.2a) can be taken into account by a geometry factor calculated according to

$$G(\vec{r}) = \frac{\int_V \frac{\rho(\vec{r}')dV'}{|\vec{r} - \vec{r}'|^2}}{\int_V \rho(\vec{r}')dV'} \tag{7.5}$$

where V is the source core volume and $\rho(\vec{r})$ denotes radioactivity per unit volume. Since the core diameter of brachytherapy sources is usually significantly shorter than their length, the exact geometry factor of Equation 7.5 tends to that for a linear source of uniform radioactivity at distances $r > L$ from the source given by the function

$$G(r, \theta) \begin{cases} (r^2 - L^2/4)^{-1}, & \theta = 0 \\ \dfrac{\beta}{Lr\sin\theta}, & \theta \neq 0 \end{cases} \tag{7.6}$$

where β is the angle subtended to the source from point (r, θ). At distances $r > 2L$ from a source, the geometry factor tends to that of a point source [$G(r) = 1/4\pi r^2$; see Figure 7.2a].*

The attenuation of the emitted photon fluence spectrum within the source structure (compare Figures 7.1b and 7.2b) can be taken into account using the Sievert integral (Williamson 1988; Karaiskos et al. 2000; see Section 7.3.), so that an equation analogous to Equation 7.4 can still be employed for the calculation of the primary dose rate. Even if the source was poly-energetic, an equation analogous to Equation 7.4 could be summed over the emitted photon energy spectrum, or effective attenuation and mass energy absorption coefficients weighted over the emitted energy spectrum could be used, without loss of simplicity.

As with the previous example of the bare point source, however, the scattered photon fluence spectrum due to photon interactions within the source (see Figure 7.2b) and the surrounding medium (see Figure 7.2d) cannot be analytically calculated. An equation that completely determines the photon fluence

---

* This discussion about the geometry factor best describing the effect of radioactivity distribution on photon fluence at a point should not confuse the reader with regard to the line approximated geometry function used in the AAPM TG-43 formalism for practical reasons. Since this formalism is in essence a decomposition of the 2D dose distribution around a cylindrically symmetric brachytherapy source to a number of dosimetric quantities, the use of any geometry function expression is allowed as long as it is consistent (i.e., the same geometry function is used to derive the formalism's dosimetric quantities from the dose distribution and subsequent dose calculations using these quantities; see also Chapter 10).

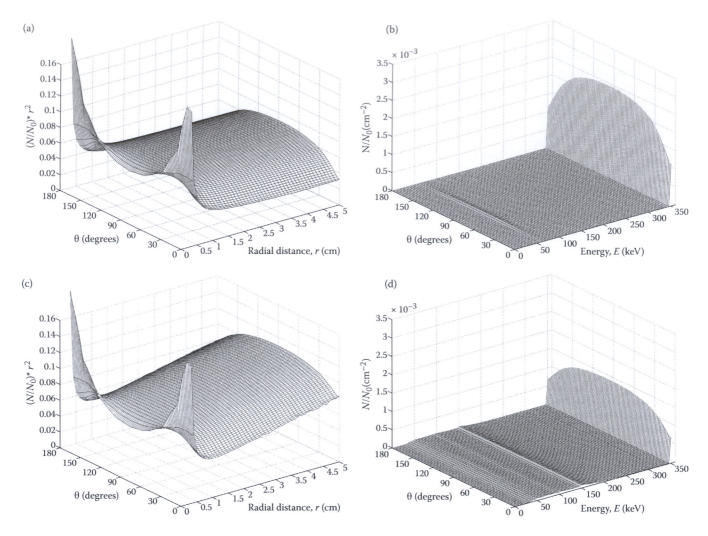

FIGURE 7.2 (a) Depiction of the photon fluence in vacuo for a hypothetical source emitting one 355-keV photon per disintegration of radioactivity uniformly distributed within its cylindrical (L = 0.5 cm, d = 0.05 cm) Ir core that is encapsulated within 0.02-cm-thick stainless steel. The number of photons per square centimeter, N, divided by the number of initially emitted photons, N_0, and multiplied by r^2 to remove the geometry factor dependence is plotted versus radial distance, r, from the source center and polar angle, θ, relative to the source longitudinal axis. The increased fluence values observed at r comparable to the source length and θ close to 0° and 180° is due to the proximity of these points to the source. The fluence is anisotropic due to the increased attenuation of photons emitted at oblique directions. At θ = 90°, the fluence is reduced by 12% relative to its value in Figure 7.1a due to the combined effect of attenuation and scatter within the source. (b) Depiction of the photon fluence spectrum in vacuo for the same source, in the form of the number of photons per square centimeter, N, reaching r = 5 cm divided by the number of initially emitted photons, N_0, plotted versus energy, E, and polar angle, θ. An energy fluence of photons that have undergone scattering in the source is discerned. The fluence of primary photons in the 345–355 keV is anisotropic due to oblique filtration within the source and reduced relative to that in Figure 7.1b (i.e., approximately 21% at θ = 90°) due to attenuation within the source. This reduction cannot be accounted for by a simple exponential as in Figure 7.1b due to the finite dimensions of the source. Still it can be calculated analytically using a Sievert integral approach (see Section 7.3). (c) Same as (a) but with the source centered in a homogeneous water sphere of 15-cm radius. In contrast to (a), since the influence of the geometry factor is removed, the fluence appears to increase with distance due to the contribution of photons scattered in water. (d) Same as (b) but with the source centered in a homogeneous water sphere of 15-cm radius. An energy fluence of photons that have undergone scattering in the source as well as water is discerned. The fluence of primary photons in the 345–355 keV energy bin is anisotropic due to oblique filtration within the source. Its value at θ = 90° is reduced relative to that in (b) by exp(–0.111*5) = 0.57, where 0.111 cm^2 g^{-1} is the linear attenuation coefficient of water for the 355 keV energy. (Courtesy of P. Papagiannis.)

spectrum in any geometry exists; it is the Boltzmann transport equation. The increased number in independent variables required to define photon fluence renders the analytical solution of this equation impossible even for the simple case of our bounded homogeneous water geometry.

Since experimental brachytherapy dosimetry presents a formidable task associated with a level of uncertainty not appropriate for treatment planning dosimetry, it is reserved for QA purposes and benchmarking of computational dosimetry results. Computational dosimetry methods used in brachytherapy can be categorized in the following three broad classes: (1) semiempirical algorithms and the numerical methods of (2) stochastic Monte Carlo simulation and (3) deterministic solvers of the linear Boltzmann equation.

7.3 Semiempirical Dose Calculation Methods

In the previous section, we have seen that Equation 7.3 gives the dose rate to water *in free space* (in vacuo) from a perfect point source. Other consideration must be taken if homogeneous, semi-infinite, water medium or heterogeneous media are considered.

7.3.1 Dose from a Point Source in a Homogeneous Water Medium

If the source is surrounded by a medium, the interplay between the attenuation of the primary photon fluence and the increased contribution of the scattered photons must be taken into account. As such, Equation 7.3 can be rewritten to give the dose rate at a distance r from a point source in medium:

$$\dot{D}_w(r) = S_K \frac{1}{r^2} \left[\mu_{en}(E) \Big/ \rho \right]_{air}^{water} f_{as,w}(r) \tag{7.7}$$

where the nomenclature of Baltas et al. (2007) is used, and $f_{as,w}(r)$ represents the correction factor for attenuation and scattering in water and can be expressed as the ratio of the dose at a distance r in medium to that at the same point in free space. It is the same as the tissue attenuation and buildup factor, $T(r)$, used in previous literature (Williamson 2005). It is also relevant to recall that in describing broad beam attenuation, the notion of buildup factor can be introduced. It is defined as the total dose in medium at a distance r (primary + scatter) to that of the primary dose alone also in the medium at the same distance (Attix 1986). As such, the relationship between $f_{as,w}(r)$ and $B(r)$ can be expressed as

$$f_{as,w}(r) = T(r) = e^{-\mu r} B(r)$$

$$= e^{-\mu r} \left\{ 1 + \frac{D_{scatter}(r)}{D_{primary}(r)} \right\} = e^{-\mu r} \{1 + \text{SPR}(r)\}. \tag{7.8}$$

Consequently, combining Equations 7.7 and 7.8, the dose rate from a point source at distance r in medium can be separated into primary and scatter components.

7.3.2 Dose from Realistic Sources and the Sievert Integral

In clinical practice, sources have finite dimensions such that they will not represent the ideal point source geometry at all distances of interest from them. In fact, most modern sources have cylindrical geometries, with or without a connecting cable at one end. Moreover, the source core and encapsulation are likely to have materials different from tissues (from a physics interaction point of view). Therefore, attenuation and scatter of the primary photons will occur within the source and fluorescence x-ray can be produced, all contributing to a change in the energy fluence, which can be angle dependent (nonisotropic) at the surface of the source capsule.

A general formulation of the geometry factor over the volume of a source was given in the previous section by Equation 7.5. Replacing the simple inverse square law in Equation 7.7, the dose rate for a realistic source can be written as

$$\dot{D}_w(r,\theta) = S_K \left[\overline{\mu_{en}(E)/\rho} \right]_{air}^{water} G(r,\theta) f'_{as,w}(r,\theta) \tag{7.9}$$

where the use of polar coordinates is justified by the cylindrical symmetry discussed above. The energy-weighted ratio of the mass-energy absorption coefficient is used to account for the poly-energetic nature of many radionuclides. Note that in Equation 7.9, $f'_{as,w}(r,\theta)$ is different from the parameter expressed in Equation 7.8; it now includes the effect of the core and encapsulation in addition to that of the absorption and scatter contribution from the medium. Note that links could be made between $T(r)$, $f'_{as,w}(r,\theta)$, and TG-43 (Appendixes A and B of the original TG-43 report; Nath et al. 1995).

Historically, the issue of sources with finite dimension has been handled using a 1D path-length model or Sievert integral (Williamson 2005). In this approach, the contribution to the dose from a source having an active core of length L is considered to be the sum of multiple smaller source elements of dimension ΔL. First assuming that attenuation and scatter of all origin can be neglected, the dose rate for a single element ΔL at a point of interest P having coordinates (x, y) is given by

$$\Delta \dot{D}_{wat}(x,y) = S_K \frac{\Delta L}{L} \frac{(\mu/\rho)_{air}^{wat}}{(x/\cos\vartheta)^2}. \tag{7.10}$$

Using polar coordinates ($r = x/\cos\vartheta$; $\Delta L = x \sec^2\vartheta\, d\vartheta$) and integrating over all elements along L, from ϑ_1 to ϑ_2 corresponding to the angle subtended by both ends of the source relative to P, as illustrated in Figure 7.3, the dose rate can be written as

$$\dot{D}_{wat}(r,\vartheta) = \frac{S_K (\mu/\rho)_{air}^{wat}}{Lx} \int_{\vartheta_1}^{\vartheta_2} d\vartheta$$

$$= \frac{S_K (\mu/\rho)_{air}^{wat}}{Lr \cos\vartheta} \Delta\vartheta. \tag{7.11}$$

If the attenuation by the encapsulation of radial thickness t is now considered, Equations. 7.10 and 7.11 become

$$\Delta \dot{D}_{wat}(x,y) = \frac{S_K}{L} (\mu/\rho)_{air}^{wat} \frac{x/\cos^2\vartheta}{(x/\cos\vartheta)^2} \frac{e^{-\mu't/\cos\vartheta}}{e^{-\mu't}} d\vartheta \tag{7.12}$$

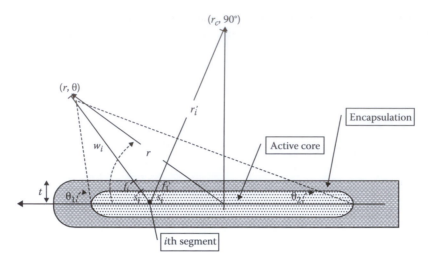

FIGURE 7.3   Geometry for the Sievert integral. (Courtesy of P. Papagiannis.)

$$\dot{D}_{\text{wat}}(r,\vartheta) = \frac{S_K(\mu/\rho)_{\text{air}}^{\text{wat}}}{Lr\cos\vartheta} e^{\mu't} \int_{\vartheta1}^{\vartheta2} e^{\mu't/\cos\vartheta} \mathrm{d}\vartheta \qquad (7.13)$$

where μ' is the effective attenuation coefficient, and t is the radius of the source. The term $e^{\mu't}$ is necessary to account for the fact that the measurement of S_K already takes into account the attenuation along the bisector axis and must be removed to avoid double correction. The integral term in Equation 7.13 is the Sievert integral. Finally, medium absorption and scatter correction can also be taken into account by combining Equations 7.8 and 7.13:

$$\dot{D}(r,\vartheta)_{\text{water}} = S_K \frac{\left(\overline{\mu_{en}/\rho}\right)_{\text{air}}^{\text{water}}}{Lr\cos\vartheta} \int_{\vartheta_1}^{\vartheta_2} e^{-\mu(r\cos\vartheta - t)\sec\vartheta}$$

$$\times\{1 + \text{SPR}[(r\cos\vartheta - t)\sec\vartheta]\} e^{-\mu't} \mathrm{d}\vartheta. \quad (7.14)$$

Further refinements can take into account the nonfinite size of the core itself (3D integration) by partitioning the 3D source core into infinitesimal point sources and integrating over the volume V:

$$\dot{D}(r,\vartheta)_{\text{water}} = S_K \frac{\left(\overline{\mu_{en}/\rho}\right)_{\text{air}}^{\text{water}}}{F(r_c)}$$

$$\times \int_V r_i^{-2} \exp(-\mu_s s_i - \mu_f f_i - \mu_w w_i)[1 + \text{SPR}(w_i)]\mathrm{d}V'$$

$$= \frac{S_K}{N} \frac{\left(\overline{\mu_{en}/\rho}\right)_{\text{air}}^{\text{water}}}{F(r_c)}$$

$$\times \sum_{i=1}^{N} \left[\exp(-\mu_s s_i - \mu_f f_i - \mu_w w_i)[1 + \text{SPR}(w_i)]/r_i^2 \right]$$

$$(7.15)$$

where V denotes the active source core volume; s_i, f_i, and w_i are the distances traversed by photons within the source core, filter material, and water, respectively. r_i is the distance between the center of the ith segment and the calculation point ($r_i = s_i + f_i + w_i$), and μ_s, μ_f and μ_w are thickness-dependent effective filtration coefficients of the source, the filter material, and water, respectively. This approach was shown to be within a few percent of Monte Carlo calculation for ^{137}Cs (Williamson 1988), but greater discrepancy was seen for low-energy photon sources (Williamson 1996). Williamson improved the accuracy of the Sievert model yielding satisfactory agreement with Monte Carlo simulation results for ^{192}Ir as well as lower energy ^{169}Yb and ^{125}I sources by using the Sievert model to calculate the primary dose rate component and approximating the scatter dose rate component by an isotropic distribution, that is,

$$\dot{D}(r,\vartheta)_{\text{water}} = \dot{D}(r,\vartheta)_{\text{prim}} + \dot{D}(r,\pi/2)_{\text{scat}} = \dot{D}(r,\vartheta)_{\text{prim}}$$

$$+ \dot{D}(r,\pi/2)_{\text{prim}} \text{SPR}(r). \qquad (7.16)$$

It should be noted that in this context, the term "primary" stands for photons that emerge from the source structure (i.e., includes photons scattered within the source), whereas the term "scatter" stands for photons that have undergone scattering in water. Karaiskos et al. (2000) further improved the isotropic scattering Sievert model of Williamson by incorporating a simple empirical correction for deviation of scatter dose from isotropy, among other things.

7.3.3 Advanced Scatter-Separation Techniques

Another approach has been proposed in which the dose distribution around a source is expressed in terms of the contributions of the primary photon dose and of that coming from the scatter particles (Russell and Ahnesjö 1996). Here, primary dose relates to all photons that leave the source capsule surface and contribute to energy deposition from the first interaction. The scatter dose is the dose that results from subsequent interactions. The total dose can be written as

$$D = D_{prim} + D_{scat}$$

$$= D_{prim} + (D_{1sc} + D_{msc}). \qquad (7.17)$$

Generally, D_{prim} and D_{scat} are obtained from Monte Carlo calculations that include detailed simulations of the geometry (in 3D). It follows that D_{prim} can be calculated without any approximation through use of ray-tracing techniques that propagate the photon energy fluence from the source through the geometry. As such, the calculation of D_{prim} will be equivalent for any model-based dose calculation algorithms (MBDCAs). D_{scat} can be modeled using exponential functions or using precalculated Monte Carlo kernels in water (Russell and Ahnesjö 1996; Carlsson and Ahnesjö 2000a). One highly efficient way of dealing with the scatter component is the collapsed-cone convolution technique (Carlsson and Ahnesjö 2000b; Carlsson Tedgren and Ahnesjö 2003; Carlsson Tedgren and Carlsson 2009), which uses a successive-scattering superposition method. Monte Carlo is used to generate the once-scatter and multiscatter dose point kernels that are fit to analytical monoexponential or biexponential functions (Carlsson and Ahnesjö 2000a). This method is likely to be included into a commercial treatment planning system from Nucletron (Nucletron BV, The Netherlands). In short, D_{prim} is calculated explicitly from the collision kerma as described above and is used to evaluate the scerma, $S(r)$, or scatter energy released per unit mass:

$$S(r) = \frac{\mu - \mu_{en}}{\mu_{en}} D_{prim}(r). \qquad (7.18)$$

$S(r)$ can then be weighted by the once-scatter point dose kernel and integrated over the whole volume to obtain D_{1sc}. The latter is then used to calculate D_{msc} proceeding in a similar fashion as for the once-scatter case using the appropriate dose point kernel. One must recall that, at this point, all calculations are done in homogeneous water geometry.

Interestingly, Russell et al. (2005) have shown that the primary-scatter separation approach can be used to derive TG-43 parameters and vice versa. Therefore, the primary-scatter separation formalism contains all the necessary information for extensive comparison of various dose calculation methods around a given source in water.

7.4 Stochastic Dose Calculation Methods: Monte Carlo Simulation

7.4.1 Basis of Monte Carlo Simulation for Dosimetry

Although the Monte Carlo method is often introduced in the context of radiological science as a stochastic alternative to

solving the Boltzmann equation, it can also be used for solving problems of nonstochastic character, for example, integral computation or solving systems of linear equations. In essence, it comprises a numerical method that relies on random sampling to produce observations on which statistical inference can be performed to extract information about a system (Lemioux 2009). It is based on the law of large numbers and the central limit theorem.

Assume that we want to calculate an unknown quantity, m, and that we can find a random variable, k, of expectation value equal to m [$E(k) = m$] with variance, b^2 [$Var(k) = b^2$].[*] If $k_1, k_2, \ldots, k_N$ are N independent values of k, then according to the law of large numbers, as N increases, the arithmetic average of these values tends to the expectation value of k. Expressed as an equation, the relationship is: $\frac{1}{N}\sum_{i=1}^{N} k_i = E(k) = m$. According to the central limit theorem, if our N values of k are randomly selected from its probability distribution, their sum is normally distributed. In other words, the probability that the sum $\sum_{i=1}^{N} k_i$ is within three standard deviations ($3b\sqrt{N}$) of its expectation value (Nm) is 0.997:

$$P\left(Nm - 3b\sqrt{N} < \sum_{i=1}^{N} k_i < Nm + 3b\sqrt{N} \right) \approx 0.997 \qquad (7.19)$$

or equivalently

$$P\left(-\frac{3b}{\sqrt{N}} < \frac{1}{N}\sum_{i=1}^{N} k_i - m < \frac{3b}{\sqrt{N}} \right) \approx 0.997. \qquad (7.20)$$

The last equation provides a summary (abstract for the moment, yet concise) of the Monte Carlo method for the estimation of an unknown quantity, as well as the associated precision. We have to come up with a random variable that has a known probability distribution and an expectation value equal to our unknown quantity, and randomly sample a sufficient number of values from its probability distribution. The arithmetic mean of our sample will approximate the unknown quantity with a precision depending only on the sample size and its variance.

---

[*] Although an effort is made throughout this chapter to refrain from mathematical nomenclature, a minimum level of familiarization to statistics and probability theory is necessary. Readers lacking it, but still interested in this chapter, should consider the following: a continuous random variable X of real values x is associated with a distribution of the probability that each value occurs. This distribution is described by a *cumulative distribution function* $F(x)$ giving $P(x' \leq x)$ as $F(x) = \int_{-\infty}^{x} f(x)dx$, where $f(x)$ is the *probability density function* giving $P(a \leq x' \leq b)$ as $f(x) = \int_{a}^{b} f(x)dx$. The *expectation value* (or average) of X is defined as $E(X) = \int_{-\infty}^{\infty} xf(x)dx$, and its *variance* is defined as $V(X) = \{E[X - E(X)]^2\}$.

In the context of computing the dose distribution around a brachytherapy source, Equation 7.1 shows how this could be generally accomplished if the photon fluence rate spectrum was known everywhere in space. The photon fluence rate spectrum consists of all possible photon trajectories in the geometry. If we could produce a random sample from the distribution of all possible trajectories, we could estimate the kerma at any point by the arithmetic mean of results obtained by an estimator of the contribution to kerma* of those trajectories encountering a finite volume element defined about the point. The problem is that we cannot sample directly from the distribution of all possible tracks. The solution is to come up with a Markov chain process corresponding to the distribution from which we want to sample.

Assume a photon trajectory is composed of sequential photon states, $S_j(\vec{r}_j, \vec{\Omega}_j, E_j)$, just before each interaction j, where $\vec{r}$, $\vec{\Omega}$, and E denote photon position, direction, and energy, respectively. The probability that each photon state j occurs depends only on the probability of occurrence of state $j-1$, which is the definition of a Markov chain. Specifically, the probability that photon state j occurs depends only on the probability that the photon interacts at r_{j-1}, the probability that a specific kind of interaction occurs, and the probability that, during this interaction, the photon is scattered in direction Ω_j with energy E_j given Ω_{j-1} and E_{j-1}. The abovementioned probability distributions, however, are known in physics.

Hence, the only component missing for performing a random walk through the Markov chain of sequential photon states, in order to produce a sample of trajectories and use them for kerma estimation, is a method to randomly sample from the known probability distributions. Numerous methods exist, and for illustrational purposes, we will consider the inversion theorem. Assume that X is a continuous random variable of values x over the interval $[a, b]$ with a probability density function $f(x)$ and a cumulative distribution function $F(x)$ that is invertible. Given a random number, r, in the interval $[0, 1]$, a value x' can be randomly selected from the distribution of X according to (see also Figure 7.4)

$$r = F(x') = \int_a^{x'} f(x)dx. \tag{7.21}$$

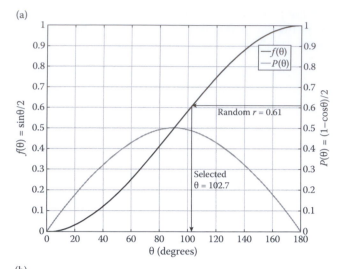

(a)

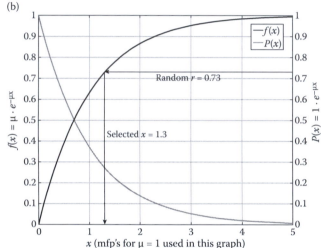

(b)

FIGURE 7.4 (a) Example of random sampling of the angle θ for the isotropic emission of photons from a point source. The left axis presents the probability density function of θ, and the right axis presents the corresponding cumulative distribution function (see text). Selecting a random number, r, equating it to the value of the cumulative distribution function, and solving for θ yield our randomly selected value. (b) Example of random sampling of the distance x traversed by photons in a material before their first interaction occurs. The left axis presents the probability density function of x (for $\mu = 1$), and the right axis presents the corresponding cumulative distribution function (see text). Selecting a random number, r, equating it to the value of the cumulative distribution function, and solving for x yield our randomly selected value in units of mean free paths since $\mu = 1$ was used in this graph. (Courtesy of P. Papagiannis.)

In case of a discrete random variable X taking N values, x_i, of probability P_i so that $\sum_{i=1}^{N} P_i = 1$, a value x' can be randomly selected as

$$x' = x_j \text{ where } j = \min\left\{j : r < \sum_{i=1}^{j} P_i\right\}. \tag{7.22}$$

---

* Examples include the photon energy loss that might occur within the volume element when a photon interaction happens within the volume, divided by its mass (commonly referred to as an analog estimator), or weighting the energy of photons whose trajectory intersects the volume element by the mass energy absorption coefficient and the distance the photon travels through the volume according to Equation 7.1, since fluence and total photon path length per unit volume are equivalent, and dividing by the volume element mass (commonly referred to as a track length estimator). See Williamson (1987) and Section 7.4.2.7 on variance reduction techniques.

Putting everything together in a simple example of a mono-energetic point photon source centered in a bounded water phantom, a Monte Carlo code would have to generate photon trajectories within the phantom by first simulating the emission of a photon. The emission is isotropic, and in spherical coordinates, the probability density function of emission into a solid angle element $d\Omega$ is $f(\theta,\phi)d\theta d\phi = \dfrac{d\Omega}{4\pi} = \dfrac{\sin\theta d\theta d\phi}{4\pi}$ so that $f(\theta)d\theta = \dfrac{\sin\theta d\theta}{2}$ and $f(\phi)d\phi = \dfrac{d\phi}{2\pi}$ since θ and ϕ are independent. Using the inversion theorem, given random numbers r, r' in the unitary interval, a value for θ can be randomly sampled as (see also Figure 7.4)

$$r = P(\vartheta') = \int_0^{\theta'} \frac{\sin\vartheta d\vartheta}{2} \Rightarrow r = \frac{1-\cos\theta'}{2} \Rightarrow \cos\theta' = 1 - 2r$$

and a value for ϕ can be randomly sampled as

$$r' = P(\phi') = \frac{1}{2\pi}\int_0^{\phi'} d\phi \Rightarrow \phi' = 2\pi r'.$$

The next step would be to decide randomly on the site of the first interaction. Employing a random number and the inversion theorem once again, we obtain (see also Figure 7.4)

$$r = P(x) = \int_0^{x'} \mu \exp(-\mu x)dx \Rightarrow r = -\exp(-\mu x') + 1 \Rightarrow$$

$$x' = -\frac{1}{\mu}\ln(1-r) = -\frac{1}{\mu}\ln(r).$$

The interaction type is a discrete random variable of i values so that $P_i = i_i/i_{\text{total}}$, and inversion in the form of Equation 7.22 can be employed. Assuming the outcome is an incoherent scattering event, our code would have to randomly sample a new energy and a new direction for the photon using the appropriate differential cross section $\dfrac{d\sigma}{d\Omega}(E,\theta,Z)$. What follows would be a new random sampling for the site of the next interaction as above, and the procedure would continue until the photon underwent photoelectric absorption or escaped the geometry, in which case the procedure would be repeated with the emission of a new photon.

The aim of this oversimplified discussion is to show that the main elements of a Monte Carlo code are a random number generator, a set of routines for sampling the probability distributions of physical processes underlying photon transport, and a set of input data. For brachytherapy source dosimetry, photoelectric effect, coherent and incoherent scattering, and characteristic x-ray production must be considered (Rivard et al. 2004). Besides information for the geometry and materials, the input data set includes linear attenuation coefficients, partial interaction cross sections, atomic form factors, and incoherent scattering factors.

7.4.2 Practical Aspects of Monte Carlo for Brachytherapy Dosimetry

7.4.2.1 Choice of Monte Carlo Code

Why not develop your own code? A linear congruential random number generator can be easily employed and tested (see the next subsection), and techniques for sampling from the necessary probability distributions and scoring are available in the literature (Morin 1988). Additionally, geometry description or photon crossing of material boundaries can be coded relatively easily, at least for our simple geometry of a brachytherapy source centered in a homogeneous water phantom, especially if generality or time efficiency is not of major concern.

Although the effort required to develop a code might seem as the main deterrent, maintaining the code and relevant documentation might prove even harder. In the long run, issues like flexibility and ease of use will push users toward one of the general-purpose Monte Carlo simulation packages developed and maintained by groups and collaborations at institutes like Los Alamos National Laboratory (LANL), National Research Council Canada (NRC), Nuclear Energy Agency (NEA), and Conseil Européen pour la Recherche Nucléaire (CERN, the European Laboratory for Particle Physics).

It is recommended (Rivard et al. 2004) that investigators obtain brachytherapy dosimetry data for clinical use by utilizing well-benchmarked Monte Carlo codes like PTRAN (Williamson 1987), EGS (Kawrakow and Rogers 2001), and MCNP (Briesmeister 2000). GEANT (Agostinelli et al. 2003) and PENELOPE (Sempau et al. 1997) have also been extensively used in the literature for brachytherapy dosimetry. These codes differ in several aspects, for example, original intended use, whether they are open source or not, operating system, data output format and analysis, underlying programming language, and associated required technical skill level. Besides these differences that might make them more appealing to a particular user, they are all appropriate for radiation therapy, including brachytherapy applications. Due to their general-purpose nature, they all share a number of invaluable features including extensive documentation, powerful geometry packages, geometry visualization applications, and a variety of built-in estimators (tallies) and variance reduction techniques. It is also due to the general-purpose nature of these codes, however, that there is a learning curve to them, and that minor bugs may go undetected in spite of rigorous benchmarking procedures. It is therefore sensible to follow the user community forum/mailing list for announcements and patches. The recommendation to benchmark a code new to a user, a new feature of a code that a user is familiar with, or a new code version, by reproducing results for at least one widely used brachytherapy source before reporting data for clinical use (Rivard et al. 2004) cannot be stressed enough. An example is provided in Figure 7.5.

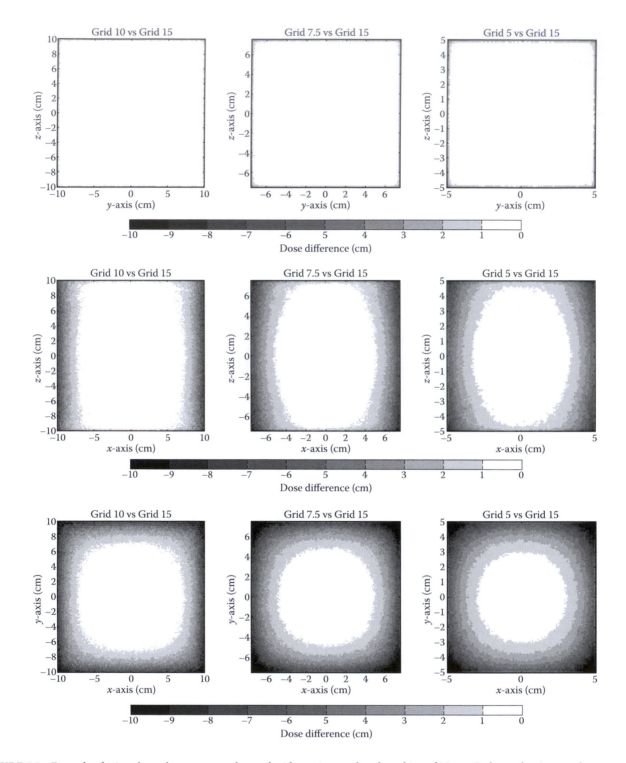

FIGURE 7.5 Example of minor bugs that may go undetected without rigorous benchmarking of Monte Carlo results. A general-purpose code was used to score water kerma around a point source emitting the photon spectrum of ^{192}Ir, centered in a water phantom of given dimensions. Different scoring grids were used, and as their dimension decreases, significant differences from results in a grid equal to the phantom dimensions are found on two of the three central orthogonal planes shown here. (Courtesy of P. Papagiannis.)

It should also be mentioned that new Monte Carlo codes have been introduced as candidate dose calculation algorithms in routine clinical brachytherapy practice (Taylor et al. 2007; Chibani and Williamson 2005; Pratx and Xing 2011) to account for effects ignored by the currently utilized TG-43 formalism, such as patient inhomogeneity or interseed attenuation in permanent implants of low-energy seed sources. While many of them are based on well-benchmarked general-purpose codes, potential users should pay attention to the special implementation characteristics necessary to achieve the time efficiency demands of a clinical setting.

7.4.2.2 Random Number Generator

Often overlooked by general-purpose code users, the performance of the random number generator forms the basis of the Monte Carlo method. Due to the finite accuracy of computers in the representation of real numbers, the sequence of random real numbers uniformly distributed in the unitary interval, r_n, needed for Monte Carlo simulation is obtained by a sequence of integers, x_n, uniformly distributed between 0 and m, according to $r_n = x_n/m$. The integer sequence is provided by a pseudorandom number generator. Although not the most robust, the oldest and most commonly employed type is that of linear congruential generators due to their simplicity, speed, and limited memory requirements. Integer x_{n+1} in the sequence is given by the remainder of $(ax_n + c)$ divided by m [i.e., $x_{n+1} = (ax_n + c)$ mod m]. The linear congruential generator is defined by four integers: the modulus m (>0), the multiplier a ($0 \leq a < m$), the increment c ($0 \leq c < m$), and the starting number or seed x_0 ($0 \leq x_0 < m$). m is usually equal to 2^N, where N is often related to the number of bits allocated for the storage of x_n.

The choice of (m,a,c,x_0) is critical for the quality of the generator. The choice of m and c affects both speed and period of the generator, the latter reaching 2^m at best (Zeeb and Burns 1999). Since large simulations involve 10^9 or more particle histories necessitating about 10^5 random numbers per history (Brown and Nagaya 2002), the reuse of a portion of the random number sequence that could lead to an artificial decrease of result variance is possible. All linear congruential generators also exhibit serial correlation, that is, correlation of the variable with itself over successive estimation periods leading to the formation of lattice structures in result plots (Marsaglia 1968). Nevertheless, the random number generators in general-purpose codes are characterized by increased periods and negligible serial correlation, and they are subjected to meticulous testing (Marsaglia 1995). End users should not be particularly concerned. In case the user is presented with an option of alternative generators, the choice should not be made lightly.

7.4.2.3 Input Data

Information about the source that is to be dosimetrically characterized using Monte Carlo simulation, or any other computational dosimetry method, includes its detailed geometry as well as the elemental composition and density of materials used.

Apparently, the precision of the reproduction of this information in the simulation model for the source as well as the uncertainty associated with this information are more crucial to the dosimetry of low-energy photon-emitting sources ($E \leq 50$ keV; Rivard et al. 2004) due to the predominance of the photoelectric effect. Information from the source manufacturer comes with tolerance statements. Apart from the source active core length, these do not affect the dosimetry of high-energy photon-emitting sources significantly (Wang and Sloboda 1998; Papagiannis et al. 2002). For low-energy sources, however, not only uncertainty in source design but also potential mobility of active parts within the source capsule may significantly affect the dose distribution and have to be taken into account (DeWerd et al. 2011; Rivard et al. 2004). Regardless of source energy, experimental verification of manufacturer design is recommended (Rivard et al. 2010; Pérez-Calatayud et al. 2012), and results should be used in the simulation source model since the dosimetric investigation pertains to the actual sources for clinical use.

With regard to the radiation spectra emitted by the radionuclide contained in the source, sources of precise information exist (e.g., Sonzogni 2005), and while it is acceptable to introduce a cutoff energy (usually 10 keV) under which photons are not simulated in order to increase efficiency, investigators should include all high-energy emissions of low probability (Rivard et al. 2010; Pérez-Calatayud et al. 2012).

As mentioned above, partial and total photon cross sections [$\sigma(E,Z)$ or $\mu(E,Z)$], as well as differential cross sections $d\sigma(E, \theta, Z)/d\Omega$ including atomic form factors $F(x, Z)$ and incoherent scatter factors $S(x, Z)$, and mass energy absorption coefficients are required. These cross sections must be complete (in terms of E, Z) and self-consistent. Since cross sections are data dynamically reevaluated through experiment and theory (Hubbell 2006), it is important that they are up-to-date. Current up-to-date cross sections are the EPDL97 compilation (LLNL) in the ENDL and ENDF/B-VI, or the DLC-146 format (ORNL) and XCOM database (NIST) (IAEA 2011). Mass energy-absorption coefficients for water by Seltzer and Hubbell (1995) (available through NIST) should be used.

The simulation geometry is another important input data element. Dose rate estimations will be increasingly affected by the lack of backscatter as distance from the geometry boundary decreases. Studies for the dosimetric characterization of brachytherapy sources most commonly employ a spherical, 15-cm radius, water phantom with the source centered in it. Although patients are not infinite, the TG-43 dosimetric formalism (Rivard et al. 2004) stipulation to perform source dosimetric characterization in a phantom resembling full scatter conditions translates to at least 5 cm of phantom material beyond the last point where dose rate is reported for low-energy sources (Rivard et al. 2004) and raises phantom radius to 40 cm for ^{192}Ir, ^{137}Cs, and [169]Yb sources, and double that for [60]Co (Pérez-Calatayud et al. 2012; Melhus and Rivard 2006). The geometry of the phantom also affects scatter conditions at points relatively close to its boundaries (Granero et al. 2008).

7.4.2.4 Scoring

All codes offer estimators for energy deposition within volume elements defined in the geometry as well as photon energy fluence at a point or averaged over a surface or over a volume element. The former is a direct estimation of dose. The latter requires multiplication by the appropriate mass energy absorption coefficient to arrive at an estimation of kerma, which can be considered equivalent to dose in our context except for distances a few millimeters from sources of relatively high photon energy emissions (Rivard et al. 2010). The kerma estimator used as well as the resolution of mass energy absorption data and the interpolation method within that data are important considerations. The interested reader should refer to code documentations and the classic publication by Williamson (1987).

The resolution of the scoring geometry should be sufficient to resolve the sharp dose gradient around brachytherapy sources (Rivard et al. 2004; Pérez-Calatayud et al. 2012). Given this sharp dose gradient, volume-averaging effects may ensue from the finite dimensions of scoring elements. Volume-averaging effects have to be kept within 0.1% (DeWerd et al. 2011) through appropriate choice of estimator and voxel size.

7.4.2.5 Air-Kerma Strength Calculations

If intended for clinical use, Monte Carlo simulation results of dose rate (usually in units of energy per gram per simulated photon) must be converted to dose rate per unit source air-kerma strength. Hence, the process of Monte Carlo simulation for source dosimetric characterization consists of two simulations for the same source model: one in water as discussed above and one for the determination of air-kerma strength per simulated photon. According to the definition of air-kerma strength, the latter is defined by the product of the air-kerma rate scored at a distance along the transverse bisector of the source in vacuo multiplied by the distance squared. If the simulation is performed in air, the results must be corrected for attenuation and scatter in air (Hedtjarn et al. 2000; Karaiskos et al. 2001). In either case, photon emissions of energy lower than 5 keV (including x-rays following photoelectric absorption of emitted photons by the source materials) must be supressed since they would contribute substantially to the air-kerma result, while their contribution to the dose in tissue is negligible at distances greater than 1 mm.

Another point that requires attention is the scoring geometry. In some low-energy sources, a part of the spatial radioactivity distribution might be screened by high-Z core materials such that it would not contribute to air-kerma strength results using a point detector on the transverse source bisector. For example, the emissions from radioactivity distributed on the edges of a radio-opaque cylindrical core would be screened (Karaiskos et al. 2001). In the NIST primary standard for air-kerma strength calibration, however, the Wide-Angle Free-Air Chamber (WAFAC) averages photon fluence over a 7.6° half-angle cone about the source transverse bisector. Hence, air-kerma rate in simulations for air-kerma strength per photon should be scored using the exact WAFAC geometry or at least a scoring geometry replicating the abovementioned photon fluence averaging (Rivard et al. 2004).

7.4.2.6 Uncertainty

Just as with experiment, a simulation result is meaningless without a statement of uncertainty.

Uncertainty can be subdivided in components falling in two broad classes: that evaluated by statistical methods (type A, commonly referred to as random error in the past) and that evaluated by other means (type B, commonly referred to as systematic error in the past).

Recalling Equation 7.20: $P\left(-\dfrac{3b}{\sqrt{N}} < \dfrac{1}{N}\sum_{i=1}^{N} k_i - m < \dfrac{3b}{\sqrt{N}} \right) \approx$ 0.997, it follows that the expanded type A uncertainty of Monte Carlo simulation output (i.e., its precision or the confidence interval about our estimate, $\dfrac{1}{N}\sum_{i=1}^{N} k_i$, within which our unknown quantity, m, resides due to type A uncertainty components) is equal to $\dfrac{3b}{\sqrt{N}}$ (for a coverage factor of 3 in this case). This leaves two strategies to reduce type A uncertainty: increasing the number of histories simulated, N, or decreasing the variance of our estimator, b, as this is quantified using our sample of results, k_i, for a given number of histories. The first strategy, commonly referred to as "brute force," is computationally intense, and therefore, investigators often resort to reducing the sample variance for the same number of histories. Several variance reduction techniques have been developed, ranging from simple problem truncation to elegant sampling and scoring techniques (see the next subsection). Some general-purpose codes provide output statistics that should not only be used for uncertainty reporting but also carefully reviewed to ensure the validity of this uncertainty statement.

An uncertainty statement is not complete without taking components of type B uncertainty into account. In our context, these include the user (due to inexperience or erroneous input), the code itself, and the uncertainty of input data. The first two components are difficult to include in the uncertainty budget. Measures, such as benchmarking tests, have to be taken against them, and the key features of simulation methodology must be communicated along with results (see Rivard et al. 2004 for a list). Uncertainty due to radionuclide spectra, cross sections, source geometry, and scoring geometry must be considered. The reader is referred to the work of DeWerd et al. (2011) and Rivard et al. (2004) for a discussion on the order of magnitude of these type B uncertainties, recommendations on type A uncertainty limits, and the method to combine them in a total uncertainty budget of simulation results.

7.4.2.7 Efficiency: Variance Reduction

The Monte Carlo method is often taken as a slow calculation approach because it needs a large number of independent repetitions (or particles) N to achieve a small statistical uncertainty of a quantity of interest. The efficiency, ε, of an MC calculation is given by (Sheikh-Bagheri et al. 2006)

$$\varepsilon = \frac{1}{\sigma^2 T} \tag{7.23}$$

where T is the CPU time and σ^2 is the estimated variance of the quantity of interest. Since T is proportional to N and σ^2 decreases as $1/N$, the denominator of Equation 7.23 is a constant for a given algorithm (using a certain set of parameters). Thus, the efficiency of one MC code over another for a given calculation resides in differences between the transport algorithms, geometry handling, etc. In addition, techniques can be implemented that increase the efficiency by simulating more independent particles for a given CPU time; these are called variance reduction techniques. The American Association of Physicists in Medicine Monograph 32 has a whole section dedicated to Monte Carlo, including variance reduction or approximate efficiency improvement techniques (Sheikh-Bagheri et al. 2006). In brachytherapy, the most obvious and used speed-up techniques are the approximation of dose by using kerma and the use of track-length estimators (Williamson 1987; 2005). The kerma approximation assumes charged-particle equilibrium. This condition is satisfied for clinically used voxelized geometries (1-mm^3 voxels or larger) at ^{192}Ir photon energy or lower. As a result, electron transport can be omitted, and their energy can be considered to be deposited locally at the interaction point, enabling a direct increase in N per unit of CPU time. In the track estimators, all photons whose trajectories traverse scoring voxels are used to calculate the kerma whether or not collisions occur within the voxels. This approach greatly improves the scoring statistics and thus the variance. It leads to an increase in efficiency by one to two orders of magnitude over the analog scoring method (Williamson 1987; Hedtjärn et al. 2002). These approaches have been implemented in brachytherapy-specific MC codes such as MCPT (Hedtjärn et al. 2002), PTRAN (Williamson 1987), MCPI (Chibani and Williamson 2005), and BrachyDose (Taylor et al. 2007). Other approaches, such as particle recycling in which photons emitted by one seed are used for every seed in a multiseed implant, have also been shown to be highly effective (Thomson et al. 2008).

7.5 Explicit Deterministic Dose Calculation Methods: Grid-Based Solvers of Boltzmann Equation

As discussed above, Monte Carlo methods solve the linear Boltzmann transport equation (LBTE) "inexplicitly" by simulating many particle behaviors to stochastically infer the average behavior of all particles in the transport medium (using the central limit theorem). In this section, deterministic methods to solve the LBTE "explicitly" are briefly described. Deterministic methods obtain the average particle behavior using the LBTE differential form to yield approximate solutions that converge to the true continuous LBTE solution in the limit of very fine phase-space mesh spacing. Because these methods are based on all phase-space discretization, such methods are referred to as the grid-based Boltzmann equation solvers (GBBSs). In general,

the method-of-characteristics, spherical harmonics, and discrete ordinates are classified as deterministic, where the latter is most commonly applied to medical physics applications (Shapiro et al. 1976; Nigg et al. 1991; Borgers 1998; Daskalov et al. 2000, 2002; Gifford et al. 2006, 2008, 2010; Vassiliev et al. 2008; Mikell and Mourtada 2010; Han et al. 2011). These methods solve the LBTE by discretizing spatial (via finite difference or element meshes), angular (via discrete ordinates, spherical harmonics, etc.), and energy variables (via the multigroup method), which results in a linear system of equations that are iteratively solved. Major differences between deterministic and Monte Carlo solvers include the following:

1. Deterministic methods are nonstochastic; hence, solution errors arise from systematic sources rather than statistical.
2. Deterministic methods provide full solution for the entire space rather than for specific regions (or tally location) done in Monte Carlo.
3. Deterministic solutions can be more efficient than Monte Carlo once derived for similar problems solved previously, that is, similar brachytherapy sources and patient volumes.

For brachytherapy, the most common LBTE deterministic solver investigated is discrete ordinates (Lewis and Miller 1984) and is the main focus of this section. Although the name specifically refers to the angular differencing, it is frequently used to describe a class of solvers that discretize in energy, angle, and space. The differential form of the transport equation is then iteratively solved. Discrete-ordinate methods solve for the phase-space solution everywhere in the computational domain. While relatively unknown in the radiotherapy community, discrete-ordinate methods have been shown to be useful for neutral particle applications such as neutron-capture therapy (Nigg et al. 1991; Moran et al. 1992) and brachytherapy (Shapiro et al. 1976). They have also been used in a variety of shielding applications, where large attenuations considerably lengthen Monte Carlo computational times. Traditional discrete-ordinate solvers have only been applicable for neutral particle transport (neutrons and photons). The main discussion here will be about the 2D DANTSYS (Alcouffe et al. 1995) and the 3D general-purpose Attila and Acuros solvers (Wareing et al. 2000, 2001), which were developed at LANL (Los Alamos, NM) and Transpire Inc (Gig Harbor, WA), respectively.

7.5.1 Basic Description of Differential LBTE

The general GBBS equations used for neutral particles transport are presented. GBBS solutions for coupled photon-electron radiation transport using the Boltzmann–Fokker–Plank (BFP) equation (Lewis and Miller 1984) will not be discussed here, since the focus is brachytherapy photon emitters where secondary CPE is a valid approximation.

GBBS solves the 3D LBTE, a six-variable integrodifferential governing equation for radiation transport (Lewis and Miller

1984). Briefly, for volume, V, with surface, δV, the LBTE, along with vacuum boundary conditions, is given by

$$\hat{\Omega}\cdot\vec{\nabla}\Psi(\vec{r},E,\hat{\Omega})+\sigma_t(\vec{r},E)\Psi(\vec{r},E,\hat{\Omega})$$

$$=Q^{scat}(\vec{r},E,\hat{\Omega})+Q^{ex}(\vec{r},E,\hat{\Omega}),\ \vec{r}\in V,\ \ (7.24)$$

$$\Psi(\vec{r},E,\hat{\Omega})=0,\vec{r}\in\delta V,\hat{\Omega}\cdot\vec{n}<0. \quad (7.25)$$

Here $\Psi(\vec{r},E,\hat{\Omega})$ is the angular flux at position $\vec{r}=(x,y,z)$, energy E, and direction $\hat{\Omega}=(\mu,\eta,\xi)$, and $\vec{n}$ is the normal vector to surface δV. The first term on the left-hand side of Equation 7.24 is termed the *streaming operator*. The second term on the left-hand side of Equation 7.25 is termed the *collision operator*, whereas $\sigma_t(\vec{r},E)$ is the macroscopic total cross section. The right-hand side of Equation 7.24 includes the *source* terms, where $Q^{scat}(\vec{r},E,\hat{\Omega})$ is the *scattering source* and $Q^{ex}(\vec{r},E,\hat{\Omega})$ is the *extraneous source*. The scattering source is explicitly given as

$$Q^{scat}(\vec{r},E,\vec{\Omega})=\int_0^\infty dE'\int_{4\pi}\sigma_s(\vec{r},E'\to E,\hat{\Omega}\cdot\hat{\Omega}')\Psi(\vec{r},E',\hat{\Omega}')d\hat{\Omega}', \quad (7.26)$$

where $\sigma_s(\vec{r},E'\to E,\hat{\Omega}\cdot\hat{\Omega}')$ is the macroscopic differential scattering cross section. For most deterministic transport methods, it is customary to expand the macroscopic differential scattering cross section in Legendre polynomials and to expand the angular flux appearing in the scattering source in spherical harmonics. Further details on the expansion operators are found in the literature (Lewis and Miller 1984; Wareing et al. 1998, 2001).

Once the solution to the transport equation has been obtained, any reaction rate, such as kerma rate or absorbed dose rates to medium, can be obtained from

$$\dot{R}(\vec{r})=\int_0^\infty\left(\frac{\sigma_{kerma/a}(\vec{r})}{\rho}\right)\phi_{0,0}(\vec{r})dE. \quad (7.27)$$

Typically $\phi_{0,0}(\vec{r})$ is referred to as the "scalar flux" and often times is just given by $\phi(\vec{r})$.

As discussed, when deterministically solving the LBTE (Equation 7.24), one must discretize all variables: energy (E), angle ($\hat{\Omega}$), and space ($\vec{r}$). A brief discussion of the discretization methods is presented in Section 7.5.1.1.

7.5.1.1 Energy Discretization

Traditionally, deterministic codes apply the multigroup method for the energy discretization using precalculated multigroup cross-section libraries (Lewis and Miller 1984). The multigroup method uses a division of the particle energy range of interest, that is, $E_{min}\le E\le E_{max}$, into a finite number, G, of intervals separated by the energies, E_g, where $g=1,2,3,\ldots,G$, as shown in

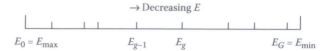

FIGURE 7.6 Division of energy range into G energy groups. (Courtesy of F. Mourtada.)

Figure 7.6. Each energy interval is called a group, and the order of number is such that as g increases, the energy decreases. Energy discretization errors result in the LBTE solution biases over a large region in the solution domain. For brachytherapy sources, several authors studied the impact of the number of energy groups on both accuracy and speed of convergence (Daskalov et al. 2000, 2002; Gifford et al. 2006, 2008).

Cross sections produced by CEPXS4 are typically used (Lorence et al. 1989). CEPXS includes all photon interactions with the exception of Rayleigh scatter, the effect of which is insignificant for dose distributions at energies produced by brachytherapy sources such as ^{192}Ir.

7.5.1.2 Angular Discretization

The discretization in angle is often referred to as the S_n method. The discrete-ordinate method consists of requiring the transport equation given by the LTBE (Equation 7.24) to hold only for a finite number of angles, $\hat{\Omega}_n$. These discrete angles are chosen from an angular quadrature set that also serves to integrate the angular integrals in the scattering source.

Angular discretization errors typically result in "ray effects" or other nonphysical angular oscillations in the solution (see example in Figure 7.7). This is more pronounced for radiation transport problems with a localized source (i.e., near a point source) in weakly scattering media and results in solution anomalies. A nonphysical buildup in flux along the discrete angles results in nonphysical oscillations in the solution. Ray effects are most pronounced at far distances from localized sources. As the quadrature order is increased (adding more angles), these ray effects are mitigated. Hence, initial careful settings of the spatial discretization parameters for brachytherapy sources must be performed and fixed for well-defined scenarios such as within the boundary of patient skin as derived from CT. Although ray effects are mitigated with increasing quadrature order, this is often prohibited by the increase in computational times. To mitigate these ray effects, many discrete-ordinate codes employ some type of first collision source technique. These techniques, using either stochastic (Monte Carlo) or semiempirical methods for the solution of the uncollided flux and hence the first collision source, have been limited, however, to rectangular meshes. Another approach is a semianalytic first collision source method, which is used in the Attila discrete ordinate code (Gifford et al. 2006). Attila performs semianalytic ray tracing from a predetermined number of uniformly distributed points within the source. Ray tracing from these source points is tracked to four Gaussian spatial integration points on each tetrahedral element, instead of the corner nodes or cell centers. This approach provides a highly accurate linear representation of the first-collided source within

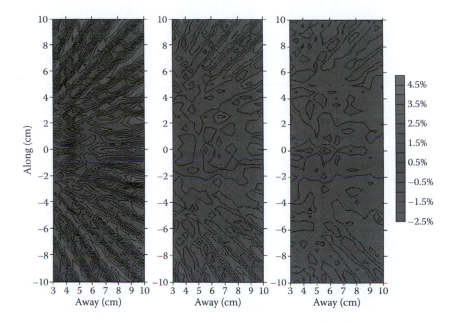

FIGURE 7.7 **(See color insert.)** Left to right: S14 (224 angles), S18 (360 angles), and S22 (528 angles). S18 dose variation due to ray effects is less than 1.5% out to 6 cm from an ^{192}Ir source in water. Based on data used in the work of Mourtada et al. (2004) using the Attila code. (Courtesy of F. Mourtada.)

each element, from which the full scattering calculation is performed. The source volumes and materials are explicitly modeled, which ensures that secondary scattering and attenuation are correctly accounted for within sources. Ray tracing time is independent of the source spectrum.

7.5.1.3 Spatial Discretization

Spatial discretization schemes are described in the literature for several GBBS deterministic transport codes. In this section, we provide details on the spatial discretization in Attila and Acuros as examples of modern GBBS codes. Attila solves the multigroup discrete-ordinate equations on fully unstructured tetrahedral elements using a high-order Galerkin-based linear discontinuous finite element method (DFEM) (Lewis and Miller 1984). The Acuros solver uses a novel adaptive Cartesian element, which is spatially variable, where the local element size is adapted based on the local anatomical and applicator material properties and gradients in the scattered photon fluence. Acuros also used DFEM for the spatial discretization. The material properties are assumed to be constant within each computational finite element. A unique feature of this discretization scheme is that, through solving multiple spatial degrees of freedom within each computational element, it calculates a linear solution spatially within each element. Hence, the angular fluence is defined everywhere in each element, not just at the element center or element nodes.

With the fully discretized DFEM multigroup discrete-ordinate equations, there are $4 \times N_{elements} \times N_{angles} \times N_{groups}$ unknowns that need to be calculated. All discrete-ordinate codes use source iteration (SI) to iteratively solve the resulting system of equations (Lewis and Miller 1984). SI can converge slowly for problems where scattering is dominant, a known problem for

discrete-ordinate methods. An efficient diffusion synthetic acceleration (DSA) algorithm has been applied to greatly reduce the number of iterations required for convergence and, hence, can significantly reduce the CPU time (Wareing et al. 2001).

It should be noted that due to the discontinuous nature of the DFEM spatial differencing, spatial discretization errors would typically be manifested by local solution over/under shooting and caused by the requirement to fit a linear solution in each element while conserving particle balance.

7.5.2 General Aspects of GBBS Implementation for Brachytherapy Dosimetry in Water

Daskalov et al. (2000, 2002) applied a discrete-ordinate solver (DANTSYS) for the dosimetric modeling of ^{125}I and ^{192}Ir brachytherapy sources. 2D comparisons were made with the EGS4 Monte Carlo code, and in both cases, discrete-ordinate was found to provide good agreement (less than 5% root mean square difference) with Monte Carlo. For the ^{125}I source, it was found that high accuracy could be achieved even when the number of energy groups was reduced from 210 down to 3. The three-group calculation was 80 times more efficient than EGS4 in 2D. Daskalov concluded, "the results strongly support the hypothesis that deterministic solutions of the linear Boltzmann transport equations, which have been largely neglected by the radiotherapy community, have significant potential as treatment planning tools for brachytherapy and other treatment modalities utilizing low-energy photon fields."

However, Daskalov also highlighted limitations of the DANTSYS implementation. To mitigate ray effects, which are caused by solving the transport equation with a discrete number of angles, Monte Carlo was used to calculate the first-collided source. Using this approach, differences as large as 14% were found using S20 quadrature (440 angles solved for each spatial element) for the ^{125}I source. When the number of groups was reduced, the time required for the Monte Carlo ray tracing emerged as a limiting factor to achieving further efficiency gains. Another drawback is the requirement by DANTSYS and other traditional discrete-ordinate solvers to use uniformly defined Cartesian or polar elements. While this was mitigated in their study through an axisymmetric polar mesh, in 3D, these symmetries do not exist, and a large number of phase-space variables would result. To increase the computational efficiency, Daskalov recommended the development of higher-order spatial differencing methods and variably sized mesh capabilities. These conclusions were consistent with those of Borgers (1998), who suggested that discrete-ordinate methods would be highly attractive for clinical treatment planning under these conditions.

Zhou and Inanc (2003) developed a GBBS, which they applied to low-energy brachytherapy, based on direct evaluation of the integral LBTE, expanded in orders of scattering. Mourtada et al. (2004) evaluated the performance of a GBBS method using the 3D Attila radiation transport code for the dosimetry of a pulsed–dose-rate (PDR) ^{192}Ir source in water. The ray tracing was performed from a predetermined number of uniformly distributed points; in this case, eight points consisting of four circumferential positions at two axial positions within the PDR source were used. Ray tracing from these source points was tracked to four Gaussian spatial integration points on each tetrahedral element, instead of the corner nodes or cell centers. This approach provided an accurate linear representation of the first-collided source within each element, which was directly coupled with the discontinuous finite element spatial differencing within Attila. The source volumes and materials were explicitly modeled, which ensures that secondary scattering and attenuation within the PDR source are correctly accounted for. The Attila calculation parameters included 37 energy groups (CEPXS cross sections), P2 scattering order, S22 angular quadrature (528 angles solved), and 100,000 elements (see Figure 7.8). MCNPX and Attila simulations assumed that the

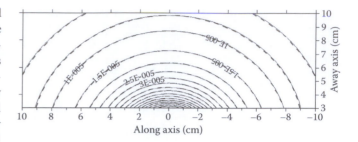

FIGURE 7.9 Isodose contours (MeV/g) for the ^{192}Ir PDR 1-mm source in water estimated by MCNPX (solid line) and Attila (dashed line). (Courtesy of F. Mourtada.)

source is inside a 30-cm water sphere. Details of the ^{192}Ir source description and MCNPX model validation are found in the literature (Mourtada et al. 2004; Gifford et al. 2008). For MCNPX, the relative error of the estimate is kept to within 1%, and the energy cutoff for photons is set to 2 keV. For both codes, the primary photon spectrum for ^{192}Ir was that of Glasgow and Dillman (1979). Results were compared with calculations performed with the MCNPX Monte Carlo code. An excellent agreement between Attila and MCNPX is obtained throughout clinically relevant distances (see Figures 7.8 and 7.9).

Further refinements of the Attila code for brachytherapy were performed by Gifford et al. (2008) for the Nucletron microSelectron mHDRv2 source. The number of energy groups, S_n (angular order), P_n (scattering order), and mesh elements were varied in addition to the method of analytic ray tracing to assess their effects on the deterministic solution. It was shown that a 5-energy-group cross-section set calculated results to within 0.5% of a 15-group cross-section set. S_{12} was sufficient to resolve the solution in angle. P_2 expansion of the scattering cross section was necessary to compute accurate distributions. A computational mesh with 55,064 tetrahedral elements in a 30-cm diameter phantom resolved the solution spatially. An efficiency factor of 110 with the above parameters was realized in comparison to MC methods. The Attila code provided an accurate and efficient solution of the LBTE for the mHDRv2 ^{192}Ir source.

More recently, Transpire Inc has created an optimized radiation-therapy specific version of the Attila GBBS called Acuros for brachytherapy and then for external beam (Zourari et al. 2010; Petrokokkinos et al. 2011; Mikell and Mourtada 2010; Han et al. 2011). The origin of Acuros is the general-purpose Attila solver discussed above (Wareing et al. 2000, 2001), which was developed at LANL.

7.6 TG-43 Source Dosimetric Characterization, and Beyond

This chapter has been limited to single source dosimetry in homogeneous water environments for illustrative purposes. Notwithstanding their variation in simplicity, speed, and generality, all three broad classes of computational dosimetry methods discussed (semiempirical algorithms, stochastic Monte Carlo

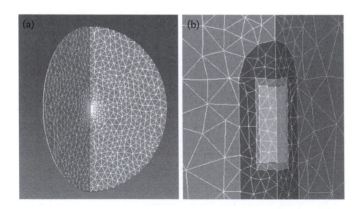

FIGURE 7.8 (See color insert.) (a) Full computational domain, with ^{192}Ir PDR source in the center of a 30-cm water sphere. (b) Computational mesh around source [zoom image in (a)]. (Courtesy of F. Mourtada.)

simulation, and deterministic solvers of the LBTE) are capable of producing accurate results in this context. Following its experimental benchmarking some two decades ago, Monte Carlo simulation has been widely accepted as the reference method for single source dosimetry. It has hence played a major role in promoting the accuracy of dosimetry in brachytherapy since current conventional treatment planning practice is largely based on Monte Carlo data in the widespread TG-43 formalism format.

Brachytherapy is currently shifting toward individualized patient dosimetry through the introduction of advanced dose calculation algorithms that account for applicator materials and patient inhomogeneity, as well as actual scatter conditions. In this process, Monte Carlo simulation still presents an attractive alternative using optimization/variance reduction (Taylor et al. 2007; Chibani and Williamson 2005) or general-purpose computing on graphics processing units (Pratx and Xing 2011; Hissoiny et al. 2011) in an effort to increase computational efficiency and allow calculation times suitable for the clinical setting. Computational methods relying on precalculated data that are application specific (Poon and Verhaegen 2009) have also been proposed. The generally applicable method of kernel superposition/convolution using the collapsed cone algorithm (Carlsson and Ahnesjö 2000a,b) is currently under implementation, and a GBBS has been incorporated in a commercially available treatment planning system (Zourari et al. 2010; Petrokokkinos et al. 2011). Furthermore, it has been recently shown that precalculated Monte Carlo 3D dose kernels could be coupled with inverse planning to generate plans for any single source afterloading-type treatments (D'Amours et al. 2011). A literature review of the available methods and their potential and, most importantly, aspects regarding their clinical implementation is presented in Chapter 11.

References

Agostinelli, S., Allison, J., Amako, K. et al. 2003. G4—A simulation toolkit. *Nucl Instrum Methods Phys Res A* 506:250–303.

Alcouffe, R.E. et al. 1995. *DANTSYS: A Diffusion Accelerated Neutral Particle Transport Code System*. Los Alamos National Laboratory, Los Alamos.

Attix, F.H. 1986. *Introduction to Radiological Physics and Radiation Dosimetry*. Wiley-Interscience, New York.

Baltas, D., Sakelliou, L., and Zamboglou, N. 2007. *The Physics of Modern Brachytherapy for Oncology*. Taylor & Francis Group, Boca Raton.

Borgers, C. and Larsen, E.W. 1996. On the accuracy of the Fokker-Planck and Fermi pencil beam equations for charged particle transport. *Med Phys* 23:1749–59.

Borgers, C. 1998. Complexity of Monte Carlo and deterministic dose-calculation methods. *Phys Med Biol* 43:517–28.

Briesmeister, J.F. 2000. MCNP—A General Monte Carlo N-Particle Transport Code System, Version 4C, LA-13709-M, Los Alamos National Laboratory.

Brown, F.B. and Nagaya, Y. 2002. The MCNP5 random number generator. Tech. rep. LA-UR-02-3782, Los Alamos National Laboratory.

Carlsson, Å.K. and Ahnesjö, A. 2000a. Point kernels and superposition methods for scatter dose calculations in brachytherapy. *Phys Med Biol* 45:357–82.

Carlsson, Å.K. and Ahnesjö, A. 2000b. The collapsed cone superposition algorithm applied to scatter dose calculations in brachytherapy. *Med Phys* 27:2320–32.

Carlsson Tedgren, Å.K. and Ahnesjö, A. 2003. Accounting for high Z shields in brachytherapy using collapsed cone superposition for scatter dose calculation. *Med Phys* 8:2206–17.

Carlsson Tedgren, A. and Carlsson, G.A. 2009. Influence of phantom material and dimensions on experimental 192Ir dosimetry. *Med Phys* 36:2228–35.

Chibani, O. and Williamson, J.F. 2005. MCPI: a sub-minute Monte Carlo dose calculation engine for prostate implants. *Med Phys* 32:3688–98.

D'Amours, M., Pouliot, J., Dagnault, A., Verhaegen, F., and Beaulieu, L. 2011. Patient-specific Monte Carlo-based dose-kernel approach for inverse planning in afterloading brachytherapy. *Int J Radiat Oncol Biol Phys* 81:1582–9.

Daskalov, G.M., Baker, R.S., Rogers, D.W., and Williamson, J.F. 2000. Dosimetric modeling of the microselectron high-dose rate 192Ir source by the multigroup discrete ordinates method. *Med Phys* 27:2307–19.

Daskalov, G.M., Baker, R.S., Rogers, D.W., and Williamson, J.F. 2002. Multigroup discrete ordinates modeling of 125I 6702 seed dose distributions using a broad energy-group cross section representation. *Med Phys* 29:113–24.

Dewerd, L.A., Ibbott, G.S., Meigooni, A.S. et al. 2011. A dosimetric uncertainty analysis for photon-emitting brachytherapy sources: Report of AAPM Task Group No. 138 and GEC-ESTRO. *Med Phys* 38:782–801.

Gifford, K.A., Horton, J.L., Wareing, T.A., Failla, G., and Mourtada, F. 2006. Comparison of a finite-element multigroup discrete-ordinates code with Monte Carlo for radiotherapy calculations. *Phys Med Biol* 51:2253–65.

Gifford, K.A., Price, M.J., Horton, J.L. Jr., Wareing, T.A., and Mourtada, F. 2008. Optimization of deterministic transport parameters for the calculation of the dose distribution around a high dose-rate 192Ir brachytherapy source. *Med Phys* 35:2279–85.

Gifford, K.A., Wareing, T.A., Failla, G. et al. 2010. Comparison of a 3-D multi-group SN particle transport code with Monte Carlo for intracavitary brachytherapy of the cervix uteri. *J Appl Clin Med Phys* 11:3103.

Glasgow, G.P., and Dillman, L.T. 1979. Specific gamma-ray constant and exposure rate constant of 192Ir. *Med Phys* 6:49–52.

Granero, D., Pérez-Calatayud, J., Pujades-Claumarchirant, M.C. et al. 2008. Equivalent phantom sizes and shapes for brachytherapy dosimetric studies of ^{192}Ir and ^{137}Cs. *Med Phys* 35:4872–7.

Han, T., Mikell, J.K., Salehpour, M., and Mourtada, F. 2011. Dosimetric comparison of Acuros XB deterministic radiation transport method with Monte Carlo and model-based convolution methods in heterogeneous media. *Med Phys* 38:2651–64.

Hedtjarn, H., Carlsson, G.A., and Williamson, J.F. 2000. Monte Carlo-aided dosimetry of the Symmetra model I25.S06 125I, interstitial brachytherapy seed. *Med Phys* 27:1076–85.

Hedtjärn, H., Alm Carlsson, G., and Williamson, J.F. 2002. Accelerated Monte Carlo based dose calculations for brachytherapy planning using correlated sampling. *Phys Med Biol* 47:351–76.

Hissoiny, S., Ozell, B., Despres, P., and Carrier, J.F. 2011. Validation of GPUMCD for low-energy brachytherapy seed dosimetry. *Med Phys* 38:4101–7.

Hubbell, J.H. 2006. Review and history of photon cross section calculations. *Phys Med Biol* 51:R245–62.

IAEA. 2011. International Atomic Energy Agency. Nuclear Data Guide. Catalog of nuclear data information available - online or off-line from the IAEA Nuclear Data Section. Available online at: http://www-nds.iaea.org/indg_intro.html.

Karaiskos, P., Sakelliou, L., Sandilos, P., and Vlachas, L. 2000. Limitations of the point and line source approximations for the determination of geometry factors around brachytherapy sources. *Med Phys* 27:124–8.

Karaiskos, P., Papagiannis, P., Sakelliou, L., Anagnostopoulos, G., and Baltas, D. 2001. Monte Carlo dosimetry of the selectSeed 125I interstitial brachytherapy seed. *Med Phys* 28:1753–60.

Kawrakow, I. and Rogers, D.W.O. 2001. The EGSnrc code system: Monte Carlo simulations of electron and photon transport. NRCC Report PITS-701, National Research Council of Canada, Ottawa.

Lemioux, C. 2009. *Monte Carlo and Quasi-Monte Carlo Sampling*. Springer, New York.

Lewis, E.E. and Miller, W.F. 1984. *Computational Methods of Neutron Transport*. Wiley, New York.

Lorence, L., Morel, J., and Valdez, G. 1989. Physics guide to CEPXS: A multigroup coupled electron-photon cross section generating code. Sandia National Laboratory Report No. SAND89-1685.

Marsaglia, G. 1968. Random numbers fall mainly in the planes. *Proceedings of the National Academy of Science* 61:25–8.

Marsaglia, G.S. 1995. The DIEHARD Battery of Tests of Randomness. Available online at http://stat.fsu.edu/pub/diehard.

Melhus, C.S. and Rivard, M.J. 2006. Approaches to calculating AAPM TG-43 brachytherapy dosimetry parameters for 137Cs, 125I, 192Ir, 103Pd, and 169Yb sources. *Med Phys* 33:1729–37.

Mikell, J.K. and Mourtada, F. 2010. Dosimetric impact of an 192Ir brachytherapy source cable length modeled using a grid-based Boltzmann transport equation solver. *Med Phys* 37:4733–43.

Moran, J.M., Nigg, D.W., Wheeler, F.J., and Bauer, W.F. 1992. Macroscopic geometric heterogeneity effects in radiation dose distribution analysis for boron neutron capture therapy. *Med Phys* 19:723–32.

Morin, R.L. (ed.). 1988. *Monte Carlo Simulation in the Radiological Sciences*. CRC Press Inc., Boca Raton, FL.

Mourtada, F., Wareing, T., Horton, J. et al. 2004. *A Deterministic Dose Calculation Method with Analytic Ray Tracing for Brachytherapy Dose Calculations*. AAPM, Pittsburg, PA.

Nath, R., Anderson, L.L., Luxton, G. et al. 1995. Dosimetry of interstitial brachytherapy sources—Recommendations of the AAPM Radiation-Therapy Committee Task Group No 43. *Med Phys* 22:209–34.

Nigg, D.W., Randolph, P.D., and Wheeler, F.J. 1991. Demonstration of three-dimensional deterministic radiation transport theory dose distribution analysis for boron neutron capture therapy. *Med Phys* 18:43–53.

Papagiannis, P., Angelopoulos, A., Pantelis, E. et al. 2002. Dosimetry comparison of 192Ir sources. *Med Phys* 29:2239–46.

Pérez-Calatayud, J., Ballester, F., Das, R.K. et al. 2012. Dose calculation standards for photon-emitting brachytherapy sources with average energy higher than 50 keV: Report of the AAPM and ESTRO. *Med Phys* 39(5):2904–29.

Petrokokkinos, L., Zourari, K., Pantelis, E. et al. 2011. Dosimetric accuracy of a deterministic radiation transport based 192Ir brachytherapy treatment planning system. Part II: Monte Carlo and experimental verification of a multiple source dwell position plan employing a shielded applicator. *Med Phys* 38:1981–92.

Poon, E. and Verhaegen, F. 2009. A CT-based analytical dose calculation method for HDR 192Ir brachytherapy. *Med Phys* 36:3982–94.

Pratx, G. and Xing, L. 2011. GPU computing in medical physics: A review. *Med Phys* 38:2685–97.

Rivard, M.J., Coursey, B.M., DeWerd, L.A. et al. 2004. Update of AAPM Task Group No. 43 Report: A revised AAPM protocol for brachytherapy dose calculations. *Med Phys* 31:633–74.

Rivard, M.J., Granero, D., Pérez-Calatayud, J., and Ballester, F. 2010. Influence of photon energy spectra from brachytherapy sources on Monte Carlo simulations of kerma and dose rates in water and air. *Med Phys* 37:869–76.

Russell, K.R. and Ahnesjö, A. 1996. Dose calculation in brachytherapy for a 192Ir source using a primary and scatter dose separation technique. *Phys Med Biol* 41:1007–24.

Russell, K.R, Carlsson Tedgren, Å.K., and Ahnesjö, A. 2005. Brachytherapy source characterization for improved dose calculations using primary and scatter dose separation. *Med Phys* 32:2739–52.

Seltzer, S.M. and Hubbell, J.H. 1995. Tables and graphs of mass attenuation coefficients and mass energy-absorption coefficients for photon energies 1 keV to 20 MeV for elements Z = 1 to 92 and some dosimetric materials. ISSN 1340-7716. Japanese Society of Radiological Technology, Tokyo.

Sempau, J., Acosta, E., Baro, J., Fernandez-Varea, J.M., and Salvat, F. 1997. An algorithm for Monte Carlo simulation of coupled electron-photon transport. *Nucl Instrum Methods Phys Res B* 132:377–90.

Shapiro, A., Schwartz, B., Windham, J.P., and Kereiakes, J.G. 1976. Calculated neutron dose rates and flux densities from implantable californium-252 point and line sources. *Med Phys* 3:241–7.

Sheikh-Bagheri, D., Kawrakow, I., Walters, B., and Rogers, D.W.O. 2006. Monte Carlo simulations: Efficiency improvement techniques and statistical considerations. *Integrated New Technologies into the Clinic: Monte Carlo and Image-Guided Radiation Therapy   Proceedings of the 2006 AAPM Summer School*.

Sonzogni, A.A. 2005. NuDat 2.0: Nuclear structure and decay data on the Internet. *AIP Conf Proc* 769:574–77.

Taylor, R.E.P., Yegin, G., and Rogers, D.W.O. 2007. Benchmarking BrachyDose: Voxel based EGSnrc Monte Carlo calculations of TG-43 dosimetry parameters. *Med Phys* 34:445–57.

Thomson, R.M., Taylor, R.E.P., and Rogers, D.W.O. 2008. Monte Carlo dosimetry for 125I and 103Pd eye plaque brachytherapy. *Med Phys* 35:5530–43.

Vassiliev, O.N., Wareing, T.A., Davis, I.M. et al. 2008. Feasibility of a multigroup deterministic solution method for three-dimensional radiotherapy dose calculations. *Int J Radiat Oncol Biol Phys* 72:220–7.

Wang, R. and Sloboda, R.S. 1998. Influence of source geometry and materials on the transverse axis dosimetry of 192Ir brachytherapy sources. *Phys Med Biol* 43:37–48.

Wareing, T.A., Morel, J.E., and McGhee, J.M. 2000. Coupled electron-photon transport methods on 3-D unstructured grids. *Trans Am Nucl Soc* 83:240–2.

Wareing, T.A., Morel, J.E., Parsons, D.K. et al. 1998. A first collision source method for Attila, an unstructured tetrahedral mesh discrete ordinates code. Proceedings of the ANS Radiation Protection and Shielding Division Topical Conference, Nashville, TN.

Wareing, T.A., McGhee, J.M., Morel, J.E., and Pautz, S.D. 2001. Discontinuous finite element S_n methods on three-dimensional unstructured grids. *Nucl Sci Eng* 138(2):256–68.

Williamson, J.F. 1987. Monte Carlo evaluation of kerma at a point for photon transport problems. *Med Phys* 14:567–76.

Williamson, J.F. 1988. Monte Carlo and analytic calculation of absorbed dose near 137Cs intracavitary sources. *Int J Radiat Oncol Biol Phys* 15:227–37.

Williamson, J.F. 1996. The Sievert integral revisited: evaluation and extension to 125I, 169Yb, and 192Ir brachytherapy sources. *Int J Radiat Oncol Biol Phys* 36(5):1239–50%U http://www.ncbi.nlm.nih.gov/pubmed/8985050.

Williamson, J.F. 2005. Semiempirical dose-calculation models in brachytherapy. In: *AAPM Summer School. Brachytherapy Physics*. B.R. Thomadsen, M.J. Rivard, and W.M. Butler (eds.). Medical Physics Publishing. pp. 201–32.

Zeeb, C.N. and Burns, P.J. 1999. Random Number Generator Recommendation, 1999: Colorado State Univ. Available online at http://www.colostate.edu/~pburns/monte/rngreport.pdf.

Zhou, C. and Inanc, F. 2003. Integral-transport-based deterministic brachytherapy dose calculations. *Phys Med Biol* 48:73–93.

Zourari, K., Pantelis, E., Moutsatsos, A. et al. 2010. Dosimetric accuracy of a deterministic radiation transport based 192Ir brachytherapy treatment planning system. Part I: Single sources and bounded homogeneous geometries. *Med Phys* 37:649–61.

III

Brachytherapy Treatment Planning and Imaging

Historical Development of Predictive Dosimetry "Systems"

Ali S. Meigooni
*Comprehensive Cancer
Centers of Nevada*

Margaret Bidmead
*Royal Marsden NHS
Foundation Trust and Institute
of Cancer Research*

Jack L. M. Venselaar
Instituut Verbeeten

8.1 Introduction

The use of radium in the treatment of gynecological cancer was first reported in 1903 (Meredith 1967), a few years after the discovery of the radioactivity by Henri Becquerel. During the early years of the introduction of brachytherapy as a treatment modality for cancer patients, determination of source strengths and source arrangements in relevant geometric patterns was a challenging subject. In clinical procedures, relative location of the treatment volume and the surrounding normal tissues/organs had to be considered in order to make these determinations. For example, in the treatment of cervical cancer patients having treatment using intracavitary brachytherapy, careful placement of sources is required in order to minimize the dose to bladder and rectum. It is interesting to observe that in the early days of development of brachytherapy, clinical observation was used as the first method of investigation, followed by the development of rules and regulations for a radiation therapy procedure (Paterson and Parker 1934, 1952; Parker 1938; Paterson 1948).

Developments of logistics and procedures for the distribution of radioactivity by different investigators have led to several differing techniques, which are known as *systems* or *schools*.

8.2 Commonly Used Brachytherapy Systems or Schools

A brachytherapy dosimetric system consists of a set of rules for the arrangement of a specific set of radioactive sources in order to deliver a prescribed dose to a designated point or volume of interest. The most commonly used brachytherapy systems for intracavitary and interstitial brachytherapy procedures are described in the next sections.

8.2.1 Intracavitary Implants

8.2.1.1 Stockholm System

The traditional Stockholm system of cervical brachytherapy treatments was based on a fractionated course of radiotherapy delivery using ^{226}Ra sources over a period of 1 month (Heymann 1935; Kottmeier 1964; Ryberg et al. 1990). In this treatment technique, usually two to three applications were used for each patient, with each application lasting a time period of 20–30 h. In this system, the applicators consisted of intravaginal applicators, which were made of lead or gold, and also an intrauterine tube, which was made of flexible rubber. The Stockholm system advocated an unequal loading for the uterine and vaginal radium sources. In this system, in order to deliver the prescribed radiation dose, normally, 30–90 mg of radium (1 mg of ^{226}Ra is equivalent to an activity of 1 mCi, 37 MBq) was placed inside the uterus, while 60–80 mg sources were placed inside the vagina. Originally, the prescription of the treatment in this system was reported as the product of the amount of source loading in terms of the milligram of radium content and the number of hours of treatment duration (i.e., the milligram-hours concept of prescription) (Heymann 1935; Kottmeier 1964; Ryberg et al. 1990; Tod and Meredith 1938). Usually, a total "dose" of 6500–7100 mg-h was prescribed for the treatment of a cervical cancer patient, from which 4500 mg-h was contributed by the vaginal applicator and the remainder by the intrauterine tube.

8.2.1.2 Manchester System

Historically, the Manchester system was developed by Tod and Meredith (Tod and Meredith 1938, 1953; Meredith 1967), and it was one of the most commonly used brachytherapy systems until computer-based dosimetry systems became available. It was the first system to attempt to define dosimetry to particular reference points. These points were based on a geometrical arrangement in order to standardize dosimetry from patient to patient. This system was designed for cervical brachytherapy in terms of prescription of dose to a point, known as *Point A*. Selection of this point enabled the investigators to perform the following:

- Standardize the treatment of one patient with another.
- This point was located in a relatively low dose gradient area. Therefore, dose to this point was not very sensitive to small alteration in applicator position.
- Correlate the dose to point A with the clinical results.

In order to achieve this goal, a set of applicators and loading schemes (i.e., the amount and distribution pattern of radium in the applicator) were designed to create the same dose rate independent of the applicator arrangement. Point A corresponds to the paracervical triangle at the medial edge of the broad ligament where the uterine vessels cross the ureter. Geometrically, this point was defined by drawing a line connecting the superior aspects of the vaginal ovoids and measuring 2 cm superiorly along the tandem and then 2 cm along the perpendicular direction to the tandem (Figure 8.1). This point was then revised to 2 cm above the external cervical os and 2 cm lateral to midline, or 2 cm above the distal end of the lowest source in the tandem and 2 cm lateral to the tandem. In addition to Point A, the Manchester system uses another point (Point B), representing the pelvic nodes. This point was originally defined as 5 cm lateral from the midline at the same level as Point A. It is worth noting that this is only a geometrical measurement representing the position of the pelvic side wall nodes. The true position of these nodes varies from patient to patient. The Manchester system created a set of rules regarding the activity, relationship, and positioning of the radium sources in the uterine tandem and the vaginal ovoids, which resulted in the desired dose rate. Normally, the prescribed dose to Point A was 80 Gy, delivered in two applications with approximately a total application time of 144 h, in the absence of any external beam. In this system, applicators consisted of a rubber tandem and two ellipsoid "ovoids" with diameters 2, 2.5, or 3 cm. The ovoids were often separated by a rubber spacer in an attempt to keep a constant separation between the ovoids. Initially, there was no shielding present in the ovoids. Therefore, a generous packing anteriorly and posteriorly was needed to reduce the dose to bladder and rectum to acceptable levels. The loading in the uterine tubes was 10+10+15, 10+15, and 20 mg of Ra for long, medium, and short tubes, respectively. The ovoids were loaded with 22.5, 20, and 17.5 mg of Ra for large, medium, and small ovoids, respectively. Standard insertion times were calculated for fixed applicator configurations and not adjusted between patients unless the measured rectal dose was greater than the chosen tolerance value. The designated Manchester system plan for cervical cancer patients included the following:

- Point A dose rate was approximately 0.53 Gy/h for all allowed applicator loadings. This dose rate applies for an ideal insertion, and often the extent of the disease does not allow perfect positioning of the applicator. However, even when the sources were displaced, the dose rate at Point A for the "ideal" insertion is that which determined the treatment time (Wilkinson et al. 1983).
- Vaginal contribution to Point A was limited to 40% of the total dose.
- The rectal dose should be 80% or less of the Point A dose.

NB. The policy in Manchester system was to measure the rectal dose with a probe. The tolerance dose to the rectum was taken to be 65% of the dose to Point A.

8.2.1.3 Paris System

In the Paris system for intracavitary applications, a single application of radium brachytherapy was specified for treatment of cervical cancer patients (Pierquin 1964; Pierquin et al. 1978). Unlike the Stockholm system, the Paris system utilized equal amounts of radium in the vaginal applicator and the intrauterine tube (Figure 8.2). In this system, two cylindrical cork colpostats were utilized along with a silk rubber intrauterine tube to deliver a "dose" of 7000–8000 mg-h of radium over a period of 5 days. The intrauterine sources contained three radioactive sources, with source strengths in the ratio of 1:1:0.5. The source strengths in the colpostats were the same as the source strength of the topmost uterine source.

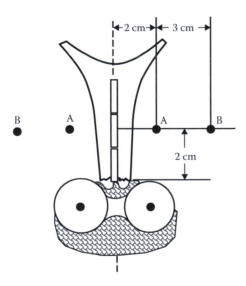

FIGURE 8.1 Schematic diagram of points "A" and "B" in the classical Manchester system. (Courtesy of A.S. Meigooni. Modified from the International Commission on Radiation Units and Measurements (ICRU). 1985. Dose and volume specification for reporting and recording intracavitary therapy in gynecology. Report 38 of ICRU, ICRU Publications, Bethesda, MD.)

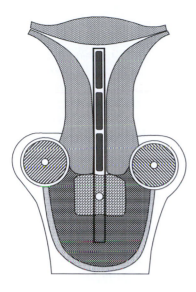

FIGURE 8.2 Schematic diagram of the applicator used in the Paris system for treatment of cervix carcinoma. This applicator typically consisted of three individualized vaginal radium sources (one in each lateral fornix and one centrally placed in front of the cervical os, and one intrauterine source composed of three radium tubes) in the so-called tandem position. (Courtesy of A.S. Meigooni. Modified from Report 38 of ICRU (ICRU 1985).)

8.2.1.4 ICRU 38

In 1985 the International Commission on Radiation Units and Measurements published its Report #38 (ICRU 38) on *Dose and volume specification for reporting and recording intracavitary therapy in gynecology* (ICRU 1985). Although this report is not a "system" for source arrangements and dose prescription itself, it is comparable to the ones discussed in the previous sections. ICRU Report 38 provided guidelines for recording and reporting dose points within adjacent normal tissues (bladder and rectum), and representative points were defined for the lymphatic system during the use of either low-dose-rate or high-dose-rate intracavitary therapy for the treatment of cervical cancer. This report provided a rational basis for dose calculation, recording, and reporting for the treatment of cancer of the cervix and other gynecologic malignancies.

ICRU 38 discourages the use of Points A and B because the exact meaning and their definitions have not always been interpreted in the same way in different centers and even in the same center over a period of time. These differences may provide significantly different values for the calculation of dose rate to Point A. Therefore, if the prescribed dose to Point A is used to calculate the total time of application, different values of time will be obtained for different methods used to assign the prescription point. The major thrust of this report was to identify a volume encompassed by an absorbed dose level of 60 Gy as the appropriate reference dose level for low-dose-rate (LDR) treatments, resulting in the requirement to specify the dimensions (width, height, and thickness) of the pear-shaped 60-Gy isodose reference volumes (Figure 8.3). If there is an

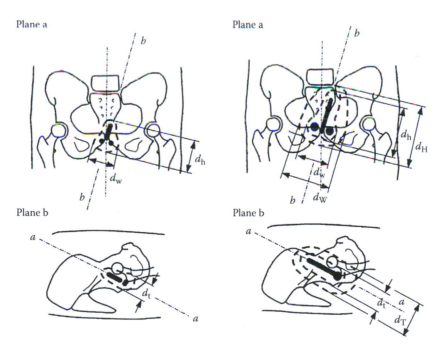

FIGURE 8.3 Schematic diagram of 60-Gy dose on the AP (planes a) and lateral (planes b) views from an implant patient. Width, length, and thickness of the 60-Gy dose isodose volumes are measured taking into account the contribution from external dose: the dimensions d_h, d_w, and d_t refer to the brachytherapy application only (left part of the figure). When a combination of brachytherapy and external beam therapy is taken into account (right part), these dimensions are symbolized by d_H, d_W, and d_T, respectively. (Reproduced from Gerbaulet, A. et al., *The GEC-ESTRO Handbook of Brachytherapy*. ESTRO, Brussels, www.estro.org, 2002. With permission from ESTRO.)

external beam component in addition to the brachytherapy, then the brachytherapy isodose must be selected to allow for the external beam dose. For example, 50 Gy external beam treatment means 10 Gy equivalent isodose from the brachytherapy distribution.

ICRU 38 recommended reporting the following information for brachytherapy treatment of gynecological patients:

- Description of the treatment technique (source, applicator)
- Total reference air-kerma rate
- Time dose pattern
- Description of the reference volume
- Dose at reference points (bladder, rectum, lymphatic trapezoid, pelvic wall)

Report 38 was written in an era of 2D image-based dosimetry, mainly using orthogonal x-rays for the geometrical reconstructions. Therefore, the concepts expressed in that report no longer reflect the possibilities of modern medical technology, in which the translation has taken place from 2D or 2.5D toward 3D techniques using CT and MR image-based brachytherapy, although some of the recommendations are still relevant. (For further reading, see the Chapters 21 and 24.)

8.2.2 Interstitial Implants

8.2.2.1 Manchester System or Patterson Parker System

Prior to 1930, a uniform distribution of the radioactivity in the interstitial implants was most commonly used in order to produce a uniform dose distribution. However, in 1930s, Paterson and Parker developed a system known as the Manchester system or Paterson and Parker system, in which the use of nonuniform distribution of the radioactivity was recommended to achieve a uniform dose distribution (Paterson and Parker 1934, 1952). In order to obtain homogenous dose distributions, normally sources with three different linear activities (0.66, 0.50, and 0.33 mg/cm) were utilized. Depending on the shape (linear, planar, and volume implant) and size of the implant, the use of a specific pattern of distribution of radioactivity was recommended. Paterson and Parker developed a tabulation system for the required total source strength in terms of milligram-hours (mg-h), along with some recommendations for source distribution within each implant type, in order to achieve 1000 cGy (with the unit "rad" at the time) at the prescription point. For example, in a planar implant, in addition to the total milligram-hours, the recommended ratio of the source strength on the peripheral and central aspect of the implant was given (Table 8.1).

TABLE 8.1 Paterson and Parker Recommendation of Distribution of Activity for Planar Implant

Area (cm²)	Fraction of Activity Used in the Periphery	Fraction of Activity in the Middle
Less than 25	2/3	1/3
25–100	1/2	1/2
Over 100	1/3	2/3

Source: Modified from Heymann, J., *Acta Radiol* 16, 129–48, 1935.

Moreover, it was recommended that the source strength be adjusted if the implant needles were not closed-ended or the shape of the implant was not square. The aim of the Patterson–Parker dosimetry system was to plan and deliver a uniform dose (within ±10% from the prescribed or stated dose) throughout the treatment area or volume. Moreover, Paterson and Parker recommended the use of a single-plane implant for every 1 cm thick slab of tissue, with the dose prescribed to a 0.5 cm distance from the plane of the implant. Therefore, for a 2 cm thick target, use of two planes (double-plane implant) was recommended, with 1 cm spacing between the two planes and the dose prescribed halfway between the two planes.

Some investigators have introduced a few corrections to the tables of Paterson and Parker, but they left the format of the tabulated data unchanged. These corrections included the change of gamma-constant of ^{226}Ra from 8.4 R cm^2 mg^{-1} h^{-1} to 8.25 R cm^2 mg^{-1} h^{-1} (with the unit R for the exposure rate in roentgen) and also the incorporation of a factor (0.957 cGy/R) to convert exposure to dose (Meredith 1967). The tables of Paterson and Parker are still utilized either as a primary method of determination of distribution of source strength or a secondary method for a second check in brachytherapy implants.

8.2.2.2 Quimby System

When the Manchester system was developed, the 0.66 and 0.33 mg Ra/cm radium sources were not available in the United States. Therefore, the Quimby system was developed at Memorial Hospital in New York City based on one linear source activity of ^{226}Ra sources that was available in the United States (Glasser et al. 1952; Quimby and Castro 1953; Quimby 1944). This system was mostly concentrated on planar implants. Similar to the Manchester system, tabulated data have been generated to provide the total source strength in terms of milligram-hours to provide 1000 cGy to the prescription point, for a uniform distribution of the source activity. The dose specification is in terms of the minimum dose, which occurs in the actual implanted region. However, unlike the Manchester system, a higher dose rate (60–70 cGy/h versus 40 cGy/h) has been used with the Quimby system for patient treatments. In addition, dose homogeneity is not a factor in this system. Initially, Quimby investigated that 50–60 Gy total doses in 3–4 days would provide biologically equivalent dose to 60–80 Gy in 6–8 days recommended in the Manchester system. However, in 1952, Quimby and Castro (1953) concluded that they have to use a linear activity of 0.5 mg Ra/cm instead of 1 mg Ra/cm in order to obtain 60 Gy in 3–4 days. It is worth noting that for an equivalent dose delivery to a point along the perpendicular direction of a planar implant, the total milligram-hours recommended by the Quimby system is less than the values recommended by the Patterson–Parker system.

8.2.2.3 Paris System

The Paris system was developed by Pierquin, Dutreix, and Chassagne for ^{192}Ir wire implants in 1960s and 1970s (Dutreix and Marinello 1987; Leung 1990; Pierquin et al. 1987). This system utilizes a uniform distribution of activity and is based on reference dose and basal dose for prescription (Figure 8.4). Iridium wires became available in the period 1950–1960 and, compared with the use of rigid radium needles, offered another type of flexibility for interstitial implants. These wires were available at various lengths and 0.3 mm in diameter and range of activities suitable for use in pre-implanted catheters and needles. Primarily, the system is intended for removable or temporary single and double-plane implants, and it does not address the other types of volume implants, although it can be adapted to fit larger differentially shaped volumes. In this system, the general rules for the single and double-plane implants are as follows:

- Sources are linear and their placements are parallel.
- The linear source strength (activity/cm) is uniform and identical for all sources.
- Adjacent sources must be equidistant from each other. The spacing between sources should be wider when using long sources.
- The prescription dose is made to the "central plane," which is perpendicular to the direction of the sources, at the midpoint of the implant.
- The sources are normally extended beyond the treated volume by 20% to 30%.

In volume implants, wires should be arranged so that the cross-sectional source distribution forms a series of equilateral triangles or squares. With regard to the volume

- The length, *l*, of treated volume is parallel to the wires and is the minimum distance across the reference isodose line (surface).
- The width, *w*, of treated volume is defined for single-plane implants only, it is the width of reference isodose in central transverse plane.
- The lateral margin, *d*, is used for multiplane implants to describe the distance reference isodose that extends

beyond physical boundaries of the implant; it is the minimum distance in the central transverse plane between the reference isodose and a line joining two outermost wires.

- The thickness, *t*, is the minimum distance between two parallel lines that are tangent to reference isodose line in the central transverse plane.

Dose (or dose rate) is calculated at the points in the central plane (which is defined as the plane perpendicular to the sources) in between the triangular or square geometries between the needles: these are the points with the minimum dose rate between a pair or group of sources: the basal dose points (BD). The average of all basal dose rate values is defined as 100% in the relative dose distribution. The value of the BD at the different calculation points across the implant should be within ±10% of the mean value for acceptable homogeneity. The clinical dose specification to the target volume is based on an isodose surface called the reference isodose surface. In practice, its value is 85% of the (average) basal dose rate. Within the limits of source geometry, this value leads to an acceptable compromise between too steep a dose gradient in passing from the margin of the treatment volume toward the middle and too great a ripple of the contour of the treatment envelope. The value of the reference isodose surface is often called the minimum peripheral dose (MPD).

The rules of the Paris system are valid also in modern applications as they provide guidance for pre-implant planning (volume determination, placement of needles, distance, and parallelism). As real implants are seldom perfect, miniaturized high-dose-rate (HDR) or pulsed-dose-rate (PDR) [192]Ir sources can be planned with optimized dwell times and dwell positions in order to achieve the desired dose coverage to the target volume. A summary of the rules for determining the active length of the sources, the spacing between the sources, and the lateral margin for single-plane implants and double-plane implants is shown in Table 8.2. All of these relationships were established assuming that all lines have the same active length, but they are also valid for lines of slightly unequal length. Although primarily established for a single-plane implantation, these rules have been extended to rectilinear and parallel sources arranged in

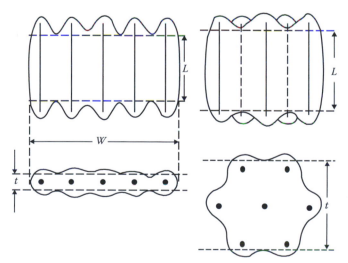

FIGURE 8.4 Schematic diagram of the implant geometry for Paris system. (Courtesy of A.S. Meigooni. Modified from Pierquin et al. 1978.)

TABLE 8.2 Relationships between Treated Volume and Implant Geometry

Ratio	Two Lines	Planar Implant	Squares	Triangles
Treatment length/ active length	0.69–0.81	0.69–0.78	0.66–0.73	0.65–0.76
Treated thickness/ separation	0.54–0.58	0.57–0.69	1.55–1.59	1.25–1.35
Safety margin/ separation			0.26–0.28	0.18–0.21
Lateral margin/ separation	0.37	0.33		

Source: Modified from Gerbaulet, A. et al., *The GEC-ESTRO Handbook of Brachytherapy.* ESTRO, Brussels, www.estro.org, 2002.

TABLE 8.3 Limits of Source Separations Used in Paris System

Active Length (mm)	Minimum Separation (mm)	Maximum Separation (mm)
10 to 40	8	15
50 to 90	10	20
≥100	15	25

Source: Modified from Gerbaulet, A. et al., *The GEC-ESTRO Handbook of Brachytherapy*. ESTRO, Brussels, www.estro.org, 2002.

TABLE 8.4 Comparison between Manchester and Paris Systems for "Typical" Interstitial Implant

Manchester	Paris
Parallel with crossing sources	Parallel sources
Activity distribution is nonuniform and depends on area; using standard distribution rules	Equal source strength; flexible dosimetry rules
Spacing between sources may vary by the height of the prescription point and area of the implant	Spacing depends on target volume thickness
Limited number of standard sources immediately available at any one institution	Any activity available at a week's notice
Active lengths usually fall within target volume	Active lengths of sources are greater than target volume length
Dose rate defined at a plane parallel to plane of implant, normally at 0.5 cm from it	Basal dose rate is the average of the dose minima; reference dose rate is 85% of basal dose rate

Source: Courtesy of A.M.B. Bidmead.

squares and triangles, to sources arranged on curved surfaces, and to loops and hairpins.

The method advocated for predicted dosimetry is to evaluate the target volume to be implanted carefully and define its three dimensions. The number of source planes to be used depends on the thickness. If the thickness exceeds 12 mm, there must be at least two source planes. The arrangement of sources in triangles or squares depends on the shape of the target volume. With the help of the ratio of treated thickness to spacing, corresponding to the chosen arrangement, the minimum spacing between the sources is next determined and the number of sources required is deduced. Finally, the radioactive length of the source is calculated from the ratio of treated length to active length.

Spacing between sources as a function of the active length of the source in the Paris system is shown in Table 8.3. As shown in this table, source separations for a 10 to 40 mm long source can never be less than 8 mm or exceed 15 mm. Table 8.4 gives a summary of the differences between the Manchester and Paris systems for a "typical" interstitial implant.

8.2.2.4 ICRU 58 Recommendations

Most brachytherapy treatments utilize some modification of a historical treatment system. As computer dosimetry has become more prevalent, dose prescription has become less dependent on a system and more dependent on the computer-generated dose distributions. Frequently, the set of rules in

the system is modified by the clinician. This may compromise the acceptable dose uniformity that was permitted by the rules of the system. The ICRU has attempted to address this problem in its report *Dose and Volume Specification for Reporting Interstitial Therapy* (ICRU 1997). In this ICRU 58 report, parameters are recommended for reporting, which are closely related to those of the Paris system. Unlike the ICRU 38, this report considers the 3D-dose distribution evaluation of an implant. This report encourages users not to change their practices for performing implants or prescribing the dose, but, instead, to adopt the ICRU terminology and recommendations to report the results in an interstitial brachytherapy implant. The ICRU 58 report lists the various items of interest for reporting interstitial brachytherapy and gives a definition of the main terms. The concepts used are in full agreement with Report 38 on intracavitary therapy in gynecology (ICRU 1985) and with the Report 18 (ICRU 1970) on prescribing, recording, and reporting photon beam therapy.

The recommendations cover the following items: description of volumes; description of sources; description of technique, source pattern, and time pattern; total reference air kerma; and description of dose and dose distribution, prescribed dose, minimum target dose (MTD), mean central dose (MCD), and high and low dose regions. The volumes of tissue defined are similar to those defined for external beam therapy. The planning target

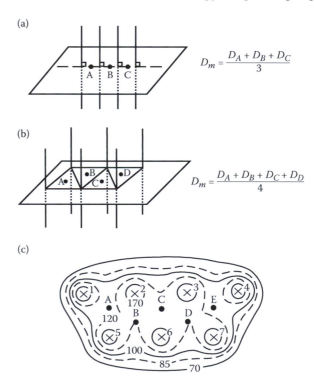

FIGURE 8.5 Geometric location of the minimum dose values (D_m) used for implants with a different number of sources. (a) Single-plane implant. (b) Two-plane implant. (c) Transversal view of a two-plane implant. (Courtesy of A.S. Meigooni. Modified from Pierquin, B. and Marinello, G. *A Practical Manual of Brachytherapy*. Medical Physics Publishing. Madison, WI, 1997.)

volume (PTV) does not need the same margin for patient motion required for external beam therapy. So it is, in general, identical to the clinical target volume (CTV). The treatment volume is the volume of tissue based on the actual implant that is encompassed by the isodose corresponding to the MTD. This isodose should, ideally, entirely encompass the CTV. Thus, reported doses are defined primarily in the central plane as in the Paris system, and the basal dose and reference dose have been renamed the MCD and the MTD, respectively. In this report, the implants are assumed to be linear sources, and the minimum dose (D_m) values between linear source values are defined in a "central plane" that is perpendicular to the midpoint of the linear sources, as shown in Figure 8.5.

The prescribed dose is normally 85% of D_m. Two uniformity parameters are suggested as measures of dose uniformity using central and MTD: (1) the spread in the individual minimum doses, averaged to get the mean value, and (2) the ratio of the MTD to the MCD. The report recommends priorities for the items to be reported depending on the type of technique used and on the level of computation available (Table 8.5).

8.2.2.5 Memorial System

The 3D imaging techniques enabled more accurate assessment of brachytherapy radiation therapy to be made. Using the value of the implanted volume, a nomograph was developed at Memorial Sloan-Kettering Cancer Center (Hilaris et al. 1988) correlating

TABLE 8.5 Hierarchy of Dosimetric Information to Be Reported for Interstitial Implant as Recommended by ICRU 58 (1997)

Hierarchy of Dosimetric Information	Priority	Computing Procedure
Description of clinical target volumes	1	1
Description of sources and techniques	1	1
Description of implant time	1	1
Total reference air kerma (TRAK)	1	1
Description of dose:	1	2
• Prescribed dose, including point or surface of prescription		
• Reference doses in central plane		
• MCD		
• MTD		
Description of high and low dose regions	2	2
Uniformity indices	3	3
Alternative representation of dose distribution [dose-volume histogram (DVH), etc.]	3	3
Dose rates	3	1

the total seed strength to the average target dimension for delivery of a matched peripheral dose (MPD) of 160 Gy with ^{125}I source. Figure 8.6 shows a sample of this nomograph for ^{125}I source (Thomadsen et al. 1997). This system is an extension of the Quimby system. It is characterized by complete dose distributions around lattices of point sources of uniform strength spaced 1 cm apart and is based on computer-generated distributions.

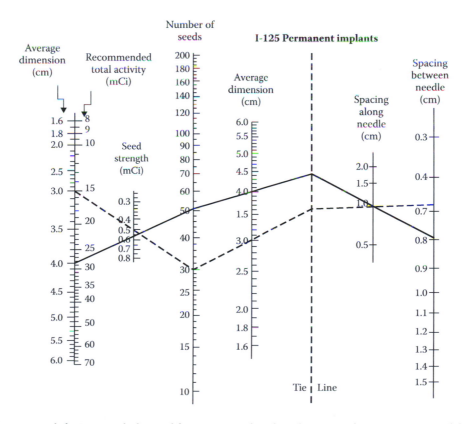

FIGURE 8.6 Sample nomograph for I-125 seeds designed for prostate seed implant. (Courtesy of A.S. Meigooni. Modified from Thomadsen et al. 1997.)

This information was then extended to other sources such as [103]Pd source (Anderson et al. 1993). The nomograph includes additional scales to guide needle spacing for a given spacing of seeds along the needle track. These nomographs are intended as intraoperative planning guides and should not be substituted for more definitive planning (e.g., for prostate implants) that uses 3D images.

8.3 Single Source Dosimetry

8.3.1 Sievert Integral

The "systems" described in the previous sections allowed the users to administer a safe dose to the patient for the treatment of the tumor by providing the rules for arrangement and placement of the sources including the treatment duration of the application. Clinical experience played a major role in their developments. The advent of other source types and radionuclides necessitated the development of more individualized dosimetry based on the dose distribution around a single source. The superposition principle assumed the total dose distribution to be a summation of the dose contributions of all individual sources in an application. The traditional single source was based on the exposure rate from a bare source with point source approximation. However, this approximation was not suitable for dose calculations in the close vicinity of a linear source with core and capsule materials that have significant effect on the radiation emitted by the source. The impact of the capsule filtrations of a linear source on the exposure rate and dose rate distribution around a brachytherapy source was calculated by Rolf Sievert (1930). Later, this technique has been more widely used to calculate exposure and dose rate distributions about different filtered linear brachytherapy sources (Shalek and Stovall 1969; Krishnaswamy 1972; Young and Batho 1964). The accuracy of this technique has been examined by comparison of the calculated dose rate around a source with the measured (Diffey and Klevenhagen 1975) and Monte Carlo simulated data (Williamson 1983). Figure 8.7 shows the schematic diagram of the source geometry and algorithm used with Sievert integral for dose calculations to a point p.

A more detailed description of the methodology of the Sievert integral is presented in Chapter 7 and is therefore not further elaborated in this paragraph. For clear reasons, the practice of using the Sievert integral for a dose calculation system of individual implants has been abandoned and replaced by the approach described in the report of AAPM task group 43 (Nath et al. 1995) and slightly modified in later updates (Rivard et al. 2004, 2007). This latter work is presented in Chapter 10.

References

Anderson, L.L., Moni J.V., and Harrison L.B. 1993. A nomogram for permanent implants of Paladium-103 seeds. *Int J Radiat Oncol Biol Phys* 27:129–35.

Diffey, B.L. and Klevenhagen, S.C. 1975. An experimental and calculated dose distribution in water around CDC K-type Caesium-137 sources. *Phys Med Biol* 20:446–54.

Dutreix, A. and Marinello, G. 1987. The Paris System. In: *Modern Brachytherapy*. B. Pierquin, J.F. Wilson, and D. Chassagne (eds.). *Masson Publishing*, U.S.A., pp. 25–42.

Gerbaulet, A., Pötter, R., Mazeron, J.-J., Meertens, H., and Van Limbergen, E. 2002. *The GEC-ESTRO Handbook of Brachytherapy*. ESTRO, Brussels. www.estro.org.

Glasser, O., Quimby, E.H., Taylor, L.S., and Weatherwax, A.J.L. 1952. *Physical Foundations of Radiology*, 2nd ed. Paul B. Boeber, New York.

Heymann, J. 1935. The so-called Stockholm method and the results of treatment of uterine cancer at Radiumhemmet. *Acta Radiol* 16:129–48.

Hilaris, B.S., Nori, D., and Anderson, L.L. 1988. *Atlas of Brachytherapy*. Macmillan Publishing Company, USA.

International Commission on Radiation Units and Measurements (ICRU). 1970. Specification of High Activity Gamma-Ray Sources. Report 18 of ICRU, ICRU Publications, Bethesda, MD.

International Commission on Radiation Units and Measurements (ICRU). 1985. Dose and volume specification for reporting and recording intracavitary therapy in gynecology. Report 38 of ICRU, ICRU Publications, Bethesda, MD.

International Commission on Radiation Units and Measurements (ICRU). 1997. Dose and Volume Specification for Reporting Interstitial Therapy. Report 58 of ICRU, ICRU Publications, Bethesda, MD.

Kottmeier, H.L. 1964. Surgical and radiation treatment of carcinoma of the uterine cervix. Experience by the current individualized Stockholm Technique. *Acta Obstet Gynecol Scand* 43 (Suppl. 2):1–48.

Krishnaswamy, V. 1972. Dose distribution about 137Cs sources in tissue. *Radiology* 105:181.

Leung, S. 1990. Perineal template techniques for interstitial implantation of gynecological cancers using the Paris system of dosimetry. *Int J Radiat Oncol Biol Phys* 19:769–74.

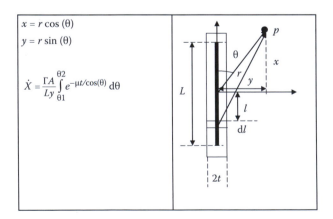

FIGURE 8.7 Schematic diagram of the source geometry used for Sievert integral. (Courtesy of A.S. Meigooni.)

Meredith, W.J. 1967. *Radium Dosage: The Manchester System.* Livingston Inc., Edinburgh.

Nath, R., Anderson, L.L., Luxton, G., Weaver, K.A., Williamson, J.F., and Meigooni, A.S. 1995. Dosimetry of interstitial brachytherapy sources: Recommendations of the AAPM Radiation Therapy Committee Task Group No. 43. American Association of Physicists in Medicine. *Med Phys* 22:209–34.

Parker, H.M. 1938. A dosage system for interstitial radium therapy. Physical aspects. *Br J Radiol* 11:252–66.

Paterson, J.R. 1948. *The Treatment of Malignant Disease by Radium X-rays Being a Practice of Radiotherapy.* Edward Arnold Publ., London.

Paterson, J.R. and Parker, H.M. 1934. A dosage for gamma-ray therapy. *Br J Radiol* 7:592–612.

Paterson, J.R. and Parker. H.M. 1952. A dosage system for interstitial radium therapy. *Br J Radiol* 25:505–16.

Pierquin, B. 1964. *Précis de Curiethérapie. Endocuriethérapie et Plésiocuriethérapie.* Masson, Paris.

Pierquin, B. and Marinello, G. 1997. *A Practical Manual of Brachytherapy.* Medical Physics Publishing. Madison, WI.

Pierquin, B., Dutreix, A., Paine, C.H., Chassagne, D., Marinello, G., and Ash, D. 1978 The Paris system in interstitial radiation therapy. *Acta Radiol Oncol* 17(1):33–48.

Pierquin, B., Wilson, F., and Chassagne, D. 1987. *Modern Brachytherapy.* Masson Publishing, New York.

Quimby, E.H. 1944. Dosage tables for linear radium sources. *Radiol* 43:572–7.

Quimby, E.H. and Castro, V. 1953. The calculation of dosage in interstitial radium therapy. *Am J Roentgenol* 70:739–49.

Rivard, M.J., Coursey, B.M., DeWerd, L.A. et al. 2004. Update of AAPM Task Group No. 43 Report: A. revised AAPM protocol for brachytherapy dose calculations. *Med Phys* 31:633–74.

Rivard, M.J., Butler, W.M., DeWerd, L.A. et al. 2007. Supplement to the 2004 update of the AAPM Task Group No. 43 Report. *Med Phys* 34:2187–205.

Ryberg, M., Björkholm, E., Båryd, I. et al. 1990. Adaptation of the Stockholm cervical cancer treatment technique to Selectron afterloading. A preliminary report. Brachytherapy Working Conference. 6th International Selectron User's Meeting. Florence, Italy.

Shalek, R.J. and Stovall M. 1969. In: *Radiation Dosimetry*, Vol. 111. F.H. Attix and E. Tochlin (eds.). Academic Press, New York, p. 743.

Sievert, R.M. 1930. Die γ-Strahlungsintensitat an der Oberflache und in der nachsten Umgebug von Radiumnadeln. *Acta Radiol* 11:249–301.

Thomadsen, B., Ayyangar, K., Anerson, L., Luxton, G., Hansen, W., Wilson, J.F. 1997. Brachytherapy treatment planning. In: *Principles and Practice of Brachytherapy.* Subir Nag (ed.). Futura Publishing Company, Inc., USA, pp. 127–199.

Tod, M. and Meredith, W.J. 1938. A dosage system for use in the treatment of cancer of the uterine cervix. *Br J Radiol* 11: 809–24.

Tod, M. and Meredith, W.J. 1953. Treatment of cancer of the cervix uteri—A revised Manchester method. *Br J Radiol* 26: 252–7.

Wilkinson, J.M., Moore, C.J., Notley, H.M., and Hunter, R.D. 1983. The use of Selectron afterloading equipment to simulate and extend the Manchester System for intracavitary therapy of the cervix uteri. *Br J Radiol* 56:409–14.

Williamson, J.F. 1983. Monte Carlo evaluation of the Sievert integral for brachytherapy dosimetry. *Phys Med Biol* 28:1021–32.

Young, M.E.J. and Batho, H.F. 1964. Dose tables for linear radium sources calculated by an electronic computer. *Br J Radiol* 37–8.

Two-, Three-, and Four-Dimensional Brachytherapy

Jean Pouliot
University of California

Ron Sloboda
Cross Cancer Institute

Brigitte Reniers
Maastro Clinic

9.1 Transition from 2D to 3D Image-Based Planning

9.1.1 Overview

The use of 3D imaging for brachytherapy treatment planning has dramatically increased over the past decade. Today, planning methods based on the availability of 3D images from various modalities are progressively superseding those developed early in the second half of the last century for 2D radiographs and subsequently refined since then. This section describes the transition from 2D to 3D image-based planning separately for intracavitary and superficial/intraluminal/interstitial brachytherapy. For each main type of therapy, a brief historical introduction, a summary of traditional 2D image-based planning methods and their limitations, and a synopsis of emergent 3D image-based methods are presented. This section concludes by identifying some key issues associated with the transition from 2D to 3D imaging for brachytherapy treatment planning.

9.1.2 Intracavitary Therapy

9.1.2.1 Development of "Systems"

Intracavitary therapy is commonly used in combination with external beam therapy to treat cancers of the cervix, uterine body, rectum, and nasopharynx, among others. Here we focus on cervix carcinoma with the understanding that similar considerations apply at the other sites.

According to ICRU Report 38 (ICRU 1985), three fundamental systems based on extensive clinical experience have been developed for the treatment of cervix carcinoma: the Stockholm system, the Paris system, and the Manchester system. The term "system," as discussed in Chapter 8, refers to a set of rules for obtaining a clinically acceptable dose distribution over a specified treatment volume and a corresponding set of recommendations for reporting key treatment parameters. The rules are formulated to account for anticipated variations in clinical administration, most importantly the radiation source strength, geometry, and time-dose pattern. The Manchester system, introduced in 1938, is derived from the original Paris system and has seen the most widespread use (Tod and Meredith 1938). Hence, in the following, we describe the evolution of image-based treatment planning in that context.

9.1.2.2 Traditional 2D Image-Based Planning

The Manchester system was originally designed to deliver, for ideal applicator geometry, a constant low dose rate (LDR) of ~54 cGy·h⁻¹ to dose point A near the cervix, regardless of the size and shape of the uterus and the vagina (Figure 9.1). The original applicator was made of rubber and consisted of an intrauterine tandem and two intravaginal ovoids holding radium tubes; modern applicators are rigid and adjustable and may be fully compatible with 3D imaging systems (Figure 9.2). In principle, the Manchester system could be utilized to deliver brachytherapy without imaging (assuming ideal applicator geometry). However, in practice, imaging was introduced early on to verify applicator placement and, later, to enable patient-specific dose determination.

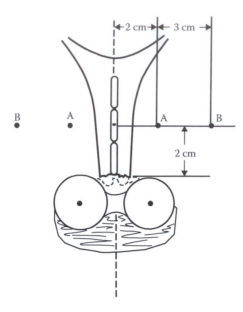

FIGURE 9.1 Definition of points "A" and "B" in the classical Manchester system. Point A is defined as being 2 cm lateral to the central canal of the uterus and 2 cm up from the mucous membrane of the lateral fornix, in the axis of the uterus. Point B is defined as being in the transverse axis through point A, 5 cm from the midline. In clinical practice where dose calculations are made from 2D or 3D images, point A is usually taken 2 cm up from the flange of the intrauterine source and 2 cm lateral from the central canal. (From International Commission on Radiation Units and Measurements (ICRU). *Dose and Volume Specification for Reporting and Recording Intracavitary Therapy in Gynecology.* Report 38 of ICRU, ICRU Publications, Bethesda, MD, 1985. With permission).

Calculated dosimetry for individual patients was typically performed using radiographs of the applicator and surrounding pelvic anatomy obtained from two complementary views (Gerbaulet et al. 2002). Nonradioactive replica or dummy sources were used to simulate the actual source placement (Figure 9.3). First, source locations in 3D were reconstructed from their radiographic coordinates, and then a 3D dose distribution was calculated and displayed

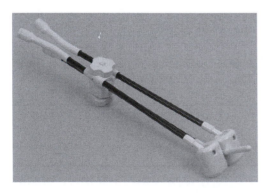

FIGURE 9.2 Modern rigid, adjustable Manchester-style applicator fabricated from composite fiber tubing. The composite material mostly eliminates the distortions seen with traditional stainless-steel tubing in CT and MR images acquired to plan cervix brachytherapy treatments. (Courtesy of Nucletron, Veenendaal, The Netherlands.)

with respect to the applicator in orthogonal planes. Points deemed to have clinical significance such as A and B (Figure 9.1) were similarly reconstructed and corresponding doses calculated. ICRU Report 38 (ICRU 1985) recommended a set of standard parameters for reporting intracavitary therapy. These included a description of the technique used, the total reference air kerma, the dose level and dimensions of the reference (treatment) volume, the absorbed dose at a number of reference points related to tissues at risk in the pelvis, and the time-dose pattern of irradiation. Among the reference points, the ICRU rectal point and bladder point were widely adopted in clinical practice. Shortly afterward, a new system, known as the Madison system, was specifically created to supersede the Manchester system for high-dose-rate (HDR) cervix brachytherapy (Thomadsen et al. 1992).

Treatment planning using 2D images suffers from several fundamental limitations made plainly evident by the introduction of CT in cervix brachytherapy (Lee et al. 1980; Gebara et al. 1998; Martel and Narayana 1998). Briefly, they are as follows:

- The extent of clinical target volume (CTV) coverage by the reference volume is not known quantitatively.
- There is limited opportunity for adaptation of the dose distribution to the CTV for individual patients, as the CTV is not visualized.
- The maximum doses received by organs at risk (OARs) are not accurately known, as geometrically derived point doses are an insufficient representation of these doses.

Making use of CT, MR, and, in some cases, PET 3D imaging, numerous investigators have gone on to quantify these shortcomings of the 2D image-based approach to planning (Gebara et al. 1998; Lin et al. 2007; Kim et al. 2007; Pötter et al. 2008a). Their findings include underdosing the CTV especially for advanced tumors, underestimating maximum doses received by rectum and bladder, not accounting for the dose delivered to other critical structures such as sigmoid colon and small bowel, and having limited potential for adapting the dose distribution to individual patient anatomy. Such shortcomings confound attempts to correlate doses calculated at the time of treatment planning with subsequently observed tumor responses and treatment-related morbidities.

9.1.2.3 Emergent 3D Image-Based Planning

Treatment planning based on 3D imaging, in combination with remotely afterloaded brachytherapy delivery, promises several opportunities to improve patient outcomes compared with traditional 2D methods. First, by using magnetic resonance imaging (MRI) or possibly PET-CT to localize the extent of disease (Lin et al. 2007; Pötter et al. 2008a), the CTV can be visualized in 3D, and planning target volume (PTV) dose coverage can be made more conformal for each patient. This is accomplished by adjusting applicator source positions and dwell times appropriately and adding concurrent interstitial therapy when needed (Pötter et al. 2008a). An early comparative study found CT to be less useful for PTV definition than MRI as it overestimated treatment volume width and thus would lead to irradiation of more normal tissue (Viswanathan et al. 2007). Second, delineation of OARs

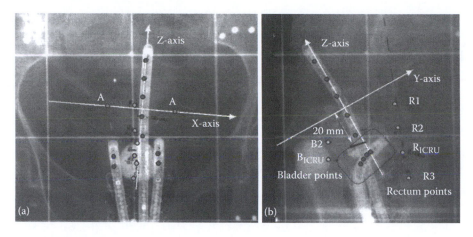

FIGURE 9.3 Orthogonal radiographs for a typical treatment of cancer of the cervix with a Manchester-style applicator, illustrating active source positions (darker-filled circles) and bladder and rectum dose calculation points (lighter-filled circles). (a) AP projection; (b) lateral projection. (From Gerbaulet, A., Pötter, R., Mazeron, J.-J., Meertens, H., and Van Limbergen, E. (eds.), 2002. *The GEC-ESTRO Handbook of Brachytherapy*. ESTRO, Brussels. With permission.)

in 3D enables calculation of dose-volume (D-V) data for these structures, which is expected to make correlation of dosimetric parameters with patient outcomes considerably more reliable and accurate than before. Third, the use of structure-based forward and inverse planning methods designed to optimize the therapeutic ratio for individual patients is made feasible (Lessard et al. 2002). At present, planners pursue this objective by distributing physical dose; in future, they may work directly with radiobiological response data as well. Finally, 4D planning—adaptive therapy for multiple brachytherapy insertions/fractions—can be realized by performing 3D image-based planning prior to each irradiation (Lin et al. 2007; Pötter et al. 2008a).

9.1.3 Superficial, Intraluminal, and Interstitial Therapy

9.1.3.1 Introduction

Superficial, intraluminal, and interstitial brachytherapy are used to treat cancers at many disease sites including skin, connective tissue, head and neck, bronchus, esophagus, breast, cervix, prostate, and rectum. Although these three modes of treatment are obviously different, the physical factors that govern their dose distributions and the parameters currently used to describe their dosimetry are quite similar. Consequently, we focus on interstitial brachytherapy and make reference to the other two modes only occasionally.

ICRU Report 58 (ICRU 1997) observes that classical systems of interstitial brachytherapy, among them the well-known Manchester, Quimby, and Paris systems, were developed fairly early in the last century (Chapter 8). Just as for intracavitary therapy, each system comprises a set of rules specifying the strength and placement of sources that yield an appropriate dose distribution over a specified treatment volume. Clinical experience with these systems has been gained over many decades using manually placed, LDR sources. By comparison, current practice increasingly relies on 3D image-based, computerized

planning to guide remotely afterloaded, HDR sources. Such technological advancements have led to a departure from the classical system approaches toward a new paradigm for interstitial brachytherapy planning.

9.1.3.2 Traditional 2D Image-Based Planning

The Manchester (or Paterson–Parker) system was designed to deliver a relatively uniform dose to interstitial lesions of various shapes and sizes using ^{226}Ra needles or ^{222}Rn seeds arranged in planes or 3D geometrical forms such as cylinders or spheres. Two different linear activity densities are utilized to achieve good dose uniformity. The Paris system was developed for afterloading interstitial implants consisting of one, two, or even more source planes of ^{192}Ir wires having a single lineal activity density. Wire spacing and length are adjusted to have the dose distribution match the boundaries of the assumed target volume. Both systems can be used with any "radium-equivalent" radionuclide source (i.e., one emitting photons with energies predominantly >200 keV). Figure 9.4 illustrates their application to a single plane implant. The Manchester system can also be used to treat superficial lesions with sources placed on the outer surface of a conformal mold. As in intracavitary therapy, radiographs were employed to verify source placement and perform patient-specific dose calculations.

With the widespread adoption of remote afterloaders in the 1990s, a number of variations of the classical implant dosimetry systems were developed to enable specific improvements. These included geometrical optimization (Edmundson 1990; see further in Chapter 12), which provided better dose uniformity for clinically realized source arrangements, and the stepping source dosimetry system (van der Laarse 1994), which was designed to minimize irradiation of the surrounding normal tissue.

In 1997, ICRU Report 58 (ICRU 1997) identified and updated core dosimetry concepts exemplified by the classical systems and recommended a standard set of parameters for reporting interstitial therapy. These included a detailed description of clinical

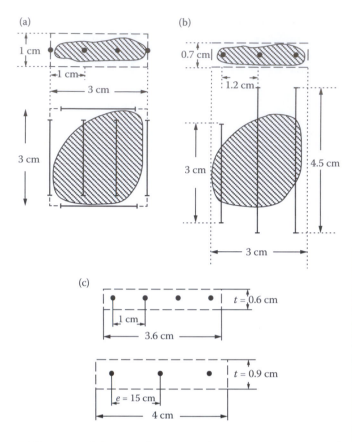

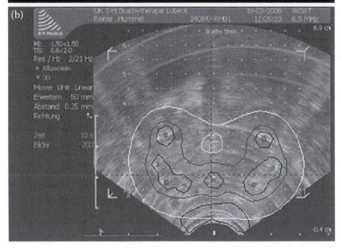

9.1.3.3 Emergent 3D Image-Based Planning

Just as for intracavitary therapy, 3D image-based planning for superficial, intraluminal, and interstitial therapy uses detailed anatomical and functional information to optimally conform the dose distribution to the target volume. In conjunction with remote afterloading for temporary treatments, or image-guided source placement for permanent implants, this approach has the potential to improve patient outcomes and reduce treatment-related morbidity.

We illustrate with some examples. For accelerated partial breast irradiation (APBI) delivered via HDR interstitial implant, Major et al. (2005) showed that the mean value of PTV D_{90} for $n = 17$ consecutively treated patients could be increased from

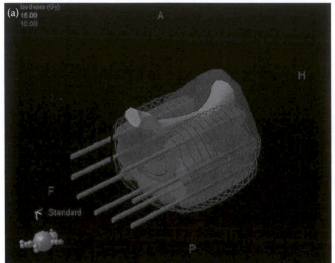

FIGURE 9.4 Schematic illustration of source arrangements and target volume coverages for a single plane implant in the Manchester (a) and Paris (b) systems. Source strength lineal density is greater for peripheral sources than for interior sources in the Manchester system, but equal for all sources in the Paris system. The thickness of the treated slab of tissue is fixed at 0.5 cm in the Manchester system, but varies proportionally to the separation e between the radioactive sources in the Paris system (c) as $t = 0.6 \times e$. (From Gerbaulet, A., Pötter, R., Mazeron, J.-J., Meertens, H., and Van Limbergen, E. (eds.), 2002. *The GEC-ESTRO Handbook of Brachytherapy*. ESTRO, Brussels. With permission.)

FIGURE 9.5 (See color insert.) (a) Intensity-modulated HDR brachytherapy 3D dose distribution for prostate boost treatment created using volume-optimized inverse planning. Red: prostate; yellow: urethra; magenta: needles; green: 15-Gy isodose surface; blue: 10-Gy isodose surface. (b) Corresponding 2D dose distribution on an axial TRUS slice. Green: prostate contour; dark yellow: urethra contour; red: 15.0 Gy reference isodose line; blue: 22.5 Gy isodose line; cyan: 30.0 Gy isodose line; yellow: 7.5 Gy isodose line. (From Kovács, G. et al., *Brachytherapy* 6, 142–8, 2007. With permission.)

volumes (using terminology previously introduced for external beam therapy), radiation sources, technique and source arrangement, time pattern of irradiation, total reference air kerma, and the dose distribution itself. For the latter, the mean central dose and minimum target dose were identified as key reporting parameters; additional data that should be recorded when available included low and high dose volume extents, dose uniformity metrics, and D-V histograms (DVHs). Parameters for reporting superficial and intraluminal therapies were also recommended. A notable feature of ICRU Report 58 was the introduction of a three-level reporting hierarchy based on different levels of dose calculation sophistication, mirroring the wide variation in capabilities of treatment planning systems in use at that time.

The use of 2D images for interstitial treatment planning entails limitations very similar to those identified in Section 9.1.2.2 for cervix brachytherapy. For example, see studies comparing 2D and 3D approaches for breast (Major et al. 2005) and prostate (Nickers et al. 2005) brachytherapy.

64% to 94% if CT-based conformal planning was used instead of the Paris system. In a similar study of $n = 166$ prostate implant patients comparing Paris dosimetry for ^{192}Ir wires to CT-based, inverse-planned, conformal HDR dosimetry, Nickers et al. (2005) reported that the proportion of satisfactory treatment plans (as defined using multiple criteria) could be increased from 82% to 95%. Using computer-optimized 3D planning and a highly stable implant delivery system, Yamada et al. (2006) found excellent disease control with minimal significant toxicity for $n = 105$ patients with localized prostate cancer receiving an HDR boost. The 5-year prostate-specific-antigen (PSA) relapse-free survival rate for high-risk patients in this study group was 92%. A similar technique has been described in detail by Kovács et al. (2007) (Figure 9.5). Improvements in dosimetry and outcomes directly attributable to 3D image-based planning have also been reported for other treatment sites including base of tongue (Hayes et al. 1992), scalp (Liebmann et al. 2007), and eye (Astrahan 2005).

9.1.4 Adaptive and Image-Guided Brachytherapy

In external beam radiotherapy (EBRT), the imaging acquisitions performed for the purpose of treatment planning (e.g. CT to define electron densities from Hounsfield units) are usually not seen as part of the image-guided treatment. In EBRT, image guidance usually refers to the image acquisitions performed during the treatment to verify several aspects of it or to adapt the treatment. In brachytherapy, often treatment planning and the actual source implant procedure are performed much closer together in time or simultaneously. Therefore, image guidance in brachytherapy generally refers to both imaging needed for treatment planning and for treatment verification.

Similar to EBRT, intrafraction and interfraction motion in brachytherapy need to be captured by the utilized imaging modality. Tumor motion, a problem of great significance for EBRT (e.g., lung motion during breathing), is of little significance for brachytherapy, as long as the applicator is solidly positioned close to the tumor. Motion of tumors does occur, but as long as the source/applicator moves with it, no large changes in tumor dose are expected in brachytherapy. In brachytherapy, it is mostly the relative position of the OARs versus the target and geometry changes (e.g., variable bladder filling, prostate edema) that cause dose distributions to differ during a treatment and between different fractions. The concept of treatment margins is also different in EBRT and brachytherapy (Pötter et al. 2008b). Examples of intrafraction and interfraction movements in brachytherapy are given in Chapter 16 (on clinical uncertainties) and further discussed in Chapter 17 (on the concept of margins in brachytherapy). Due to the different nature and consequences of geometrical changes during EBRT and brachytherapy, one needs to be aware that image guidance and treatment adaptation may require different measures.

In 2008, Williamson warned that guidelines for brachytherapy imaging are largely lacking (Williamson 2008). Recently, the American Association of Physicists in Medicine (AAPM) TG-137 report (Nath et al. 2009) was issued on "Dose Prescription and Reporting Methods for Permanent Interstitial Brachytherapy for Prostate Cancer," which discusses briefly the imaging modalities used in conjunction with this brachytherapy technique. However, comprehensive guidelines are still lacking.

Single fraction, permanent LDR seed implant is one of the most frequently performed image-guided brachytherapy techniques. Usually, transrectal ultrasound (TRUS) imaging is used to visualize the prostate, to perform treatment planning, and to implant the seeds. The TRUS-guided LDR brachytherapy technique has been described extensively in the literature; see, for example, the reference list of Nath et al. (2009). Once implanted, the seeds are not sufficiently clearly visualized, hence the need for another imaging modality for the dose assessment. Both the ESTRO/EAU/EORTC (Ash et al. 2000) and the ABS (Nag et al. 2000) recommend using TRUS imaging for implantation, which is also endorsed by the more recent AAPM TG-137 (Nath et al. 2009). For dose assessment, either CT (Nath et al. 2009; Ash et al. 2000; Nag et al. 2000) or MRI (Nath et al. 2009) imaging is recommended, and these images are usually acquired weeks after implantation. Recently, it has been shown that the agreement between dosimetry at time of implant and post-implant is poor (Acher et al. 2010) begging questions as to when the dosimetry should be performed.

In the very near future, there will be other considerations that come into play when considering imaging modalities. The AAPM TG-186 has been mandated to issue recommendations regarding dose calculation to real tissues, rather than to water as is now common practice according to the AAPM TG-43 recommendations (Rivard et al. 2004), further addressed in Chapter 11. To extract the required tissue characteristics to enable accurate dose calculations, in particular, for low-energy photon emitters such as ^{125}I, it may be necessary to investigate novel imaging modalities such as dual-energy CT (Bazalova et al. 2008; Landry et al. 2011).

9.1.5 Issues Arising in 3D Image-Based Planning

Brachytherapy treatment planning continues to evolve away from its historical dependence on classical rule-based dosimetry systems developed in the radiographic era, and toward a growing reliance on assessment of patient-specific dose distributions calculated in anatomical and functional volumes defined from 3D images. This transition has been particularly evident over the past decade and raises several important issues that, in many instances, can only be fully addressed through ongoing, site-specific clinical investigation. Some key questions are as follows:

1. Which imaging modalities, either alone or in combination, are sufficient/preferred for 3D image-based planning? How can access to required equipment be secured?
2. What are the clinically important features of classical dose distributions that ought to be preserved in 3D image-based planning? How should the transition from 2D to 3D planning be managed in the clinic?

3. How should dose to PTV be prescribed and doses to OARs limited to provide effective treatment while minimizing treatment-related morbidity? Are dose-volume parameters alone sufficient for this purpose and if so, which parameters are most appropriate for a given treatment? Are dose conformality and dose homogeneity indices useful?

4. Is it necessary to develop new dose calculation algorithms to better account for tissue and applicator boundaries and composition?

5. What planning methods (e.g., forward planning, inverse planning) are optimal, and how can they be standardized between clinics?

6. How can dosimetry and clinical outcomes be compared when different time–dose patterns [i.e., HDR, pulsed dose rate (PDR), LDR] are used to deliver treatment?

These issues are dealt with to varying degrees in other sections of this book describing current 3D planning practices; however, definitive guidance is not always available. In particular, much important work correlating dosimetry with clinical outcomes remains to be done.

9.2 3D Image-Based Planning

9.2.1 Introduction

The use of CT images for brachytherapy treatment planning was first reported around 1980 for both cervix insertions (Lee et al. 1980) and interstitial implants (Elkon et al. 1981). Early proponents realized its potential for localizing disease, planning and reconstructing radionuclide source positions, and calculating 3D dose distributions indexed to patient anatomy. However, widespread clinical adoption did not begin until much later when increased access to scanners, smaller CT slice thicknesses more appropriate for brachytherapy applications, CT-compatible applicators, applicator reconstruction methods yielding accuracies similar to those achieved with orthogonal films, and suitable treatment planning software became commonly available. In 1985, ICRU Report 38 (ICRU 1985) mentioned CT only very briefly as an alternate means of locating dosimetric reference points; by 1997, ICRU Report 58 (ICRU 1997) had identified D-V data as additional information that should be reported when available. Today, commercial brachytherapy treatment planning systems support CT-based applicator/source localization, 3D anatomical structure definition, and D-V analysis. A 2007 American Brachytherapy Society (ABS) survey of U.S. gynecologic brachytherapy practitioners found that a majority used CT for dose specification, although three-quarters still prescribed to point A rather than a 3D target volume (Viswanathan and Erickson 2010). In a patterns-of-care survey for brachytherapy performed in Europe, it was found that the use of CT-based dosimetry increased substantially in a group of "high resource countries" from 33% of centers in 2002 to 61% in 2007 (Guedea et al. 2010).

An early account of the use of MRI to plan ^{252}Cf brain implants was related by Maruyama et al. (1984). The authors noted that compared with CT, MRI allowed better 3D localization of tumor and enhanced visualization of normal anatomy and peritumoral edema. Yet despite such advantages and the capacity to form images directly in nonaxial planes (that can be oriented with respect to an applicator, for instance), MRI integration into radiation therapy planning, in general, and brachytherapy planning, in particular, has proceeded slowly because of fundamental technical issues and limited access to equipment. Among the former, image distortion and assignment of tissue/applicator composition for accurate dose calculation, identified some time ago by Fraass et al. (1987), continue to require attention. Specifically for brachytherapy, applicator and source position identification may require a special procedure. Consequently, MRI has not yet been widely adopted for brachytherapy planning and is frequently used in combination with CT (Krempien et al. 2003). The ABS survey cited earlier revealed that only 2% of respondents employed MRI to specify dose to the cervix (Viswanathan and Erickson 2010). By comparison, MRI-based dosimetry in "high resource" European countries accounted for 16% of brachytherapy procedures in 2007 versus 5% in 2002 (Guedea et al. 2010).

However, a recent study (Aubry et al. 2010) showed that the distortions measured in the presence of MRI-compatible applicators were small enough to validate the use of MRI for GYN brachytherapy treatment planning and verification. Other studies have shown that MRI alone can be used for treatment planning, therefore eliminating registration error from CT/MRI fusion (Kubicky et al. 2008; Pérez-Calatayud et al. 2009). Both the European Gynecological GEC-ESTRO Working Group and the American Image-Guided Brachytherapy Working Group have proposed guidelines for target delineation and recommend MRI as the imaging modality of choice for GYN brachytherapy (Haie-Meder et al. 2005).

In spite of the relatively slow but steady transition, significant efforts were deployed in the last 5 years to accelerate the clinical introduction of CT and MRI in brachytherapy. Inherently, 3D images provide more anatomical information. Most soft tissue is not visible on x-ray films. The tumor definition, its relation to the surrounding anatomy and the OARs, as well as the position of the brachytherapy applicator or catheters, relative to the anatomy, are better assessed with 3D imaging.

The limitations and uncertainties introduced using 3D imaging (CT and MRI or MRS) must be understood and taken into account. However, one must keep in mind that the additional knowledge from 3D imaging for target definition, OAR identification, catheter and applicator reconstruction, dose planning, quality assurance, and safety carries significantly more weight in helping to achieve the overall treatment quality.

9.2.2 3D Treatment Planning Process

9.2.2.1 Overview

Brachytherapy can be delivered using a variety of irradiation schedules: continuous for LDR, fractionated for HDR, and hyperfractionated for PDR. At the same time, 3D imaging modalities, mainly ultrasound and CT, but also MRI for prostate (Ménard et al. 2004; Pouliot et al. 2004) or GYN (Haie-Meder et al. 2005; Kubicky et al. 2008) and functional imaging such as MRS (Pouliot et al. 2004) or PET (Guha et al. 2008; Grigsby 2009) and SPECT (Wahl et al. 2011), are increasingly being used for dose planning in brachytherapy. These imaging modalities are used either alone or in combination. Given such diversity, the treatment planning process might be expected to vary with each particular treatment approach. While this is true to some extent, it is nonetheless possible to identify a common workflow that forms the basis for all 3D planning, and this is done in Section 9.2.2.2. Section 9.2.2.3 describes the general aspects of plan optimization in greater detail, as this procedure, perhaps more than any other, is at the heart of 3D planning. Clinical applications by disease site follow in Section 9.2.3.

It should be mentioned that 3D image-based treatment planning can be a resource-intensive process, particularly if multimodality images and numerous anatomical structures are involved. Hence, the continuing development of software tools for efficient image fusion and structure delineation is seen as critically important in facilitating broad acceptance of this approach. On the other hand, 3D image-based planning combined with new inverse optimization tools can offer significant time savings by eliminating steps that are no longer necessary or by replacing them with more efficient ones. Examples of those are automatic catheter reconstruction on CT, determination of active dwell positions, and anatomy-based dose distribution optimization tools. More details are presented in Section 9.2.2.2.

9.2.2.2 Process Workflow

9.2.2.2.1 Imaging Prior to Applicator Placement

The acquisition of images is usually performed before the placement of the applicators. Typically, these images are acquired a long time before the implant for diagnostic or staging purposes. The evolution of the tumor shape and volume makes these images unsuitable for planning purpose. Images are sometimes also acquired just prior to the implant to provide information and guidance for the insertion of the applicator and catheters. These images, however, are also of little use for dose planning itself. Effectively, the presence of the applicator, the insertion of the catheters, and the swelling that results from the procedure may substantially change the anatomy. For this reason and others, it is usually required to acquire a complete 3D image data set immediately after the insertion of the applicator prior to dose delivery.

9.2.2.2.2 Imaging with Applicators in Position

The acquisition of planning 3D images implies the use of applicators compatible with the imaging modality. Metallic applicators, with or without shielding, create strong artifacts on CT images, rendering them unusable for applicator reconstruction and target delineation. Fortunately, CT- and MRI-compatible applicators are now commercially available. The increase in demand should contribute to reduce the cost and improve the availability of the applicators.

The image acquisition is typically performed after the patient recovered from the OR procedure (for applicator insertion).

The challenge is to ensure the position stability of the implant between the imaging session and the dose delivery. Modern brachy-suites are equipped with imaging devices such as fluoroscopy and cone-beam CT that can be used to verify the implant integrity. Some brachy-suites are also shielded for HDR brachytherapy, allowing one to image and treat without moving the patient. More routinely, however, CT or MR images are acquired first, and the patient is brought to the treatment room for dose delivery after a dose plan was prepared. Extreme care must be taken to minimize the chance of applicator displacement. Interstitial implants are usually considered more stable (Foster et al. 2011).

9.2.2.2.3 Registration and Fusion of Multimodality Images

One of the main advantages of 3D planning is the possibility of combining information from different sources to improve the target definition. This is inherently difficult with 2D or film-based planning. The new generation of planning systems allows 3D planning, by definition based on a primary 3D image data set, usually CT or MR. By using the registration tools, information from other images can be made available in the coordinate system of the primary image set. PET/CT images can be registered to planning CT. Another example is the different MR image sequences acquired (T1, T2, spin echo, etc.) during the same imaging session (Kubicky et al. 2008). For HDR prostate, MR spectroscopy can be combined with CT-based planning to boost the dominant intraprostatic lesion (Pouliot et al. 2004). A magnetic resonance spectroscopic imaging (MRSI) to MRI/CT alignment protocol was developed to exploit the high specificity of combined MRI/MRSI for detecting and localizing prostate cancer within the prostate and to accurately transfer this information to the planning images (Reed et al. 2011). In this example, it was necessary to model the probe-induced prostatic distortion (see Figure 9.9). CT images from previous fractions are sometimes combined with the current CT image to determine the stability of the implant for the current fraction (Foster et al. 2011). New functionalities are also being developed to make available information on the planning image. Positive biopsy locations can be displayed on the ultrasound image used for permanent prostate implant. Gaudet et al. (2010) proposed to use this information to perform dose escalation to the dominant intraprostatic lesion (DIL) defined by sextant biopsy in a permanent prostate ^{125}I implant.

The tools found in planning systems are generally based on rigid-body registration algorithms, manual or automatic (mutual information, fiducials, etc.). It is rare that the anatomy of a patient does not change shape in time. Therefore, the precision of the alignment, as well as the residual position error of an anatomical feature after registration, depends on the point of alignment. One must be aware of the limitations of precision when using these tools. Deformable registration is a complicated task due to organ motion and the deformation of the tissue due to the insertion of the applicators or needles necessary for brachytherapy. In the present state, the total dose is evaluated by adding DVH parameters, taking into account the difference in biological effect due to dose rate when combining with external beam. In this approach, as the used DVH parameters are always the maximum doses to OARs, ignoring the deformation can only lead to a global overestimation of the doses (Pötter et al. 2006). To obtain a correct estimate of the total dose for multiple fractions or for brachytherapy + EBRT, it is necessary to use nonrigid image registration. Even for EBRT, those algorithms are still at the very early stage of clinical use, and validation is still difficult (Wang et al. 2005). For brachytherapy, the deformations involved are very often much larger than for EBRT with the addition/subtraction of material (needles, applicators) making the problem much more complicated. Indeed, a major assumption in deformable image registration is that, if two images are being registered, every point of one image corresponds appropriately to some point in the other. This is not the case when combining brachytherapy with external beam. Moreover, the most common brachytherapy applications are in tumors of the prostate and cervix, both situated in the pelvic region where the CT image contrast is low. This low contrast adds to the difficulty even in the case of interfraction registration. Solutions to overcome this problem could be the use of organ contours to guide the deformation (Christensen et al. 2001) or the use of MRI in place of CT (Yu et al. 2010). However, before the implementation of the deformation algorithms can reach its full potential, further work is needed to verify the accuracy of these algorithms.

9.2.2.2.4 Structure Definition: PTVs and OARs

Prior to the availability of 3D imaging, information on the dose delivered to the anatomy was limited to a few dose points. A good example is the report of the dose to points A in gynecology. If those points had the benefit of insuring a level of consistency in applicator placement, very little, if any, correlation was ever shown between complication rates and dose to point A.

Similarly to external-beam radiation therapy, the anatomic structures of interest are contoured and used to define PTVs and OARs. Here PTVs and CTVs are often considered the same in brachytherapy. A discussion to differentiate them is beyond the scope of this chapter. Clearly, the availability of 3D imaging improves tumor targeting and OAR avoidance and allows for a more precise reporting of dose delivered and dose-response curves.

9.2.2.2.5 Dose Prescription

The passage from 2D to 3D image-based planning also calls for an evolution on the methodology of prescribing dose. If dose to a point was the standard (for instance, see discussion on point A in Section 9.1.2.2), the ability to contour the anatomy requires a more complete specification of the dose distribution (Kim et al. 2004). The new generation of 3D planning systems can generate DVHs. As a result, the dose distribution of any contoured organ can be described by its DVH. In other words, the intended outcome becomes the input. To be practical, one can derive a few volumetric dose indices from the DVH. However, it is always desirable to evaluate the DVH over its complete dose range and use volumetric prescription parameters. For the moment, the

available forward planning approaches require that a plan be prepared, that is, that a dose distribution is obtained before the DVH could be evaluated. The process is therefore highly iterative where the DVH-extracted dosimetric indices are compared with intended values. The difference observed is then used to modify the plan, and a new evaluation is performed until the difference is clinically negligible. New inverse planning optimization tools will soon have the ability to directly use the dosimetric indices as input parameters to determine the dose distribution.

9.2.2.2.6 D-V Objective Specification

V_n is the *volume* of the region of interest (ROI) that receives n% of the prescribed dose. Generally, V is reported as the percentage of the ROI volume. For instance, V_{100} (%) represents the percentage of the ROI volume receiving at least 100% prescription dose. Similarly, D_m is the minimum *dose* that covers m% of the ROI volume. D_m can be expressed in percentage of the prescribed dose or simply in Gy. D_{90} (Gy) would therefore represent the minimum dose that covers 90% of the ROI volume.

The ABS has published general guidelines for different anatomical sites (Nag et al. 1999) that consist of a set of dose limit specifications. Other agencies have since proposed guidelines. Although they will be introduced in the site-specific sections later in the text, a few examples are given here.

For prostate cancer, Stock et al. (1998) have demonstrated that the dose delivered to 90% of the gland (D_{90}) had a critical impact on the subsequent risk of PSA diagnosed recurrence. D_{90} has therefore become an important prescription value for prostate implants. For HDR prostate brachytherapy, a multicenter study, Radiation Therapy Oncology Group (RTOG) 0321 protocol (Hsu et al. 2010), completed in 2007, was tasked with developing a quality assurance process for HDR prostate brachytherapy. The RTOG 0321 recommendations are as follows: (1) deliver the prescription dose to at least 90% of the primary target volume ($V_{Prostate100} > 90$%), (2) restrict the volume of bladder and rectum tissue receiving 75% of the prescription dose to less than 1 cc ($V_{Rectum75}$ and $V_{Bladder75} < 1$ cc), and (3) restrict the volume of urethra receiving 125% of the prescription dose to less than 1 cc ($V_{Urethra}125 < 1$ cc). It is interesting to note here that because HDR planning is generally CT-based, while PPI is ultrasound-based, more organs can be contoured while performing HDR, allowing more dosimetric indices to be specified.

9.2.2.2.7 Applicator Definition

Although ingenious approaches were developed over the years, applicator coordinate reconstruction from two or more 2D images was a delicate and time-consuming practice. The introduction of 3D imaging has greatly facilitated the process of obtaining the relevant coordinates of some areas of an applicator relative to the anatomy. CT-based planning allows for some automation in the anatomy delineation and catheter definition. For instance, automatic catheter reconstruction in prostate HDR can easily be achieved. The uncertainty in the localization of the longitudinal catheter-tip positions caused by the discrete slice thickness results in some dose uncertainty. The trade-off between the treatment planning time and the number of CT slices employed in HDR brachytherapy treatment planning is changing as planning systems become capable of handling an increasing number of slices. The passage of 2D to 3D planning and the visualization of applicators have been reviewed in the literature. More discussions on the precision and uncertainty of 3D imaging can be found in Chapter 13 of *Leibel and Phillips Textbook of Radiation Oncology* (Pouliot and Beaulieu 2010).

9.2.2.2.8 Plan Evaluation (Clinical)

Perhaps one of the most important benefits of the introduction of 3D imaging and planning is the ability to verify the spatial correspondence of the dose distribution with the anatomy. The user can look at the isodose on each image and verify the conformity and the avoidance of the dose distribution. For instance, the presence and location of hotspots, if any, can be assessed through a simple visualization. Quantitatively, dose distributions are usually evaluated using a limited number of dosimetric indices derived from the DVH. Modern planning systems now have the capability to update the DVH automatically after each dose modification and perform comparisons with expected values.

9.2.2.2.9 Plan QA

The delivery of a high dose calls for the verification of a specific dose plan performed by an independent physicist who was not involved in the generation of that plan. This step is essential, and is often mandatory, depending on the local regulations. Items to be verified include the following:

General
- Confirmation of patient name and identification number on plan printout
- Number of fractions for this plan, and in total
- Dose (cGy) for each fraction

Catheters
- Consistency of catheter numbering between digital plan and catheter map
- Number of catheters, size of the applicator, numbering scheme
- Verification of catheter geometry using 3D tools (check for catheter crossover)

Dose distribution
- Optimization method
- Prescription

Source activity
- Source strength checked against decay tables [air-kerma strength, or reference air-kerma rate (RAKR), or eventually in terms of the quantity activity GBq (or Ci)]

9.2.2.3 Dose Calculation, Source Loading, and Plan Optimization

9.2.2.3.1 Dose Calculation

The AAPM TG-43 dose calculation formalism (see Chapter 10) is considered the standard and is used in all commercial brachytherapy treatment planning systems. The TG-43 formalism was originally intended for LDR, low-energy seeds used in permanent seed implants, but it is now widely used as well for HDR and PDR brachytherapy. The formalism is based on a simple geometry, using a water sphere of 15 cm diameter with the radioactive source placed at its center. The characterization of new sources is generally performed by both measurements and Monte Carlo dose calculations. This systematic process has not yet been officially implemented for HDR and PDR sources at this time. As such, the source registry is currently limited to iodine and palladium sources. However, the practice of performing experimental measurements and Monte Carlo simulations of any type of brachytherapy source can almost be considered a standard procedure in the scientific community. New joint reports of AAPM Brachytherapy Subcommittee and ESTRO are forthcoming, both on the use of high-energy brachytherapy dosimetry (HEBD) sources and on model-based dose calculation algorithms (MBDCAs).

Beyond these recommendations, it is the responsibility of the medical physicist to review periodically the scientific literature for published consensus or new dosimetric parameters for the sources used in clinics and to ensure proper entry of the parameters in the planning system.

Further details are given in Chapter 10. Material considerations (other than water) are absent, and the dose calculation assumes a homogeneous material. This formalism provides a very good dose calculation approximation for almost all scenarios encountered in clinic.

However, some studies have shown that dose calculation engines, such as Monte Carlo or Boltzmann solver, can take heterogeneities into account and potentially improve treatment quality. A review of potential algorithms has recently been published (Rivard et al. 2009a). A new optimization framework that combines an inverse planning approach, namely, inverse planning with simulated annealing (IPSA), with the Monte Carlo method to perform inverse planning dose optimization for realistic geometries and tissue heterogeneities has recently been proposed (D'Amours et al. 2011).

9.2.2.3.2 Source Loading

Before the introduction of 3D imaging, it was important to plan implants that would be reproducible from one patient to another and that would minimize the risk of hot spots. This has led to the development of numerous systems described in Chapter 8 and more briefly earlier in this chapter. As it was discussed previously in this chapter, 3D imaging allows the visualization of the catheter location relative to the anatomy, the dose calculation everywhere in the patient, the evaluation of the locations and amplitudes of hot spots inherent to brachytherapy, and the computation of dosimetric indices. Those considerably limit the need for a rigid system and open new degrees of freedom to allow the physician to customize the dose distribution to the specific anatomy of each patient. In other words, although the fundamentals of the old systems remain valid, their application in the face of the new information available is superseded.

9.2.2.3.3 Plan Optimization

In permanent seed implants, a large number of seeds are used. The proper number and their best positions in the treatment volume are the results of the planning process. In HDR brachytherapy, the dose is controlled by altering the source dwell times—the time spent at points along the already implanted catheters. Plan optimization is often referred to as a manual modification of the dose distribution after a system was used. However, plan optimization nowadays refers more to the use of a computer algorithm to change or optimize parameters. As such, forward planning optimization falls in the category of modifying a dose distribution, whereas inverse planning will use computer algorithms to maximize an objective function that guides the dose distribution.

9.2.2.3.4 Forward Planning

Many tools are available in current planning systems that greatly facilitate the modification of the dose distribution. A seed location is rapidly determined and can be moved, eliminated, or dwell time-modified if the local dose is inadequate. A motion of the mouse on the computer screen can modify a given isodose. In that case, a mathematical model is used to revise the dwell times in such a way that it will produce a change in isodose in the direction of the mouse. Caution must be exercised as it is difficult for the user to correctly assess the changes in dwell times resulting from dragging an isodose. This is amplified by the fact that the user follows the isodose on a 2D plane, while the dose distribution is affected in the whole volume.

9.2.2.3.5 Inverse Planning

The inverse planning approach can be described as a method of radiation treatment planning in which one starts with the desired dose distribution, or clinical objectives, and then determines the treatment parameters that will achieve it. This is opposed to the conventional forward planning approach in which the treatment parameters are first chosen, and then the resulting dose distribution is calculated and evaluated. Because inverse treatment planning begins with the description of the desired dose distribution, it represents a change of paradigm in the planning process. CT or MRI contours of the CTVs and OARs are used not only to define the anatomy for visual assessment and DVH calculation but also to guide the dose distribution optimization process.

IPSA is an example of inverse planning optimization systems. It is based on simulated annealing and is now widely available (Lessard and Pouliot 2001). The IPSA optimization tool has been designed to produce plans for any type of brachytherapy automatically and in a few seconds. These treatment plans are specially optimized for each single patient because the whole routine is controlled by the patient's specific anatomy. IPSA determines the active dwell positions and their optimal dwell times based on the digitized catheters, the contoured anatomy, and the dose objective parameters. More details and discussion of other clinically available inverse planning optimization tools are available in Chapter 12.

9.3 Applications

9.3.1 GYN

Moving away from the still used point-based 2D brachytherapy for the definitive treatment of cervical cancer, volume-based CT- and MRI-guided 3D brachytherapy is considered the most promising approach and is therefore increasingly being used in many centers. The results reported so far are very promising, although mainly obtained from single institution registration studies. The major advantage of this technique, with dwell time optimization and optionally using interstitial needles in combination with the tandem-ring or tandem-ovoids applicator sets, is the possibility of conforming the dose given by brachytherapy with regard to both volume (3D) and time (4D).

For gynecological brachytherapy, both European (GEC-ESTRO) and American (ABS) recommendations have been published regarding the use of imaging. They both acknowledge that the visualization of the tumor on CT is very difficult, which makes MRI necessary. In their view, ideally each brachytherapy implant should be followed by MRI imaging with the applicator in situ and a new dose plan. The GEC-ESTRO recommendations (Haie-Meder et al. 2005; Pötter et al. 2005, 2006) have found a strong basis in several professional societies and provide the necessary guidance for users to adapt their treatment to imaging tools that gradually become more easily accessible to them. These recommendations and the volume specifications are discussed in more detail in Chapter 21 on brachytherapy for gynecological cancers and Chapter 24 on the transition from ICRU-38 to image-guided brachytherapy. For example, the gross tumor volume (GTV) and CTV definitions of the GEC-ESTRO recommendations are shown in Figures 24.1 and 24.2.

While CT imaging-based treatment planning represents a substantial step forward relative to orthogonal film-based treatment planning, it still has limitations. For example, Viswanathan et al. (2007) noted that CT imaging overestimated tumor volumes resulting in increased dose delivered to adjacent normal tissue. From such experiences, there is common agreement in a growing number of reports on the superiority of MR as compared to CT for determining both the initial extent of cervical tumor (Pötter et al. 2006, 2008a; Viswanathan et al. 2007; Dimopoulos et al. 2009; Nag et al. 2004; Haie-Meder et al. 2005; Mayr et al. 2010) and response to treatment as seen at the time of brachytherapy and beyond (Pötter et al. 2008a; Dimopoulos et al. 2009; Haie-Meder et al. 2005; Mayr et al. 2010). MRI is capable of assessing tumor size with an accuracy of ±0.5 cm and is also capable of assessing parametrial extension correctly 77%–96% of the time (Pötter et al. 2008a). In this regard, MRI is clearly superior to CT or other clinical or radiographic processes. Assessment of nodal metastases remains poor with either CT or MRI (Pötter et al. 2008a).

It was hypothesized that better imaging would lead to more accurate brachytherapy dosimetry and ultimately better treatment outcomes as a result of improved local control (resulting from dose escalation and/or improved dose distribution) and/or reduced toxicity (from lower maximal dose to OAR and/or smaller volumes of OAR receiving large doses). A report on 145 patients with locally advanced cervical cancer suggests that dose escalation would be possible without exceeding normal tissue tolerance (Pötter et al. 2007). In 2011, the same Vienna group reported on 165 patients treated with 3D conformal radiotherapy ± chemotherapy plus image (MRI) guided adaptive intracavitary brachytherapy including needle insertion in advanced disease with local control rates of 95%–100% at 3 years in limited/favorable (IB/IIB) and 85%–90% in large/poor response (IIB/III/IV) cervix cancer patients associated with a moderate rate of treatment-related morbidity (Pötter et al. 2011). Compared with the historical Vienna series, there is relative reduction in pelvic recurrence by 65%–70% and reduction in major morbidity. The local control improvement seems to have impact on cancer-specific survival and overall survival.

As the OARs (bladder, sigmoid, and bowel) can be seen on CT, CT can replace MRI if more than one fraction is applied. On the MRI images, the target, composed of the GTV, the whole cervix plus suspected residual extra cervical disease, can be delineated at the time of brachytherapy, which can be different from the target prior to EBRT and chemotherapy. This allows taking into account tumor regression. The review of the plan is based on very clear rules for the DVH parameters (Haie-Meder et al. 2005; Pötter et al. 2006). Based on the experience collected so far, the image-based brachytherapy approach significantly improves the DVH parameters, and the improved dose delivered seems to have a major impact on the clinical outcome with a concomitant decrease in the rates of both local failure and morbidity.

In order to move forward and collect conclusive clinical data, results of multicenter studies are required. EMBRACE (http://www.embracestudy.dk) is a prospective observational study on MRI-guided brachytherapy in locally advanced cervical cancer, which is now running by introducing MRI-based brachytherapy in a multicenter setting with the aim of correlating image-based DVH parameters for the CTV and for OAR with outcome. Based on these results, the investigators hope to develop prognostic and predictive statistical models for clinical outcomes including volumetric, dosimetric, clinical, and biological risk factors as well as radiobiological parameter estimates that will allow a precise risk assessment in individual patients and aid in further development of new treatment protocols. The interested reader is referred to the Web site and the underlying literature.

9.3.2 Prostate

9.3.2.1 LDR Permanent Seed Implants

Permanent implantation of ^{125}I, ^{103}Pd, or ^{131}Cs seeds is used to treat low-risk and low-tier intermediate-risk prostate cancer and can also be utilized as a boost in combination with external beam therapy to treat higher-risk disease. Treatment planning is usually based on an axial set of 5-mm-thick TRUS images, onto which are drawn a CTV consisting of the whole prostate plus margin and (optionally) intraprostatic urethra and proximal rectum contours. The TRUS images can be obtained several weeks prior to implant or in the operating room immediately beforehand. A preplanned seed distribution usually follows one of four manually generated patterns (uniform loading, peripheral loading, modified uniform loading, and modified peripheral loading), with 75% of brachytherapists using a modified peripheral loading approach (Merrick et al. 2005). When created intraoperatively, the seed distribution is normally obtained using inverse planning software to save time (Rubens et al. 2006) (see Figure 9.6). In a very few centers, interventional MRI is used instead of TRUS to guide needle placement and perform intraoperative planning based on observed needle locations (Rubens et al. 2006).

The GEC-ESTRO recommendations (Ash et al. 2000; Salembier et al. 2007) and the AAPM TG-137 report (Nath et al. 2009) provide guidance on dose prescription, recording, and reporting for permanent seed implants, including recommended treatment margins and D-V constraints. By themselves, the latter are insufficient to uniquely determine the seed distribution. This is well illustrated in a multi-institutional planning study conducted by Merrick et al. (2005), who reported that eight brachytherapists planning treatment for the same patient with ^{125}I and ^{103}Pd seeds used a wide range of seed strengths and arrangements. Although all of these plans met the TG-137 D-V constraints excluding CTV $V_{150} < 50\%$, dose uniformity within and conformity to the PTV varied widely. Such diversity in the seed distributions used clinically contrasts sharply with the standardized source arrangements promoted by classical implant planning systems and complicates comparison of clinical outcomes between institutions. Therefore, treatment planning practice for permanent implants would benefit from the introduction of additional, broadly applicable, evidence-based constraints.

TG-137 recommends CT images be used for dosimetric evaluation of the implant as the standard of care and identifies fused CT and MR images acquired on the same day as ideal (Nath et al. 2009). Figure 9.7 shows an example of the latter utilizing T2-weighted MRI, which has been found to provide good soft tissue contrast (McLaughlin et al. 2004). As is readily appreciated, seeds are best visualized on CT and anatomical structures on MRI.

9.3.2.2 HDR Temporary Implants

Temporary implants using HDR ^{192}Ir sources are used to treat localized prostate cancer at all levels of disease risk, either alone or in combination with external beam therapy. HDR implants are particularly suitable for the treatment of advanced disease, as catheters (rigid needles) can be placed in periprostatic tissue and source dwell times adjusted to more accurately irradiate a larger volume of tissue than could be irradiated with a permanent seed implant, while sparing tissues at risk. The basic technique for advanced disease (Vicini et al. 2003) involves (1) an initial HDR boost to the gland in combination with 4–5 weeks of external beam therapy delivering 40–50 Gy to the prostate and periprostatic tissues; (2) dose escalation via a further HDR boost; and (3) often, neoadjuvant or adjuvant androgen deprivation therapy. Boost doses range from 16 to 30 Gy and fraction sizes from 3 to 15 Gy, delivered before, during, or after external beam therapy

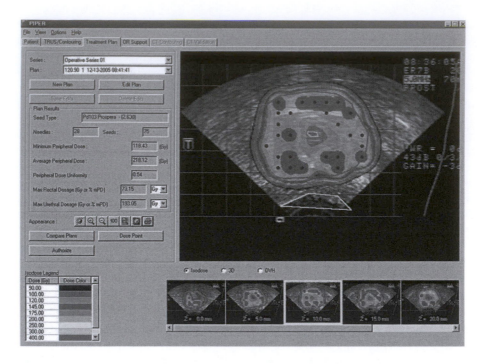

FIGURE 9.6 (See color insert.) Axial ultrasound image of a prostate at midgland with ¹⁰³Pd seed dosimetry superimposed. Needle positions are indicated by blue dots; needles containing seeds at this level by red dots. The prostate margin as originally traced on the intraoperative plan is outlined in pink, the urethra in blue, and the rectum in yellow. The remaining colored regions represent the isodose delivered to each area of the prostate and to the surrounding tissues as indicated in the color code table at the lower left of the figure. The minimum peripheral dose to the entire prostate is 118 Gy. (From Rubens, D.J. et al., *Radiol Clin N Am* 44, 735–48, 2006. With permission.)

in one or two implants done 1–2 weeks apart. As for gynecologic brachytherapy, dosimetry and outcomes for different boost schedules can be compared by calculating biologically effective dose using the linear–quadratic model of tissue response to irradiation (Yamada et al. 2006). Technique and outcome details for nine centers with extensive experience in conformal HDR prostate brachytherapy can be found in the work of Vicini et al. (2003). General guidelines for dose specification, optimization, and reporting of HDR prostate brachytherapy have been published by the ABS (Rodriguez et al. 2001) and GEC-ESTRO (Kovács et al. 2005). The toxicity results of the first prospective, multi-institutional trial of computed tomography-based HDR brachytherapy and EBRT (RTOG-0321) have been recently published (Hsu et al. 2010).

TRUS imaging and a perineal template are commonly used to plan and guide needle placement. Treatment planning can be done postoperatively using TRUS or CT images, but is sometimes done intraoperatively using live TRUS images. As well, it has been demonstrated that needle placement and intraoperative treatment planning can be accomplished using a standard (closed bore) 1.5-T MRI scanner and an endorectal coil (Susil et al. 2004) (see Figure 9.8). Advantages of this novel approach that may benefit future clinical practice include better visualization of patient anatomy and ease of fusion of any spectroscopic or other image data obtained concurrently with the endorectal coil. Disadvantages include a relatively long scanner access time (~2 h) and increased cost.

A key clinical observation is that locally advanced prostate cancer patients treated with radiotherapy do significantly better as the dose to the gland is escalated. For HDR brachytherapy, high gland doses and excellent outcomes are achievable through a combination of 3D computer-optimized treatment planning and a stable implant delivery system, as demonstrated by Yamada et al. (2006). In that work, the CTV was delineated from CT images as whole prostate plus extensions to areas of concern for extracapsular disease, and PTV was defined as CTV plus margin, typically 5 mm. Urethra (identified by a Foley catheter) and rectum were also contoured and implanted catheters identified, providing an exact 3D model of the implant in relation to pelvic structures of interest. Source loading was determined using a genetic algorithm that was given dose constraints specifying desired PTV dose coverage, OAR dose limits, and a dose conformity index, and that gave dwell times for all dwell positions in the implant. This type of inverse planning approach has generally been found to provide superior PTV dose coverage along with adequate OAR sparing.

The advent of functional imaging modalities such as positron emission tomography (PET; discussed in Section 9.4) or combined MRI/MRSI provides new tools for better cancer-validated tumor targeting in radiation therapy. An HDR brachytherapy clinical protocol currently under way uses a combination of MRI and MRSI to define DILs. The inverse planning algorithm is then used to increase the dose to the prostatic lesions defined by MRI/MRSI while providing the usual dose coverage of the

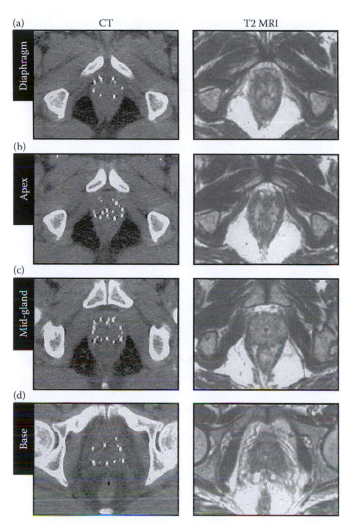

FIGURE 9.7 Comparison of T2-weighted MRI registered to CT for permanent implant evaluation. Prostate region: (a) GU diaphragm; (b) apex; (c) midgland; (d) base. (From McLaughlin, P.W. et al., *Brachytherapy* 3, 61–70, 2004. With permission.)

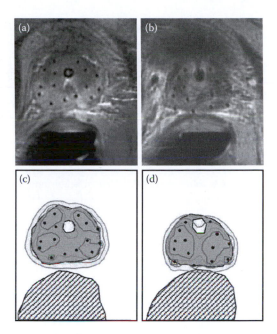

FIGURE 9.8 **(See color insert.)** MRI-guided HDR prostate brachytherapy catheter placement and isodose maps. Brachytherapy was performed at the beginning (a) and end (b) of a 5-week course of external beam radiation therapy. Needle locations are visible as dark dots inside the prostate in these axial FSE images. (c, d) Radiation isodose maps, corresponding to (a) and (b), indicate 150% (red), 125% (orange), 100% (green), and 75% (blue) of the prescribed radiation dose (1050 cGy). The prostate (gray-filled region), urethra (white region inside the prostate), and rectum (hatched region) are also shown. Note that the green, 100% dose contour conforms well to the prostate margin, while overdose of the urethra and rectum is avoided. (From Susil, R.C. et al., *Magn Reson Med* 52, 683–7, 2004. With permission.)

prostate and the protection to the urethra, rectum, penile bulb, and bladder (Pouliot et al. 2004; Kim et al. 2008) (see Figure 9.9).

9.3.3 Breast

Breast cancer patients often receive radiotherapy after lumpectomy. This most frequently consists of external photon beam treatment of the whole breast followed by a boost with a narrow field electron beam. It is known, however, that the electron dose distributions may significantly miss the target (Fraser et al. 2010). To alleviate this, radiotherapy may also be performed in the form of brachytherapy, such as in ^{103}Pd LDR seed implants, electronic brachytherapy, or with an ^{192}Ir source in a MammoSite applicator.

A recent particular image-guided breast brachytherapy system is the AccuBoost (Nucletron, Veenendaal, The Netherlands). Its concept and dosimetry have been described in the literature (Rivard et al. 2009b; Yang and Rivard 2009). The technology integrates mammographic breast imaging with HDR brachytherapy with a standard ^{192}Ir afterloader (Figure 9.10a). The compressed breast is imaged with the mammography system to locate the target. With the breast still under compression, round or D-shaped applicators are then positioned on the skin, in a parallel-opposed fashion. The afterloader is then used to step the HDR source through channels in the applicator to deliver noninvasive brachytherapy. The applicator is made of tungsten and serves to collimate the photon beam. A range of different sizes of applicators is available. Dose delivery can be done with both applicators in a horizontal position (Figure 9.10b) resulting in a craniocaudal irradiation or in a vertical position with a mediolateral exposure. This way, a fairly uniform dose may be delivered under image guidance, while keeping the dose to the lungs, heart, and contralateral breast to a minimum. Due to the collimator, the dose distributions are not typical brachytherapy isodoses but resemble more external photon beam doses. Dose distributions may be obtained from functions fitted to Monte Carlo simulations (Rivard et al. 2009b).

Das et al. (2004) gave a detailed account of their center's forward planning process for HDR APBI, which utilizes Paris dosimetry, geometrical optimization, manual optimization,

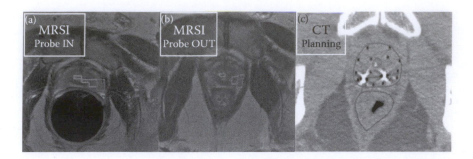

FIGURE 9.9 **(See color insert.)** (a) Original acquisition image with the suspicious regions delineated. (b) Axial probe-out volume with the rotated, translated, and warped spectral delineations. (c) MRSI scores of 4 and 5 are delineated on the planning CT. (From Reed, G. et al., *J Contemp Brachytherapy* 1, 26–31, 2011. With permission.)

and a dose homogeneity index. In contrast, the more innovative approach of using anatomical D-V parameters directly to guide the planning process, which is standard practice for inverse-planned prostate implants, is still at the developmental stage for APBI (Berger et al. 2008).

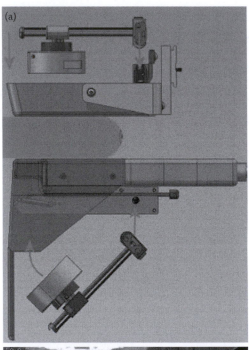

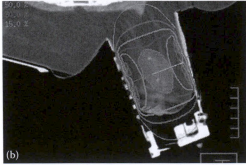

FIGURE 9.10 **(See color insert.)** Integrated breast HDR brachytherapy/mammography unit, AccuBoost (a). Dose distribution in the compressed breast (b). (Courtesy of Nucletron, Veenendaal.)

Remotely afterloaded interstitial breast implants can be planned from CT images in several different ways. For 50 consecutively treated patients, Das et al. (2004) employed a combination of geometric and manual optimization methods to deliver 3.4 Gy in 10 HDR fractions or 4 Gy in 8 fractions to Paris system basal dose points, while achieving a median target volume coverage of 96%. For each plan, they found that source strength multiplied by treatment time for every unit of prescribed dose agreed within ±7% with the Manchester system volume implant table corrected for modern units. For 17 patients, Major et al. (2005) conducted a planning study comparing average target volume coverage and dose homogeneity for the Paris system and a CT-based HDR system, among others. For the latter, dose was optimized to a set of points placed on the surface of the PTV. As expected, target coverage improved from 61% to 87% when 3D structure-based planning was used; however, the dose homogeneity index fell from 0.78 to 0.37. The authors observed that preimplant volumetric imaging would be beneficial in devising catheter arrangements that better conform to the treatment volume than the traditional geometrical arrangements required by the Paris system. Finally, for 25 patients treated with PDR to 50.4 Gy (0.8 Gy per 1-h pulse), Berger et al. (2008) conducted a retrospective planning study in an effort to identify D-V parameters that could be used to reliably and reproducibly spare skin, lung, and ribs. Dose was prescribed according to the Paris system, and both geometrical and manual optimization of dwell weights were applied in order to obtain good CTV coverage while protecting OARs. For skin, a virtual bolus was drawn to minimize the influence of an assumed skin thickness. The authors succeeded in identifying several dose parameters for absolute OAR volumes that they believe are good candidates for ongoing clinical study.

9.3.4 Other Treatment Sites

As with cervix, breast, and prostate, 3D planning for other treatment sites relies primarily on CT scanning, with MRI currently utilized in only a few centers. For many of these sites, an incremental approach to making the transition from 2D to 3D image-based planning is described in the literature.

The flexibility afforded by 3D structure-based planning with CT and MRI enables classical treatment techniques to be improved and new techniques to be developed. Almost two decades ago, Hayes et al. (1992) reported that CT planning permits more objective definition of head-and-neck implant parameters important in multi-institutional clinical trials analysis. More recently, Liebmann et al. (2007) described a novel, CT-based mold technique for uniform HDR irradiation of the whole scalp, which was successfully used to treat chronic lymphatic leukemia infiltrates of the skin. In the foreseeable future, we expect CT to continue as the clinical standard for 3D image-based planning and MRI to increasingly be integrated into clinical practice, particularly for cervix brachytherapy (Pötter et al. 2008a).

9.4 Potential for PET Imaging

9.4.1 Overview of PET Imaging for Cancer Radiotherapy

PET is a molecular imaging technique capable of creating a "visual representation, characterization and quantification of biological processes at the cellular and subcellular level" (Massoud and Gambhir 2003). The application of PET imaging in oncology requires a fundamental understanding of biochemistry and molecular biology and aims to image particular traits of malignant cells in vivo such as the activation or inhibition of tumor development pathways. At present, several molecular probes (or markers) are available to image a range of tumor expression parameters including glucose and fatty acid metabolism, proliferation, hypoxia, angiogenesis, and apoptosis (Schöder and Ong 2008). Table 9.1 lists some of these markers and their clinical applications (Guha et al. 2008).

The steps involved in PET imaging are the following (Kapoor et al. 2004): selection and production of a suitable positron-emitting

TABLE 9.1 Tumor Expression Parameters Suitable for PET Imaging

Tumor Expression Parameter	PET Marker	Clinical Application
Glucose metabolism	^{18}F-FDG	General tumor imaging
Proteins/amino acids	^{11}C-methionine	Brain tumors
	^{11}C-choline	Prostate tumors
	^{18}F-DOPA	Carcinoid
	^{18}F-methyltyrosine	Musculoskeletal tumors
Proliferation (DNA)	^{18}F-thymidine	Radiation response
Apoptosis	^{18}F-annexin V	Treatment response
Hypoxia	^{18}F-misonidazole	Radiation planning
Receptor binding (avidity)	^{18}F-estradiole	Breast cancer imaging
Angiogenesis/blood flow/perfusion	^{18}F-galacto RDG	Integrin $\alpha v \beta 3$ binding
Membrane/lipid synthesis	^{18}F/^{11}C-acetate	Proliferation
Bone turnover	^{18}F	Skeletal disease

Source: Adapted from Guha, C. et al. *Semin Nucl Med* 38, 105–13, 2008. With permission.

radionuclide; combination of the radionuclide with a biologically active molecule to create a desired marker; administration of the marker to the patient (most often by venous injection); and determination of the 3D radionuclide distribution in the patient at an appropriate time after marker administration. The latter step is accomplished using a commercial PET or PET/CT scanner, which detects pairs of 511-keV coincidence photons emitted nearly back to back during positron annihilations subsequent to positron emissions from radionuclide decays occurring in the patient (see Figure 9.11). An image reconstruction algorithm groups recorded coincidence events into projection images called sinograms, which are analogous to CT scanner projections, and from them creates a 3D image of the radionuclide distribution. In so doing, a snapshot of the underlying functional process that created the radionuclide distribution is indirectly mapped. A "physiological map" of functional process dynamics can also be created by repeated imaging at different time points after marker administration; however, this procedure is resource-intensive and logistically challenging.

PET images must be interpreted carefully as their fidelity in depicting the spatial extent and intensity of molecular processes is limited by the physics of image formation (Kapoor et al. 2004). First, the statistical quality of PET projection data is relatively poor. A typical PET acquisition consists of millions of counts, whereas a CT acquisition can reach a few billion counts. Second, a point-like positron-emitting source is not imaged as such, but rather as a distributed source with reduced intensity because of blurring caused by positron travel prior to annihilation, variability in the angle between the annihilation photons, photon scattering in the patient and scanner, photon penetration of multiple detectors, and so-called random events. Finally, to enable quantitation of the radionuclide distribution, PET images must be corrected for the effects of photon attenuation in the patient. This correction is approximate and occasionally introduces image artifacts. Generally speaking, PET images have a

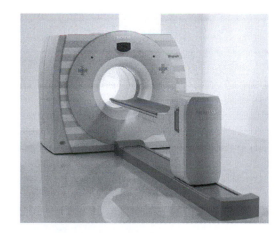

FIGURE 9.11 Contemporary PET-CT scanner in which the PET and CT detectors are mounted on a single gantry. An alternative design places separate PET and CT gantries in close proximity with each other. (From http://www.Siemens.Medical.com, accessed November 1, 2010. With permission.)

spatial resolution of no better than ~4 mm and display very little anatomic detail. For these reasons, PET-CT scanners that provide coregistered functional and anatomic images are used in radiotherapy applications in order to facilitate differentiation of normal from abnormal radionuclide uptake.

PET-CT is fast becoming an accepted clinical tool in radiotherapy. To prove valuable over the long term, PET-CT must continue to establish its usefulness for staging cancers, assessing the extent of disease, delineating target volumes for radiotherapy planning, and assessing and predicting treatment response including the capacity to differentiate between treatment sequelae and recurrent disease (Schöder and Ong 2008). Its future role in brachytherapy will ultimately depend on how well it performs in each of these key areas.

9.4.2 Incorporating PET-CT Imaging into Brachytherapy Treatment Planning

Traditional radiotherapy PTV design based on anatomical structures and probabilistic models of cancer spread has limitations and can lead to overtreatment of normal tissues and undertreatment of disease. The concept of a 3D or 4D biological target volume (BTV) that also considers functional processes that affect the radiation response, such as cancer metabolism and proliferation, has therefore been proposed as an improvement (Schöder and Ong 2008). It is estimated that approximately 60% of patients who receive functional imaging have changes made to their PTV or to planned dose distribution parameters (Guha et al. 2008).

PET-CT imaging is capable of providing the information required for BTV definition. Once defined, a clinical BTV can be incorporated into the treatment planning process in much the same way as a PTV or used to design simultaneously

integrated boost treatment (see Section 9.4.2.2). On the other hand, PET-CT–based BTV delineation is subject to uncertainties associated with scanner variability, image blurring, and limited sensitivity and specificity of the molecular marker. Image interpretation introduces additional uncertainties arising from a lack of image standardization (realized for CT through the use of Hounsfield units) and from the difficulty in choosing appropriate signal thresholds to identify regions of abnormal uptake (Chiti et al. 2010). As a result, BTV definition using PET-CT and target volume comparison with alternative imaging modalities such as MRI, functional MRI, and functional CT continue to be active areas of investigation. Widespread adoption of PET-CT in radiotherapy has been limited by a paucity of data unequivocally demonstrating clinical benefit and by the relatively high cost of PET radionuclide production and transportation.

In brachytherapy, PET-CT–based planning could prove beneficial for a number of sites including breast, cervix, head and neck, lung, and prostate. We illustrate this by briefly describing early applications to planning cervix and prostate treatments.

9.4.2.1 Adaptive Treatment Planning for Cervical Cancer

Grigsby and colleagues reported the use of ^{18}F FDG PET for BTV definition in a theoretical study of adaptive treatment planning for 11 cervix cancer patients who underwent a total of 31 intracavitary treatments (Lin et al. 2007). Their overall clinical experience with PET-CT has been summarized in a recent review article (Grigsby 2009). The author cites three advantages of FDG PET over MRI for this site, namely, the absence of need for special applicators, improved sensitivity for detection of advanced stage disease, and sustained sensitivity to the presence of disease as brachytherapy progresses.

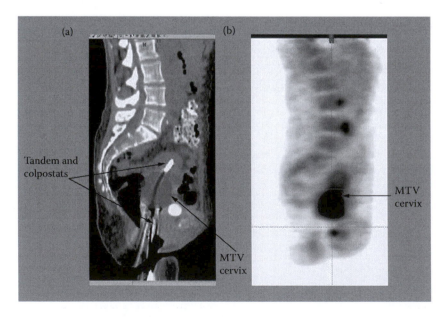

FIGURE 9.12 Sagittal images near midline of a tandem and colpostats insertion imaged with FDG PET-CT. (a) CT showing pelvic anatomy and the applicator; (b) PET image showing the metabolic (biological) tumor volume (MTV) and other regions of increased FDG uptake. (From Grigsby, P.W., *Future Oncol* 5, 953–8, 2009. With permission.)

A pretreatment FDG PET scan was transferred to the treatment planning system, and the cervical tumor volume was contoured using an image intensity threshold of 40% of the peak value. Subsequent scans were acquired and contours drawn in conjunction with the first, middle, and final brachytherapy fractions for HDR or with each of two insertions for LDR. For all scans, intravenous hydration and diuretic agents were administered to minimize FDG accumulation in the urinary tract and to enhance visualization of disease (Figure 9.12). No margin for microscopic disease was added to the BTV, as it was presumed to include all such disease. Treatment plans were done for a standard source distribution delivering 6.5 Gy to point A at time of initial treatment, and for an optimized source distribution covering at least 80% of the BTV with 6.5 Gy while keeping bladder $D_{2cc} < 7.5$ Gy and rectum $D_{2cc} < 5$ Gy, and then compared.

The mean tumor size as determined by FDG PET was found to be 56 cm^3 (range 7–137 cm^3) for the first implant, and 17 cm^3 (range 2–38 cm^3) for the mid and last implants. BTV dose coverage with and without optimization was 73% and 68% ($p = 0.21$) for the first implant, and 83% and 70% ($p = 0.02$) for the mid and last implants, respectively. This work illustrates the potential of FDG PET-CT–based adaptive brachytherapy to treat cervical cancer.

9.4.2.2 Localized Boost Therapy Planning for Prostate Cancer

Unlike cervix cancers, prostate cancers cannot generally be visualized using ^{18}F FDG PET because of their relatively low glycolytic activity and the predominantly urinary excretion of FDG leading to high bladder activity that can mask prostate uptake. Instead, prostate avid markers with low bladder excretion such as ^{11}C-acetate, ^{11}C-choline, and anti-1-amino-3-18F-fluorocyclobutane-1-carboxylic acid have demonstrated potential to localize prostate tumors (Turkbey et al. 2009).

Seppälä et al. (2009) demonstrated the feasibility of ^{11}C-acetate PET-CT to delineate intraprostatic lesions (IPLs) for the purpose of BTV boost planning in EBRT. Twelve men with intracapsular prostate carcinoma were imaged. A BTV was delineated on the PET image using a standard uptake value threshold of 2.0 (Figure 9.13), and a PTV consisting of the whole prostate gland plus 6-mm margin was delineated on CT. A series of treatment plans designed to deliver 72 Gy to the PTV and simultaneously integrated boost (SIB) doses of 72, 77.9, 81, 84, 87, and 90 Gy to the BTV were generated. Tumor control probabilities (TCPs) were calculated using a modified Zaider–Minerbo model, and normal tissue complication probabilities (NTCPs) for bladder and rectum were calculated using a Lyman–Kutcher–Burman model. The authors found for all patients that TCP increased with SIB without increasing NTCP for bladder or rectum. The probability of uncomplicated control increased on average by 28% with SIB doses above 72 Gy.

Niyazi et al. (2010) developed a simple TCP model based on the Poisson distribution to assess potential effects of dose escalation in prostate subvolumes identified with ^{11}C-choline PET-CT. Their focus was on determining the influence of relevant

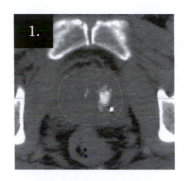

FIGURE 9.13 Distribution of ^{11}C-acetate in a cancer patient's prostate (thin contour) in an axial plane near midgland, overlaid on a concurrently acquired CT slice. A standard uptake value of 2.0 was used for BTV delineation (thicker contour). (From Seppälä, J. et al., *Radiother Oncol* 93, 234–40, 2009. With permission.)

variables on TCP: sensitivity of ^{11}C-choline PET, linear–quadratic model parameters prostate α and α/β, slope of cell killing curve γ50, whole prostate dose, SIB dose, and dose per fraction. They observed that dose escalation to SIBs can potentially increase TCP, but that no firm conclusions can be made because of the lack of exact values for PET sensitivity and prostate α/β values.

Although the above studies were conducted for SIBs delivered using intensity-modulated radiotherapy (IMRT), their findings are also relevant for SI brachytherapy boosts. The treatment delivery challenges are quite different, however: for IMRT, patient setup error and prostate motion require attention; for brachytherapy, source placement error must be minimized.

References

Acher, P., Puttagunta, S., Rhode, K. et al. 2010. An analysis of intraoperative versus post-operative dosimetry with CT, CT-MRI fusion and XMR for the evaluation of permanent prostate brachytherapy implants. *Radiother Oncol* 96:166–71.

Ash, D., Flynn, A., Battermann, J., de Reijke, T., Lavagnini, P., and Blank, L. 2000. ESTRO/EAU/EORTC recommendations on permanent seed implantation for localized prostate cancer. *Radiother Oncol* 57:315–21.

Astrahan, M. 2005. Improved treatment planning for COMS eye plaques. *Int J Radiat Oncol Biol Phys* 61:1227–42.

Aubry, J.F., Beaulieu, L., Hsu, I.-C., and Pouliot, J. 2010. Investigation of geometric distortions on MR and CBCT images used for planning and verification of HDR brachytherapy cervical cancer treatment. *Brachytherapy* 9:266–73.

Bazalova, M., Carrier, J.F., Beaulieu, L., and Verhaegen, F. 2008. Dual-energy CT-based material extraction for tissue segmentation in Monte Carlo dose calculations. *Phys Med Biol* 53:2439–56.

Berger, D., Kauer-Dorner, D., Seitz, W., Pötter, R., and Kirisits, C. 2008. Concepts for critical organ dosimetry in three-dimensional image-based breast brachytherapy. *Brachytherapy* 7:320–6.

Chiti, A., Kirienko, M., and Grégoire, V. 2010. Clinical use of PET-CT data for radiotherapy planning: What are we looking for? *Radiother Oncol* 96:277–9.

Christensen, G.E., Carlson, B., Chao, K.S. et al. 2001. Image-based dose planning of intracavitary brachytherapy: Registration of serial-imaging studies using deformable anatomic templates. *Int J Radiat Oncol Biol Phys* 51:227–43.

D'Amours, M., Pouliot, J., Dagnault, A., Verhaegen, F., and Beaulieu, L. 2011. Patient-specific Monte Carlo based dose-kernel approach for inverse planning in HDR brachytherapy. *Int J Radiat Oncol Biol Phys* 85:1582–9. On-line doi:10.1016/j.ijrobp.2010.09.029.

Das, R.K., Patel, R., Shah, H., Odau, H., and Kuske, R.R. 2004. 3D CT-based high-dose-rate breast brachytherapy implants: Treatment planning and quality assurance. *Int J Radiat Oncol Biol Phys* 59:1224–8.

Dimopoulos, J.C., Schirl, G., Baldinger, A., Helbich, T.H., Potter, R. 2009. MRI assessment of cervical cancer for adaptive radiotherapy. *Strahlenther Onkol* 185:282–7.

Edmundson, G.K. 1990. Geometry based optimization for stepping source implants. 1990. In: *Brachytherapy HDR and LDR.* A.A. Martinez, C.G. Orton, and R.F. Mould (eds.). Nucletron Corp., Columbia.

Elkon, D., Kim, J.A., and Constable, W.C. 1981. CT scanning and interstitial therapy. *J Comput Tomogr* 5:268–72.

Foster, W., Cunha, A., Hsu, I.-C., Weinberg, V., Krishnamurthy, D., and Pouliot, J. 2011. Dosimetric impact of inter-fraction catheter movement in HDR prostate brachytherapy. *Int J Radiat Oncol Biol Phys* 80(1):85–90.

Fraass, B.A., McShan, D.L., Diaz, R.F. et al. 1987. Integration of magnetic resonance imaging into radiation therapy treatment planning: I. Technical considerations. *Int J Radiat Oncol Biol Phys* 13:1897–908.

Fraser, D.J., Wong, P., Sultanem, K., and Verhaegen, F. 2010. Dosimetric evolution of the breast electron boost target using 3D ultrasound imaging. *Radiother Oncol* 96:185–91.

Gaudet, M.,Vigneault, E., Aubin, S. et al. 2010. Dose escalation to the dominant intraprostatic lesion defined by sextant biopsy in a permanent prostate I-125 implant: A prospective comparative toxicity analysis. *Int J Radiat Oncol Biol Phys* 77:153–9.

Gebara, W.J., Weeks, K.J., Hahn, C.A., Montana, G.S., and Anscher, M.S. 1998. Computed axial tomography tandem and ovoids (CATTO) dosimetry: Three dimensional assessment of bladder and rectal doses. *Radiat Oncol Invest* 6:268–75.

Gerbaulet, A., Pötter, R., Mazeron, J.-J., Meertens, H., and Van Limbergen, E. (eds.) 2002. *The GEC-ESTRO Handbook of Brachytherapy.* ESTRO, Brussels.

Grigsby, P.W. 2009. PET/CT imaging to guide cervical cancer therapy. *Future Oncol* 5:953–8.

Guedea, F., Venselaar, J., Hoskin, P. et al. 2010. Patterns of care for brachytherapy in Europe: Updated results. *Radiother Oncol* 97:514–20.

Guha, C., Alfieri, A., Blaufox, M.D., and Kalnicki, S. 2008. Tumor biology-guided radiotherapy treatment planning: Gross tumor volume versus functional tumor volume. *Semin Nucl Med* 38:105–13.

Haie-Meder, C., Pötter, R., Van Limbergen, E. et al. 2005. Recommendations from Gynaecological (GYN) GEC-ESTRO Working Group (I): Concepts and terms in 3D image based 3D treatment planning in cervix cancer brachytherapy with emphasis on MRI assessment of GTV and CTV. *Radiother Oncol* 74:235–45.

Hayes, J.K., Moeller, J.H., Leavitt, D.D., Davis, R.K., and Harnsberger, H.R. 1992. Computed tomography treatment planning in Ir-192 brachytherapy in the head and neck. *Int J Radiat Oncol Biol Phys* 22:181–9.

Hsu, I.-C., Bae, K., Shinohara, K. et al. 2010. Phase II Trial of Combined High Dose Rate Brachytherapy and External Beam Radiotherapy for Adenocarcinoma of the Prostate: Preliminary Results of RTOG 0321. *Int J Radiat Oncol Biol Phys* 78:751–8, on-line DOI: 10.1016/j.ijrobp.2009.08.048.

International Commission on Radiation Units and Measurements (ICRU). 1985. *Dose and Volume Specification for Reporting and Recording Intracavitary Therapy in Gynecology.* Report 38 of ICRU, ICRU Publications, Bethesda, MD.

International Commission on Radiation Units and Measurements (ICRU). 1997. *Dose and Volume Specification for Reporting Interstitial Therapy.* Report 58 of ICRU, ICRU Publications, Bethesda, MD.

Kapoor, V., McCook, B.M., and Torok, F.S. 2004. An introduction to PET-CT imaging. *Radiographics* 24:523–43.

Kim, Y., Hsu, I.-C., Lessard, E., Vujic, J., and Pouliot, J. 2004. Dose uncertainty due to computed tomography CT slice thickness in CT-based high dose rate brachytherapy of the prostate. *Med Phys* 31:2543–8.

Kim, R.Y., Shen, S., and Duan, J. 2007. Image-based three-dimensional treatment planning of intracavitary brachytherapy for cancer of the cervix: Dose-volume histograms of the bladder, rectum, sigmoid colon, and small bowel. *Brachytherapy* 6:187–94.

Kim, Y., Hsu, I.C.J., Lessard, E. et al. 2008. Class solution in inverse planned HDR prostate brachytherapy for dose escalation of DIL defined by combined MRI/MRSI. *Radiother Oncol* 88:148–55.

Kovács, G., C. Melchert, C., Sommerauer, M., and Walden, O. 2007. Intensity modulated high-dose-rate brachytherapy boost complementary to external beam radiation for intermediate- and high-risk localized prostate cancer patients—How we do it in Lübeck/Germany. *Brachytherapy* 6:142–8.

Kovács, G., Pötter, R., Loch, T. et al. 2005. GEC/ESTRO-EAU recommendations on temporary brachytherapy using stepping sources for localised prostate cancer. *Radiother Oncol* 74:137–48.

Krempien, R.C., Daeuber, S., Hensley, F.W., Wannenmacher M., and Harms, W. 2003. Image fusion of CT and MRI data enables improved target volume definition in 3D-brachytherapy treatment planning. *Brachytherapy* 2:164–71.

Kubicky, C.D., Yeh, B., Lessard, E. et al. 2008. Inverse planning simulated annealing for magnetic resonance imaging-based intracavitary high-dose-rate brachytherapy for cervical cancer. *Brachytherapy* 7:242–7.

Landry, G., Granton, P.V., Reniers, B. et al. 2011. Simulation study on potential accuracy gains from dual energy CT tissue segmentation for low-energy brachytherapy Monte Carlo dose calculations. *Phys Med Biol* 56:6257–78.

Lee, K.R., Mansfield, C.M., Dwyer, S.J. III, Cox, H.L., Levine, E., and Templeton, A.W. 1980. CT for intracavitary radiotherapy planning. *AJR* 135:809–13.

Lessard, E. and Pouliot, J. 2001. Inverse planning anatomy-based dose optimization for HDR brachytherapy of the prostate using fast simulated annealing algorithm and dedicated objective function. *Med Phys* 28:773–9.

Lessard, E., Hsu, I.-C., and Pouliot, J. 2002. Inverse planning for interstitial gynecologic template brachytherapy: Truly anatomy-based planning. *Int J Radiat Oncol Biol Phys* 54:1243–51.

Liebmann, A., Pohlmann, S., Heinicke, F., and Hildebrandt, G. 2007. Helmet mold-based surface brachytherapy for homogeneous scalp treatment: A case report. *Strahlenther Onkol* 183:211–4.

Lin, L.L., Mutic, S., Low, D.A. et al. 2007. Adaptive brachytherapy treatment planning for cervical cancer using FDG-PET. *Int J Radiat Oncol Biol Phys* 67:91–6.

Major, T., Fodor, J., Takácsi-Nagy, Z., Ágoston, P., and Polgár, C. 2005. Evaluation of HDR interstitial breast implants planned by conventional and optimized CT-based dosimetry systems with respect to dose homogeneity and conformality. *Strahlenther Onkol* 181:89–96.

Martel, M.K. and Narayana, V. 1998. Brachytherapy for the next century: Use of image-based treatment planning. *Radiat Res* 150 (Suppl):S178–88.

Maruyama, Y., Chin, H.W., Young, A.B. et al. 1984. CT and MR for brain tumor implant therapy using Cf-252 neutrons. *J Neuro-Oncol* 2:349–60.

Massoud, T.F. and Gambhir, S.S. 2003. Molecular imaging in living subjects: Seeing fundamental biological processes in a new light. *Genes Dev* 17:545–80.

Mayr, N.A., Wang, J.Z., Lo, S.S. et al. 2010. Translating response during therapy into ultimate treatment outcome: A personalized 4-dimensional MRI tumor volumetric regression approach in cervical cancer. *Int J Radiat Oncol Biol Phys* 76:719–27.

McLaughlin, P.W., Narayana, V., Kessler, M. et al. 2004. The use of mutual information in registration of CT and MRI datasets post permanent implant. *Brachytherapy* 3:61–70.

Ménard, C., Susil, R., Choyke, P. et al. 2004. MRI-guided HDR prostate brachytherapy in a standard 1.5T scanner. *Int J Radiat Oncol Biol Phys* 59:1414–23.

Merrick, G.S., Butler, W.M., Wallner, K.E. et al. 2005. Variability of prostate brachytherapy preimplant dosimetry: A multi-institution analysis. *Brachytherapy* 4:241–51.

Nag, S., Beyer, D., Friedland, J., Grimm, P., and Nath, R. 1999. American Brachytherapy Society (ABS) recommendations for transperineal permanent brachytherapy of prostate cancer. *Int J Radiat Oncol Biol Phys* 44:789–99.

Nag, S., Bice, W., De Wyngaert, K., Prestidge, B., Stock, R., and Yu, Y. 2000. The American Brachytherapy Society recommendations for permanent prostate brachytherapy post-implant dosimetric analysis. *Int J Radiat Oncol Biol Phys* 46:221–30.

Nag, S., Cardenes, H., Chang, S. et al. 2004. Proposed guidelines for image-based intracavitary brachytherapy for cervical carcinoma: Report from the image-guided brachytherapy working group. *Int J Radiat Oncol Biol Phys* 60:1160–72.

Nath, R., Bice, W.S., Butler, W.M. et al. 2009. AAPM recommendations on dose prescription and reporting methods for permanent interstitial brachytherapy for prostate cancer: Report of Task Group 137. *Med Phys* 36:5310–22.

Nickers, P., Lenaerts, E., Thissen, B., and Deneufbourg, J.-M. 2005. Does inverse planning applied to Iridium 192 high dose rate prostate brachytherapy improve the optimization of the dose afforded by the Paris system? *Radiother Oncol* 74:131–6.

Niyazi, M., Bartenstein, P., Belka, C., and Ganswindt, U. 2010. Choline PET based dose-painting in prostate cancer—Modelling of dose effects. *Radiat Oncol* 5:23.

Pérez-Calatayud, J., Kuipers, F., Ballester, F. et al. 2009. Exclusive MRI-based tandem and colpostats reconstruction in gynaecological brachytherapy treatment planning. *Radiother Oncol* 91:181–6.

Potter, R., Dimopoulos, J., Kirisits, C. et al. 2005. Recommendations for image-based intracavitary brachytherapy of cervix cancer: the GYN GEC ESTRO Working Group point of view: in regard to Nag et al. (*Int J Radiat Oncol Biol Phys* 2004; 60:1160–1172). *Int J Radiat Oncol Biol Phys* 62:293–5; author reply 295–6.

Pötter, R., Haie-Meder, C., Van Limbergen, E. et al. 2006. Recommendations from gynecologic (GYN) GEC-ESTRO working group (II): Concepts and terms in 3D image-based treatment planning in cervix cancer brachytherapy—3D dose volume parameters and aspects of 3D image-based anatomy, radiation physics, radiobiology. *Radiother Oncol* 78:67–77.

Pötter, R., Dimopoulos, J., Georg, P. et al. 2007. Clinical impact of MRI assisted dose volume adaptation and dose escalation in brachytherapy of locally advanced cervical cancer. *Radiother Oncol* 83:148–55.

Pötter, R., Fidarova, E., Kirisits, C., and Dimopoulos, J. 2008a. Image-guided adaptive brachytherapy for cervix carcinoma. *Clin Oncol* 20:426–32.

Pötter, R., Kirisits, C., Fidarova, E.F. et al. 2008b. Present status and future of high-precision image guided adaptive brachytherapy for cervix carcinoma. *Acta Oncol* 47:1325–36.

Pötter, R., Georg, P., Dimopoulos, J.C. et al. 2011. Clinical outcome of protocol based image (MRI) guided adaptive brachytherapy combined with 3D conformal radiotherapy with or without chemotherapy in patients with locally advanced cervical cancer. *Radiother Oncol* 100:116–23.

Pouliot, J., Kim, Y., Lessard, E., Hsu, I.-C., Vigneron, D., and Kurhanewicz, J. 2004. Inverse planning for HDR prostate brachytherapy used to boost dominant intraprostatic lesions defined by magnetic resonance spectroscopy imaging. *Int J Radiat Oncol Biol Phys* 59:1196–207.

Pouliot, J. and Beaulieu, L. 2010. Modern principles of brachytherapy physics: from 2D to 3D to dynamic planning and delivery. In: *Leibel and Phillips Textbook of Radiation Oncology*. R. Hoppe, T. Phillips, and M. Roach (eds.). Elsevier, Philadelphia. 3rd ed., pp. 224–244.

Reed, G., Cunha, J.A., Noworolski, S. et al. 2011. Interactive, multi-modality image registrations for combined MRI/MRSI-planned HDR prostate brachytherapy. *J Contemp Brachytherapy* 1:26–31.

Rivard, M.J., Coursey, B.M., DeWerd, L.A. et al. 2004. Update of AAPM Task Group No. 43 Report: A revised AAPM protocol for brachytherapy dose calculations. *Med Phys* 31:633–74.

Rivard, M.J., Venselaar, J.L.M., and Beaulieu, L. 2009a. The evolution of brachytherapy treatment planning. *Med Phys* 36:2136–53.

Rivard M.J., Melhus, C.S., Wazer, D.E., and Bricault, R.J. Jr. 2009b. Dosimetric characterization of round HDR 192Ir AccuBoost applicators for breast brachytherapy. *Med Phys* 36:5027–32.

Rodriguez, R.R., Nag S., and Mate, T.P. 2001. High dose rate brachytherapy for prostate cancer: Assessment of current clinical practice and recommendations of the American Brachytherapy Society. *J Brachy Int* 17:265–82. Updated recommendations available at http://www.americanbrachytherapy.org/guidelines/HDRTaskGroup.pdf, accessed October 12, 2010.

Rubens, D.J., Yu, Y., Barnes, A.S., Strang, J.G., and Brasacchio, R. 2006. Image-guided brachytherapy for prostate cancer. *Radiol Clin N Am* 44:735–48.

Salembier, C., Lavagnini, P., Nickers, P. et al. On behalf of the PROBATE group of GEC ESTRO. 2007. Tumour and target volumes in permanent prostate brachytherapy: A supplement to the ESTRO/EAU/EORTC recommendations on prostate brachytherapy. *Radiother Oncol* 83:3–10.

Schöder, H. and Ong, S.C. 2008. Fundamentals of molecular imaging: Rationale and applications with relevance for radiation oncology. *Semin Nucl Med* 38:119–28.

Seppälä, J., Seppänen, M., Arponen, E., Lindholm, P., and Minn, H. 2009. Carbon-11 acetate PET/CT based dose escalated IMRT in prostate cancer. *Radiother Oncol* 93:234–40.

Stock, R.G., Stone, N.N., Tabert, A., Iannuzzi, C., and DeWyngaert, J.J. 1998. A dose-response study for I-125 prostate implants. *Int J Radiat Oncol Biol Phys* 41:101–8.

Susil, R.C., Camphausen, K., Choyke, P. et al. 2004. System for prostate brachytherapy and biopsy in a standard 1.5 T MRI scanner. *Magn Reson Med* 52:683–7.

Thomadsen, B.R., Shahabi, S., Stitt, J.A. et al. 1992. High dose rate brachytherapy for carcinoma of the cervix: The Madison system: II. Procedural and physical considerations. *Int J Radiat Oncol Biol Phys* 24:349–57.

Tod, M.C. and Meredith, W.J. 1938. A dosage system for use in the treatment of cancer of the uterine cervix. *Br J Radiol* 11:809–24.

Turkbey, B., Albert, P.S., Kurdziel, K., and Choyke, P.L. 2009. Imaging localized prostate cancer: Current approaches and new developments. *AJR* 192:1471–80.

Van der Laarse, R. 1994. The stepping source dosimetry system as an extension of the Paris System. In: *Brachytherapy from Radium to Optimization*. A.A. Martinez, C.G. Orton, and R.F. Mould (eds.). Nucletron International BV, Veenendaal.

Vicini, F.A., Vargas, C., Edmundson, G., Kestin, L., and Martinez, A. 2003. The role of high-dose rate brachytherapy in locally advanced prostate cancer. *Semin Radiat Oncol* 13:98–108.

Viswanathan, A.N. and Erickson, B.A. 2010. Three-dimensional imaging in gynecologic brachytherapy: A survey of the American Brachytherapy Society. *Int J Radiat Oncol Biol Phys* 76:104–9.

Viswanathan, A.N., Cormack, R., Holloway, C.L. et al. 2006. Magnetic resonance-guided interstitial therapy for vaginal recurrence of endometrial cancer. *Int J Radiat Oncol Biol Phys* 61:91–9.

Viswanathan, A.N., Dimopoulos, J., Kirisits, C., Berger, D., and Pötter, R. 2007. Computed tomography versus magnetic resonance imaging–based contouring in cervical cancer brachytherapy: Results of a prospective trial and preliminary guidelines for standardized contours. *Int J Radiat Oncol Biol Phys* 68:491–8.

Wahl, R.L., Herman, J.M., and Ford, E. 2011. The promise and pitfalls of positron emission tomography and single-photon emission computed tomography molecular imaging-guided radiation therapy. *Semin Radiat Oncol* 21(2):88–100.

Wang, H., Dong, L., O'Daniel, J. et al. 2005. Validation of an accelerated 'demons' algorithm for deformable image registration in radiation therapy. *Phys Med Biol* 50:2887–905.

Williamson, J.F. 2008. Current brachytherapy quality assurance guidance: does it meet the challenges of emerging image-guided technologies? *Int J Radiat Oncol Biol Phys* 71:18–22.

Yamada, Y., Bhatia, S., Zaider, M. et al. 2006. Favorable clinical outcomes of three-dimensional computer-optimized high-dose-rate prostate brachytherapy in the management of localized prostate cancer. *Brachytherapy* 5:157–64.

Yang Y. and Rivard, M.J. 2009. Monte Carlo simulations and radiation dosimetry measurements of peripherally applied HDR 192Ir breast brachytherapy D-shaped applicators. *Med Phys* 36:809–15.

Yu, Z., Richardson, S., Zhang, Y., Klopp, A., Dong, L., and Mourtada, F. 2010. Using MRI-based deformable image registration to accumulate 3D total dose distribution from intracavitary brachytherapy fractions of cervical cancer. *Med Phys* 37:3392 (abstract).

TG-43 Dose Calculation Formalism: Development, Insights, and Modernization

Mark J. Rivard
Tufts University School of Medicine

Ali S. Meigooni
Comprehensive Cancer Centers of Nevada

Ravinder Nath
Yale University School of Medicine

10.1 TG-43 Report Background

As of the year 2012, the international standard for brachytherapy dose calculations is the TG-43 formalism. This formalism was developed by the American Association of Physicists in Medicine (AAPM) Task Group No. 43 and published in 1995 (Nath et al. 1995). Prior to the introduction of TG-43 dose calculation algorithm, brachytherapy dosimetry formalisms were based on calculating the exposure rate in air at the point of interest and making a correction for the attenuation of photons in water. The simplest version of these protocols assumed a point isotropic source resulting in a 1D formulation such as

$$\dot{D}(r) = \frac{\Gamma A}{r^2} \cdot \left(\frac{\mu_{en}}{\rho} \right)_{air}^{water} ,$$ where the product of source activity and

the exposure rate constant was simply corrected by ratio of the mass-energy absorption coefficients for water and air. This was improved through accounting for dose falloff based on physical properties of medium and capsule, such as exponential attenuation, by the Sievert integral

$$\dot{D}(r,\theta) = S_K \cdot e^{\mu' t} \cdot \frac{\left(\overline{\mu_{en}/\rho} \right)_{air}^{water} \Gamma A}{Lr\cos\theta} \cdot \int_{\theta_1}^{\theta_2} e^{-\mu' t \sec\phi} T[(r\cos\theta - t)\sec\phi] d\phi$$

as described in Appendix B of the 1995 TG-43 report, which provides a detailed description of this process (see also Chapter 7). TG-43 formalism is instead based upon the dose rate at a reference point in water from the actual source geometry, which is then corrected for relative attenuation in water at all other points using factors such as radial dose function and anisotropy function. Deviation from the physics-oriented model of the Sievert integral toward the formalism recommended in the TG-43 report was a bold move. Given that the goals of a brachytherapy dose calculation formalism are accuracy and consistency of clinical implementation, this was the right decision to overcome the limitations of the Sievert integral to model radiological interactions and needless complication and specification of various terms. This is discussed in greater detail in Chapter 7.

The 1995 report of TG-43 was updated by the AAPM (TG-43U1) in 2004 (Rivard et al. 2004) and includes the following:

- A revised definition of air-kerma strength
- Elimination of apparent activity A_{app} for specification of source strength
- Elimination of the anisotropy constant in favor of the distance-dependent 1D anisotropy function
- Guidance on extrapolating tabulated TG-43 parameters to longer and shorter distances
- Correction for minor inconsistencies and omissions in the original protocol and its implementation

In this chapter, we present a description of the latest version of the TG-43 formalism, the improvements made by TG-43 in dosimetry for brachytherapy sources emitting low-energy photons, and methods to maintain quality and consistency of dose calculation accuracy in clinical implementation.

10.1.1 2D Dose Calculation Formalism

The dose rate in water at the point of interest $P(r, \theta)$ is shown on a polar coordinate system in Figure 10.1.

Using this coordinate system, Equation 10.1 depicts the 2D dose calculation formalism for linear photon-emitting brachytherapy sources, which is given by

$$\dot{D}(r,\theta) = S_K \cdot \Lambda \cdot \frac{G_L(r,\theta)}{G_L(r_0,\theta_0)} \cdot g_L(r) \cdot F(r,\theta) \quad (10.1)$$

where the dose rate at any point $\dot{D}(r,\theta)$ is the product of the following brachytherapy dosimetry parameters:

S_K the brachytherapy source air-kerma strength measured on the source transverse plane, with units U where $1\ U = 1\ \text{cGy cm}^2\ \text{h}^{-1}$. This quantity is specified in terms of traceable quantities measured at primary standard dosimetry laboratories (PSDLs) or national metrology institutes (NMIs) in Europe. In the United States, this quantity is traceable to the National Institute of Standards and Technology (NIST) wide-angle free-air ionization chamber (WAFAC) for low-energy photon-emitting sources. An energy threshold δ is used to remove low-energy photons that do not significantly contribute to absorbed dose in the patient. This energy threshold is set to 5 keV for low-energy photon-emitting sources like ^{103}Pd, ^{125}I, and ^{131}Cs. In Europe, reference air-kerma rate (RAKR) is the traceable quantity for source strength specification, differing from S_K only through specification of a measurement distance. The calibration methods of brachytherapy

sources in Europe and the United States are discussed in greater detail in Chapter 5. Regardless of the calibration method, treatment planning system (TPS) vendors should uniformly embrace traceable quantities for source strength specification and should remove usage of antiquated terms such as A_{app} immediately as has been disapproved by professional societies (AAPM Report 21 1987).

Λ the dose rate constant, having unit of cGy h^{-1} U^{-1}, which has dimensions of cm^{-2} in the SI system. This quantity is the ratio of the dose rate to water in water at the reference position $\dot{D}(r_0,\theta_0)$ to the air-kerma strength S_K. It is determined either through measurements or Monte Carlo (MC) methods to simulate radiation transport. Measurements are frequently performed in machined plastic phantom materials using thermoluminescent dosimeters (TLDs). There is debate in the field as to whether the definition should be based on the ratio of dose rate and air-kerma strength strictly on the transverse plane, or accounting for the volume averaging intrinsic to physical measurement systems that include some degrees of polar angle averaging. This concern is pertinent for brachytherapy sources in which there is significant variation of air-kerma strength as a function of polar angle, such as for sources having right cylindrical geometry with end-face contributions.

$G_L(r, \theta)$ the geometry function, specified as a function of r and θ, using a line-source approximation of the physical distribution of radiation emissions. The reference geometry function $G_L(r_0, \theta_0)$ is calculated for the special case of the reference position. The line-source approximation is based on $G_L(r,\theta) = \dfrac{\beta}{Lr\sin(\theta)}$ using the angle β (see Figure 10.1) with units radians, subtended from the point of interest to the two ends of the specified active length L with units cm. Along the source long axis, $G_L(r,0) = \dfrac{1}{r^2 - (L/2)^2}$.

For large distances (i.e., $r \gg L$), the geometry function using the line-source approximation may be readily approximated as the inverse-square law or point-source approximation. The 2004 AAPM TG-43 report presents the complete details of this definition. For sources not having an easily specified active length, such as if there are several radioactive pellets inside the source capsule, an effective length L_{eff} may be calculated as $L_{eff} = s \times n$, where s is the spacing (units cm) between source pellets, and n is the number of active pellets. Should L_{eff} exceed the capsule outer length, the maximum separation between the proximal and distal aspects of the source distribution should be used as L_{eff}.

$g_L(r)$ the radial dose function using the line-source approximation. It is a dimensionless function pertinent to positions on the transverse plane where $\theta = 90°$ and takes a value of unity at r_0. The line-source radial dose function is the ratio of dose rate falloff (mainly due to radiation attenuation) as a function of r, relative to the dose rate at the reference position. This parameter characterizes

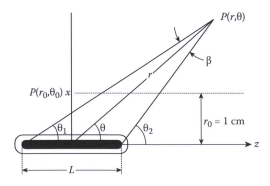

FIGURE 10.1 Polar coordinate system for 2D dose calculations in the vicinity of a photon-emitting brachytherapy source as a function of radial distance r and polar angle θ at any point $P(r,\theta)$. A special reference position is defined at $P(r_0,\theta_0)$ where $r_0 = 1$ cm and $\theta_0 = 90°$. If the dose distribution is not symmetric on the transverse plane bisecting the source center along the z-axis, then θ will range from 0° to 180° instead of 0° to 90°. (Courtesy of M. Rivard.)

dose rate falloff expected due to solid angle as accounted for by the line-source geometry function as shown by $g_L(r) = \dfrac{\dot{D}(r,\theta_0)}{\dot{D}(r_0,\theta_0)} \dfrac{G_L(r_0,\theta_0)}{G_L(r,\theta_0)}$. In brachytherapy TPSs, $g_L(r)$ is sometimes introduced as the fifth-order polynomials fits, as remnant of pre–TG-43 dose calculations in which third-order polynomial fitting functions (commonly known as Meisberger coefficients) (Meisberger et al. 1968) were more popular, instead of high-resolution specification of brachytherapy dosimetry parameters in a tabular form.

$F(r,\theta)$ the 2D anisotropy function, used to correct for the self-absorption, radiation emission distribution, and any remaining dosimetric effects that may have not been accounted for by the other dosimetry parameters. This quantity is the ratio of dose rates and geometry functions for a given r value as a function of θ as shown by $F(r,\theta) = \dfrac{\dot{D}(r,\theta)}{\dot{D}(r,\theta_0)} \dfrac{G_L(r,\theta_0)}{G_L(r,\theta)}$. $F(r,\theta)$ values are unity on the transverse plane, where $\theta = 90°$, and typically decrease as polar angles change toward 0° or 180° due to radiation attenuation by the capsule and the source core. However, $F(r,\theta)$ can take values greater than unity for $60° < \theta < 90°$

due to dose contributions from the end faces of cylindrical line sources as described for Λ. For a given θ value, $F(r,\theta)$ generally increases as r increases beyond $r = 0.5$ cm because the proportion of radiation scatter relative to the primary radiation increases with increasing distance (Figures 10.2 and 10.3). For poly-energetic photon sources such as ^{103}Pd, there is significant dose anisotropy at small distances due to the low-energy photon emissions, whereas there is negligible anisotropy at larger distances due to the high-energy emissions. The quantitative behavior of $F(r,\theta)$ and the distance at which the low-energy anisotropy effects give way to the high-energy effects are radionuclide dependent.

Equation 10.1 depicts the 2D dose calculation formalism, which is applicable for photon-emitting brachytherapy sources that have a finite length and width. It assumes that the dose distributions are axially symmetric. In order to use this equation for dose calculation in a patient, the linear orientation of the brachytherapy source within the patient anatomy must be known.

10.1.2 1D Dose Calculation Formalism

When the source orientation is not known, the TG-43 1D dose calculation formalism is recommended by the AAPM. This is a simplification of Equation 10.1 by integrating the values over the polar angle. On the other hand, the AAPM TG-43U1 report recommended using a line-source geometry function within the 1D formalism for improved accuracy for $r < L$, as shown in Equation 10.2:

$$\dot{D}(r) = S_K \cdot \Lambda \cdot \frac{G_L(r,\theta_0)}{G_L(r_0,\theta_0)} \cdot g_L(r) \cdot \phi_{an}(r). \qquad (10.2)$$

All clinical TPSs use some form of Equation 10.3 for 1D dose calculations:

$$\dot{D}(r) = S_K \cdot \Lambda \cdot \left(\frac{r_0}{r}\right)^2 \cdot g_P(r) \cdot \phi_{an}(r). \qquad (10.3)$$

This is because, in part, it is not obvious to employ the line-source geometry function as needed in Equation 10.2 for $G_L(r,\theta)$ and $g_L(r)$.

Differences between Equations 10.1 and 10.3 include the following terms:

$\dfrac{1}{r^2}$ the point-source geometry function. As required for the line-source geometry function, division by the reference geometry function is needed for this quotient to be dimensionless. When dividing $1/r^2$ by $1/r_0^2$, the r_0 term moves to the numerator of the quotient as shown in Equation 10.3.

$g_P(r)$ the radial dose function using the point-source approximation. It is a dimensionless function and takes a value

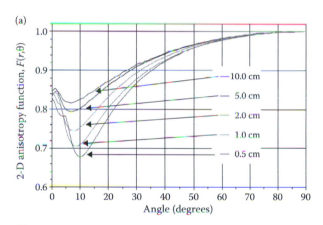

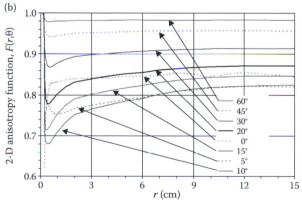

FIGURE 10.2 2D anisotropy functions for a representative ^{125}I brachytherapy source. The upper graph (a) depicts the conventional means of plotting $F(r,\theta)$ versus θ. However, the lower graph (b) readily permits assessment of $F(r,\theta)$ as a functional of r. (Courtesy of M. Rivard.)

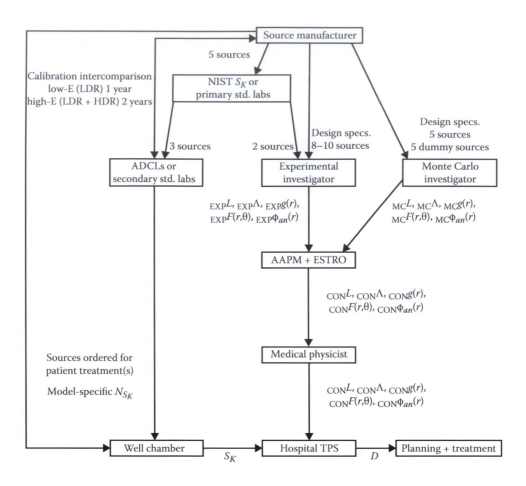

FIGURE 10.3 Dosimetry parameter data trail and associated dosimetric uncertainties. (Courtesy of M. Rivard.)

of unity at r_0. The point-source radial dose function is the ratio of dose rate falloff as a function of r relative to the dose rate at the reference position, accounting for dose rate falloff expected due to the solid angle as accounted for by the point-source geometry function as shown by

$$g_P(r) = \frac{\dot{D}(r)}{\dot{D}(r_0)} \left(\frac{r}{r_0} \right)^2.$$

$\phi_{an}(r)$ the 1D anisotropy function and is identical to the anisotropy factor defined in the 1995 TG-43 report. For a given r value, $\phi_{an}(r)$ is the ratio of the solid angle-weighted dose rate, averaged over the entire 4π steradian space, to the dose rate at the same r distance on the transverse plane as shown by

$$\phi_{an}(r) = \frac{\int_0^\pi \dot{D}(r,\theta)\sin(\theta)d\theta}{2\dot{D}(r,\theta_0)}.$$

This equation is based on the cylindrical symmetry of the source dose distribution. One should integrate dose rate and not $F(r,\theta)$ values over the polar angle to determine $\phi_{an}(r)$. For sources exhibiting high $F(r,\theta)$ gradients, especially near the transverse plane, derivation of $\phi_{an}(r)$ can

be a strong function of the angular bin spacing. While 10° increments are commonly used, 1° increments will better approximate the above integration method if spherical sampling bins are not directly evaluated.

For traditional dose calculation formalisms, the variation of the 1D anisotropy function with radial distance was ignored, and only one factor known as anisotropy constant was used. The TG-43U1 report recommended not using the simplified 1D equation (Equation 10.4) where the 1D anisotropy function was reduced to the anisotropy constant $\bar{\phi}_{an}$:

$$\dot{D}(r) = S_K \cdot \Lambda \cdot \left(\frac{r_0}{r} \right)^2 \cdot g_P(r) \cdot \bar{\phi}_{an}. \tag{10.4}$$

This recommendation to use Equation 10.4 improves the dose calculation accuracy in the vicinity of the source where dose rates are highest and usually of greater clinical importance. While TPS vendors are advised to take the lead on this recommendation, the clinical medical physicist can adapt noncompliant TPS as an interim solution by integrating $\phi_{an}(r)$ behavior into $g_P(r)$ through using modified dosimetry parameters, using $g_P(r) \cdot \phi_{an}(r)$ in place of $g_P(r)$, and setting $\bar{\phi}_{an} = 1$.

Not all brachytherapy source dose distributions are readily fit to the TG-43 formalism. This is especially true for short-ranged radiations such as beta particles. Chapter 26 provides a summary of beta dosimetry methods specific to intravascular brachytherapy.

10.2 Dose Calculation Improvements Obtained by TG-43 Formalism

The 1995 TG-43 report was an improvement over prior brachytherapy dose calculation formalisms for the following reasons:

(1) Explicit recommendations to clinical medical physicists, source and TPS vendors, and calibration laboratories to use traceable source strength quantities measured at NIST, PSDLs, or NMIs

(2) Elimination of the exposure rate constant in favor of the dose rate constant, again due to requirements for traceable source strength quantities instead of activity

(3) Elimination of exposure-to-dose conversion factors

(4) Elimination of the tissue attenuation factor $T(r)$ in favor of the radial dose function $g(r)$ due to its dependence on exposure

(5) Elimination of $\bar{\phi}_{an}$ in favor of $\phi_{an}(r)$

(6) Introduction of the 2D anisotropy function to account for the self-absorption of the source and physical distribution of radiation emissions

While the naming of $T(r)$ implies some relationship to human tissue, radiological differences between water and tissue are not significant for high-energy photon-emitting brachytherapy sources such as ^{192}Ir, ^{137}Cs, and ^{60}Co. However, differences in dose falloff and absorbed dose between water and tissue become more pronounced as photon energy decreases (as for low-energy photon-emitting brachytherapy sources such as ^{103}Pd, ^{125}I, and ^{131}Cs) and as distance from the source(s) increases (Luxton 1994; Luxton and Jozsef 1999). Further, experimental measurements in liquid water phantoms for determining brachytherapy dosimetry parameters have been performed only recently (Tailor et al. 2008). Thus, it may have been more suitable to establish a standardized medium other than water that has machining compatibility for experimental dosimetry measurements and dosimetric equivalence to tissue. This is discussed in greater detail in Chapter 6.

An additional improvement of the TG-43 formalism over other approaches was explicit standardization of the phantom features such as size, shape, and composition. While assignment of these features was certainly better than without their specification, the differences between clinical dose distributions and those obtained using the TG-43 formalism soon became apparent. Pérez-Calatayud et al. (2004) observed 10%–30% dose differences for low- and high-energy photon-emitting brachytherapy sources between spherical phantoms of different size. These findings were supported by other observations (Melhus and Rivard 2006), producing quantitatively similar results and

further investigating the general influence of phantom composition as a function of photon energy. In collaboration (Granero et al. 2008), these investigators examined the influence of phantom shape, correlating apparently disparate dosimetry results in the literature when accounting for their differing phantom shapes. Though these three phantom features can manifest errors in dose distributions between the TG-43 formalism and more sophisticated means such as MC methods, the TG-43 formalism is the current dose calculation standard upon which clinical prescriptions (and related patient outcomes) are based. Potential advantages of more sophisticated brachytherapy dose calculation methods are discussed in Chapter 11.

Due to the success of the TG-43 brachytherapy dose calculation formalism to provide accurate results in a simple context with widespread TPS appeal, there was interest to further simplify the formalism by eliminating the geometry function and employing a high-resolution grid for all the dosimetry parameters. This was discussed in the literature (Rivard 1999; Kouwenhoven et al. 2001; Meli 2002; Rivard et al. 2002; Song et al. 2003) and indicates that the simplicity of specifying L in concert with using the geometry function is superior to employing a high-resolution dose grid without utilizing the brachytherapy dosimetry parameters. Additionally, several fitting functions have been examined to improve the TPS dose calculation accuracy given the availability of reference data. These functions include the historically based nth-order polynomials dating from the era of Meisberger coefficients (Meisberger et al. 1968) and an assortment of other fitting functions either explicitly based on physical properties or chosen to approximate the observed trend (Furhang and Anderson 1999; Moss 2000; Furhang and Wallace 2000; Fung 2007; Taylor and Rogers 2008; Wu et al. 2009; Pujades-Claumarchirant et al. 2010). In general, these fitting functions do not provide significant (greater than 2%) improvements in dose calculation accuracy in comparison to the log–linear and linear–linear methods in current TPS used for low-energy and high-energy brachytherapy dose calculations.

10.3 Methods to Maintain Quality and Consistency of Dose Calculation Accuracy

10.3.1 Dosimetry Parameter Evaluation

The process for derivation, review, and clinical implementation of brachytherapy dosimetry parameters into TPS has been examined by the joint AAPM/ESTRO (European Society for Radiotherapy and Oncology) Task Group 138 (DeWerd et al. 2011) and is shown in Figure 10.3 as adapted for this chapter.

Currently, the AAPM/RPC Brachytherapy Source Registry (http://rpc.mdanderson.org/RPC/home.htm) at the Radiological Physics Center at the M.D. Anderson Cancer Center includes low- and high-energy photon-emitting sources that have met the AAPM Dosimetric Prerequisites (Williamson et al. 1998; Li et al. 2007; Pérez-Calatayud et al. 2012). It is recommended that all brachytherapy TPSs utilize these dosimetry parameters for the clinical applications. However, the dosimetry parameters and dose rate distribution tables are not in a convenient, spreadsheet-oriented format. The ESTRO Web site (http://www.estro.org/estroactivities/Pages/TG43HOMEPAGE.aspx) has postings of brachytherapy source dosimetry papers and dosimetry parameters compiled into spreadsheet, but the review process for posting is currently not as robust as the AAPM/RPC Registry. Neither society nor their Web sites have in place a process for updating recommended dosimetry parameters in light of new publications or findings that augment previously recommended dosimetry parameters. Under the direction of David Rogers, Carleton University similarly has a Web-based database listing brachytherapy dosimetry parameters for a variety of low- and high-energy brachytherapy sources (http://www.physics.carleton.ca/clrp/seed_database/), but there is no information on methods to update these data in light of new findings. Until the completion of the tasks listed in the following paragraph, the user is advised to verify independently the accuracy of dosimetry data presented in these resources.

In part due to need for development of long-term medical infrastructure, the AAPM and ESTRO have approved (late 2011) a new Working Group for ongoing evaluation of brachytherapy dosimetry parameters. The leadership and membership for this joint AAPM/ESTRO Working Group is shared between both professional societies. Membership to this Working Group by other professional societies with expertise in brachytherapy dosimetry is possible. In addition to posting and ongoing evaluation of brachytherapy dosimetry parameters onto a joint AAPM/ESTRO Registry, the Working Group is charged with the following tasks:

(1) Develop a limited number (approximately 5) of well-defined test case plans and perform model-based dose calculations and dosimetric comparisons.
(2) Identify the best venue for housing the reference plans/data, and put in place in collaboration with identified partners of the Registry.
(3) Propose to the brachytherapy community well-defined prerequisites for test case plans to be submitted to the Registry.
(4) Develop a review process for evaluation as new reference data meeting the prerequisites; this may include Digital Imaging and Communications in Medicine (DICOM)-RT import and export for quantitative dose comparisons.
(5) Engage the brachytherapy source and TPS vendors and manufacturers to promote uniformity of practice and to assign dimensional tolerances for sources and applicators.

10.3.2 Dose Calculation Accuracy Evaluation

One aspect to developing the field of brachytherapy dosimetry is to consider advanced dose comparison metrics beyond the AAPM TG-56 and TG-43U1 rules for 2% at any point. This approach examines only specific user-defined points and does not cover all potential positions toward the physicist commissioning a region/space. This is pertinent since large distances

may possibly be of concern due to water/media nonequivalence; small distances may possibly be of concern due to dose/kerma nonequivalence or neglect of beta or other short-range contributions. Further, it is likely that different criteria are needed for a variety of dose gradient regions, clinical applications, and radiation energies (Venselaar et al. 2001). In addition to percentage agreement, there is a need to expand the comparison metric beyond a simple criterion (percentage agreement) and to consider distance to agreement (DTA). The technique of using combinations of percentage agreement and DTA for a gamma-index comparison has been performed with great success for the heterogeneous dose distributions inherent to intensity-modulated radiotherapy fields (Low and Dempsey 2003; Low et al. 2011). Preliminary investigations into the applicability of gamma-index comparison have been made for brachytherapy (Yang et al. 2011; Petrokokkinos et al. 2011; Bannon et al. 2011) using agreement criteria as low as 1%/1 mm to as large as 2%/5 mm. There can be different reasons for applying DTA within the gamma-index comparison method (Low et al. 2011), such as correlating the various dose grids from MC results, TPS physics characterization, TPS dose calculation, TPS dose/fluence export, and within the gamma-index comparison analysis software. This is a ripe area for future research and subsequent societal guidelines.

10.3.3 Need for Reducing Brachytherapy Dosimetry Investigator Uncertainties

From the uncertainty analyses presented in the joint AAPM/ESTRO TG-138 report (DeWerd et al. 2011) and Chapter 15, the limiting dosimetric uncertainties may be subdivided into the following categories:

(1) Basics physics data and related uncertainties such as atomic and molecular radiological cross sections, μ/ρ, μ_{en}/ρ, radiation energy spectra, and radiation source half-lives.
(2) MC-related uncertainties such as comparisons of physics models among different codes (with subsequent evaluation of the appropriate dosimetric uncertainties), tally development for user convenience and clear/consistent implementation, and capability for DICOM and/or DICOM-RT input for characterizing simulation environments.
(3) Experimental dosimetry-related uncertainties such as detector characteristics (Williamson and Rivard 2009) and phantom influence on dosimetry such as size, shape, and composition. Related to this last item, some materials such as PMMA have larger medium corrections than specially designed plastics that are designed for the purpose of matching radiological water equivalence for a given photon energy range. The poly(methyl methacrylate) (PMMA) uncertainty corrections may be smaller than for the specially designed plastic due to increased uncertainties in their compositions in comparison to high-purity PMMA.

As the majority of the aforementioned issues are beyond the control or capabilities of clinical medical physicists, there is consequently a need to embrace the basic physics community to work collaboratively to research and minimize brachytherapy dosimetric uncertainties. Reducing the uncertainties further would require modifications in dose calculation formalisms.

10.4 Summary

The TG-43 brachytherapy dose calculation formalism is an evolution of methods in use for over four decades. This most recent dose calculation technique is used worldwide and has reached across conventional boundaries separating clinical medical physicists, brachytherapy source and TPS vendors, and source calibration laboratories. Through the simplification of prior methods and generalization to a wide variety of brachytherapy sources, the TG-43 formalism permits consistent and widespread brachytherapy dose calculations with reasonable accuracy under most clinical circumstances.

References

AAPM Report No. 21. 1987. Recommendations of AAPM Task Group 32: Specification of Brachytherapy Source Strength. American Institute of Physics, New York.

Bannon, E.A., Yang, Y., and Rivard, M.J. 2011. Accuracy of the superposition principle for evaluating dose distributions of elongated and curved ^{103}Pd and ^{192}Ir brachytherapy sources. *Med Phys* 38:2957–63.

DeWerd, L.A., Ibbott, G.S., Meigooni, A.S. et al. 2011. A dosimetric uncertainty analysis for photon-emitting brachytherapy sources: Report of AAPM Task Group No. 138 and GEC-ESTRO. *Med Phys* 38:782–801.

Fung, A.Y. 2007. Comment on "Functional fitting of interstitial brachytherapy dosimetry data recommended by the AAPM Radiation Therapy Committee Task Group 43" [*Med Phys* 1999; 26: 153–60] and "Fitting and benchmarking of dosimetry data for new brachytherapy sources" [*Med Phys* 2000; 27: 2302–06]. *Med Phys* 28:400.

Furhang, E.E. and Anderson, L.L. 1999. Functional fitting of interstitial brachytherapy dosimetry data recommended by the AAPM Radiation Therapy Committee Task Group 43. *Med Phys* 26:153–60.

Furhang, E.E. and Wallace, R.E. 2000. Fitting and benchmarking of dosimetry data for new brachytherapy sources. *Med Phys* 27:2302–6.

Granero, D., Pérez-Calatayud, J., Pujades-Claumarchirant, M.C., Ballester, F., Melhus, C.S., and Rivard, M.J. 2008. Equivalent phantom sizes and shapes for brachytherapy dosimetric studies of ^{192}Ir and ^{137}Cs. *Med Phys* 35:4872–7.

Kouwenhoven, E., van der Laarse, R., and Schaart, D.R. 2001. Variation in interpretation of the AAPM TG-43 geometry factor leads to unclearness in brachytherapy dosimetry. *Med Phys* 28:1965–6.

Li, Z., Das, R.K., DeWerd, L.A. et al. 2007. Dosimetric prerequisites for routine clinical use of photon emitting brachytherapy

sources with average energy higher than 50 keV. *Med Phys* 34: 37–40.

Low, D.A. and Dempsey, J.F. 2003. Evaluation of the gamma dose distribution comparison method. *Med Phys* 30:2455–64.

Low, D.A., Moran, J.M., Dempsey, J.F., Dong, L., and Oldham, M. 2011. Dosimetry tools and techniques for IMRT. *Med Phys* 38:1313–38.

Luxton, G. 1994. Comparison of radiation dosimetry in water and in solid phantom materials for I-125 and Pd-103 brachytherapy sources: EGS4 Monte Carlo study. *Med Phys* 21:631–41.

Luxton, G. and Jozsef, G. 1999. Radial dose distribution, dose to water and dose rate constant for monoenergetic photon point sources from 10 keV to 2 MeV: EGS4 Monte Carlo model calculation. *Med Phys* 26:2531–8.

Meisberger, L.L., Keller, R.J., and Shalek, R.J. 1968. The effective attenuation in water of gamma rays of gold 198, iridium 192, cesium 137, radium 226, and cobalt 60. *Radiol* 90: 953–7.

Melhus, C.S. and Rivard, M.J. 2006. Approaches to calculating AAPM TG-43 brachytherapy dosimetry parameters for ^{137}Cs, ^{125}I, ^{192}Ir, ^{103}Pd, and ^{169}Yb sources. *Med Phys* 33:1729–37.

Meli, J.A. 2002. Let's abandon geometry factors other than that of a point source in brachytherapy dosimetry. *Med Phys* 29:1917–8.

Moss, D.C. 2000. Technical note: Improved analytical fit to the TG-43 radial dose function, g(r). *Med Phys* 27:659–61.

Nath, R., Anderson, L.L., Luxton, G., Weaver, K.A., Williamson, J.F., and Meigooni, A.S. 1995. Dosimetry of interstitial brachytherapy sources: Recommendations of the AAPM Radiation Therapy Committee Task Group No. 43. *Med Phys* 22:209–34.

Pérez-Calatayud, J., Granero, D., and Ballester, F. 2004. Phantom size in brachytherapy source dosimetric studies. *Med Phys* 31:2075–81.

Pérez-Calatayud, J., Ballester, F., Das, R.K. et al. 2012. Dose calculation for photon-emitting brachytherapy sources with average energy higher than 50 keV: Report of the AAPM and ESTRO. *Med Phys* 39:2904–29.

Petrokokkinos, L., Zourari, K., Pantelis, E. et al. 2011. Dosimetric accuracy of a deterministic radiation transport based ^{192}Ir brachytherapy treatment planning system. Part II: Monte Carlo and experimental verification of a multiple source dwell position plan employing a shielded applicator. *Med Phys* 30:1981–92.

Pujades-Claumarchirant, M.C., Granero, D., Pérez-Calatayud, J., Ballester, F., Melhus, C.S., and Rivard, M.J. 2010. Evaluation of interpolation methods for TG-43 dosimetric parameters based on comparison with Monte Carlo data for high-energy brachytherapy sources. *J Contemp Brachyther* 2:28–32.

Rivard, M.J. 1999. Refinements to the geometry factor used in the AAPM Task Group Report No. 43 necessary for brachytherapy dosimetry calculations. *Med Phys* 26:2445–50.

Rivard, M.J., Coursey, B.M., DeWerd, L.A. et al. 2002. Comment on: Let's abandon geometry factors other than that of a point source in brachytherapy dosimetry. *Med Phys* 29:1919–20.

Rivard, M.J., Coursey, B.M., DeWerd, L.A. et al. 2004. Update of AAPM Task Group No. 43 Report: A revised AAPM protocol for brachytherapy dose calculations. *Med Phys* 31:633–74.

Song, H., Luxton, G., and Hendee, W.R. 2003. Calculation of brachytherapy doses does not need TG-43 factorization. *Med Phys* 30:997–9.

Tailor R., Ibbott, G., Lampe, S., Warren, W.B., and Tolani, N. 2008. Dosimetric characterization of a ^{131}Cs brachytherapy source by thermoluminescence dosimetry in liquid water. *Med Phys* 35:5861–8.

Taylor, R.P.E. and Rogers, D.W.O. 2008. More accurate fitting of ^{125}I and ^{103}Pd radial dose functions. *Med Phys* 35:4242–50.

Venselaar, J., Welleweerd, H. and Mijnheer, B. 2001. Tolerances for the accuracy of photon beam dose calculations of treatment planning systems. *Radiother Oncol* 60:191–201.

Williamson, J.F., Coursey, B.M., DeWerd, L.A., Hanson, W.F., and Nath, R. 1998. Dosimetric prerequisites for routine clinical use of new low energy photon interstitial brachytherapy sources. *Med Phys* 25:2269–70.

Williamson, J.F. and Rivard, M.J. 2009. Thermoluminescent detector and Monte Carlo techniques for reference-quality brachytherapy dosimetry. In: *Clinical Dosimetry for Radiotherapy: AAPM Summer School*. D.W.O. Rogers and J.E. Cygler (eds.). Medical Physics Publishing, Madison, WI, pp. 437–99.

Wu, X., Brezovich, I.A., and Fiveash, J.B. 2009. Bi- and tri-exponential fitting to TG-43 radial dose functions of brachytherapy sources based on a genetic algorithm. *Brachytherapy* 8:361–6.

Yang, Y., Melhus, C.S., Sioshansi, S., and Rivard, M.J. 2011. Treatment planning of a skin-sparing conical breast brachytherapy applicator using conventional brachytherapy software. *Med Phys* 38:1519–25.

11

On the Introduction of Model-Based Algorithms Performing Nonwater Heterogeneity Corrections into Brachytherapy Treatment Planning

Åsa Carlsson Tedgren
Linköping University and the Swedish Radiation Safety Authority

Frank Verhaegen
Maastro Clinic

Luc Beaulieu
Centre Hospitalier Universitaire de Québec and Université Laval

11.1 Introduction

Radiotherapy aims to cure cancer by delivering absorbed doses in a narrow interval (the "therapeutic window") that balances tumor control probability and normal tissue complications. Accurate delivery of absorbed dose is crucial since dose response curves for tumors and normal tissues can be steep. Dose calculation methods used in treatment planning constitute one link in the dosimetric chain. Results from studies using Monte Carlo (MC) simulations in realistic brachytherapy (BT) geometries show differences with results from calculations using the American Association of Physicists in Medicine (AAPM) Task Group 43 report (TG-43) formalism in use today (Nath et al. 1995; Rivard et al. 2004). These differences are pronounced at low photon energies (<50 keV) but are more moderate at higher energies.

Algorithms accounting for heterogeneities are common practice in external beam radiation therapy (EBRT) but not so in BT. A reasonable requirement on calculation methods for use in treatment planning is that they should be accurate enough as to not contribute significantly to the overall uncertainty in dose delivery. This is demonstrated in a summary of contributions to the total

uncertainty for external photon beams in Table 2 of Ahnesjö and Aspradakis (1999). The total uncertainty in dose delivery by BT has been estimated to be 5% to 10% depending on the application and can be divided into parts of physical and of clinical origin (Nath et al. 1997). Dosimetric uncertainties of nonclinical origin in BT have been recently reported by the AAPM Task Group 138 (DeWerd et al. 2011), and a project to estimate uncertainties of clinical origin has been started by the BT physics group of ESTRO, BRAPHYQS (Kirisits et al. 2011). See also Section IV on "Uncertainties in Clinical Brachytherapy" (Chapters 15–17) of this book.

The TG-43 method, described in detail in Chapter 10, derives absorbed dose values in a water volume substituting the patient through superimposing precalculated single-source dose distributions. The TG-43 formalism provides a parameterization of prederived (measured and/or calculated) water data accounting for the full source geometry. The TG-43 approach to calculate patient doses is far more simplistic than methods routinely used in external beam treatment planning. Over many years, planning of EBRT has evolved from adding data from premeasured fields in water into calculating absorbed dose individually for each patient using anatomical information obtained from CT images and verified models of linear accelerator-produced treatment beams. Calculation methods of various complexities, ranging from 1D pencil beam scaling to advanced 3D methods like collapsed cone (CC) kernel superposition and MC simulation, exist. For a review of external beam photon dose calculations, see Ahnesjö and Aspradakis (1999).

There are several reasons why heterogeneity corrections have not been routinely incorporated into BT treatment planning. The inverse square falloff in primary fluence with distance from the source dominates BT dose distributions, such that heterogeneity corrections are a second-order effect in comparison. Furthermore, BT was an early modality for radiation therapy and has a successful tradition based on surgical implant skills and empirical knowledge. As long as a successful and reproducible system is unchanged, it can be safely handled by empirical knowledge. However, modern BT is in an era of rapid evolution, including use of image guidance, introduction of new radiation sources, and application of new techniques, among other developments (Williamson 2006; Rivard et al. 2009). Accurate dosimetry is of utmost importance when clinical knowledge obtained with one system is to be transferred into another.

Interest in the research community to study the presence of heterogeneities in calculated dose distributions has resulted in implementation of model-based dose calculation algorithms (MBDCAs) in clinical treatment planning systems (TPSs) (Petrokokkinos et al. 2011; van Veelen 2011). Accounting for nonwater heterogeneities in dose calculation for BT treatment planning implies several differences compared to today's practice, with the largest differences seen in the low energy region (<50 keV). Anticipating an emerging need for guidelines, the AAPM formed Task Group 186 in 2009, named "Model-Based Dose Calculation Techniques in Brachytherapy: Status and Clinical Requirements for Implementation Beyond AAPM TG-43." A report from TG-186 can be expected during 2012.

The main steps of the dose calculation process using the TG-43 method and MBDCA are presented in Figure 11.1. Different dose calculation algorithms are used, and differences are also evident in the calculated doses (the output) and in the data required about the radiation source and the individual patient (the input).

This chapter starts in Section 11.2 with a short introduction to the main characteristics of photon interaction at BT energies. Insight into the photon physics of the BT energy region will be helpful in understanding the assumptions made by BT MBDCAs, the magnitudes of and major reasons for differences in absorbed doses calculated with TG-43 and MBDCAs, as well as why Hounsfield numbers (HUs) obtained from computed tomography (CT) imaging of the patient are well suited to describe the material composition of patients in megavoltage (MV) EBRT beams and high-energy BT dose calculations but not for low-energy BT.

An overview of MBDCAs available for BT follows in Section 11.3. This section provides a brief overview of the capabilities, advantages, and limitations of correction-based dose calculation methods and MC simulations, including analytical grid-based Boltzmann transport equation solvers (GBBS) and CC

(a)

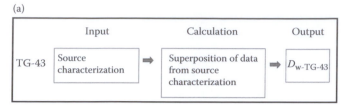

(b)

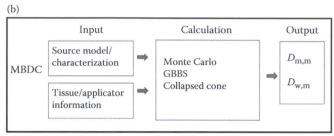

FIGURE 11.1 Main steps of the dose calculation process for treatment planning using the TG-43 method (a) and model-based dose calculation algorithms (b). The output in terms of calculated doses relies on the calculation algorithms, and the type of input includes the radiation source (source model/characterization) and the geometry in which calculations are performed (i.e., tissue and applicator information for the actual patient). Definitions of the output quantities $D_{\text{w,w-TG-43}}$, $D_{\text{w,m}}$, and $D_{\text{m,m}}$ are given in Table 11.2. (Courtesy of Å. Carlsson Tedgren.)

superposition methods. Aspects related to MBDCA input in the form of radiation source models and individual patient models (e.g., geometry and atomic composition) will be covered in Sections 11.4 and 11.5, respectively. Section 11.6 will provide an introduction to the topic of dose reporting in terms of "dose to medium in medium, ($D_{\text{m,m}}$)" or "dose to water in medium, ($D_{\text{w,m}}$)." Most importantly from the clinical perspective are differences between dose distributions derived by the TG-43 method on which clinical experience rests and MBDCAs, that is, differences in the output between traditional and new methods. This will be covered in Section 11.7 that provides a review of dose differences in clinical situations between TG-43 and MBDCAs accounting for tissue and applicator heterogeneities, presence of shields, and finite patient dimensions in various BT energy regions.

11.2 Interaction Characteristics for Photons 20–1250 keV

BT can be divided into a low ($E < 50$ keV) and a high ($E > 50$ keV) energy region. Different photon interaction modes dominate in these regions. Figure 11.2 shows the relative contributions of photoelectric absorption and coherent and incoherent scatter to the total mass attenuation coefficient in water as a function of photon energy.

As can be seen from Figure 11.2, there are substantial differences in the main photon interaction mode with energies over the range of interest to BT. Coherent and incoherent scattering dominates in the high energy region (>50 keV). Therefore, at these energies, the dose is sensitive to changes in the 3D

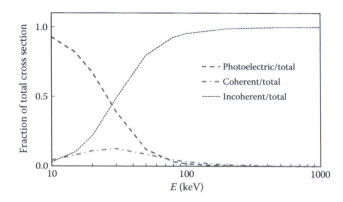

FIGURE 11.2 Proportion of the photon mass attenuation coefficient in water according to the contributions from photoelectric absorption, coherent, and incoherent scattering as a function of photon energy E. (Courtesy of Å. Carlsson Tedgren).

scattering environment, for example, finite patient dimensions. Dosimetry for low-energy BT sources (<50 keV) is less sensitive to changes in the 3D scatter environment but very sensitive to small changes in atomic composition of the transport medium since the dominating photoelectric cross section is approximately proportional to Z^3, where Z is the atomic number. The main photon interaction modes as a function of photon energy explain why doses calculated by MBDCAs and TG-43 vary in magnitude with photon energy in nonwater tissues.

The quantity $1 - \mu_{en}/\mu$, where μ_{en} and μ are the linear energy absorption and linear attenuation coefficients, respectively, is a measure of the average amount of energy that is transferred to the secondary photon in a single photon interaction (Table 11.1). It takes on high values in materials of low atomic numbers in the high energy range (>50 keV). Another important aspect of dose calculations in most of the BT photon energy range is the short continuous slowing down approximation (CSDA) range of secondary electrons; see Table 11.1, which also includes the photon mean free paths at the selected energies.

If the maximum range of the released secondary electrons is short compared to the corresponding mean free path of the photon, charged particle equilibrium (CPE) prevails at a point provided that the medium surrounding the point is homogeneous up to a distance equal to the maximum range of the secondary electrons (Alm Carlsson 1985). In a low atomic number medium

TABLE 11.1 CSDA Range of Electrons, Photon Mean Free Path, $\dfrac{1}{\mu}$, and Fraction $1 - \mu_{en}/\mu$ of Energy Transferred to Secondary Photons per Photon Interaction in Water as a Function of Energy

Energy (keV)	CSDA-Range Electrons (g/cm²)	$\dfrac{1}{\mu}$ (cm)	$1 - \mu_{en}/\mu$
30	1.8E-03	2.5	0.56
100	1.2E-02	5.6	0.85
350	1.1E-01	8.9	0.71
1000	4.4E-01	14	0.58

Note: Here μ_{en} is the linear energy absorption coefficient, and μ is the linear attenuation coefficient.

such as water, CPE is a good approximation for photon energies < 1 MeV, and the "kerma approximation" can be used to estimate the absorbed dose, that is, absorbed dose is considered equivalent to the medium collision kerma. If the electron ranges are short compared to the dimensions of voxels used in treatment planning (on the order of 1 mm³), CPE can be assumed to exist throughout the voxel, that is, interface effects between voxels of different atomic compositions can be neglected, and the kerma approximation used to derive values of absorbed dose except for voxels very close to sources (Yu et al. 1999) and in the relatively rare use of the radionuclide ^{60}Co (Figure 11.3).

The absorbed dose contributed by primary photons, "the primary dose," can now be obtained using 1D ray-tracing and analytical calculations. The relationship between absorbed dose in a point P from primary photons, D_{prim}, and collision kerma, $K_{col,prim}$, for a monoenergetic point source is given by

$$D_{prim}(P,E) = K_{col,prim}(P,E)$$

$$= \frac{1}{4 \cdot \pi \cdot r^2} \cdot \frac{\mu_{en}(P,E)}{\rho} \cdot \Psi_{prim}(P,E) \cdot e^{-\int_0^P \mu(E,l) \cdot l \cdot dl} \qquad (11.1)$$

where E is the photon energy, r is the distance between the source and P, μ_{en} and μ are the linear energy absorption and linear attenuation coefficients, ρ is the density of the medium, and Ψ is the energy fluence of primary photons. The absorbed dose contributed by subsequent generations of photons, that is, "the scatter dose," depends on the full 3D environment surrounding the calculation point and requires use of 3D calculation methods for full accuracy.

At 60–350 keV, around 85% to 70% of the energy of the interacting photon is transferred to the secondary photon in low

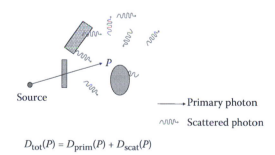

$$D_{tot}(P) = D_{prim}(P) + D_{scat}(P)$$

FIGURE 11.3 When CPE prevails, as at most energies and situations of interest to BT, the primary dose (the absorbed dose contributed by charged secondaries to primary photons) in the point P, $D_{prim}(P)$, can be calculated by 1D ray-tracing using the kerma approximation for absorbed dose (see text for explanation). The scatter dose in P, $D_{scat}(P)$ (i.e., the dose contributed by photons that interact for the second- or higher-order time outside the BT source), depends on the full 3D geometry surrounding P and must, for the highest accuracy, be derived with methods capable of 3D scatter integration. (Courtesy of Å. Carlsson Tedgren.)

atomic number media (see Table 11.1 for water). This is caused by the scattering process being relatively elastic (i.e., with small fractional energy loss) in this energy interval. The relatively elastic scattering process reduces the calculation efficiency of MC simulations since many interactions take place before the photon has lost a considerable fraction of its initial energy. On the other hand, the secondary electron range is considerably shorter than the photon mean free path and can be considered to deposit their energy locally under CPE assumption. Thus, there is no need to simulate electron transport which increases the efficiency of MC calculations.

11.3 Principles of MBDCAs

While the first principle approaches (Section 11.3.2) are the most accurate, the semiempirical methods reviewed in Section 11.3.1 can be useful when increased computational efficiency is needed, for example, in optimizing source placement and dwell times.

11.3.1 Semiempirical Approaches

11.3.1.1 Ray-Tracing with Scaling

Methods that apply 1D ray-tracing with scaling to correct total doses, due to the presence of high atomic number shields around ^{137}Cs sources, have been developed (van der Laarse and Meertens 1984; Weeks and Dennett 1990) and are also in use for ^{192}Ir sources. The approach is correct for the dose contributed by primary photons but not for scattered photons. It can be a reasonable approximation at short distances from ^{137}Cs and ^{192}Ir sources, where doses from primary photons dominate.

11.3.1.2 Applicator-Kernel Superposition

Superposition of MC precalculated applicator-specific dose kernels to account for shields of high atomic numbers has been tested for ^{192}Ir sources (Markman et al. 2001; Yan et al. 2008). This approach is less feasible at low energies where the impact of patient-specific heterogeneities is more pronounced.

11.3.1.3 Scatter Separation Methods

Several methods that separate the primary and scatter dose components have been developed using 1D ray-tracing calculations for the primary dose and approaches of different complexity for the scatter dose. A scatter separation method that accounts for the location of heterogeneities but not for their lateral extent was developed for ^{137}Cs and ^{226}Ra sources (Williamson 1990). An "effective path-length" method with additional empirical corrections for the scatter component was developed to account for individual lead shields in radiotherapy using ^{241}Am sources that emit 60 keV photons (Nath et al. 1990). The scatter-to-primary ratio in low-Z materials at 60 keV is around 80%. The individual lead shields and bulky high-Z ^{241}Am sources presented a challenging and truly 3D problem that would clearly benefit from being solved by 3D first-principle methods. A BT version of the scatter-subtraction method (Lulu and Bjärngard 1982) has

been developed and applied for calculations around ^{192}Ir sources (Williamson et al. 1993). It was later extended to perform 2D scatter integration in 3D space (Kirov and Williamson 1997) and complemented with an analytical approach to account for the atomic number of the heterogeneities (Daskalov et al. 1998). An analytical method using precalculated scatter-to-primary ratios was used to account for limited phantoms and heterogeneities around ^{192}Ir sources (Anagnostopoulos 2003). A method separating the calculation of dose from primary, once-scattered, and multiple-scattered photons has recently been presented (Wang and Sloboda 2007a,b). Poon and Verhaegen developed correction-based methods intended for use with ^{192}Ir sources to take into account lack of scatter in breast implants (Poon and Verhaegen 2009a) and also the presence of heterogeneities (Poon and Verhaegen 2009b).

11.3.2 First-Principle Methods

The first-principle methods represent general approaches that perform transport calculations in the actual media and account for the dependence of the scatter dose on the 3D geometry. MC simulation solves the linear Boltzmann transport equation (LBTE) by random sampling and is the current state of the art in computational dosimetry. The description below concentrates on applications in treatment planning.

11.3.2.1 Point Kernel Superposition/Convolution Methods and CC Method

Superposition/convolution methods are based on superposing the distribution of energy released by a group of photons with a kernel describing the dose distribution around an interaction site of these photons. Ray-tracing operations are used to scale the kernels, initially determined for a homogeneous medium, for heterogeneities leading to dose distributions involving no principal approximations up to and including dose from once-scattered photons. This approach has been extensively investigated for EBRT (Ahnesjö and Aspradakis 1999) and applied in BT (Williamson et al. 1991; Carlsson and Ahnesjö 2000a). The CC algorithm is a fast method for kernel superposition designed for treatment planning applications with the calculation speed stemming from an efficient kernel ray-tracing (Ahnesjö 1989) provided through "collapsing" the transport of energy released onto cone axes (angular discretization) in combination with suitable lattices of transport lines (computational grid) over the calculation volume. CC algorithms are in clinical use for EBRT, and they compare well with full MC simulations (Fogliata et al. 2007). A BT version of CC has been developed (Carlsson and Ahnesjö 2000b). It calculates the primary dose analytically under the CPE assumption and the dose from scattered photons in two subsequent steps using a "successive scattering" approach (Carlsson and Ahnesjö 2000a) designed to minimize systematic errors due to kernel discretization and approximations in multiple-scatter dose. The CC method has been used to calculate dose for sources in the 30- to 662-keV energy range (Carlsson and Ahnesjö 2000b), around clinical sources using primary-scatter separation (PSS) source data precalculated with MC (Russell et al. 2005), and to account for high-atomic number heterogeneities, including the effects of characteristic x-ray production (Carlsson Tedgren and Ahnesjö 2003). The BT CC algorithm is currently being integrated to the Oncentra Brachy TPS from Nucletron BV (Veenendaal, the Netherlands) (van Veelen 2011).

11.3.2.2 Deterministic Solutions to LBTE

Principles of analytical solvers to the LBTE, also referred to as GBBS, have been described in detail in Chapter 7 of this book and will not be covered here. Analytical solvers have been tested successfully for BT (Daskalov et al. 2000, 2002; Zhou and Inanc 2003; Gifford et al. 2006). A rationale for their use in treatment planning is that they can be faster than MC simulations. A GBBS named Acuros and developed by Transpire, Inc. (Gig Harbor, WA) has recently been implemented in the Varian Medical Systems Inc. (Palo Alto, CA) TPS BrachyVision for ^{192}Ir dose calculations and has been benchmarked against MC and measurements (Petrokokkinos et al. 2011).

11.3.2.3 MC Simulations

Principles of the MC method for photon and electron transport have been covered in Chapter 7 and will not be repeated here. A large number of examples on the use of MC codes to investigate BT treatment planning will be provided in Section 11.7. Practical application in treatment planning requires fast calculations and that the dose is derived in all voxels of interest in a 3D region around the target. Point dose estimators, which can be fast in calculating dose to a single point (Williamson 1987), are not efficient in treatment planning applications. MC simulations for BT can mostly assume CPE and disregard electron transport, allowing the use of "track-length estimators" that score the photon energy fluence differential in energy and calculate absorbed dose by combining this estimate with mass energy absorption coefficients. Each photon passing a voxel contributes to the track-length estimator's score of photon fluence, which is why these are more efficient than analogue estimators to which only photons interacting in a voxel contribute (Williamson 1987). Also due to the prevalence of CPE, MC can be used for the scatter dose part only, leaving calculation of the primary dose to analytical 1D ray-tracing methods. The high and intermediate energy regions of BT are among the most demanding for MC since scattering with relatively little energy loss is the dominating interaction mode. The situation is less demanding in the low energy range. Several codes aimed at being sufficiently fast for use in BT treatment planning have been developed (Chibani and Williamson 2005; Taylor et al. 2007). Effort to increase the efficiency of MC simulations by finding new variance reduction techniques such as correlated sampling can be noted in the literature (Hedtjärn et al. 2002).

Rivard et al. (2009) published a review of dose-calculation algorithms in a Vision 20/20 article, and a separate review can also be found in Chapter 14 of the 2005 AAPM BT Physics Summer School (Williamson 2005).

11.4 Source Characterization

Precalculated information on a radiation source, that is, source characterization, is an important preprocessing step to treatment planning using both MBDCA and TG-43.

The TG-43 formalism enables accurate reproduction of the dose distribution determined in a large water phantom surrounding a single source, initially derived by either MC simulations and/or by measurements using lithium fluoride (LiF) thermoluminescent dosimeters (TLDs). For specific details regarding obtaining calculated or measured source data, see Section V in the TG-43U1 report (Rivard et al. 2004). Source component configuration influences dosimetry significantly, and a reliable source characterization is of the utmost importance (see, e.g., Rivard 2001).

Although some MBDCAs (e.g., MC and GBBS) are capable of explicitly modeling the source, remodeling a source of identical design many times and in every patient is unnecessary. Computational power is better applied to the patient geometry during run time calculations for treatment planning. Therefore, a method for precharacterizing the source that can be utilized by MBDCAs is needed. For MC simulations, "phase space" files have been used, for example, see Poon and Verhaegen (2009b). A phase space file for a BT source is a file that has been derived in an MC simulation of the source geometry by scoring the energy, angle, and exit position of each photon emitted from the capsule of the source. Phase space files are large and thus difficult to distribute and commission.

11.4.1 Primary and Scatter Separation— Method for Source Characterization

An alternative source characterization method, which is potentially useful for all kinds of MBDCAs and has the advantage of being backward compatible with and very similar to the current method recommended by TG-43, has been suggested under the name of "primary scatter separation" (PSS) (Russell and Ahnesjö 1996; Russell et al. 2005). The PSS concept has hitherto been used for clinical BT sources with the CC algorithm (Russell et al. 2005) and with a 2D scatter separation method (Kirov and Williamson 1997). The PSS concept has the potential to serve as a source characterization method also for MC and GBBS methods, as outlined in Figure 11.4.

Since BT is performed with photon energies low enough for CPE to prevail, it can be shown that the primary dose around the source in any homogenous media is a fingerprint of its internal geometry and energy characteristics. To generate PSS data, there is only one additional requirement to the standard TG-43 recommendations for MC source modeling: that the dose in water outside the source should be separated into its primary* and scatter constituents during scoring; such results would be useful as source input to MBDCAs. CC superposition for BT has utilized

---

* In the PSS source characterization, the primary dose is the dose from photons interacting for the first time outside the source active material and its encapsulation.

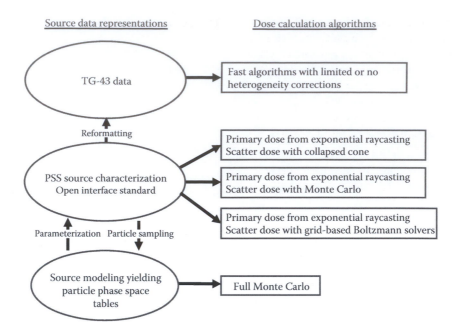

FIGURE 11.4 Illustration of the basic principles of the PSS source characterization approach. Data from PSS source characterization have the potential to drive several kinds of dose calculation methods, although this has not yet been explicitly tested. PSS data are preferably derived following the already existing TG-43 recommendations for source characterization. (Courtesy of A. Ahnesjö. With permission.)

PSS data combined with ray-tracing and 1D scaling in the actual arbitrary heterogeneous environment to derive the primary dose and to initiate scatter dose calculations in the patient geometry (Russell et al. 2005). When CPE prevails, the primary dose can be calculated correctly by 1D methods, and several MC codes in use for low-energy photon transport can calculate the primary dose analytically with actual MC simulations performed for only the scattered dose (Williamson 1987; Sandborg et al. 1994).

The PSS approach allows storage of source data in tabular form, either with data parameterization for primary and scatter dose following the TG-43 formalism or through fitting of data to exponential functions as suggested by Russell et al. (2005) and Russell and Ahnesjö (1996). An effective source data set, which is useful as input to algorithms of different complexity, is highly desired for a TPS that supports several types of dose calculation algorithms. Differences in single source dose distributions obtained with diverse calculation algorithms within the same TPS originating from variations in the initial source modeling (Mikell and Mourtada 2010) would be avoided. Several recent publications on BT source characterization have, in addition to TG-43 data, also provided the absorbed dose to water separated into its primary and scatter components in preparation for the PSS concept (Taylor and Rogers 2008; Meigooni et al. 2009; Granero et al. 2011).

11.5 Tissue and Applicator Considerations

In principle, some MBDCAs are capable of calculating doses in arbitrary configurations of material and geometry. A rationale

of using MBDCAs in treatment planning is to calculate dose distributions for each patient accounting for their individual anatomic configuration and the distribution of applicators and sources. To utilize this capability, input in the form of density and material composition of applicators and patient anatomy is required. Methods for obtaining this information with sufficient accuracy to achieve high precision in the calculated doses are essential. Computerized tomography (CT) imaging is the method for tissue segmentation in using MBDCA for EBRT. Neither magnetic resonance (MR) nor ultrasound (US) imaging can, at present, identify the tissue types in the way required for dose calculations. These imaging methods are important for other clinical purposes such as improved contouring due to better soft tissue contrast (MR) and allowing image guidance during needle and source implantation (US). The reason why CT is successfully used in EBRT is that the Hounsfield units (HU), which make up a CT image, are proportional to the electron density of tissues (Papanikolaou et al. 2004). The cross section for Compton scattering is proportional to the electron density and hence HU provides the information required for MBDCAs at photon energies where Compton scattering is the dominant interaction mode. Compton interaction dominates in the high energy range of BT to which the common radionuclide ^{192}Ir belongs, see Figure 11.2. However, in the low energy BT range (<50 keV) to which ^{125}I and [103]Pd sources (commonly referred to as seeds) and electronic BT x-ray sources belong, a large fraction of the photon interactions take place through the photoelectric effect. Because the interaction probabilities for the photoelectric effect are approximately proportional to Z^3, low-energy photon transport calculations are very sensitive to material composition. Therefore, conventional CT does not provide material composition with sufficient accuracy in the low energy range of BT (Verhaegen and Devic 2005; Furstoss et al. 2009; Landry et al. 2010; Afsharpour et al. 2010). Research on the use of dual-energy CT to increase accuracy of tissue segmentation in the low energy BT range is currently of high interest and a topic that BT shares with proton and ion beam therapy. These external beam modalities also benefit from more detailed tissue composition information due to the strong relationship between nuclear interaction cross sections and atomic material composition (Unkelbach et al. 2009; Yang et al. 2011). An interim solution for the low energy range would be to define tissues based on organ contouring and using standard compositions of various human tissues, such as those provided in ICRU Report No. 44 (ICRU 1989). In this way, dosimetric consistency between institutions, existing with the current TG-43 approach, could be preserved through use of recommended tissue compositions for organ types. Numerical values of calculated doses would be closer to true ones than those calculated by the TG-43 approach; however, one would still not obtain truly individual dose values that account for dosimetric effects due to individual variations in tissue composition, for example, calcifications of prostate and breast tissues (Landry et al. 2010).

Streak artifacts in CT images taken after permanent seed implant can cause problems when used as input for dose calculation, and actions need to be taken to remove imaging artifacts as otherwise erroneous information is supplied to the MBDCA (Xu et al. 2011). Applicators for BT can be CT/MR compatible (Barillot et al. 2006; Viswanathan et al. 2007). For applicators with shields, unless the shields are loaded into the applicator after imaging, with density and material composition information added into the MBDCA during a separate step, they would also contribute artifacts in the images (Price et al. 2009; Reniers and Verhaegen 2011). Furthermore, it has been shown that the differences between soft tissues and water are within a few percent for ^{192}Ir (Poon and Verhaegen 2009a), which would make it possible to use MR or US as an imaging modality at this energy, assuming tissue to be water-equivalent and using MBDCA to account only for the effect of finite patient dimensions and applicator components/shields.

11.6 Dose-Reporting Medium

First-principle MBDCAs are used for transport calculations in the local medium; however, for reporting, either the absorbed dose to the local medium or to a small water cavity in that medium can be chosen. The topic has its roots in EBRT and, as will be seen below, becomes increasingly complicated for BT, especially at energies <50 keV. For high-energy BT sources, the problem is much less, with a situation close to that encountered in EBRT. There is, at the time of writing this chapter, a lack of published investigations on the topic at BT energies, and the aim of this section is to provide the current background for the problem under discussion.

11.6.1 EBRT Background

When MBDCAs were introduced in EBRT, values of absorbed dose to media were converted into absorbed dose to water cavities in the local medium using unrestricted stopping power ratios between the medium and water. This approach reports dose to a small (Bragg–Gray) water cavity in the medium and was chosen since values of dose to water were closer to the values derived by earlier simpler algorithms for which clinical experience existed. Since that time, a debate has arisen in the EBRT community with arguments in favor of both quantities (Liu and Keall 2002). The AAPM TG-105 report on photon MC treatment planning (Chetty et al. 2007) does not take a standpoint for either quantity, but states that it is important to report which one is calculated and to provide data for a conversion of one quantity into the other. Differences between dose to medium and dose to water in medium at MV photon and electron beams are around 1% to 3% for soft tissues and up to 10% for bone (Walters et al. 2010). At MV energies, the conversion is made using ratios of stopping powers under the assumption that the cavity is a small Bragg–Gray cavity of water in the medium. Also, in proton therapy, the issue of converting doses to medium into doses to small water cavities while using MC methods for treatment planning is of interest (Paganetti 2009).

11.6.2 Problem Description for BT

In BT, choice of dose-reporting medium has not been addressed until very recently. The majority of published MC studies in patient geometries compare dose to medium in medium calculated by MBDCAs to dose to water as derived using the TG-43 formalism. A few recent BT MC studies have compared values of dose to medium in medium with dose to water in medium. Values of dose to water in medium were, in these studies, derived through applying mass-energy absorption coefficients for either water or the local medium to the values of energy fluence obtained using track-length estimators in the MC simulations (Landry et al. 2010, 2011; Rivard et al. 2010). Thus, differences between values of dose to medium and dose to water reported in these studies equal the ratios of mass-energy absorption coefficients between water and medium, that is, they have been obtained under assumption of CPE, thus assuming the water cavity to be large in comparison with the ranges of secondary electrons.

One argument in favor of dose to water in medium stems from viewing the cell, or part thereof, as the radiation target and the assumption that this structure has, in general, a composition closer to water than the averaged bulk tissue (Liu and Keall 2002). This leads to an issue of concern at low BT energies considering that ranges of secondary electrons (Table 11.1) are comparable to the dimensions of cellular components. At low BT energies, it is not immediately obvious how dose to the local medium, calculated in a millimeter-sized voxel grid, should be converted into values of dose to a water cavity of cellular/subcellular dimensions (see Figure 11.5).

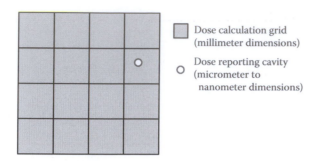

FIGURE 11.5 Cavity theory conversion for dose reporting is a process in which absorbed doses to the bulk medium, defined on a macroscopic grid (dimensions in millimeters), are converted into the absorbed dose valid for a "cavity" of water of cellular/subcellular dimensions (micrometers to nanometers) located in that medium. As a result, the issue of dose reporting does not affect the initial transport calculations for treatment planning being performed in a grid with voxel dimensions of at least 1 mm³ and the average bulk medium composition. (Courtesy of Å. Carlsson Tedgren.)

The ranges of secondary electrons in MV photon and ^{192}Ir beams are long compared to cellular dimensions, and ratios of unrestricted stopping powers in the medium and water can be used to toggle between dose to medium and dose to water. At low-energy BT, however, the ranges of secondary electrons are such that a cavity of cellular dimensions cannot be assumed either large or small. The conversion of dose to medium in medium into dose to water in medium at low-energy BT is thus best investigated using full MC simulations to assumed cavity sizes. Such investigations have been initiated, and preliminary results report a complicated pattern that depends heavily on assumptions of material compositions and with no obvious advantage of either quantity (Enger et al. 2010, 2011a; Thomson et al. 2011). Enger et al. (2012) used MC and monoenergetic photons 20–300 keV to irradiate a 1 cm³ tissue box in which a spherical cavity of diameter 14 micrometer was centered and filled with either (1) the surrounding tissue (breast, prostate, muscle or adipose), (2) water, or (3) any out of several different atomic compositions available in literature for cell nuclei. Hypothesizing the cell nuclei (n) to be the radiation target, the aim was to study which of $D_{w,m}$ and $D_{m,m}$ come closest to $D_{n,m}$. Under the investigated assumptions, $D_{w,m}$ was found to be a good substitute for $D_{n,m}$ for most of the investigated materials. Further studies on the topic are warranted and can be expected in the close future.

Table 11.2 summarizes the various quantities available for dose reporting in BT. The notation for absorbed dose estimation at location r is $D_{x,y-Z}(\mathbf{r})$, where x denotes the choice of dose-specification medium (either water or local medium at r); y denotes the voxel composition used for the radiation transport calculations (where voxel composition can either be homogeneous water, all biological tissue approximated by water except for applicators and air–tissue interfaces, or applicators and air–tissue interfaces with actual tissue compositions and densities); and Z denotes the choice of algorithm (either MBDCA or TG-43). The superscript "ct" seen on the dose to water in medium is to

TABLE 11.2 Quantities for Dose Reporting

Notation	Dose Calculation Method	Comment
$D_{w,w-TG-43}$	Superposition of precalculated absorbed dose-to-water distributions around single sources (the AAPM TG-43 formalism)	The quantity on which current clinical experience rests
$D_{m,m-MBDC} = D_{m,m}$	MBDCAs calculating dose to the media comprising the calculation volume (part of patient, sources, applicators)	Requires knowledge of the composition of the calculation volume (from CT); calculated by MBDCA (MC, CC, or GBBS)
$D_{w,m-MBDC}^{ct} = D_{w,m}^{ct}$	MBDCA radiation transport as above, but the dose to bulk medium is converted into dose-to-water using cavity theory	Default in external beam therapy. Concerns relating to the conversion method may arise at low BT energies (i.e., <50 keV), where cellular components have dimensions similar to or smaller than ranges of secondary electrons and where calculated doses are sensitive to small differences in atomic composition

indicate the cavity theory used for conversion. The notation follows that put forth by AAPM TG-186 (Beaulieu et al. 2012).

In summary, it can be said that the topic of dose reporting is similar to that in EBRT for high-energy BT but becomes complicated at low BT energies. At low energies, calculated doses are sensitive to small differences in atomic composition, and ranges of secondary electrons are comparable to cellular dimensions. The topic would definitely benefit from further investigations. Furthermore, previously published studies (Landry et al. 2010, 2011; Rivard et al. 2010) have investigated the difference between the quantities for dose reporting using ratios of mass-energy absorption coefficients to convert dose-to-medium into dose-to-water. More knowledge on the differences between dose-to-medium in medium and dose-to-water in medium under various assumptions of the size of the water cavity would be valuable.

11.7 Effects of Heterogeneity Corrections on Calculated Doses—Clinical Implications

In a Vision 20/20 article, Rivard et al. (2010) generated an overview of the anticipated effect of heterogeneities for various treatment sites. This manuscript presented the results in terms of high (e.g., ^{192}Ir) and low (e.g., ^{125}I) photon energy. The same information can be reformatted in terms of major physical effects on the dose relative to a water-only geometry (Table 11.3). The following definitions of AAPM and ESTRO will be used in the following

TABLE 11.3 Summary of Major Contributors to Dose Departure from TG-43 Based on Table 1 from Rivard et al. (2009)

Energy Range	Physical Effect
High:	Scatter condition
^{192}Ir	Shielding (applicator related)
Low:	Absorbed dose (μ_{en}/ρ)
^{103}Pd/^{125}I/eBx	Attenuation (μ/ρ)
	Shielding (applicator, source)

Source: Rivard, M.J. et al., *Med Phys* 36, 2136–53, 2009.

discussion, namely, "low energy sources" corresponds to photon energy below 50 keV [this includes electronic BT (eBx) sources (Rivard et al. 2005)], while all sources above 50 keV are classified as "high energy," including experimental sources such as ^{169}Yb (Lymperopoulou et al. 2005; Mason et al. 1992; Piermattei et al. 1995) and ^{170}Tm (Ballester et al. 2010; Enger et al. 2011b). In addition to the review presented in the following, the interested reader can find further information in the work of Thomadsen et al. (2008) and Rivard et al. (2009).

11.7.1 Low Energy Region (<50 keV)

In the energy range below 50 keV, photon mean free paths are short, the radial dose function decreases quickly, and the contribution of primary photons becomes less important compared with the contribution from scattered photons as close as 2 cm from the source. For most clinical applications, the scatter conditions are not the most important factor influencing the dose deposition relative to TG-43 (Melhus and Rivard 2006; Pérez-Calatayud et al. 2004), except for the case of lung BT (Chen et al. 1999; Johnson et al. 2007; Santos et al. 2003; Voynov et al. 2005; Yang and Rivard 2011). Taking into account the importance of the photoelectric effect as depicted in Figure 11.2, one can quickly estimate that any departure from a water-equivalent medium (both in terms of μ_{en}/ρ and μ/ρ) will lead to significant differences between various possible ways to score the dose, namely, $D_{w,w-TG-43}$, $D_{w,m}$, and $D_{m,m}$. Therefore, low-energy photon sources are not very forgiving of any type of heterogeneities, as illustrated in Figure 11.6. This would include the effect of photon attenuation by other sources in multiple source implants, that is, shielding due to interseed attenuation.

Note that the three most common clinical uses of low energy sources are permanent seed prostate implants followed by eBx source for breast BT and eye plaque treatments with seeds. Permanent breast seed implants (Pignol et al. 2006) and lung mesh BT (Santos et al. 2003) have also been reported. The effect of considering heterogeneities rather than homogeneous water was initially studied in the early 1990s and showed deviations up to 50% for adipose tissues (Huang et al. 1990; Meigooni and Nath 1992). Similarly, Reniers et al. (2004) studied the impact of the medium on the radial dose function. Their results were expected based on fundamental physics cross-section behavior. However, these were for single source configurations in geometries where uniform water was replaced by a uniform medium

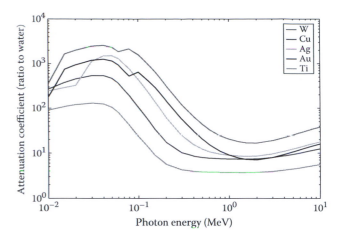

FIGURE 11.6 Attenuation coefficients relative to that of water as a function of photon energy for various materials found in applicators and source compositions. (Courtesy of Luc Beaulieu. From Hubbell, J.H. and Seltzer, S.M. Tables of X-Ray Mass Attenuation Coefficients and Mass Energy-Absorption Coefficients (version 1.4). Available at http://physics.nist.gov/xaamdi. National Institute of Standards and Technology, Gaithersburg, MD, 2004. With permission.)

of interest. More recently, clinical configurations have been studied using increasing medium complexity. In the following sections, a treatment site-specific review is given.

11.7.1.1 Permanent Prostate Implants

One of the characteristics of permanent prostate seed implants is the use of multiple seeds, from a low number of 30 to as many as 130. While the attenuation coefficients of most high-Z materials exhibit a strong departure from water for all energies used in radiation therapy, the size of the sources, typically 0.8 mm in diameter and 4.5 to 5 mm in length, used in permanent implants limits the amount of materials in the path of photons. As such, the AAPM TG-64 report considered the effect of shielding from one source by another, that is, the interseed attenuation (ISA) effect, to be negligible (Yu et al. 1999). Still Meigooni and Nath (1992) reported a possible dosimetric impact of up to 6% on the minimum peripheral dose, while Mobit and Badragan (2004) studied the ISA for a cubic arrangement of seeds, giving a decrease in the minimum peripheral dose of 10% or less. This was followed by characterization of ISA using clinical seed configurations. ISA was shown to reduce the clinical D_{90} metric (the dose received by 90% of the CTV) from TG-43 dose calculations by 2% to 5% depending on the seed configuration (Carrier et al. 2007), on the number of seeds used (i.e., the seed density), which depends on the seed activity (Carrier et al. 2006), and, to a weaker extent, on the radionuclide used (i.e., comparing ^{103}Pd to ^{125}I sources) (Chibani et al. 2005). Similarly, it was shown that for a given source/patient configuration, the ISA will lead to D_{90} variations of 2% to 3% depending on the seed model used, with some seeds being more water-like than others. A new, water-equivalent seed model was proposed in the same work (Afsharpour et al. 2008). Of course, these are average variations over the whole organ. Local dose variations in close proximity to seeds are much higher at tens of percent (Chibani et al. 2005; Afsharpour et al. 2008).

The effect of heterogeneous tissue composition for permanent prostate geometries has also been studied by Chibani and Williamson (2005). Deviation of a few percent was found between $D_{w,w\text{-}TG\text{-}43}$ and $D_{m,m}$. Retrospective dose calculations of a clinical cohort by Carrier et al. (2007) confirmed an average difference of 3% to 4% on D_{90} due to tissue composition and density being only different than water. Furthermore, these variations were patient dependent, that is, based on variation in the CT datasets leading to variation in tissue densities, and differences of the D_{90} metric up to 14% lower were obtained with MC simulations combining both the effect of ISA and tissue heterogeneities. The effect of calcification within the prostate gland was investigated by Chibani and Williamson (2005). They showed that a calcium content of only a few percent diffused within the prostate could change the clinical dose parameters, such as D_{90}, by up to 37% relative to TG-43. An improved characterization of the effect of localized calcification from clinical image sets is needed. In addition, a study on the impact of metal artifacts around the seeds on CT image sets is warranted (Xu et al. 2011).

11.7.1.2 Breast BT

In the low energy regime, breast BT has two primary modalities: high-dose-rate (HDR) BT using eBx and permanent low-dose-rate (LDR) seed implants. The average energy for both is approximately 21 keV, although eBx is an x-ray source and has a broad, filtered spectrum (Liu et al. 2008). eBx is currently used exclusively with balloon applicators in breast cancer treatments. Contrast agents, such as iodine-based saline solution or barium sulfide, used in balloon-type applicators for high-energy HDR BT would clearly significantly attenuate the photon fluence of an eBx source. As such, only the balloon wall is filled with contrast agent, which leads to an attenuation of about 6% (Mille et al. 2010).

The effect of tissue heterogeneity is significant in breast (Furstoss et al. 2009; Landry et al. 2010; Afsharpour et al. 2010; Mille et al. 2010). This is because the breast composition, a mix between adipose and glandular tissues, is significantly different from water at these low energies. As such, dose differences around a single seed embedded in breast tissue can lead to dose differences above 30% to 40% (Landry et al. 2011). Furthermore, the overall level of differences between TG-43 and MC depends on which quantity, either $D_{w,m}$ or $D_{m,m}$, is computed, both being significantly different than the $D_{w,w\text{-}TG\text{-}43}$ distribution (Landry et al. 2011). For clinical breast multiseed implant configurations, taking into account the tissue composition impact on clinical parameters was shown to lead to a decrease of 4% to 35% depending on the proportion of glandular and adipose tissues (Afsharpour et al. 2010). Similarly, a breast composed only of adipose tissue would lead to increased dose to the skin by as much as 25% due to the lower attenuation of the adipose tissue. Since breast composition was shown to be age-dependent (Yaffe et al. 2009), Landry et al. (2010) conducted a sensitivity analysis

and concluded that the expected variation within the population remains smaller than the effect of going from a water-based dose calculation to a mean population breast composition. Still, CT imaging might enable us to devise better tissue segmentation schemes for breast (Afsharpour et al. 2011; Sutherland et al. 2011).

The reader should note that contrary to permanent prostate seed implants, ISA in breast remains at the level of 2% to 3% and is a negligible effect for this procedure. This is because tissue heterogeneity is the dominant effect in breast for low-energy photons (Landry et al. 2010).

11.7.1.3 Eye Plaques

The gold-alloy backing (Modulay) and silicon polymer seed carrier (Silastic) of eye plaque applicators have a large effect on the dose distributions. Early TLD measurements and MC dose calculations for a single seed showed that the presence of these high-Z materials decreased the dose on the central axis of the applicators by 8% to 10% at 1 cm and up to 15% at 2 cm or off-axis (Chiu-Tsao et al. 2007; Cygler et al. 1990; Waever 1986; Wu et al. 1988). This decrease is due to the reduced contribution from backscattered radiation relative to a homogeneous water geometry. The gold backing itself was shown to produce a dose increase in the first few millimeters from the applicator due to the increased fluence from low-energy fluorescent photons and secondary electrons. However, these are absorbed by the Silastic holder (Cygler et al. 1990). Dose to critical ocular structures was also shown to be 16% to 50% less than expected compared with the TG-43 dose calculation (Astrahan 2005).

Simulations of realistic 3D configurations using fully loaded plaques, from 10 to 22 mm in diameter, have shown that the dose on the central axis is around 11% (at 5 mm) to 15% less (at 10 mm) when ^{125}I seeds are used (Melhus and Rivard 2008; Thomson et al. 2008, Thomson and Rogers 2010). These values are radionuclide-dependent with [103]Pd producing larger differences from 20% to 30% due to its lower-energy photon spectrum, while [131]Cs with its higher average photon energy gives a reduction of about 9% at 5 mm relative to a full water environment. However, the off-axis dose can be reduced by as much as 90% (Thomson et al. 2008). In contrast, ISA for eye plaque simulations is of a similar magnitude to that for permanent prostate implants at about 2% (Thomson et al. 2008).

The change to the dose delivered to various points of interest (critical structures) has been extensively studied by Thomson et al. (2008) and Rivard et al. (2011) and was shown to yield dose reduction of the order of 20% to 30% for most of them and up to 92% dose reduction for the lacrimal gland. Furthermore, Rivard et al. (2011) have shown significant differences in the 85-Gy prescription dose at 5-mm depth, including a 9-Gy reduction for ^{125}I and an 18-Gy decrease for [103]Pd relative to the water-only geometry (Nag et al. 2003). Note that for most of the abovementioned works, the eye was considered to consist of water, and the heterogeneous environment referred to the gold backing, the Silastic (a silicon polymer that is used as a seed carrier), and the eye–air interface. Only the work by Thomson and Rogers (2010) looked at the effect of using realistic eye tissue compositions. They found that the eye fluid yields a dose difference of about 2% to 3% relative to water, and the eye lens dose is about 7% to 9% lower. The AAPM TG-129 report will provide specific recommendations for the clinical medical physicists for this treatment modality (Chui-Tsao et al. 2012).

11.7.1.4 Lung Seed (Mesh) Implants

Lung mesh seed implants were first used at the end of the 1990s and initially described by Chen et al. (1999) for patients having received sublobar resection in stage I nonsmall cell lung cancer. It consists of ^{125}I seeds embedded in a synthetic suture material mesh, namely, polyglyconate (Vicryl). The seeds are weaved into the mesh along lines to constitute a 2D array. The seed array is set to cover the staple line plus a 2-cm margin and to deliver a prescription dose of 100 to 120 Gy at 0.5 cm above and below the implant plane. The array can be as large as 50 cm^2 (Johnson et al. 2007; Santos et al. 2003; Voynov et al. 2005).

Another BT option is called double suture (Lee et al. 2003; Powell et al. 2009). The seeds are placed 1 cm apart in a strand made of similar suture materials as above. Two of such strands are used and affixed 0.5 cm on each side of the resection margin.

In both approaches, the dosimetry is performed using the TG-43 formalism. However, the immediate implant environment would clearly be of interest to advanced dose calculation methods due to conditions that are clearly different from an all-water geometry (Rivard et al. 2011). This was further illustrated in a retrospective dose calculation study of a few clinical cases using MC. Both $D_{w,m}$ and $D_{m,m}$ were compared to TG-43. Differences over 30% are seen on key clinical parameters such as D_{90}. (Sutherland et al. 2012).

11.7.2 High Energy Region (>50 keV)

For high-energy photon sources used in BT, the effect of the medium in terms of attenuation or dose deposition relative to water is negligible for the most part. However, any use of high-Z materials such as shields or metallic applicators would introduce significant differences in the photon attenuation cross sections, as shown in Figure 11.6. Furthermore, high-energy photons have a longer radiation path length in water (or tissues). As such, for a finite geometry, the dose at the boundary will be affected by changes in the scatter conditions. Breast, head and neck, and superficial BT are among the procedures affected. Others, like prostate or gynecological treatments (with unshielded applicators), will see little benefit in moving from water-based tissue to heterogeneous composition (Melhus and Rivard 2006).

11.7.2.1 Breast BT

For ^{192}Ir HDR BT, Pantelis et al. (2005) showed that the finite size of the calculation phantom does not notably affect the high dose region but can reduce the skin dose by at least 5%. On a patient-by-patient basis, differences could be even larger depending on the configuration. This result was recently confirmed by

experimental measurements using TLD to validate MC dose calculations (Raffi et al. 2010). At lower energy, namely, that of ^{169}Yb, Lymperopoulou et al. (2005) have shown skin and lung dose differences of 15% to 30% for an interstitial implant based on an idealized configuration. Doses to the breast–air and breast–lung interfaces have also been simulated by Cazeca et al. (2010) for balloon BT for both radionuclides discussed above and found dose reduction of up to 16%.

Breast balloon BT comes in many forms, from a single catheter in the middle of the balloon to multi-strut configurations (see, e.g., Chapter 22). Some leave air in the balloon, while others recommend contrast agents for planning purposes. Kassas et al. (2004) have studied the effect of iodine-based saline solution as a contrast agent for the MammoSite balloon (Hologic Inc., Bedford, MA). The difference relative to a water-only geometry showed variations depending on the balloon diameter (4, 5, and 6 cm) and iodine content (5%, 10%, 15%, 20%, and 25%), with a maximum of about 6% for the 6-cm-diameter balloon with 25% contrast agent (Kassas et al. 2004). For the same balloon, Cazeca et al. (2010) found an even more pronounced effect than the previous studies using a different geometry. More importantly, they showed that the contrast has a larger effect on lower energy ^{169}Yb source than on ^{192}Ir.

Richardson and Pino (2010) studied the effect of having an air cavity in the SAVI balloon applicator (Cianna Medical, Aliso Viejo, CA) for ^{192}Ir sources. The dose was calculated using MC at various distances from 1 to 10 cm away from the device and comparing various configurations from single to multidwell positions. At the largest distances, the difference between water and air (within the balloon) is negligible. However, it can reach up to 9% at the prescription point, 1 cm from the surface of the balloon.

11.7.2.2 Head and Neck

Poon and Verhaegen (2009a) studied a case of head and neck ^{192}Ir HDR BT comparing TG-43, MC, and an analytical method. Doses to the skin, brain stem, nasal cavities, and bone were compared. TG-43 was found to be only marginally different than MC for the cases studied. Further comparison on a greater cohort of patients is needed to confirm this.

11.7.2.3 Shielded Applicators

Shielding is usually composed of high-Z materials. The effect of such materials will introduce a major perturbation relative to water-only geometry (Williamson et al. 1993). Sureka et al. (2006) have nicely illustrated this effect for GYN applicators using a 90°, 180°, and 270° shields within a vaginal cylinder. Poon et al. (2008) used the MC technique to calculate the dose around a shielded rectal applicator for 40 consecutive patients. They found that the dose to the contralateral normal tissues was decreased on average by 24% and also decreased the target dose by an average of 3%. More importantly, the changing of the position of the shield relative to the target and organs at risk (in this case, changes due to rotation) leads to significant modification of the dose distribution.

11.8 Summary and Conclusions

This chapter has presented an overview and assessment of model-based dose calculation algorithms available for use in BT that are able to calculate dose in heterogeneous (nonwater) geometries. Other challenges arising from the introduction of new calculation methods have also been introduced, such as new requirements for characterization of the radiation sources, uncertainties in the information available about nonwater media that can cause uncertainty in calculated doses, and alternative dose-reporting quantities. The current literature on the effects of heterogeneity corrections on calculated doses and the associated clinical implications have been summarized. The introduction of model-based algorithms into BT treatment planning is an active research-and-development field, and the interested reader should follow the literature for updates and professional society recommendations.

References

Afsharpour, H., D'Amours, M., Cote, B., Carrier, J.F., Verhaegen, F., and Beaulieu, L. 2008. A Monte Carlo study on the effect of seed design on the interseed attenuation in permanent prostate implants. *Med Phys* 35:3671–81.

Afsharpour, H., Pignol, J.-P., Keller, B. et al. 2010. The influence of breast composition and interseed attenuation in dose calculations for post-implant assessment of permanent breast ^{103}Pd seed implant. *Phys Med Biol* 55:4547–61.

Afsharpour, H., Landry, G., Reniers, B., Pignol, J.P., Beaulieu, L., and Verhaegen, F. 2011. Tissue modeling schemes in low energy breast brachytherapy. *Phys Med Biol* 56(22): 7045–60.

Ahnesjö, A. 1989. Collapsed cone convolution of radiant energy for photon dose calculation in heterogeneous media. *Med Phys* 16:577–92.

Ahnesjö, A. and Aspradakis, M.M. 1999. Review: Dose calculations for external photon beams in radiotherapy. *Phys Med Biol* 44:R99–155.

Alm Carlsson, G. 1985. Theoretical basis for dosimetry. In: *The Dosimetry of Ionizing Radiation*. K. Kase, B.E. Bjärngard, and F.H. Attix, (eds.). Academic, New York, pp. 1–75.

Anagnostopoulos, G., Baltas, D., Karaiskos, P., Pantelis, E., Papagiannis, P., and Sakelliou, L. 2003. An analytical dosimetry model as a step towards accounting for inhomogeneities and bounded geometries in ^{192}Ir brachytherapy treatment planning. *Phys Med Biol* 48:1625–47.

Astrahan, M. 2005. Improved treatment planning for COMS eye plaques. *Int J Radiat Oncol Biol Phys* 61:1227–42.

Ballester, F., Granero, D., Perez-Calatayud, J., Venselaar, J.L.M., and Rivard, M.J. 2010. Study of encapsulated ^{170}Tm sources for their potential use in brachytherapy. *Med Phys* 37:1629.

Barillot, I. and Reynaud-Bougnoux, A. 2006. The use of MRI in planning radiotherapy for gynaecologic tumours. *Cancer Imag* 6:100–6.

Beaulieu, L., Carlsson Tedgren, Å., Carrier, J.F. et al. 2012. Report TG-186 of the AAPM, ESTRO, and ABG on model-based dose calculation techniques in brachytherapy: Status and clinical requirements for implementation beyond the TG-43 formalism. *Med Phys* (in preparation).

Carlsson, Å.K. and Ahnesjö, A. 2000a. Point kernels and superposition methods for scatter dose calculations in brachytherapy. *Phys Med Biol* 45:357–82.

Carlsson, Å.K. and Ahnesjö, A. 2000b. The collapsed cone superposition algorithm applied to scatter dose calculations in brachytherapy. *Med Phys* 27:2320–32.

Carlsson Tedgren, Å.K. and Ahnesjö, A. 2003. Accounting for high Z shields in brachytherapy using collapsed cone superposition for scatter dose calculation. *Med Phys* 8:2206–17.

Carrier, J.F., Beaulieu, L., Therriault-Proulx, F., and Roy, R. 2006. Impact of interseed attenuation and tissue composition for permanent prostate implants. *Med Phys* 33:595–604.

Carrier, J.F., D'Amours, M., Verhaegen, F. et al. 2007. Postimplant dosimetry using a Monte Carlo dose calculation engine: A new clinical standard. *Int J Radiat Oncol Biol Phys* 68:1190–8.

Cazeca, M.J., Medich, D.C., and Munro III, J.J. 2010. Effects of breast-air and breast-lung interfaces on the dose rate at the planning target volume of a MammoSite catheter for Yb-169 and Ir-192 HDR sources. *Med Phys* 37:4038–45.

Chen, A., Galloway, M., Landreneau, R. et al. 1999. Intraoperative 125I brachytherapy for high-risk stage I non-small cell lung carcinoma. *Int J Radiat Oncol Biol Phys* 44:1057–63.

Chetty, I.J., Curran, B., Cygler, J.E. et al. 2007. Report of the AAPM Task Group No. 105: Issues associated with clinical implementation of Monte Carlo-based photon and electron external beam treatment planning. *Med Phys* 34:4818–53.

Chibani, O. and Williamson, J.F. 2005. MCPI: A sub-minute Monte Carlo dose calculation engine for prostate implants. *Med Phys* 32:3688–98.

Chibani, O., Williamson, J.F., and Todor, D. 2005. Dosimetric effects of seed anisotropy and interseed attenuation for 103Pd and 125I prostate implants. *Med Phys* 32:2557–66.

Chiu-Tsao, S.-T., Schaart, D.R., Soares, C.G., and Nath, R. 2007. Dose calculation formalisms and consensus dosimetry parameters for intravascular brachytherapy dosimetry: Recommendations of the AAPM Therapy Physics Committee Task Group No. 149. *Med Phys* 34:4126–57.

Chui-Tsao, S.T., Astrahan, M.A., Finger, P.T. et al. 2012. Dosimetry of ^{125}I and ^{103}Pd eye plaques for COMS intraocular tumors: Report of Task Group 129 by the AAPM and ABS. *Med Phys* (submitted).

Cygler, J., Szanto, J., Soubra, M., and Rogers, D. 1990. Effects of gold and silver backings on the dose rate around an 125I seed. *Med Phys* 17:172–8.

Daskalov, G.M., Kirov, A.S., and Williamson, J.F. 1998. Analytical approach to heterogeneity correction factor calculation for brachytherapy. *Med Phys* 25:722–32.

Daskalov, G.M., Baker, R.S., Little, R.C., Rogers, D.W.O., and Williamson, J.F. 2000. Two-dimensional discrete ordinates photon transport calculations for brachytherapy dosimetry applications. *Nucl Sci Eng* 134:121–34.

Daskalov, G.M., Baker, R.S., Rogers, D.W.O., and Williamson, J.F. 2002. Multigroup discrete ordinates modeling of ^{125}I 6702 seed dose distributions using a broad energy-group cross section representation. *Med Phys* 29:113–24.

DeWerd, L.A., Ibbott, G. S., Meigooni, A. S. et al. 2011. A dosimetric uncertainty analysis for photon-emitting brachytherapy sources: Report of AAPM Task Group No. 138 and GEC-ESTRO. *Med Phys* 38:782–801.

Enger, S.A., Ahnesjö, A., and Beaulieu, L. 2010. Dose to medium or dose to a water cavity embedded in medium? *Brachytherapy* 9:S1–35.

Enger, S.A., Ahnesjö, A., Verhaegen, F., and Beaulieu, L. 2011a. Dose to medium or dose to a water cavity embedded in medium? Proceedings of the International Workshop on Recent Advances in Monte Carlo Techniques in Radiation Therapy, McGill University, Montreal, Canada.

Enger, S.A., D'Amours, M., and Beaulieu, L. 2011b. Modeling a hypothetical ^{170}Tm source for brachytherapy applications. *Med Phys* 38:5307–10.

Enger, S.A., Ahnesjö, A., Verhaegen, F., and Beaulieu, L. 2012. Dose to tissue medium or water cavities as surrogate for the dose to cell nuclei at brachytherapy photon energies. *Phys Med Biol* 57:4489–500.

Fogliata, A., Vanetti, E., Albers, D. et al. 2007. On the dosimetric behaviour of photon dose calculation algorithms in the presence of simple geometric heterogeneities: Comparison with Monte Carlo calculations. *Phys Med Biol* 52:1363–85.

Furstoss, C., Reniers, B., Bertrand, M.-J. et al. 2009. Monte Carlo study of LDR seed dosimetry with an application in a clinical brachytherapy breast implant. *Med Phys* 36:1848–58.

Gifford, K.A., Horton, J.L.J., Wareing, T.A., Failla, G., and Mourtada, F. 2006. Comparison of a finite-element multipgroup discrete-ordinates code with Monte Carlo for radiotherapy calculations. *Phys Med Biol* 51:2253–65.

Granero, D., Vijande, J., Ballester, F., and Rivard, M.J. 2011. Dosimetry revisited for the HDR 192Ir brachytherapy source model mHDR-v2. *Med Phys* 38:487–94.

Hedtjärn, H., Alm Carlsson, G., and Williamson, J.F. 2002. Accelerated Monte Carlo based dose calculations for brachytherapy planning using correlated sampling. *Phys Med Biol* 47:351–76.

Huang, D.Y.C., Schell, M.C., Weaver, K.A., and Ling, C.C. 1990. Dose distribution of ^{125}I sources in different tissues. *Med Phys* 17:826–32.

Hubbell, J.H. and Seltzer, S.M. 2004. Tables of X-Ray Mass Attenuation Coefficients and Mass Energy-Absorption Coefficients (version 1.4). Available at http://physics.nist.gov/xaamdi. National Institute of Standards and Technology, Gaithersburg, MD.

ICRU. 1989. Tissue Substitutes in Radiation Dosimetry and Measurements, ICRU Report No. 44. International Commission on Radiation Units and Measurements, Bethesda, MD.

Johnson, M., Colonias, A., Parda, D. et al. 2007. Dosimetric and technical aspects of intraoperative I-125 brachytherapy for stage I non-small cell lung cancer. *Phys Med Biol* 52:1237–45.

Kassas, B., Mourtada, F., Horton, J.L., and Lane, R.G. 2004. Contrast effects on dosimetry of a partial breast irradiation system. *Med Phys* 31:1976–9.

Kirisits, C., Rivard, M.J., Ballester, F. et al. 2011. Dosimetric uncertainties in the practice of clinical brachytherapy. *Brachytherapy* 10:S32.

Kirov, A.S. and Williamson, J.F. 1997. Two-dimensional scatter integration method for brachytherapy dose calculations in 3D geometry. *Phys Med Biol* 42:2119–35.

Landry, G., Reniers, B., Murrer, L. et al. 2010. Sensitivity of low energy brachytherapy Monte Carlo dose calculations to uncertainties in human tissue composition. *Med Phys* 37:5188–98.

Landry, G., Reniers, B., Pignol, J., Beaulieu, L., and Verhaegen, F. 2011. The difference of scoring dose to water or tissues in Monte Carlo dose calculations for low energy brachytherapy photon sources. *Med Phys* 38:1526–33.

Lee, W., Daly, B., DiPetrillo, T. et al. 2003. Limited resection for non-small cell lung cancer: Observed local control with implantation of I-125 brachytherapy seeds. *Ann Thor Surg* 75:237–42.

Liu, H.H. and Keall, P. 2002. Point/counterpoint: D_m rather than D_w should be used in Monte Carlo treatment planning. *Med Phys* 29:922–4.

Liu, D., Poon, E., Bazalova, M. et al. 2008. Spectroscopic characterization of a novel electronic brachytherapy system. *Phys Med Biol* 53:61–75.

Lulu, B.A. and Bjärngard, B.E. 1982. Batho's correction factor combined with scatter summation. *Med Phys* 9:372–7.

Lymperopoulou, G., Papagiannis, P., Sakelliou, L., Milickovic, N., Giannouli, S., and Baltas, D. 2005. A dosimetric comparison of 169Yb versus 192Ir for HDR prostate brachytherapy. *Med Phys* 32:3832–42.

Markman, J., Williamson, J.F., Dempsey, J.F., and Low, D.A. 2001. On the validity of the superposition principle in dose calculations for intracavitary implants with shielded vaginal colpostats. *Med Phys* 28:147–55.

Mason, D., Battista, J., Barnett, R., and Porter, A. 1992. Yb-169 - Calculated physical-properties of a new radiation source for brachytherapy. *Med Phys* 19:695–703.

Meigooni, A.S. and Nath, R. 1992. Tissue inhomogeneity correction for brachytherapy sources in a heterogeneous phantom with cylindrical symmetry. *Med Phys* 19:401–7.

Meigooni, A.S., Wright, C., Koona, R.A. et al. 2009. TG-43 U1 based dosimetric characterization of model 67-6520 Cs-137 brachytherapy source. *Med Phys* 36:4711–19.

Melhus, C.S. and Rivard, M.J. 2006. Approaches to calculating AAPM TG-43 brachytherapy dosimetry parameters for 137Cs, 125I, 192Ir, 103Pd, and 169Yb sources. *Med Phys* 33:1729–37.

Melhus, C.S. and Rivard, M.J. 2008. COMS eye plaque brachytherapy dosimetry simulations for ^{103}Pd, ^{125}I, and ^{131}Cs. *Med Phys* 35:3364.

Mikell, J.K. and Mourtada, F. 2010. Dosimetric impact of an 192Ir brachytherapy source cable length modeled using a grid-based Boltzmann transport equation solver. *Med Phys* 37:4733–43.

Mille, M.M., Xu, X.G., and Rivard, M.J. 2010. Comparison of organ doses for patients undergoing balloon brachytherapy of the breast with HDR 192Ir or electronic sources using monte carlo simulations in a heterogeneous human phantom. *Med Phys* 37:662–71.

Mobit, P. and Badragan, I. 2004. Dose perturbation effects in prostate seed implant brachytherapy with I-125. *Phys Med Biol* 49:3171–8.

Nag, S., Quivery, J., Earle, J., Followill, D., Fontanesi, J., and Finger, P. 2003. The American Brachytherapy Society recommendations for brachytherapy of uveal melanomas. *Int J Radiat Oncol Biol Phys* 56:544–55.

Nath, R., Park, C.H., King, C.R., and Muench, P. 1990. A dose computation model for [241]Am vaginal applicators including the source-to-source shielding effects. *Med Phys* 17:833–42.

Nath, R., Anderson, L.L., Luxton, G., Weaver, K.A., Williamson, J.F., and Meigooni, A.S. 1995. Dosimetry of interstitial brachytherapy sources: Recommendations of the AAPM radiation therapy committe task group No. 43. *Med Phys* 22:209–36.

Nath, R., Anderson, L., Meli, J., Olch, A., Stitt, J., and Williamson, J. 1997. Code of practice for brachytherapy physics: Report of the AAPM Radiation Therapy Committee Task Group No. 56. *Med Phys* 24:1557–98.

Paganetti, H. 2009. Dose to water versus dose to medium in proton beam therapy. *Phys Med Biol* 54:4399–421.

Papanikolaou, N., Battista, J.J., Boyer, A. et al. 2004. *AAPM Report No. 85: Tissue inhomogeneity corrrections for megavoltage photon beams*. AAPM REport No 85: Report of Task Group No. 65 of the Radiation Therapy Committee of the American Association of Physicists in Medicine. Medical Physics Publishing, Madison, WI.

Pantelis, E., Papagiannis, P., Karaiskos, P. et al. 2005. The effect of finite patient dimensions and tissue inhomogeneities on dosimetry planning of ^{192}Ir HDR breast brachytherapy: A Monte Carlo dose verification study. *Int J Radiat Oncol Biol Phys* 61:1596–602.

Pérez-Calatayud, J., Granero, D., and Ballester, F. 2004. Phantom size in brachytherapy source dosimetric studies. *Med Phys* 31:2075–81.

Petrokokkinos, L., Zourari, K., Pantelis, E. et al. 2011. Dosimetric accuracy of a deterministic radiation transport based ^{192}Ir brachytherapy treatment planning system. Part II: Monte Carlo and experimental verification of a multiple source dwell position plan employing a shielded applicator. *Med Phys* 38:1981–92.

Piermattei, A., Azario, L., and Montemaggi, P. 1995. Implantation guidelines for [169]Yb seed interstitial treatments. *Phys Med Biol* 40:1331–8.

Pignol, J.-P., Keller, B., Rakovitch, E., Sankreacha, R., Easton, H., and Que, W. 2006. First report of a permanent breast 103Pd seed implant as adjuvant radiation treatment for early-stage breast cancer. *Int J Radiat Oncol Biol Phys* 64:176–81.

Poon, E., Williamson, J.F., Vuong, T., and Verhaegen, F. 2008. Patient-specific Monte Carlo dose calculations for high-dose-rate endorectal brachytherapy with shielded intracavitary applicator. *Int J Radiat Oncol Biol Phys* 72:1259–66.

Poon, E. and Verhaegen, F. 2009a. A CT-based analytical dose calculation method for HDR ^{192}Ir brachytherapy. *Med Phys* 36:3982–94.

Poon, E., and Verhaegen, F. 2009b. Development of a scatter correction technique and its application to HDR ^{192}Ir multi-catheter breast brachytherapy. *Med Phys* 36:3703–13.

Powell, J.W., Dexter, E., Scalzetti, E.M., and Bogart, J.A. 2009. Treatment advances for medically inoperable non-small-cell lung cancer: Emphasis on prospective trials. *Lancet Oncol* 10:885–94.

Price, M.J., Jackson, E.F., Gifford, K.A., Eifel, P.J., and Mourtada, F. 2009. Development of prototype shielded cervical intracavitary brachytherapy applicators compatible with CT and MR imaging. *Med Phys* 36:5515–24.

Raffi, J.A., Davis, S.D., Hammer, C.G. et al. 2010. Determination of exit skin dose for 192Ir intracavitary accelerated partial breast irradiation with thermoluminescent dosimeters. *Med Phys* 37:2693–702.

Reniers, B. and Verhaegen, F. 2011. 3D image guided brachytherapy using cone beam CT. *Med Phys* 38:2762–7.

Reniers, B., Verhaegen, F., and Vynckier, S. 2004. The radial dose function of low-energy brachytherapy seeds in different solid phantoms: Comparison between calculations with the EGSnrc and MCNP4C Monte Carlo codes and measurements. *Phys Med Biol* 49:1569–82.

Richardson, S.L. and Pino, R. 2010. Dosimetric effects of an air cavity for the SAVI™ partial breast irradiation applicator. *Med Phys* 37:3919.

Rivard, M.J. 2001. Monte Carlo calculations of AAPM Task Group Report No. 43 dosimetry parameters for the MED3631-A/M125I source. *Med Phys* 28:629–37.

Rivard, M.J., Coursey, B.M., DeWerd, L.A. et al. 2004. Update of AAPM Task Group No. 43 Report: A revised AAPM protocol for brachytherapy dose calculations. *Med Phys* 31:633–74.

Rivard, M.J., DeWerd, L.A., and Zinkin, H.D. 2005. Brachytherapy with miniature electronic X-ray sources. In: *Brachytherapy Physics*, 2nd ed. B.R. Thomadsen, M.J. Rivard, and W.M. Butler (eds.). Medical Physics Publishing, Madison, Wisconsin, pp. 889–900.

Rivard, M.J., Venselaar, J., and Beaulieu, L. 2009. The evolution of brachytherapy treatment planning. *Med Phys* 36:2136–53.

Rivard, M.J., Beaulieu, L., and Mourtada, F. 2010. Enhancements to commissioning techniques and quality assurance of brachytherapy treatment planning systems that use model-based dose calculation algorithms. *Med Phys* 37:2645–58.

Rivard, M.J., Chiu-Tsao, S.-T., Finger, P.T. et al. 2011. Comparison of dose calculation methods for brachytherapy of intraocular tumors. *Med Phys* 38:306.

Russell, K.R. and Ahnesjö, A. 1996. Dose calculation in brachytherapy for a ^{192}Ir source using a primary and scatter dose separation technique. *Phys Med Biol* 41:1007–24.

Russell, K.R., Carlsson Tedgren, Å.K., and Ahnesjö, A. 2005. Brachytherapy source characterization for improved dose calculations using primary and scatter dose separation. *Med Phys* 32:2739–52.

Sandborg, M., Dance, D.R., Persliden, J., and Alm Carlsson, G. 1994. A Monte Carlo program for the calculation of contrast, noise and absorbed dose in diagnostic radiology. *Comp Methods Programs Biomed* 42:167–80.

Santos, R., Colonias, A., Parda, D. et al. 2003. Comparison between sublobar resection and 125Iodine brachytherapy after sublobar resection in high-risk patients with stage I non–small-cell lung cancer. *Surgery* 134:691–7.

Sureka, C.S., Aruna, P., Ganesan, S., Sunny, C.S., and Subbaiah, K.V. 2006. Computation of relative dose distribution and effective transmission around a shielded vaginal cylinder with 192Ir HDR source using MCNP4B. *Med Phys* 33:1552–61.

Sutherland, J.G.H., Thomson, R.M., and Rogers, D.W.O. 2011. Changes in dose with segmentation of breast tissues in Monte Carlo calculations for low-energy brachytherapy. *Med Phys* 38:4858–65.

Sutherland, J.G.H., Furutani, K.M., Garces, Y.I., and Thomson, R.M. 2012. Model-based dose calculations for 125I lung brachytherapy. *Med Phys* 39: 4365–77.

Taylor, R.E.P., Yegin, G., and Rogers, D.W.O. 2007. Benchmarking BrachyDose: Voxel based EGSnrc Monte Carlo calculations of TG-43 dosimetry parameters. *Med Phys* 34:445–57.

Taylor, R.E.P. and Rogers, D.W.O. 2008. EGSnrc Monte Carlo calculated dosimetry parameters for ^{192}Ir and ^{169}Yb sources. *Med Phys* 35:4933–44.

Thomadsen, B.R., Williamson, J.F., Rivard, M.J., and Meigooni, A.S. 2008. Anniversary paper: Past and current issues, and trends in brachytherapy physics. *Med Phys* 35:4708.

Thomson, R.M., Taylor, R.E.P., and Rogers, D.W.O. 2008. Monte Carlo dosimetry for ^{125}I and ^{103}Pd eye plaque brachytherapy. *Med Phys* 35:5530–43.

Thomson, R.M. and Rogers, D.W.O. 2010. Monte Carlo dosimetry for I and Pd eye plaque brachytherapy with various seed models. *Med Phys* 37:368.

Thomson, R.M., Carlsson Tedgren, Å., and Williamson, J.F. 2011. Comparison of bulk medium dose descriptors and cellular absrobed doses for brachytherapy. Proceedings of the International Workshop on Recent Advances in Monte Carlo Techniques in Radiation Therapy, McGill University, Montreal, Canada:73.

Unkelbach, J., Bortfeld, T., Martin, B.C., and Soukop, M. 2009. Reducing sensitivity of IMPT treatment plans to setup errors and range uncertainties via probabilistic treatment planning. *Med Phys* 36:146–63.

van der Laarse, R. and Meertens, H. 1984. An algorithm for ovoid shielding of a cervix applicator. In 8th International Conference on the Use of Computers in Radiation Therapy. J.R. Cunningham, D. Ragan, and D. Van DyK, (eds.). IEEE Computer Society, Los Angeles, CA, Toronto, Canada.

van Veelen, B. 2011. Collapsed cone dose calculations for model based brachytherapy utilizing Monte Carlo source characterization data. Proceedings of the International Workshop on Recent Advances in Monte Carlo Techniques in Radiation Therapy, McGill University, Montreal, Canada:40.

Verhaegen, F. and Devic, S. 2005. Sensitivity study for CT image use in Monte Carlo treatment planning. *Phys Med Biol* 50:937–46.

Voynov, G., Heron, D.E., Lin, C.J. et al. 2005. Intraoperative I125 Vicryl mesh brachytherapy after sublobar resection for high-risk stage I nonsmall cell lung cancer. *Brachytherapy* 4:278–85.

Viswanathan, A., Dimopoulos, J., Kirisits, C., Berger, D., and Pötter, R. 2007. Computed tomography versus magnetic resonance imaging-based contouring in cervical cancer brachytherapy: Results of a prospective trial and preliminary guidelines for standardized contours. *Int J Radiat Oncol Biol Phys* 68:491–8.

Waever, K. 1986. The dosimetry of ^{125}I seed eye plaques. *Med Phys* 13:78–83.

Walters, B.R.B., Kramer, R., and Kawrakow, I. 2010. Dose to medium versus dose to water as an estimator of dose to sensitive skeletal tissue. *Phys Med Biol* 55:4535–46.

Wang, R. and Sloboda, R.S. 2007a. Brachytherapy scatter dose calculation in heterogeneous media: I. A microbeam ray-tracing method for the single-scatter contribution. *Phys Med Biol* 52:5619–36.

Wang, R. and Sloboda, R.S. 2007b. Brachytherapy scatter dose calculation in heterogeneous media: II. Empirical formulation for the multiple-scatter contribution. *Phys Med Biol* 52:5637–54.

Weeks, K.J. and Dennett, J.C. 1990. Dose calculation and measurements for a CT compatible version of the Fletcher applicator. *Int J Radiat Oncol Biol Phys* 18:1191–8.

Williamson, J.F. 1987. Monte Carlo evaluation of kerma at a point for photon transport problems. *Med Phys* 14:567–76.

Williamson, J.F. 1990. Dose calculations about shielded gynecological colpostates. *Int J Rad Biol Phys* 19:167–78.

Williamson, J.F. 2005. Semiempirical dose-calculation models in brachytherapy. In: *AAPM Summer School: Brachytherapy Physics.* B.R. Thomadsen, M.J. Rivard, and W.M. Butler (eds.). Medical Physics Publishing, Madison, Wisconsin, pp. 201–32.

Williamson, J.F. 2006. Brachytherapy technology and physics practice since 1950: A half-century of progress *Phys Med Biol* 51:R303–25.

Williamson, J.F., Baker, R.S., and Li, Z. 1991. A convolution algorithm for brachytherapy dose computations in heterogeneous geometries. *Med Phys* 18:1256–65.

Williamson, J.F., Li, Z., and Wong, J.W. 1993. One-dimensional scatter-subtraction method for brachytherapy dose calculation near bounded heterogeneities. *Med Phys* 20:233–44.

Wu, A., Sternick, E., and Muise, D. 1988. Effect of gold shielding on the dosimetry of an 125I seed at close range. *Med Phys* 15:627–8.

Xu, C., Verhaegen, F., Laurendeau, D., Enger, S.A., and Beaulieu, L. 2011. An algorithm for efficient metal artifact reductions in permanent seed implants. *Med Phys* 38:47–56.

Yaffe, M.J., Boone, J.M., Packard, N. et al. 2009. The myth of the 50-50 breast. *Med Phys* 36:5437–43.

Yan, X., Poon, E., Reniers, B., Vuong, T., and Verhaegen, F. 2008. Comparison of dose calculation algorithms for colorectal cancer brachytherapy treatment with a shielded applicator. *Med Phys* 35:4824–30.

Yang, Y. and Rivard, M.J. 2011. Evaluation of brachytherapy lung implant dose distributions from photon-emitting sources due to tissue heterogeneities. *Med Phys* 38:5857.

Yang, M., Virshup, G., Clayton, J., Zhu, X.R., Mohan, R., and Dong, L. 2011. Theoretical variance analysis of single- and dual-energy computed tomography methods for calculating proton stopping power ratios of biological tissues. *Phys Med Biol* 55:1343–62.

Yu, Y., Anderson, L.L., Li, Z. et al. 1999. Permanent prostate seed implant brachytherapy: Report of the American Association of Physicists in Medicine Task Group No. 64. *Med Phys* 26:2054–76.

Zhou, C. and Inanc, F. 2003. Integral-transport-based deterministic brachytherapy dose calculations. *Phys Med Biol* 48:73–93.

12

Optimization and Evaluation

Dimos Baltas
Klinikum Offenbach GmbH

Inger-Karine K. Kolkman-Deurloo
Erasmus MC–Daniel Den Hoed Cancer Center

12.1 Optimization

12.1.1 Introduction

The objectives of brachytherapy treatment planning are to deliver a sufficiently high dose in the cancerous tissue and to protect the surrounding normal tissue (NT) and organs at risk (OAR) from excessive radiation. The problem is to determine the position and number of sources (permanent implants) or source dwell positions (SDPs), number of catheters, and the dwell times (temporary implants), such that the obtained dose distribution is as close as possible to the desired dose distribution. Additionally, the stability of solutions can be considered with respect to possible movements of the sources or SDPs. The planning includes clinical constraints such as a realistic range of catheters and sources and their positions and orientations. The determination of an optimal number of catheters and sources or SDPs is a very important aspect of treatment planning, as a reduction of the number of catheters and sources simplifies the treatment plan in terms of time and complexity. It also reduces the possibility of treatment errors and is less invasive for the patient.

As analytical solutions cannot be determined, the solution is obtained by *inverse optimization* or *inverse planning*. The term *inverse planning* is used considering this as the opposite of the forward problem, that is, the determination of the dose distribution for a specific set of sources or SDPs and corresponding dwell times. If the positions and number of sources or of catheters and the SDPs are given after the implantation of the catheters, we term the process *postplanning*. Then, the optimization process to obtain an optimal dose distribution is called *dose optimization*. Dose optimization can be considered as a special type of inverse optimization where the positions and number of sources or catheters and the SDPs are fixed.

The term *inverse problem* is a well-known mathematical term dating at least from the 19th century, and there are several mathematical journals devoted to this topic, including, for example, *Inverse Problems*, a journal published by the Institute of Physics.

Inverse optimization or *inverse planning* in brachytherapy is therefore a specific case within the general spectrum of inverse problems. Or in plain terms: first you know the ideal answer, then you take into account any constraints, and finally, you mathematically determine the optimum parameter values to provide the ideal answer. In other words, you start with the wished result and the inverse problem (inverse planning or optimization) is to determine the cause of this result.

With low-dose-rate (LDR) brachytherapy, only permanent seed applications are considered, with radionuclides such as ^{125}I, ^{103}Pd, and ^{131}Cs, because the earlier methods of LDR manual afterloading of linear sources with radionuclides such as ^{137}Cs, ^{182}Ta, and ^{192}Ir are not much used anymore. With high-dose-rate (HDR) brachytherapy, a single ^{192}Ir or ^{60}Co source is used in the remote-controlled afterloading unit, which steps through various dwell positions in the catheters.

Inverse planning has to consider many objectives and is thus a multiobjective (MO) optimization problem (Miettinen 1999). We have a set of competing objectives. Increasing the dose in the planning target volume (PTV) will increase the dose outside the PTV and in the OARs. A trade-off between the objectives exists as we have never had a situation in which all the objectives can, in a best possible way, be satisfied simultaneously. One solution of this MO problem is to convert it into a specific single objective (SO) problem by combining the various objective functions with different weights into an SO function (aggregation). Optimization and analysis of the solutions are repeated with different sets of weights until a satisfactory solution is obtained, as the optimal weights are a priori unknown. In MO optimization, a representative set of all possible so-called *nondominated solutions* is obtained, and the best solution is selected from this set. The optimization and decision processes are decoupled. The set provides a coherent global view of the trade-offs between the objectives necessary to select the best possible solution, whereas the SO approach is a trial–and-error method in which optimization and decision processes are coupled. Due to its complexity,

MO has found, until now, limited application in the clinical treatment planning and optimization process.

Until the early 1990s, there was no real 3D anatomical information widely available for the routine treatment planning in brachytherapy. CT-based planning has been established in the late 1990s and MR-based planning only first at the beginning of the twenty-first century. Ultrasound (US) based planning, even if it was previously available, was limited to 2D imaging and to prostate permanent implants. With missing detailed 3D anatomy information, sources and catheter placement was based on simple rules ("systems," nomograms, or geometrical figures; see Chapter 8) relating tumor dimensions, source strength, and source distribution within the tumor.

With the introduction of HDR afterloading technology with ^{192}Ir sources in the late 1980s and with the availability of the first PCs, 3D acquisition and reconstruction of catheter geometry using x-ray projectional methods were developed, and a computerized calculation of adequate sources or SDPs became possible. Due to missing 3D anatomical information, this was based on the assumption that the implanted catheter geometry follows the target morphology, something that the physician should have to ensure based on his or her experience and how his or her hands feel. If this is the case, then the catheter geometry and consequently the geometry of SDPs themselves could be used for optimally adjusting the shape of the dose distribution. A representative of that era of catheter-based dose adjustment is the geometrical optimization (GO) method as suggested by Edmundson (1990).

12.1.2 Catheter-Based Methods

HDR and pulsed-dose-rate (PDR) brachytherapy are usually performed with single-stepping source afterloaders. These afterloaders contain a small, mostly ^{192}Ir, source mounted at the end of a flexible steel wire, which is transported under computer control in previously implanted applicators. The machine is programmed to position the source at predefined positions—the so-called dwell positions—in the applicators. The time the source spends at each dwell position—the so-called dwell time—can freely be chosen. This enables optimization of the dose distribution by optimization of the dwell times over all dwell positions in the implant.

In this section, the following catheter-based optimization techniques are discussed:

- Dose point optimization
- Geometric optimization

A detailed description of the 3D dose calculation methodology for a single source (position) according to AAPM TG-43 (Nath et al. 1995) and to its updates (Rivard et al. 2004, 2007) is presented and further discussed in Chapter 10. In order to be able to explain the basic dose optimization algorithms and for the sake of simplicity, we neglect the radial dose function and the anisotropy function and assume a point source approximation (see Equation 10.1). Furthermore, the parameters air-kerma

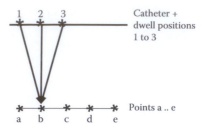

FIGURE 12.1 Implant geometry consisting of one catheter, including dwell positions 1, 2, and 3, and several arbitrary points a to e. (Courtesy of I.-K.K. Kolkman-Deurloo.)

strength S_K and dose rate constant Λ are omitted from the following equations.

Then, the dose contribution (in arbitrary units) in point $P(r)$ at the radial distance r from the source is approximated by

$$D(r) \approx T \cdot \frac{1}{r^2} \qquad (12.1)$$

Let us suppose an implant according to Figure 12.1. The definition of dwell time optimization is to determine the unknown dwell times T_1, T_2, T_3, etc. from equations similar to Equation 12.2:

$$D_b \approx \left(T_1 \cdot \frac{1}{r_{1b}^2} \right) + \left(T_2 \cdot \frac{1}{r_{2b}^2} \right) + \left(T_3 \cdot \frac{1}{r_{3b}^2} \right) \qquad (12.2)$$

with D_j the dose in (dose) point j, T_i the dwell time at dwell position i, and r_{ij} the distance between dwell position i and dose point j.

12.1.2.1 Dose Point Optimization

If the target volume is characterized by a (sufficient) number of reference points (the so-called dose points), the dose distribution can be optimized such that the prescribed reference dose is obtained in these points (the so-called optimization on dose points).

Assume a geometry according to Figure 12.2, that is, two catheters indicated by the thick lines, with dwell positions 1, 2, and 3 indicated, in combination with anatomical points a, b, c,..., indicated by the crosses, serving as dose points that should receive the prescribed dose.

According to Equation 12.2, one can calculate the dose in points a, b, and c as follows:

$$D_b \approx \left(T_1 \cdot \frac{1}{r_{1b}^2} \right) + \left(T_2 \cdot \frac{1}{r_{2b}^2} \right) + \left(T_3 \cdot \frac{1}{r_{3b}^2} \right) \qquad (12.3)$$

$$D_a \approx \left(T_1 \cdot \frac{1}{r_{1a}^2} \right) + \dots \qquad (12.4)$$

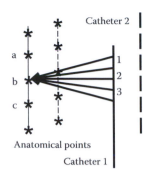

FIGURE 12.2 Implant geometry consisting of two catheters, including dwell positions (1, 2, and 3), and anatomical points (a, b, and c). (Courtesy of I.-K.K. Kolkman-Deurloo.)

$$D_c \approx \left(T_1 \cdot \frac{1}{r_{1c}^2} \right) + \ldots \qquad (12.5)$$

Furthermore, 100% of the reference dose is prescribed at points *a*, *b*, and *c*:

$$D_a = D_b = D_c = \ldots = 100\% \qquad (12.6)$$

From Equations 12.3 to 12.6, the dwell times T_i can be calculated by simple mathematics in case the number of equations (= number of dose points) is equal to the number of variables (= number of dwell positions/times).

However, a solution including negative dwell times can result from these equations. This is solved by the introduction of a dwell time gradient factor. Extra equations are added limiting the gradient of (or difference between) neighboring dwell times, enabling determination of a solution. Furthermore, the number of points will generally not be equal to the number of dwell positions/times (an overdetermined or underdetermined math problem), which can result in either an infinite number of solutions or no solution at all. This situation is mathematically solved by considering the sum of squares of dwell times and the dwell time gradient factor in order to find a unique solution. However, these solutions will not fulfill all requirements exactly but present the best approximation to all objectives.

This procedure can be extended by differentiating between target points and NT points and requiring that the dose in these NT points should be below a chosen tolerance level.

12.1.2.2 Geometric Optimization

In geometric optimization, the dwell times are determined such that the dwell time in each dwell position is inversely proportional to the dose contribution of neighboring source positions. When looking at Figure 12.1 and Equations 12.3 to 12.5 for this case, the points *a*, *b*, and *c* are replaced by dwell positions in which the dose contribution from dwell positions 1, 2, and 3 is calculated (see also Figure 12.3):

$$D_b \approx \left(T_1 \cdot \frac{1}{r_{1b}^2} \right) + \left(T_2 \cdot \frac{1}{r_{2b}^2} \right) \ldots \qquad (12.7)$$

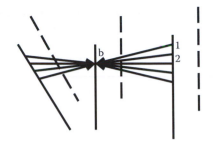

FIGURE 12.3 Implant geometry consisting of six catheters, including a dwell position b in one catheter in which the dose contribution from dwell positions 1 and 2 in another catheter is calculated. (Courtesy of I.-K.K. Kolkman-Deurloo.)

The dwell time in dwell position *b* is made inversely proportional to the dose contribution from the other dwell positions 1, 2, etc. by

$$T_b := \left(\frac{1}{D_b} \right) \qquad (12.8)$$

At the start of the calculation, the dwell times *T* are all set to 1; the new dwell times are iteratively calculated using Equations 12.7 and 12.8 until a stable solution is achieved. It is easily recognized that if the total dose contribution D_b in dwell position *b* from other dwell positions is large, the dwell time T_b will become small.

12.1.2.3 Different Modes in Geometric Optimization

The dwell time in a dwell position is inversely proportional to the dose contribution of neighboring source positions. For these neighbors, one can take all other positions into account (Figure 12.4a)—this is called the *distance mode*—or only the positions in other catheters than the one considered (Figure 12.4b)—this is called the *volume mode*.

The volume mode is developed in order to suppress the catheter itself, as the contribution from the dwell positions in this catheter will be dominant in case the spacing between neighboring catheters is larger than the dwell position separation. The volume mode is used to fill in the region between the catheters

(a)   (b)

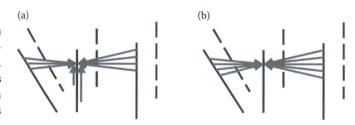

FIGURE 12.4 (a) Geometric optimization taking into account *all other* dwell positions (distance mode). (b) Geometric optimization taking into account only the dwell positions *in other catheters* (volume mode). (Courtesy of I.-K.K. Kolkman-Deurloo.).

and "compensate" for diverging catheters. The distance mode is used to obtain a homogeneous dose distribution along the catheters.

12.1.2.4 Pitfalls in Geometric Optimization

The differences in results between the distance and volume modes in geometric optimization are very well illustrated by the example shown in Figure 12.5 (Edmundson 1994). For an equal dose along the catheter(s), the distance mode results in homogeneous dose delivery. The volume mode fills in the gap in between the catheters. Which to use between the two modes depends on the clinical decision.

12.1.2.5 Result of Equal Dwell Times, Dose Point Optimization, and Geometric Optimization

The resulting dose distributions along a single catheter in case of equal dwell times, geometric optimization, and dose point optimization are illustrated in Figure 12.6. The differences at the edge of the catheter, that is, the coverage of the dose points by the reference isodose, are well recognized. The corresponding relative dwell times at the edge of the catheter can be found in Table 12.1.

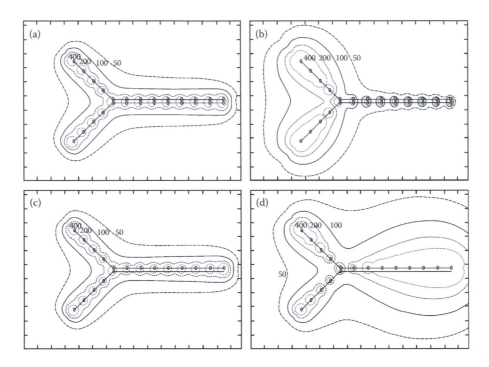

FIGURE 12.5 Example of two catheters partly running parallel and partly diverging. (a) and (c) distance mode geometric optimization; (b) and (d) volume mode geometric optimization. (a) and (b) both catheters loaded over entire lentgh. (c) and (d) one catheter loaded over entire length and second catheter partly loaded. (Courtesy of G. Edmundson).

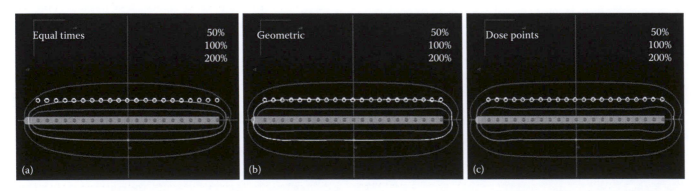

FIGURE 12.6 A single catheter of 10 cm length with 21 positions spaced 5 mm apart and 21 (dose) points. Shown are the 50%, 100% and 200% isodose lines in case of equal dwell times (a), geometric optimization (b) and dose point optimization (c). (Courtesy: I.-K.K. Kolkman-Deurloo.)

TABLE 12.1 Relative Dwell Times at the Edge of the Catheter Shown in Figure 12.6 in Case of Equal Dwell Times, Geometric Optimization, and Dose Point Optimization

Dwell Position	Relative Dwell Time		
	Equal Dwell Times	Geometric Optimization	Dose Point Optimization
1	1.7	3.3	3.9
2	1.7	2.0	2.7
3	1.7	1.9	1.7
4	1.7	1.8	1.5
5	1.7	1.8	1.7
6	1.7	1.7	1.8
7	1.7	1.7	1.8
8	1.7	1.7	1.7
…	..	..	..

12.1.2.6 Clinical Examples of Geometric and Dose Point Optimization

12.1.2.6.1 Dose Point Optimization

Dose point optimization can be used in many occasions. Illustrating examples in intracavitary brachytherapy can be found in the works of Niël et al. (1994) and Levendag et al. (1997). Niël et al. (1994) adapt a gynecological applicator consisting of two ovoids and an intrauterine tube in order to treat patients with a cervix carcinoma also extending to the vaginal wall and the parametria. By using dose point optimization, a smooth dose distribution is obtained with the prescribed dose at point A as well as at 5-mm depth of the vaginal wall. Levendag et al. (1997) demonstrate the use of dose point optimization for endocavitary treatment of the nasopharynx; the dose distribution is optimized such that points describing the target receive the prescribed dose while other points representing surrounding NTs receive a dose as low as possible. Dose point optimization is also very well suited to optimize the dose distribution in intraoperative brachytherapy using catheters inserted in a flat template positioned over the surgical bed (Beddar et al. 2006; Kolkman-Deurloo et al. 2004). A slightly different application of dose point optimization is its use in interstitial volume implants in the so-called stepping source dosimetry system (SSDS) (Thomadsen et al. 1994; Major et al. 2000, 2002). In this case, the dose distribution is optimized using dose points placed in between the catheters along the whole active lengths, that is, through the whole target volume. In this way, the active lengths can be chosen shorter than in the Paris system (Dutreix et al. 1982) while keeping the treated area comparable.

12.1.2.6.2 Geometric Optimization

The use of geometric optimization for interstitial volume implants has been extensively studied. Some examples of comparisons of optimized versus nonoptimized dose distributions can be found in the works of Major et al. (2000), Kolkman-Deurloo et al. (1994), Anacak et al. (1997), Pieters et al. (2001),

and Berns et al. (1997). These examples mostly concern the so-called regular implants containing parallel catheters/needles, for example, (two-plane) breast implants. However, Berns et al. (1997) also consider single-plane surface molds, and Kolkman-Deurloo et al. (1994) include irregular volume implants as used in the treatment of base-of-tongue tumors. Regarding the regular implants, all studies clearly indicate an increase in treated volume (TV) and in dose homogeneity due to geometric optimization. Therefore, it is concluded for this case that either a shorter loading of the catheters/needles can be used or a different isodose for dose prescription can be chosen, for example, 90% instead of 85% as is customary in the Paris system (Dutreix et al. 1982). The study of Kolkman-Deurloo et al. (1994) shows that, in case of the irregular volume implants, the TV does not increase after geometric optimization; however, the increase in dose homogeneity remains present. When comparing this application of geometric optimization to SSDS, it is clear that a significant advantage of geometric optimization is the absence of the necessity of placing all the dose points along the catheters.

12.1.3 Anatomy-Based Methods

An ideal dose distribution $D(\mathbf{r})$ with a specific prescription dose D_{ref} regarding the PTV is

$$D(\mathbf{r}) = D(x, y, z) = \begin{cases} D_{ref} & \text{if } \mathbf{r} \in \text{PTV} \\ 0 & \text{else} \end{cases} \qquad (12.9)$$

This dose distribution is not obtainable since, due to inverse square law, radiation cannot be confined to the PTV only. Part of the radiation will traverse the OARs and the surrounding NT, and thus, dose will be always delivered to volumes outside the PTV (OARs and NT).

Optimization is the process of determining or approximating the parameter sets that maximize (e.g., for the PTV) or minimize (e.g., for OARs) a given number of objective functions. A distinction is made between *global* and *local* optimization (local search). Optimal solutions are defined with respect to a given neighborhood in the search space. For global optima, the neighborhood includes the whole search space.

The objective, goal, or score function is the function to be optimized, depending on the parameters (also referred to as object parameters or search space parameters, e.g., SDPs and dwell times for temporary implants). The input is the set of the parameter values. The output is the objective value representing the evaluation/quality of the solution. In a multicriteria optimization problem where many objectives are considered, the optimization produces a vector of objective values. The objective value is often referred to as *fitness*.

Out of all possible dose distributions, the problem is to obtain the optimal dose distribution without any a priori knowledge of the physical restrictions. Optimality requires quantifying the quality of a dose distribution. This can be achieved with a similarity measure as a distance function between a given dose distribution and the ideal one.

Quality indices can be defined for a dose distribution such as the conformity index. For the ideal dose distribution, a quality index would give an optimum value. Another method is to use the dose-volume histogram (DVH) specifying constraints that have to be satisfied by a dose distribution.

Conformal anatomy-based dose optimization considers the dose in the PTV and OARs, including the surrounding NT. The aim of conformal anatomy-based dose optimization is the complete coverage of the PTV with a dose at least equal to the prescription dose and to simultaneously avoid dose values above some critical value. These values are specific for each OAR and the surrounding NT.

With anatomy-based optimization, we also mean that the user can define the dosimetric targets for the PTV and the dosimetric limitations to the OARs. In contrast to the catheter-based methods discussed previously, anatomy-based dose optimization requires modern 3D imaging-based brachytherapy for the complete 3D description of the relevant anatomy and of the 3D implant geometry.

12.1.3.1 Distance-Based Objective Functions

A natural measure quantifying the similarity of a dose distribution at N sampling points with dose values d_i to the corresponding optimal dose values d_i^* is a distance measure. A common measure is the L_p norm:

$$L_p = \left(\sum_{i=1}^{N} \left(d_i - d_i^* \right)^p \right)^{\frac{1}{p}} \quad (12.10)$$

For $p = 2$, that is, L_2, we have the Euclidean distance.

The dose optimization problem is transformed into an optimization problem by introducing as an objective the minimization of the distance between the ideal and the achievable dose distribution. These objectives can be expressed, in general, by the objective functions $f_L(\mathbf{x})$ and $f_H(\mathbf{x})$:

$$f_L(\mathbf{x}) = \frac{1}{N} \sum_{i=1}^{N} \Theta \left(D_L - d_i(\mathbf{x}) \right) \left(D_L - d_i(\mathbf{x}) \right)^p \quad (12.11)$$

$$f_H(\mathbf{x}) = \frac{1}{N} \sum_{i=1}^{N} \Theta \left(d_i(\mathbf{x}) - D_H \right) \left(d_i(\mathbf{x}) - D_H \right)^p \quad (12.12)$$

where $d_i(\mathbf{x})$ is the dose at the ith sampling point that depends on parameter x such as source dwell times, p is a parameter defining the distance norm (see Equation 12.10), N is the number of dose sampling points, and D_L and D_H are the low and high dose limits, respectively, for the specific anatomical volume of interest (VOI). These are used if dose values below D_L and above D_H are to be considered as expressed by the step function $\Theta(x)$.

For $p = 2$, we obtain the quadratic type of objectives (Milickovic et al. 2002), and for $p = 1$, a linear form (Lessard and Pouliot 2001). For $p = 0$ (Lahanas et al. 2003a), the DVH-based objective as the DVH value at the dose D_H is given by

$$DVH \left(D_H \right) = \frac{100}{N} \sum_{i=1}^{N} \Theta \left(d_i - D_H \right) \quad (12.13)$$

The difference between various dosimetric-based objective functions is the norm used for defining the distance between the ideal and actual dose distribution, on how the violation is penalized, and what dose normalization is applied. For $p = 2$ or for variance-based objective functions, dose values above or below a critical dose value are penalized quadratically. For $p = 1$ there is a linearly dependency of the penalty of the dose value, but for $p = 0$ this penalty is independent.

12.1.3.2 DVH-Based Objectives

DVH-derived quantities can also be used to specify properties of the desired dose. These values are used as optimization objectives specifying that a dose distribution should result to a specific DVH value at a given set of dose values for different anatomical volumes of interest (e.g., PTVs, OARs).

Dose-volume constraints specify requirements for a clinically acceptable dose distribution. These can be used for a constraint dose optimization. Such constraints could specify upper bounds for the fraction of the volume of a region that can accept a dose larger than a specific level, or a lower bound for the fraction that should have a dose at least larger than a specific value. If $V(D)$ is the volume of a region that receives a dose greater than or equal to D, then an upper dose-volume constraint V_{max} for a dose D' is an inequality of the form $V(D') < V_{max}$, and a lower dose-volume constraint V_{min} for a dose value D'' is an inequality of the form $V(D'') > V_{min}$.

Different dose distributions can be obtained if only DVH-based objectives are considered, as the dose distributions are only required to satisfy some integral properties. This could be a benefit if we want to obtain a large range of possible dose distributions. It can also be a reason that various optimization algorithms cannot be used for such a dose optimization. Objectives that use variances and try to minimize the distance in terms of some metric can fix the range of possible dose distributions that can be obtained significantly. The most commonly commercially implemented metrics are for $p = 1$ (linear) and for $p = 2$ (quadratic or variance-based), as explained in the discussion for Equations 12.11 and 12.12.

12.1.3.3 Multicriteria Optimization and Multiobjective Problem

As explained above, the problem of dose optimization is transformed into a problem of minimization of the values of several objective functions with respect to targets, OARs and NT. In the majority of the cases, the minimization of one of the objective functions (achievement of a goal of planner) is in conflict with one or more of the other objective functions and planner goals. This is a typical situation in the multicriteria and multiobjective optimization (Miettinen 1999). Efforts have been made in the past to introduce multiobjective optimization techniques for the brachytherapy dose optimization (Lahanas et al. 1999, 2001, 2003a,b, 2004; Milickovic et al. 2001, 2002). The complexity and the runtime demands of the required evolutionary optimization methods have limited these efforts to scientific and research activities.

The only practical approach established over the last two decades is to transform the multiobjective problem to a single-objective one (Karabis et al. 2005; Trnková et al. 2009), where several methods are available for the objective function minimization.

12.1.3.4 Single-Objective Approach

For reducing the multiobjective problem to a single-objective optimization problem, the following aggregate form out of M different objective functions $f_i(\mathbf{x})$ is considered:

$$f(\mathbf{x}) = \sum_{m=1}^{M} w_m f_m(\mathbf{x}). \qquad (12.14)$$

$\mathbf{x}$ is the vector of the optimization parameters, for example, source dwell times for the temporary implants. The aggregation form given in Equation 12.14 defines an SO function $f(\mathbf{x})$ by using a weighted sum of all considered objectives such as those defined in Equations 2.11 and 2.12.

The weights w_m are also known as *importance factors* (Lahanas et al. 1999, 2001, 2003a,b; Karabis et al. 2005) or *penalties* (Lessard and Pouliot 2001; Lessard et al. 2006) and are considered as a measure of the significance of each objective in the optimization process.

One could consider this as an a priori multiobjective optimization, but this requires some knowledge of the *importance factors* and their influence on the results. Even if the solution obtained is a global optimal solution, that is, the best possible for the aggregate SO function, it is possible that by using other importance factors, other better solutions can be obtained. This often requires repeating the optimization with other importance factors. If the result for some of the objectives is not satisfactory, then the corresponding *importance factor* is increased. As this has an effect on the other importance factors, the result for another objective could deteriorate.

In a trial-and-error method, the optimization is repeated with different *importance factors* until the treatment planner considers that the optimization result is acceptable. While for two objectives a solution very close to the optimal can be found, the calculation is more difficult as more objectives are considered. The clinical acceptance criteria are commonly defined as a set of dose-volume constraints for PTV and OARs, which have to be satisfied by the dose distribution (see also Section 12.2). The objective functions described above represent the following:

(1) The maximum dose limit D_H for an anatomical volume represents the dose-volume constraint $V(D_H) = 0$; the volume of the organ receiving a dose of at least D_H should be 0.

(2) The low dose limit D_L for an anatomical volume represents the dose constraint $V(D_L) = 1.0$; the volume of the organ receiving a dose of at least D_L should be 1 or 100%.

Practically and based on the experience accumulated with the available optimization engines in the commercial treatment planning systems, some default sets for the dose limit values in Equations 12.11 and 12.12 as well as for the *importance factors* or *penalties* in Equation 12.14 are provided for different tumor localizations. Such sets of values are commonly called *protocols* (Karabis et al. 2005) or class *solutions* (Lessard et al. 2006).

12.1.3.5 Optimization Algorithms

There is no optimization algorithm that is optimal for all optimization problems. For stochastic algorithms such as genetic algorithms or simulated annealing, the *no-free-lunch theorem* (Wolpert and Macready 1997) has proven that without incorporation of special knowledge, optimization problems exist for these algorithms that do not perform more effectively than a random blind search. Deterministic algorithms, although very fast, can easily be trapped in local minima. The best algorithm depends on the set of objectives used.

The most important single-objective optimization algorithms used for optimization of dose distributions in brachytherapy are given below, separated into two main categories of optimization algorithms.

- *Deterministic*: Optimization methods that do not use any random elements during the optimization. Linear and quadratic programming algorithms belong to this class. For continuous parameters such as the source dwell times and for continuous differentiable objective functions, gradient-based optimization algorithms can be used. For quadratic or nonlinear functions, nonlinear optimization algorithms are required. Most common are constraint-free gradient-based algorithms that are based on line search (Lahanas et al. 2003b; Milickovic et al. 2001, 2002).
- *Stochastic*: Optimization algorithms that use random elements produced by the use of random numbers. Such algorithms are evolutionary algorithms (Goldberg 1989; Koza 1992; Lahanas et al. 1999, 2001, 2003a, 2004) and simulated annealing (Kirkpatrick et al. 1983; Tsallis and Stariolo 1996; Ingber 1989, 1993, 1996; Szu and Hartley 1987; Lessard et al. 2001, 2006).

12.2 Evaluation

12.2.1 Introduction

There have been several concepts and parameters defined and proposed for the evaluation of the 3D dose distribution in brachytherapy.

ICRU report 58 (ICRU 1998) offers an extended summary of the classical parameters that could be used for evaluating the dose distribution, but it is mainly focused on the system-based treatment planning in interstitial brachytherapy. This report, on the other hand, introduces for the first time anatomy-oriented parameters for evaluation as well as the volume definitions as already known in the external beam treatment planning: *gross tumor volume* (GTV), *clinical target volume* (CTV), *planning target volume* (PTV), and *treated volume* (TV).

In this report, it is recommended to use the *minimum target dose* (MTD), defined as the minimum dose at the periphery of the CTV, and the *mean central dose* (MCD), which is taken as the arithmetic mean of the local minimum doses between sources or catheters in the central plane, for reporting and evaluating. The latest is according to the Paris system of dosimetry in interstitial LDR brachytherapy (see Chapter 8). Finally, the *high dose volume*, defined as the volumes encompassed by the isodose surface corresponding to 150% of the MCD, and the *low dose volume*, defined as the volume within the CTV encompassed by the 90% isodose surface, are considered.

In the past, efforts have been made for establishing some parameters or figures to describe the homogeneity of the dose distribution in brachytherapy. All of these have been based on a dose calculation space without accounting for any anatomic detail. In other words, these efforts have considered the dose distribution simply around the catheters. The method that found a wide application at least in the field of LDR brachytherapy is that of *natural dose-volume histogram* (NDVH) introduced by Anderson (1986). Other published recommendations (Ash et al. 2000; Nag et al. 1999, 2001; Pötter et al. 2002, 2006; Kovács et al. 2005; Salembier et al. 2007) proposed to use DVH-based parameters based on the cumulative DVHs for the evaluation and documentation of the dose distribution. These evaluation tools are discussed in more detail in the following subsection.

12.2.2 Dose-Volume Histogram

The DVHs were first introduced by Goitein and Verhey in a publication by Shipley et al. (1979) and have been in use for over 30 years in radiotherapy treatment planning (See for more information Nioutsikou et al. 2005). They have proved to be an invaluable tool for summarizing a 3D dose distribution in a 2D graph (Nioutsikou et al. 2005).

Generally for a DVH, the dose distribution within a VOI is considered. The VOI is either the volume of an OAR or target or, in case of brachytherapy, could be also a volume defined simply in relation to the implant itself. This 3D dose distribution is considered for the purposes of DVH in a discrete form (discretization of 3D dose distribution). The dose is then calculated on a regular 3D grid or at a set of sampling points generated in such a way and number to cover adequately and with a homogeneous distribution the VOI. DVHs are the result of a statistical analysis of the discrete dose values at the dose sampling points (3D grid or randomly distributed dose sampling points) with respect to volume.

12.2.2.1 Differential DVH

The differential DVH is the frequency distribution of dose values within the VOI under investigation. To calculate this, the range of dose values at the sampling points within the VOI is divided into a specific number of bins, Nb. These bins are of an arbitrary size. The number of sampling points with dose values within each of the bins is counted. The result, the frequency distribution, is usually presented as a continuous line or bar graph. The horizontal axis represents the dose expressed either in absolute values (cGy, Gy) or in relative values (percentage) normalized to the prescription dose (100%). The vertical axis represents that part of the VOI receiving dose values within a specific bin. This volume is calculated depending on the methodology followed for the generation of dose sampling points. In the case of a regular 3D dose grid, it is the sum of volumes of all voxels with dose values being within that bin. If a random sampling technique is used (this is very common in brachytherapy to account for the very high dose gradients existing close to catheter and applicators), this volume is calculated by the number of sampling points accounted to have dose values within that bin multiplied by the assumed volume element for each sampling point. The latter is calculated by dividing the VOI by the total number of generated sampling points within that VOI.

The volume is expressed either in absolute units (mm³ or cm³) or as percentage of the total VOI. The narrower the peak in the differential DVH of an implant, the more homogeneous the dose distribution within that implant geometry (see Figure 12.7).

12.2.2.2 Cumulative DVH

The cumulative DVH is extracted out of the previously calculated differential DVH. Similarly to the differential DVH, the *x*-axis represents the dose range, and it is based on the bins used for the calculation of the differential DVH. The *y*-axis represents the VOI that receives at least the dose of the corresponding dose bin (see Figure 12.7):

$$DVH(D) = \int_D^{\infty} dDVH(x)dx = 1 - \int_0^D dDVH(x)dx. \qquad (12.15)$$

In this equation, DVH(*D*) is the value of the cumulative DVH at the dose value *D*. dDVH(*x*) is the value of the differential DVH at the dose level *x*. dDVH(*x*) is calculated at that dose bin, which contains the dose level *x*. The cumulative DVH demonstrates the coverage of the VOI by a given dose value.

12.2.3 Dose-Volume Parameters

DVHs represent the statistical evaluation of the dose value distribution in the 3D space as differential or cumulative frequency distribution and have found a wide acceptance in the field of radiation oncology treatment planning and evaluation. Due to the complexity of the DVH curves and the fact that, in each plan, several VOIs are considered (target volumes GTVs, CTVs, and PTV and different OARs), it is practically impossible to consider these curves for comparing plans and for looking for correlations to clinical outcomes and biological effects. It has been established, however, in the above-cited recommendations

(a)

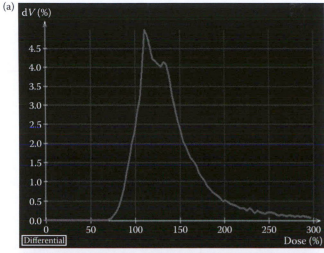

(b)

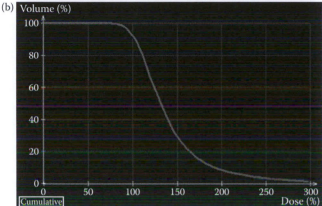

FIGURE 12.7 Example of DVHs for the target volume for an HDR prostate implant. (a) Differential DVH. (b) Cumulative DVH. Both axes are in relative units; *x*-axis in percentages of the prescription dose and *y*-axis in percentages of the PTV volume. (Courtesy of D. Baltas.)

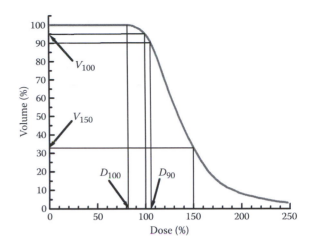

FIGURE 12.8 Graphical demonstration of the definition of the D_{100}, D_{90}, V_{100}, and V_{150} dosimetric parameters for an imaging-based 3D brachytherapy treatment planning based on the cumulative DVH of a target volume (GTV, CTV, PTV). Dose is normalized relative to the prescription dose (= 100%). (Courtesy of D. Baltas.)

Of course, statistical values like the *mean dose value* (D_{mean}) and the *standard deviation* of the dose value distribution in the PTV can also be used.

12.2.3.2 OAR Orientated Parameters

For the OARs, there are no widely established dosimetric parameters for recording and reporting. Exceptions are the values considered to be representative for the irradiation of bladder and rectum for the primary intracavitary brachytherapy of the cervix carcinoma as proposed by ICRU 38 (ICRU 1985) and the gynecological (GYN) GEC-ESTRO working group (Pötter et al. 2006). The same was proposed for prostate (Kovács et al. 2005; Salembier et al. 2007).

Pötter et al. (2002) proposed to use the maximum dose for the OARs at least for the case of primary intracavitary brachytherapy of the cervix carcinoma, where the maximum doses are considered to be the doses received in a volume of at least 2 cm³ (D_{2cc}) and 5 cm³ (D_{5cc}) for bladder and rectum, respectively. In the latest GEC-ESTRO recommendations (Pötter et al. 2006), this is recommended for the OARs (rectum, sigmoid, bladder) if the organ walls are contoured. Otherwise, it is recommended in this report that the minimum dose in the most irradiated 0.1 ($D_{0.1cc}$), 1 (D_{1cc}), and 2 cm³ (D_{2cc}) of the OAR be stated. These values are also suggested by the GEC-ESTRO-EAU recommendations on temporary HDR brachytherapy of the prostate (Kovács et al. 2005) and by the ESTRO-EAU-EORTC recommendations for permanent LDR prostate brachytherapy (Salembier et al. 2007). $D_{0.1cc}$ and D_1, the highest dose covering 1% of the OAR, are also suggested for urethra in the temporary brachytherapy report, where Salembier et al. (2007) suggest the use of D_5 (highest dose covering 5% of the OAR volume), D_{10} (highest dose covering 10% of the OAR volume), and D_{30} (highest dose covering 30% of the OAR volume) for urethra. Figure 12.9 demonstrates these parameters on DVH graphs of the different OARs.

to consider specific dose-volume pair values, the dose-volume indices, which define representative points on the DVH curve (cumulative DVH curve of a specific VOI), for reporting.

12.2.3.1 Target Orientated Parameters

Firstly, dose-volume indices defined for the different types of target volumes, GTVs or CTVs or PTV, will be listed.

> D_{100}: the dose that covers 100% of the target volume, which is equivalent to the MTD proposed by ICRU report 58 (ICRU 1998); for the case, we consider CTV.
> D_{90}: the dose that covers 90% of the target volume.
> V_{100}: the percentage of the target volume that receives at least the prescribed dose, which is set to 100%.
> V_{150}: the percentage of the target volume, which receives at least 150% of the prescribed dose.

The definition of these dose-volume parameters for the target volumes is graphically shown in Figure 12.8.

(a)

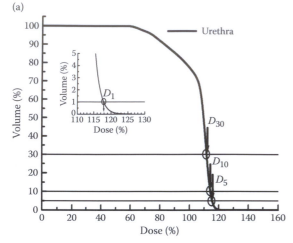

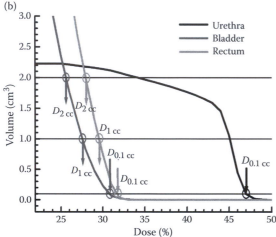

(b)

FIGURE 12.9 Graphical demonstration of the definition of the several recommended dosimetric parameters for OARs. (a) Definition of the D_1 (figure insert), D_5, D_{10}, and D_{30} dose volume parameters on DVH graph with the volume as percentage of the total contoured organ volume. (b) Definition of the $D_{0.1cc}$, D_{1cc}, and D_{2cc} parameters on DVH graphs with the volume of the organ in cubic centimeters. The dose in both figures is expressed as percentage of the prescription dose (= 100%). (Courtesy of D. Baltas.)

12.2.3.3 Conformity Indices

Modern brachytherapy is a 3D conformal radiotherapy delivery technique, with an extreme intensity modulation, especially when utilizing HDR or PDR single-stepping source afterloading devices. It is based on 3D medical imaging techniques, thus enabling delineation of target volumes and critical healthy tissues. Often, more than one solution can be found using one or more optimization strategies for the given objectives of a plan. Clinical judgment will then allow a better adaptation of the dose distribution to the tumor volume, while limiting irradiation of healthy tissues. The dosimetric indices described in the previous section are based on the calculated DVHs for the VOI. But these planning parameters are not necessarily sufficient or suited to simply evaluate a treatment plan or to compare alternatively

calculated plans. One plan could be better regarding D_{90} and V_{100} but worse regarding V_{150} and possibly for one or another OAR.

To overcome these difficulties, indices have been developed aiming to offer a measure of the conformity of the achieved 3D dose distribution. In this way, conformity indices can be considered as complementary tools to DVH parameters that attribute a score to a treatment plan or that can compare several treatment plans calculated for the same implant.

12.2.3.3.1 Conformal Index

Baltas et al. (1998) introduced a utility function as a measure of the implant quality, the Conformal Index (COIN), which has been later expanded to include OARs (Milickovic et al. 2002). The COIN takes into account all relevant patient VOIs: the PTV, the surrounding normal tissue (NT) and the OARs. The COIN for the prescription reference dose value D_{ref} (prescribed dose) is defined as

$$COIN = c_1 \cdot c_2 \cdot c_3 \quad (12.16)$$

where

$$c_1 = \frac{PTV_{ref}}{PTV} \quad c_2 = \frac{PTV_{ref}}{V_{ref}} \quad c_3 = \prod_{i=1}^{N_{OAR}}\left(1 - \frac{V_{OAR}^i\left(D > D_{crit}^i\right)}{V_{OAR}^i}\right)$$

The coefficient c_1 is the fraction of the PTV (PTV_{ref}) that receives dose values of at least D_{ref}. The coefficient c_2 is the fraction of the reference isodose volume V_{ref} that is within the PTV (see also Figure 12.10).

V_{OAR}^i is the volume of the ith OAR, and $V_{OAR}^i\left(D > D_{crit}^i\right)$ is the volume of the ith OAR that receives a dose that exceeds the critical dose value D_{crit}^i defined for that OAR. The product in the equation for c_3 is calculated for all N_{OAR} OARs included in the treatment planning.

In the case where an OAR receives a dose D above the critical value defined for that structure, the conformity index will be reduced by a fraction that is proportional to the volume that exceeds this limit. The ideal situation is COIN = $c_1 = c_2 = c_3 = 1.0$.

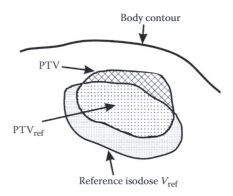

FIGURE 12.10 Schematic diagrams of volumes necessary for computation of the COIN for the prescription reference dose D_{ref}. (Courtesy of D. Baltas.)

The COIN assumes in this form that the PTV, the OARs, and the surrounding NT are of the same importance. When several types of target volumes are considered, such as GTVs or CTVs and PTV, then all corresponding c_1 and c_2 coefficients have to be considered in the COIN formula according to Equation 12.16 for the corresponding specified prescription doses for each one of those target volumes.

COIN values can be calculated for any dose level of a 3D dose distribution. It is important to mention that the c_1 and c_2 coefficients in Equation 12.16 depend on the selected dose value, whereas the c_3 coefficient depends only on the predefined critical dose values D_{crit}^i for the different OARs considered in the treatment planning process.

When the COIN value is calculated for every dose value D according to Equation 12.16, then the conformity distribution or equivalently the COIN histogram is generated. Figure 12.11 demonstrates the conformity distribution for an example of an HDR prostate implant. A good implant is that where the maximum COIN value is observed exactly at or close to the prescribed reference dose V_{ref} (100%).

12.2.3.3.2 Conformation Number and External Index

The product $c_1 \cdot c_2$ in Equation 12.16 has originally been introduced as the conformation number (*CN*):

$$CN = c_1 \cdot c_2 \qquad (12.17)$$

by van't Riet et al. (1997). The *CN* defines the quality of coverage of the target and the degree of conformation of the prescription reference dose to the target volume (not extended to NTs). As it is also the case for the COIN, considering the product of the separate conformity components regarding target volume and regarding NT, a total conformity of, for example, *CN* = 0.72 could be the result of a high coverage of the target volume, c_1 = 0.90, and an intermediate degree of saving of NT, c_2 = 0.80, or of the opposite combination, c_1 = 0.80 and c_2 = 0.90. When evaluating or comparing plans for a specific implant, it is useful to consider the separate conformity components such as c_1 and c_2 and, for COIN, also c_3 in order to identify the reason for the eventually low conformity, that is, low *CN* or COIN values.

Meertens et al. (1994) defined the external volume index (EI) to be the amount of NT volume that receives a dose equal to or greater than the prescribed reference dose. Originally, the volume should be expressed as a percentage of the target volume. Based on Equation 12.16 and Figure 12.10, the external volume index EI is given by the following equation:

$$EI = \left[\frac{c_1}{c_2} - c_1 \right] \cdot 100(\%) \qquad (12.18)$$

Avoiding the multiplication by 100 in the above equation, *EI* is then expressed as a *fraction* of the target volume. Depending on the value of the c_2 coefficient, it can appear that the resulting

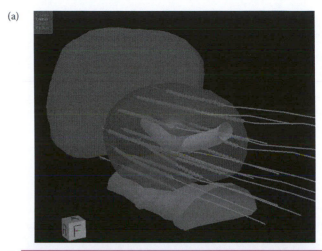

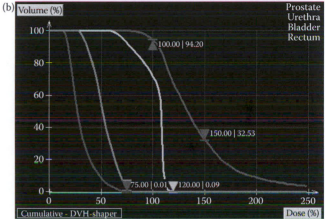

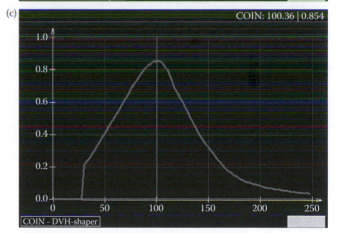

FIGURE 12.11 **(See color insert.)** Example of an HDR prostate implant. (a) 3D view of the contoured prostate as PTV, urethra, rectum, and bladder and the 15 implanted catheters with a total of 282 active SDPs at a step of 2.5 mm. (b) Cumulative DVHs for PTV and the OARs urethra, rectum, and bladder. The critical dose values or dose limits for the three OARs are 120% for urethra and 75% of the reference dose (100%) for rectum and bladder. (c) Conformity distribution (COIN distribution) calculated according to Equation 12.16 and based on the above DVHs, including all three OARs, demonstrates a COIN value of 0.854 for the 100% reference isodose line. The maximum COIN value is 0.857 obtained for the 102.5% isodose value that is very close to the reference dose for that implant and treatment plan. (Courtesy of D. Baltas.)

value for *EI* is higher than 1.0 (i.e., for low c_2 values, $c_2 \ll 1.0$), which can never be the case for COIN or *CN*.

12.2.3.3.3 *Natural DVH*

In implant-based brachytherapy, a frequently used evaluation method consists of calculating differential DVHs for a volume encompassing the implant. These DVHs aim at evaluating the (in)homogeneity of the dose distribution in and around the implant and the choice of the reference isodose in relation to the homogeneously irradiated volume. These histograms do not assess the target coverage or the dose in OARs.

A limitation of a differential DVH, depicting d*V*/d*D* as a function of *D*, is the fact that the curve is largely influenced by the inverse square law behavior of the brachytherapy irradiation. Due to this inverse square law, the change of volume with dose, d*V*/d*D*, near the point source, that is, in the high dose region, is small; at a large distance from the source, that is, in the low dose region, d*V*/d*D* is large. This effect obscures local effects due to the implant geometry. In the case of an ideal point source, with *V* changing as a function of r^3 and *D* as a function of r^{-2}, one can simply derive that d*V*/d*D* is a function of $D^{-5/2}$.

By defining as differential DVH the graph of d*V*/d*u* versus *u*, with $u = D^{-3/2}$, the so-called NDVH is obtained (Anderson 1986). For an ideal point source, the NDVH shows up as a straight line at $(4/3)\pi S^{3/2}$, with *S* being the source strength. In this way, the dependence of the inverse square law is removed. However, the volume receiving at least a given dose is still kept proportional to the area under the curve. If an implant is delivering dose values within a narrow range to a large part of the target volume, the NDVH deviates from this constant value and will show a peak for that range of dose values. Thus, a peak in the histogram represents a large volume with a low dose gradient. For volume implants, it follows that a better dose uniformity is characterized by a high and narrow peak. The position of the peak [indicated by the parameters *LD* and *HD* (see Figure 12.12), being the doses at half-maximum of the peak at, respectively, the low dose side and the high dose side] relative to the actual prescription dose, *PD*, is a measure of the choice of the reference isodose in relation to the relatively homogeneous irradiated volume. Ideally, *PD* should coincide with *LD*. Choice of a prescription dose, *PD*, with *PD* < *LD* produces a relatively low uniformity index. On the other hand, a prescription dose, *PD*, with *PD* > *LD* indicates a large volume with a low dose gradient, receiving lower doses than the reference dose.

The uniformity index, *UI*, which can be extracted from the NDVH (Anderson 1986), is defined as

$$UI = \frac{V(PD-HD)}{V(PD)} * \frac{u(PD)}{u(PD)-u(HD)} \qquad (12.19)$$

with *PD* being the prescription dose, *HD* the dose at half-maximum of the peak at the high dose side, *V(PD-HD)* the volume receiving a dose between *PD* and *HD*, *V(PD)* the volume receiving at least the prescription dose *PD*, and $u(D) = D^{-3/2}$.

This uniformity index, thus, depends not only on the uniformity of the dose distribution but also on the choice of the reference isodose. It expresses the amount of the treatment volume that is concentrated between the prescription dose, *PD*, and the high dose side of the peak, *HD*.

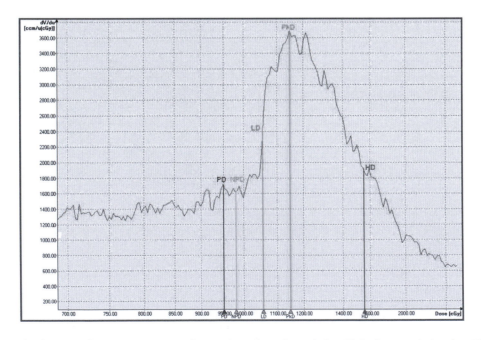

FIGURE 12.12 Example of NDVH of an HDR prostate implant. Indicated are the peak dose Pk*D*, the prescription dose *PD* (also referred to as treatment dose *TD*) being 950 cGy, the dose at half-maximum of the peak at the high and low dose side, *HD* and *LD*, respectively, and the natural prescription dose *NPD*. (Courtesy of I.-K.K. Kolkman-Deurloo.)

Therefore, a second index, the quality index, *QI*, was defined, which suppresses the influence of the chosen reference dose:

$$QI = \frac{V(LD-HD)}{V(LD)} * \frac{u(LD)}{u(LD)-u(HD)} \qquad (12.20)$$

with *LD* being the dose at half-maximum of the peak at the low dose side, *V(LD–HD)* the volume receiving a dose between *LD* and *HD*, and *V(LD)* the volume receiving at least dose *LD*.

A different way to assess the ideal prescription dose is presented by Moerland et al. (2000). They explain that when this prescription dose is positioned at the base of the peak at the low dose side, the peak of the NDVH, that is, the volume of the implant irradiated with a low dose gradient, receives at least the prescription dose, while, at the same time, the volume receiving a dose lower than the prescription dose experiences a steep dose decrease due to the inverse square law. Moerland et al., therefore, call this dose the natural prescription dose, *NPD*. However, in image-based brachytherapy planning, one tends to decide on a prescription dose *PD* leading to a predefined target coverage. Any mismatch between the *PD* and the *NPD*, thus the mismatch between coverage and homogeneity, is scored by the natural dose ratio *NDR*, which is defined as

$$NDR = \frac{NPD}{PD}. \qquad (12.21)$$

Ideally, *NDR* = 1; the target is overdosed if *NDR* > 1 and underdosed if *NDR* < 1.

The NDVH and the quality indices derived from it are valuable, although not frequently used, tools for evaluation of volume implants. Its use for head–and–neck brachytherapy is demonstrated in the works of Kolkman-Deurloo et al. (1994), evaluating optimized versus nonoptimized dose distributions for base-of-tongue implants, and Teguh et al. (2009), evaluating the quality of life in relation to these indices for patients who have had a base-of-tongue implant. The merits of geometric dose optimization for surface mold brachytherapy based on an analysis of the NDVH and its indices are presented in the work of Berns et al. (1997). The value of *NDR* for permanent prostate seed implants is demonstrated by Moerland et al. (2000). An application of *NDR* for HDR brachytherapy can be found in the work of Shwetha et al. (2010) using it for evaluation of different dwell time optimization techniques for interstitial cervix implants.

12.3 Conclusion

Optimization tools and technology have been developed over the last two decades enabling the automatic generation of source positions and dwell time patterns in order to fulfill clinical criteria. The critical point here is the fact that evaluation criteria not expressed in numerical expressions cannot be considered by such tools. Furthermore, the different algorithms take a specific subset of such criteria into consideration. It depends thus on the user needs and demands to select the appropriate optimization tools to achieve his or her targets.

DVHs have been proven to be useful tools in helping planners evaluate and compare treatment plans. Dose-volume and quality parameters can be derived from such DVHs. The predictive value of those parameters still has to be proven in general. Although DVHs and evaluation parameters are useful tools, inspection of the full 3D dose distribution remains an essential step for approval of a final treatment plan.

References

Anacak, Y., Esassolak, M., Aydin, A., Aras, A., Olacak, I., and Haydaroglu, A. 1997. Effect of geometrical optimization on the treatment volumes and the dose homogeneity of biplane interstitial brachytherapy implants. *Radiother Oncol* 45:71–6.

Anderson, L.L. 1986. A "natural" volume-dose histogram for brachytherapy. *Med Phys* 13:898–903.

Ash, D., Flynn, A., Battermann, J., de Reijke, T., Lavagnini, P., and Blank, L. 2000. ESTRO/EAU/EORTC recommendations on permanent seed implantation for localized prostate cancer. *Radiother Oncol* 57:315–21.

Baltas, D., Kolotas, C., Geramani, K. et al. 1998. A conformal index (COIN) to evaluate implant quality and dose specification in brachytherapy. *Int J Radiat Oncol Biol Phys* 40: 515–24.

Beddar, A.S., Krishnan, S., Briere, T.M. et al. 2006. The optimization of dose delivery for intraoperative high-dose-rate radiation therapy using curved HAM applicators. *Radiother Oncol* 78:207–12.

Berns, C., Fritz, P., Hensley, F.W., and Wannenmacher, M. 1997. Consequences of optimization in PDR brachytherapy — is a routine geometrical optimization recommendable? *Int J Radiat Oncol Biol Phys* 37:1171–80.

Dutreix, A., Marinello, G., and Wambersie, A. 1982. *Dosimétrie en Curiethérapie*. Masson, Paris, France.

Edmundson, G.K. 1990. Geometry-based optimization for stepping source implants. In: *Brachytherapy HDR and LDR*. A.A. Martinez, C.G. Orton, R.F. Mould, and M.D. Columbia (eds.). Nucletron International BV, Veenendaal, The Netherlands, pp. 184–92.

Edmundson, G.K. 1994. Volume optimization: An American viewpoint. In: *Brachytherapy from Radium to Optimization*. R.F. Mould, J. Battermann, A.A. Martinez, and B.L. Speiser (eds.). Nucletron International BV, Veenendaal, The Netherlands, pp. 314–8.

Goldberg, D.E. 1989. *Genetic Algorithms in Search, Optimisation and Machine Learning*. Addison Wesley, Reading, MA.

Ingber, L. 1989. Very fast simulated re-annealing. *J Math Comput Modeling* 12:967–73.

Ingber, L. 1993. Simulated annealing: practice versus theory. *J Math Comput Modeling* 18:29–57.

Ingber, L. 1996. Adaptive simulated annealing (ASA): Lessons learned. *J Control Cybernet* 25: 33–54.

International Commission on Radiation Units and Measurements (ICRU). 1985. *Dose and Volume Specification for Reporting and Recording Intracavitary Therapy in Gynecology*. Report 38 of ICRU, ICRU Publications, Bethesda, MD.

International Commission on Radiation Units and Measurements (ICRU). 1998. *Dose and Volume Specification for Reporting Interstitial Therapy*. Report 58 of ICRU, ICRU Publications, Bethesda, MD.

Karabis, A., Giannouli, S., and Baltas, D. 2005. HIPO: A hybrid inverse treatment planning optimization algorithm in HDR brachytherapy. *Radiother Oncol* 76(Supplement 2):29.

Kirkpatrick, S., Gelatt, C.D., and Vecci, M.P. 1983. Optimisation by simulated annealing. *Science* 220:671–80.

Kolkman-Deurloo, I.-K.K., Visser, A.G., Niël, C.G.J.H., Driver, N., and Levendag, P.C. 1994. Optimization of interstitial volume implants. *Radiother Oncol* 31:229–39.

Kolkman-Deurloo, I.-K.K., Nuyttens, J.J., Hanssens, P.E.J., and Levendag, P.C. 2004. Intraoperative HDR brachytherapy for rectal cancer using a flexible intraoperative template: standard plans versus individual planning. *Radiother Oncol* 70:75–9.

Kovács, G., Pötter, R., Loch, T. et al. 2005. GEC/ESTRO-EAU recommendations on temporary brachytherapy using stepping sources for localised prostate cancer. *Radiother Oncol* 74:137–48.

Koza, J.R. 1992. *Genetic Programming: On the Programming of Computers by Means of Natural Selection*. MIT, Cambridge, MA.

Lahanas, M., Baltas, D., and Zamboglou, N. 1999. Anatomy-based three-dimensional dose optimisation in brachytherapy using multiobjective genetic algorithms. *Med Phys* 26:1904–18.

Lahanas, M., Milickovic, N., Baltas, D., and Zamboglou, N. 2001. Application of multiobjective evolutionary algorithms for dose optimisation problems in brachytherapy. In: *Proceedings of the First International Conference, EMO 2001, Zurich, Switzerland*. E. Zitzler, K. Deb, L. Thiele, C.C.A. Coello, and D. Corne (eds.). Lecture Notes in Computer Science Vol. 1993, Springer, Berlin, pp. 574–87.

Lahanas, M., Baltas, D., and Zamboglou, N. 2003a. A hybrid evolutionary multiobjective algorithm for anatomy based dose optimisation algorithm in HDR brachytherapy. *Phys Med Biol* 48:399–415.

Lahanas, M., Baltas, D., Giannouli, S., and Zamboglou, N. 2003b. Global convergence analysis of fast multiobjective gradient based dose optimisation algorithms for high dose rate brachytherapy. *Phys Med Biol* 48:599–617.

Lahanas, M., Karouzakis, K., Giannouli, S., Mould, R.F., and Baltas, D. 2004. Inverse planning in brachytherapy from radium to high dose rate 192-Iridium afterloading. *J Oncol* 54:535–46.

Lessard, E. and Pouliot, J. 2001. Inverse planning anatomy-based dose optimisation for HDR-brachytherapy of the prostate using fast simulated annealing and dedicated objective functions. *Med Phys* 28:773–9.

Lessard, E., Kwa, S.L.S., Pickett, B., Roach, III, M., and Pouliot, J. 2006. Class solution for inversely planned permanent prostate implants to mimic an experienced dosimetrist for pre and real-time treatment planning. *Med Phys* 33:2773–82.

Levendag, P.C., Peters, R., Meeuwis, C.A. et al. 1997. A new applicator design for endocavitary brachytherapy of cancer in the nasopharynx. *Radiother Oncol* 45:95–8.

Major, T., Polgar, C., Somogyi, A., and Nemeth, G. 2000. Evaluation of the dose uniformity for double-plane high dose rate interstitial breast implants with the use of dose reference points and dose non-uniformity ratio. *Radiother Oncol* 54:213–20.

Major, T., Polgar, C., Fodor, J., Somogyi, A., and Nemeth, G. 2002. Conformality and homogeneity of dose distributions in interstitial implants at idealized target volumes: A comparison between the Paris and dose-point optimized systems. *Radiother Oncol* 62:103–11.

Meertens, H., Borger, J., Steggerda, M., and Blom, A. 1994. *Evaluation and Optimization of Interstitial Brachytherapy Dose Distributions*. R.F. Mould, J.J. Battermann, A.A. Martinez, and B.L. Speiser (eds.). Nucletron International BV, Veenendaal, The Netherlands, pp. 301–6.

Miettinen, K.M. 1999. *Nonlinear Multiobjective Optimisation*. Kluwer Academic Publisher, Boston.

Milickovic, N., Lahanas, M., Baltas, D., and Zamboglou, N. 2001. Comparison of evolutionary and deterministic multiobjective algorithms for dose optimisation in brachytherapy. In: *Proceedings of the First International Conference, EMO 2001, Zurich, Switzerland*. E. Zitzler, K. Deb, L. Thiele, C.A. Coello Coello, D. Corne (eds.). Lecture Notes in Computer Science Vol. 1993, Springer, Berlin, pp. 167–80.

Milickovic, N., Lahanas, M., Papagiannopoulou, M., Zamboglou, N., and Baltas, D. 2002. Multiobjective anatomy-based dose optimization for HDR-brachytherapy with constraint free deterministic algorithms. *Phys Med Biol* 47:2263–80.

Moerland, M.A., van der Laarse, R., Luthmann, R.W., Wijrdeman, H.K., and Battermann, J.J. 2000. The combined use of the natural and the cumulative dose volume histograms in planning and evaluation of permanent prostatic seed implants. *Radiother Oncol* 57:279–84.

Nag, S., Beyer, D., Friedland, J., Grimm, P., and Nath, R. 1999. American Brachytherapy Society (ABS) recommendations for transperineal permanent brachytherapy of prostate cancer. *Int J Rad Oncol Biol Phys* 44:789–99.

Nag, S., Cano, E., Demanes, J., Puthawala, A., and Vikram, B. 2001. The American Brachytherapy Society recommendations for high-dose-rate brachytherapy for head-and-neck carcinoma. *Int J Rad Oncol Biol Phys* 50:1190–8.

Nath, R., Anderson, L.L., Luxton, G., Weaver, K.A., Williamson, J.F., and Meigooni, A.S. 1995. Dosimetry of interstitial brachytherapy sources: Recommendations of the AAPM Radiation Therapy Committee Task Group No. 43. *Med Phys* 22:209–34.

Niël, C.G.J.H., Koper, P.C.M., Visser, A.G., Sipkema, D., and Levendag, P.C. 1994. Optimizing brachytherapy for locally advanced cervical cancer. *Int J Radiat Oncol Biol Phys* 29:873–7.

Nioutsikou, E., Webb, S., Panakis, N., Bortfeld, T., and Oelfke, U. 2005. Reconsidering the definition of a dose-volume histogram. *Phys Med Biol* 50:L17–9.

Pieters, B.R., Saarnak, A.E., Steggerda, M.J., and Borger, J.H. 2001. A method to improve the dose distribution of interstitial breast implants using geometrically optimized stepping source techniques and dose normalization. *Radiother Oncol* 58:63–70.

Pötter, R., van Limbergen, E., and Wambersie, A. 2002. Reporting in brachytherapy: Dose and volume specification. In: *The GEC ESTRO Handbook of Brachytherapy*. A. Gerbaulet, R. Pötter, J.J. Mazeron, H. Meertens, and E. van Limbergen (eds.). ESTRO, Brussels, pp. 153–215.

Pötter, R., Haie-Meder, C., van Limbergen, E. et al. 2006. Recommendations from gynaecological (GYN) GEC ESTRO working group (II): Concepts and terms in 3D image-based treatment planning in cervix cancer brachytherapy—3D dose volume parameters and aspects of 3D image-based anatomy, radiation physics, radiobiology. *Radiother Oncol* 78:67–77.

Rivard, M.J., Coursey, B.M., DeWerd, L.A. et al. 2004. Update of AAPM Task Group No. 43 Report: A revised AAPM protocol for brachytherapy dose calculations. *Med Phys* 31:633–74.

Rivard, M.J., Butler, W.M., DeWerd, L.A. et al. 2007. Supplement to the 2004 update of the AAPM Task Group No. 43 Report. *Med Phys* 34:2187–205.

Salembier, C., Lavagnini, P., Nickers, P. et al. 2007. Tumour and target volumes in permanent prostate brachytherapy: A supplement to the ESTRO/EAU/EORTC recommendations on prostate brachytherapy. *Radiother Oncol* 83:3–10.

Shipley, W.U., Tepper, J.E., Prout, G.R., et al. 1979. Proton radiation as boost therapy for localized prostatic carcinoma. *JAMA* 241:1912–5.

Shwetha, B., Ravikumar, M., Katke, A. et al. 2010. Dosimetric comparison of various optimization techniques for high dose rate brachytherapy of interstitial cervix implants. *J Appl Clin Med Phys* 11:225–30.

Szu, H. and Hartley, R. 1987. Fast simulated annealing. *Phys Lett A* 122:157–62.

Teguh, D.N., Levendag, P.C., Kolkman-Deurloo, I.-K.K., van Rooij, P., and Schmitz, P.I.M. 2009. Quality of life of oropharyngeal cancer patients treated with brachytherapy. *Curr Oncol Rep* 11:143–50.

Thomadsen, B.R., Houdek, P.V., Edmundson, G., van der Laarse, R., Kolkman-Deurloo, I.-K.K., and Visser, A.G. 1994. Treatment planning and optimization. In: *High Dose Rate (HDR) Brachytherapy: A Textbook*. S. Nag (ed.). Futura Publishing Company, Amonk, NY, pp. 79–145.

Trnková, P., Pötter, R., Baltas, D. et al. 2009. New inverse planning technology for image-guided cervical cancer brachytherapy: Description and evaluation within a clinical frame. *Radiother Oncol* 93:331–40.

Tsallis, C. and Stariolo, D.A. 1996. Generalized simulated annealing. *Physica A* 233:395.

Van't Riet, A., Mak, A.C., Moerland, M.A. et al. 1997. A conformation number to quantify the degree of conformality in brachytherapy and external beam irradiation (application to the prostate). *Int J Radiat Oncol Biol Phys* 37:731–6.

Wolpert, D.H. and Macready, W.G. 1997. No free lunch theorems for optimization. *IEEE Trans Evol Comput.* 1:67–82.

13

Quality Management of Treatment Planning

Ali S. Meigooni
*Comprehensive Cancer
Centers of Nevada*

Jack L. M. Venselaar
Instituut Verbeeten

13.1 Introduction

Advancement of the brachytherapy treatment modalities can be attributed to the technological development of the equipment, improvement of treatment planning techniques and software, and enhancement of our knowledge in dosimetric and biological aspects of radiation. In a detailed publication, Rivard et al. (2009) have described the evolution of the brachytherapy treatment planning systems (TPSs) in the past century. At the early stage of developing the brachytherapy procedures, the planning technique was simply the amount of milligram-hour (mg-h) of radium used in the treatments (Paterson and Parker 1938). However, these techniques have been expanded at the present time to the use of 3D computerized planning systems (Martin et al. 2000; Das et al. 2004) and image-guided radiation therapy (D'Amico et al. 1998; Rubens et al. 2006). The clinical application of these computerized planning systems requires quality assurance (QA) tests to verify their accuracy for patient dosimetry. Several task groups from the American Association of Physicists in Medicine (AAPM) have published recommendations on QA of the TPSs (Kutcher et al. 1994; Nath et al. 1997; Kubo et al. 1998; Fraass et al. 1998). Other reports were published by the European Society for Therapeutic Radiology and Oncology in the ESTRO Booklet No. 8 (Venselaar and Pérez-Calatayud 2004) and by the International Atomic Energy Agency in IAEA-TECDOC-1540 (IAEA 2007). These recommendations

included the required QA procedures and the frequency of tests, from acceptance testing, characterization, and commissioning to routine QA of clinical system use. In addition, these tests included the verification of the algorithm, source data entry, geometric modeling of the implant target, and source localization mechanism.

The goals of this chapter are to review some of the existing protocols and recommendations for QA of the brachytherapy planning systems and also to present the consensus of these protocols for daily clinical applications.

13.2 Review of Existing Publications

A review of the recommendations by different AAPM and other task group reports on QA of the TPSs that focused on brachytherapy is presented below. AAPM task group reports are ordered here according to their TG numbers, being not necessarily the same as the order of the publications.

13.2.1 Recommendations Published by the American Association of Medical Physicists

13.2.1.1 AAPM TG-40

This recommendation of 1994 is a detailed QA for general radiation therapy field and includes a section (Section V) for brachytherapy (Kutcher et al. 1994). According to this report, one goal of QA is to achieve a desired level of accuracy and precision in the delivery of dose. It was noted that as compared to external beam therapy where it is generally accepted that the dose should be delivered to within 5% limits, for intracavitary and other forms of brachytherapy procedures, an uncertainty of ±15% in the delivery of prescribed dose is a more realistic level. Radiation oncology physicists are primarily and professionally engaged in the evaluation, delivery, and optimization of radiation therapy. Their role has clinical, research, and educational components. In addition to their advanced degree, these individuals will have received instruction in concepts and techniques of applying physics to medicine and practical training in radiation oncology physics. A major responsibility is to provide a high standard of clinical physics service and supervision. The task group has recommended the following roles and responsibilities (without copying the detailed descriptions from this report) for the radiation oncology physicist regarding QA procedures:

- Calibration of radiation oncology equipment
- Specifications of therapy equipment
- Acceptance testing, commissioning, and QA
- Measurement and analysis of data
- Tabulation of beam data for clinical use
- Establishment of dose calibration procedures
- Establishment of treatment planning and treatment procedures
- Treatment planning
- Establishment of QA procedures
- Education

13.2.1.2 AAPM TG-43

The TG-43 report (followed by updates in 2004 and 2007) on brachytherapy dose calculation of 1995 was a landmark in setting the standard dosimetric procedure and dose calculation algorithm used in all current brachytherapy TPS (Nath et al. 1995; Rivard et al. 2004, 2007). Although this report was primarily for low–dose-rate (LDR) low-energy brachytherapy sources of—mostly—^{125}I and ^{103}Pd, the TG-43 formalism has been widely used and is also internationally accepted for high-energy high-dose-rate (HDR) and pulsed–dose-rate (PDR) sources used in remote afterloading systems. Essentially, the formalism is open for use with other radionuclides than ^{192}Ir, for example, for ^{169}Yb, ^{170}Tm, and ^{60}Co HDR or PDR sources. In the older formalisms, dose calculations were based on the dosimetric parameters such as apparent activity (A_{app}), equivalent mass of radium, exposure-rate constants, tissue attenuation coefficients, and anisotropy factors (although considered obsolete, at some places in this chapter, some old quantities are maintained for historical reasons), which were in part determined in air and depended only on the applied radionuclide. The algorithm usually did not account for differences in active core construction and the encapsulation design of the sources. The TG-43 recommendations introduced and incorporated dose rate constants and other dosimetric parameters of each specific brachytherapy

source design. TG-43 is a consistent formalism, simple to implement, and based on a small number of parameters/quantities that can be easily extracted from Monte Carlo (MC) calculated dose rate distributions around the sources in a water phantom, or from measurements in water or water-equivalent medium. This increases the accuracy of calculations that are carried out in the clinic, which are always for water medium and not in free space. The basic concept of the TG-43 dosimetry protocol is to define a clear formalism expressed in mathematical formalism and incorporate parameters and quantities that enable users to accurately and consistently calculate dose and dose rate distributions around common designs of radioactive sources in the routine clinical procedures. In addition, this enables the use of common and consistent data and databases for commercially available source designs. In its design, however, the TG-43 formalism has several limitations, as it presumes homogeneity of the medium, and it does not take into account other influences (e.g., source-to-source shielding). These limitations are further discussed by Rivard et al. (2009).

The original and updated TG-43 report and its updates proposed no specific recommendations for TPS QA [TG-43U1 in Rivard et al. (2004) and TG-43U1S1 in Rivard et al. (2007)].

13.2.1.3 AAPM TG-53

Four years after publication of TG-40 and one year after TG-56 (see below), the TG-53 report was published on QA for treatment planning of external beam and brachytherapy sources (Fraass et al. 1998). In the AAPM TG-53 recommendation, the acceptance tests have been defined as the tests that are performed to confirm that the RTP system performs according to its specifications. This recommendation indicates that one should have a rigorous and careful design of the specifications for acquisition of an radiotherapy treatment planning (RTP) system if one wants to (1) know how the RTP system should perform in various situations, and (2) be able to design and perform a formal acceptance test to verify that the system works as specified. This task group placed a disclaimer in their report indicating that creation of specifications for a modern 3D RTP system is a large task, and it is beyond the scope of their report. While not prescribing specific QA tests, they recommend that each organization use their guidelines to create an individualized framework for the QA of their program to address the issues on their planning systems. In addition, they have noted that satisfaction of specifications should not be dependent on clinic-specific beam data. TG-53 identifies the items suitable for specifications; the task group has divided these items into three broad categories:

- Computer hardware: CPU and all the peripheral devices that are part of the RTP system, such as the display monitor(s), printer, plotter, and tape drive.
- Software features and functions: Many software feature specifications will be of the yes/no or exists/does-not-exist type, rather than quantitative.
- Benchmark tests: Performance on benchmark tests indicates the accuracy of the dose calculation algorithm under very specific circumstances with specific beam data. Calculation times can also be measured.

TG-53 recommends that if these tests are performed by the vendor, the user may want to repeat some or all of the tests to verify the results. It is also recommended to perform dose calculation verification tests independent of the TPS results. TG-53 briefly mentions the complexity of dosimetrically accounting for material heterogeneities, but users are left on their own to "understand the implications of those approximations" without any specific guidelines provided. One dose calculation issue pertinent to this report was, "Verify correct behaviour of dose calculations, sometimes including tissue multiple scattering and attenuation, at selected distances from the source." With no references or elaboration on methods to perform this verification, the physicist is left feeling concerned without recourse. The main outline of the TG-53 report is guiding for the further discussions in this chapter.

13.2.1.4 AAPM TG-56

The TG-56 report "Code of Practice for Brachytherapy Physics" (Nath et al. 1997) was published one year earlier than the comprehensive TG-53 report and was specifically dedicated to brachytherapy. The TG-56 did not significantly add to the specific recommendations beyond those presented in TG-53. Section VI.C of TG-56 focused on QA for brachytherapy TPS, but mainly addressed tests of system functionality instead of specific tests of dose calculation accuracy.

13.2.1.5 AAPM TG-59

The TG-59 report focused on HDR ^{192}Ir brachytherapy QA (Kubo et al. 1998). In addition to the previously listed items to assess during TPS commissioning, a list was provided of 13 QA items to be reviewed preceding treatment delivery for individual patients. The scope of these items generally agreed with those outlined in prior reports such as from TG-53. Example nomograms and formulae were provided to permit the physicist to estimate expected treatment times to supplement calculations independent from the TPS.

13.2.1.6 AAPM TG-64

Permanent prostate seed implant brachytherapy was covered in TG-64 (Yu et al. 1999). The report focused on technical innovations and treatment delivery techniques. Treatment planning for this LDR approach was discussed from a practical perspective, with a quantitative benchmark of calculating total dose as a function of distance following complete decay from three low-energy LDR sources. No new TPS QA criteria were established, and the authors endorsed the AAPM TG-40 recommendations.

13.2.2 Other Publications on Quality Management of Brachytherapy Dose Planning

13.2.2.1 Van Dyk et al. (1993)

These authors were the first to describe specific and quantitative recommendations for QA of brachytherapy dose calculations. It was acknowledged that uncertainty estimates for brachytherapy

calculations are much more difficult to assess due to the short treatment distances and the resulting sharp dose gradients that can yield large uncertainties in both the calculated and measured data. For the testing of computerized calculations, the best approach should be to assume that the source specification is known accurately. Calculations should then be performed both manually and by computer for a specified number of configurations. It was recommended to use published dose distributions around individual sources as benchmark data. A criterion of acceptability of 5%, quoted as a percentage of the local dose value, was considered to be reasonable for points at distances of 0.5–5.0 cm from the source (for linear sources: along lines normal to the source axis).

Uncertainties associated with brachytherapy shielding and tissue inhomogeneity calculations were recognized as more difficult to quantify, and no further criteria were suggested. For brachytherapy calculations, the authors suggested for developers of software a goal of 3% at distances of 0.5 cm or more at any point from any source.

As initial and regular tests to be done by the radiation oncology physicist, inspection of data entered into the TPS was suggested [exposure (dose) rate constant; parameters for tissue attenuation and scatter, half-life, source wall information, activity (strength), and decay for short-live isotopes]. For source decay, it should be inspected that the decay is handled correctly within 0.1%. Comparison of calculated data should be performed against published data for a few single- and multiple-source distributions, both for seed and line source calculations. Rotation and translation of the coordinate system of the plane of calculation were considered important at the time. The constancy checks should be thoroughly performed at least at the minimum frequency indicated in the paper or after any modification of data files, software, or hardware. A minimum frequency was suggested as semiannually for a subset (e.g., 3 points near the source) of the tests for single- and linear-source distributions.

13.2.2.2 Venselaar and Pérez-Calatayud (2004)

Chapter 9 of the ESTRO Booklet No. 8 (Venselaar and Pérez-Calatayud 2004) largely follows the previously published documents in the frame of a discussion on the role of the responsible physicist at commissioning and continued use of the TPS. There is much emphasis on the understanding of the calculational models, available data sets, and system files implemented in the TPS, including the tasks for proper documentation. Recommendations for tolerances and frequencies are similar to the TG-56 report (Nath et al. 1997) with 2% tolerance level, a thorough inspection initially and with each new source and software updates or changes. Any deviation found to be >5% should be investigated in detail. A number of suggestions were provided to the readers with regard to dose-volume histogram (DVH) calculations and the use of optimization routines. An important part of the chapter dealt with verification of the individual clinical treatment plan. An inspection of the integrity of the data transfer from TPS to remote afterloader and the use of an indication of the reasonableness of the final plan are discussed. For this, the use of certain indices and the total reference air kerma (TRAK) is explained. Education and continued training of the members of the brachytherapy team are considered key.

13.2.2.3 IAEA-TECDOC-1540 (IAEA 2007)

A working group of the International Atomic Energy Agency prepared a document that was published as IAEA-TECDOC-1540 (IAEA 2007) with the purpose of proposing guidelines on how to implement acceptance and commissioning procedures for newly purchased TPSs. Although the document is largely focusing on external beam therapy, many aspects also refer to the similar situation in brachytherapy. The report uses the IEC 62083 standard as a basis for defining specifications and tests of TPSs, and gives formal descriptions of both Type Tests, to be performed by the manufacturer to establish compliance with specified criteria, and Site Tests, to be used after installation on an individual device to establish compliance with specified criteria ("acceptance test"). This distinction is important for external beam therapy planning, but is also more relevant in brachytherapy, as the complexity of the software is rapidly increasing, for example, with the upcoming advanced model-based dose calculation algorithms (MBDCAs). Therefore, this document provides interesting conceptual material that needs to be translated from external beam into brachytherapy approaches.

13.2.2.4 Rivard et al. (2010)

In this article, the authors discuss the transition from the current standard for brachytherapy dose calculation based on the TG-43 formalism (Nath et al. 1995) to the new MBDCAs. This MBDCA approach requires a paradigm shift toward new QA standards. The current societal recommendations should be augmented to consider the consequences for dose specification (see Chapter 11), and also be modified to address a new infrastructure to safely and uniformly introduce these new algorithms. In the meantime, manufacturers are making progress in developing these MBDCAs and in introducing new sources, source types, and delivery systems that go beyond the current societal QA guidelines. The article discusses the options for benchmarking using reference phantom configurations, with compiled reference datasets for each brachytherapy source to validate the capability of the MBDCA to account for scatter conditions and material inhomogeneities. Some possibilities are discussed for using the concepts of percentage of agreement and the distance to agreement or a combination of the two, that is, the gamma index. The tolerances may depend on the distance to the phantom periphery and may be specific for given material heterogeneities and photon energy of the considered source. It is noted that such tolerance recommendations propagate into the overall uncertainties in brachytherapy dosimetry, a topic discussed thoroughly in the work of the AAPM task group 138 (DeWerd et al. 2011; also see Chapter 15). The authors end with a suggestion to use a reference brachytherapy source registry, which includes a database of the reference sets of quantitative data in an easily accessible electronic format. If the repository includes well-described test cases, which vendors should include in their

commissioning tests (in its approach much alike the test set for external beam TPSs as described in IAEA-TECDOC-1540 discussed above), extensive international societal efforts should take place to establish an augmented acceptance testing procedure.

13.3 Acceptance and Periodic Clinical Tests for TPSs

After installing or upgrading a TPS for brachytherapy procedures, medical physicists must perform a series of tests to assure the proper function of the system as described in the agreement with the vendor (Acceptance Tests) and to prepare the system for use with the clinical applications of the institution (Commissioning). The results of these tests should be carefully documented, along with any variation from the defined procedures, and kept as long as the TPS is being used in the department. The following section presents the details of commonly used QA and acceptance tests, recommended within the above-noted publications (Kutcher et al. 1994; Nath et al. 1995, 1997; Fraass et al. 1998; Kubo et al. 1998; Yu et al. 1999; Van Dyk et al. 1993; Venselaar and Pérez-Calatayud 2004; IAEA 2007; Rivard et al. 2010), which might be included in this procedure for brachytherapy TPSs. Medical physicist may add his or her own tests upon their institutional needs and any new features of the planning systems that are not listed here until the consensus of such a test becomes available.

13.3.1 Dose Calculation Algorithm

Various dose calculation algorithms have been utilized in different brachytherapy TPSs. The most commonly used algorithms were composed of one of the following methodologies, which must be tested during the acceptance test of the system, prior to their clinical applications:

- Sievert integral (see, e.g., Chapter 7)
- Traditional linear and point source approximation formalism using exposure rate constant (see, e.g., Baltas et al. 2007)
- TG-43 1D dose calculation algorithms (point source approximation; Chapter 10)
- TG-43 2D dose calculation formalism (linear-source approximation; Chapter 10)

For new MBDCAs, a similar testing requirement holds. The medical physicist must find the exact formalism used in his or her planning system and verify the accuracy of the corresponding dosimetric parameters and their units (Table 13.1). In these tests, care must be taken to assure that the possible error of one parameter is not affecting the evaluation of other parameters. For instance, in evaluating the accuracy of the algorithms, one must eliminate the possibility of the source-positioning error by keyboard entry of their coordinates. In addition, the accuracy of the source dosimetry parameters must be examined by comparison with the published data (Section 13.3.3).

As a first step in the acceptance, the dose calculation formalism can be tested by performing single-source dose calculations

TABLE 13.1 Acceptance Testing of Dose Calculation Algorithm for Brachytherapy TPS

Item	Material	Frequency*
Dose calculation algorithm	• Verify the accuracy of the algorithm in the program.	• A, B
	• Verify the accuracy of point source versus linear-source calculations.	• A, B
	• Verify the accuracy of corresponding source parameters (e.g., TG-43 parameters).	• A, B, C
	• Verify the capability and accuracy of the 2D versus 3D dose calculations.	• A, B
	• Verify the anisotropy or orientation-dependent features of the dose distribution for each type of source. If anisotropy is being neglected, it should be so noted in the dose distribution documentation.	• A, B, C
	• Use a series of plans, source strength, etc., to verify that all dose output methods are in agreement. Consider total dose initial dose rate, average dose rate at time of implant, permanent implant total dose, and any other methods of dose display/specification that are available.	• A, B

Note: Frequencies are recommended for initial/acceptance, annual (or with each update), and daily (which comes with each clinical case review) testing.

*A = initial acceptance tests, B = annual QA or after any update of the system, C = daily clinical test.

at various points relative to the source for each source type. The results of these calculations will be compared with the published data or values obtained in one spreadsheet calculation using the accurate algorithm and source dosimetric parameters. Then, dose calculations for multisource implants can be performed for some easily reproducible source geometries, such as the tandem of sources for ^{137}Cs or ^{192}Ir sources or a planar-array of sources for ^{192}Ir, ^{125}I, and ^{103}Pd sources. As per AAPM TG-138 recommendation (DeWerd et al. 2011), uncertainty of equal to or less than 5% is acceptable for dose delivery at any point in an implant, in which the contribution of the calculation steps is merely a part.

The user must also examine the capability of TPS for dose calculation with either point source approximation or linear-source approximation. The point source approximation is used for implants for which seed orientations are random or cannot be identified in the images, such as prostate seed implants. However, linear-source approximation is used for implants where source orientations are controlled using an applicator, such as GYN implants. It should be noted that some of the source dosimetric parameters are different between the point and linear-source approximations. For example, with the TG-43 formalism, one requires either anisotropy constant or 1D anisotropy function to perform the dose calculations. However, with linear-source approximation, the 2D anisotropy function will be needed.

Finally, the users must examine the capability and the accuracy of the dose calculations for typical temporary and permanent implants. For these verifications, one can calculate the integrated dose and dose rates around a single-source configuration. The final outcome of these tests shall be documented in a logbook or binder, throughout the time that the TPS is in use in the department.

13.3.2 Source Input and Geometric Accuracy

The accuracy of dose calculations in a brachytherapy implant is strongly dependent on the accuracy of the source localization within the implanted area or volume. Source localizations are commonly performed using a digitizer and orthogonal or stereo-shift films, or a 3D reconstruction technique using CT imaging procedure. In the first technique, digitization of the film, the accuracy of the system must be tested (Table 13.2) by creating images on a graph paper or film with well-defined source positions (i.e., 2D array of sources with 1-cm spacing in between). Figure 13.1 shows a sample of these arrangements for source location test with a digitizer. The coordinates of these seeds must be reproduced with an accuracy of ±1 mm. It should be noted that one could use the center of one of the seeds as the origin for these digitization processes.

For a 3D reconstruction, a set of CT images from a phantom embedded with well-defined source geometry would be required to examine the accuracy of the source localization. 3D seed coordinate representation after entry should be confirmed. Figure 13.2 shows a commercially available phantom, known as the "Baltas phantom" (Pi Medical Ltd., Karneadou 40-42 Str., 106 76 Kolonaki, Athens, Greece), that can be used for the purpose of checking the geometric reconstruction techniques of an institution (Baltas 1993; Venselaar and Pérez-Calatayud 2004).

TABLE 13.2 Testing Accuracy of Source Geometric Input for Brachytherapy TPS

Item	Material	Frequency*
Source input and geometric accuracy	• For source location entry using a digitizer and orthogonal or stereo-shift films, check should be made of the data entry software, the film acquisition process, source identification, and other associated activities. 3D seed coordinate representation after entry should be confirmed.	• A, B
	• For automatic seed identification with the software, a test with well-defined implant geometry should be performed.	• A, B
	• For applicator trajectory identification, the appropriate tests should be performed.	• A, B, C

*A = initial acceptance tests, B = annual QA or after any update of the system, C = daily clinical test.

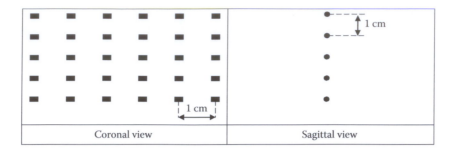

Coronal view	Sagittal view

FIGURE 13.1 Sample source arrangement for source localization test of digitizer. (Courtesy of A.S. Meigooni.)

The results of these tests shall be documented in the logbook for the TPS.

13.3.3 Source Information

Some of the source dosimetric parameters and information entered in a TPS are dependent on the formalism used for the calculations, and some are independent. For example, if a planning system is using the TG-43 formalism, it requires dose rate constant, radial dose function, and 2D and 1D anisotropy functions. However, half-life of the source is usually independent of the formalism. It is very crucial to know the exact parameters needed and also the proper units for each unit. Although vendors may provide the information for different brachytherapy sources, it is the responsibility of the users to verify their accuracies by comparison with the published data. Table 13.3 presents the commonly recommended acceptance tests for the source information in a brachytherapy planning system. In addition, the users should be able to add or delete any new source or update the dosimetric information of existing sources at any time during the use of the system. For TG-43–based dose calculation algorithm, the required input data of AAPM-approved sources can be found from the references at the MD Anderson RPC

Web site–based repository. Away-and-along tables are useful for direct comparison, and these are published for many source types, both in print (see, e.g., Venselaar and Pérez-Calatayud 2004; Baltas et al. 2007) and on the Internet. See, for example, http://www.physics.carleton.ca/clrp/seed_database/ and http://www.estro.org/estroactivities/Pages/TG43HOMEPAGE.aspx.

13.3.4 Image Input

The relative location of the brachytherapy sources to the treatment volume in the present brachytherapy planning systems is

TABLE 13.3 Testing Accuracy of Source Information in Brachytherapy TPS

Item	Material	Frequency*
Source information	• For each source and source type, check specification of source strength**: • Reference air-kerma rate • Air kerma strength • Apparent activity (mCi) • Apparent activity (MBq) • Equivalent mass of radium (mgRaEq)	• A, B, C
	• Verify all conversions between source strength specifications of source suppliers and the RTP system. Should be done for each source type individually.	• A, B, C
	• For each source type, check specification of decay constant, half-life, average life, dose constants, and other related parameters.	• A, B, C
	• Verify that source strength decay calculations work correctly, for each source type individually. Determine at what time during the implant (e.g., beginning, midpoint) the source strength is specified.	• A, B, C

*A = initial acceptance tests, B = annual QA or after any update of the system, C = daily clinical test.

**Recommended source strength specifications are reference air-kerma rate or air-kerma strength. Activity in terms of MBq or GBq may be required for licensing purposes, but other specification quantities (Ci, mgRaEq) are considered obsolete in the most recent protocols.

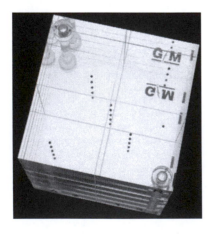

FIGURE 13.2 Commercially available phantom, known as "Baltas phantom," used for verification of the geometric reconstruction techniques used in brachytherapy (http://www.pi-medical.gr/products/tools/matrix-phantom). (Courtesy of Olympios Marneros, pi-Medical Ltd.)

based either on orthogonal films for 2D dose calculations or CT/MRI/Ultrasound scanning images for 3D dose calculations. The method of incorporating images in the dose calculations and the accuracy of the images transferred into the planning system must be examined before its clinical application (Table 13.4). For a 2D dose calculation, a set of orthogonal x-ray images are sufficient for geometrical reconstructions. The information from these orthogonal films is extracted using either a standard digitizer or a film scanning device. To examine the accuracy of this methodology, a set of orthogonal films should be taken from a phantom with embedded applicator and x-ray markers. Using several fiducial markers, one could identify the points of interests, for example, corresponding to points A and B for a GYN implant. The medical physicists must be able to use either a digitizer or scanner and extract all of the required information in terms of source position as well as the points of interests into the TPS.

For the 3D planning systems, image transfer in a hospital is often performed via Digital Imaging and Communications in Medicine (DICOM). A single DICOM file normally contains the header, which stores information about the patient's name, the type of scan, image dimensions, etc. In addition, the files contain the 3D image data for the patient as well as the brachytherapy sources. The TPS should be able to accurately import this information from imaging devices such as CT and MR. The accuracy of this process must be examined by creating a phantom with an embedded applicator and scanned with one of the imaging techniques. Choice of modes and settings of the imaging devices

FIGURE 13.3 Commercially available CT phantom with embedded heterogeneity sections (http://www.pnwx.com/Accessories/Phantoms/Dosimetry/CIRS/ElectronDensity/). (Courtesy of CIRS.)

(CT and MR) for image acquisition is important, as it can affect, to a large extent, the result of the reconstructions (e.g., Siebert et al. 2005, 2007). Discussion, examples, and references for further reading are provided in Chapter 14. Both the demographic information of the file and relative positions of the applicator and sources must be examined. Medical physicists must be able to examine the accuracy of the demographic information of the file and the possibility of contouring different organs represented on these images, and localize the brachytherapy sources for dose calculations. For TPSs with heterogeneity correction capability, the accuracy of the CT numbers on the images must be examined using the images from commercially available phantom materials with embedded components with different chemical compositions and densities (Figure 13.3). Results must be logged.

13.3.5 Dose Display and DVHs

The outcomes of a TPS are normally presented in graphical form (isodose curves, DVHs) and also text data. Medical physicists should be able to examine the accuracy of both graphical and textual dose display and DVH for their TPSs. In particular, the calculated isodose curves for different anatomical structures, in 3D images, are desirable for a better assessment of the treatment delivery. The dose display should include the patient's demographic information (name, birthday, etc.), source information (name, type, etc.), and implant type (interstitial versus intracavitary, permanent versus temporary).

Cumulative DVHs are recommended for evaluation of the complex dose heterogeneity. The 3D planning systems are normally equipped with DVHs to evaluate the results of dose calculations and optimization. The accuracy of DVH calculation cannot easily be verified in an independent way. In some instances, a DVH calculation on a second TPS can be performed, and the results of both systems can be compared. Only for isotropic point sources can the DVH be calculated analytically (Van der Laarse and Luthmann 2001; see Chapter 12 for further reading). A study was performed by Kirisits et al. (2007)

TABLE 13.4 Verification of Accuracy of Incorporating Images for 2D or 3D Dose Calculations in Brachytherapy TPSs

Item	Material	Frequency*
Image input	• For 2D calculations, create a set of orthogonal film from a phantom with embedded applicator and x-ray markers. Use several fiducial markers to identify the points of interests, such as points A and B for a GYN implant. Digitize the films and verify the accuracy of the points of interests relative to the sources.	• A, B
	• For 3D calculations, create an anatomical description based on a standard set of CT scans from a commercially available humanoid phantom, in the format that will be employed by the user. Both the demographic information of the file and relative positions of the applicator and sources must be examined.	• A, B
	• For treatment planning with heterogeneity correction capability, the accuracy of the CT numbers on the images must be examined.	• A, B

*A = initial acceptance tests, B = annual QA or after any update of the system.

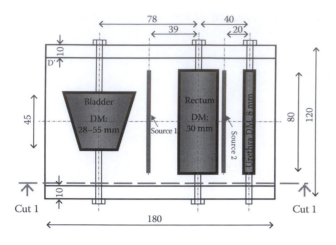

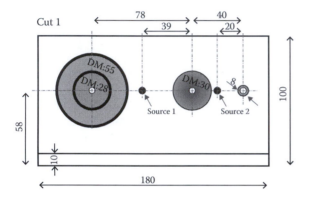

FIGURE 13.4 Model of a phantom used by Kirisits et al. (2007) in a study to compare the accuracy of DVH calculations by different brachytherapy TPSs. (Courtesy of C. Kirisits and D. Berger.)

comparing seven different planning systems with a phantom containing three different volumes, scanned with MR and CT (Figure 13.4). They found that differences in reconstructed volumes resulted from the methods of calculation in the first and last slices of a contoured structure. For this situation and in case of high dose gradients inside analyzed volumes, high uncertainties were observed.

Lessons learned from the study were that the use of DVH parameters in clinical practice should take into account the method of calculation and the possible uncertainties, and that the test of the accuracy of DVH calculation is far from a simple task. This work is typically suited for cooperative user-group testing, while the individual medical physicist should be aware of the published reports and should create for his or her own department a derivative in the form of a simple example for consistency checks (Table 13.5).

13.3.6 Print, Plot, and Export of Implant Information

Final results of a treatment are normally stored as printed materials or electronic charting. This plan normally includes the patient's demographic information in addition to the treatment

TABLE 13.5 Verification of Accuracy of Incorporating Images for 2D or 3D Dose Calculations in Brachytherapy TPSs

Item	Material	Frequency*
Dose display, DVHs	• Verify the accuracy of dose display for some sample plans of different types (temporary, permanent, intracavitary, interstitial, etc.) and verify their accuracies in terms of the information that must be on them (i.e., patient's demographics, source information, treatment date and time, source type, source activity, implant type, prescription dose, dose at different points of interests, physician's name, physicist/dosimetrist name, etc.).	• A, B
	• Verify the capability and accuracy of display of isodose lines, DVHs, BEVs, DRRs, 3D views, for the above-noted plans and verify their content and accuracies relative to the display on the computer as well as prescription by the physicians.	• A, B

*A = initial acceptance tests, B = annual QA or after any update of the system.

technique, source information, treatment date and time, dose prescription and dose distributions to the treatment target volume, and also dose to the surrounding normal tissue. The users should make sure that they can obtain complete and correct hardcopy and/or electronic copy of the treatment plan for patient documentations (Table 13.6). The plan identification or label should be recorded on each relevant page, as more than one plan version may have been calculated for a patient case. Moreover, the TPS normally produces graphical data of the geometrical and dosimetric aspects of the treatment plan such as plots with isodose lines, DVHs, beam's eye views (BEVs), digitally reconstructed radiographs (DRRs), 3D views of patient anatomy, beam setup, and dose distributions. This information will be used for evaluation, quality control, and implementation on the treatment machine.

Some institutions use the digital output of the brachytherapy TPS for export of the calculated data file to a second and independent calculation device. In this way, by using a spreadsheet program, it is possible to verify the outcome of the TPS and to identify irregularities, data base corruption, etc. An example of such an approach was published by Carmona et al. (2010).

13.3.7 Contouring Tools

The TPSs with 3D planning capability should have contouring tools for identifying the treatment volume as well as the surrounding organs at risks. The identified organs will be used for optimization of the plan, and the DVH for each organ can be used for determination of the quality of the treatment plan. These contouring tools should enable the users to contour different geometric shapes as fast as possible. Interpolation between

TABLE 13.6 Verification of Accuracy of Graphical and Texture Print as Well as Digital Export of Results of Brachytherapy TPSs

Item	Material	Frequency*
Print, plot, and export of the implant information	• Print the output of some sample plan for different types (temporary, permanent, intracavitary, interstitial, etc.) and verify their accuracies in term of the information that must be on them (i.e., patient's demographics, source information, treatment date and time, source type, source activity, implant type, prescription dose, dose at different points of interests, physician's name, physicist/dosimeterist name, etc.).	• A, B, C
	• Plot a copy of the isodose lines, DVHs, BEVs, DRRs, 3-D views, for the above noted plans and verify their content and accuracies relative to the display on the computer as well as prescription by the physicians.	• A, B, C
	• Create a digital copy of text as well as the graphical display of the sample plans for various implant types (temporary, permanent, intracavitary, interstitial, etc.).	• A, B
	• Verify the export and import capability of the plan (e.g., for Radiation Therapy Oncology Group (RTOG) protocols, one should be able to export the plan with patient identification masked).	• A, B
	• Verify the possibility of creating composite plan from different implants.	• A, B

*A = initial acceptance tests, B = annual QA or after any update of the system, C = daily clinical test

the contours at different image slices and also Boolean logic will enable the users to achieve their results as fast as possible. The correctness of this interpolation should be checked, especially in regions with large contour variations.

Scanning of the structure of known volumes (such as blocks or spheres) may be useful in estimating the uncertainties in volume reconstruction. It is noted that this section on contouring

TABLE 13.7 Suggested Tests Regarding Contouring Tools in Brachytherapy TPSs

Item	Material	Frequency*
Contouring tools	• Verify the accuracy and user-friendliness of the contouring tools in the planning system.	• A, B
	• Verify the accuracy of the contouring template that may exist in the TPS system.	• A, B

*A = initial acceptance tests, B = annual QA or after any update of the system.

TABLE 13.8 Suggested Tests Regarding Evaluation of Optimization Techniques of Brachytherapy TPS

Item	Material	Frequency*
Optimization and evaluation	• Test automated brachytherapy optimization tools, such as automatic determination of dwell position and time to yield a specific dose distribution with an afterloader unit. Test designs should be very dependent on the algorithm used.	• A, B
	• Test other standard tools such as DVHs.	• A, B

*A = initial acceptance tests, B = annual QA or after any update of the system.

tools is not different from the requirements for TPSs used in external beam therapy (Table 13.7).

13.3.8 Optimization Techniques

Presently, the optimization techniques for brachytherapy are only applicable for treatments with HDR and PDR remote afterloading stepping source systems. In these systems, dwell positions and dwell times of a single source are determined for each catheter or applicator, based on the dose limits provided by the users. There are different optimization techniques (volume optimization, applicator optimization, inverse planning anatomy-based dose optimization, etc.) in various brachytherapy TPSs. The users must identify the method of optimization in their TPS and also the limitations of their planning system. For example, in a balloon brachytherapy treatment with BrachyVision (Varian), volume optimization is being used with the upper and lower dose limits to planning target volume (PTV) or other organs that are being selected. In contrast, in the PLATO and Brachy-OMP (Oncentra Master Plan, Nucletron) TPS, the generation of dose points is an explicit requirement for a dose prescription. These dose points are associated with a structure or with the applicator, and they can be used for the optimization step. See also Chapter 12 for further explanation of optimization strategies (Table 13.8).

13.4 Dose Calculation Commissioning

Commonly, the TPSs utilize a superposition technique to calculate the dose for a multiseed implant from dosimetric information of a single source. Although many software vendors supply their codes with the brachytherapy source dosimetric information, they are not responsible for the codes' accuracy, and also, they may not provide the dosimetric information for all of the commercially available sources. Therefore, prior to the clinical application of newly installed treatment planning software, the users may want to add the dosimetric information of the sources that is not already included within the available planning system. After verification and clarification of the possibilities for any license limitation for adding such source information to their system, the users should include the dosimetric information to the system and perform some appropriate tests to verify their accuracies. These tests may include the following.

13.4.1 Single-Source Dosimetry Verifications

The medical physicist should verify the accuracies of the dosimetric parameters of all the sources in their planning system by comparing them with the published references. If any new source information has been added or edited in the TPS, he or she should verify the accuracy of this entry by performing some tests, as discussed in Section 13.3.3. The original data and the printed final data must be documented for any inspection or future spot check throughout the time that the software is being used.

13.4.2 Source Position Accuracy

In case of adding any new source design to the existing software, or adding a new methodology for introducing the source position into the implant (i.e., CT versus radiograph), the physicist must verify the accuracy of the source positioning following the same procedure that has been described in Section 13.3.2. The results of these tests must be kept in a logbook, as long as the treatment planning is in use and the new source/source positioning technique is being utilized for the clinical procedures.

13.4.3 Multisource Dosimetry Accuracy

The check of the accuracy of a multisource implant with any new source design and geometric information must be performed using a simple and reproducible geometry such as a linear or planar implant. These geometries normally can be created by keyboard entry, similar to the one shown in Figure 13.1. The medical physicist should create an implant geometry, perform the dose calculations in a superposition manner, and compare these results with either published or spreadsheet data obtained by the same group, using the accurate dose calculation algorithm. The agreement between the results should be within ±5%. The information should be documented throughout the use of the TPS.

13.5 Periodic Clinical Treatment Planning QA

13.5.1 Inspection of Individual Brachytherapy Treatment Plan

Prior to each treatment, the several items on the plan as well as the treatment console must be examined for accurate treatment delivery. Table 13.9 lists a number of aspects that should be tested by the physicist to detect any possible errors in individual treatment plan with brachytherapy treatment. Many issues follow from the pretreatment QA report for the treatment planning program, such as the use of the correct version of the algorithm

TABLE 13.9 Error Detection Review of Plan

Item	Material
Patient identification	All documents, films, prints, plots provided for a treatment, should include the patient's name, date of birth, and hospital or clinical number.
Dose prescription	Delivered dose versus prescribed dose Evaluation of uniformity of the dose Location of prescribed dose Type of the implant (i.e., intracavitary, interstitial, HDR, LDR, etc.) Dose distribution/differentials in dose Begin and end positions of the active source along each catheter (i.e., active length) as well as the number of catheters.
Implant date and time	The physicists must verify the accuracy of the implant date, time, and treatment duration.
Dose to normal structures	Location of high dose areas, location of normal structures, constraints fulfilled.
Program identification	Correct algorithm used (i.e., point source versus linear source), version number, shielding, correction factors.
Source information	Correct use of radionuclide, source strength, step size (source arrangement for LDR implants), tip length (distance from the HDR or PDR remote afterloader to the first dwell position).
Transfer of data	Correct transfer of dwell positions, dwell time, total time, and channels from TPS to the patient or treatment console (i.e., HDR and PDR cases)..
Independent verification of treatment dose	Using an independent software or hand calculations, the physicist has to verify the accuracy of the dose calculations for each case.

Note: See also ESTRO Booklet No. 8 (Venselaar and Pérez-Calatayud 2004).

and the selection of the correct source. This is followed by the inspection of high dose areas (i.e., 150% or 200% isodose line) and organ-at-risk doses. For HDR and PDR treatments, very often, the overall treatment time is used for inspection, but this has its limitations. The historical and obsolete concept of using milligram-hour or milligrams radium equivalent-hours was used by many investigators who had utilized it in the Manchester system for dose and treatment prescription in gynecological treatments with ^{226}Ra and later ^{137}Cs and ^{192}Ir sources, that is, with the long-lived and LDR sources. A very similar quantity (i.e., Ci × s), namely, the product of the source activity content in Ci and the duration of the treatment in seconds, can be used for HDR and PDR treatments. The number of milligrams radium equivalent-hours or millicurie seconds is often within a small bandwidth of values seen over many patients, even if some individualization of prescription is applied. Therefore, it can give the user confidence in the result of a treatment plan calculation. The quantities recommended for use nowadays, such as the TRAK, which is recommended for reporting in ICRU Reports 38 (ICRU

1985) and 58 (ICRU 1997), serve well for the same purpose and can be used as an indicator of consistency for these types of treatment.

In a practical procedure between the creation of the treatment plan and the start of the treatment, the medical physicist is often allowed to use only a few minutes to inspect a plan for "reasonableness." Such an inspection should be performed to identify serious errors at the very least. For this purpose, the goal suggested in the AAPM TG-56 reportable rule for brachytherapy clinical dose delivery is an actionable threshold of 5%–10% deviation (Nath et al. 1997). The aim should be to have more precise checks, but the 10% rule can be used as a minimum demand. At least, the physics steps in the dosimetry chain should aim at considerably lower levels (DeWerd et al. 2011).

13.5.2 Dose Index Test

One of the techniques for verification of the reasonableness of dose calculation for an implant is to add some extra dose points outside of the implant volume (i.e., at distances such as 5 or 10 cm away from the center of the implant) as a standard in the calculation. Dose values to these points can be independently verified assuming the entire implant geometry as a point source (see Figure 13.5).

With optimized plans, such a manual calculation still may be too complex, but then, a comparison can be made with data from a precalculated table of dose values at such distances for sources with unit strength (Venselaar et al. 1994). Ezzell (1994) followed a similar approach for planar implants and Rogus et al.

2nd Phys. Check for a Contura Multi-channel Brachytherapy with HDR
(VariSource)

Patient Name:

Patient ID: Date:

TG-43 Formula for point source calculation:

$$Time\ (hr) = \frac{Dose(cGy)*\gamma^2(cm^2)}{DoseRateConst(cGy/hr/U)*SourceStrength(U)*F(\gamma,\theta)*g(\gamma)}$$

Dose Rate Constant (VariSource Ir-192) = 1.10 cGy/hr/U

1 mCi = 4.035 U

$F(r,\theta) = 1.0$ $g(r) = 1.0$

$$Time\ (Sec) = \frac{3.27*10^3 * Dose(cGy)*\gamma^2(cm^2)}{SourceStrength(U)}$$

V-Reference = Volume of PTV_Eval + Volume of Balloon
$r_{ref} = (3*V\text{-}Reference/4*x)^{1/4}$

Prescription Dose =	340	cGy	
V-Reference =	141	CC	
r_{ref} =	3.23	cm	
Source Activity (Ci) =	5.776	Ci	Source Strength (U) = 23.28*10³ U
Time (Sec) =	499	Sec	
Expected Value (Plane) =	522.4	Sec	
% difference	−4.5%		Acceptable Limit = ±5%

Physicist   Date

FIGURE 13.5 Point source calculation for a multisource implant using a point source approximation. (Courtesy of A.S. Meigooni.)

(1998) for single catheter HDR applications. They used the following expression and compared the obtained "dose index" with an expected range of values:

$$dose\ index = \frac{100*[D(+10\,cm)+D(-10\,cm)]}{strength\ of\ the\ source*total\ time} \quad (13.1)$$

The method may work very well to detect 5%–10% deviations and can be applied for both planar and volume implants. As pointed out in the work of Nag (1994), the dose-at-distance method will not detect all types of errors; notably, it will not detect whether or not the implant covers the target volume. For example, consider a situation in which the planner intended to place active dwell positions at 5-mm step size but inadvertently used the same number of steps at 2.5-mm step size. The dose distribution in the implant is then considerably different and the catheters would be loaded to only half the intended length, but the dose-at-distance method would not detect this error. It is therefore important to understand the limitations of such methods.

Other methods that go into more details have been developed. These may sometimes be more accurate in prediction of the "reasonableness" but may also be quite specific for a given technique. Most can be performed within a few minutes using a programmable hand calculator. Only a few references are given here that may be useful if a user wants to develop such methods; see, for example, Kubo (1992), Kubo and Chin (1992), Ezzell (1994), and Rogus et al. (1998).

13.5.3 Total Time Index Test

A "total time index" for intracavitary applications was proposed by Williamson et al. (1994) as

$$total\ time\ index = \frac{sum\ of\ dwell\ time*strength\ of\ the\ source}{prescription\ dose*number\ of\ dwell\ positions} \quad (13.2)$$

If used in the proper quantities, this expression gives the average TRAK over all dwell positions per unit dose. Thomadsen (2000) stated that the total time index is related to the integral dose to the patient.

13.5.4 Tip Time Index Test

An additional index was suggested to verify that the dose near the tip of the intrauterine tandem falls within a certain range of values for similar patients. In the technique used in the Wisconsin department, the dwell time 1 cm inferior to the tip of the tandem tends to be relatively stable. A "tip time index" analogous to the total time index is characteristic of the cervical cancer implant, indicating the variation dwell time for this dwell position (i.e., 1 cm from the tip of the tandem):

$$tip\ time\ index = \frac{dwell\ time\ 1\,cm\ from\ tip*strength\ of\ the\ source}{prescription\ dose} \quad (13.3)$$

As reference data, tables with average values obtained from grouped clinical cases were derived in the department, to which future individual cases can be compared. Techniques like those described here may be somewhat department dependent, but similar ideas can easily be developed for other clinics.

13.5.5 Summary of Clinical Plan Checks

In the clinical application of this technique, for tests such as those described in this section, it is obvious that the user must first gain experience over a number of cases. Often, if the method works well, a certain confidence level can be defined. It is noted that, quite often, the individual case value of such indices exceeds 10%, as certain (patient-specific) aspects are not taken into account. Then, if a procedure falls beyond this level, it may need further inspection to try to find an explanation for the deviation.

It is good practice for a second person to perform the check. Sometimes it is necessary that the whole treatment plan is repeated independently from scratch by this second person. The results of both plans should then be discussed by all team members, not just by the physicists. A serious and unexplained discrepancy between the two plans can be a reason for aborting the treatment, rather than continuing with the execution of a possibly erroneous plan.

In addition to such methods, the use of well-designed treatment and planning protocols, the use of standardized forms, the inspection of the result of the treatment planning by a second and independent expert, and a proper training of all brachytherapy team members are keys to the success of the quality management program (Table 13.10).

TABLE 13.10 Reasonableness, Clinical Aspects

Item	Material	Frequency
Test of outcome of LDR, PDR, and HDR calculation	Apply the methods that were developed at the institution for specific clinical cases: • Dose at distance method • TRAK • Others, for example, time indices	Each patient
Protocols	All types of applications should be described in detail.	Each patient
Standard forms	To be developed for each application.	Each patient
Independent check	Ensure that, as often as possible, a second person with expert knowledge checks the work of the first planner.	Each patient
Training	Adhere strictly to the designed training program.	Each patient

Note: See also ESTRO Booklet No. 8 (Venselaar and Pérez-Calatayud 2004).

13.6 Model-Based Dose Calculation Algorithms

Essentially, all sections of this chapter refer to the present-day practices, imaging tools, and treatment planning facilities, while it is clear that we are at the upfront of a new era in which MBDCAs will form a new standard for dose calculation for our plans. CT-based 3D patient information will allow us to include tissue information (structures, densities, outer contours, patient-specific radiation scatter conditions) and the presence of shielding material into the treatment planning procedure. It will replace the parameter-based TG-43 approach with its assumptions for a virtually unbounded homogeneous water environment.

Consistency in the application of treatment planning using TG-43 has been considered one of the key points for the success of brachytherapy development during the past 15 years. This technique has become a worldwide standard and forms the basis of the current practice. As MBDCA becomes more commonly available, medical physicists will depart from "water-only dosimetric material," which have been assumed until now. There is clearly a need for societal efforts to provide guidance for its introduction into the clinic, as there is the strong possibility that inter- and intra-institutional variability will occur. As stated in a Vision 20/20 paper by Rivard et al. (2010), intrinsically, MBDCAs have the potential to offer treatment plans with more accurate dosimetric depiction, but they also are highly dependent on the accuracy of information provided as input. The lower the energy of the radiation emitted by the source, the larger the effect. These authors suggest a similar process of guidance to be performed by the professional societies and users for validating their MBDCA-based TPS systems. At the very least, any MBDCA-based TPS system should reproduce the results obtained using the TG-43 formalism in the appropriate water geometry. Further, phantom configurations should be made available for the users, with compiled reference datasets for each brachytherapy source that could suffice the AAPM TG-43 dosimetric prerequisites. This information should enable the medical physicist to independently validate the capability of the new MBDCA to account for scatter conditions and material heterogeneities. A repository is suggested with a reliable content and an easy access for users to derive and develop their own test procedure, logically connected to the steps, checks, and tests described in this chapter (Rivard et al. 2010).

Joint databases for MC-based algorithms will form the backbone of the new generation of commercial TPSs, in combination with a new validation prerequisite for those new algorithms. Initiatives have been taken by AAPM Brachytherapy Subcommittee and GEC-ESTRO (Rivard et al. 2010) to define the contours of the societal infrastructure for this societal guidance.

References

Baltas, D. 1993. Quality assurance in brachytherapy with special reference to the microSelectron-HDR. *Activity International Selectron Brachytherapy Journal, Special Report No. 2.* Nucletron-Veenendaal.

Baltas, D., Sakelliou, L., and Zamboglou, N. 2007. *The Physics of Modern Brachytherapy for Oncology.* Taylor and Francis Group, Boca Raton.

Carmona, V., Perez-Calatayud, J., Lliso, F. et al. 2010. A program for the independent verification of brachytherapy planning system calculations. *J Contemp Brachyther* 2:129–33.

D'Amico, A.V., Cormack, R., Tempany, C.M. et al. 1998. Real-time magnetic resonance image-guided interstitial brachytherapy in the treatment of select patients with clinically localized prostate cancer. *Int J Radiat Oncol Biol Phys* 42:507–15.

Das, R.K., Patel, R., Shah, H. et al. 2004. 3D CT-based high-dose-rate breast brachytherapy implants: treatment planning and quality assurance. *Int J Radiat Oncol Biol Phys* 59:1224–8.

DeWerd, L.A., Ibbott, G.S., Meigooni, A.S. et al. 2011. A dosimetric uncertainty analysis for photon-emitting brachytherapy sources: Report of AAPM Task Group No. 138 and GEC-ESTRO. *Med Phys* 38:782–801.

Ezzell, G.A. 1994. Quality assurance of treatment plans for optimized high dose rate brachytherapy planar implants. *Med Phys* 21:659–61.

Fraass, B., Doppke, K., Hunt, M. et al. 1998. American Association of Physicists in Medicine Radiation Therapy Committee Task Group 53: Quality assurance for clinical radiotherapy treatment planning. *Med Phys* 25:1773–829.

International Atomic Energy Agency. 2007. *Specification and Acceptance Testing of Radiotherapy Treatment Planning Systems.* IAEA-TECDOC-1540. IAEA, Vienna.

International Commission on Radiation Units and Measurements (ICRU). 1985. *Dose and Volume Specification for Reporting and Recording Intracavitary Therapy in Gynecology.* Report 38 of ICRU, ICRU Publications, Bethesda, MD.

International Commission on Radiation Units and Measurements (ICRU). 1997. *Dose and Volume Specification for Reporting Interstitial Therapy.* Report 58 of ICRU, ICRU Publications, Bethesda, MD.

Kirisits, C., Siebert, F.-A., Baltas, D. et al. 2007. Accuracy of volume and DVH parameters determined with different brachytherapy treatment planning systems. *Radiother Oncol* 84:290–7.

Kubo, H.D. 1992. Verification of treatment plans by mathematical formulas for single catheter HDR brachytherapy. *Med Dosimetry* 17:151–5.

Kubo, H.D. and Chin, R.B. 1992. Simple mathematical formulas for quick-checking of single catheter high dose rate brachytherapy treatment plans. *Endocurie Hypertherm Oncol* 8:165–9.

Kubo, H.D., Glasgow, G.P., Pethel, T.D. et al. 1998. High dose-rate brachytherapy treatment delivery: Report of the AAPM Radiation Therapy Committee Task Group No. 59. *Med Phys* 25:375–403.

Kutcher, G.J., Coia, L., Gillin, M. et al. 1994. Comprehensive QA for radiation oncology: Report of AAPM Radiation Therapy Committee Task Group 40. *Med Phys* 21:581–618.

Martin, T., Hey-Koch, S., Strassmann, G. et al. 2000. 3D interstitial HDR brachytherapy combined with 3D external beam radiotherapy and androgen deprivation for prostate cancer. Preliminary results. *StrahlentherOnkol* 176:361–7.

Nag, S. 1994. In: *High Dose Rate Brachytherapy: A Textbook*. S. Nag (ed.), Futura Publishing Company, Inc., Armonk, New York.

Nath, R., Anderson, L.L., Luxton, G. et al. 1995. Dosimetry of interstitial brachytherapy sources: Recommendations of the AAPM Radiation Therapy Committee Task Group No. 43. American Association of Physicists in Medicine. *Med Phys* 22:209–34.

Nath, R., Anderson, L.L., Meli, J.A. et al. 1997. Code of practice for brachytherapy physics: AAPM Radiation Therapy Committee Task Group 56. *Med Phys* 24:1557–98.

Paterson, R. and Parker, H.M. 1938. A dosage system for interstitial radium therapy. *Br J Radiol* 1:252–340.

Rivard, M.J., Coursey, B.M., DeWerd, L.A. et al. 2004. Update of AAPM Task Group No. 43 Report: A revised AAPM protocol for brachytherapy dose calculations. *Med Phys* 31:633–74.

Rivard, M.J., Butler, W.M., DeWerd, L.A. et al. 2007. Supplement to the 2004 update of the AAPM Task Group No. 43 Report. *Med Phys* 34:2187–205.

Rivard, M.J., Venselaar, J.L.M., and Beaulieu, L. 2009. The evolution of brachytherapy treatment planning. *Med Phys* 36:2136–53.

Rivard, M.J., Beaulieu, L., and Mourtada, F. 2010. Enhancements to commissioning techniques and quality assurance of brachytherapy treatment planning systems that use model-based dose calculation algorithms. *Med Phys* 37:2645–58.

Rogus, R.D., Smith, M.J., and Kubo, H.D. 1998. An equation to QA check the total treatment time for single-catheter HDR brachytherapy. *Int J Radiat Oncol Biol Phys.* 40:245–8.

Rubens, D.J., Yu, Y., Barnes, A.S., and Strang, J.G. 2006. Image-guided brachytherapy for prostate cancer. *Radiol Clin North Am* 44:735–48.

Siebert, F.-A., Kohr, P., and Kovács, G. 2005. The design and testing of a solid phantom for the verification of a commercial 3D seed reconstruction algorithm. *Radiother Oncol* 74:169–75.

Siebert, F.-A., De Brabandere, M., Kirisits, C., Kovács, G., and Venselaar, J. 2007. Phantom investigations on CT seed imaging for interstitial brachytherapy. *Radiother Oncol* 85:316–23.

Thomadsen, B.R. 2000. *Achieving Quality in Brachytherapy*. Medical Science Series, Institute of Physics Publishing, Bristol and Philadelphia.

Van der Laarse, R. and Luthmann, R.W. 2001. Computers in brachytherapy dosimetry. In: *Principles and Practice of Brachytherapy Dosimetry Using Afterloading Systems*. C.A.F. Joslin, A. Flynn, and E.J. Hall (eds.), Arnold, London, pp. 49–80.

Van Dyk, J., Barnett, R.B., Cygler, J.E., and Shragge, P.C. 1993. Commissioning and quality assurance of treatment planning computers. *Int J Radiat Oncol Biol Phys* 26:261–73.

Venselaar, J.L.M., Bierhuizen, H.W.J., and Klop, R. 1994. A method to check treatment time calculations in Ir-192 high dose rate volume implants. *Med Phys* 22:1499–500.

Venselaar, J.L.M. and Pérez-Calatayud, J. 2004. *A Practical Guide to Quality Control of Brachytherapy Equipment*. ESTRO Booklet No. 8. European Society for Therapeutic Radiology and Oncology, Brussels.

Williamson, J.F., Ezzell, G.A., Olch, A., and Thomadsen, B.R. 1994. Quality assurance for high dose rate brachytherapy. In: *High Dose Rate Brachytherapy: A Textbook*. S. Nag (ed.). Futura Publishing Company Inc., Armonk, New York.

Yu, Y., Anderson, L.L., Li, Z. et al. 1999. Permanent prostate seed implant brachytherapy: Report of the American Association of Physicists in Medicine Task Group No. 64. *Med Phys* 26:2054–76.

Practical Use, Limitations, and Quality Control of Imaging in Brachytherapy

Frank-André Siebert
Clinic of Radiotherapy (Radiooncology), UKSH

Taran Paulsen Hellebust
Oslo University Hospital – The Norwegian Radium Hospital

14.1 Introduction

Imaging has probably been the most important driving force for the development of brachytherapy treatments during the past several years. Through 3D imaging, brachytherapy became a highly accurate and reliable treatment option in radiotherapy. Tumor volumes and organs at risk (OARs) are visibly good in many cases, and applicators can be reconstructed in three dimensions. Moreover, intraoperative techniques, in particular, transrectal ultrasound (US) for prostate, are nowadays standard in most clinics. This chapter discusses the possibilities and limitations of various imaging methods for brachytherapy. It contains a short physical explanation of each methodology followed by a description of the practical use in different body sites.

14.2 Radiography Imaging

14.2.1 Physics of Radiography

Radiographic imaging has been known since Roentgens' discovery of x-rays in 1885 and continues to be of great importance for medical science and thus, in the present context, for brachytherapy (Gerbaulet et al. 2002). The principle of radiographic imaging is relatively simple. In a typical x-ray tube, bremsstrahlung and characteristic radiation are produced in a continuous energy spectrum, which is normally composed for the purpose of medical imaging in an energy range between 20 and 150 kVp. Usually, a set of orthogonal transmission radiographic images are produced from implanted patients for dosimetric purposes. Transmission radiography means that the patient is positioned between the x-ray tube and the detector (i.e., a film or an image intensifier). The transit photons interact with the film by photoelectric absorption and Compton scattering interactions, thus creating the image of the patient. It is very easy to identify high and low density as well as the high-Z structures, such as bone and metallic structures of the brachytherapy sources and applicators, and the cavities in these images. However, due to the fact that the absorption coefficients of different soft tissues are quite similar, they often cannot easily be differentiated on a regular radiograph.

Various film types have been developed for radiographic imaging procedures, for example, conventional Kodak radiographic films and self-developing GAFCHROMIC films. Moreover, image intensifiers are often used on conventional radiotherapy simulators and C-arms in order to allow the low-intensity x-rays to be converted to a visible light and create images at low x-ray dose. However, some investigators have shown that the use of image intensifiers may also lead to distortions in the images (Liu et al. 2003). This should be taken into account when image intensifier data are being used for the purpose of fusion or applicator reconstruction.

The main application of radiographs in brachytherapy is the reconstruction of applicators or permanent-implanted sources for more accurate radiation dosimetry. Visualization of target volumes and critical organs enables the user to create an optimum plan for the patient's treatment. For example, using a set of isocentric orthogonal films and knowing the target-to-film and target-to-isocenter distances, a computerized software-based dosimetry can determine the magnification factors of the source and applicators needed for reconstruction purposes. Figure 14.1 shows a schematic diagram of a reconstructed point of interest $P(x,y,z)$ on an applicator using a set of orthogonal radiographs.

If the radiographic films are produced by using a non-isocentric device such as a C-arm system, a reconstruction jig

can be used to obtain the relevant geometric parameters (Fung 2002). This device consists of specific markers with known geometrical information and is placed in the x-ray field. The projected images of these markers can be used to obtain the required information for the reconstruction purpose. The accuracy of reconstruction techniques depends on the image quality and the angles of the projecting beams, in addition to the accuracy of the geometric parameters of the jig. Reconstruction with a set of perpendicular beams has shown to be the most accurate technique. Moreover, it is of importance that patient's movements between radiographs be minimal.

Occasionally, the orthogonal film may not be suitable for the brachytherapy dosimetry procedure due to the shielding of the brachytherapy source by structures such as bone or prosthesis. For example, shielding of the prostate seeds by the femoral bone or femoral prosthesis in the lateral view may not allow one to create a useful set of orthogonal films. In these cases, alternatives to the orthogonal film technique can be used such as the stereoshift method. With this method, two or more films are produced by shifting the x-ray tube of the patient along the longitudinal axis, in between the films (van der Laarse and Luthman 2001).

Despite its simplicity, the orthogonal film-based reconstruction technique is error prone. Specifically, the influence of the chosen magnification factors from the applicator plane to the film plane is significant in the reconstruction results. A proper commissioning and quality control procedure with a phantom test is therefore strongly recommended. With changes of equipment, the commissioning tests should be repeated.

In the traditional reconstructions, the 3D positions of implanted seeds with orthogonal films were composed of manual identification of the sources on both images. This technique was more practical for implants with smaller seed/source numbers. However, for the interstitial implant with a large number of sources, identification of all sources on both films would be very cumbersome, time consuming, and sometimes impossible as the projections of some seeds may overlap. This technique was particularly inefficient for interstitial prostate permanent seed implant or interstitial vaginal implants with many sources in a small volume with possibility of the images of several sources falling at the same points. Rosenthal and Nath (1983) recommended the use of the three-film technique (e.g., with beams incident at AP/PA, Lat, and 45°) for these implants. Using a sophisticated mathematical algorithm, computerized software could associate the images of the sources from the three films by using at least two or three reference non-coplanar points. Since the introduction of this technique, several other investigators have developed solutions for the problem of overlapping seed images in the radiographs (e.g., Su et al. 2004). Presently, the application of computed tomography (CT) has resolved the issues of the multiseed implant.

14.2.2 Clinical Application: Prostate Implants (HDR, LDR)

Treatment planning of high-dose-rate (HDR) prostate implants themselves is nowadays mainly based on full 3D US or CT

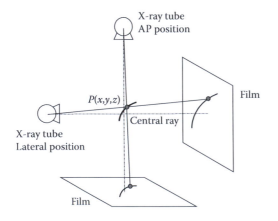

FIGURE 14.1 Reconstruction of a point of interest $P(x,y,z)$ using orthogonal radiographs. (Courtesy of F.-A. Siebert.)

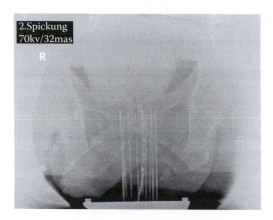

FIGURE 14.2 AP radiograph taken from a C-arm for an interstitial HDR implant. Visible are the bony structures, the implant needles, the contrast-filled bladder, and a fivefold diode for in vivo dosimetry in the rectum. (Courtesy of U.K. S-H, Campus Kiel, Clinic of Radiotherapy.)

imaging techniques. In these techniques, not only the brachytherapy sources and applicators are identified, but also the treatment target volume as well as the surrounding normal tissues can be included in the dose calculation procedures to achieve an optimum treatment. Therefore, radiographic images do not provide enough information on soft tissue resolutions for these treatment techniques. Nevertheless, the use of radiographic images can be a reasonable addition to the procedure, as suggested in the GEC-ESTRO recommendations describing the use of a fluoroscopy device to perform CT imaging in treatment position for needle control (Kovács et al. 2005). Many radiation oncologists use x-ray images to verify the seed implant procedure. Furthermore, the position of implanted needles in HDR procedures can be checked using an x-ray device or a C-arm device prior to the treatment,

and one can compare the results with 3D US or CT images. An example of an AP C-arm radiograph of an interstitial vaginal implant is given in Figure 14.2. Especially in HDR monotherapy, the verification of the needle positions after the creation of an optimized treatment planning and before administration of the treatment, or just before the start of new pulses in a fractionated schedule, is very useful (Hoskin et al. 2003).

For institutions where a CT scanner is not available, a radiographic film technique can be used for seed reconstructions. Figure 14.3 shows the schematic of a three-film technique, used for reconstruction of the source positions in a multiseed implant with an accuracy of approximately 0.2 mm (Siebert et al. 2007a). Some radiography-based reconstruction algorithms even offer the possibility of detecting the individual seed orientation (Siebert et al. 2007a). Because of the high contrast of the implanted seeds in pelvic radiographs, they can be used for verifying the number of implanted seeds in the prostate, which is of clinical interest due to the risk of seed migration. Sometimes, depending on the seed model, overlapping seeds are not easy to discern. In such cases, a second radiograph taken with a tilt of the incident beam of only 5°–10° may be useful. Images of overlapping sources will then move a little apart. In most cases, it is advantageous for image quality to avoid lateral images. Due to the fact that more patient tissue must be passed by the roentgen photons in this direction, the radiographs are often blurred and seeds are more difficult to detect.

14.2.3 Clinical Application: Breast Implants

14.2.3.1 Interstitial Multicatheter Implant

Modern treatment planning systems (TPSs) for breast implants are nowadays mostly based on 3D CT datasets. Nevertheless,

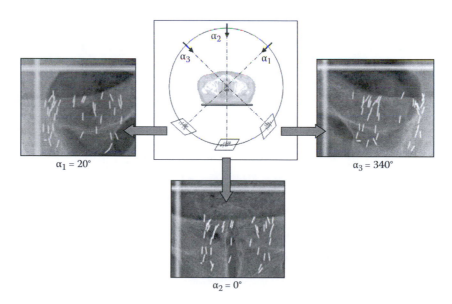

FIGURE 14.3 Illustration of a three-film reconstruction technique. Radiographs from three different gantry angles α_i are taken. With known distances and gantry angles, seeds can be reconstructed in 3D. Depending on the reconstruction algorithm, the user has some degrees of freedom to choose the gantry angles. (Courtesy of F.-A. Siebert.)

radiography remains an important imaging modality for use with brachytherapy breast implants. C-arm imaging is widely used to support the implant of the catheters in the operation room. Guided by fluoroscopy, the optimal implant plane can be located, and the physician can control the position and angles of insertion of the needles or the catheters. By rotating the gantry, a 3D impression of the needle locations can be obtained during the implant. Tumor extensions are normally identified by using radio-opaque clips at the superior/inferior and AP/PA directions. In case the procedure is performed with a radioactive seed type of an implant, the appearance of these clips is distinctly different from the brachytherapy seeds, in order to avoid any confusion during seed counts. To distinguish different catheters in the radiographs, dedicated marker wires with a distinct sign for each wire (e.g., for wire marker no. 2, there will be a larger marker on seed no. 2, and for wire no. 3, there will be a larger marker at position 3) are helpful. Commercial products are available for this purpose from several vendors, known as (dedicated) x-ray catheters.

Entrance and exit points of the catheters can be identified using the fixation buttons, which are used to keep the catheters at the skin surface. These will clearly appear on the x-ray images.

Note that full 3D reconstruction of a multicatheter breast implant for treatment planning is nowadays commonly performed by performing CT scanning of the patient.

14.2.3.2 Balloon Brachytherapy

Another option for partial breast treatments is the balloon technique. A deflated double-lumen balloon is placed in the lumpectomy cavity and then inflated to fill the cavity size. This treatment is normally performed with an HDR remote afterloading system. With radiographs, US, or eventually CT scanning techniques, the position and the durability of the balloon filling can be examined prior to each fraction or pulse of a fractionated treatment schedule. Treatment planning itself is always carried out using 3D CT. Kassas et al. (2004) showed that the use of radio-opaque contrast fillings (normally 1% to 5%, depending on the resolution of the scanning system) in the balloon for a better visualization does not influence the dose rate significantly.

14.2.4 Clinical Application: Gynecological Implants

For many years, the radiography technique was the only imaging modality used for treatment planning of the gynecological brachytherapy procedures. Two orthogonal projection images were used to identify the orientation and position of the sources relative to the patients' anatomy (Gerbaulet et al. 2002). Despite the poor tissue contrast in x-ray images, reference points, like points A and B, could be defined as these are defined in direct relation to the applicators (flange); see Chapter 24 on ICRU 38. However, with the new technological advancements of the field and development of the CT- and magnetic-resonance-imaging (MRI)-compatible applicators instead of the traditional stainless-steel applicators, the 3D treatment planning can be performed

using CT or MRI images. With this improvement, not only the dose to the treatment volume can be determined and recorded more accurately, but also the dose to the normal tissues such as bladder and rectum can be better controlled. In a further step, the application of the images from a PET scanning system could allow a better delineation of the target volume and a better optimized treatment planning for cervical patients (Lin et al. 2007).

14.3 US Imaging

14.3.1 Physics of US

US is the application of an alternating pressure in a medium with frequencies above 20 kHz (Webb 1988). Medically used US is in the range of 1 to 20 MHz. The generation and detection of US are performed by using a piezoelectrical crystal device known as the transducer. If alternating voltage is applied at the transducer, it oscillates and initiates alternating pressure into a medium in the form of ultrasonic sound waves. The same device will convert the received US waves into detectable electrical signals. Therefore, after the emission of an US pulse, the transducer would detect the echo signals. The time difference between sending the US waves and receiving the echo represents the distance between the transducer and the reflecting object. This device will show the amplitude of the detected signals (A-mode) or its intensity in the form of brightness (B-mode). Distances between the peaks of the signals in the A-mode or different levels of brightness in the B-mode represent the distances between the transducer and the reflecting object or other change of the medium in the pathway of the US. Medical US systems are able to emit the US in a specific direction, either mechanically or by electronic focusing. Modern US devices often use arrays of transducers consisting of many transducer elements. A time delay between sending impulses to the outer elements and the inner ones can change the focal position of the sound. Thus, it is possible to create a 2D image with several profiles from different directions.

14.3.2 Clinical Application of US in Brachytherapy: Transrectal US for Prostate Implants

The most common use of US in brachytherapy is in the treatment of prostate cancer. With transrectal US (TRUS), probes are composed of different arrays of transducers that would enable the users to obtain biplane US images (i.e., transversal and longitudinal) of the prostate. Mostly the TRUS probes are mounted on a stabilizer device along with a stepping device in order to move the probe in a defined and reproducible way. 3D reconstruction is possible when images are acquired in different stepper positions. Typical step sizes for transversal images are 5 or 2.5 mm. US offers very accurate imaging of the prostate in terms of discerning of prostate tissue from surrounding tissues such as bladder and rectum. In some cases, the zonal anatomy of the prostate can be visualized, but there is no reliable differentiation between tissue and hyperechoic cancer.

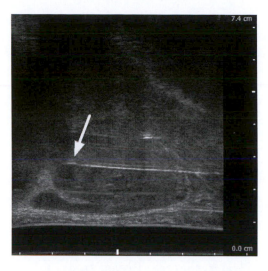

FIGURE 14.4 Longitudinal view of a prostate (bladder left side) in a TRUS image using 9 MHz. One single seed (arrow) in the most distal position (zero plane) was just ejected from the implant needle. (Courtesy of F.-A. Siebert.)

One of the important advantages of the modern TRUS-based implant procedure compared to the methodologies used in the early 1980s is that it can be carried out as an interactive treatment planning technique (Nath et al. 2009). This technique allows for a stepwise refinement of the treatment plan due to image-based needle position feedback to the TPS. After the start of the technique and the descriptions given, for example, by Holm et al. (1983), who published the use of TRUS guidance for transperineal prostate seed [low dose rate (LDR)] implants, this technique was further developed by different groups. Much more detailed descriptions of developments toward the present-day more-dynamic dose calculation approach were given by Nag et al. (2001) and Polo et al. (2010). In these methods, the dose is "painted" to the 3D target volume that might change during the implant procedure. Dynamic dose calculation relies on full 3D adaptive dose planning with permanent feedback of the actual seed locations. In Figure 14.4, a typical longitudinal view is depicted. Beneath the prostate gland and the seminal vesicles, the implant needle and one seed that just dropped out the needle are clearly visible.

For postplanning purposes, however, it seems that the advantage of US-based treatment planning is limited. Han et al. (2003) showed in a study using stored US images that the average percent of the seeds allegedly identified per patient ranged from 51% to 83%. To date, postplanning with US images only is not a recommended technique; see, for example, the AAPM TG-137 report (Nath et al. 2009).

14.3.3 Clinical Application of US in Brachytherapy: Breast

US can be used to measure the distance between the surface of the balloon and the skin of the patient. Esthappan et al. (2009) described the possibility of checking the balloon diameter and integrity either by orthogonal radiographs or US.

14.3.4 Clinical Application of US in Brachytherapy: Gynecological Brachytherapy

For gynecological brachytherapy treatment, CT and MRI scanning are the most common imaging modalities. Nevertheless, several authors have described the advantage of using transabdominal US for cervical brachytherapy treatment (van Dyk et al. 2009; Davidson et al. 2008). Davidson et al. (2008) described an intraoperative US approach for image-guided insertions of the tandem applicator in 35 consecutive insertions. After the applicator placement, a planning CT was performed. They were able to decrease the implant time and reported improved selection of tandem length and angle as well as a decrease in out-of-department consultations due to perforations. In a publication by van Dyk et al. (2009), it was shown that transabdominal US offers an accurate, quick, and cost-effective method of conformal brachytherapy planning. In comparison with MRI, they computed target equivalent doses ($\alpha/\beta = 10$) of 80.8 Gy for US-based and 78.1 Gy for MRI-based treatment plans, whereas the standard radiography-based plan had a dose of 87.5 Gy.

14.4 Computed Tomography

14.4.1 Physics of CT

X-ray transmission CT is a very important image modality in external beam radiotherapy as well as in brachytherapy. An x-ray source rotates around the patient while the scanner couch with the patient moves in longitudinal direction, and the resulting attenuated intensities of the beams are detected and stored. Today, in most cases, filtered back-projection algorithms are in use to compute the 3D dataset of a patient. Data are stored as Hounsfield units, which are, after transfer to the TPS, converted into electron densities. As long as inhomogeneity correction is not a common feature of the brachytherapy TPS as it is for external beam therapy, the accuracy of this conversion has no influence on the resulting dose calculation. This may change in the future, with the upcoming treatment planning with model-based dose calculation algorithms. This will have consequences for the commissioning and quality control of this part of the brachytherapy dosimetry.

CT images are, at present, used for organ delineation and catheter reconstruction. The existence of metal in a patient can cause very disturbing artifacts in the images. Different methods are under investigation to reduce these artifacts. A good overview on CT principles and reconstruction techniques is given in the works of Kalender (2000) and Herman (2009).

14.4.2 Clinical Application of CT in Brachytherapy: Prostate Imaging

The implantation procedure of HDR prostate brachytherapy is generally performed under US guidance. The additional use of CT for treatment planning has also been described in the literature (e.g., Hoskin 2000). Catheter reconstruction can be best performed using a small slice thickness and table index. As with all catheter reconstruction techniques, it must be considered that the physical tip of the implant needle is not the first dwell position. Dimensions of the dead space end of the needle or source lengths must be properly taken into account. Nevertheless, the soft tissue contrast in CT imaging is limited so that the delineation of the prostate volume is not straightforward.

CT is considered as the "gold standard" for postplanning with LDR seeds. After the US-based implant, the postplanning procedure starts in most cases, either at day zero or more commonly at 3–4 weeks after the implant, when the prostate swelling has already decreased.

The location of the urethra can be visualized by inserting a catheter. It was shown that small slice thicknesses and table indices result in a better reconstruction of the seeds (Siebert et al. 2007b). In a phantom reconstruction, accuracies of better than 1 mm can easily be reached when slice thickness and table index are taken equal to smaller than 3 mm. Seeds can be detected automatically by software. This works fine in most cases, but caution must be taken when seeds are close together. In these cases, a manual intervention of the user of the planning software is often required.

14.4.3 Clinical Application of CT in Brachytherapy: Gynecological Imaging

Due to its 3D capabilities and high availability, CT is often used for gynecological imaging in brachytherapy. Visualization of bladder, rectum, sigmoid, bowel, and vagina with CT is quite similar in comparison to MRI. However, CT images are limited for the delineation of gynecological target volumes (Wang et al. 2009).

In the work of Fellner et al. (2001), the dosimetric benefit of 3D CT treatment planning for cervical cancer over point dose dosimetry methods based on 2D radiographs was demonstrated. Similar results have been published by Kim and Pareek (2003). They found for a cohort of 15 patients an overestimation of tumor dose in radiography-based conventional treatment planning, in particular, for patients with more advanced tumors.

Despite its general advantages for 3D imaging in gynecological brachytherapy, the limitation for tumor volume determination is apparent in comparison to MR. In a prospective trial for cervical cancer patients, statistically significant differences in the volume treated with the prescription dose or greater were found between CT and MR datasets, because CT contouring can overestimate the tumor width yielding to significant differences in the D_{90} and D_{100} of the target dose (Viswanathan et al. 2007). In their study, Viswanathan et al. report also on limitations in CT-based delineation of the superior border of the cervix and the lateral border of the parametria, as well as difficulties with contouring the OARs.

Next to the reconstruction of target and organ volumes, applicator reconstruction is another important issue for brachytherapy. To reconstruct applicators, different approaches exist: direct manual reconstruction in the CT images, reconstruction in multiplanar reconstructed images, and application of library

plans. The smallest uncertainties were found for library-based reconstruction (Hellebust et al. 2007). But all methods provide an uncertainty of less than 3.7% in dedicated dose points, which is low in comparison with other uncertainties in brachytherapy. Moreover, it was found that the orientation of the applicator relative to the image plane does not influence the overall accuracy of the treatment planning process. After optimization of scan protocols and parts of the x-ray tube, a cone beam CT (CBCT) was found to be an appropriate imaging modality for gynecological brachytherapy cases. Reniers and Verhaegen (2011) addressed the advantages of a CBCT over conventional CT: acquisition can be performed in the treatment room and carried out in patient treatment position due to more space between the patient and the CBCT tube.

14.4.4 Clinical Application of CT in Brachytherapy: Breast Imaging

CT offers a good imaging modality for brachytherapy multicatheter and balloon breast techniques (Aristei et al. 2009). After the implant with a multicatheter technique, the patient is positioned on the CT couch. If necessary, for example, when using plastic catheters, thin metal wires can be inserted to improve the visualization on CT. Special markers (buttons) can be used for visualization of the entrance points of the catheters at the skin surface. Since the breast is a mobile part of the body, the setup on the CT must be identical to the treatment position. Moreover, fast CT scan protocols should be used for brachytherapy breast techniques to minimize breathing movements of the breast during the scan. A slice width of 3–5 mm is typical. Scanderbeg et al. (2009) reported the use of a breast board to reduce interfraction patient movements and recommended CT scans using a breath-hold technique. The reconstruction of the catheters can be performed either manually or using automatic catheter detection algorithms. Care should be taken with the order of the catheters. Numbering of digitally reconstructed catheters must be taken with care to avoid interchanging catheters. Sometimes surgical clips after lumpectomy are placed in the breast to aid the delineation process for the target volume. Without the use of clips, higher uncertainties in contouring of the clinical target volume (CTV) are expected (Landis et al. 2007). Wong et al. (2006) analyzed the impact on contouring guidelines for partial breast radiotherapy and found a lower interobserver variability when guidelines were followed.

In a study by Major et al. (2005), it was shown that using CT-based dose planning, the coverage of the target is improved compared with conventional techniques: mean target coverage 70% for conventional and 87% for conformal planned cases. Similar results were published by Das et al. (2004). For 50 consecutive accelerated partial-breast irradiation (APBI) patients, they showed coverage of 100% of the lumpectomy cavity in all cases. The target volume was covered in median with 96%.

Also for breast balloon techniques, the CT is accepted as being the adequate imaging device (Edmundson et al. 2002). To take into account different shapes of target volumes and lumpectomy cavities, various balloon designs were developed. For an increased contrast of the balloon and the catheter, these are filled with radiopaque contrast solution. Typically, the target volume includes the lumpectomy cavity with a margin of 1–2 cm. In CT images, the balloon diameter can easily be detected. This is useful when ensuring the stability of the balloon filling over the treatment period. Subsequent daily checks of balloon size can be performed by x-ray.

14.4.5 Clinical Application of CT in Brachytherapy: Others (Head and Neck)

CT scans are of great use in head-and-neck brachytherapy treatment planning. Soft tissue and bony structures are well visualized, and CT is superior to MRI in the evaluation of lymph nodes (Mazeron et al. 2009). Reconstruction of catheters is possible on CT basis. In some cases, when using loops, additional x-ray imaging can be of help in identifying the applicators. After the reconstruction, the loop diameter can be easily determined in the TPS and checked if the minimum curvature for the afterloader model was reached and if problems might occur when the plan is to be delivered.

14.5 Magnetic Resonance Imaging

14.5.1 Physics of MRI

14.5.1.1 Basic Principles

MR exploits the magnetic properties of atoms with an odd number of nuclei (Bushong 1996). Every nucleus has an associated quantity called spin, and this spin is quantized into units of half-integer values, called the spin quantum number. The spin quantum of hydrogen nuclei is 1/2, and only two spin states are allowed: +1/2 and –1/2. Since the nucleus is a spinning charged particle, a magnetic field is associated with it, the so-called nuclear magnetic moment. In a static external magnetic field, the positively charged nuclei will have different energy levels depending on their spin quantum, and they will orient themselves in restricted orientation. A collection of hydrogen nuclei will align their spinning movement (precession) either parallel or antiparallel to the external magnetic field, and precess with their rotational frequency (the Larmor frequency) around the axis of the magnetic field according to the following equation:

$$\omega = \gamma B_0 \tag{14.1}$$

where ω is the frequency of precession, γ is the gyromagnetic ratio, and B_0 is the strength of the external magnetic field. At equilibrium, more nuclei will align parallel to the external magnetic field, and therefore, a net magnetization, M_0, along B_0 will be seen. It is not possible to measure M_0 at equilibrium. By applying a radiofrequency (RF) pulse (excitation) tuned to the nuclei's Larmor frequency, the magnetization vector can be flipped into another direction to have nonzero component in the transversal plane, M_T (see Figure 14.5).

The component in the longitudinal direction, M_L, will be smaller than M_0. M_T will rotate about the longitudinal axis, and therefore, a current will be induced if an MR detector coil is positioned in the transversal plane. The signal in such a detector will be proportional to the magnitude of M_T.

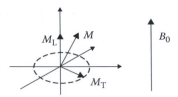

FIGURE 14.5 After an RF pulse, the magnetization vector M is turning from a straight longitudinal direction to a transverse direction. Thus, the magnetization vector M will have a longitudinal (M_L) and a transversal (M_T) component. (Courtesy of T.P. Hellebust.)

Following the termination of the RF excitation pulse, nuclei will return to equilibrium. To arrive at equilibrium, the nuclei must give up the energy gained from the RF pulse. This energy is partly released by interaction between the individual hydrogen nuclei (spin) and the surrounding molecule (lattice) or by interaction between the hydrogen nuclei. The rest of the energy is emitted in the form of RF that can be detected in a receiver coil, referred to as echo. The time constant that describes the rate at which M_L returns to M_0 is the T1 relaxation time. The time constant that describes the rate at which M_T returns to zero is the T2 relaxation time. T1 is also called the longitudinal or spin–lattice relaxation time, whereas T2 is called the transverse or spin–spin relaxation time. T1 and T2 will depend on the tissue and give rise to contrast in an MR image. To receive a proper signal from the patient, more than one RF signal has to be applied, the so-called pulse sequence. A pulse sequence is typically a combination of several 90° and 180° flipping RF pulses. One sequence is not sufficient to generate an image; it has to be repeated many times. The time between the sequences is called repetition time (TR). The time between the application of the RF pulse and the peak of the detected echo is called echo time (TE). In MRI, it is common to manipulate different signal parameters, as TR and TE, to maximize contrast and obtain better visualization. A short TR and TE will result in a T1-weighted image, while a long TR and TE will lead to a T2-weighted image. In a T1-weighted image, fat tissue will appear bright and tissue containing a lot of water will be darker, whereas the opposite is true in a T2-weighted image.

To obtain spatial information in the MR signal, three orthogonal magnetic field gradients are applied. A magnetic field that changes linearly with position makes the nuclei process with different Larmor frequency along the direction of the gradient (Equation 14.1). An applied RF pulse will flip the magnetization vector of nuclei in a narrow section of the body (or slice) having Larmor frequency matching the RF pulse frequency. Spatial encoding in the transversal plane is obtained by sequentially turning on and off the remaining gradient fields. In the work of Hashemi et al. (2010), more detailed information on MRI basic physics is described.

14.5.1.2 MR Pulse Sequences

The pulse sequences are wave forms of the gradients and RF pulses applied to create the MR image. There are two fundamental types of MR pulse sequences: spin echo (SE) and gradient echo (GRE). Other sequences are variations of these two.

In a SE sequence, a 90° RF pulse is followed by one or more 180° pulses at some later time (Bushong 1996). A ghost echo signal, called a spin echo, will be generated and detected. Conventional SE sequences have very long acquisition time. However, modern MRI technology enables a reduced acquisition time by using the so-called fast SE (FSE). Many different sequences are available, and often, various vendors are using different trade names (Bitar et al. 2006). A more detailed description is not within the scope of this chapter.

14.5.1.3 Geometrical Distortion, Field Strength, and Lack of Electron Density

MRI could be prone to geometrical distortion, and these distortions are influenced by the choice of MR pulse sequence, the field strength, as well as the construction of the MR machine and the environments. In radiotherapy, SE sequences are usually used since others are more prone to geometrical distortions. In brachytherapy, large dose gradients exist, and it is therefore important to be aware of the magnitude of the distortions. In general, the spatial accuracy decreases with the distance from the magnet isocenter (Fransson et al. 2001), that is, in pelvic brachytherapy (prostate and gynecology), the distortions are smallest in the area of the implant. If appropriate sequences are used for a field strength of 1.5 T together with distortion correction algorithms provided by the manufacturers, the geometrical distortion is considered to be small (<2 mm) in the region of interest (ROI) in pelvic brachytherapy (Wills et al. 2010).

Published clinical experiences with MR and brachytherapy are mainly based on MRI with field strength of 0.2–1.5 T. Stronger field strength (e.g., 3 T) will give a better signal-to-noise ratio (SNR) and thus higher image contrast (Kataoka et al. 2007). Additionally, the acquisition speed is faster (Hussain et al. 2006). However, stronger field strength is also associated with larger geometrical uncertainties and should be cautiously used (Kim et al. 2011).

Today, most brachytherapy TPSs do not take into account tissue and material heterogeneities. However, new model-based dose calculation algorithms are developed accounting for such effects (Rivard et al. 2009). To perform dose calculations including inhomogeneity correction, electron density information is necessary. Unfortunately, there is no direct link between tissue electron density data and the MR signal as is the case for CT imaging. According to Rivard et al. (2009), the significance of performing inhomogeneity correction in prostate and gynecological implants using high-energy sources (e.g., ^{192}Ir) is limited. However, such correction might be important using low-energy photon-emitting sources (e.g., ^{125}I and ^{103}Pd).

14.5.1.4 Functional MRI

Functional MRI (fMRI), such as dynamic contrast-enhanced MR (DCEMR), diffusion weighted imaging (DWI), and MR spectroscopy (MRS), allows to image microenvironmental characteristics of a tumor that could be important for treatment planning. DCEMR imaging visualizes differences between tissues in the uptake of an MR contrast medium as a function of time (Zahra et al. 2007). The most frequently used method is to administer a paramagnetic agent that will alter the local magnetic environment, for example, Gd-DTPA (a gadolinium-based contrast agent). By looking at the dynamic properties of the contrast enhancement in different tissues, it will be possible to distinguish between tissues with large blood flow and those with low blood flow. In diffusion weighted MRI, the translational mobility of water molecules is explored (Luypaert et al. 2001). The mobility of water will be reduced in dense tissue with low fraction of extracellular space compared with that in tissue with a large fraction. In MRS, chemical information about the tissue is imaged (De Graaf 2002). This is obtained by exploiting the principle that different chemicals containing the same nucleus have different characteristic resonance frequencies, depending on their molecular environment and their interaction with other chemical structures. In in vivo MRS, the signal from water overwhelms that emitted from other metabolites. By suppressing the water signal, the signal from relevant metabolites can be measured. However, clinical limitations restrict in vivo MRS to some high-concentration metabolites, such as lipids, choline, creatine, and lactate.

To perform fMRI, high sensitivity of the system is required and field strengths ≥1.5 T should be used.

14.5.2 Clinical Application of MRI: Prostate Imaging

For radiological staging of prostate cancer, MRI is considered to be the optimal image modality (Carey 2005), and the best possible image quality is achieved using an endorectal coil in conjunction with a pelvic phased-array coil on a mid- to high-field strength MR machine (Hricak et al. 2007). The role of MRI in prostate brachytherapy has increased during the last decade, both for treatment planning and post-implant evaluation. MRI provides excellent visualization of the prostate gland and the surrounding anatomy, making it appropriate both for preplanning (Tanaka et al. 2008) and intraoperative treatment planning (Simnor et al. 2009; Ménard et al. 2004; D'Amico et al. 1998). In the latter method, the treatment planning is based on the true position of the needles and seeds in the prostate, allowing a direct comparison between the implanted versus planned catheter position (Polo et al. 2010). The pretreatment plan based on images without needles could then be adjusted accordingly.

Furthermore, MRI offers a unique possibility of biologic imaging, such as MRS and DCEMR, to image molecular properties (Kurhanewicz and Vigneron 2008; Hricak et al. 2007; Choi et al. 2007; Hoskin et al. 2007). Specific areas within the gland with a high burden of disease or with biological characteristics indicating radioresistance may be targeted for higher dose delivery (Kim et al. 2008; Zelefsky et al. 2000). Hoskin et al. (2007) used blood oxygen level-dependent (BOLD) MRI and showed that such imaging is a potential noninvasive technique for mapping prostatic tumor hypoxia. Kim et al. (2008) delineated intraprostatic lesion (DIL) defined by MRS (elevation of the choline peak and reduction of the citrate, creatine and polyamine peak) in 15 patients and showed that it was feasible to significantly escalate the dose to the tumor while maintaining the risk of complications to an acceptable level.

Although the needle placements usually are guided by TRUS, MR-guided prostate brachytherapy has been reported both with an open low field MR scanner (Ares et al. 2009) and with a "closed-bore" 1.5-T scanner (Ménard et al. 2004).

In permanent prostate brachytherapy, a post-implant evaluation is usually performed some 4 weeks after the implantation. Optimal imaging will include T2-weighted MRI to improve the definition of the CTV (Salembier et al. 2007). However, seed localization using MRI is usually considered to be challenging due to distortion and susceptibility artifacts. On the other hand, De Brabandere et al. (2006) showed in a phantom study that it was feasible to localize the seeds with an acceptable accuracy using T1-weighted images, while localization was impossible using T2-weighted images. They found that the seed localization accuracy increased when the slice thickness increased from 3 to 5 mm. Bloch et al. (2007) performed a study on 13 patients where they compared enumeration and localization of seeds between contrast-enhanced T1-weighted 3D MRI (CEMR), T2-weighted MRI, and CT. The results showed that using CEMR in combination with T2-weighted MR images was more accurate than using CT or T2-weighted MRI alone (Bloch et al. 2007).

14.5.3 Clinical Application of MRI: Gynecological Imaging

MRI is well established as a diagnostic image modality for cervical cancer and appears to discriminate soft tissue and tumor better than CT imaging (Bipat et al. 2003; Hricak et al. 1988; Subak et al. 1995). This means that the CTV and the critical anatomic structure can be appropriately defined. Concepts for image guided brachytherapy (IGBT) in cervical cancer have been developed by the Gynecological (GYN) GEC-ESTRO Working Group, and T2-weighted MRI is the preferred modality (Haie-Meder et al. 2005; Pötter et al. 2006). The number of centers worldwide that are implementing this technique clinically is increasing, and several institutional reports have been published (Pötter et al. 2007; Lindegaard et al. 2008; De Brabandere et al. 2008; Chargari et al. 2009; Jurgenliemk-Schulz et al. 2009).

In IGBT, the treatment planning is based upon an image series with the applicator in situ at the time of brachytherapy. The gross tumor volume (GTV) on these images obtained after external beam radiation includes tumor mass with high signal intensity on T2-weighted images. Gray zones of intermediate signal intensity in the parametria represent pathological tissue, which topographically corresponds to tumor extension at diagnosis. These "gray zones" are assumed to be at high risk for residual cancer and should be included in the high-risk-CTV (Haie-Meder et al. 2005). The definition of "gray zones" is challenging and difficult even for an experienced radiologist or radiotherapist. Therefore, Dimopoulos et al. (2006) published a way of systematic interpretation and quantification of MR findings.

The use of functional MRI might guide the definition of target volume in cervical cancer brachytherapy in the future. Mayr et al. (2010) found that patients with persistently low dynamic contrast enhancement throughout the first part of the radiotherapy had a higher risk of treatment failure. They conclude that these findings likely reflect reoxygenation and that this may have potential for image-guided radiotherapy. Haack et al. (2010) used DWI and found that the apparent diffusion coefficient was significantly different in Gross tumor volume (GTV), High risk clinical target volume (HR-CTV), and Intermediate risk clinical target volume (IR-CTV) in 15 cervical cancer patients. They conclude, however, that to evaluate the role of DWI in target contouring, further studies are needed.

The applicator reconstruction should ideally be performed in the same image series as the contouring, that is, T2-weighted images (Hellebust et al. 2010). In intracavitary gynecological brachytherapy, usually solid plastic applicators are used. The applicator and its lumen appear as a black area in images, and it is impossible to visualize the source path without using an MR marker string (Hellebust et al. 2010). The marker string has to be produced from a catheter containing a fluid material, for example, a $CuSo_4$ solution (Haack et al. 2009; Bengtsson et al. 2008), water (Pérez-Calatayud et al. 2009), or glycerine (Chargari et al. 2009). The challenge is, however, that many applicators have a narrow entrance diameter limiting the volume of the fluid, and thus limiting the signal in T2-weighted images. When a combined intracavitary/interstitial applicator is used, the unused

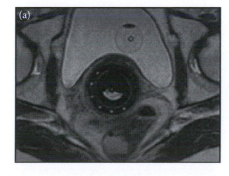 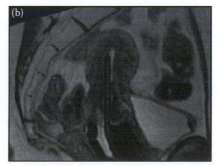

FIGURE 14.6　Para-transversal (a) and para-sagittal (b) views of Vienna ring applicator set with three plastic needles. T2-weighted 3D sequence with isotropic voxel size (1 mm) is used. Catheters with $CuSo_4$ solution are positioned inside the ring and the tandem applicator to visualize the source path. The needles are positioned in the upper left holes of the ring. The wholes without needles are clearly visible. (Courtesy of T.P. Hellebust.)

needle holes will be filled by fluid (Figure 14.6) and can be used as reference structures for applicator reconstruction as long as the locations of these holes in relation to the dwell positions are known (Berger et al. 2009). Modern TPS often contains an applicator library where the position of the source path is defined as well as the physical outer dimensions of the rigid applicator. Since the applicator appears as a black area, the applicator file can be imported and rotated/translated until it fits these black areas. Such a procedure is fast and simple, and there will be a very small risk of introducing incorrect applicator geometry (Hellebust et al. 2010).

The choice of MR sequence is important for the accuracy of the applicator reconstruction. Para-transversal images (aligned according to the applicator) should preferably be used if the TPS accepts this (Hellebust et al. 2010). A small slice thickness is recommended (≤5 mm), but too long acquisition times should be avoided in order to minimize the patient movements (Hellebust et al. 2010). 3D sequence with isotropic voxel size gives an excellent visualization of both the applicator and the anatomy (Petric et al. 2008). Figure 14.6 shows an example of such acquisition.

14.6 Positron Emission Tomography

14.6.1 Physics of PET

Positron emission tomography (PET) is an imaging technique where functional processes in the body are imaged (Phelps 2004). A positron-emitting radionuclide (tracer) is introduced into the body, and the imaging device detects pairs of gamma rays produced from annihilation of the emitted positrons. The tracer is chemically incorporated into a biologically active molecule, and it is important that this incorporation does not change the biological properties of the molecule in question. By removing a small group of atoms and replacing it with an isotope with similar atomic weight and electron structure, such a requirement is fulfilled. The emitted positron will interact with an electron in an annihilation process, and a pair of gamma rays is produced. Most of the electron–positron interactions result in two 511-keV gamma photons being emitted at almost 180° to each other. The gamma rays are detected by scintillation detectors organized in a circular design surrounding the patient. By detecting two 511-keV photons at 180° (within a short time span), it is possible to localize their source of origin along a straight line. This line is called a line of response (LOR), and the two detected events are referred to as coincidence events. Under ideal circumstances, the photons hit the detectors simultaneously, but this is not always the case. Therefore, a time window is needed. The time window to test for coincidence depends on the scintillator material used and is typically in the order of a few nanoseconds (Phelps 2004). Using a too wide window will lead to an increased chance of recording a random coincidence.

Photons originating from deep in the body must traverse more tissues compared with photons originating close to the body surface. Consequently, the photons are attenuated differently. This means that structures deep in the body will be reconstructed as having false lower tracer uptake. Therefore, an attenuation correction should be performed.

The assumption is made that the positron annihilates at the site of the biologically active molecule. This is not exactly true since the emitted positron will have a component of kinetic energy. The positron is therefore able to travel for a certain distance before it interacts with an electron. The distance will depend on the radionuclide used, but is typically less than 1 mm (Phelps 2004). This effect causes image blurring and limits the spatial resolution that could theoretically be possible using PET imaging.

Since both the positron and the electron have some kinetic energy, the scatter angle between the resulting photons is not exactly 180°. It will be a roughly Gaussian-shaped distribution around a 180° angle. This means that the true origin of the interaction is not localized on a straight line between the coincidence events. The blurring due to this effect depends on the diameter of the PET scanner (Phelps 2004).

When the gamma ray is deposited in the scintillator, a light is generated that is proportional to the deposited energy. The light is channeled toward a photomultiplier tube (PMT) that transforms it to a proportionally amplified electrical signal. Today, usually a block detector is used, and this design was originally proposed by Casey and Nutt (1986). A relatively large rectangular or square block of the scintillator detector (typically 4 × 4 cm in area by 3 cm deep) is cut into multiple smaller detector elements (typically 8 × 8), and the scintillator block is connected to four PMTs (Phelps 2004). The cuts are filled with white-reflecting material to optically isolate the elements. By using different depths of the cuts, each detector element will produce a unique distribution of light that can be detected by the PMTs. Such a design allows for the use of a reduced number of PMTs compared to a nonblock detector design.

The raw data from a scan are usually stored in a 2D matrix where each element reflects the number of coincidence events along one specific LOR given by a particular angle and a radial offset from the center of the scanner. Such a 2D matrix is called a sinogram since a point source will appear as a sinusoidal path in the matrix. When a sinogram is registered during a patient scan, an image can be reconstructed using a back-projection algorithm (Phelps 2004). However, the most frequently used method today is the iterative reconstruction method. This method starts with an initial guess of the image distribution, and the corresponding projection data are calculated. These data are compared with the actual measured projection data; the initial guess will be adjusted accordingly, and the whole process is repeated until the estimated image converges toward the true image. This latter method will usually lead to better image quality than the (filtered) back-projection method (Phelps 2004).

The most widely used oncology tracer is ^{18}F-fluoro-deoxyglucose ([18]FDG). This tracer is a glucose analog and will be phosphorylated by hexokinase in the cells. Glucose hypermetabolism may indicate malignant tissue, and FDG-PET could therefore be used to differentiate active tumors from, for example, edema. However, FDG-PET has some limitations with regard to distinguishing tumors from inflammation (Morris et al. 2002) and necrotic tissue, especially in lower-grade, less metabolic tumors. Other tracers are currently being investigated clinically. ^{11}C-methionine will differentiate tumor from normal tissue due to elevated protein synthesis (Jager et al. 2001). This tracer has a half-time of only 20 min, and this means that it can only be used in facilities close to a cyclotron where the tracer is produced.

The standardized uptake value (SUV) is a semiquantitative measure of describing PET data. SUV in its simplest form is the specific activity in a given region divided by the total injected activity per body mass (Phelps 2004). If the tracer is uniformly distributed in the body, the SUV will be equal to 1. SUV could be used to define a ROI in radiotherapy, either as a given threshold (e.g., 2.5; Hong et al. 2007; Schinagl et al. 2011) or as a percentage of the maximum SUV value (e.g., 40% or 50%; Hong et al. 2007; Schinagl et al. 2011; Ma et al. 2011). However, it is pointed out by Westerterp et al. (2007) that the SUVs depend on acquisition, reconstruction, and ROI parameters, and such parameters should be standardized in a multicenter trial setting.

In a PET scan, functional processes in the body are imaged, however, with a lack of anatomical information. Modern PET equipment is often combined with an anatomical image modality, usually CT. Since these two image modalities are integrated in the same unit, the image acquisition could be performed sequentially, allowing a rather precise image registration between the two image sets. In a PET/CT scanner, the CT detector is utilized in the attenuation correction process, and this will speed up the acquisition of the PET scan by approximately 30% (Morris et al. 2002).

A good overview of the principles in PET imaging is given in the work of Phelps (2004).

14.6.2 Clinical Application of PET: Prostate Imaging

Disappointingly low sensitivity has been reported for FDG-PET for tumor detection in primary prostate cancer (Hricak et al. 2007). One of the problems has been the streak artifacts appearing from the FDG-filled bladder if the back-projection algorithm is used. Turlakow et al. (2001) showed in a study that the iterative reconstruction method significantly improved the image quality compared to the back-projection method and may help to eliminate this problem. Another problem is that prostate cancer has a low metabolic rate since it is a slow-growing tumor with a reduced rate of glycolysis. ^{11}C-choline and ^{11}C-acetate appear as an attractive alternative (John et al. 2008). However, the application of these tracers is hampered by short half-time and, as mentioned above, can only be used in facilities close to a cyclotron where the tracer is produced.

No clinical studies have been published using brachytherapy in combination with PET imaging. However, a potential of localizing the tumor within the prostate gland with PET/CT imaging and boosting this volume with brachytherapy has been described by John et al. (2008).

14.6.3 Clinical Application of PET: Gynecological Imaging

Several groups have investigated whether FDG-PET of the primary tumor could be used as an indicator of prognosis for cervical cancer (Xue et al. 2006; Nakamura et al. 2010). Xue et al. analyzed data from 96 patients and found a significant difference in the 5-year disease-free survival for patients with maximum SUV < 10.2 compared with patients with maximum SUV ≥ 10.2 (71% versus 52%, respectively; $p = 0.0289$) within the primary tumor (Xue et al. 2006). They point out that high FDG uptake may be useful in identifying patients who might require more aggressive initial therapy. Nakamura et al. (2010) investigated the correlation between the maximal SUV of the primary tumor as well as serum squamous cell carcinoma antigen and clinicopathological characteristics and survival rates for 52 cervical cancer patients. Their results indicate that high maximal SUV plus lymph node metastasis may be associated with poor prognosis.

Studies using PET imaging in combination with gynecological brachytherapy are limited. The group at Mallinckrodt Institute of Radiology in St. Louis has published several studies where FDG-PET is used to localize the target volume in cervical cancer brachytherapy (Malyapa et al. 2002; Lin et al. 2007). Malyapa et al. (2002) compared 3D FDG-PET–based treatment planning with conventional 2D x-ray–based planning, and they concluded that FDG-PET is a feasible and accurate method for treatment planning of cervical cancer. The same group used this 3D FDG-PET method to optimize the dose distribution to the PET-defined target volume and showed that they were able to improve the target coverage without significant increase in the dose to the bladder and rectum for 11 patients (Lin et al. 2007). In a recent publication, MRI- and FDG-PET–defined GTV was compared for 47 patients with cervical cancer carcinoma (Ma et al. 2011). They found that the MR and FDG-PET tumor volumes were similar, whereas the location varied, especially for tumors <14 cm^3. It is suggested that further improvements in both imaging techniques could reduce the observed differences.

14.7 Image Registration and Fusion

14.7.1 Goal of Image Registration

Multimodal imaging plays a central role in management of external beam radiotherapy including initial diagnosis and treatment planning, as well as in follow-up (Kessler 2006). As is evident from the previous sections, multimodal imaging has, in the last 10 to 15 years, also become important for brachytherapy. Several image studies using different modalities provide complementary information in assessing the target volumes and OARs. In brachytherapy, one image modality could be optimal for visualization of the applicator, while another modality is optimal for the delineation process. The information from both modalities should be combined to utilize the potential of different image modalities. Imaging in brachytherapy is often performed prior to each fraction. All the information in these image studies acquired at different points in time should ideally be combined to assess the total treatment delivered to the patient.

To be able to combine different image series, the data from each series have to be geometrically registered to a common coordinate system; this process is formally called image registration. The process of mapping the complementary data in each study is called image fusion (Kessler 2006).

14.7.2 Techniques of Image Registration and Fusion

The image registration process usually involves three general components: the transformation model, the metric used to measure how well the images are registered, and the optimizer including the optimization scheme used to bring the imaging data into alignment (Kessler 2006).

The image conditions, the clinical site, and the particular application will be important for the type of transformation model that is chosen. The simplest model will be an identity transformation where the coordinates for all the points in the two image studies are similar, that is, the patient is positioned in an identical orientation for both image studies and the scale and center of the coordinate system coincide. This situation most closely exists in a dual image modality device such as in a PET/CT scanner or when different image orientations in an MR uptake are acquired (e.g., transversal and coronal). However, usually such ideal situations do not exist, and a more sophisticated model is needed.

If the position and orientation of the anatomy are defined in a rigid system (e.g., the brain), a simple rotate–translate model can be used. A linear transformation could be performed to map points from each dataset by specifying three rotational angles and three translations in the *x*, *y*, and *z* direction. Affine transformation model is a more general linear transformation where collinearity is preserved ("parallel lines remain parallel") (Kessler 2006). Even if it is not possible to achieve an acceptable alignment of the full anatomy using rigid or affine transformation, local rigid motion in some subvolumes could be assumed. For example, the prostate may be considered rigid even if it moves relative to the bony anatomy in the pelvis.

When an organ or a structure changes in size and shape, a nonrigid or deformable model should be used. Such models range in complexity from an affine transformation extension with a few parameters to models using a number of parameters exceeding the number of voxels considered. It is beyond the scope of this chapter to describe them in detail.

In a registration algorithm, a metric is either maximized or minimized to measure the similarity or the dissimilarity of two image series. The metric can be either geometry-based or intensity-based. The former method involves the use of points, lines, or surfaces, whereas the latter method uses the numerical grayscale information in the images. Fiducial markers or anatomical landmarks are examples of points that can be used in a geometry-based image registration. An algorithm using mutual information metric is an example of an intensity-based registration procedure (Kessler 2006).

A good overview of the principles of image registration and fusion is given in the work of Kessler (2006).

14.7.3 Clinical Application: Prostate Imaging

Post-implant dosimetry after permanent seed implantation is important to evaluate and report the quality of the implant. Currently, CT-based post-implant analysis is usually performed. However, MRI has better soft-tissue contrast compared to CT (Carey 2005; Hricak et al. 2007), and a lower interobserver reproducibility in contouring has been reported (Dubois et al. 1998). The post-implant analysis is very sensitive to uncertainties in target definitions. Polo et al. (2004) demonstrated that an expansion of the target volume of only a few millimeters would result in an 18% change in the D_{90} for a typical prostate implant. On the other hand, the seed localization procedure is easier, faster, and more accurate using CT imaging compared to MRI (De Brabandere et al. 2006). By combining these two image modalities using image fusion, the contouring and the seed localization can be based on MRI and CT imaging, respectively (Polo et al. 2004; Tanaka et al. 2006; Crook et al. 2004).

Polo et al. (2004) compared CT- and CT/MRI fusion-based post-implant dosimetry for 52 patients and reported that the latter method allows accurate determination of the prostate size and significantly improves the dosimetric evaluation. Crook et al. (2004) used MRI/CT fusion for 29 patients to evaluate the effect of prolonged edema on dosimetry after permanent prostate brachytherapy. CT/MRI fusion-based post-implant dosimetry is, however, a labor-demanding and costly procedure and might therefore not be able to be implemented in many hospitals.

Other groups suggest MR/x-ray fusion (Miquel et al. 2006), CT/TRUS fusion (Steggerda et al. 2005), and CBCT/TRUS fusion (Ng et al. 2008) for post-implant dosimetry. Reynier et al. (2004) suggest MR/TRUS data fusion for intraoperative treatment planning of permanent brachytherapy implants. They point out that the prostate apex and base are sometimes difficult to visualize in transverse TRUS images and have demonstrated that MR/TRUS fusion will influence the segmentation of these regions (Reynier et al. 2004).

14.7.4 Clinical Application: Gynecological Imaging

In image-guided gynecological brachytherapy, T2-weighted MRI is the recommended image modality for target delineation (Haie-Meder et al. 2005; Pötter et al. 2006). The applicator reconstruction procedure is feasible but challenging using T2-weighted MRI, especially with interstitial implants. Haack et al. (2009) showed that the largest uncertainty was found for titanium applicators in the longitudinal direction, most probably due to difficulties in interpretation of susceptibility artifacts. They suggest acquiring an additional CT scan to increase the precision of defining the end of the applicators, especially for free oblique needles. Figure 14.7 shows an example of a geometry-based image registration of MR and CT images. From

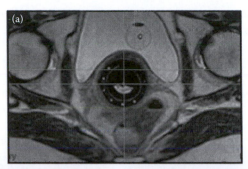

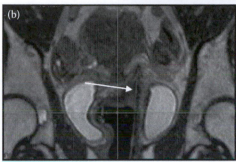

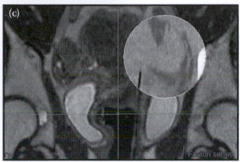

FIGURE 14.7 Para-transversal (a) and para-coronal (b) and (c) views of Vienna ring applicator set with three plastic needles. T2-weighted 3D sequence with isotropic voxel size (1 mm) is used. It is difficult to define the tip of the needles (arrow) using MR images. In image (c), CT information is shown inside the circle. The tip of the needle is easily defined using this information. Geometry-based image registration is used with two needle holes and the tip of the tandem as landmarks. (Courtesy of T.P. Hellebust.)

these images, it is evident that it is much easier to define the tip of the needles using CT images compared to MR images.

Ma et al. (2011) manually registered MR and PET/CT image studies using pelvic and vertebral bony anatomy. They found that the target volume deviated between these two image modalities for small tumors (<14 cm³).

Kirisits et al. (2006) showed that the use of only one treatment plan for several applications resulted in higher dose to the target and OARs. They suggested that individual treatment planning should be performed for every application. In many hospitals, it is logically challenging to perform MRI prior to all brachytherapy treatments. Usually, CT imaging is a less expensive and readily available alternative. A compromise would therefore be to perform MRI prior to the first brachytherapy fraction and CT for the subsequent fractions. To guide the target delineation in the CT-based fractions, an image registration between the MR and CT images could be performed using well-defined points in the applicator as landmarks. Such an approach requires that the same applicators are used for subsequent fractions and that the fractions are not too much separated in time.

14.8 Quality Control for Imaging in Brachytherapy

14.8.1 Introduction

Several reports have been published to provide guidance on quality control procedures and setting up quality assurance (QA) programs for medical devices used for imaging in radiation oncology. Only little specific attention was paid to its use in brachytherapy. Another observation that we make is that only recently the QA of imaging devices has become a part of the training programs of the medical physicist with specialty in radiation oncology. The AAPM Report 66 on the CT simulation process (Mutic et al. 2003) and a more recent overview paper by Cormack (2008) on the QA issues of CT-, US-, and MRI-guided brachytherapy are useful references for the development of a departmental QA program with emphasis on brachytherapy.

14.8.2 Modality-Specific Concerns

14.8.2.1 Radiographic Imaging

In brachytherapy, radiographic imaging is mainly used for reconstruction purposes. This means that the geometric accuracy of the radiography devices has to be checked carefully.

Roué et al. (2006) presented a European study in which measurements were performed with the Baltas phantom for 152 brachytherapy reconstructions according to clinically used imaging protocols. The use of this so-called Baltas phantom is further discussed in Chapter 31 in the frame of quality audits. Using this phantom, distances between marker points were analyzed of which 86% were found to be in the acceptance level, whereas 14% showed mean deviations greater than 1 mm. The deviations in the reconstructions checked in this study included uncertainties of the x-ray imaging and the reconstruction algorithm of the treatment planning software. For radiographic systems, the cause of all deviations was consistently shown to be related to the use of erroneous user-definable magnification factors in the software. Once detected, these could easily be corrected.

When using imaging with an image intensifier, the spatial distortion of the image should be taken into account. Due to the projection of the x-rays on the curved image intensifier surface, electron optics, influences of (earth) magnetic fields, and the video component, images are distorted. (Cosby and Leszczynski 1998) addressed this topic in their publication. These authors presented mean and maximum residual errors (measured in the plane of the isocenter) of 0.4 and 1.0 mm, respectively, after the correction of the images.

An overview on basic quality control checks of simulators used in radiotherapy was published in the IAEA Report 2008 and the AAPM TG-66 Report (Mutic et al. 2003). For brachytherapy, relevant parts of the described procedures, as well as frequencies and tolerance levels, are depicted in Table 14.1.

14.8.2.2 Computed Tomography

CT provides good spatial resolution, as well as the ability to discern objects on the basis of the electron density. The reconstruction of the CT when used for brachytherapy can be checked in the commissioning phase. The experiences reported by Roué et al. (2006) using the Baltas phantom on CT reconstructions were very favorable with no systematic errors detected.

Some users may wish to employ thin metal marker wires to enhance visualization of implanted catheters in the CT images. This can be checked in a phantom experiment in order to have a compromise between the artifacts of the metal wire and the enhanced visibility. It is specifically noted that the introduction of such devices in the imaging volume can cause significant artifacts that will hamper the ability to discern the anatomy of interest.

The use of applicator libraries in modern brachytherapy TPSs can be checked, and this should be performed in similar phantom experiments; moreover, the accurate reconstruction of each department-specific set of applicators should be validated.

It has been shown in the literature that seed models may have different appearance on CT imaging. This can affect seed-finding algorithms as was shown in a study by Siebert et al. (2007b). It seems advantageous to validate the imaging of a given seed type before using it clinically. For example, scanning several dummy seeds positioned in a gel phantom may help in getting a proper feeling for the seed imaging using CT scanning and, moreover, facilitates the test of any seed-finding software.

Electron density has, until now, not been a real issue in brachytherapy dosimetry, as dose calculation algorithms have ignored tissue inhomogeneity corrections, for example, in the most used TG-43 formalism. Application of any new model-based dose calculation algorithms, however, will require that the user identifies the conversion of Hounsfield numbers of the CT scanner into electron densities. The quality control procedure is then identical as required with external beam treatment planning. Still, the required accuracy of Hounsfield units for model-based dose calculation algorithms in brachytherapy is under investigation.

Basic CT quality control checks were published in the AAPM Task Group Report 66 (Mutic et al. 2003). In Table 14.2, relevant CT checks are summarized.

14.8.2.3 Ultrasonography

US is mainly used for LDR or HDR prostate techniques in brachytherapy. The number of vendors of TRUS equipment worldwide is rather limited, but there are—in contrast—a huge variety of vendors of the other parts of the system used in (perineal prostate) implant procedures, such as the stepper, implant needles, templates, and radioactive seeds. The manufacturing details influence largely the US visibility of needles and seeds, and this may have an effect on the placement of the seeds or the temporary implanted needles for HDR brachytherapy. Therefore, it is recommended to include in a QA program the evaluation of the coincidence between the user's image-based planned needle or source placement and the measurement of the reality of the source or needle placement. Manual image acquisition is often used to generate a 3D US dataset. Volumetric images for TRUS-guided implants are generally produced with the aid of a manual stepper and data import. Offsets and accurate capture of the template coordinates (scaling) need to be checked on a regular basis using a system-related calibration protocol, for example, performed quarterly and at any repair.

TABLE 14.1 Quality Checks of Simulators

Check	Frequency	Tolerance
Gantry/collimator angle indicators	Monthly	1°
Cross-hair centering	Monthly	2 mm
Fluoroscopic image quality	Monthly	baseline
Collimator rotation isocenter	Annually	2-mm diameter
Gantry rotation isocenter	Annually	3-mm diameter

Source: International Atomic Energy Agency (IAEA). 2008. *Setting Up a Radiotherapy Programme: Clinical, Medical Physics, Radiation Protection and Safety Aspects.* Vienna. With permission.

TABLE 14.2 Overview of Tests for CT for Use in Brachytherapy

Check	Frequency	Tolerance
Table vertical and longitudinal motion	Monthly	±1 mm over the range of table motion
Table indexing and position	Annually	±1 mm over the scan range
CT number accuracy	Monthly	For water, 0±5 HU (*)
Electron density to CT number conversion	Annually	Consistent with commissioning results
Visibility of clinically applied seeds, tubes, catheters	At commissioning	
Reconstruction of applicators	At commissioning	

Note: Most of the checks taken from Mutic et al. (2003).

*This value is for external beam planning algorithms.

The 3D data are susceptible to geometric errors arising from data entry errors. The US probe is sometimes repositioned in the procedure, and it follows that a new stepping process may result in each image being acquired as slightly different in the patient anatomy as the previous one. Equipment manufacturers have attempted to reduce this source of error by automated readout of the stepper location or automated control of the US beam using either electronic or mechanical methods.

In 2008, the AAPM TG-128 Report (Pfeiffer et al. 2008) was published on QA tests for prostate brachytherapy US systems. In this report, several tests are described in detail to check the TRUS device with usage of phantoms. Some of the QA checks from this report are included in Table 14.3. Mutic et al. (2000) introduced a simple technique how to test the alignment of the implant needles with the template coordinates achieving a reproducibility of better than 1 mm. A volume test can be used to compare a scanned well-known phantom volume with the volume reconstructed in the TPS. Discrepancies should be smaller than 5% of the volume.

In prostate treatment, the TRUS is normally used with a stepping unit. It moves the probe in the craniocaudal direction and, for biplane probes, supports rotations. Movements (back, forth, and rotations) are detected electronically and transferred to the TPS. Hence, it is of great importance to check the tracking stability of the stepper and the readouts. In the longitudinal direction, for example, four clicks back with the stepper should result in a distance of 2 cm in the TPS if every click moves the probe 5 mm. Probe rotations can be checked by using template grid points: the angle between two grid points and the center of the TRUS probe can be calculated manually. This value is then compared against the digital readout when rotating the probe from one grid point to the other. Longitudinal tracking accuracy should be within 0.5 mm, and the rotational accuracy should be within 0.5°. When using biplane probes, the relation between the two image planes must be taken into account. Due to the fact that the transversal and longitudinal transducer array of the TRUS probe is not exactly on the same place on the probe, an offset between the two viewing planes sometimes exists. This can be checked in a water tank, such as the one described in the work of Siebert et al. (2009). According to the knowledge of the authors, this offset can also vary slightly between TRUS probes of the same model.

14.8.2.4 Magnetic Resonance Imaging

The use of certain metal devices in an MRI environment requires that the devices be safe in the strong magnetic field at the site of the scanner. Equipment should be cleared for use in the MRI scanner, and if any doubts exist, this should be checked. Additionally, the equipment must be MRI compatible. Equipment that is safe can still produce large artifacts on the images and can even obscure the entire imaging volume. MRI may suffer from geometric distortions, but this is usually minimal with modern scanner types and especially rather limited in the central part of the scan volume. Distortions can be checked with a dedicated phantom with known dimensions or a phantom containing a known grid or pattern of holes, often as part of the regular service and maintenance procedure by the vendor (Jackson et al. 2010). For a test, images should be analyzed in all three main viewing planes. The medical physicist should discuss with the service engineers the importance of minimizing the geometric distortions. Different pulse sequences in MRI can be used to enhance certain anatomic features.

Checking the slice thickness ensures a constant spatial resolution because slice thickness variations result in partial volume effects that degrade the spatial resolution. Another important issue is the spatial relation between the image artifact and applicator location, as well as any shifts in artifact location that arise from changes in pulse sequences. Reconstruction of gynecological applicators in MRI is, for example, described in the recommendations of GEC-ESTRO (Hellebust et al. 2010). When using MRI for LDR postplanning, phantom investigations help in assessing the visualization of the sources and in optimizing the sequences. De Brabandere et al. (2006) have studied this in great detail. In their gel phantom, seed visualization was very similar to that in patients.

Table 14.4 contains some checks that could be performed when using MRI for brachytherapy.

TABLE 14.3 Tests for Quality Control of TRUS Including Frequency and Tolerances

Check	Frequency	Tolerance
Needle template alignment	Annually	<3 mm
Volume test	Annually	<5%
Longitudinal tracking	Quarterly	<0.5 mm
Rotational tracking	Quarterly	<0.5°
Offset between longitudinal and transversal view	Annually	<1 mm

Note: First two tests taken from Pfeiffer et al. (2008); the rest is taken from the authors (personal communications).

TABLE 14.4 Brachytherapy-Specific Checks of MRI

Check	Frequency	Tolerance
Image distortion in all three orthogonal main viewing planes		<2%
Slice thickness test		Within 10% for SE sequences for 5 mm or thicker slices
Check MRI safety of applicator materials	Before first use	
Check of visibility of the clinically applied seeds, tubes, and/or catheters, marker wires	Before first use	
Check of reconstruction of the applicator according to library definition (differences in assumed library data and reality; if not done so for CT)	Before first use	

Note: Details of tests can be found in the recent AAPM update report by Jackson et al. (2010).

14.8.3 Multimodal Image Registration to Intraoperative Imaging

It is generally accepted that MRI is the best modality to image the soft-tissue structures of the pelvis, and thus for both prostate implants and gynecologic brachytherapy. Intraoperative imaging using MRI is very uncommon due to the limited availability of, for example, an open MRI system at the brachytherapy department. At best, recent preoperative MR images are made available, which are used to guide the actual implant procedure in which operative imaging is performed using a different modality. The placement of seeds, applicators, or needles is then guided by suboptimal imaging, but this is still the best achievable method nowadays. This naturally leads to attempts to fuse the high-quality pre-procedure diagnostic image sets with the lower-quality images used to guide the intervention. In many cases, the fusion will be performed with the preoperative MR images on a post-implant CT image set. If such fusion is desired, this implies that some features of concern will not be well visible on the CT images, and registration must be done using other features visible on both datasets.

The evaluation of registration quality is generally visual and 3D, and, as such, it is much more complex than the choice of reference plane in TRUS-guided implants, with no objective criteria for the evaluation of quality. Registrations in the brachytherapy setting are subject to all the problems previously indicated in the Section 14.7.2 on tissue deformation. Because registrations might have a significant effect on structure definition, a method to evaluate the consequences of registration uncertainty on the treatment plan should be available. Evaluation of uncertainties in image fusion is not easy for patient data because the "true" fusion is unknown. Maybe it does not even exist due to the deformations or anatomical changes between the time of data acquisitions.

Analysis of image fusion can be performed in two steps: first, evaluation using a phantom is useful. This shows the uncertainties based on technical deficiencies. Moreover, errors like swapping of data can easily be detected. In a second step, fusion of real patient data can be analyzed. Results of this step can be compared against a "standard," and interobserver and intraobserver uncertainties can be analyzed. A study that compared MRI and CT image fusion for LDR postplanning purposes was presented by De Brabandere et al. (2011). In this evaluation, it was found that contouring and fusion seem to be the "weak links" in the postplanning procedure. Image fusions of CT and T2-weighted MRI datasets showed a mean interobserver error of 16% of the $D_{prostate90}$. The reason for this quite large variability can be the lack of appropriate landmarks in the T2 images. Sources give no signal in MRI but appear as voids with expanded dimensions. In the future, automated fusion software may help in reducing the interobserver variability.

Although brachytherapy was not part of the discussion, the publication of Sharpe et al. (2008) forms a recommended further reading for the interested medical physicist.

14.9 Conclusions

This chapter on imaging in brachytherapy shows the possibilities and limits of modern 3D visualization for brachytherapy treatment planning. At the moment, there is no gold standard for imaging—the imaging modalities all have advantages and disadvantages. Usage of more than one modality by image fusion is a further promising approach. QA is an important factor in brachytherapy imaging and helps in maintaining quality and avoid errors in treatment planning. Novel dummy markers could help reduce the challenge of reconstructing source locations and provide a framework for imbedding additional information in the images. Additionally, improving the resolution of scanners could reduce the uncertainty in position resulting from slice thickness. More image-friendly applicators are essential. A limited number are currently available, leaving much room for improvement. Shielding is generally incompatible with imaging. Providing a method to add shielding after the placement and imaging of an applicator would be an obvious improvement in the applicators. Independent checks of crucial inputs are a vital part of any QA process. An RF localizer moving through the applicator or needles could provide an independent verification of the implant's geometry and offer a way to measure any image distortion that the applicator creates in the image.

Improvements in the importing tools and displays of planning systems could greatly reduce the simplest data entry errors. Moreover, objective contouring tools could help reduce interuser variability and provide a rational framework for correlation of dosimetric quantities with outcome data. Further developments of automated fusion algorithms are of interest: they will minimize fusion uncertainties, and, when using novel nonrigid fusion methods, they could improve the overall quality of 3D image fusion.

References

Ares, C., Popowski, Y., Pampallona, S. et al. 2009. Hypofractionated boost with high-dose-rate brachytherapy and open magnetic resonance imaging-guided implants for locally aggressive prostate cancer: A sequential dose-escalation pilot study. *Int J Radiat Oncol Biol Phys* 75:656–63.

Aristei, C., Tarducci, R., Palumbo, I. et al. 2009. Computed tomography for excision cavity localization and 3D-treatment planning in partial breast irradiation with high-dose-rate interstitial brachytherapy. *Radiother Oncol* 90:43–7.

Bengtsson, E., Vargas, R., Nordell, B., and Lundell, M. 2008. Markers to be used with MR compatible gynaecological brachytherapy applicators. *Radiother Oncol* 88(S2):S427.

Berger, D., Dimopoulos, J., Pötter, R., and Kirisits, C. 2009. Direct reconstruction of the Vienna applicator on MR images. *Radiother Oncol* 93:347–51.

Bipat, S., Glas, A.S., van der Velden, J. et al. 2003. Computed tomography and magnetic resonance imaging in staging of uterine carcinoma: a systematic review. *Gynecol Oncol* 91:59–66.

Bitar, R., Leung, G., Perng, R. et al. 2006. MR Pulse Sequences: What every radiologist wants to know but is afraid to ask. *RadioGraphics* 26:513–37.

Bloch, B.N., Lenkinski, R.E., Helbich, T.H. et al. 2007. Prostate postbrachytherapy seed distribution: comparison of high-resolution, contrast-enhanced, T1- and T2-weighted endorectal magnetic resonance imaging versus computed tomography: initial experience. *Int J Radiat Oncol Biol Phys* 69:70–8.

Bushong, S.C. 1996. *Magnetic Resonance Imaging. Physical and Biological Principles*, 2nd ed. Mosby, St. Louis, USA.

Carey, B.M. 2005. Imaging for prostate cancer. *Clin Oncol* 17: 553–9.

Casey, M.E. and Nutt, R. 1986. A multicrystal two dimensional BGO detector system for positron emission tomography. *IEEE Trans Nucl Sci* 33:460–3.

Chargari, C., Magne, N. Dumas, I. et al. 2009. Physics contribution and clinical outcome with 3D-MRI-based pulsed dose rate intracavitary brachytherapy in cervical cancer patients. *Int J Radiat Oncol Biol Phys* 74:133–9.

Choi, Y.J., Kim, J.K., Kim, N. et al. 2007. Functional MR imaging of prostate cancer. *RadioGraphics* 27:63–77.

Cormack, R.A. 2008. Quality assurance issues for computed tomography-, ultrasound-, and magnetic resonance imaging-guided brachytherapy. *Int J Radiat Oncol Biol Phys* 71(1 Suppl):S136–41.

Cosby, N.S., Leszczynski, K.W. 1998. Computer-aided radiation therapy simulation: image intensifier spatial distortion correction for large field of view digital fluoroscopy. *Phys Med Biol* 43: 2265–78.

Crook, J., McLean, M., Yeung, I., Williams, T., and Lockwood, G. 2004. MRI-CT fusion to assess postbrachytherapy prostate volume and the effects of prolonged edema on dosimetry following transperineal interstitial permanent prostate brachytherapy. *Brachytherapy* 3:55–60.

D'Amico, A.V., Cormack, R., Tempany, C.M. et al. 1998. Real-time magnetic resonance image-guided interstitial brachytherapy in the treatment of selected patients with clinically localized prostate cancer. *Int J Radiat Oncol Biol Phys* 42: 507–15.

Das, R., Patel, R., Shah, H. et al. 2004. 3D CT–Based High-Dose-Rate Breast Brachytherapy Implants: Treatment Planning and Quality Assurance. *Int J Radiat Oncol Biol Phys* 59:1224–8.

Davidson, M.T.M., Yuen, J., D'Souza, D.P. et al. 2008. Optimization of high-dose-rate cervix brachytherapy applicator placement: The benefits of intraoperative ultrasound guidance. *Brachytherapy* 7:248–53.

De Brabandere, M., Kirisits, C., Petters, R., Haustermans, K., and Van den Heuvel, F. 2006. Accuracy of seed reconstruction in prostate postplanning studied with a CT- and MRI-compatible phantom. *Radiother Oncol* 79:190–7.

De Brabandere, M., Mousa, A.G., Nulens, A., Swinnen, A., and Van Limbergen, E. 2008. Potential of dose optimization on MRI-based PDR brachytherapy of cervix carcinoma. *Radiother Oncol* 88:217–26.

De Brabandere, M., Haustermans, K., van den Heuvel, F. et al. 2011. Prostate post-implant dosimetry: Interobserver variability in seed localisation, contouring and fusion. *Radiother Oncol* 99(Supplement 1):S86–7.

De Graaf, R.A. 2002. *In Vivo NMR Spectroscopy. Principles and Techniques*. John Wiley & Sons Ltd, West Sussex, England.

Dimopoulos, J.C., Schard, G., Berger, D. et al. 2006. Systematic evaluation of MRI findings in different stages of treatment of cervical cancer: Potential of MRI on delineation of target, pathoanatomic structures, and organs at risk. *Int J Radiat Oncol Biol Phys* 64:1380–8.

Dubois, D.F., Prestige, B.R., Hotchkiss, L.A. et al. 1998. Intraobserver and interobserver variability of MR imaging- and CT-derived prostate volumes after transperineal interstitial permanent prostate brachytherapy. *Radiology* 207:785–9.

Edmundson, G.K., Vicini, F.A., and Chen, P.Y. 2002 Dosimetric Characteristics of the Mammosite RTS, a new Breast Brachytherapy Applicator. *Int J Radiat Oncol Biol Phys* 52:1132–9.

Esthappan, J., Santanam, L., Yang, D. et al. 2009. Use of serial CT imaging for the quality assurance of MammoSite therapy. *Brachytherapy* 8:379–84.

Fellner, C., Pötter, R., Knocke, T.H. et al. 2001. Comparison of radiography- and computed tomography-based treatment planning in cervix cancer in brachytherapy with specific attention to some quality assurance aspects. *Radiother Oncol* 58:53–62.

Fransson, A., Andreo, P., and Pötter, R. 2001. Aspects of MR image distortions in radiotherapy treatment planning. *Strahlenther Onkol* 177:59–73.

Fung, A.Y.C. 2002. C-Arm imaging for brachytherapy source reconstruction: Geometrical accuracy. *Med Phys* 29:724–6.

Gerbaulet, A., Pötter, R, Mazeron, J.-J., Meertens, H., and Van Limbergen, E. 2002. *GEC ESTRO Handbook of Brachytherapy*. A. Gerbaulet, R. Pötter, and J.J. Mazeron (Eds.). ESTRO, Brussels.

Haack, S., Nielsen, S.K., Lindegaard, J.C., Gelineck, J., and Tanderup, K. 2009. Applicator reconstruction in MRI 3D image-based dose planning of brachytherapy for cervical cancer. *Radiother Oncol* 91:187–93.

Haack, S., Pedersen, E.M., Jespersen, S.N. et al. 2010. Apparent diffusion coefficient in GEC ESTRO target volumes for image guided adaptive brachytherapy of locally advanced cervical cancer. *Acta Oncol* 49:978–83.

Haie-Meder, C., Pötter, R., Van Limbergen, E. et al. 2005. Recommendations from Gynaecological (GYN) GEC-ESTRO Working Group (I): Concepts and terms in 3D imaging based 3D treatment planning in cervix cancer brachytherapy with emphasis on MRI assessment of GTV and CTV. *Radiother Oncol* 74:235–45.

Han, B.H., Wallner, K., Merrick, G. et al. 2003. Prostate brachytherapy seed identification on post-implant TRUS images. *Med Phys* 30:898–900.

Hashemi, R.H., Bradley, W.G., Lisanti, C.J. 2010. In: *MRI: The Basics*, 3rd ed. Lippincott Williams & Wilkins, Philadelphia April 6, 2010.

Hellebust, T.P., Tanderup, K., Bergstrand, E.S. et al. 2007. Reconstruction of a ring applicator using CT imaging: Impact of the reconstruction method and applicator orientation. *Phys Med Biol* 52:4893–904.

Hellebust, T.P., Kirisits, C., Berger, D. et al. 2010. Recommendations from Gynaecological (GYN) GEC-ESTRO Working group: Considerations and pitfalls in commissioning and applicator reconstruction in 3D image-based treatment planning of cervix cancer brachytherapy. *Radiother Oncol* 96:153–60.

Herman, G.T. 2009. *Advances in Computer Vision and Pattern Recognition,* 2nd ed. Springer, London. September 24, 2009.

Holm, H.H., Juul, N., Pedersen, J.F. et al. 1983. Transperineal 125iodine seed implantation in prostatic cancer guided by transrectal ultrasonography. *J Urol* 130:283–6.

Hong, R., Halama, J., Bova, D., Sethi, A., and Emami, B. 2007. Correlation of PET standard uptake value and CT window-level thresholds for target delineation in CT-based radiation treatment planning. *Int J Radiat Oncol Biol Phys* 67:720–6.

Hoskin, P.J. 2000. High dose rate brachytherapy boost treatment in radical radiotherapy for prostate cancer. *Radiother Oncol* 57:285–8.

Hoskin, P.J. Bownes, P.J., Ostler, P. et al. 2003. High dose rate afterloading brachytherapy for prostate cancer: catheter and gland movement between fractions. *Radiother Oncol* 68: 285–8.

Hoskin, P.J., Carnell, D.M., Taylor, N.J. et al. 2007. Hypoxia in prostate cancer: correlation of BOLD-MRI with pimonidazole immunohistochemistry—Initial observations. *Int J Radiat Oncol Biol Phys* 68:1065–71.

Hricak, H., Lacey, C., Sandles, L. et al. 1988. Invasive cervical carcinoma: Comparison of MR imaging and surgical findings. *Radiology* 166:623–31.

Hricak, H., Choyke, P.L., Eberhardt, S.C., Leibel, S.A., and Scardino, P.T. 2007. Imaging in prostate cancer: A multidisciplinary perspective. *Radiology* 243:28–53.

Hussain, S.M., van der Bos, I.C., Oliveto, J.M. et al. 2006. MR imaging of the female pelvis at 3T. *Magn Reson Imaging Clin N Am* 14:537–44, vii.

International Atomic Energy Agency (IAEA). 2008. *Setting Up a Radiotherapy Programme: Clinical, Medical Physics, Radiation Protection and Safety Aspects.* IAEA, Vienna.

Jackson, E.F., Bronskill, N.J., Drost, D.J. et al. 2010. Acceptance Testing and Quality Assurance Procedures for Magnetic Resonance Imaging Facilities. Report of MR subcommittee Task Group I, AAPM Report 100.

Jager, P.L., Vaalburg, W., Pruim, J. et al. 2001. Radiolabeled amino acids: basic aspects and clinical applications in oncology. *J Nucl Med* 42:432–45.

John, S.S., Zietman, A.L., Shipley, W.U., and Harisinghani, M.G. 2008. Newer image modalities to assist with target localization in the radiation treatment of prostate cancer and possible lymph node metastases. *Int J Radiat Oncol Biol Phys* 71(Suppl):S43–7.

Jurgenliemk-Schulz, I., Tersteeg, R.J., Roesink, J.M. et al. 2009. MRI-guided treatment planning optimisation in intracavitary or combined intracavitary/interstitial PDR brachytherapy using tandem ovoid applicators in locally advanced cervical cancer. *Radiother Oncol* 93:322–30.

Kalender, W.A. 2000. *Computed Tomography: Fundamentals, System Technology, Image Quality, Applications.* Publicis Corporate Publishing, Erlangen.

Kassas, B., Mourtada, F., Horton, J.L. et al. 2004. Contrast effects on dosimetry of a partial breast irradiation system. *Med Phys* 31:1976–9.

Kataoka, M., Kido, A., Koyama, T. et al. 2007. MRI of the female pelvis at 3T compared to 1.5T: Evaluation on the high-resolution T2-weighted and HASTE images. *J Magn Reson Imaging* 25:527–34.

Kessler, M.L. 2006. Image registration and data fusion in radiation therapy. *Br J Radiol* 79(1):S99–108.

Kim R.Y., Pareek, P. 2003. Radiography-based treatment planning compared with computed tomography CT)-based treatment planning for intracavitary brachytherapy in cancer of the cervix: Analysis of dose-volume histograms. *Brachytherapy* 2:200–6.

Kim, Y., Hsu, I.C.J., Lessard, E., Kurhanewicz, J., Norworolski, S.M., and Pouliot, J. 2008. Class solution in inverse planned HDR prostate brachytherapy for dose escalation of DIL defined by combined MRI/MRSI. *Radiother Oncol* 88:148–55.

Kim, Y., Muruganandham, M., Modrick, J.M., and Bayouth, J.E. 2011. Evaluation of artefacts and distortions of titanium applicators on 3.0 Tesla MRI: Feasibility of titanium applicators in MRI-guided brachytherapy for gynaecological cancer. *Int J Radiat Oncol Biol Phys* 80:947–55.

Kirisits, C., Lang, S., Dimopoulos, J., Oechs, K., Georg, D., and Pötter, R. 2006. Uncertainties when using one MRI-based treatment plan for subsequent high-dose-rate tandem and ring applications in brachytherapy of cervix cancer. *Radiother Oncol* 81:269–75.

Kovács, G., Pötter, R., Loch, T. et al. 2005. GEC/ESTRO-EAU recommendations on temporary brachytherapy using stepping sources for localised prostate cancer. *Radiother Oncol* 74:137–48.

Kurhanewicz, J. and Vigneron, D.B. 2008. Advances in MR spectroscopy of the prostate. *Magn Reson Imaging Clin North Am* 16:697–710.

Landis, D.M., Luo, W., Song, J. et al. 2007. Variability among Breast Radiation Oncologists in Delineation of the Postsurgical Lumpectomy Cavity. *Int J Radiat Oncol Biol Phys* 67:1299–308.

Lin, L.L., Mutic, S., Low, D.A. et al. 2007. Adaptive brachytherapy treatment planning for cervical cancer using FDG-PET. *Int J Radiat Oncol Biol Phys* 67:91–6.

Lindegaard, J.C., Tanderup, K., Nielsen, S.K., Haack, S., and Gelineck, J. 2008. MR-guided 3D optimization significantly improves DVH parameters of pulsed-dose-rate brachytherapy in locally advanced cervical cancer. *Int J Radiat Oncol Biol Phys* 71:756–64.

Liu, L., Bassano, D.A., Prasad, S.C. et al. 2003. On the use of C-arm fluoroscopy for treatment planning in high dose rate brachytherapy. *Med Phys* 30:2297–302.

Luypaert, R., Boujraf, S., Sourbron, S., and Osteaux, M. 2001. Diffusion and perfusion MRI: basic physics. *Eur J Radiol* 38:19–27.

Ma, D.J., Zhu, J.M., and Grigsby, P.W. 2011. Tumor volume discrepancies between FDG-PET and MRI for cervical cancer. *Int J Radiat Oncol Biol Phys* 98:139–42.

Malyapa, R., Mutic, S., Low, D.A. et al. 2002. Physiologic FDG-PET three-dimensional brachytherapy treatment planning for cervical cancer. *Int J Radiat Oncol Biol Phys* 54:1140–6.

Mayr, N.A., Wang, J.Z., Zhang, D. et al. 2010. Longitudinal changes in tumor perfusion pattern during the radiation therapy course and its clinical impact in cervical cancer. *Int J Radiat Oncol Biol Phys* 77:502–8.

Mazeron, J.J., Ardiet, J.-M., and Haie-Méder, C. 2009. GEC-ESTRO recommendations for brachytherapy for head and neck squamous cell carcinomas. *Radiother Oncol* 91:150–6.

Ménard, C., Susil, R., Choyke, P. et al. 2004. MRI-guided HDR prostate brachytherapy in standard 1.5T scanner. *Int J Radiat Oncol Biol Phys* 59:1414–23.

Miquel, M.E., Rhode, K.S., Acher, P.L. et al. 2006. Using combined x-ray and MR imaging for prostate I-125 post-implant dosimetry: phantom validation and preliminary patient work. *Phys Med Biol* 51:1129–37.

Morris, M.J., Akhurst, T., Osman, I. et al. 2002. Fluorinated deoxyglucose positron emission tomography imaging in progressive metastatic prostate cancer. *Urology* 59:913–8.

Mutic, S., Low, D.A., Nussbaum, G.H. et al. 2000. A simple technique for alignment of perineal needle template to ultrasound image grid for permanent prostate implants. *Med Phys* 27:141–3.

Mutic, S., Palta, J.R., Butker, E.K. et al. 2003. Quality assurance for computed-tomography simulators and the computed tomography-simulation process: Report of the AAPM Radiation Therapy Committee Task Group No. 66. *Med Phys* 30:2762–92.

Nag, S., Ciezki, J.P., Cormack, R. et al. 2001. Intraoperative planning and evaluation of permanent prostate brachytherapy: report of the American Brachytherapy Society. *Int J Radiat Oncol Biol Phys* 51:1422–30.

Nakamura, K., Okumura, Y., Kodama, J., Hongo, A., Kanazawa, S., and Hiramatsu, Y. 2010. The predictive value of measurement of SUVmax and SCC-antigen in patients with pretreatment of primary squamous cell carcinoma of cervix. *Gynecol Oncol* 119:81–6.

Nath, R., Bice, W., Butler, W. et al. 2009. AAPM Recommendations on dose prescription and reporting methods for permanent interstitial brachytherapy for prostate cancer. *Med Phys* 36:5310–22.

Ng, A., Beiki-Ardakan, A., Tong, S. et al. 2008. A dual modality phantom for cone beam CT and ultrasound image fusion in prostate implant. *Med Phys* 35:2062–71.

Pérez-Calatayud, J., Kuipers, F., Ballester, F. et al. 2009. Exclusive MRI-based tandem and colpostats reconstruction in gynaecological brachytherapy treatment planning. *Radiother Oncol* 91:181–6.

Petric, P., Hudej, R., Rogelj, P., and Logar, B.Z. 2008. 3D T2-weighted fast recovery fast spin echo sequence MRI for target contouring in cervix cancer brachytherapy. *Brachytherapy* 7:109–10.

Pfeiffer, D., Sutlief, S., Feng, W. et al. 2008. AAPM Task Group 128: Quality assurance tests for prostate brachytherapy ultrasound systems. *Med Phys* 35:5471–89.

Phelps, M.E. 2004. *PET Molecular Imaging and Its Biological Applications*. Springer, New York.

Polo, A., Cattani, F., Vavassori, A. et al. 2004. MR and CT image fusion for postimplant analysis in permanent prostate seed implants. *Int J Radiat Oncol Biol Phys* 60:1572–9.

Polo, A., Salembier, C., Venselaar, J., and Hoskin, P. 2010. Review of intraoperative and planning techniques in permanent seed prostate brachytherapy. *Radiother Oncol* 94:12–23.

Pötter, R., Haie-Meder, C., Van Limbergen, E. et al. 2006. Recommendations from Gynaecological (GYN) GEC_ESTRO Working group (II): Concepts and terms in 3D imaging based 3D treatment planning in cervix cancer brachytherapy: Aspects of 3D imaging, radiation physics, radiobiology, and 3D dose volume parameters. *Radiother Oncol* 78:67–77.

Pötter, R., Dimopoulos, J.C., Georg, P. et al. 2007. Clinical impact of MRI assisted dose volume adaptation and dose escalation in brachytherapy of locally advanced cervix cancer. *Radiother Oncol* 83:148–55.

Reniers, B. and Verhaegen, F. 2011. Cone beam CT imaging for 3D image guided brachytherapy for gynecological HDR brachytherapy. *Phys Med* 38:2762–7.

Reynier, C., Troccaz, J., Fourneret, P. et al. 2004. MRI/TRUS data fusion for prostate brachytherapy. Preliminary results. *Med Phys* 31:1568–75.

Rivard, M.J., Venselaar, J.L., and Beaulieu, L. 2009. The evolution of brachytherapy treatment planning. *Med Phys* 36:2136–53.

Rosenthal, M. and Nath, R. 1983. An automated seed identification technique for interstitial implants using three isocentric radiographs. *Med Phys* 10:475–9.

Roué, A., Ferreira, I.H., Dam, J.V. et al. 2006. The EQUAL-ESTRO audit on geometric reconstruction techniques in brachytherapy. *Radiother Oncol* 78:78–83.

Salembier, C., Lavagnini, P., Nickers, P. et al. 2007. Tumour and target volumes in permanent prostate brachytherapy: A supplement to the ESTRO/EAU/EORTC recommendations on prostate brachytherapy. *Radiother Oncol* 83:3–10.

Scanderberg, D.J., Yashar, C., Rice, R. et al. 2009. Clinical implementation of a new HDR brachytherapy device for partial breast irradiation. *Radiother Oncol* 90:36–42.

Schinagl, D.A.X., Span, P.N., Oyen, W.J., and Kaanders, J.H.A.M. 2011. Can FDG PET predict radiation treatment outcome in head and neck cancer? Results of a prospective study. *Eur J Nucl Med Mol Imaging* 38:1449–58.

Sharpe, M. and Brock, K. 2008. Quality assurance of serial 3D image registration, fusion, and segmentation. *Int J Radiat Oncol Biol Phys* 71(Supplement):S33–7.

Siebert, F.A., Srivastav, A., Kliemann, L. et al. 2007a. Three-dimensional reconstruction of seed implants by randomized rounding and visual evaluation. *Med Phys* 34:967–75.

Siebert, F.A., De Brabandere M., Kirisits, C. et al. 2007b. Phantom investigations on CT seed imaging for interstitial brachytherapy. *Radiother Oncol* 85:316–23.

Siebert, F.A., Hirt, M., Niehoff, P. et al. 2009. Imaging of implant needles for real-time HDR-brachytherapy prostate treatment using biplane ultrasound transducers. *Med Phys* 36:3406–12.

Simnor, T., Li, S., Lowe, G. et al. 2009. Justification for interfraction correction of catheter movement in fractionated high dose-rate brachytherapy treatment of prostate cancer. *Radiother Oncol* 93:253–8.

Steggerda, M., Schneider, C., van Herk, M., Zijp, L., Moonen, L., and van der Poel, H. 2005. The applicability of simultaneous TRUS-CT imaging for the evaluation of prostate seed implants. *Med Phys* 32:2262–70.

Su, Y., Davis, B., Herman, M. et al. 2004. Prostate brachytherapy seed localization by analysis of multiple projections: Identifying and addressing the seed overlap problem. *Med Phys* 31:1277–87.

Subak, L.L., Hricak, H., Powell, C.B., Azizi, L., and Stern, J.L. 1995. Cervical carcinoma: Computed tomography and magnetic resonance imaging for preoperative staging (Review). *Obstet Gynecol* 86:43–50.

Tanaka, O., Hayashi, S., Matsuo, M. et al. 2006. Comparison of MRI-based and CT/MRI fusion-based postimplant dosimetric analysis of prostate brachytherapy. *Int J Radiat Oncol Biol Phys* 66:597–602.

Tanaka, O., Hayashi, S., Matsuo, M. et al. 2008. MRI-based pre-planning in low-dose-rate prostate brachytherapy. *Radiother Oncol* 88:115–20.

Turlakow, A., Larson, S.M., Coakley, F. et al. 2001. Local detection of prostate cancer by positron emission tomography with 2-fluorodeoxyglucose: Comparison of filtered back projection and iterative reconstruction with segmented attenuation correction. *Q J Nucl Med* 45:235–44.

Van der Laarse, R. and Luthman, R.W. 2001. Computers in brachytherapy dosimetry. In: *Principles and Practise of Brachytherapy Using Afterloader Systems*. C.A.F. Joslin, A. Flynn, and E.J. Hall (eds.). Arnold Publ., Cambridge, pp. 49–80.

Van Dyk, S., Narayan, K., Fischer, R. et al. 2009. Conformal brachytherapy planning for cervical cancer using transabdominal ultrasound. *Int J Radiat Oncol Biol Phys* 75:64–70.

Viswanathan, A.N., Dimopoulos, J., and Kirisits, C. 2007. Computed tomography versus magnetic resonance imaging based contouring in cervical cancer brachytherapy: Results of a prospective trial and preliminary guidelines for standardized contours. *Int J Radiat Oncol Biol Phys* 68:491–8.

Wang, B., Kwon, A., Zhu, Y. et al. 2009. Image-guided intracavitary high-dose-rate brachytherapy for cervix cancer: A single institutional experience with three-dimensional CT-based planning. *Brachytherapy* 8:240–7.

Webb, S. 1988. *The Physics of Medical Imaging. Medical Sciences Series*. Adam Hilger, IOP Publishing Ltd., Bristol and Philadelphia.

Westerterp, M., Pruim, J., Oyen, W. et al. 2007. Quantification of FDG PET studies using standardised uptake values in multi-centre trails: effects of image reconstruction, resolution and ROI definition parameters. *Eur J Nucl Med Mol Imaging* 34:392–404.

Wills, R., Lowe, G., Inchley, D. et al. 2010. Applicator reconstruction for HDR cervix treatment planning using images from 0.35 T open MR scanner. *Radiother Oncol* 94:346–52.

Wong, E.K., Truong, P.T., Kader, H.A. et al. 2006. Consistency in seroma contouring for partial breast radiotherapy: Impact of guidelines. *Int J Radiat Oncol Biol Phys* 66:372–6.

Xue, F., Lin, L.L., Dehdashti, F., Miller, T.R., Siegel, B.A., Grigsby, P.W. et al. 2006. F-18 fluorodeoxyglucose uptake in primary cervical cancer as an indicator of prognosis after radiotherapy. *Gynecol Oncol* 101:147–51.

Zahra, M.A., Hollingsworth, K.G., Sala, E., Lomas, D.J., and Tan, L.T. 2007. Dynamic contrast-enhanced MRI as a predictor of tumour response to radiotherapy. *Lancet Oncol* 8:63–74.

Zelefsky, M.J., Cohen, G., Zakain, K.L. et al. 2000. Intraoperative conformal optimization for transperineal prostate implantation using magnetic resonance spectroscopic imaging. *Cancer J* 6:249–55.

IV

Uncertainties in Clinical Brachytherapy

15

Uncertainties Associated with Brachytherapy Source Calibrations and Dose Calculations

Larry A. DeWerd
University of Wisconsin

Mark J. Rivard
Tufts University School of Medicine

Hans-Joachim Selbach
Physikalisch-Technische Bundesanstalt

15.1 Introduction: Formation of Uncertainties and Their Determination

Uncertainties as a topic have been receiving much more attention recently. The use and investigation of uncertainties in medical physics have grown to include a number of publications on uncertainty determination in radiation dosimetry for both external beam and brachytherapy applications (Mitch et al. 2009; DeWerd et al. 2011). The chapter (Mitch et al. 2009) in the 2009 *American Association of Physicists in Medicine (AAPM) Summer School* addresses this topic for high-dose-rate (HDR), pulsed-dose-rate (PDR), and low-dose-rate (LDR) sources. The recent publication (DeWerd et al. 2011) elaborates on LDR sources. For both HDR and LDR applications, the division of the topic involves the uncertainty in the source, the uncertainty in the calibration and measurement, and finally the uncertainty involved in the source parameters, which would include Monte Carlo (MC) calculations. This chapter will only deal with photon-emitting sources.

15.1.1 Calibration Uncertainties

Primary standard dosimetry laboratories (PSDLs) provide calibrations with the smallest uncertainty possible to the conventional true value, from which the uncertainty chain begins. The chain continues through the secondary laboratories and then the transfer of these uncertainties through the clinical calibration process. Separately, determination of the dosimetric parameters is performed by measurement or MC calculations. However, this chain involves the calibration of the source as well, either from a primary or a secondary laboratory. It is preferable that the parameter is traceable to a primary standard.

Reference dosimetry is, by definition, directly traceable to a primary laboratory. This traceability is usually transferred to the clinic through a primary laboratory, an accredited dosimetry calibration laboratory (ADCL) in the United States, or through a secondary standard dosimetry laboratory (SSDL). In this chapter, the propagation of uncertainty from the primary calibration standard to the ADCL and finally to the clinic is detailed according to the recommendations of the International Organization for Standardization (ISO) 17025:2005 report (Honsa and McIntyre 2003).

This chapter also addresses uncertainties pertaining to photon-emitting brachytherapy source dosimetry, determined by MC-estimated and experimentally measured values. The 2004 AAPM TG-43U1 report considered these uncertainties in a cursory manner (Rivard et al. 2004a). Before publication of the TG-43U1 report, estimates of dosimetry uncertainties for brachytherapy were limited. Most investigators using MC techniques presented only statistical uncertainties; only recently have other MC uncertainties been examined. The 2004 AAPM TG-43U1 report presented a generic uncertainty analysis specific to calculations of brachytherapy dose distributions. This analysis included dose calculations based on simulations using MC methods and experimental measurements using thermoluminescent dosimeters (TLDs). These simulation and measurement uncertainty analyses included components toward developing an uncertainty budget.

15.1.2 Methodology of Uncertainty Estimation

All measurements and calculations in radiation dosimetry have an associated uncertainty that takes into account possible factors that may affect determination of the final quantity. This uncertainty can be thought of as a defining interval or probability distribution, which is believed to contain the true value of the quantity, called the conventional true value, within a certain level of confidence. The difficult part of uncertainty analysis is to account for all possible influences on the determination of the quantity of interest.

The current approach to evaluating uncertainty in measurements is based on the recommendations of the Comité International des Poids et Mesures (CIPM) (Giacomo 1981). This brief CIPM document was expanded by an ISO working group into the *Guide to the Expression of Uncertainty in Measurement*, first published in 1993 and subsequently updated in 2010; the original version and its supplements can be found at the Bureau International des Poids et Mesures (BIPM) link (BIPM 2011). This formal method of assessing, evaluating, and reporting uncertainties in measurements was presented in a succinct fashion in NIST Technical Note 1297, *Guidelines for Evaluating and Expressing the Uncertainty of NIST Measurement Results* (Taylor and Kuyatt 1994). Other primary laboratories also have versions of uncertainty documents, for example, by the Physikalisch-Technische Bundesanstalt (PTB) (Kochsiek 2001). The basic concepts of uncertainty in radiation dosimetry are given in the work of Siebert (2006). More recently, the overall uncertainty methodology has been outlined in detail in Chapter 22 of the 2009 AAPM Summer School proceedings (Mitch et al. 2009). The relevant points of these references are summarized below.

The terminology used for uncertainty analysis categorizes the components of uncertainty according to the method used to determine them (Type A or Type B) rather than using the term *error* to describe all sources of uncertainty. Components of measurement uncertainty may be classified into two types: those evaluated by statistical methods (Type A) and those evaluated by scientific judgment (Type B). In addition, the terms *accuracy* and *precision* are still maintained but with slightly different definitions. *Accuracy* is defined as the proximity of the result to the conventional true value (albeit unknown) or a reference value and is an indication of the correctness of the result. *Precision* is defined as a measure of the reproducibility or repeatability of the result. Accuracy is the goal of the primary laboratories and precision that of the ADCL or SSDL.

The general approach for uncertainty propagation is to combine observed and estimated uncertainties at the level of one standard deviation and apply a coverage factor, k, to express differing levels of confidence in the determined value (68% at $k = 1$, 95% at $k = 2$ for a normal distribution). It is standard practice for primary and secondary calibration laboratories to report expanded uncertainties with a coverage factor of $k = 2$ per the ISO 17025:2005 report (Honsa and McIntyre 2003).

Each component of uncertainty is described by an estimated standard deviation known as the *standard uncertainty*, u. For the ith Type A component of uncertainty, u_i, the standard uncertainty is the statistically estimated standard deviation (evaluated as the standard deviation of the mean of a series of measurements). The jth Type B component of uncertainty, u_j, is an estimate of the corresponding standard deviation of an assumed probability distribution (e.g., normal, rectangular, or triangular) based on scientific judgment, experience with instrument behavior, and/or the instrument's specifications. These assumed probability distributions are considered to cover 99% of the values. For a triangular distribution, all values fall within limits ($\pm M$ in %) with the values more weighted toward the central value. Therefore, the Type B uncertainty can be estimated by

$$\mu_{iB} = \frac{M}{\sqrt{6}}. \tag{15.1}$$

For the rectangular distribution, all values have an equal chance between the limits ($\pm L$ in %), and the Type B uncertainty can be estimated by

$$\mu_{jB} = \frac{L}{\sqrt{3}}. \tag{15.2}$$

Historical data in the form of control charts from a given measurement process or manufacturer claims may be used to evaluate Type B components of uncertainty.

The *combined standard uncertainty*, u_c, represents the estimated standard deviation of a measurement result and is calculated by taking the square root of the sum of the squares (referred to as the RSS method or quadratic summation) of the Type A and Type B components. In this chapter, uncertainty propagation is accomplished by quadratic summation of the relative (%) uncertainties at $k = 1$ for each step of a measurement traceability chain. If the probability distribution characterized by the measurement result y is approximately normal, then $y \pm u_c$ gives an interval within which the true value is believed to lie with a 68% level of confidence. The expanded uncertainty, $U = ku_c$, where k is the coverage factor, is typically reported to express higher confidence in a measured result and is applied only to the combined uncertainty, not at each stage of an evaluation. Assuming an approximately normal distribution, $U = 2u_c$ ($k = 2$) defines an interval with a 95% level of confidence, and $U = 3u_c$ ($k = 3$) defines an interval with a 99% level of confidence.

15.2 Uncertainty in Brachytherapy Source Dosimetry Characterization

There are a number of uncertainties involved in brachytherapy dosimetry measurements and MC calculations. These measurements and calculations are usually performed at research facilities outside of the clinic. Dosimetry investigators should quantify all these uncertainties and specify them in their publications. Traceability to the primary standards is crucial for these measurements, and therefore, the calibration material later in this chapter should be considered for this section as well.

15.2.1 Other Source Parameters Involved in Dosimetry Characterization

Inherent characteristics of the source and devices used for dosimetric measurements include knowledge of the source activity distribution and source-to-detector positioning. These characteristics contribute to dosimetric uncertainties, often specific to the model of the source and the detector.

An uncertainty in source activity distribution on the internal substrate components becomes a systematic uncertainty, propagating to all measurements. Most brachytherapy sources are assumed to be uniform about the circumference of the long axis due to their cylindrical symmetry. However, in reality, the vast majority of LDR sources demonstrate variations of 20% to 2% in the intensity of emissions about the long axis for low- and high-energy photon emitters. Such variations are reflected in the statistical uncertainty of measurements if measurements are made at numerous circumferential positions around the source, and the results are averaged (Rivard et al. 2004b, 2005). Variations around the LDR source have been demonstrated in the calibrations performed at NIST (Mitch and Seltzer 2007).

15.2.2 Measurements for Dosimetry Characterization

Generally most measurements of TG-43 dosimetry parameters are done using TLDs. There are unique challenges to measuring radiation dose in the presence of either a high-dose gradient or a very low dose rate, particularly for low-energy photon-emitting sources. The major consideration is the need for a detector with a wide dynamic range, flat energy response, small geometric dimensions, and adequate sensitivity. Radiation measurement

devices in general use for brachytherapy source dosimetry are LiF TLDs, which fulfill many of the above criteria. The accuracy of the results is subject to the uncertainties due to volume averaging, self-attenuation, and absorbed-dose sensitivity. At the small source–detector distances of brachytherapy, detector size can influence self-attenuation and volume averaging.

Several types of uncertainty arise from the relative positions of the source and the detector and depend on the phantom material and the detector. If TLD is used, the shape of the detector (TLD rods, chips, or capsules of powder) may lead to different uncertainties in the location of the detector relative to the source. For measurements of some parameters, such as dose rate constant Λ and radial dose function $g(r)$, the source is positioned normal to the detector plane. A Type A dosimetric uncertainty of detector distance from the source relative to the mean detector distance appears as an uncertainty in detector reading. However, a Type B uncertainty in the mean distance of a group of detectors must be considered in the analysis of Λ and $g(r)$. For measurements of the 2D anisotropy function $F(r,\theta)$, the uncertainty in the distance of each detector from the source must be determined. In addition, the uncertainty in the angle from the source long axis must be considered. Tailor et al. (2008a) determined the uncertainty ($k = 1$) in mean distance to the detectors in a water phantom to be 0.09 mm. However, Tailor et al. (2008b) claimed a seed-to-TLD positioning uncertainty of 0.05 mm ($k = 3$) for a 0.3% Type B component dosimetric uncertainty at $r = 1$ cm. More typical Type B values obtained by a routine investigator would fall around 0.5 mm ($k = 1$).

TLDs have been the main dosimeter used for measurement of brachytherapy source dose. Typically, these measurements have been made in solid–water phantoms composed of plastics having radiological characteristics similar to water. Kron et al. (1999) provided characteristics that should be reported each time a TLD measurement is made. A calibration of the TLDs to a known energy and dose is necessary to perform dosimetry. Two major sources of uncertainty are the annealing regime used by different investigators and the intrinsic energy dependence $k_{Bq}(Q)$, which is per unit of activity (i.e., becquerel). Depending on the temperatures and cooling for the materials, the uncertainty can increase drastically, from 1% to 5%. The uncertainty is reduced when meticulous care is used in the handling, reading, and irradiation conditions. The other large source of uncertainty is the variation in the TLD absorbed dose sensitivity between the energy used for calibration and that of the brachytherapy source. This is the uncertainty in the relation of the energy dependence of the absorbed dose sensitivity relative to that in the beam quality used for calibration. Each reading regime should be the same to reduce the variation. The characteristics that affect thermoluminescence are elaborated upon in Chapter 24 of the 2009 AAPM Summer School text (DeWerd et al. 2009). The use of TLDs for various measurements of high-energy sources can be less problematic than for low-energy sources. If care is taken in each of the regimes, an overall estimate of the expanded uncertainty ($k = 2$) to measure absorbed dose would be less than 6% (Raffi et al. 2010).

15.2.3 MC Dosimetric Uncertainties

MC methods may have both obvious and hidden uncertainties associated with the brachytherapy source dosimetry process. For large numbers of histories where Poisson statistics applies, the uncertainty in the estimated results decreases by the square root of the number of particle histories. This uncertainty is referred to as the Type A uncertainty for MC methods and should be kept to < 0.1% when feasible so as to be negligible in comparison to other computational uncertainties. In many cases, it is unfeasible to simulate additional histories due to processing power and time constraints. While variance reduction techniques are sometimes used to diminish Type A uncertainties, careful benchmarking is required for radiation transport codes and their individual features and subroutines. An example of an MC uncertainty table for LDR sources is given in the TG-138 report (DeWerd et al. 2011).

Characterization of brachytherapy dose rate distributions for MC calculation of source parameters starts with a full understanding of the source construction. In general, brachytherapy sources contain radionuclides that are sealed in a single capsule. HDR sources usually have the capsule attached to a delivery cable used to position the individual source at multiple locations within the patient. PDR sources are similar to HDR, but the treatment is applied in a protracted manner. LDR sources may be described as individual entities and do not utilize a delivery cable. However, they may be contained within metal or plastic cylinders or a surgical suture material as is the case for stranded seeds. With the current TG-43 dosimetry formalism based on superposition of individual sources within a 30-cm diameter water phantom to provide full-scatter conditions for $r \leq 10$ cm for low-energy sources, characterization of the active radionuclide and the source capsule is all that is required (Rivard et al. 2004a).

As shown by Rivard (2001), the internal components within the capsule may change position. At distances of a few millimeters from some sources, the dose rate can change more than a factor of two upon varying capsule orientation (Rivard 2001). Since most low-energy sources do not have their internal components rigidly attached to the encapsulation, it is possible that the internal components may move about based on the source orientation. Especially for a low-energy photon-emitting source containing radio-opaque markers for localization, such dynamic aspects may be of clinical relevance under certain circumstances. While this effect can be observed experimentally when the source orientation is rotated 180°, this behavior is readily assessable using MC methods, but more challenging with experimental techniques where localization of the internal components may be unknown. To ascribe MC dosimetric uncertainties to this component, the full range of motion should be considered, along with possibilities for configuring internal components if multiple items are free to move and subtend different geometries upon settling within the capsule.

Presently, the TG-43 dosimetry formalism does not account for material heterogeneities and recommends liquid water as the reference media for specification of in vivo dose rate distributions. Being a simple and readily available material, it is not challenging to simulate the composition (H_2O) and mass density ($\rho = 0.998$ g/cm³ at 22°C) of liquid water. However, care must be taken when the dosimetry investigator aims to simulate the geometry of a physical experiment. Here, the setup will often include a plastic medium in place of liquid water. Due to the variable nature in fabricating these plastic media, the dosimetry investigator is advised to determine the composition and mass density independently and assign uncertainties to this assessment. Furthermore, these uncertainties directly impact the resultant dosimetric uncertainties, which should be assigned to the phantom composition. In contrast to phantom size, the MC dosimetric uncertainties due to phantom composition generally increase with decreasing photon energy and increase with increasing radial distance.

The use of MC simulations for these higher energy sources involves the same parameters as given in the TG-138 report. The phantom setup, cross section, and other uncertainties should be <1% for all classes (HDR/LDR and low/high energy) of brachytherapy sources. The number of particle histories used for the estimated results determines the Type A uncertainty for MC methods. An appropriate number of histories should be used such that the uncertainty should be kept to <0.1% when feasible so as to be negligible in comparison to other computational uncertainties. In many cases, it is infeasible to simulate additional histories due to processing power and time constraints.

All MC codes use approximations and assumptions when simulating radiological interactions. For example, generation of multiple-photon emissions following characteristic x-ray production may be simplified to the most probable photons, some MC codes ignore electron binding effects, and electron transport is often reduced to a multigroup algorithm or ignored entirely. Though molecular form factors can be used in some codes, there is no significant dosimetric effect when using an independent-atom approximation for coherent scattering form factors (Taylor and Rogers 2008a). Specific to use of radiation transport codes for determining brachytherapy dose rate distributions, there is a practical energy limit for simplification to a photon-only transport technique at the exclusion of coupled photon–electron transport, and high-energy photon-emitting radionuclides such as ^{192}Ir and ^{137}Cs may not be simulated accurately when close to the source. Electron contributions to the dosimetric uncertainty could be negligible given accurate transport equations, empirically derived atomic form factors, and proper implementation of the code by the dosimetry investigator. However, dosimetric differences within 1 mm of an ^{192}Ir source capsule between photon-only and coupled photon-electron transport may exceed 15% (Taylor and Rogers 2008b; Wang and Li 2002; Ballester et al. 2009). Estimates of $k = 1$ dosimetric uncertainties due to the physics implementation within MC radiation transport algorithms at $r = 1$ cm are 0.3% and 0.2% for low-energy and high-energy sources, respectively, and 0.7% and 0.3% at $r = 5$ cm (Rivard 2007; Rivard et al. 2010).

With the computational geometry established, progression of radiation transport is governed by atomic and nuclear cross sections that dictate the type and frequency of radiological interactions. These cross sections are organized into libraries that are maintained by international agencies such as the NNDC (NUDAT 2011). Uncertainties in the cross-sections within the source affect radiation emitted in the phantom. These cross-sections are typically calculated and compared with experimental cross-sections, determined at discrete energies. Given the physics model used to characterize the element and radiological interaction, a fitting function (such as a log-log fit) is used by the radiation transport code to interpolate between reported cross section values. Since the interpolation fit may not be robust for all element and energy possibilities, it is recommended to use recently derived cross section libraries with high resolution in energy. Sensitivity of dosimetric results on cross section libraries was illustrated by DeMarco et al. (2002).

MC-based radiation transport codes utilize μ_{en}/ρ towards calculating dose rates, and are separated from μ/ρ as, for example, one could determine dose to muscle in water instead of dose to water in water. Here, the μ/ρ and μ_{en}/ρ values for water and muscle would be used, respectively. Thus, the uncertainties ($k = 1$) of both μ/ρ and μ_{en}/ρ are of concern, and are about 1.2% and 1.0% for low-energy and high-energy sources, respectively (Cullen et al. 1997; Seltzer 1993). The influence of the cross section uncertainties on absorbed dose is a function of distance from the source with larger distances subject to larger dosimetric uncertainties. For low-energy sources, the dosimetric uncertainties at 0.5 and 5 cm are about 0.08% and 0.76%, respectively, and for high-energy sources, the dosimetric uncertainties are 0.01% and 0.12% for the same distances (Rivard 2007; Rivard et al. 2010). Further research on a modern assessment of cross-section uncertainties is needed.

All the prior steps set the simulation framework in which the calculations are performed. The dosimetry investigator must select the scoring algorithm used to determine the dose rate distributions. While some estimators are more appropriate than others (Williamson 1987), none will truly represent the desired output resultant from the dosimetry calculations. Typically, some form of volume averaging or energy-weighted modification will be used to determine dose rate at a given location within the calculation phantom. These uncertainties should be <0.1% for all classes (HDR/LDR and low/high energy) of brachytherapy sources. For path-length estimators used to determine collisional kerma, decreases in voxel thickness along the radial direction will diminish volume averaging within the voxel without significant influence on the Type A uncertainties (Rivard et al. 2009). However, MC estimators based on energy deposition within the voxel will have Type A uncertainties inversely proportional to the square root of the voxel volume, and are thus influenced by voxel thickness along the radial direction. Derivation of brachytherapy dosimetry parameters such as Λ, $g(r)$, $F(r,\theta)$, and $\phi_{an}(r)$ using MC methods involves the summation of results over various tallied voxels, weighting results based on solid angle, or taking ratios of simulated dose rates. Since all brachytherapy dosimetry parameters are ratios of dose rates, except for Λ, it is often straightforward to simply take ratios of

the raw simulated results. Systematic uncertainties in postsimulation processing may arise when energy thresholds δ (Taylor and Rogers 2008a), intentional volume averaging, or tally energy modifiers are employed. Further research on these uncertainties is needed.

Drawing from the TG-138 report (DeWerd et al. 2011) and the MC-based uncertainty analyses by Rivard and colleagues (Rivard 2007; Rivard et al. 2010; Granero et al. 2011), values for individual uncertainty components are examined. The influence of the aforementioned uncertainty components described in this subsection is shown in Figure 15.1 for an LDR [131]Cs source and an HDR ^{192}Ir source.

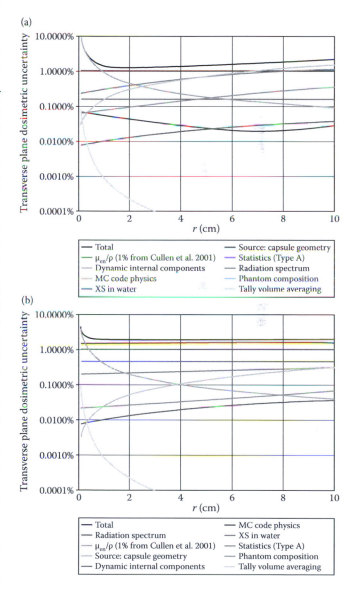

FIGURE 15.1 Transverse plane dosimetric uncertainties for [131]Cs (a) and [192]Ir (b) as representative low- and high-energy photon-emitting brachytherapy sources. The total of the uncertainty components is depicted as the top curve of each image. (Courtesy of M.J. Rivard.)

The uncertainties on the transverse plane at 0.1, 0.2, 0.5, 1, 2, 5, and 10 cm were fit to curves for depicting general trends. The volume-averaging results are truncated for better visualization of the other uncertainty components. However, the uncertainty decreased to 0.00001% at $r = 9.5$ cm.

For the ^{131}Cs source, the potential for dynamic internal components dominates the overall dosimetric uncertainty on the transverse plane for $r < 0.9$ cm. For $0.9 \leq r < 6.5$ cm, the cross-section uncertainty for absorbed dose calculations becomes most predominant. For $r \geq 6.5$ cm, the cross-section uncertainty in the medium (water) dominates, followed closely by the dosimetric uncertainty associated with the MC code.

For the ^{192}Ir source, there is less potential for dynamic internal component motion, with this only being of concern for $r < 0.3$ cm. Beyond 0.3 cm, dosimetric uncertainties in the ^{192}Ir photon spectrum and the cross-section uncertainty for absorbed dose calculations are dominant. Notably, the influence of manufacturer tolerances for the source–capsule geometry is more important for ^{192}Ir than ^{131}Cs largely due to the high-Z components.

These dosimetric uncertainty components may be categorized according to areas of responsibility. Table 15.1 illustrates the role of the medical physicist dosimetry investigator(s), source manufacturer, and basic scientists.

From the two graphs in Figure 15.1, it is evident that the choice of 0.003-cm-thick tally cells and 10^9 simulation histories by the dosimetry investigator produced dosimetric uncertainties much less than the other dosimetry components. Had the tally thickness been 10 times larger, the volume averaging would have been 100 times larger. Had the number of simulation histories been 100 times less, the Type A statistical uncertainties would have been 10 times larger. Thus, these uncertainty components would have no longer become negligible (as they should be) in comparison to the other dosimetric uncertainty components.

Dosimetric uncertainties due to dynamic internal components and design tolerances are the manufacturer's responsibility and play a large role in the overall dosimetric uncertainty. However, the largest observed dosimetric uncertainties arise from the factor out of control of either the manufacturer or dosimetry investigator. Efforts are needed by the basic science community to improve the dosimetric uncertainties (or their estimates) for absorbed dose calculations, MC code physics, radionuclide radiation spectra, and medium cross sections. It is these four dosimetric uncertainty components that are beyond the control of the dosimetry investigator, yet dominate the overall dosimetric uncertainties by approximately an order of magnitude.

15.3 Primary Laboratory Brachytherapy Calibration Uncertainties

Calibrations of brachytherapy sources need to be traceable to international standards, maintained by primary laboratories around the world. Radium nuclides will not be discussed in this chapter. The standard for brachytherapy photon sources is the quantity air-kerma strength, S_K, or reference air-kerma rate (RAKR). Air-kerma strength has a distance in its unit, whereas the RAKR is specified at 1 m. Both quantities are the same at 1 m. There is research being done to establish an absorbed dose to water quantity for brachytherapy sources. The uncertainty is not much different, although the measurement of a dose rate constant is not necessary with an absorbed dose to water, so that the uncertainty would be less. Details of the establishment of these quantities are given in Chapter 5 and will only be reviewed slightly here to discuss the uncertainties involved. This section will be divided into low-energy LDR standards and high-energy LDR and HDR source standards. There are ^{192}Ir seeds and wires and ^{137}Cs needles in use, which have a standing history of calibrations for the past 40 years.

Modern brachytherapy source standards were first established in the United States, but a number of other primary laboratories have established standards within the past 10 years. In particular, other primary laboratories, such as the PTB in Braunschweig, Germany, the National Physical Laboratory (NPL) in the United Kingdom, and the Laboratoire National Henri Becquerel (LNHB) in France, perform brachytherapy source calibrations, with each measurement system having an associated uncertainty budget. Other primary laboratories are developing standards but do not offer them as yet. It should be noted that many of these uncertainties, as well as those from the secondary laboratories given later, affect source parameters before use in the clinic, and the clinical medical physicist has no control over them.

15.3.1 Low-Energy LDR Brachytherapy Sources

The uncertainty involved in low-energy LDR sources has been extensively covered by the TG-138 report (DeWerd et al. 2011). The U.S. national primary standard of air-kerma strength ($S_{K,NIST}$) for low-energy (≤ 50 keV) photon-emitting brachytherapy sources, containing the radionuclide ^{103}Pd, ^{125}I, or ^{131}Cs, is realized using the NIST wide-angle free-air chamber (WAFAC)

TABLE 15.1 Responsible Party for Brachytherapy Dosimetric Uncertainties

Dosimetric Uncertainty Component	Dosimetry Investigator	Source Manufacturer	Basic Scientists
Source–capsule geometry		X	
Dynamic internal components		X	
Source radiation spectrum			X
MC code physics			X
Phantom composition			X
Medium cross sections			X
Absorbed dose calculation			X
Tally volume averaging	X		
Tally statistics	X		

(Seltzer et al. 2003). The WAFAC is an automated, free-air ionization chamber with a variable volume. As of June 2011, over 1000 sources of 42 different designs from 19 manufacturers have been calibrated using the WAFAC since 1999. The uncertainty $S_{K,NIST}$ consists of the standard deviation of the mean of replicate measurements (Type A) and the quadrature sum of all Type B components of uncertainty. The uncertainty in the NIST WAFAC measurement is 0.8% ($k = 1$), or the expanded uncertainty is 1.6% ($k = 2$) (Mitch and Soares 2009). Following the $S_{K,NIST}$ measurement, the responses of several well-type ionization chambers of different designs and emergent photon spectra are measured with a high-purity germanium spectrometer at NIST (Soares et al. 2009). The first primary standard device in Europe for calibration of low-energy photon sources was the large-volume extrapolation chamber (GROVEX) built at the PTB (Figure 15.2).

The procedures for the GROVEX are, in principle, the same as those at NIST (Selbach et al. 2008). However, instead of using the difference of the charges collected in two volumes, as it is the principle of the WAFAC, the PTB standard allows a continuous change in the collecting volume. This method eliminates the wall effect of the electrodes by measuring the increment of ionization per increment of the collecting volume for electrode spacings greater than the range of secondary electrons originating from the walls. The stated uncertainty of the reference air-kerma rate realized by the GROVEX is comparable to the uncertainty of the WAFAC and amounts to 0.9% ($k = 1$), or the expanded uncertainty is 1.8% ($k = 2$). For each seed type (not necessarily for each individual seed of the same type), the spectral photon distribution to obtain the spectrum-dependent correction factors for air attenuation, scattering, etc., is determined. Details are given in the work of Selbach et al. (2008). Using a sensitive scintillation detector free-in-air at 1 m, polar and azimuthal anisotropy is measured for each individual seed to be calibrated.

The results of the anisotropy measurements are part of the calibration certificate. The NPL also provides air-kerma rate calibrations of ^{125}I sources using their secondary-standard radionuclide calibrator, a well-type ionization chamber for which the calibration coefficient is traceable to the NIST primary air-kerma standard (Mitch and Soares 2009).

The expanded uncertainty ($k = 2$) for low-energy LDR sources is 1.8% for a primary calibration. This value covers all the primary laboratories doing these calibrations. It should be noted, however, that this value may have to be increased due to the special conditions of the individual source, for example, insufficient dose rate (low signal-to-background ratio) or variation of dose rate dependence on source orientation. For example, at PTB, the calibration consists of at least eight measurements, four with the source up and four with the source down. The standard deviation of the mean value of all measurements is then taken into account for the final combined uncertainties stated in the calibration certificate.

15.3.2 High-Energy LDR Brachytherapy Sources

The U.S. national primary standard of $S_{K,NIST}$ for LDR high-energy gamma-ray-emitting brachytherapy sources containing the radionuclide ^{192}Ir is realized using a spherical graphite-wall cavity chamber that is open to the atmosphere (Loftus 1980). Since arrays of approximately 50 sources were required to perform the cavity chamber measurement due to low detector sensitivity, the $S_{K,NIST}$ of individual sources is determined by using a spherical-Al re-entrant chamber working standard with a ^{226}Ra source to verify the stability of the re-entrant chamber over time. The expanded uncertainty ($k = 2$) in $S_{K,NIST}$ for LDR ^{192}Ir sources is 2%. Well-chamber response is not as sensitive to small changes in source construction due to manufacturing variability for high-energy photon emitters in comparison to low-energy sources (Li et al. 2007).

Similar to ^{192}Ir, the U.S. national primary standard of $S_{K,NIST}$ for LDR high-energy photon-emitting brachytherapy sources containing ^{137}Cs is also realized using a spherical graphite-wall cavity chamber that is open to the atmosphere (Loftus 1980). For routine calibrations, a spherical-Al cavity chamber with several ^{137}Cs working standard sources is used. The expanded uncertainty ($k = 2$) in $S_{K,NIST}$ for LDR ^{137}Cs sources is 2%. As is the case with LDR ^{192}Ir sources, well-chamber response is relatively insensitive to small changes in source construction.

At NPL, air-kerma rate calibrations are performed for ^{192}Ir wires and pins using the secondary standard radionuclide calibrator, which is traceable to the NPL air-kerma primary standard. The expanded uncertainty ($k = 2$) for an ^{192}Ir air-kerma rate measurement is stated to be 1.5%.

PTB offers reference air-kerma rate calibrations for ^{192}Ir wires (1 to 10 GBq) with an uncertainty of 2.5% ($k = 2$). This is slightly higher than for high-energy HDR sources because of the wire dimensions (2–7 cm length) and due to the noncollimated

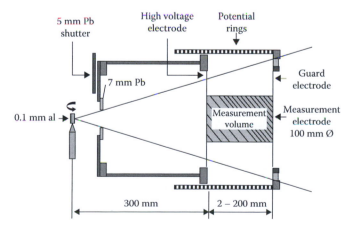

FIGURE 15.2 Schematic of the GROVEX measurement system. By varying the distance between the high voltage electrode and the measurement electrode, the measurement volume can change between 16 and 1600 cm². (Courtesy of H.-J. Selbach.)

calibration arrangement, which causes greater corrections for air scatter.

The expanded uncertainty ($k = 2$) for high-energy LDR sources is 2.5% for a primary calibration. This value covers all the primary laboratories doing these calibrations.

15.3.3 High-Energy HDR Brachytherapy Sources

There are a number of high-energy sources, but the major radioactive nuclide in use is ^{192}Ir, which is the radionuclide discussed here. All of the HDR ^{192}Ir afterloaders when measured by a seven-distance technique (Goetsch et al. 1991; Stump et al. 2002) or by some variation of it result in the same calibration factor for a well chamber within ±1% (Rasmussen et al. 2011). The first analysis for HDR ^{192}Ir sources was published by Stump et al. (2002). The value presented for the uncertainty of the primary calibration of the source using the seven-distance technique was 2.15% ($k = 2$).

NIST traceability for the measurement of air-kerma strength for HDR ^{192}Ir sources is based on the interpolation of air-kerma calibration coefficients of a secondary standard ionization chamber (Goetsch et al. 1991). The weighted average energy of these sources is 397 keV, and thus, an interpolated value between the calibration points of ^{137}Cs and 250-kVp x-rays is used. However, more rigorous methodologies for the ionization chamber ^{192}Ir air-kerma calibration coefficient have been suggested (Mainegra-Hing and Rogers 2006; van Dijk et al. 2004), which use the inverse of N_K with or without additional calibration points for the chamber. For a chamber having a calibration factor with little energy response, the results agree within 0.5%, which falls within the 2.15% uncertainty ($k = 2$).

There are two techniques to measure S_K using an ionization chamber calibrated as above: the shadow shield method and the seven-distance technique. Air-kerma strength for the seven-distance technique is thus given as

$$S_K = \frac{N_{S_K}^{Ir-192}(M_d - M_S)(d + c)^2}{\Delta t} \qquad (15.3)$$

where $N_{S_K}^{Ir-192}$ is the air-kerma calibration coefficient for ^{192}Ir; M_d is the direct measurement including the primary beam, scatter M_S, distance to the source center d, setup distance error c, and irradiation time Δt. The value of S_K is then transferred to a well-type ionization chamber.

HDR ^{192}Ir air-kerma standards are established at LNHB, PTB, and NPL (Bidmead et al. 2010). An intercomparison of the University of Wisconsin ADCL calibration standard with the LNHB calibration standard showed agreement for a specific HDR ^{192}Ir source within 0.3% (Douysset et al. 2005). Intercomparisons done between NPL and LNHB demonstrated agreement to within 0.3% to 0.5% (Douysset et al. 2008). When

uncertainty analysis is performed for all other HDR ^{192}Ir source models and intercomparisons, the overall expanded uncertainty ($k = 2$) for S_K is 2.15% (van Dijk et al. 2004; Stump et al. 2002). LNHB achieves an HDR ^{192}Ir calibration uncertainty ($k = 2$) of 1.3% for well-type ionization chambers (Douysset et al. 2005). At PTB, calibrations of HDR ^{192}Ir and ^{60}Co sources in terms of reference air-kerma rate are performed with an overall expanded uncertainty of 1.8% ($k = 2$). The lowest PSDL uncertainty offered can be seen in the Calibration and Measurement Capabilities (CMC) list from the BIPM (http://kcdb.bipm.org/appendixC/). Given the assortment of HDR high-energy sources and variety of calibration methods used at the various primary standard labs, the aforementioned calibration uncertainties are not necessarily indicative for other sources or other labs. So in summary, the expanded uncertainty ($k = 2$) for high-energy HDR sources is 2.2% for a primary calibration. This value covers all the primary calibrations.

15.4 Secondary Laboratory Brachytherapy Calibration Uncertainties and Transfer to Clinic Well Chamber

Europe does not yet have the same scale of infrastructure for low-energy brachytherapy source calibrations as does the United States. Therefore, the program described here would be that of the United States, although it certainly could transfer to other secondary laboratories. The AAPM ADCLs are responsible for transferring a traceable calibration coefficient to the clinics. Therefore, the ADCLs maintain secondary air-kerma strength standards using well-type ionization chambers, which are directly traceable to NIST to a great precision and add about 0.1% to the uncertainty budget. The AAPM Calibration Laboratory Accreditation subcommittee monitors this traceability. ADCLs establish their on-site secondary standard by measuring the response of a well chamber to a NIST-calibrated source. The ratio of air-kerma strength, S_K, to current, I, yields a calibration coefficient for a given source type. The ADCLs use their calibrated well chamber and manufacturer-supplied sources to calibrate well chambers for clinics. Calibrations of electrometers and instruments monitoring atmospheric conditions are also necessary to complete the system. Intercomparisons amongst ADCLs and proficiency tests with NIST ensure that each ADCL is accurate in its dissemination and that the calibrations from different ADCLs are equivalent.

15.4.1 Low-Energy LDR Brachytherapy Sources

For LDR low-energy photon-emitting brachytherapy sources, the NIST air-kerma strength standard for each new source model is initially transferred to all ADCLs that are accredited by the AAPM to perform brachytherapy source calibrations by sending a batch of three WAFAC-calibrated sources in turn to each ADCL. To ensure that the NIST-traceable standard at each ADCL remains consistent over time with the initial baseline values, subsequent batches of three sources of each model are calibrated by NIST and circulated among all ADCLs at least annually (DeWerd et al. 2004). Based on data collected by NIST and the ADCLs over many years, it appears that the accuracy achievable in a secondary standard is not the same for all source models. Variations in emergent spectrum and spatial anisotropy of emissions influence well-chamber-to-WAFAC response ratios, and how well such variations are minimized during source fabrication affects the magnitude of variability in well-chamber measurements for sources of supposedly identical construction. As can be seen by reference to the TG-138 report (DeWerd et al. 2011), the uncertainty added by the secondary laboratory results in a value from the well chamber of 1.1% or an expanded uncertainty ($k = 2$) of 2.2% for low-energy LDR sources. When the clinic well chamber is calibrated, the uncertainty is 1.3%, with the expanded uncertainty to the clinic of 2.6% for low-energy LDR brachytherapy sources.

15.4.2 High-Energy LDR Brachytherapy Sources

A NIST-traceable air-kerma strength standard for both high-energy gamma-ray-emitting brachytherapy sources (i.e., ^{192}Ir and ^{137}Cs) has been available from all ADCLs for many years. The continued accuracy of the secondary standards is verified through the performance of periodic Measurement Quality Assurance tests. Recommendations have been published specifying that a check of the accuracy of manufacturer source or equipment calibrations be verified by either NIST or an AAPM-accredited ADCL on an annual basis (DeWerd et al. 2004).

When a well chamber is calibrated using the value determined for the source, the total uncertainty is 2.56% ($k = 2$) for high-energy LDR sources. Consideration of the primary calibration should include an additional value, since the original manufacturer's model is no longer in production; this is true for cesium sources as well as low-dose ^{192}Ir sources. Applying the calibration values to new sources would then increase the uncertainties to 2.2% with well-chamber calibrations to 3.0% at $k = 2$ for both radionuclides. For details on determination of uncertainties and the methodology, we ask the reader to refer to the TG-138 report (DeWerd et al. 2011). Therefore, the clinic well chamber has a calibration value for high-energy LDR sources of 3.0% ($k = 2$).

15.4.3 High-Energy HDR Brachytherapy Sources

The uncertainty for the secondary laboratory would be 2.2% ($k = 2$). When the well chamber is calibrated for the clinical use, the uncertainty would increase to 2.4% ($k = 2$). Thus, the clinic well chamber has a calibration value for high-energy HDR brachytherapy sources at 2.7% ($k = 2$).

15.5 Brachytherapy TG-43 Dosimetry Parameters

The AAPM has developed the TG-43 methodology for brachytherapy sources. The AAPM TG-43 update (Rivard et al. 2004a) has set forth data and the methodology for sources using the equation and parameters as given below to determine the dose rate:

$$\dot{D}(r,\theta) = S_K \cdot \Lambda \cdot \frac{G_L(r,\theta)}{G_L(r_0,\theta_0)} \cdot g_L(r) \cdot F(r,\theta) \qquad (15.4)$$

where S_K is the air kerma strength, determined from a calibrated well chamber, Λ is the dose rate constant, $G_L(r,\theta)$ is the geometry function at point $P(r,\theta)$, $G_L(r_0,\theta_0)$ is the reference geometry function at the reference position, $r_0 = 1$ cm, $\theta_0 = 90°$, $g_L(r)$ is the radial dose function, and $F(r,\theta)$ is the anisotropy function.

Each of the above parameters is determined experimentally or by MC calculations. The update paper provides consensus data to be used for each source type. The purpose here is to give the uncertainties for each parameter of Equation 15.4. The uncertainty determination is elaborated in the TG-138 report (DeWerd et al. 2011).

The air-kerma strength is determined using the clinic calibrated well chamber, with the associated uncertainty as given in Section 15.4. The dose rate constant varies depending on the source type. Generally, the consensus value is an average of the experimentally determined value (using TLDs) and the MC determined value. The TLD value would be approximately 2.8%, and the MC value would be at 1.2%. Thus, since this is a combined value, the dose rate constant would have an uncertainty of 3.0%. The other parameters in Equation 15.4 are generally determined by MC calculations but verified by experimental techniques. Each of those parameters would have an uncertainty of 1.2% and each would vary depending on what type of source is under consideration as given above. Therefore, an overall uncertainty for a typical source for the dose rate, $\dot{D}(r,\theta)$, is given in Table 15.2 to be 3.9% at $k = 1$ or an expanded uncertainty of 7.8% at $k = 2$.

TABLE 15.2 Overall Dose Rate Uncertainty of LDR Brachytherapy Source for TG-43 Formalism

TG-43 Dosimetry Parameter	Relative Propagated Uncertainty (%)
Air-kerma strength from calibrated well chamber	1.3
Dose rate constant	3.27
Geometry function	3.48
Radial dose function	3.68
Anisotropy function	3.87
Uncertainty in dose rate ($k = 1$)	3.9
Expanded uncertainty ($k = 2$)	7.8

15.6 Clinical Dosimetry Uncertainties

The above information is the beginning of the uncertainty involved in the clinic. The uncertainty of dose calculated by a treatment planning system (TPS) will be based on the combination of uncertainties of NIST-traceable S_K and the dose rates determined by the dosimetry investigator. However, there are additional uncertainties introduced by the TPS.

Commissioning of the brachytherapy source for dose calculations requires the physicist or other responsible person to install source characterization data into the TPS computer. Therefore, the question becomes, What additional uncertainty is associated with the installation of source characterization data, and the use of those data in the TPS, to calculated dose distributions? When dosimetry parameters are entered, the frequency and spacing of the data are key since the TPS performs interpolation on the entered data. Unless spacing varies in inverse proportion to the contribution of a parameter, the uncertainty is likely to be different at different distances. When fits to experimental- or MC-derived dosimetry parameters are entered, the uncertainty relates to the quality of the fit. The fit approach and model used will affect the uncertainty. Further, the TPS dose calculation uncertainty depends on the implementation of the algorithm, the calculation matrix spacing, and the veracity of the output mechanisms. Consequently, it is impossible to determine explicitly the uncertainty introduced by model fitting and interpolation.

Uncertainty analyses should include all dosimetric properties of clinical brachytherapy sources and follow a common set of guidelines and principles, analogous to TG-43 parameters for brachytherapy sources. To minimize uncertainties, clinical medical physicists should use the consensus brachytherapy dosimetry data. Use of nonconsensus data would lead to a mistake rather than an increase in uncertainties. The primary aspects under control by the clinical medical physicist are measurements of S_K and TPS data entry. For the first aspect, the clinical medical physicist should follow the 2008 AAPM brachytherapy source calibration recommendations (Butler et al. 2008; Li et al. 2007).

15.7 Conclusions

Throughout the current report, the AAPM and GEC-ESTRO have refined clinical expectations of brachytherapy dosimetric uncertainty. Uncertainties are involved in all aspects of the dosimetry process. Every aspect of the process results in a greater uncertainty in the estimation of patient dose. The end result for consideration is the uncertainties involved in patient treatments.

References

Ballester, F., Granero, D., Pérez-Calatayud, J., Melhus, C.S., and Rivard, M.J. 2009. Evaluation of high-energy brachytherapy source electronic disequilibrium and dose from emitted electrons. *Med Phys* 36:4250–6.

Bidmead, A.M., Sander, T., Locks, S.M. et al. 2010. The IPEM code of practice for determination of the reference air kerma rate for HDR ^{192}Ir brachytherapy sources based on the NPL air kerma standard. *Phys Med Biol* 55:3145–59.

Bureau International des Poids et Mesures (BIPM). 2011. Available at http://www.bipm.org/en/publications/guides/gum.html, accessed July 11, 2011.

Butler, W.M., Bice, W.S. Jr., DeWerd, L.A. et al. 2008. Third-party brachytherapy source calibrations and physicist responsibilities: Report of the AAPM Low Energy Brachytherapy Source Calibration Working Group. *Med Phys* 35:3860–5.

Cullen, D.E., Hubbell, J.H., and Kissel, L. 1997. EPDL97: The Evaluated Photon Data Library, '97 Version. Lawrence Livermore National Laboratory Report No. UCRL-50400, Vol. 6, Revision 5. September 19, 1997.

DeMarco, J.J., Wallace, R.E., and Boedeker K. 2002. An analysis of MCNP cross-sections and tally methods for low-energy photon emitters. *Phys Med Biol* 47:1321–32.

DeWerd, L.A., Huq, M.S., Das, I.J. et al. 2004. Procedures for establishing and maintaining consistent air-kerma strength standards for low-energy, photon-emitting brachytherapy sources: Recommendations of the Calibration Laboratory Accreditation Subcommittee of the American Association of Physicists in Medicine. *Med Phys* 31:675–81.

DeWerd, L.A., Bartol, L.J., and Davis, S.D. 2009. Thermoluminescence dosimetry. In: *Clinical Dosimetry for Radiotherapy: AAPM Summer School*. D.W.O. Rogers and J.E. Cygler (eds.). Medical Physics Publishing, Madison, WI, pp. 815–40.

DeWerd, L.A., Ibbott, G.S., Meigooni, A.S. et al. 2011. A dosimetric uncertainty analysis for photon-emitting brachytherapy sources: Report of AAPM Task Group No. 138 and GEC-ESTRO. *Med Phys* 38:782–801.

Douysset, G., Gouriou, J., Delaunay, F., DeWerd, L., Stump, K., and Micka, J. 2005. Comparison of dosimetric standards of USA and France for HDR brachytherapy. *Phys Med Biol* 50:1961–78.

Douysset, G., Sander, T., Gouriou, J., and Nutbrown, R. 2008. Comparison of air kerma standards of LNE-LNHB and NPL for ^{192}Ir HDR brachytherapy sources: EUROMET project no 814. *Phys Med Biol* 53:N85–97.

Giacomo, P. 1981. News from the CIPM. *Metrologia* 17:69–74.

Goetsch, S.J., Attix, F.H., Pearson, D.W., and Thomadsen, B.R. 1991. Calibration of ^{192}Ir high-dose-rate afterloading systems. *Med Phys* 18:462–7.

Granero, D., Vijande, J., Ballester, F., and Rivard, M.J. 2011. Dosimetry revisited for the HDR ^{192}Ir brachytherapy source model mHDR-v2. *Med Phys* 38:487–94.

Honsa, J.D. and McIntyre, D.A., 2003. ISO 17025: Practical benefits of implementing a quality system. *AOAC Int* 86:1038–44.

Kochsiek, M. 2001. *PTB-Mitteilungen* 111, Heft 3, Themenschwerpunkt: Messunsicherheiten.

Kron, T., DeWerd, L.A., Mobit, P. et al. 1999. A checklist for reporting of thermoluminescence dosimetry (TLD) measurements. *Phys Med Biol* 44:L15–9.

Li, Z., Das, R.K., DeWerd, L.A. et al. 2007. Dosimetric prerequisites for routine clinical use of photon emitting brachytherapy sources with average energy higher than 50 keV. *Med Phys* 34:37–40.

Loftus, T.P. 1980. Standardization of iridium-192 gamma-ray sources in terms of exposure. *J Res Natl Bur Stand* 85:19–25.

Mainegra-Hing, E. and Rogers, D.W.O. 2006. On the accuracy of techniques for obtaining the calibration coefficient N_K of ^{192}Ir HDR brachytherapy sources. *Med Phys* 33:3340–7.

Mitch, M.G. and Seltzer, S.M. 2007. Model-specific uncertainties in air-kerma strength measurements of low-energy photon-emitting brachytherapy sources. *Med Phys* 34:2337 (abstract).

Mitch, M.G., DeWerd, L.A., Minniti, R., and Williamson, J.F. 2009. Treatment of uncertainties in radiation dosimetry. In: *Clinical Dosimetry for Radiotherapy: AAPM Summer School.* D.W.O. Rogers and J.E. Cygler (eds.). Medical Physics Publishing, Madison, WI, pp. 723–57.

Mitch, M.G. and Soares, C.G. 2009. Primary standards for brachytherapy sources. In: *Clinical Dosimetry for Radiotherapy: AAPM Summer School.* D.W.O. Rogers and J.E. Cygler (eds.). Medical Physics Publishing, Madison, WI, pp. 549–65.

NUDAT 2.5. 2011. National Nuclear Data Center, Brookhaven National Laboratory, Upton, NY, USA. http://www.nndc.bnl.gov/nudat2/index.jsp, accessed July 11, 2011.

Raffi, J.A., Davis, S.D., Hammer, C.G. et al. 2010. Determination of exit skin dose for ^{192}Ir intracavitary accelerated partial breast irradiation with thermoluminescent dosimeters. *Med Phys* 37:2693–702.

Rasmussen, B.E., Davis, S.D., Schmidt, C.R., Micka, J.A., and DeWerd, L.A. 2011. Comparison of air-kerma strength determinations for HDR ^{192}Ir sources. *Med Phys* 38:6721–9.

Rivard, M.J. 2001. Monte Carlo calculations of AAPM Task Group Report No. 43 dosimetry parameters for the MED3631-A/M ^{125}I source. *Med Phys* 28:629–37.

Rivard, M.J. 2007. Brachytherapy dosimetry parameters calculated for a ^{131}Cs source. *Med Phys* 34:754–62.

Rivard, M.J., Coursey, B.M., DeWerd, L.A. et al. 2004a. Update of AAPM Task Group No. 43 Report: A revised AAPM protocol for brachytherapy dose calculations. *Med Phys* 31:633–74.

Rivard, M.J., Melhus, C.S., and Kirk, B.L. 2004b. Brachytherapy dosimetry parameters calculated for a new ^{103}Pd source. *Med Phys* 31:2466–70.

Rivard, M.J., Kirk, B.L., and Leal, L.C. 2005. Impact of radionuclide physical distribution on brachytherapy dosimetry parameters. *Nucl Sci Eng* 149:101–6.

Rivard, M.J., Melhus, C.S., Granero, D., Pérez-Calatayud, J., and Ballester, F. 2009. An approach to using conventional brachytherapy software for clinical treatment planning of complex, Monte Carlo-based brachytherapy dose distributions. *Med Phys* 36:1968–75.

Rivard, M.J., Granero, D., Pérez-Calatayud, J., and Ballester, F. 2010. Influence of photon energy spectra from brachytherapy sources on Monte Carlo simulations of kerma and dose rates in water and air. *Med Phys* 37:869–76.

Selbach, H.-J., Kramer, H.-M., and Culberson, W.S., 2008. Realization of reference air-kerma rate for low-energy photon sources. *Metrologia* 45:422–8.

Seltzer, S.M. 1993. Calculation of photon mass energy-transfer and mass energy-absorption coefficients. *Rad Res* 136:147–70.

Seltzer, S.M., Lamperti, P.J., Loevinger, R., Mitch, M.G., Weaver, J.T., and Coursey, B.M. 2003. New national air-kerma-strength standards of I-125 and Pd-103 brachytherapy seeds. *J Res Natl Inst Stand Technol* 108:337–58.

Siebert, B.R.L. 2006. Uncertainty in radiation dosimetry: Basic concepts and methods. *Rad Prot Dos* 121:3–11.

Soares, C.G., Douysset, G., and Mitch, M.G. 2009. Primary standards and dosimetry protocols for brachytherapy sources. *Metrologia* 46:S80–98.

Stump, K.E., DeWerd, L.A., Micka, J.A., and Anderson D.R. 2002. Calibration of new high dose rate ^{192}Ir sources. *Med Phys* 29:1483–8.

Tailor, R.C., Ibbott, G. S., and Tolani, N. 2008a. Thermoluminescence dosimetry measurements of brachytherapy sources in liquid water. *Med Phys* 35:4063–9.

Tailor, R., Ibbott, G., Lampe, S., Bivens-Warren, W., and Tolani, N. 2008b. Dosimetric characterization of a ^{131}Cs brachytherapy source by thermoluminescence dosimetry in liquid water. *Med Phys* 35:5861–8.

Taylor, B.N. and Kuyatt, C.E. 1994. Guidelines for evaluating and expressing the uncertainty of NIST measurement results. NIST Technical Note 1297 (U.S. Government Printing Office, Washington, DC, 1994). http://physics.nist.gov/Pubs/guidelines/contents.html, accessed July 11, 2011.

Taylor, R.E.P. and Rogers, D.W.O. 2008a. An EGSnrc Monte Carlo-calculated database of TG-43 parameters. *Med Phys* 35:4228–41.

Taylor, R.E.P. and Rogers, D.W.O. 2008b. EGSnrc Monte Carlo calculated dosimetry parameters for ^{192}Ir and ^{169}Yb brachytherapy sources. *Med Phys* 35:4933–44.

van Dijk, E., Kolkman-Deurloo, I.-K.K., and Damen, P.M.G. 2004. Determination of the reference air kerma rate for ^{192}Ir brachytherapy sources and the related uncertainty. *Med Phys* 31:2826–33.

Wang, R. and Li, X.A. 2002. Dose characterization in the near-source region for two high dose rate brachytherapy sources. *Med Phys* 29:1678–86.

Williamson, J.F. 1987. Monte Carlo evaluation of kerma at a point for photon transport problems. *Med Phys* 14:567–76.

16

Uncertainties Associated with Clinical Aspects of Brachytherapy

Peter Hoskin
Mount Vernon Hospital

Alfredo Polo
Ramon y Cajal University Hospital

Jack L. M. Venselaar
Instituut Verbeeten

16.1 Introduction and General Considerations

The degree of accuracy clinically required in radiotherapy for dose delivery to a patient has received great attention over the years. These considerations are generally based on the difference in radiation response between tumors and normal tissues. In radiation oncology, the tolerance of normal tissue is often the limiting factor for the dose that can be delivered to the patient. The relation between the dose and the biological effect is described by dose–effect curves. For normal tissue complications, the dose–effect curve is generally steeper than for local tumor control, and the same level of biological response is usually found at a higher dose level for normal tissue than for tumors. The precise form and steepness of these curves depend on many factors, including the intrinsic radiosensitivity of the tissue, the treatment modality, and the fractionation scheme as discussed in Chapter 19 on radiobiology. Considerations are essentially the same for external beam radiotherapy and brachytherapy.

The term "accuracy" usually is taken separately for clinical aspects and physics aspects. This often is associated with geometrical miss and dosimetrical deviations, respectively, although the connection between the two is sometimes made in more generic approaches as in the application of type of gamma-index analysis (e.g., 2 mm/3% correspondences in treatment plan comparisons).

From clinical studies, it was concluded by the International Commission on Radiation Units and Measurements (ICRU Report 24, 1976) that the available evidence for certain types of tumor points to the need for an accuracy of ±5% in delivery of an absorbed dose to a target volume if the eradication of the primary tumor is sought. Closer limits were considered virtually unachievable at the time of writing of that report. In the ICRU report, no indication is given about the confidence level of this 5% value, that is, whether it concerns one or two standard deviations or an action level.

One decade later, Mijnheer et al. (1987) discussed the clinical observations of normal tissue reactions. These authors concluded that an increase in the absorbed dose of ±7% can result in observable and unacceptable normal tissue complication probabilities. If information from one radiotherapy department is transferred to another, unacceptable risks are involved if the overall uncertainty in the absorbed dose is larger than this ±7%. According to the authors, this value should be interpreted as twice the standard deviation of the absorbed dose. Therefore, they concluded that a total uncertainty of ±3.5% (1 SD) in the absorbed dose at the dose specification point is desirable and should be strived for in routine clinical practice (Mijnheer et al. 1987).

Similarly, Brahme et al. (1988) concluded that a relative standard deviation in tumor control probability of less than ±10%, and preferably less than ±5%, is needed to have a reasonable probability of distinguishing the outcome from comparable studies with patient groups of a few hundred persons. To maintain a high quality of treatment, the loss in tumor control probability due to dose variations should not be more than 5% and preferably less than 3% to the target. When transformed into a recommended tolerance level for accuracy in dose delivery, a value of 3% relative standard deviation in the absorbed dose was proposed. The action level, above which it is recommended to work to improve the accuracy in dose delivery, is at a relative standard deviation of 5% (Brahme et al. 1988).

To achieve a 3.5% (1 SD) accuracy in the physical dose delivery requires accurate and reproducible calibration of the treatment delivery machine and a high precision in the dose calculation procedures. These considerations form the basis of the many national and international recommendations for quality assurance of equipment and audit procedures for radiation oncology departments. For brachytherapy, these have been discussed in depth in Chapter 4 on QA of equipment and in Chapter 15, following the outline of the recent comprehensive AAPM TG-138 report (DeWerd et al. 2011).

There are also significant patient-specific clinical uncertainties related to interfraction and intrafraction movement of both target and adjacent organs at risk (OARs). Individual patients will also exhibit variable radiation response characteristics related to both tumor and normal tissues.

A final consideration and perhaps the greatest variable in clinical practice relates to the target and OAR delineation by the clinician with considerable interobserver and intraobserver variations seen.

When the tumor or parts of it are not irradiated or not treated with the intended dose, the success of the therapy will be reduced drastically. In particular, in brachytherapy, it should be considered that treatments are often given in one or a very low number of fractions in comparison to external beam radiation therapy (EBRT). For the dose delivery to a brachytherapy target volume, uncertainties will be encountered in the contouring (interobserver and intraobserver variabilities), afterloader performance (both with spatial and temporal uncertainties), imaging (reconstruction, volume interpolation, fusion), and dose calculation in inhomogeneous media, and at the patient level, there will be intrafraction and interfraction movements, organ motion, and swelling. It is the intention of this chapter to discuss these conditions and their consequences for variations in dose and geometry in the following paragraphs.

16.2 Process of Clinical Brachytherapy

Brachytherapy should be understood as a process comprising a series of activities that are performed sequentially, leading to an acceptable dose distribution around an applicator positioned precisely in the target anatomy. This will depend initially upon the implant geometry for which careful preimplant planning is important. This process includes the following considerations:

(1) Definition of volumes to be irradiated, that is, the volumes within which the dose is to be prescribed to the tumor and the dose limits of the healthy tissues must be determined.

(2) Preimplant planning should be performed to localize the target volume and then to decide the technique and optimal distribution of the applicators.

(3) Implantation is performed with the aid of imaging and intraoperative navigation tools.

(4) Post-implant imaging is performed, a 3D reconstruction is created, and the volumes of interest are identified; in this process, image artifacts, distortions, and applicator suitability for imaging should all be considered; imaging technique and sequences should be optimized and the use of appropriate contrast included.

(5) The dose distribution is calculated using the 3D reconstruction; this may be affected by imperfections in the algorithm, which are related to, for example, tissue boundary effects, tissue inhomogeneities, effects of applicator shields, and interseed shielding.

(6) Imperfect implant geometry may be modified by dosimetric optimization, implementing clinical dose constraints and physical criteria such as dose homogeneity and conformation; the dose distribution may also be biologically optimized, while defining the overdose volumes, altering fractionation, or modifying the normalization isodose level.

(7) Treatment delivery takes place using the afterloading equipment under clinical conditions, where interfraction and intrafraction movements, organ movement, or swelling is present.

(8) Imperfections in technical treatment delivery must be minimized using a quality management system, which is used to guarantee the whole process; in vivo dosimetry is sometimes used to provide confidence in the correct course of the treatment.

The degree of sophistication of the planning of the procedure depends on the area to be treated. For example, a simple and straightforward treatment such as an esophageal endocavitary palliative irradiation is distinct from a complex combination of intracavitary and interstitial brachytherapy for a cervical cancer. The clinical context can be completely different as with primary radiation versus re-irradiation. The tools in each of the planning phases are selected in accordance with these criteria.

Similarly, the workflow will depend on the situation. Thus, a gynecological relapse to be salvaged using interstitial brachytherapy may require exploration under anesthesia in the treatment position, a preimplant simulation by means of MR scanning for target and OAR definition, a post-implant CT scan for volumetric reconstruction, and the calculation of the dose distribution, possibly combined (fused) with a separate post-implant MR scan.

The eight points mentioned above will be elaborated in the next sections and illustrated with a few examples of specific published studies.

16.3 Definition, Localization, and Delineation of Volumes of Interest

Knowledge of the volumes to be irradiated is essential in the clinical planning process. This will include the definition of the target volume (*what do we want to irradiate?*), its intraoperative localization (*where is the target during the operation?*), and its delineation on the axial images used for dosimetric calculation (*how is it viewed in the planning images?*).

- *Definition*: the description of the structures and margins of the target volume to be radiated. This will be based on our knowledge of the natural history of the disease, information from clinical and radiological examination, and pathological features from biopsy. A suitable definition of the volumes of interest in brachytherapy can be found in the ICRU 58 report (ICRU 1997). The clinical target volume (CTV) is the clinical volume that contains the gross tumor volume and any subclinical microscopic malignant disease that has to be eliminated. The planning target volume (PTV) represents the CTV with an expansion to accommodate organ movement and set up error in the external beam treatment; in brachytherapy, organ movement will, in general, be considered as minimal; however, in low–dose-rate or pulsed-dose-rate (PDR) delivery, edema may occur during the period of irradiation. The addition of margins will be largely related to estimated geometric uncertainties that may occur between fractions (applicator displacement, movement of internal organs) and with variations in delineation. In general, however, in brachytherapy, the required expansion from CTV to PTV tends to be less than in EBRT, and in many circumstances, no expansion will be used, that is, CTV = PTV. Greater appreciation of the uncertainties in volume definition and dose delivery brings this zero expansion concept into question. Margins of 3–5 mm to account for uncertainties in treatment delivery have been suggested (Han et al. 2000; Butzbach et al. 2001; Merrick et al. 2007). However, expansion of a volume of 50 cm^3 with a 2-mm larger margin in all directions increases the volume to be treated to 64 cm^3, which is an increase of 28% in the volume. This may influence PTV cover as illustrated in a study by Crook et al. (2010), who investigated a total of 131 men with early-stage prostate cancer treated with iodine-125 brachytherapy monotherapy. Postplan assessment was performed at 1 month using magnetic resonance (MR)–computed tomography (CT) fusion. The prostate V_{100} and D_{90} were calculated with 2-, 3-, and 5-mm margins. Satisfactory coverage of a 2- and 3-mm periprostatic margin was obtained with the described planning approach, but the target coverage fell off significantly by 5 mm.
- *Localization*: the identification of the target volumes previously described using appropriate clinical examination and imaging including planar x-ray, US, CT, or MRI. Target volume localization may be undertaken intraoperatively; however, all cases will require post-implant localization and/or treatment planning. Use of ultrasound, CT, MR, and PET-CT may lead to different volumes. A study of 82 patients who underwent ^{103}Pd prostate brachytherapy using real-time intraoperative transrectal US-guided

treatment planning (Nag et al. 2008) performed postoperative CT-based dosimetry a few hours later. The study was set up to compare the two imaging modalities while there was a correlation between the US- and CT-based dosimetry, but also large variations in the implant-quality parameters of the two modalities were found. It was concluded that dosimetry using intraoperative US-based planning does not accurately reflect the postoperative CT-based dosimetry.

- *Delineation*: the action required to present the required structures to the dosimetric treatment planning system (TPS) on which the calculation of the absorbed doses will be made. In modern brachytherapy, volume delineation is no longer dependent on reference points based on bony structures or the applicator as seen on orthogonal imaging but utilizes 3D volume determination from cross-sectional images from CT scans, MRI, and US. Interobserver variation studies for volume definitions have been published by numerous groups, both for external beam therapy and for brachytherapy. An interoperator study was performed by Wills et al. (2010) to assess the variability of applicator reconstruction between operators when CT or MR images were used alone to reconstruct the applicators. This showed variability in applicator reconstruction between operators with both MR and CT (median MR/CT 1.3 mm/0.9 mm, range 0–3.6 mm/0–3.3 mm). Interoperator variation in PTV V_{100} and PTV D_{90} for MR/CT was 6.1%/3.0% and 7.4%/6.3%, respectively, and D_{2cc} OAR doses varied by up to 1.0 Gy between operators for both MR and CT.

A study of 19 clinical cervical cancer patients treated with image-guided brachytherapy (IGBT) from two radiation oncology centers compared the outlines defined independently by two radiation oncologists (Dimopoulos et al. 2009) according to the GYN GEC-ESTRO recommendations (Haie-Meder et al. 2005; Pötter et al. 2006). The absolute, common, and encompassing volumes and their conformity indices (CIs) were assessed for the GTV, high-risk CTV, and intermediate-risk CTV. D_{90} and D_{100} for each volume were assessed. Significant differences were observed only for the mean volumes of the IR CTV of both centers ($p < 0.05$). CIs ranged from 0.5 to 0.7. Dose-volume histogram (DVH) parameter analyses did not reveal any statistical differences, except for D_{100} for the GTV at one center and D_{90} for the IR CTV at the other center, with $p < 0.05$. Underlying reasons for interobserver differences included image contrast adjustment and neglecting to consider anatomical borders.

Xue et al. (2006) studied the results of three observers after volume definition in 20 preplanned peripherally loaded ^{125}I prostate implants. Multiple pair-wise comparisons showed that the prostate volumes delineated by observer 3 differed significantly from those of observers 1 and 2 ($p < 0.003$). The volumes of observers 1 and 2 were not significantly different ($p > 0.5$). The mean values of D_{90} ranged from 124.2 to 171.1 Gy (median: 154.7 Gy), having SDs that ranged from 0.6% to 24.4% of the mean D_{90} (median: 7.8%). The mean values of V_{100} ranged from 82.3% to 95.1% (median: 92.8%), having SDs that ranged from 0.4% to 11.2% of the mean V_{100} (median: 4.0%). The values of both D_{90} and V_{100} calculated from the volumes of observer 3 were significantly ($p < 0.003$) different from those of observers 1 and 2, which did not differ significantly ($p > 0.5$). Significant interobserver differences in delineating the prostate volume on post-implant TRUS images were observed; however, these differences were less than generally reported for post-implant CT images.

16.4 Preimplant Planning of a Procedure

In some cases, there will be a standard applicator insertion for which minimal preimplant planning is required, for example, with vaginal vault brachytherapy. In other cases, preimplant planning must be based on diagnostic imaging using, for example, MR images to determine whether or not interstitial needles are required for an adequate coverage of a cervical tumor with parametrial spread. In some instances, for example, with ^{125}I prostate implants, detailed preimplant and interactive perimplant dosimetry will be required to optimize the implant procedure and identify patients where pubic arch interference may be a problem.

16.5 Intraoperative Planning

Intraoperative planning refers more precisely to the possibility of localizing the target volume and of knowing the dose distribution associated with the implant (or an estimate thereof) at the time of execution. It requires an imaging modality and a calculation system that can be accessed in the operating room. This is used to optimize the geometry of the implant during execution. In the same way, dose distribution can be optimized as the implant progresses, modifying the initial plan according to the clinical criteria that emerge during execution (anatomical characteristics that condition the appearance of hot or cold areas, interposed critical organs, etc.).

Real-time navigation refers to the use of continuous intraoperative imaging during the execution of the implant. Multiple combined imaging modalities may be used (US + fluoroscopy). This allows one to control the geometry by showing both the target volume and the surrounding critical organs. The clearest example may be a uterovaginal application, in the case of cervical cancer or interstitial breast brachytherapy in which imaging (generally ultrasound) is used to guarantee accurate intraoperative localization of the applicator. The treatment of inoperable gynecological relapse (centropelvic or paravaginal) by means of intraoperative navigation techniques is addressed elsewhere in this book. Currently, technological limitations prevent connecting intraoperative planning to real-time navigation in most

clinical localizations, with the exception of prostate brachytherapy (Polo et al. 2010).

The ideal intraoperative procedure will deliver treatment in the implantation position avoiding the additional errors introduced in patient movement during which applicators may be displaced and OARs assume a different configuration. In this setting, all other steps including imaging delineation, dose calculation and the afterloader must be available in the operating area. This requires full radiation protection in case of application of a high-dose–rate (HDR) high-energy source, a condition that is much more easily fulfilled when low-energy sources such as ^{125}I or [103]Pd are used (and, maybe in the future, [169]Yb) compared to the high photon energy from the ^{192}Ir source types. The brachytherapy suite should preferably be equipped with the 3D imaging device, ultrasound, and an open MR system currently representing state of the art. However, open MRI scanners typically have lower magnetic field strengths than used in standard imaging departments (e.g., 0.35 T versus 1.5 T). This means that there is a lower signal-to-noise ratio unless compensated for by reducing resolution. Open scanners tend to have greater image distortion, especially toward the edge of the field, but this is likely to be acceptable (<1 mm) for the normally deep-seated and centrally located areas under consideration (Wills et al. 2010).

16.6 Post-Implant Planning and Definitive Dosimetry

As indicated above, the intraoperative technique requires a dedicated brachytherapy suite, a facility that is not widely available throughout the world. Therefore, most procedures rely on an offline, that is, a post-implant, imaging procedure, reconstruction, and definitive dosimetry before radiation delivery.

The use of planar radiography for the creation of the treatment plan has, in most centers, gradually been replaced by axial CT and MR imaging. However, radiography is still often used for guidance and verification during the implant procedure itself. From the planar x-rays, even when taken from two or three or more directions, information only on the position of the applicators can be achieved, not on tumor volume definition. What results in such a case is an applicator-based dosimetry result, which is, in most cases, considered as being outdated for modern brachytherapy. The only role for radiography is in those cases where the relation between the applicator and the volume is well defined, for example, for those locations where the clinical observation by the radiation oncologist would allow this approach (e.g., skin tumors).

As discussed elsewhere, MR scanning has advantages over CT in terms of better tissue contrast to distinguish between tumor and healthy tissue. CT is more widely available and therefore used in many centers. CT imaging provides information on tissue inhomogeneity, which is relevant when used with model-based advanced dose calculation algorithms. A combination of MR and CT is often considered the optimal imaging information on which to base dosimetry in brachytherapy. This requires a

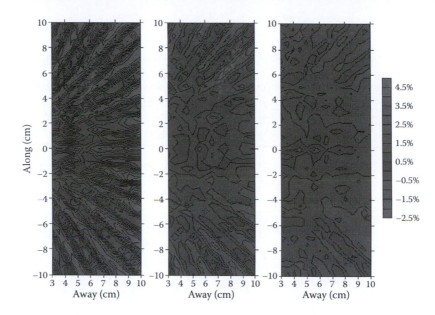

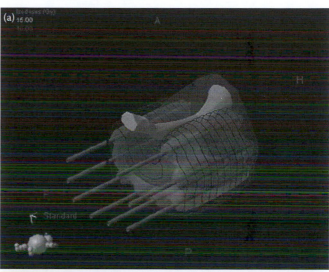

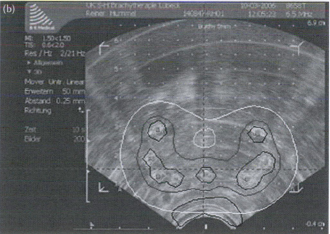

FIGURE 7.7 Left to right: S14 (224 angles), S18 (360 angles), and S22 (528 angles). S18 dose variation due to ray effects is less than 1.5% out to 6 cm from an ^{192}Ir source in water. Based on data used in the work of Mourtada et al. (2004) using the Attila code. (Courtesy of F. Mourtada.)

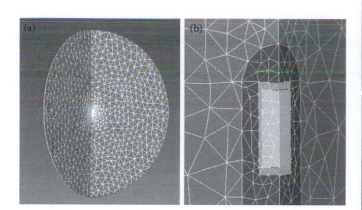

FIGURE 7.8 (a) Full computational domain, with ^{192}Ir PDR source in the center of a 30-cm water sphere. (b) Computational mesh around source [zoom image in (a)]. (Courtesy of F. Mourtada.)

FIGURE 9.5 (a) Intensity-modulated HDR brachytherapy 3D dose distribution for prostate boost treatment created using volume-optimized inverse planning. Red: prostate; yellow: urethra; magenta: needles; green: 15-Gy isodose surface; blue: 10-Gy isodose surface. (b) Corresponding 2D dose distribution on an axial TRUS slice. Green: prostate contour; dark yellow: urethra contour; red: 15.0 Gy reference isodose line; blue: 22.5 Gy isodose line; cyan: 30.0 Gy isodose line; yellow: 7.5 Gy isodose line. (From Kovács, G. et al., *Brachytherapy* 6, 142–8, 2007. With permission.)

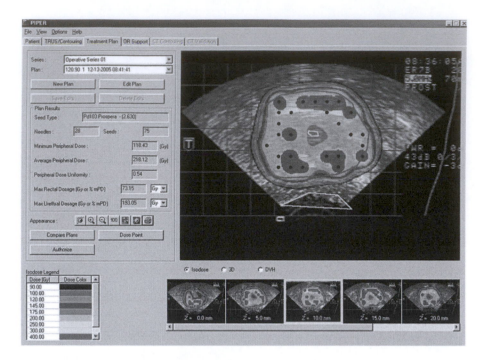

FIGURE 9.6 Axial ultrasound image of a prostate at midgland with ^{103}Pd seed dosimetry superimposed. Needle positions are indicated by blue dots; needles containing seeds at this level by red dots. The prostate margin as originally traced on the intraoperative plan is outlined in pink, the urethra in blue, and the rectum in yellow. The remaining colored regions represent the isodose delivered to each area of the prostate and to the surrounding tissues as indicated in the color code table at the lower left of the figure. The minimum peripheral dose to the entire prostate is 118 Gy. (From Rubens, D.J. et al., *Radiol Clin N Am* 44, 735–48, 2006. With permission.)

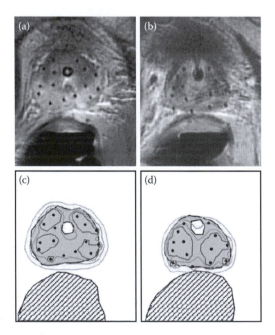

FIGURE 9.8 MRI-guided HDR prostate brachytherapy catheter placement and isodose maps. Brachytherapy was performed at the beginning (a) and end (b) of a 5-week course of external beam radiation therapy. Needle locations are visible as dark dots inside the prostate in these axial FSE images. (c, d) Radiation isodose maps, corresponding to (a) and (b), indicate 150% (red), 125% (orange), 100% (green), and 75% (blue) of the prescribed radiation dose (1050 cGy). The prostate (gray-filled region), urethra (white region inside the prostate), and rectum (hatched region) are also shown. Note that the green, 100% dose contour conforms well to the prostate margin, while overdose of the urethra and rectum is avoided. (From Susil, R.C. et al., *Magn Reson Med* 52, 683–7, 2004. With permission.)

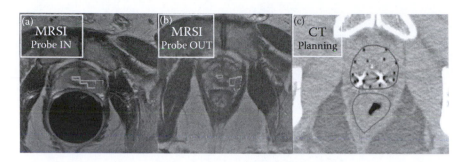

FIGURE 9.9   (a) Original acquisition image with the suspicious regions delineated. (b) Axial probe-out volume with the rotated, translated, and warped spectral delineations. (c) MRSI scores of 4 and 5 are delineated on the planning CT. (From Reed, G. et al., *J Contemp Brachytherapy* 1, 26–31, 2011. With permission.)

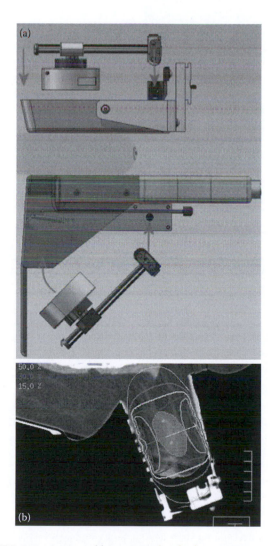

FIGURE 9.10   Integrated breast HDR brachytherapy/mammography unit, AccuBoost (a). Dose distribution in the compressed breast (b). (Courtesy of Nucletron, Veenendaal.)

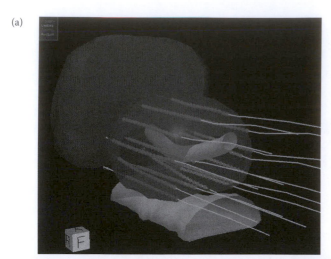

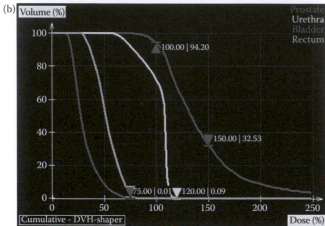

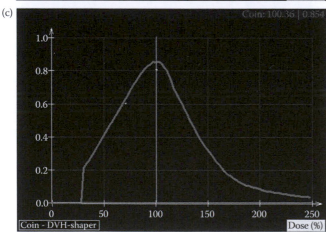

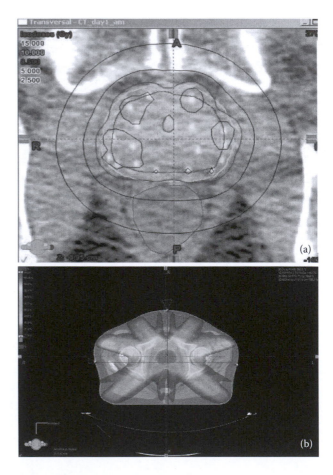

FIGURE 12.11 Example of an HDR prostate implant. (a) 3D view of the contoured prostate as PTV, urethra, rectum, and bladder and the 15 implanted catheters with a total of 282 active SDPs at a step of 2.5 mm. (b) Cumulative DVHs for PTV and the OARs urethra, rectum, and bladder. The critical dose values or dose limits for the three OARs are 120% for urethra and 75% of the reference dose (100%) for rectum and bladder. (c) Conformity distribution (COIN distribution) calculated according to Equation 12.16 and based on the above DVHs, including all three OARs, demonstrates a COIN value of 0.854 for the 100% reference isodose line. The maximum COIN value is 0.857 obtained for the 102.5% isodose value that is very close to the reference dose for that implant and treatment plan. (Courtesy of D. Baltas.)

FIGURE 18.3 Comparison of an HDR prostate brachytherapy implant (a) with a modern intensity-modulated radiotherapy distribution (b) for prostate cancer in which both plans were designed to deliver a high dose to a CTV encompassing the prostate capsule with a 3-mm margin. (Courtesy of P. Hoskin.)

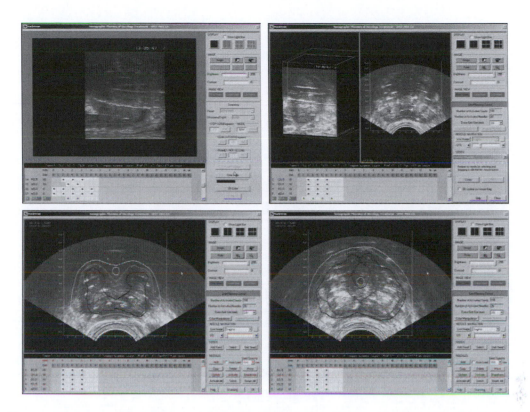

FIGURE 20.2 Ultrasound appearance of an implant in real time. Shown here is the technique of interactive planning. Individual needles can be adjusted and a new plan generated while in the operating room. (Courtesy of A. Polo.)

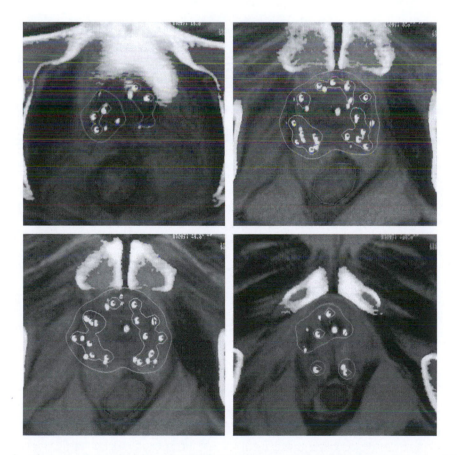

FIGURE 20.3 Post-implant dosimetry images based on CT–MRI fusion. Prostate capsule definition is performed on MRI and the pelvic bones on the CT. The margin effect is shown and cold spots are viewed at the base and the apex of the prostate. (Courtesy of A. Polo.)

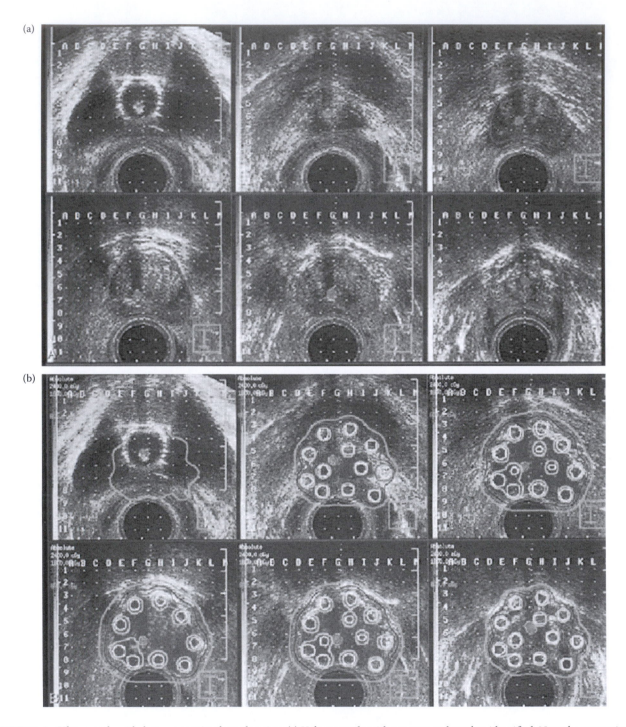

FIGURE 20.4 Ultrasound-guided permanent implant planning. (a) Volume study with prostate and urethra identified. Note the eccentric position of the urethra. (b) Isodose distribution of the same patient. Red color wash is 100%; blue line indicates 75%. (Courtesy of M. Ghilezan.)

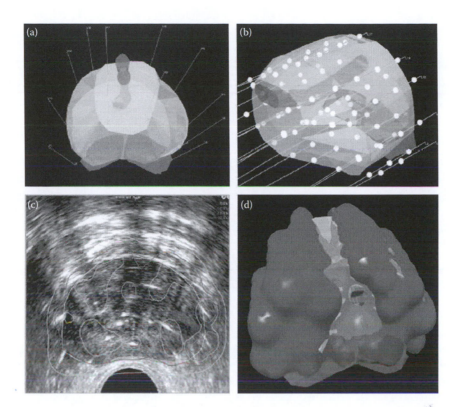

FIGURE 20.5 HDR intraoperative implant using the Nucletron SWIFT guidance system. (a) 3D reconstruction of the prostate gland, urethral trajectory, and needle orientation. (b) Anatomic relationship of the prostate, the urethra, and the needles with the selected dwell positions on each needle. (c) Final TRUS-based intraoperative dosimetry and coverage of the PTV. (d) Dosimetric rendering of the prostate coverage by the 100% isodose cloud in red with urethral sparing from modulating the dwell times and dwell positions. (Courtesy of M. Ghilezan.)

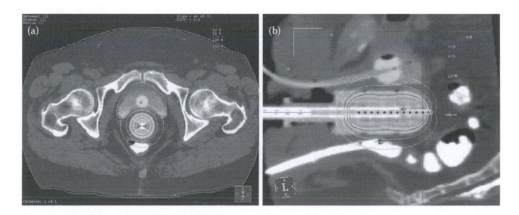

FIGURE 21.5 Axial (a) and sagittal (b) CT images with domed vaginal applicator in good contact with the vaginal walls. Opacification of the bladder and rectum is shown for critical organ dose assessment. (Courtesy of B. Erickson.)

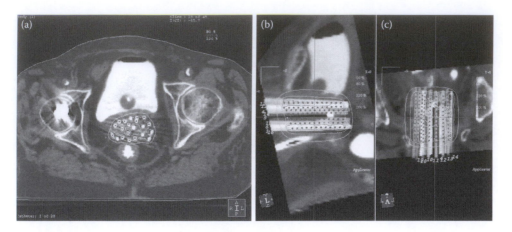

FIGURE 21.6 CT-based transperineal BT dosimetric plan used in the treatment of a vaginal cancer in (a) axial, (b) sagittal, and (c) coronal views with needles and associated isodose curves. Note the relative sparing of the opacified rectum and bladder. (Courtesy of B. Erickson.)

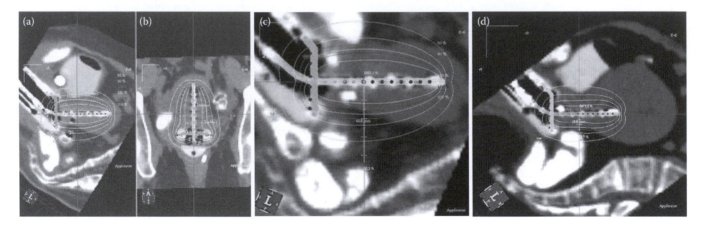

FIGURE 21.7 CT-based dose distribution system in sagittal (a) and coronal (b) views showing the pear-shaped dose distribution of the tandem and ring. Sagittal views (c) of one implant with the sigmoid far from the associated dose distribution and in close proximity to the high dose region of the pear (d). Such loops may go unrecognized with film-based dosimetry. (Courtesy of B. Erickson.)

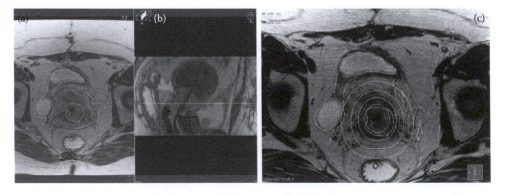

FIGURE 21.8 Axial (a) and sagittal (b) MR images with contours of the GTV, HR CTV, and IR CTV following insertion of a tandem and ring applicator. The HR CTV is not adequately encompassed by the 100% isodose curve due to the volume of the remaining disease (c). Such a patient might benefit from the addition of interstitial techniques. (Courtesy of B. Erickson.)

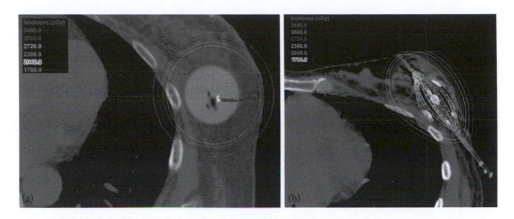

FIGURE 22.8 (a) Comparison of dose distributions for APBI, 34 Gy in 10 fractions, delivered with Contura catheter, and (b) SAVI catheter. (Courtesy of K.L. Leonard.)

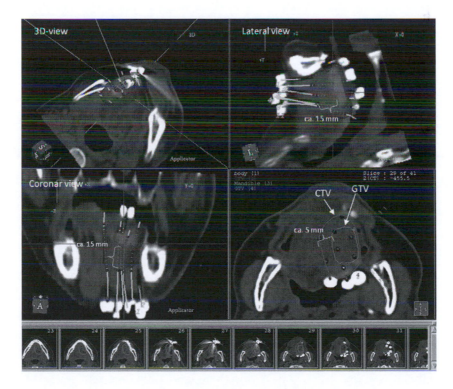

FIGURE 23.3 Examples of the contouring rules for cancer of the mobile tongue (refer to the text for more details). (Courtesy of V. Strnad, University Hospital Erlangen.)

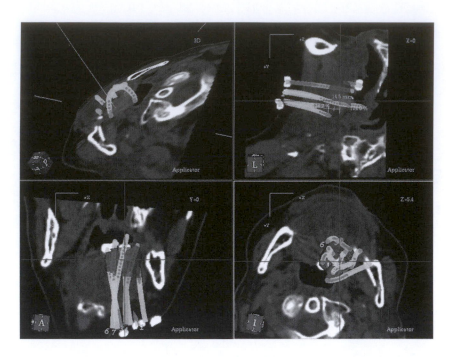

FIGURE 23.4 Typical example of an implant of a tumor in the region of the base of tongue and tonsillar fossa. (Courtesy of V. Strnad, University Hospital Erlangen.)

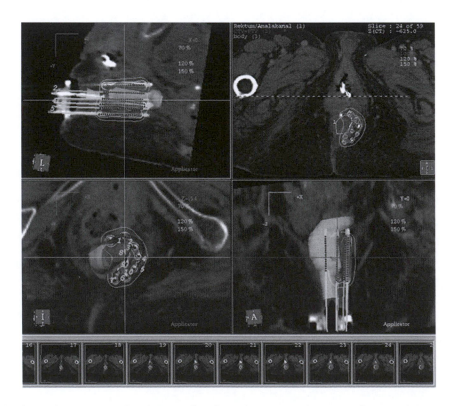

FIGURE 23.8 Example of 2D and 3D imaging of a volume implant for anal carcinoma. (Courtesy of V. Strnad, University Hospital Erlangen.)

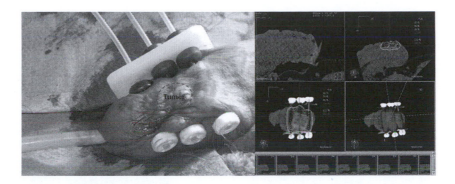

FIGURE 23.10 Example of interstitial brachytherapy of small cancer of the penis using plastic catheters and resulting dose distribution. (Courtesy of V. Strnad, University Hospital Erlangen.)

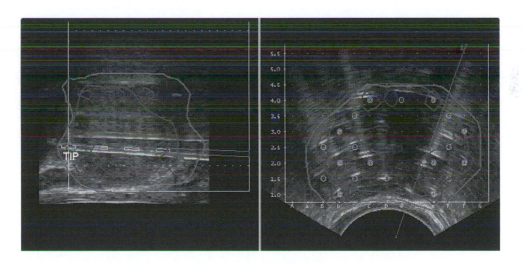

FIGURE 27.4 Illustration of sagittal needle insertion guidance and augmented reality. An axial view from the planning system is shown on the right. The related live US image with augmented reality elements is displayed on the left. See text for details. The figure is a screen capture from the SPOT system (Nucletron BV—an Elekta company, Veenendaal, The Netherlands.) (Courtesy of L. Beaulieu.)

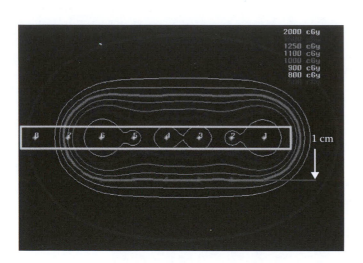

FIGURE 28.5 Dose distribution in the axial plane. Catheters 7 and 8 are not loaded. The blue rectangle represents the applicator. A single dose of 10 Gy is prescribed to 1 cm from the surface of the applicator. (Courtesy of J.J. Nuyttens.)

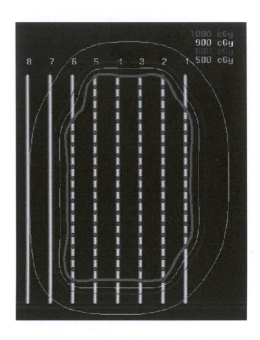

FIGURE 28.6 Dose distribution in the sagittal plane. (Courtesy of J.J. Nuyttens.)

fusion of the image sets, which will again introduce an inherent error due to the variation in patient and organ position between scans and the limitations of fusion software. In a study of MR imaging for reconstruction of HDR cervix treatment planning while distortion was minimal within a defined volume, typically within the center area of the scanner, larger distortions may be found further off from the center (Wills et al. 2010).

An Equal-ESTRO quality audit (Roué et al. 2006) used a mailed phantom geometric check procedure of reconstruction techniques used in brachytherapy TPSs; 152 routinely used reconstructions were checked with orthogonal x-ray systems, simulators, and CT scanning. Eighty-six percent of the results were within an acceptance level after the first check. For the remaining 14%, a second check had to be performed; 98% of the rechecks were within the defined acceptance level. This external audit revealed some errors previously undiscovered in clinical quality control routines. Most errors were systematic, caused by the use of erroneous magnification factors in the reconstruction algorithms for the orthogonal x-ray methods. No errors were identified with the use of CT.

Siebert et al. (2007) investigated the quality of CT- and x-ray–based seed reconstruction procedures using the Kiel phantom equipped with a test configuration composed of 17 nonradioactive seeds. The quality of seed reconstruction CT measurements with varying CT parameters and different seed models was evaluated in six centers; the results showed that when slice thickness or the table index reaches 4 or 5 mm, the accuracy of the CT seed reconstruction decreases in the longitudinal direction: <1.0 mm for 2–3 mm scans; <1.4 mm for 4–5 mm scans.

The accuracy of seed reconstruction for post-implant dosimetry of prostate seed implants has also been investigated comparing CT and MR (De Brabandere et al. 2006) using a 60-seed CT- and MRI-compatible prostate phantom. After implantation of the seeds in the phantom, CT and MRI scans with 3-, 4-, and 5-mm slice thickness were performed. The seed locations were reconstructed in the TPS and compared with the known reference positions. The observed mean reconstruction uncertainties were, in general, smaller for CT than for MRI. For the clinical sequences of both CT and MRI, the mean deviations of the reconstructed seed positions were all within 2.3 mm, which was considered acceptable for clinical use. CT systematically resulted in the lowest D_{90} values, both for the reference and mean observer values. This is explained by the fact that prostate CTV volumes are generally overestimated on CT, resulting in target areas that are not fully covered by the 145 Gy isodose.

Section 16.3 provided the definition of volumes to be delineated from the information obtained with 3D imaging procedures. The radiation oncologist must contour the CTV and OARs on these images. The presence of applicators in the images should not lead to major distortions, for example, by metal artifacts or shielding. In MR images, the rigid applicators (or seed sources) have no water content and therefore show up as—sometimes—large voids in the images. Where possible, CT- and MR-compatible applicators must be used for the insertions. The presence of contrast media for better visualization should be limited as these can similarly lead to artifacts in the images when still present in too large concentrations.

When the radiation oncologist delineates the target and determines the volumes, recent preimplant diagnostic scans can be very helpful in this process. The use of standardized protocols is strongly recommended, for which several internationally accepted recommendations have been published. A full discussion on this for different target areas can be found in the respective clinical chapters of this book. Despite this extensive guidance, however, and even with experienced clinicians, there is considerable variation in the definition of CTV between individuals. This is demonstrated in a study that looked at uncertainties introduced in the postplanning procedure (De Brabandere et al. 2011). Three techniques were compared: (1) CT, (2) CT + T2-weighted MR, and (3) T1 + T2. A variant of the third technique was added to this list: (4) CT + T1$_{int}$ + T2, in which T1-weighted images are used in an intermediate step to register CT with T2. There was a large variation in CT and T2 volume estimation reflected in considerable interobserver variability in D_{90}, with SD values of 23% for CT and 17%–18% for the other techniques using T2 for contouring. Remarkably, the D_{90} values for the reference contour were often higher than the observer mean D_{90} values. Analysis of the contours suggests that, on average, the observers extended the CTV-P cranially outside the reference volume. The study presents data on how much the interobserver variability influences the value of generally accepted reporting parameters such as D_{90} in these cases of prostate implants.

16.7 Dose Calculation

Modern brachytherapy uses reconstruction of the anatomy and geometry of the implant using CT and MR scanning. It is important to be aware of and optimize the geometrical accuracy achieved by these modalities. Proper commissioning and verification at regular intervals of the imaging steps should ensure this quality aspect. The dose calculation itself will depend upon the dosimetry software. Most commonly, at present, dose calculations in brachytherapy are determined by the AAPM recommendations comprised in the TG-43 formalism (Nath et al. 1995; Rivard et al. 2004).

The TG-43 formalism has several advantages over the calculation methods used before its clinical introduction. First of all, it is widely adopted by all main vendors of brachytherapy TPSs and forms the standard for most brachytherapy dose calculation systems at this moment. The numerical data that should be used are available for essentially all source types, and they are contained in easily accessible consensus databases on the Internet. The TG-43 data used clinically for brachytherapy dose calculations are determined with at least two independent methods, one experimental (often using TLD as the measuring method) and the other based on Monte Carlo (MC) calculation. Recommended published values for specific sources are based on a weighted average of these methods. The conditions under which these data should be obtained by investigators are described in more detail in Chapter 10 and further analyzed for their influence on the overall uncertainty of dosimetry in Chapter 16 (DeWerd et al. 2011).

Algorithms currently available for clinical use, however, do not employ MC modeling and do not take into account missing tissue, air cavities, or the influence of shields present in the applicator. A well-known effect that is not taken care of is the interseed shielding effect in a prostate implant containing low-energy photon-emitting sources like ^{125}I, ^{103}Pd, or ^{131}Cs seeds. Thus, the dose defined from a planning computer, based on dose to water rather than soft tissue, will have a number of uncertainties associated with it that may be considerable especially at low energies. The dosimetric concerns with the TG-43 formalism have recently been summarized in a Vision 20/20 article in *Medical Physics* (Rivard et al. 2009). Absorbed dose in water is about –4% and +2% compared to tissue for low- and high-energy photons, respectively. Per centimeter, the attenuation of high-energy photons is about the same between water and tissue. However, there are significant differences in attenuation for low-energy photons, increasing as photon energy decreases. The presence of high-Z materials (shields, interseed effects) can substantially alter dose distributions for low-energy photons. Dose differences of >5% are possible for high-energy sources within 5 cm of the skin. Equivalence of dose and kerma within 5% does not hold true within a few millimeters of high-energy photon-emitting sources. Dosimetric contributions from beta emissions at these distances are ignored in the current TG-43 formalism. See the work of Rivard et al. (2009) for further reading.

16.8 Dosimetry Tools, DVHs, and Radiobiology Modeling

Modern advanced IGBT aims at achieving a high dose to the PTV and defines dose constraints for OARs. Optimization of the treatment plan may help to achieve these goals. It is important to distinguish physical optimization of dwell positions and dwell times of a miniature stepping source technique and biological optimization in search for an optimal time–dose pattern.

Physical optimization is the subject of Chapter 12, where several methods are discussed in detail: catheter-based methods for optimization on dose points or on geometry and anatomy-based methods that belong more to the domain of 3D volume-based advanced brachytherapy. Optimization of dwell positions and dwell times should ideally lead to a physical dose distribution that fulfills the requirements set by the radiation oncologist. Understanding of the different optimization strategies offered by modern TPS is a prerequisite for its proper application in clinical routine. The accuracy of determination of volume and DVH parameters calculated by different brachytherapy TPSs has been studied using a phantom with three different volumes scanned with CT and MRI (Kirisits et al. 2007). An interobserver analysis was based on contouring performed by five persons. The volume of a standard contour set was calculated using seven different TPSs. For five systems, a typical brachytherapy dose distribution was used to compare DVH determination. The interobserver variability (1 SD) was 13% for a small cylindrical

volume, 5% for a large cylinder, and 3% for a conical shape. A standardized volume for a 4-mm CT scan contoured on seven different TPSs varied by 7%, 2%, and 5% (1 SD). Use of smaller slice thickness reduced the variations. A treatment plan with the sources between the large cylindrical shape and the cone showed variations for D_{2cc} of 1% and 5% (1 SD), respectively. Deviations larger than 10% were observed for a smaller source to cylinder surface distance of 5 mm. Modern TPSs minimize the volumetric and dosimetric calculation uncertainties. These are comparable to interobserver contouring variations found in this study.

However, perhaps one of the greatest uncertainties lies in the biological variation that will be encountered within any patient population. In Chapter 19 of this book, there is an extensive discussion on equivalent dose concepts that depend upon parameters such as alpha/beta ratio, half-time of repair, and proliferation rate. While these may be defined in the experimental setting, in clinical practice, they can only be derived around a population mean with each individual having their unique characteristics relating to both normal tissue and tumor response. In the future, genetic fingerprinting may allow us to predict more accurately an individual's radiation response characteristics, but at present, it remains a variable that is only defined by post hoc observation.

Brachytherapy is a rapidly moving field with considerable potential, but also great uncertainties in the final clinical outcome. Strategies for localized high dose deposition to target subregions defined with functional imaging using positron emission tomography (PET), single-photon emission computed tomography (SPECT), dynamic contrast-enhanced MRI (DCE-MRI), magnetic resonance spectroscopic imaging (MRSI) are under investigation. However, before clinical implementation of the new and challenging strategy of dose painting, several issues must be addressed. The clinical and biological rationale for inhomogeneous dose prescription must be further investigated and refined. There are considerable uncertainties in the radiobiological models currently available, discussed in more detail in Chapter 19. Introduction of multiple imaging techniques also leads to a multitude of uncertainties related to image distortion and registration. The physiological changes with time during radiotherapy within a tumor are largely unknown, and further investigations into the kinetics of relevant radiobiological features and strategies for incorporating time dependence in the treatment planning process are required. Finally, new paradigms of dose definition will be required to encompass variable prescription levels within a single plan.

16.9 Clinical Uncertainties in Treatment Delivery

Treatment delivery is the final link in the complex chain of events in modern brachytherapy. Unless meticulous attention is paid to this final aspect, then all the preceding efforts to minimize uncertainties will be to no avail. Movement of the patient from implant to treatment delivery can result in changes in implant and OAR positions; verification prior to radiation delivery is therefore a vital feature of any brachytherapy process. This is further compounded in fractionated high-dose-rate treatments or prolonged pulsed-dose-rate treatments. Implant and OAR movements are largely unpredictable and vary from patient to patient; a single solution such as a fixed CTV to PTV expansion is therefore not appropriate. Careful measurement of catheter position and pre-exposure imaging with corrective action may help in minimizing this effect and should be part of all brachytherapy protocols. This may include ultrasound or fluoroscopic evaluation with the patient in the treatment position. CT or MR verification may enable more accurate reproduction of the implant and volume but may introduce the additional hazard of another patient transfer from scanner to treatment machine. This may be overcome in some settings by a tracked moving couch from scanner to afterloader, but such facilities are rare. In vivo dosimetry may be applied for some cases. Intrafraction movement of the target and the OARs will remain a largely insoluble problem in brachytherapy. They may be minimized by ensuring good bowel preparation and continuous bladder drainage per catheter, and target organ movement may not be critical if the implant moves with the organ, for example, an intrauterine tube or prostate seed implant, or when the organ is fixed by the implant, for example, in case of a high-dose-rate prostate implant. However, the implant procedure itself can lead to hemorrhage and edema with associated swelling of the target organ and surrounding tissues. This may then influence the relative position of the needles or the seeds and, in this way, the dosimetry of the implant.

Applicator shifts and changes in DVH parameters during PDR treatment were measured using sequential MR scans in 10 patients with the applicators in situ (De Leeuw et al. 2009). The average applicator shift relative to the pelvic structures between two scans 24 h apart was 5–6 mm into the ventral direction and 3–4 mm cranially. For a single PDR fraction, the average D_{90} (HR-CTV) on day 2 was 0.2 (SD 2.0) Gy lower than that for day 1. The average increase in D_{2cc} (bladder) was 1.0 (SD 3.0) Gy ($\alpha/\beta3$) for a single PDR fraction. If the effect of both fractions was combined, for one patient, there was a total decrease in D_{90} of 7 Gy ($\alpha/\beta10$), whereas for another patient, the total increase in bladder dose was 12 Gy ($\alpha/\beta3$). In the overall accuracy during PDR brachytherapy, the reconstruction uncertainty is of minor importance. Applicator and/or organ movements during the course of the PDR fraction produce larger uncertainties.

Similar issues can arise in fractionated HDR treatments. Catheter position and dosimetry were evaluated in 20 consecutive patients having a total of 332 catheters undergoing two HDR afterloading brachytherapy fractions over 36 h (Hoskin et al. 2003). The mean interfraction movement of catheters as measured by external length was less than 1 mm, but within the prostate on consecutive CT scans, there was a mean interfraction movement of 11.5 mm away from the prostate base. This has a significant impact on implant dosimetry as measured by D_{90} and the conformal index (COIN), unless corrected by repositioning the catheters. Whitaker et al. (2011) from the Sydney Cancer Centre investigated 48 implants as a conformal boost for treating of prostate cancer. They found that in the 1–3 h between

the planning CT scan and the HDR treatment, catheter displacement can occur. The results showed an average displacement of 7.5 mm (range –2.9–23.9 mm), with 67% of the cases having a displacement of 5 mm or greater. The authors recommended a catheter position verification immediately prior to treatment delivery (Whitaker et al. 2011).

16.10 Quality Management

Quality management of equipment has already been addressed in this book in Chapter 4. In radiation oncology departments, clinical planning must be integrated in the quality management system as its extension. An adequate definition of process maps in brachytherapy will give rise to suitable procedures and instructions and their integration in the work flow. An in-depth evaluation of procedures, preferably performed in advance of the clinical introduction of a new treatment modality or a new technique, in the form of a risk assessment and analysis is very helpful and recommended to identify the weak points and to avoid future problems. This is elaborated in Chapter 30 on the modern paradigm for quality management in brachytherapy.

The quality management in the department aims at a safe operation of the equipment and a minimization of the variations in treatment delivery. For brachytherapy equipment, this means, for example, that the source is positioned as accurately as possible, with a negligible deviation of the dwell times of the source. HDR and PDR afterloaders are delivery systems designed to position a miniaturized source with the smallest possible uncertainty for a predetermined dwell time. QC procedures must ensure that criteria are met. As discussed in several recommendations on QA, and summarized in Chapter 4 on QA of equipment, a positioning accuracy of better than 1 mm and a time resolution better than 1% are achievable for these systems. It is noted, however, that the positioning checks are performed with straight transfer tubes and catheters. In clinical circumstances, the transfer tubes are often bent, while the applicator itself may have a strong curvature; for example, with the ring applicator in gynecology, the source cable moves within the lumen of the applicator, which is wider than the diameter of the cable/source and thus does not follow a perfect curve. Up to 4-mm slack in the source cable in a ring applicator has been demonstrated (Hellebust et al. 2010), which can increase the dose to rectum/bladder by 20% in individual patients.

The effect of equipment errors related to source positioning has been investigated (Kertzscher et al. 2011) in a phantom study with two PDR gynecological treatments using tandem and ring applicators and one HDR prostate treatment using interstitial needles. Treatment errors were imposed including interchanged pairs of afterloader guide tubes and 2–20 mm source displacements. The error detection capacity was evaluated at three dose levels: dwell position, source channel, and fraction. Out of 20 interchanged guide tube errors, time-resolved analysis identified 17 while fraction level analysis identified 2. Channel and fraction level comparisons could miss 10-mm dosimeter displacement errors, but dwell position dose rate comparisons correctly identified displacements ≥5 mm.

16.11 Conclusions

An analysis of clinical uncertainties is a very complex task. Brachytherapy is performed on a wide variety of treatment sites and with different techniques and fractionation schemes. The examples provided in this chapter demonstrate that systematic variations such as an erroneous magnification factor in the applicator reconstruction and random variations such as interfraction movement affect the dosimetry of the individual patient in a totally different manner. There is a great dependence on the type of application under consideration. This makes it very difficult to provide an overall uncertainty estimate based on the process of "adding up" the uncertainties of individual steps in the procedure. First overviews of systematic analysis have been presented at conferences, such as in the work of Kirisits et al. (2010, 2011a,b). The action level, at a relative standard deviation of 5% as suggested by Brahme et al. (1988) above, which is recommended to introduce actions to improve the accuracy in dose delivery, is still difficult to achieve in the clinical practice of brachytherapy. A proper quality management system in a brachytherapy department should be directed toward identifying the systematic errors affecting all patients and minimizing the random errors due to treatment- and patient-specific issues. The consequences for the clinical treatment protocol are further and systematically discussed in Chapter 17 on the concept of applying margins to the target volumes in brachytherapy.

References

Brahme, A., Chavaudra, J., Landberg, T. et al. 1988. Accuracy requirements and quality assurance of external beam therapy with photons and electrons. *Acta Oncol* Suppl 1.

Butzbach, D., Waterman, F.M., and Dicker, A.P. 2001. Can extraprostatic extension be treated by prostate brachytherapy? An analysis based on postimplant dosimetry. *Int J Radiat Oncol Biol Phys* 51:1196–9.

Crook, J., Patil, N., Ma, C., Mclean, M., and Borg, J. 2010. Magnetic resonance imaging-define treatment margins in iodine-125 prostate brachytherapy. *Int J Radiat Oncol Biol Phys* 77:1079–84.

De Brabandere, M., Kirisits, C., Peeters, R., Haustermans, K., and Vanden Heuvel, F. 2006. Accuracy of seed reconstruction in prostate postplanning studied with a CT- and MRI-compatible phantom. *Radiother Oncol* 79:190–7.

De Brabandere, M., Hoskin, P., Haustermans, K., Van den Heuvel, F., and Siebert, F.-A. 2012. Prostate post-implant dosimetry: Interobserver variability in seed localisation, contouring and fusion. Accepted for publication *Radiother Oncol* (July 03, 2012).

De Leeuw, A.A., Moerland, M.A., Nomden, C., Tersteeg, R.H., Roesink, J.M., and Jürgenliemk-Schulz, I.M. 2009. Applicator reconstruction and applicator shifts in 3D MR-based PDR brachytherapy of cervical cancer. *Radiother Oncol* 93:341–6.

DeWerd, L.A., Ibbott, G.S., Meigooni, A.S. et al. 2011. A dosimetric uncertainty analysis for photon-emitting brachytherapy sources: Report of AAPM Task Group No. 138 and GEC-ESTRO. *Med Phys* 38:782–801.

Dimopoulos, J.C., DeVos, V., Berger, D. et al. 2009. Inter-observer comparison of target delineation for MRI-assisted cervical cancer brachytherapy: Application of the GYN GEC-ESTRO recommendations. *Radiother Oncol* 91:166–72.

Haie-Meder, C., Pötter, R., Van Limbergen, E. et al. 2005. Recommendations from Gynaecological (GYN) GEC-ESTRO Working Group (I): Concepts and terms in 3D image based 3D treatment planning in cervix cancer brachytherapy with emphasis on MRI assessment of GTV and CTV. *Radiother Oncol* 74:235–45.

Han, B., Wallner, K., Aggarwal, S. et al. 2000. Treatment margins for prostate brachytherapy. *Semin Urol Oncol* 18:137–41.

Hellebust, T.P., Kirisits, C., Berger, D. et al. on behalf of the Gynaecological (GYN) GEC-ESTRO Working Group. 2010. Recommendations from Gynaecological (GYN) GEC-ESTRO Working Group: Considerations and pitfalls in commissioning and applicator reconstruction in 3D image-based treatment planning of cervix cancer brachytherapy. *Radiother Oncol* 96:153–60.

Hoskin, P.J., Bownes, P.J., Ostler, P., Walker, K., and Bryant, L. 2003. High dose rate afterloading brachytherapy for prostate cancer: Catheter and gland movement between fractions. *Radiother Oncol* 68:285–8.

ICRU Report 24. 1976. Determination of the absorbed dose in a patient irradiated by beams of X or gamma rays in radiation therapy procedures. International Commission on Radiation Units and Measurements. Washington, DC.

ICRU Report 58. 1997. Dose and volume specification for reporting interstitial brachytherapy. International Commission on Radiation Units and Measurements. Washington, DC.

Kertzscher, G., Andersen, C.E., Siebert, F.A., Nielsen, S.K., Lindegaard, J.C., and Tanderup, K. 2011. Identifying afterloading PDR and HDR brachytherapy errors using real-time fiber-coupled Al(2)O(3):C dosimetry and a novel statistical error decision criterion. *Radiother Oncol* 100:456–62.

Kirisits, C., Siebert, F.A., Baltas, D. et al. 2007. Accuracy of volume and DVH parameters determined with different brachytherapy treatment planning systems. *Radiother Oncol* 84:290–7.

Kirisits, C., Ballester, F., Baltas, D. et al. 2010. Uncertainty in modern brachytherapy-analysis concept. *Radiother Oncol* 96(Suppl 1):S71.

Kirisits, C., Rivard, M.J., Ballester, F. et al. 2011a. Dosimetric uncertainties in the practice of clinical brachytherapy. *Brachytherapy* 10:S32–3.

Kirisits, C. Rivard, M.J., Ballester, F. et al. 2011b. Comprehensive study on dosimetric uncertainties in clinical brachytherapy. Abstr. CAMCT Conference, Braunschweig, November 1, 2011. Proceedings submitted for publication in *Metrology* 2012.

Merrick, G.S., Butler, W.M., Wallner, K.E. et al. 2007. Dosimetry of an extracapsular anulus following permanent prostate brachytherapy. *Am J Clin Oncol* 30:228–33.

Mijnheer, B.J., Battermann, J.J., and Wambersie, A. 1987. What degree of accuracy is required and can be achieved in photon and neutron therapy? *Radiother Oncol* 8:237–52.

Nag, S., Shi, P., Liu, B., Gupta, N., Bahnson, R.R., and Wang, J.Z. 2008. Comparison of real-time intraoperative ultrasound-based dosimetry with postoperative computed tomography-based dosimetry for prostate brachytherapy. *Int J Radiat Oncol Biol Phys* 70:311–7.

Nath, R., Anderson, L.L., Luxton, G., Weaver, K.A., Williamson, J.F., and Meigooni, A.S. 1995. Dosimetry of interstitial brachytherapy sources: Recommendations of the AAPM Radiation Therapy Committee Task Group No. 43. *Med Phys* 22:209–34.

Polo, A., Salembier, C., Venselaar, J., and Hoskin, P. on behalf of the PROBATE group of the GECESTRO. 2010. Review of intraoperative imaging and planning techniques in permanent seed prostate brachytherapy. *Radiother Oncol* 94:12–23.

Pötter, R., Haie-Meder, C., Van Limbergen, E. et al. 2006. Recommendations from gynaecological (GYN) GEC ESTRO working group (II): Concepts and terms in 3D image-based treatment planning in cervix cancer brachytherapy—3D dose volume parameters and aspects of 3D image-based anatomy, radiation physics, radiobiology. *Radiother Oncol* 78:67–77.

Rivard, M.J., Coursey, B.M., DeWerd, L.A. et al. 2004. Update of AAPM Task Group No. 43 Report: A revised AAPM protocol for brachytherapy dose calculations. *Med Phys* 31:633–74.

Rivard, M.J., Venselaar, J.L.M., and Beaulieu, L. 2009. The evolution of brachytherapy treatment planning. *Med Phys* 36:2136–53.

Roué, A., Ferreira, I.H., Van Dam, J., Svensson, H., Venselaar, J.L. 2006. The EQUAL-ESTRO audit on geometric reconstruction techniques in brachytherapy. *Radiother Oncol* 78:78–83.

Siebert, F.A., De Brabandere, M., Kirisits, C., Kovács, G., and Venselaar, J. 2007. Phantom investigations on CT seed imaging for interstitial brachytherapy. *Radiother Oncol* 85:316–23.

Whitaker, M., Hruby, G., Lovett, A., and Patanjali, N. 2011. Prostate HDR brachytherapy catheter displacement between planning and treatment delivery. *Radiother Oncol* 101:490–4.

Wills, R., Lowe, G., Inchley, D., Anderson, C., Beenstock, V., and Hoskin, P. 2010. Applicator reconstruction for HDR cervix treatment planning using images from 0.35 T open MR scanner. *Radiother Oncol* 94:346–52.

Xue, J., Waterman, F., Handler, J., and Gressen, E. 2006. The effect of interobserver variability on transrectal ultrasonography-based postimplant dosimetry. *Brachytherapy* 5:174–82.

17

Margin Concepts in Image-Guided Brachytherapy

Kari Tanderup
Aarhus University Hospital

Christian Kirisits
Comprehensive Cancer Center Vienna

Jacob C. Lindegaard
Aarhus University Hospital

Erik Van Limbergen
University Hospital Gasthuisberg

André Wambersie
Catholic University of Louvain

Richard Pötter
Comprehensive Cancer Center Vienna

17.1 GTV, CTV, and PTV Concepts in External Beam Radiotherapy and Brachytherapy

Gross tumor volume (GTV) and clinical target volume (CTV) are oncological volume concepts that must be treated to a certain dose in order to achieve the aim of radiotherapy (ICRU 50 1993; ICRU 62 1999). GTV and CTV are not dependent on the radiotherapy technique or on treatment planning. The actual delineation depends on the imaging technique used, the time of delineation, and the experience and judgment of the radiation oncologist. GTV and CTV are used in brachytherapy (BT) in the same way as in external beam radiotherapy (EBRT), since these volumes are general oncological concepts applicable for any radiotherapy technique. Organ movements and inaccuracies in beam setup and patient setup are potential sources of uncertainties during EBRT. It is possible to compensate for these uncertainties during dose planning by using a planning target volume (PTV), which is constructed by adding a margin to the CTV. The PTV concept was officially established and recommended in ICRU reports (ICRU 50 1993; ICRU 62 1999; ICRU 71 2004; ICRU 78 2007; ICRU 83 2010). The concept has been successfully and widely used within EBRT. During dose planning, it is aimed to cover the PTV with the prescription dose. The consequence of margin application is that the prescription dose plateau is increased in size. Therefore, the risk is reduced that geometric uncertainties will lead to under-dosage of the CTV during fractionated EBRT. In general, this is not linked to a dose escalation, but only to an increase in the treated volume.

For BT, it has previously often been assumed that the PTV equals the CTV since the radioactive source(s) is assumed to be bound or fixed together with the target (Hoskin 2008; Pötter et al. 2002; Strnad 2004; Van Limbergen 2003). However, even in this typical situation of BT, recent research has shown that there are uncertainties present in the target and organ-at-risk (OAR) topography, applicator reconstruction, dose planning, and delivery (de Leeuw et al. 2009; Simnor et al. 2009; Tanderup et al. 2010b). In principle, it would therefore be ideal to introduce a BT PTV analogue to EBRT by adding margins that could compensate for such uncertainties. However, the characteristic BT dose distribution is fundamentally different from the uniform dose plateaus that are typically used with EBRT. Due to dose gradients throughout the target, it is not generally possible to create homogeneous dose plateaus in BT. As a consequence, the impact of uncertainties on delivered dose presents in a different way in BT as compared to EBRT. Therefore, it is not straightforward to adopt a PTV concept in BT by adding margins to the CTV analogue to the concept of EBRT PTV (Tanderup et al. 2010b).

17.2 Uncertainties in BT

The concept of 3D image-guided BT involves imaging with the applicator in place and treatment planning according to the anatomy and tumor extension at the time of BT (De Brabandere et al. 2008; Fenkell et al. 2011; Jürgenliemk-Schulz et al. 2009, 2010; Pötter et al. 2006, 2008a,b; Tanderup et al. 2010a). Even with such full image guidance and adaptation, uncertainties in BT are still present. Geometrical uncertainties are related to source positioning and to reconstruction of the applicator (Hellebust et al. 2010). Internal uncertainties are caused by potential displacements of the applicator relative to the target and/or movement of OARs during BT irradiation or in between imaging and dose delivery (Kirisits et al. 2006; Simnor et al. 2009). Internal uncertainties are usually better controlled for the target in BT than in EBRT. The radiation source is placed directly in the tumor region, and the sources will, in general, move together with the target. This is somewhat different for OARs, which may move independently of the source channels.

Source positioning uncertainties are monitored with continuous QA, typically at source exchange and during applicator commissioning. The tolerance limit for source positioning uncertainties in a straight catheter is usually ±1 mm (Venselaar and Pérez-Calatayud 2004). In curved applicators—such as the ring applicator—source positioning uncertainties can be larger, and source misplacements of 2–4 mm may appear (Hellebust et al. 2007; Kohr and Siebert 2007). However, during reconstruction, it is possible to correct or partly correct for this kind of positioning error in curved rigid applicators where the source path is reproducible (Hellebust et al. 2010).

Applicator reconstruction uncertainties are dependent on image modality, image quality, and the procedure used for reconstruction. In general, CT makes it possible to arrive at an accurate definition of the source channel (Hellebust et al. 2007), whereas MRI may be associated with more uncertainties due to challenges in discrimination of the source channels on MRI (Berger et al. 2009; de Leeuw et al. 2009; Haack et al. 2009; Kirisits et al. 2005; Wills et al. 2010). The uncertainties are larger in the direction perpendicular to the slice orientation due to the slice thickness and the consequent blurring of sagittal and coronal reconstructed images. In cases where fusion of two image sequences (e.g. CT and MR) is used during the reconstruction procedure, the fusion uncertainties add to the reconstruction uncertainties. Fusion uncertainties are typically also largest in the direction perpendicular to the slice orientation.

Image distortions on MRIs may also contribute to geometric uncertainties. However, for the most common sequences that are used for MRI-guided BT, the image distortions are negligible in the center of the magnet where the applicator, the target, OARs, and the general region of interest are situated (Dimopoulos et al. 2012).

In summary, geometric uncertainties are typically larger in the direction along source catheters since reconstruction uncertainties, fusion uncertainties, and source positioning uncertainties are usually largest in this direction. In the direction perpendicular to source catheters, the uncertainties are most often less pronounced (Haack et al. 2009; Wills et al. 2010).

In image-guided BT, there may be a time delay between imaging and dose delivery due to the procedure of contouring and dose planning, which may take around 1–5 h depending on the infrastructure and on the complexity of the implant and the dose plan. The applicator and/or organs may move during this period, mainly dependent on the application technique and the anatomical site. Furthermore, there will be additional risk of movement during and in between pulses/fractions with pulsed-dose-rate (PDR) BT (de Leeuw et al. 2009), or with high dose rate (HDR) if several fractions are delivered based on one applicator insertion and imaging session (Beriwal et al. 2009; Hoskin et al. 2003; Kirisits et al. 2006; Mullokandov and Gejerman 2004). The stability of the applicator is dependent on the applicator type, the methods of fixation, and the handling of the patient.

17.3 General Considerations on Dose Distribution and Use of Margins in Afterloading BT

A BT point source delivers a dose that falls off approximately by the square of the distance to the source. When several source positions are combined in afterloading BT, the dose distribution is modulated. A number of source dwell positions along a catheter will lead to a dose distribution that is similar to that from a radioactive wire. The dose distribution is of a different nature in the direction along the axis of source catheters (longitudinal direction) and in the direction perpendicular to these (radial direction). The longitudinal direction is, for example, along interstitial needles, along the tandem in intracavitary BT, along an esophageal tube, or along a vaginal cylinder.

17.3.1 Longitudinal Direction

In the longitudinal direction, the dose distribution may be elongated when it is possible to introduce and load extra source positions just at the edge and/or outside the target. This requires that the active source lengths are longer than the CTV. In this way, a dose plateau can be created, which extends beyond the CTV. This approach will make the dose distribution more robust toward uncertainties in the direction along source catheters. The dose distribution in the central region of the implant is typically almost not affected by such a modification of dwell positions at the edge of or outside the target.

17.3.2 Radial Direction

The radial dose falloff is almost exclusively determined by physics (inverse square law) and is mainly determined by the distance to the source catheter. The radial dose gradients cannot be manipulated to become less steep by modifying the loading pattern, and it is not possible to obtain a homogeneous dose plateau in this direction. Application of margins in the radial direction and renormalization according to the expanded volume will only lead to blowing up the isodoses throughout the target volume. This is equivalent to a dose escalation. This is fundamentally different from the situation in EBRT, where the dose plateau is increased in size by application of a margin, but no dose escalation is performed (Figure 17.1).

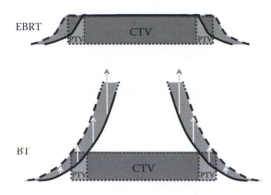

FIGURE 17.1 Effect of margins on dose distribution in EBRT and in intracavitary cervix cancer BT. In EBRT (upper panel), a PTV margin will result in an increase in the volume irradiated to a high dose. The dose plateau becomes larger in size, but the CTV dose remains unchanged. In BT (lower panel), a PTV margin into the lateral and anterior–posterior directions and a renormalization of dose according to the PTV will result in a general dose escalation. The dose throughout the CTV and OARs will systematically increase from inner to outer dose profile. (Modified from Tanderup, K. et al. *Radiother Oncol* 97(3), 495–500, 2010.)

The radial dose gradient depends strongly on the distance to the source channel. At a depth of 4 mm from a source channel (with loading similar to a line source), the dose gradient is around 50% per millimeter. This gradient decreases to around 6%–7% at a distance of 20 mm and 5% at 35 mm. Application of a radial margin could therefore result in a dose escalation of as much as 50% per millimeter of margin applied if the target has a depth of 4 mm from the source channel (e.g., in intravascular BT). The escalation becomes less dramatic but remains still very significant at larger distances to the source channel. In a typical cervical cancer case with intracavitary BT (vaginal and intrauterine sources), an application of margins of 3 and 5 mm at a source distance of 20 mm would result in dose escalations of around 25% and 40%, respectively.

Taking these general considerations into account, application of PTV margins in the radial direction is strongly discouraged since this would lead to an overall dose escalation for the patient and for the entire patient population. Therefore, it is not meaningful to systematically apply margins in all directions in BT except for nonuniform margins, which are dependent on the direction of source catheters.

17.4 Intracavitary BT in Cervix Cancer

Intracavitary BT in cervix cancer is typically administered with the use of a combination of intrauterine sources (tandem) and vaginal sources (e.g., ring, ovoids, mold). When 3D image-guided BT is performed, adaptive target volumes are defined according to the tumor extension at the time of BT [high risk CTV (HR CTV)] and according to the extension at the time of diagnosis [intermediate risk CTV (IR CTV)] (Haie-Meder et al. 2005). According to the general considerations of margins

outlined above, it is relevant to consider application of PTV margins in the longitudinal direction, which is into the uterus and maybe into the vagina, if clinically appropriate. PTV margins in the direction perpendicular to the tandem will result in dose escalation and should not be applied.

With regard to the uterine direction, the tandem is most often longer than the HR CTV. In such cases, it is possible to introduce and load extra source dwell positions just at the edge or above the target in order to obtain a margin between the HR CTV and the isodose corresponding to the prescribed dose (Figure 17.2). It is particularly important to obtain a margin in this direction because of considerable target contouring uncertainties due to the difficulties in discrimination between the cervix and uterine corpus on CT as well as on MR images. Generous loading in the longitudinal direction will make the dose distribution more robust toward such uncertainties in this direction, and most often, the additional treated volume will be inside the uterine corpus leading to no significant additional irradiation of normal tissue. This has some correspondence to the practice of traditional BT schools where the whole tandem used to be loaded, for various reasons. In cases where the bowel or sigmoid is located in close proximity to the tandem, the application of margin has to be critically considered.

Application of margins into the vaginal direction depends on whether the HR CTV extends into the vagina. Typically, the use of ring, ovoid, or vaginal mold sources produces a high dose volume that extends typically 1–2 cm below the cervix into the vagina. In cases without specific vaginal extension, there is therefore a region with ample dose below the cervical target (when ring- or ovoid-type applicators are used), and no

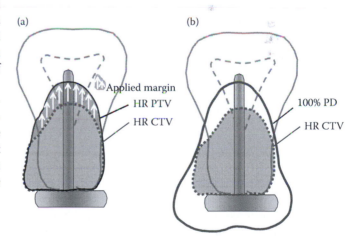

FIGURE 17.2 Schematic figure indicating ring applicator and uterus in the coronal view. HR PTV is defined as an expansion of the HR CTV in the longitudinal direction (a). The tandem is loaded in the upper part in order to maintain a margin between the HR CTV and the pear-shaped isodose in the longitudinal direction (b). The size of the margin should reflect the geometric uncertainties and the target contouring uncertainty due to difficulties in discrimination between cervix and uterine corpus. (Modified from Tanderup, K. et al. *Radiother Oncol* 97(3), 495–500, 2010.)

additional margins in this direction are needed. In cases where the HR CTV extends into the vagina, additional loading into the vagina may be required, and a caudal PTV safety margin can be used according to the same principles as a margin extending into the uterus. However, dose at the rectum has to be carefully taken into account.

17.5 Interstitial BT

Application of a PTV concept in interstitial BT has not yet been systematically addressed in the traditional BT literature, for example, for breast, head and neck, gynecological, or sarcoma BT. For prostate image-guided BT, a margin concept has been used from the very beginning (Ash 2002) and is still widely recommended (Kovacs et al. 2005; Salembier et al. 2007). A discussion of a margin concept for interstitial image-guided BT should, in our opinion, go partly along the same routes as for intracavitary BT. There are similar considerations of the difference between applying a margin along and perpendicular to source channels. Applying an isotropic PTV margin around the prostate capsule as CTV is an example of an introduction of a margin in the direction perpendicular to the source channels (lateral and anterior/posterior directions). If the source channels are inside the CTV and more than around 4 mm inside the CTV border, the application of a margin will escalate the dose in the central part of the target. The mechanism is comparable to the situation in intracavitary BT if a margin is applied perpendicular to the tandem. Dose gradients in interstitial BT are even larger than in intracavitary BT due to the close proximity between sources and target, and the impact of application of margins will be more pronounced.

In the classical Paris interstitial dosimetry system, the reference isodose is defined to be 85% of the mean central dose (MCD) (ICRU 58 1997). The reference isodose is assumed to be identical to the contour of the CTV. The lateral margins (single plane implant/square or triangle multiple plane implant) are calculated based on the specific parameters of the number of and the distance between the iridium wires related to the dimensions of the CTV and are expressed as the distance between outer sources and the reference isodose, which is assumed to be identical to the CTV contour. This is not a PTV margin concept, but a concept describing how far away from the sources the reference isodose is located, which encompasses the CTV. The size of the lateral or safety margin depends on the implant and is typically within 20%–40% of the distance between source lines, e.g., 2–4 mm for a source separation of 10 mm. Increasing the margin beyond the 85% MCD dose corresponds to prescribing to a reference isodose that is lower than 85% MCD, and this will escalate the dose to the central part of the implant. An example of this is demonstrated in a study of interstitial breast BT by Resch et al. (2002). In this study, the dose was prescribed to a distance of around 10 mm from the outer sources, and this is considerably larger than the depth of the 85% reference isodose. This increased their MCD by 30%–40% as compared to the situation where dose prescription would have been performed at 85% MCD. This strategy

means that a larger volume was treated to the prescribed dose as compared to what would be expected according to the classical Paris dosimetry system. At the same time, the dose in the central region of the implant became 30%–40% higher.

In interstitial BT, it has to be taken into account that the treated volume may also be manipulated by inserting extra needles/catheters/sources. Such preimplantation PTV can be considered as the volume that is planned to be covered through the dose distribution from the interstitial sources including the CTV and, if appropriate, an additional margin that takes into account uncertainties. In other words, the preimplantation-related PTV is a volume realized before/during implantation and not during dose planning when the needles are already in place. This is conceptually different from a post-implantation PTV, which is a static volume defined on treatment planning images by adding a margin to the CTV. The preimplantation PTV may increase the treated volume beyond the CTV by loading, for example, needles near or at the edge of the CTV or even outside, whereas a post-implantation PTV, by adding an overall margin to the CTV at the time of dose planning, may lead to dose escalation if there is no source loading available at the edge or outside the CTV border.

With the interstitial Paris system using iridium line sources, it was prescribed to extend the loading beyond the target volume in the direction of needles/catheters. This meant that the active source lengths should be 1.4 times longer than the CTV. In this way, it was possible to compensate for the dose falloff toward the longitudinal border of the target, and the target could be fully covered with the prescribed dose. For the lateral margins, there was a calculation formalism that mainly took into account the physics parameters of needle configuration (Paris system, see above). With modern possibilities of stepping source technology and dose optimization, it is possible to increase the catheter loading toward the longitudinal border of the CTV such that the active source length is more similar to the length of the CTV (Major et al. 2002; Nath et al. 1995). However, full target conformity (CTV) should not be aimed for since compensation of uncertainties in interstitial BT is very relevant in the direction along catheters (Hoskin et al. 2003) to make the dose delivered to the CTV robust to uncertainties of catheter movement and reconstruction uncertainties in this direction. If the needles/catheters are extending beyond the CTV, additional loading can be applied just outside the CTV. Special attention has to be paid in breast (image-guided) BT to avoid overdose in or close to the skin in the longitudinal direction. On the contrary, in (image-guided) head-and-neck BT, special attention has to be paid to assure sufficient dose coverage of the CTV at the mucosal surface, in particular, when using stepping source technology without the classical loops, for example, in tongue implants (Ash and Gerbaulet 2002). In case of interstitial prostate or gynecological BT, there may be consideration for nearby OARs that prevent loading beyond the CTV.

As in intracavitary and intraluminal BT, it is strongly recommended for interstitial BT to avoid application of margins after implantation in the direction perpendicular to the source channels since this will inevitably lead to dose escalation in the whole implant. Uncertainties in the perpendicular direction, if relevant, have to be taken into account before implantation of source catheters within the procedure of image-guided preplanning by adapting the source positions appropriately.

Even more complicated than geometric uncertainties in interstitial BT are uncertainties due to changes in anatomy and topography, for example, edema and tissue swelling after insertion of seeds as seen in prostate BT. The swelling has an impact on the delivered dose, and often, the consequence is that less dose is delivered to the CTV than is estimated from imaging performed just after implantation. The amount of swelling is very individual and difficult to predict, and it is recommended to perform post-implant evaluation of the dose by performing imaging at a relevant time point after implantation depending on the half-life of the isotope (Nag et al. 2004; Salembier et al. 2007). Post-implantation evaluation makes it possible to estimate the delivered dose more accurately, but it does not make it possible to compensate for uncertainties. It is currently not clear whether a PTV concept could be meaningful in the context of tissue swelling in prostate BT. A simple margin around the prostate cannot fully model the tissue swelling and the impact on dose, since swelling means that the whole organ is expanding and that the seeds are moving together with the tissue (Han et al. 2000). A more advanced approach may be of interest for the future in order to predict and compensate for the swelling of tissue and movement of seeds.

17.6 Intraluminal BT

For intraluminal BT a 3D image-based comprehensive target volume concept has not been fully elaborated. This includes the issue of setup uncertainties, which have consequently not been fully addressed yet. However, significant clinical experience has been accumulated—mainly based on 2D and only partly on 3D imaging—which underlines that, in general, there are major uncertainties in positioning intraluminal applicators appropriately in regard to the radial and longitudinal direction of the organ of interest and the extension of the GTV (Pötter et al. 2002; Pötter and Van Limbergen 2002; Van Limbergen and Pötter 2002). Furthermore, there are various uncertainties due to internal organ motion (e.g., lung/bronchus, heart). Both types of uncertainties are mainly dependent on tumor and organ site and the type of applicator used. Three-dimensional imaging technologies—including also intraluminal ultrasound devices—and their translation into treatment protocols may become important for a more comprehensive 3D image-based target volume concept. Setup uncertainties in 3D will also need to be addressed appropriately.

During the period of endovascular BT, more than one decade ago, major attention had to be paid to set up uncertainties due to clinical failures occurring frequently at the edge of the treated volume (Syeda et al. 2002). The GEC-ESTRO recommendation on intravascular BT therefore included a planning target length (Pötter et al. 2001) taking into account source positioning and internal organ motion uncertainties. After initial clinical experience with intracoronary and femoral BT, it became clear that the positioning procedure is linked to major uncertainties due to various reasons. Especially in case of coronary BT, additional applicator movement was observed in regard to the target due to the moving heart. Based on retrospective analysis, these uncertainties could be quantified, and margins for the PTL were proposed (Schmid et al. 2004). The application of such margins even showed an impact on clinical results (Syeda et al. 2002).

Deliberately, a PTL and not a PTV was introduced for intravascular BT. It seemed that uncertainties in the radial direction could not be compensated for by adding margins on top of the target depth, although there were major position uncertainties of the intraluminal catheter due to a non-optimal relation between the catheter and the vessel diameter. On top of this, an individual definition of a target depth along the length of the vessel was clinically unpractical, although intravascular US (IVUS) was partly introduced. The choice of the active source length, which was fixed to a delivery device or to the number of dwell positions, depends on the reference isodose length (RIL), which can be reached with these source configurations (Kirisits et al. 2002). The RIL was defined to reach a certain threshold (90% isodose) within the PTL at a reference depth in the vessel wall. This concept is very similar to an external beam, where margins based on retrospective analysis were derived and could be applied in daily clinical routine.

Clinical scenarios similar to intravascular BT are seen in endobronchial and endo-esophageal BT, both with regard to uncertainties of source catheter positioning and with regard to shortcomings and difficulties of definition of the target depth. In the GEC-ESTRO handbook of BT, the concept of PTL was adapted for intraluminal techniques in general (Pötter et al. 2002). Placement of intraluminal catheters in general and in particular in the bronchus and the esophagus are linked to uncertainties. A general PTV concept was described; however, this concept was only specified for the length and depth of this volume, and therefore in 2D rather than in 3D. For the PTL this concept was detailed in particular for the esophagus. No margin was recommended for the radial direction. A concept for reporting possible minimum and maximum doses was proposed as had been applied in endovascular BT and as proposed for the esophagus (Pötter et al. 2002). The same concept can also be applied for bronchus.

These considerations mainly based on 2D experience as reported and discussed in the literature should also be taken into account when moving to a volume concept. Margins as used in 2D should then also be used for 3D treatment planning. For the radial direction, this means that application of margins is strongly discouraged for similar reasons as for intracavitary BT and interstitial BT. The sharp dose falloff around the central source in intraluminal BT implies that any margin in the radial direction only leads to dose escalation within the overall volume. This also would cause significant dose escalation for the normal tissues, in particular, the usually radiosensitive mucosal structures covering the lumen of the tumor-bearing organ. For intraluminal BT, the prescription depth is even much closer to the sources as compared with intracavitary BT. The dose gradients are consequently steeper, and application of radial uncertainty margins would result in even larger dose escalation. The use of margins for geometrical uncertainties in the longitudinal direction—independent of the clinical target length determination—can/should be used in the longitudinal direction of intraluminal BT treatment plans.

17.7 Dose Reporting

In image-guided BT, it is recommended to report the target dose in terms of dose-volume histogram (DVH) parameters. Due to the high dose gradients throughout the target, it is suitable to include DVH parameters that are relevant for dose both in the outer and in the central regions of the target. With regard to the outer and coldest regions, it has been recommended to report D_{90} and D_{100} in both cervix and prostate BT (Kovacs et al. 2005; Nag et al. 2000; Nath et al. 2009; Pötter et al. 2006; Salembier et al. 2007). The D_{100} parameter may be susceptible to uncertainties since it is a point dose located in steep dose gradients, and for the future, more robust parameters like D_{95} or D_{98} could be considered. For the central and hot parts of the target parameters like D_{50}, V_{150} and V_{200} have been recommended (Kovacs et al. 2005; Nag et al. 2000; Pötter et al. 2006; Salembier et al. 2007). For OARs, absolute volumes indicating the dose to the most irradiated part of the bladder, rectum, and sigmoid are recommended: D_{2cc}, D_{1cc}, and $D_{0.1cc}$ (Pötter et al. 2006). The reported parameters are all prone to uncertainties. Relatively small geometric deviations can translate into large deviations in delivered doses due to the large dose gradients of BT. Typical changes in HR CTV dose (D_{90}, D_{100}) are below 4%–5% per millimeter of applicator displacement error (Tanderup et al. 2008). The limited extent of geometric reconstruction uncertainties means that the impact of different reconstruction methods as well as reconstruction interobserver variations has been reported to be limited in cervix BT—less than 3%–6% for HR CTV D_{90} (Berger et al. 2009; de Leeuw et al. 2009; Haack et al. 2009; Haie-Meder et al. 2005; Hellebust et al. 2007; Tanderup et al. 2008; Wills et al. 2010). Occasional applicator displacements have the potential to have much more impact on delivered doses than uncertainties in applicator reconstruction. Considerable changes in target dose may be seen in individual patients, for example, in case that an ovoid applicator changes orientation during the treatment (de Leeuw et al. 2009) or in case of needle movement in prostate BT (Pantelis et al. 2004; Simnor et al. 2009).

In EBRT, the PTV dose is representative of the CTV dose, and the PTV is regularly used for dose reporting. In BT, the dose gradients throughout the CTV and a PTV result in discrepancies between the CTV and PTV doses. In particular, a PTV expanded in the radial direction would represent a systematic underestimation of the CTV dose by ~8%–50% per millimeter of margin expansion. Even PTV margins applied only along catheters will not result in a PTV dose, which is the same as the CTV dose. There will still be dose gradients between the CTV and the PTV, although some kind of dose plateau is produced beyond the target.

In BT, the target dose may become either larger or smaller than expected from the "snapshot" images on the treatment planning system. If the CTV is contoured larger than it is in reality, the CTV dose will underestimate the true target dose. Likewise, an under-contouring will result in a reported dose that overestimates the target dose. Reconstruction uncertainties may also result in either overestimations or underestimations. It is currently not completely clear how to arrive at a dose reporting that represents the delivered dose in the best possible way. Currently, a CTV dose is clearly recommended and mandatory for reporting. However, the PTV may also have relevance for dose reporting, in particular, in estimating the minimum target dose in worst-case situations.

17.8 Discussion

Application of PTV margins should be based on a systematic evaluation of uncertainties. Currently, only a limited number of studies have been published on uncertainties in 3D image-guided BT, and further work on this is warranted. Specifically, it would be of great interest to further analyze the stability of different applicators and fixation techniques. The clinical advantage of image guidance is based on adaptation of dose to target and OARs according to the anatomical situation on treatment planning images at the time of BT (Haie-Meder et al. 2010; Lindegaard et al. 2008; Pötter et al. 2007). Significant movement of the applicators, target, or OARs may hamper the advantages of this approach. Additional movement control [e.g., imaging (Simnor et al. 2009) or in vivo dosimetry (Andersen et al. 2009)] and repositioning of catheters or replanning of source positions have to be incorporated into the BT procedure when catheter or organ movement is pronounced.

The BT dose distribution is profoundly different from that of EBRT. The dose gradients cannot, in general, be manipulated, and a homogeneous ("plateau") dose distribution cannot be obtained in or around a BT CTV. This means that BT dose planning does not allow for an overall compensation of uncertainties by application of margins. However, nonisotropic margins can be applied in selected directions whenever it is possible to expand the dose distribution by adding source positions beyond the CTV—for example, along needles or tandem. Contouring (Dimopoulos et al. 2009; Petric et al. 2008), reconstruction (Hellebust et al. 2010), and fusion uncertainties are often mostly pronounced in the direction along source catheters, and it is particularly important to maintain a margin between the target and the prescription isodose in this direction when possible. This will generate dose distributions that are robust toward geometric uncertainties and target contouring uncertainties. Application of PTV margins in the direction perpendicular to the source channels is discouraged since this would lead to an overall dose escalation for the patient and for the entire patient population.

The possibilities to compensate for uncertainties during dose planning are subject to the conditions of the realized BT implant. When the tandem does not extend beyond the HR CTV or when the interstitial needles are inside the prostate, it is not possible to apply any uncertainty margin even in the direction along source catheters. Similarly, expansion in the radial direction will blow up the isodoses and will not improve the robustness of the dose plan when there are no catheters available at the edge or beyond the CTV in the radial direction. Uncertainties must be taken into account already during the planning of the implant, so that source catheters are inserted in a way that allows for robust treatment planning, as far as this is possible. In particular, uncertainties in the perpendicular direction cannot be handled by application of margins during dose planning, but must be considered upfront. In other words, the compensation of uncertainties takes place during two steps in BT: (1) before and during the implantation of source catheters with the concept of a preimplantation PTV and (2) during dose planning with the concept of a post-implantation PTV. The possibilities during step 2—at the time of dose planning—to create a treatment plan that is robust toward uncertainties are dependent on which kind of implant is realized in step 1 during the BT application. The introduction of PTV volumes in BT has more complexity than a mechanic expansion of the CTV based on evaluation of uncertainties. It requires that both medical and physics aspects are considered. The medical part involves considerations of the possibilities for the implantation taking into account target and OAR topography, dose constraints to target and OARs, and also contouring uncertainties. Physics issues include the evaluation/consideration of geometric uncertainties and the procedures during dose planning to create a robust dose distribution by nonisotropic expansion of the isodoses in appropriate directions.

In conclusion, PTV is not equal to CTV in BT, since there are indeed uncertainties present in BT. However, the BT PTV cannot be constructed by directly expanding the CTV in all directions, because the possibilities for margin expansion depend on the conditions of the implant and on the direction of source catheters. In general, it is only possible to compensate for certain directional uncertainties by application of nonisotropic margins.

References

Andersen, C.E., Nielsen, S.K., Lindegaard, J.C., and Tanderup, K. 2009. Time-resolved in vivo luminescence dosimetry for online error detection in pulsed dose-rate brachytherapy. *Med Phys* 36(11):5033–43.

Ash, D. 2002. Prostate cancer. In: *The GEC ESTRO Handbook of Brachytherapy*. A. Gerbaulet et al. (eds.), ESTRO, Brussels. pp. 471–8.

Ash, D. and Gerbaulet, A. 2002. Oral tongue cancer. In: *The GEC ESTRO Handbook of Brachytherapy*. A. Gerbaulet et al. (eds.), ESTRO, Brussels. pp. 237–52.

Berger, D., Dimopoulos, J., Pötter, R., and Kirisits, C. 2009. Direct reconstruction of the Vienna applicator on MR images. *Radiother Oncol* 93(2):347–51.

Beriwal, S., Kim, H., Coon, D. et al. 2009. Single magnetic resonance imaging vs magnetic resonance imaging/computed tomography planning in cervical cancer brachytherapy. *Clin Oncol (R Coll Radiol)* 21(6):483–7.

De Brabandere, M., Mousa, A.G., Nulens, A., Swinnen, A., and Van Limbergen, E. 2008. Potential of dose optimisation in MRI-based PDR brachytherapy of cervix carcinoma. *Radiother Oncol* 88(2):217–26.

de Leeuw, A.A., Moerland, M.A., Nomden, C., Tersteeg, R.H., Roesink, J.M., and Jürgenliemk-Schulz, I.M. 2009. Applicator reconstruction and applicator shifts in 3D MR-based PDR brachytherapy of cervical cancer. *Radiother Oncol* 93(2):341–6.

Dimopoulos, J.C., De Vos, V., Berger, D. et al. 2009. Inter-observer comparison of target delineation for MRI-assisted cervical cancer brachytherapy: Application of the GYN GEC-ESTRO recommendations. *Radiother Oncol* 91(2):166–72.

Dimopoulos, J.A., Petrow, P., Tanderup, K., Petric, P., Berger, D., Kirisits, C., Pedersen, E.M., van Limbergen, E., Haie-Meder, C., Pötter, R. 2012 Apr. Recommendations from Gynaecological (GYN) GEC-ESTRO Working Group (IV): Basic principles and parameters for MR Imaging within the frame of image based adaptive cervix cancer brachytherapy, *Radiother Oncol* 103(1):113–22.

Fenkell, L., Assenholt, M., Nielsen, S.K. et al. 2011. Parametrial boost using midline shielding results in an unpredictable dose to tumor and organs-at-risk in combined external beam radiotherapy and brachytherapy for locally advanced cervical cancer. *Int J Radiat Oncol Biol Phys* 79(5):1572–9.

Haack, S., Nielsen, S.K., Lindegaard, J.C., Gelineck, J., and Tanderup, K. 2009. Applicator reconstruction in MRI 3D image-based dose planning of brachytherapy for cervical cancer. *Radiother Oncol* 91(2):187–93.

Haie-Meder, C., Pötter, R., Van Limbergen, E. et al. 2005. Recommendations from Gynaecological (GYN) GEC-ESTRO Working Group (I): Concepts and terms in 3D image based 3D treatment planning in cervix cancer brachytherapy with emphasis on MRI assessment of GTV and CTV. *Radiother Oncol* 74(3):235–45.

Haie-Meder, C., Chargari, C., Rey, A., Dumas, I., Morice, P., and Magne, N. 2010. MRI-based low dose-rate brachytherapy experience in locally advanced cervical cancer patients initially treated by concomitant chemoradiotherapy. *Radiother Oncol* 96(2):161–5.

Han, B., Wallner, K., Aggarwal, S., Armstrong, J., and Sutlief, S. 2000. Treatment margins for prostate brachytherapy. *Semin Urol Oncol* 18(2):137–41.

Hellebust, T.P., Tanderup, K., Bergstrand, E.S., Knutsen, B.H., Røislien, J., and Olsen, D.R. 2007. Reconstruction of a ring applicator using CT imaging: Impact of reconstruction method and applicator orientation. *Phys Med Biol* 52(16):4893–904.

Hellebust, T.P., Kirisits, C., Berger, D. et al. 2010. Recommendations from Gynaecological (GYN) GEC-ESTRO Working Group: Considerations and pitfalls in commissioning and applicator reconstruction in 3D image-based treatment planning of cervix cancer brachytherapy. *Radiother Oncol* 96(2):153–60.

Hoskin, P. 2008. High dose rate brachytherapy for prostate cancer. *Cancer Radiother* 12(6–7):512–4.

Hoskin, P.J., Bownes, P.J., Ostler, P., Walker, K., and Bryant, L. 2003. High dose rate afterloading brachytherapy for prostate cancer: Catheter and gland movement between fractions. *Radiother Oncol* 68(3):285–8.

International Commision on Radiation Units and Measurements (ICRU) Report 50. 1993. Prescribing, Recording, and Reporting Photon Beam Therapy, Bethesda, USA.

International Commision on Radiation Units and Measurements (ICRU) Report 58. 1997. Dose and Volume Specification for Reporting Interstitial Therapy, Bethesda, USA.

International Commision on Radiation Units and Measurements (ICRU) Report 62. 1999. Prescribing, Recording and Reporting Photon Beam Therapy (supplement to ICRU Report 50), Bethesda, USA.

International Commission on Radiation Units and Measurements (ICRU) Report 71. 2004. Prescribing, Recording, and Reporting Electron Beam Therapy, Bethesda, MD.

International Commission on Radiation Units and Measurements (ICRU) Report 78. 2007. Prescribing, recording, and reporting proton-beam therapy. *J ICRU* 7(2):1–210.

International Commission on Radiation Units and Measurements (ICRU) Report 83. 2010. Prescribing, recording, and reporting intensity-modulated photon-beam therapy (IMRT). *J ICRU* 10(1):41–54.

Jürgenliemk-Schulz, I.M., Tersteeg, R.J., Roesink, J.M. et al. 2009. MRI-guided treatment-planning optimisation in intracavitary or combined intracavitary/interstitial PDR brachytherapy using tandem ovoid applicators in locally advanced cervical cancer. *Radiother Oncol* 93(2):322–30.

Jürgenliemk-Schulz, I.M., Lang, S., Tanderup, K. et al. 2010. Variation of treatment planning parameters (D90 HR-CTV, D 2cc for OAR) for cervical cancer tandem ring brachytherapy in a multicentre setting: Comparison of standard planning and 3D image guided optimisation based on a joint protocol for dose-volume constraints. *Radiother Oncol* 94(3):339–45.

Kirisits, C., Georg, D., Wexberg, P., Pokrajac, B., Glogar, D., and Potter, R. 2002. Determination and application of the reference isodose length (RIL) for commercial endovascular brachytherapy devices. *Radiother Oncol* 64(3):309–15.

Kirisits, C., Pötter, R., Lang, S., Dimopoulos, J., Wachter-Gerstner, N., and Georg, D. 2005. Dose and volume parameters for MRI-based treatment planning in intracavitary brachytherapy for cervical cancer. *Int J Radiat Oncol Biol Phys* 62(3):901–11.

Kirisits, C., Lang, S., Dimopoulos, J., Oechs, K., Georg, D., and Pötter, R. 2006. Uncertainties when using only one MRI-based treatment plan for subsequent high-dose-rate tandem and ring applications in brachytherapy of cervix cancer. *Radiother Oncol* 81(3):269–75.

Kohr, P. and Siebert, F.A. 2007. Quality assurance of brachytherapy afterloaders using a multi-slit phantom. *Phys Med Biol* 52(17):N387–91.

Kovacs, G., Potter, R., Loch, T. et al. 2005. GEC/ESTRO-EAU recommendations on temporary brachytherapy using stepping sources for localised prostate cancer. *Radiother Oncol* 74(2):137–48.

Lindegaard, J.C., Tanderup, K., Nielsen, S.K., Haack, S., and Gelineck, J. 2008. MRI-guided 3D optimization significantly improves DVH parameters of pulsed-dose-rate brachytherapy in locally advanced cervical cancer. *Int J Radiat Oncol Biol Phys* 71(3):756–64.

Major, T., Polgar, C., Fodor, J., Somogyi, A., and Nemeth, G. 2002. Conformality and homogeneity of dose distributions in interstitial implants at idealized target volumes: A comparison between the Paris and dose-point optimized systems. *Radiother Oncol* 62(1):103–11.

Mullokandov, E. and Gejerman, G. 2004. Analysis of serial CT scans to assess template and catheter movement in prostate HDR brachytherapy. *Int J Radiat Oncol Biol Phys* 58(4):1063–71.

Nag, S., Bice, W., DeWyngaert, K., Prestidge, B., Stock, R., and Yu, Y. 2000. The American Brachytherapy Society recommendations for permanent prostate brachytherapy post-implant dosimetric analysis. *Int J Radiat Oncol Biol Phys* 46(1):221–30.

Nag, S., Cardenes, H., Chang, S. et al. 2004. Proposed guidelines for image-based intracavitary brachytherapy for cervical carcinoma: Report from Image-Guided Brachytherapy Working Group. *Int J Radiat Oncol Biol Phys* 60(4):1160–72.

Nath, R., Anderson, L.L., Luxton, G., Weaver, K.A., Williamson, J.F., and Meigooni, A.S. 1995. Dosimetry of interstitial brachytherapy sources: Recommendations of the AAPM Radiation Therapy Committee Task Group No. 43. American Association of Physicists in Medicine. *Med Phys* 22(2):209–34.

Nath, R., Bice, W.S., Butler, W.M. et al. 2009. AAPM recommendations on dose prescription and reporting methods for permanent interstitial brachytherapy for prostate cancer: Report of Task Group 137. *Med Phys* 36(11):5310–22.

Pantelis, E., Papagiannis, P., Anagnostopoulos, G. et al. 2004. Evaluation of a TG-43 compliant analytical dosimetry model in clinical 192Ir HDR brachytherapy treatment planning and assessment of the significance of source position and catheter reconstruction uncertainties. *Phys Med Biol* 49(1):55–67.

Petric, P., Dimopoulos, J., Kirisits, C., Berger, D., Hudej, R., and Pötter, R. 2008. Inter- and intraobserver variation in HR-CTV contouring: Intercomparison of transverse and paratransverse image orientation in 3D-MRI assisted cervix cancer brachytherapy. *Radiother Oncol* 89(2):164–71.

Pötter, R. and Van Limbergen, E. 2002. Oesophageal cancer. In: *The GEC ESTRO Handbook of Brachytherapy*, A. Gerbaulet et al. (eds.), ESTRO, Brussels. pp. 515–38.

Pötter, R., Van, L.E., Dries, W. et al. 2001. Prescribing, recording, and reporting in endovascular brachytherapy. Quality assurance, equipment, personnel and education. *Radiother Oncol* 59(3):339–60.

Pötter, R., Van Limbergen, E., and Wambersie, A. 2002. Reporting in brachytherapy: Dose and volume specification. In: *The*

GEC ESTRO Handbook of Brachytherapy, A. Gerbaulet et al. (eds.), ESTRO, Brussels. pp. 153–216.

Pötter, R., Haie-Meder, C., Van Limbergen, E. et al. 2006. Recommendations from gynaecological (GYN) GEC ESTRO working group (II): Concepts and terms in 3D image-based treatment planning in cervix cancer brachytherapy-3D dose volume parameters and aspects of 3D image-based anatomy, radiation physics, radiobiology. *Radiother Oncol* 78(1):67–77.

Pötter, R., Dimopoulos, J., Georg, P. et al. 2007. Clinical impact of MRI assisted dose volume adaptation and dose escalation in brachytherapy of locally advanced cervix cancer. *Radiother Oncol* 83(2):148–55.

Pötter, R., Fidarova, E., Kirisits, C., and Dimopoulos, J. 2008a. Image-guided adaptive brachytherapy for cervix carcinoma. *Clin Oncol (R Coll Radiol)* 20(6):426–32.

Pötter, R., Kirisits, C., Fidarova, E.F. et al. 2008b. Present status and future of high-precision image guided adaptive brachytherapy for cervix carcinoma. *Acta Oncol* 47(7):1325–36.

Resch, A., Pötter, R., Van, Limbergen, E. et al. 2002. Long-term results (10 years) of intensive breast conserving therapy including a high-dose and large-volume interstitial brachytherapy boost (LDR/HDR) for T1/T2 breast cancer. *Radiother Oncol* 63(1):47–58.

Salembier, C., Lavagnini, P., Nickers, P. et al. 2007. Tumour and target volumes in permanent prostate brachytherapy: A supplement to the ESTRO/EAU/EORTC recommendations on prostate brachytherapy. *Radiother Oncol* 83(1):3–10.

Schmid, R., Kirisits, C., Syeda, B. et al. 2004. Quality assurance in intracoronary brachytherapy. Recommendations for determining the planning target length to avoid geographic miss. *Radiother Oncol* 71(3):311–8.

Simnor, T., Li, S., Lowe, G. et al. 2009. Justification for inter-fraction correction of catheter movement in fractionated high dose-rate brachytherapy treatment of prostate cancer. *Radiother Oncol* 93(2):253–58.

Strnad, V. 2004. Treatment of oral cavity and oropharyngeal cancer. Indications, technical aspects, and results of interstitial brachytherapy. *Strahlenther Onkol* 180(11):710–17.

Syeda, B., Siostrzonek, P., Schmid, R. et al. 2002. Geographical miss during intracoronary irradiation: Impact on restenosis and determination of required safety margin length. *J Am Coll Cardiol* 40(7):1225–31.

Tanderup, K., Hellebust, T.P., Lang, S. et al. 2008. Consequences of random and systematic reconstruction uncertainties in 3D image based brachytherapy in cervical cancer. *Radiother Oncol* 89(2):156–63.

Tanderup, K., Georg, D., Potter, R., Kirisits, C., Grau, C., and Lindegaard, J.C. 2010a. Adaptive management of cervical cancer radiotherapy. *Semin Radiat Oncol* 20(2):121–9.

Tanderup, K., Potter, R., Lindegaard, J.C., Berger, D., Wambersie, A., and Kirisits, C. 2010b. PTV margins should not be used to compensate for uncertainties in 3D image guided intracavitary brachytherapy. *Radiother Oncol* 97(3):495–500.

Van Limbergen, E. 2003. Indications and technical aspects of brachytherapy in breast conserving treatment of breast cancer. *Cancer Radiother* 7(2):107–20.

Van Limbergen, E. and Pötter, R. 2002. Bronchus cancer. In: *The GEC ESTRO Handbook of Brachytherapy*, A. Gerbaulet et al. (eds.), ESTRO, Brussels. pp. 545–61.

Venselaar, J. and Pérez-Calatayud, J. 2004. *A Practical Guide to Quality Control of Brachytherapy Equipment.* ESTRO, Brussels.

Wills, R., Lowe, G., Inchley, D., Anderson, C., Beenstock, V., and Hoskin, P. 2010. Applicator reconstruction for HDR cervix treatment planning using images from 0.35 T open MR scanner. *Radiother Oncol* 94(3):346–52.

V

Clinical Brachytherapy

Clinical Use of Brachytherapy

Peter Hoskin
Mount Vernon Hospital

18.1 Historical Background

Brachytherapy has been used for the treatment of disease from the very early days of radiation discovery. ^{226}Ra provided a very stable source of gamma radiation and in the first two decades of the twentieth century was applied to many clinical settings, in particular, intracavitary treatment of gynecological tumors and surface treatment of skin and breast tumors. There was little attention paid to radiation protection or the hazards of radon release and no formal dosimetry. However, successful treatments were described in the early literature, and the use of this modality expanded. With this expansion emerged three major schools of brachytherapy in Europe based in Manchester, Paris, and Stockholm and in the United States at Memorial Hospital in New York. The "Manchester rules" published in 1938 set out requirements for brachytherapy source distributions to achieve good homogeneous dose distributions for a wide range of applications ranging from intracavitary gynecological insertions to surface molds and interstitial brachytherapy, as shown in Figure 18.1. Similarly, work in Paris, Stockholm, and Houston developed systems that were proven in clinical practice to be safe and effective and that to this day form the basis of modern brachytherapy.

Greater awareness of the potential hazards from radiation sources and a wish to use higher-energy high-dose rate (HDR) sources led to the development of afterloaders for both low-dose-rate (LDR) and HDR systems to deliver the radiation with minimal operator exposure. Alongside this, developments in the nuclear industry gave access to newer and safer isotopes, initially ^{137}Cs and ^{60}Co and more recently ^{192}Ir, which form the basis of modern afterloading techniques. Clinical experience in the introduction of these new modalities of brachytherapy led to a greater understanding of the clinical importance of dose rate and the implications for biological efficacy of even relatively modest changes in dose rate.

Technical developments in computing and imaging together with the widespread adoption of stepping source delivery systems have led to a further revolution in brachytherapy in recent years with the advent of image-guided brachytherapy. Image-based volumes are now defined based on 3D imaging with CT and MR and adopting the ICRU concepts of gross tumor volume (GTV), clinical target volume (CTV), and planning target volume (PTV), the latter being less relevant to most forms of brachytherapy where setup error and organ motion do not occur. However, 3D high-quality imaging has also shown us that applicator movement, organ swelling due to hemorrhage and edema, and changes in organs at risk may well result in significant variations in dose delivery if rigorous quality assurance is not applied in brachytherapy (Simnor et al. 2009), as shown in Figure 18.2.

Modern brachytherapy therefore has a long heritage of which it can be proud. Live source insertions are still undertaken with ^{125}I and ^{103}Pd seeds for prostate cancer, but the future will lie with modern afterloading systems using ^{192}Ir or ^{60}Co in HDR or pulsed-dose-rate (PDR) systems. The versatility created by the use of a small source and flexible catheters has greatly expanded the clinical applications where brachytherapy can have a major role in disease management. The future will no doubt see further developments with new isotopes, electronic brachytherapy, and microspheres already been seen in clinical trials. There will also be changes in the indications for brachytherapy as clinical practice evolves; thus, recent years have seen a dramatic expansion in prostate brachytherapy with the acceptance of the transperineal transrectal ultrasound-based procedure, and currently accelerated partial breast radiotherapy (APBI) is providing a further opportunity for brachytherapy development.

18.2 Clinical Advantages of Brachytherapy

The term "brachytherapy" refers to "near" therapy with radiation sources being placed directly on, in, or through the area of interest that is to receive a high radiation dose. The dose distribution in brachytherapy is determined almost exclusively by

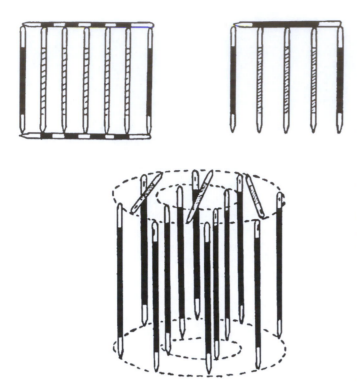

FIGURE 18.1 Manchester rules for source distribution in a cylindrical volume implant. (From Meredith, W.J. and Massey, J.B. *Fundamental Physics of Radiology*, John Wright & Sons, 1976. With permission.)

conformal dose distributions can now be achieved using IMRT techniques even when the tumor volume has an irregular shape, this still cannot rival the tight conformity and very high central dose achieved with brachytherapy. An illustration is shown in Figure 18.3 comparing a modern intensity-modulated radiotherapy distribution for prostate cancer with an HDR prostate brachytherapy implant, both of which are designed to deliver a high dose to a CTV encompassing the prostate capsule with a 3 mm margin. Other techniques that may be used include proton therapy, which will still have a significant skin entry dose even though the exit dose is eliminated. Similarly, stereotactic radiosurgery, where equivalence in terms of CTV coverage has been claimed, still delivers a low-dose body bath from the entry doses and exit doses. The implications of the increased low dose volume arising from these modern external beam techniques have yet to be manifest in terms of late follow-up, but modeling studies have suggested it may well result in a significant increase in the incidence of second malignancies (Georg et al. 2008; Fontenot et al. 2009; Kry et al. 2005).

A further clinical advantage for brachytherapy lies in the accuracy and reliability of its dose delivery. With external beam techniques, an additional margin is conventionally applied to expand the CTV to a PTV. This expansion allows for setup error due to both random and systematic errors in the translation of the external beam plan to the treatment couch and patient and also internal organ movement. In brachytherapy, there is no setup error, and organ movement is largely overcome in one of two ways. There may be fixation of the organ; for example, an HDR prostate brachytherapy implant will essentially abolish the variation in position of the prostate gland as it is transfixed by the brachytherapy applicators. In other circumstances, the organ may retain some mobility, but the brachytherapy sources will move with the organ, examples being an intrauterine insertion or an LDR ^{125}I seed implant of the prostate gland where permanent implants remain within the prostate gland, moving with the gland while decaying and delivering their radiation. Thus, radiation delivery with brachytherapy is both more accurate and more reliable than with external beam techniques.

the inverse square law. These two factors mean that the clinical result of brachytherapy is a very high conformal dose delivered to the target. This can be achieved with high accuracy, and the rapid dose falloff minimizes radiation exposure to surrounding normal tissues. Brachytherapy therefore fulfills the ideal requirements for radiation dose delivery in clinical oncology. In this respect, it has substantial advantages over the alternative approaches with external beam irradiation when it is inevitable that with the maximum energy arising some distance away from the patient, it will give a higher dose to all the tissues en route from the skin entry point to the area to be treated. While very

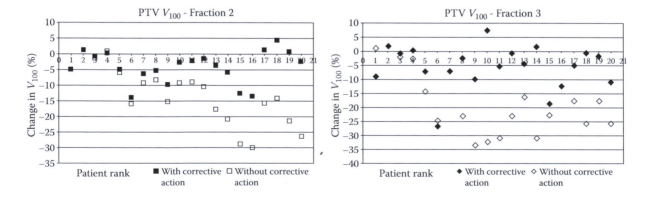

FIGURE 18.2 Dose-volume histogram (DVH) results of individual patients treated with HDR brachytherapy of the prostate. V_{100} is shown relative to fraction 1 with and without corrective action as movement correction for each fraction. A positive result indicates a higher value than that achieved in the original plan. (From Simnor, T. et al. *Radiother Oncol* 93, 253–8, 2009. With permission.)

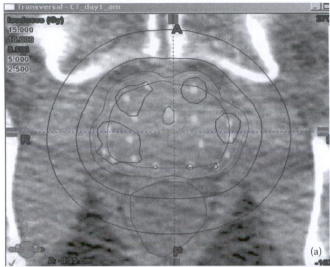

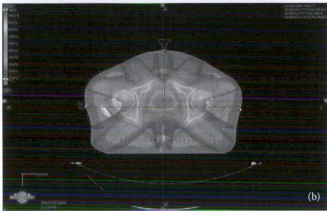

FIGURE 18.3 (See color insert.) Comparison of an HDR prostate brachytherapy implant (a) with a modern intensity-modulated radiotherapy distribution (b) for prostate cancer in which both plans were designed to deliver a high dose to a CTV encompassing the prostate capsule with a 3-mm margin. (Courtesy of P. Hoskin.)

The very high conformal dose delivery does, however, carry a stringent requirement for quality assurance. This must encompass the entire chain of treatment delivery and include source calibration, afterloader characteristics, imaging, and catheter placement as well as ensure that during radiation, delivery applicators or sources have not moved from the intended position. This is particularly critical for HDR systems where a very high dose per fraction will be delivered in a matter of minutes with no room for error. Despite this, it is clear that there are potentially significant uncertainties in brachytherapy. These relate to both physical uncertainties relating to dosimetric and geometric parameters and clinical uncertainties of which the greatest is undoubtedly related to clinician variability in contouring of both target volumes and organs at risk (De Brabandere et al. 2012).

Another clinical advantage with brachytherapy is that, in general, it will be completed within a much shorter time than external beam treatment. Even an LDR/PDR system will deliver a radical dose within 5 to 6 days. The HDR may achieve radical

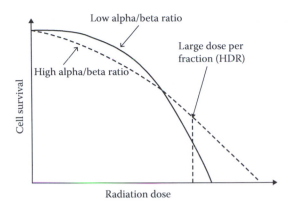

FIGURE 18.4 Cell survival curves as a function of dose for low and high alpha/beta ratios. (Courtesy of P. Hoskin.)

dose delivery within a much shorter time even though fractionated scheduling is required. Acceleration of radiation delivery may have significant advantages in tumors that have a short potential doubling time, which is a property of many common epithelial tumors treated with brachytherapy, in particular, head-and-neck cancer and cervical cancer.

A further potential advantage has emerged for HDR brachytherapy in prostate cancer and may be applicable to other primary sites. Prostate cancer, in particular, has been identified as a tumor with a low alpha/beta ratio describing its radiation response characteristics (Bentzen and Ritter 2005). This means that high doses per fraction will have relatively greater biological effect, as shown in Figure 18.4. However, this applies not only to the tumor but also to late normal tissue effects for which the alpha/beta ratio is also low. Thus, the only way to achieve a therapeutic ratio in favor of tumor cell kill with minimal normal tissue damage is by precise geometric localization of dose with sparing of organs at risk. HDR brachytherapy enables delivery of high doses per fraction, and the physical advantages of close dose conformity enable such doses to be achieved with acceptable normal tissue damage.

18.3 Clinical Disadvantages of Brachytherapy

The disadvantages of brachytherapy relate mainly to the need for an invasive procedure to place radiation applicators or sources. Perhaps the one exception to this is the use of surface molds for skin tumors. For many procedures, some form of anesthesia will be required to enable accurate placement within the limits of patient acceptability, and this may mean that an operating facility has to be accessed, increasing the complexity and cost of the procedure. It is also limited by the sites where there is easy access for source placement. Thus, surface tumors and body cavities may be readily treated as evidenced by the treatment of tumors of the gynecological tract, esophagus, and bronchus, while other sites may be accessible for interstitial treatment as in the prostate and head-and-neck region. Internal organs, however, particularly those in the abdomen, present challenges that have not been overcome in routine practice, although the developing practice of intraoperative brachytherapy has shown that even these sites are now within the remit of brachytherapy.

Brachytherapy requires specific skills and training, and in many countries, these resources are sparse. Radiation therapy training programs no longer require radiation oncologists to be competent in brachytherapy at the time of obtaining their accreditation, and only enthusiasts go on to achieve post-accreditation training. Safe and effective delivery of brachytherapy requires a skilled team including brachytherapy physicists, radiographers, and nursing staff, all of whom will require special expertise in the brachytherapy processes. Access to an operating room environment, whether integrated into the brachytherapy unit or within a surgical service, is also necessary, and skilled anesthesia and postprocedural pain control is an important aspect of patient care during brachytherapy. Specific problems relating to staff and resource utilization must also be considered when setting up a brachytherapy service. With LDR/PDR, 12- or 24-h cover may be required during the period of source exposure; in contrast, HDR demands intensive periods of intermittent activity to deliver fractionated treatment. These logistics must be accommodated within the day-to-day activity of a busy radiotherapy department.

Brachytherapy may also present specific issues related to radiation protection, which will be defined under national legislation. Apart from ^{125}I seed implants where the radiation energy is low and staff exposure negligible, live source implants are rarely undertaken, and the hazard to staff is substantially reduced or eliminated by afterloading procedures. Where LDR iridium wire implants are still undertaken, there is the hazard of radiation exposure while the implant is loaded with the patient typically in an isolated room on a hospital ward. Restrictions then have to be placed on the nursing time allowed, and visitors will be allowed for only very short limited periods of the day. HDR brachytherapy avoids these problems, but other issues arise that need further consideration and staff training, for example, the management of a failed source retraction or source contamination. Thus, it can be seen that the clinical application of brachytherapy is a complex task requiring a multidisciplinary team highly skilled in this particular form of radiation delivery.

18.4 Clinical Applications of Brachytherapy

Brachytherapy is an ideal treatment for a small confined well-demarcated tumor. There are practical limitations to the volume of tissue that can be implanted, and this will typically reflect a tumor with a maximum dimension of up to 5 cm. Larger volumes may on occasion be treated, for example, a soft tissue sarcoma of the limb. This means that brachytherapy has an important role in the treatment of localized accessible tumors but is suboptimal for prophylactic treatment of regional lymph nodes or tumors that have exceeded 5 cm or so in size. In these settings, however, it may be used in combination with external beam treatment as

a boost to enable high dose delivery to a limited primary volume while external beam treatment is used to cover a more extensive regional CTV.

18.5 Dose Rate and Clinical Brachytherapy

The definitions and biological implications of dose rate in brachytherapy are covered elsewhere in this book. In clinical practice, dose rate can be considered in two distinct groups: either low to medium dose rate (LDR/MDR/PDR) with treatment delivery over several hours or HDR in which treatment is delivered over a few minutes.

The clinical use of LDR or PDR is very different from that of HDR in terms of both logistics and dose delivery. An overview is provided in Table 18.1. LDR/PDR requires inpatient facilities with a patient undergoing a treatment that may last anywhere from 2 to 6 days with applicators in situ throughout that period. This presents important quality assurance issues with regard to avoiding or compensating for movement of applicators during treatment. There are also radiation protection considerations to allow the delivery of nursing care and procedures for source displacement. In contrast, HDR brachytherapy requires fractionated treatment with the additional challenge of either repeated implant procedures or maintaining an implant for a period during which repeated fractions are delivered through the same applicators.

The choice to install a PDR or HDR system will depend on a number of factors, not least of which are the historical experience and facilities available within a radiotherapy department. There are also well-rehearsed and often firmly held biological arguments, and increasingly, we also have to consider cost effectiveness. In a high-throughput unit, HDR brachytherapy is undoubtedly more cost effective than LDR/PDR where, depending upon the system used, only one or two patients can be treated at a time, each treatment lasting several days. With HDR, although fractionated schedules will require several visits of the patient to the unit, one or two patients an hour can be treated comfortably, resulting in a substantially greater capacity.

TABLE 18.1 Clinical Comparison of HDR and LDR/PDR Brachytherapy

	HDR (^{192}Ir)	LDR/PDR (^{192}Ir)
Dose rate	1 Gy/min	1 Gy/h
Source strength	370 GBq (10 Ci)	37 GBq (1 Ci)
Shielding required	More	Less
Treatment duration	Minutes	Days
Setting	Outpatient/day case	Inpatient
Nursing care	Minimal	Full
Capacity per unit	10 patients/week (dependent upon dose/fractionation)	2 patients/week
Clinical evidence base	Extensive	Limited

Notwithstanding the above, the original tradition of brachytherapy arose with LDR systems. More recently, cesium afterloading units became popular using dose rates of 80 to 150 cGy per hour. The majority of these units, however, are becoming obsolete and will, over the next few years, be replaced. Patterns of care studies both in Europe and Canada suggest that the predominant move is toward centers acquiring an HDR afterloader, with a minority moving to PDR (Guedea et al. 2010; Tai et al. 2004). Over the last decade or so, the concern that HDR brachytherapy could not deliver the same results as LDR has largely been abolished; for example, in gynecological brachytherapy, there are five randomized trials and a number of retrospective series comparing the two approaches and confirming equivalence in terms of clinical results, both tumor control and toxicity (see Chapter 21). The randomized trials have been submitted to a meta-analysis and confirm no difference in toxicity or control rates (Viani et al. 2009).

The arguments that prevailed when HDR was developed are once more being rehearsed for the comparison between PDR and HDR. Currently, the published data on PDR are limited, and there is no level-I evidence from randomized trials comparing this with other modalities. There is an assumption that delivering a small HDR pulse hourly to build up the total dose is equivalent to continuous LDR delivery in biological terms. Experimental data suggest that in normal tissues with a short half-time for repair, PDR will be more damaging than LDR unless the dose per pulse is kept below 1 Gy. Conversely, other modeling studies have shown that if the repair half-time is 4 h or more, then HDR may well be superior to LDR (Sminia et al. 2002). Unfortunately, there remains uncertainty over the biological parameters that should be included in these modeling exercises highlighting the uncertainty over dose equivalence between the different modalities and the need for pragmatic clinical trials to assert isoeffectiveness in terms of tumor control and toxicity.

The future, therefore, will rest with a mixture of PDR and HDR brachytherapy. PDR will remain a field of uncertainty requiring close clinical observation and comparative studies to establish the equitoxic and equieffective doses compared to the historical LDR and modern HDR brachytherapy experiences. There is no reason, given the appropriate dose corrections, that they should not be equivalent.

The following clinical chapters in this book describe the clinical brachytherapy experience for the main treatment sites.

References

Bentzen, S.M. and Ritter, M.A. 2005. The alpha/beta ratio for prostate cancer: What is it, really? *Radiother Oncol* 76:1–3.

De Brabandere, M., Hoskin, P., Haustermans, K., Van den Heuvel, F., Siebert, F-A. 2012. Prostate post-implant dosimetry: Interobserver variability in seed localisation, contouring and fusion. Accepted for publication to *Radiother Oncol* (July 03, 2012).

Fontenot, J.D., Lee, A.K., and Newhauser, W.D. 2009. Risk of secondary malignant neoplasms from proton therapy and intensity modulated x-ray therapy for early stage prostate cancer. *Int J Radiat Oncol Biol Phys* 74:616–22.

Georg, D., Kirisits, C., Hillbrand, M., Dimopoulos, J., and Pötter, R. 2008. Radiotherapy for cervix cancer: High tech external beam therapy versus high tech brachytherapy. *Int J Radiat Oncol Biol Phys* 71:1272–8.

Guedea, F., Venselaar, J., Hoskin, P. et al. 2010. Patterns of care for brachytherapy in Europe: Updated results. *Radiother Oncol* 97:514–20.

Kry, S.F., Salehpour, M., Followill, D. et al. 2005. The calculated risk of fatal secondary malignancies from intensity modulated radiation therapy. *Int J Radiat Oncol Biol Phys* 62:1195–203.

Meredith, W.J. and Massey, J.B. 1976. *Fundamental Physics of Radiology*. John Wright & Sons, Bristol.

Simnor, T., Li, S., Lowe, G. et al. 2009. Justification for interfraction correction of catheter movement in fractionated high dose-rate brachytherapy treatment of prostate cancer. *Radiother Oncol* 93:253–8.

Sminia, P., Schneider, C.J., and Fowler, J.F. 2002. The optimal fraction size in high-dose-rate brachytherapy: Dependency on tissue repair kinetics and low-dose rate. *Int J Radiat Oncol Biol Phys* 52:844–9.

Tai, P., Yu, E., Battista, J., and Van Dyk, J. 2004. Radiation treatment of lung cancer—Patterns of practice in Canada. *Radiother Oncol* 71:167–74.

Viani, G.A., Manta, G.B., Stefano, E.J., and deFendi, L.I. 2009. Brachytherapy for cervix cancer: Low-dose rate or high-dose rate brachytherapy—A meta-analysis of clinical trials. *J Exp Clin Cancer Res* 28:47.

19

Radiobiology for Brachytherapy

David J. Carlson
Yale University School of Medicine

Zhe J. Chen
Yale University School of Medicine

Peter Hoskin
Mount Vernon Hospital

Zoubir Ouhib
Boca Raton Regional Hospital

Marco Zaider
Memorial Sloan Kettering Cancer Center

19.1 Fundamentals

19.1.1 Introduction

The interaction of radiation with biological tissue is influenced by a multitude of factors. These may be related to the host, for example, intrinsic radiosensitivity, oxygenation status, proliferation rate, and damage repair capacity, or to the radiation field, for example, dose, temporal, and spatial pattern of dose delivery and radiation quality. The effect of a given radiation schedule is the result of a complex interplay among these parameters. Radiobiological models provide a mathematical link between these factors and the end point of interest.

There are two general kinds of radiobiological models: phenomenological and mechanistic. The first kind amounts to finding an analytic expression that describes with reasonable accuracy available data; as such, this is essentially an interpolation tool. As the name indicates, mechanistic models make explicit assumptions concerning the processes responsible for the radiation effect of interest. Their parameters (e.g., proliferation rates, cell loss factor, sublethal damage repair rate) should be obtainable from independent radiobiological investigations. As a matter of principle, then, only mechanistic models should be trusted for extrapolating from one kind of radiation regimen to another, although uncertainties concerning relevant parameters remain a point of concern.

The question then is, what makes two treatments equivalent? In terms of *tumor control probability* (TCP), defined as the probability of inactivating all target cells by the end of the treatment, and in the absence of net cellular proliferation (or, equivalently, when cellular proliferation during treatment may be neglected), isoeffectiveness is equivalent to equal probability of cellular survival. This is because, under these conditions,

$$\text{TCP} = [1 - S(D)]^n \approx \exp[-nS(D)] \tag{19.1}$$

where n is the initial number of malignant cells, $S(D)$ is the expected *fractional* number of clonogenic cells alive at the end of treatment, and, therefore, $nS(D)$ is the expected *number* of surviving cells. It is common to account for cellular proliferation using

$$S(T) = S_0(T)e^{(b-d)T} \tag{19.2}$$

where b and d are, respectively, the birth and spontaneous (i.e., not radiation-induced) death rates, T is the total treatment time, and $S_0(T)$ is the survival probability in the *absence* of cell proliferation. An immediate way to see that using $S(T)$ of Equation 19.2 in Equation 19.1 is erroneous is to attempt to calculate TCP for a permanent implant ($T \to \infty$).

A more important question, however, is whether this definition of TCP represents isoeffectiveness in terms of *cure*. This, and the proper expression for TCP, valid for any temporal pattern of dose delivery (permanent implants included), are discussed below.

Formulating a mechanistic description of the yield of tissue damage at the end of treatment and its subsequent healing kinetics is a substantially more complicated problem. The ability of an organ to carry out its function requires a minimum number of viable cells. The tissue (organ) consists of an operationally homogeneous population of self-maintaining cells. There are two types of tissue (organ) proliferative organization:

1. Hierarchical tissues, consisting of three compartments. *Stem cells* are capable of indefinite proliferation (first group). *Functional* or differentiated cells (third group) are not clonogenic and are derived from stem cells via a transition compartment (second group) where cells divide a finite number of times. This latter group is also referred to as the "amplification compartment."

2. The so-called *simple-duplication* tissue organization is characterized by the fact that functionally competent cells are also responsible for tissue renewal (or regeneration following damage).

Clinical data on the kinetics of time to onset and time to toxicity resolution for *properly defined* end points remain sparse. For instance, grade 2 genito-urinary (GU) toxicity associated with the treatment of prostate cancer may describe various combinations of six different symptoms (frequency, nocturia, incontinence, urgency, hematuria, and dysuria) not likely to have common mechanisms. As well, a distinction is often made between acute and late response. The demarcation line between the two is usually arbitrary; for instance, GU toxicity symptoms that occur within 3 months from the end of treatment are classified as acute.

It is reasonable to assume that the total number of cells inactivated determines the clinical outcome for tumors, but not for organs at risk, primarily because for these structures, *function* is the key clinical outcome. Because *mechanistic* normal tissue complication probability (NTCP) models remain largely unavailable, a possible option is to use empirical expressions and substitute for dose the isoeffective dose (IED).

Two different treatments can be made biologically equivalent in terms of either expected cure probability or healthy-tissue morbidity, but not both. In our view, the choice of the relative weight given in treatment planning to these (opposing) requirements should be left to the patient (Amols et al. 1997).

19.1.2 Isoeffectiveness in Terms of Cellular Survival

Brachytherapy includes procedures delivered under different temporal patterns [low dose rate (LDR), high dose rate (HDR), or combinations of both (LDR and HDR)] as well as by different radiation qualities. A general expression that describes the probability of tumor control is (Zaider and Minerbo 2000)

$$TCP(t) = \left[1 - \frac{S_0(t)e^{(b-d)t}}{1 + bS_0(t)e^{(b-d)t} \int_0^t \frac{du}{S_0(u)e^{(b-d)u}}} \right]^n \quad (19.3)$$

Here, $S_0(t)$ is the survival probability at time t, and b and d (defined above) may be calculated from the more familiar quantities T_p (clonogenic doubling time), T_{pot} (potential doubling time), and φ (cell-loss factor); thus, $b - d = \ln(2)/T_p$, $T_p = T_{pot}/(1 - \varphi)$ and $d/b = \varphi$. By comparing this expression with Equation 19.1, it is apparent that the *effective survival probability* (birth and death processes included) of each of the n clonogenic cells is given by the second term in square brackets.

In contrast to this approach, where cure is taken to mean the extinction of all clonogenic malignant cells at the end of treatment (a sufficient but not necessary condition of local control), it makes more sense to replace this quantity with the probability of *long-term, recurrence-free survival*, $G(\infty)$, which is the only *observable* quantity that actually matters. It can be shown (Zaider and Hanin 2011) that, in fact, $G(\infty) = TCP(\infty)$.

For instance, if the treatment duration is T, then

$$G(\infty) = \left[1 - \frac{1}{b \int_0^T \frac{du}{S_0(u)e^{(b-d)u}} + \frac{b}{b-d} \frac{e^{-(b-d)T}}{S_0(t)}} \right]^{n(0)}. \quad (19.4)$$

For a treatment that consists of f equal fractions delivered at times $0, \tau, 2\tau, \ldots, (f-1)\tau$, and assuming a constant cell survival probability per fraction, S_0,

$$G(\infty) = \left[1 - \left[\frac{1}{S_0} \frac{b}{b-d} \left\{ \left[1 - e^{-(b-d)\tau} \right] \frac{1 - \left[\frac{e^{-(b-d)\tau}}{S_0} \right]^{f-1}}{1 - \left[\frac{e^{-(b-d)\tau}}{S_0} \right]} \right. \right. \right.$$

$$\left. \left. \left. + \left[\frac{e^{-(b-d)\tau}}{S_0} \right]^{f-1} \right\}^{-1} \right]^{n(0)} \quad (19.5)$$

Implementing the general expression, Equation 19.4, requires analytic (or numerical) formulations of the survival probability, $S(t)$. By far, the most common choice is the so-called linear–quadratic (LQ) model (discussed below); however, the reader is encouraged to critically examine this choice and consider alternative possibilities.

19.1.3 LQ Model

The LQ model (a misnomer) has been shown to describe well cellular survival data at doses of clinical interest. Formally,

$$S(D) = \exp(-\alpha D - \beta D^2). \quad (19.6)$$

In its simplest form, the LQ model serves as a convenient way of summarizing large amounts of data with only two parameters (the coefficients of the LQ polynomial, α and β), and this is one reason for its extensive use.

The mechanistic underpinning of the LQ model is controversial and has been thoroughly debated (Sachs and Brenner 1998; Zaider 1998a,b). It rests on four claims: (1) that lesions (damage responsible for cellular inactivation) consist of lethally misrepaired double-strand breaks (DSBs) and exchange-type chromosomal aberrations; (2) that the *average* yield of such lethal lesions can be described by a second-order (LQ) polynomial in dose; (3) that a single lethal lesion is sufficient to inactivate the cell (this is nothing but the definition of a lesion); and (4) that the number of lethal lesions at dose D is Poisson distributed. Claims 1 to 3 are well accepted. The last assertion (claim 4) is known to be incorrect, except at low doses and dose rates for low linear energy transfer (LET) radiation (reviewed in Sachs et al. 1997). Then, what makes the LQ model of cellular survival attractive? By extrapolating the lethal lesion formalism to cellular survival (via claim 4), radiobiologists hoped to take advantage of two important features developed for the former model, namely, the ability to describe changes in radiation quality (microdosimetric interpretation of the ratio α/β) and in the temporal pattern of dose delivery (fractionation, low and high dose rate, dose rate decaying exponentially as a function of time). Consequently, the onus is on the LQ user to show that claim 4 above is warranted.

If Equation 19.6 is taken to be mechanistically valid (i.e., accept claim 4)), then the effect of dose rate may be accounted for by inserting a new function, $q(t)$, as follows:

$$S(D) = \exp(-\alpha D - \beta q(t)D^2). \tag{19.7}$$

The dose-rate function $q(t)$—a quantity that takes values in the interval $[0,1]$—accounts for DSB repair and its effect on intertrack exchange-type chromosome aberrations (i.e., the quadratic term in dose) (Lea and Catcheside 1942; Kellerer and Rossi 1978; Sachs et al. 1997). At very high dose rates, q is close to 1, and at very low dose rates, $q \to 0$. More precisely,

$$q(t) = \frac{\int_0^t \tau(u)h(u)du}{\int_0^t h(u)du}. \tag{19.8}$$

In this expression,

$$\tau(t) = \exp(-\mu t) \tag{19.9}$$

is the probability that a sublesion remains unrepaired t units of time after being formed, and $h(t)dt$ is the pairwise probability distribution of time intervals between pairs of sublesions. Thus,

if we denote by $I(t)$ the dose rate at time t, and if sublesion production is proportional to absorbed dose, it follows that

$$h(t) \propto \int_0^\infty I(u)I(u+t)du. \tag{19.10}$$

19.1.4 Catalog of Dose-Rate Functions

Equation 19.8 can be used to estimate the effects of protracting a single dose of radiation delivered at a constant dose rate as well as the protraction effects arising in split-dose, multifraction, and continuous LDR irradiation schemes. The biophysical interpretation of Equation 19.8 is that a sublesion, presumably the DSB, is created and, if not repaired, may interact in pairwise fashion with a second lesion produced at time t. As long as the dose rate remains constant, the irradiation time will increase with increasing dose, which means that $q(t)$ decreases with increasing dose, even though the dose rate remains constant. Even for high dose rates, dose protraction effects may have a significant impact on cell killing and, ultimately, treatment effectiveness (Wang et al. 2003; Carlson et al. 2004).

The dose rate function, $q(t)$, has been calculated for a variety of temporal patterns of dose delivery.

- For a *single dose* of radiation delivered at *constant dose rate*, $\dot{D}$, during time interval T

$$q(r) = \frac{2}{r} - \left(\frac{2}{r^2}\right)(1 - e^{-r}) \tag{19.11}$$

where $r = T/\tau$ (τ = mean sublethal damage repair time, $\tau \sim$ 1 h). When $T \gg \tau$, $q \approx 2\tau/T$.
  For a treatment consisting of f fractions, $q \approx 1/f$
- For a *temporary implant* with radioactive seeds administered over the time interval $[0,t]$,

$$q(t) = \frac{2(\lambda t)^2}{(\mu t)^2 \left(1 - \lambda^2/\mu^2\right)\left(1 - e^{-\lambda t}\right)^2}$$

$$\times \left[e^{-(\lambda+\mu)t} + \mu t\left(\frac{1 - e^{-2\lambda t}}{2\lambda t}\right) - \frac{1 + e^{-2\lambda t}}{2} \right] \tag{19.12}$$

$$D(t) = D_0(1 - e^{-\lambda t}). \tag{19.13}$$

D_0 is the total dose that would be delivered if $t \to \infty$ (i.e., a permanent implant), and μ is the fractional rate of sublethal damage repair.
- For a *permanent implant* (initial dose rate I_1, radioactive decay constant λ) to which, at time T later, a *second permanent implant* (initial dose rate I_2) is added:

For $t \in [0,T)$

$$q_{\text{LDR}}(t) = \frac{2(\lambda t)^2}{(\mu t)^2 \left(1 - \lambda^2/\mu^2\right)\left(1 - e^{-\lambda t}\right)^2}$$

$$\times \left[e^{-(\lambda+\mu)t} + \mu t \left(\frac{1 - e^{-2\lambda t}}{2\lambda t} \right) - \frac{1 + e^{-2\lambda t}}{2} \right] \quad (19.14)$$

For $t \in [T,\infty)$

$$q(t) = \frac{q_{\text{LDR}}(t)D_1^2 + q_{\text{LDR}}(t-T)D_2^2 + 2q_{12}(t)D_1 D_2}{\left(D_1 + D_2\right)^2} \quad (19.15)$$

where

$$q_{12}(t) = \frac{\lambda^2}{(\mu-\lambda)(\mu+\lambda)\left(1 - e^{-\lambda t}\right)\left(1 - e^{-\lambda(t-T)}\right)}$$

$$\left\{ \frac{\mu}{\lambda}e^{-\lambda T} - \left(1 + \frac{\mu}{\lambda}\right)e^{-\lambda(2t-T)} + e^{-(\mu+\lambda)t}\left[e^{\lambda T} + e^{\mu T}\right] - e^{-\mu T} \right\}. \quad (19.16)$$

These expressions may be further simplified by observing that for the usual implants with ^{125}I or ^{103}Pd seeds, $\lambda \ll \mu$.

- For *LDR brachytherapy followed by HDR brachytheraphy*, the function $q(t)$ can be calculated as follows:

$$q = \frac{D_{\text{LDR}}^2 q_{\text{LDR}} + D_{\text{HDR}}^2 q_{\text{HDR}} + 2D_{\text{LDR}}D_{\text{HDR}}q_{\text{LDR,HDR}}}{\left(D_{\text{LDR}} + D_{\text{HDR}}\right)^2}. \quad (19.17)$$

In this expression, D_{LDR} and D_{HDR} denote, respectively, the *total* LDR and HDR doses, and q_{LDR} and q_{HDR} are dose-rate functions for monotreatments as indicated:

$$q_{\text{LDR}}(t) = \frac{2(\lambda t)^2}{(\mu t)^2 \left(1 - \lambda^2/\mu^2\right)\left(1 - e^{-\lambda t}\right)^2}$$

$$\times \left[e^{-(\lambda+\mu)t} + \mu t \left(\frac{1 - e^{-2\lambda t}}{2\lambda t} \right) - \frac{1 + e^{-2\lambda t}}{2} \right] \quad (19.18)$$

$$q_{\text{HDR}} = \frac{1}{f} \quad (19.19)$$

$$q_{\text{LDR,HDR}}(t) = \frac{1}{f} \frac{\lambda}{1 - e^{-\lambda t}} \left[\frac{e^{-\mu T}}{\lambda - \mu} \frac{1 - e^{-\mu \Delta_1 f}}{1 - e^{-\mu \Delta_1}} - \frac{e^{-\lambda T}}{\lambda - \mu} \frac{1 - e^{-\lambda \Delta_1 f}}{1 - e^{-\lambda \Delta_1}} \right.$$

$$\left. + \frac{e^{-\lambda T}}{\lambda + \mu} \frac{1 - e^{-\lambda \Delta_1 f}}{1 - e^{-\lambda \Delta_1}} - e^{-(\lambda+\mu)t} \frac{e^{\mu T}}{\lambda + \mu} \frac{1 - e^{+\mu \Delta_1 f}}{1 - e^{+\mu \Delta_1}} \right]. \quad (19.20)$$

Here, T is the time that separates the beginning of the LDR treatment from the first HDR fraction.

19.1.5 Isoeffectiveness in Terms of NTCP

The target-cell hypothesis of Thames and Hendry (1987) states that "effects on tissue… can be described in terms of alterations of survival probability." In keeping with this doctrine, two regimens are isoeffective if their respective *effective* survivals

$$S_{\text{eff}} = \frac{S_0(t)e^{(b-d)t}}{1 + bS_0(t)e^{(b-d)t}\int_0^t \dfrac{du}{S_0(u)e^{(b-d)u}}}, \quad (19.21)$$

are the same. This extends the applicability of the formalism to phenomenological models describing NTCP. For instance, for the probability of unresolved grade 2 (or higher), urethral toxicity at 12 months:

$$\text{NTCP}_{\text{LDR}} = P_{tox,12}(\text{DU}_{20}) = \frac{1}{1 + \exp\left[-\left(\gamma + \delta \text{DU}_{20}\right)\right]} \quad (19.22)$$

with $\gamma = -2.60 \pm 0.50$ and $\delta = 0.0066 \pm 0.0016$ Gy^{-1} and where DU_{20} means that 20% of the urethral volume is treated to a dose at least that large. For rectal toxicity, another item of interest in brachytherapy, MacKay et al. (1997) have proposed the following expression to describe the probability of late injury to the rectum:

$$\text{NTCP} = \left\{ 1 - \prod_i \left[1 - \left(\frac{1}{1 + \left(D_{50}/D_i\right)^k} \right) s \right]^{\frac{v_i}{V}} \right\}^{1/s}. \quad (19.23)$$

Here, V is the total volume of the rectal wall, and v_i is a rectal wall subvolume containing cells that, upon treatment, receive the same dose (D_i). D_{50}, s, and k are empirically determined parameters. For severe late reactions, MacKay et al. (1997) recommend $k = 10.24$, $s = 0.75$, and $D_{50} = 80$ Gy—this latter being normalized to a schedule of 2 Gy per fraction.

19.1.6 What We Have Learned So Far

We have reached the end of what is (rigorously) known about deciding if two treatments are isoeffective. The formalism described here is without any doubt a simplified description of a complex reality. For instance, empirical data—against which the models are tested—represent the average response of a heterogeneous population (interpatient and intrapatient) with respect to model parameters. This means that, strictly speaking, one should include in calculations the frequency distribution of these parameters over the population under study, but information on which to base such an effort remains unobtainable. Yet there is a counterargument, based on model calculations, that says that heterogeneity of response may be less of a problem than generally assumed because radioresponse depends essentially on the most radioresistant and/or fast-proliferating moiety in the treated cell population. A more extensive discussion of other points of contention can be found in a recent review of the TCP concept (Zaider and Hanin 2011).

Many of these problems are swept under the rug known as biologically effective dose (BED), a quantity meant to condense in a single number (with units of dose) the biological effectiveness of a treatment modality. The price paid for this convenience is that one is forced to accept the LQ model. For the reader interested in this concept, we clarify in the next sections its proper usage.

19.1.7 IED and BED

From what we have discussed so far, different treatment modalities are biologically equivalent if the quantity

$$\text{IED}(t) = -\frac{1}{\alpha}\log\left[\frac{S_0(t)e^{(b-d)t}}{1 + bS_0(t)e^{(b-d)t}\int_0^t \frac{du}{S_0(u)e^{(b-d)u}}}\right] \quad (19.24)$$

is the same. This definition, *valid within the LQ model*, is such as to make IED = BED when $b = d = 0$, that is, when cell proliferation and spontaneous cell death may be neglected.

$$\text{BED} = -(1/\alpha)\log S(D). \quad (19.25)$$

If $b > d$ (the natural case) and $S_0(\infty) > 0$, then

$$\text{IED}(\infty) = \frac{1}{\alpha}\log\left[b\int_0^\infty \frac{du}{S_0(u)e^{(b-d)u}}\right]. \quad (19.26)$$

Treatments that have the same IED, IED(∞), are expected to have the same cure rate.

The real reason for using such quantities as IED or BED (this latter when appropriate) to construct equivalent treatments is that they circumvent the need to know the actual number of malignant cells, $n(0)$; see Equation 19.4. On the other hand, the applicability of BED as a substitute for IED does depend on $n(0)$ (among other factors), and this is the main reason BED should be avoided. The other reason is that BED calculated at the end of the treatment (and the same goes for IED) is often a poor predictor of IED(∞)—the quantity related to the cure rate. Unlike IED(t), however, the extension of BED to larger values of t eventually becomes meaningless (i.e., negative effective dose).

In the case of permanent implants, this well-recognized problem is artificially avoided by replacing the limit $t \to \infty$ with an "effective" treatment time, t_{eff}, defined by the provision that for $t > t_{eff}$ the treatment ceases to sterilize cells. Following Dale (1989), this quantity is given by

$$t_{eff} = -\frac{1}{\lambda}\log\left(\frac{b}{\alpha D \lambda}\right). \quad (19.27)$$

To illustrate potential differences between BED(t_{eff}) and IED(∞), consider, as an example, a permanent implant with I^{125} seeds and a prescription dose of 144 Gy (Zaider and Hanin 2007). For the target subvolumes actually treated to this dose, IED(∞) = 138.2 Gy (assuming $\alpha = 0.3$ Gy^{-1}, $\beta = 0.03$ Gy^{-2}, $b = 4.3 \times 10^{-2}$ day^{-1}, $d = 3.2 \times 10^{-2}$ day^{-1}, $T_{pot} = 16$ days, $\lambda = 0.01155$ day^{-1}, $\mu = 16.6$ day^{-1}); under the same conditions, one obtains BED(t_{eff}) = 125.3 Gy, a difference that may be important in terms of understanding the predictive value of BED (Stock et al. 2006).

19.2 Review of the Historical Use of BED in Brachytherapy

19.2.1 Introduction

For historical reasons, the BED approach has been the predominant approach in the last two decades for comparing the radiobiological effects of different external beam radiotherapy (EBRT) and brachytherapy regimes. As shown in Section 19.1.7, the use of BED depends on the applicability of LQ cell survival model. In addition, while cell proliferation is taken into account within the context of LQ model, Equation 19.2, conceptual difficulties arise whenever the condition $S(D) < 1$ is not fulfilled; see comments following Equation 19.1. A case in point is permanent implant brachytherapy. As explained in the previous section, for this situation, an "effective treatment time" was introduced in order to use the BED approach for cells that (inevitably) proliferate during treatment.

In this section, we review some of the historical applications of BED in brachytherapy and illustrate, when appropriate, the challenges posed by proliferating tissues. For clarity of presentation, uniform dose irradiation or dose to a particular point in the irradiated volume is considered. In addition, the radiobiological properties of the target tissues are modeled simply by the radiosensitivity parameters of the LQ model (α and β) with monoexponential sublethal damage repair and uniform repopulation. Issues encountered in the application of the BED approach in clinical brachytherapy, such as the intrinsic spatial heterogeneity of dose distribution within the volume of interest and the influence of cell cycle, oxygen status, tissue microenvironment, and other biological processes, will be discussed later in dedicated subsections.

The reader is strongly advised (and will be reminded every so often) to keep in mind that, unless shown otherwise, the implications of the results reviewed below may be (and often are) affected by the inapplicability of the concept of BED to proliferating tissues.

19.2.2 Radiation-Induced Cell Death and the Concept of BED

Radiation-induced cell death and the corresponding quantity of BED have often been used as surrogates for the biological effectiveness of a given ionizing radiation schedule (Fowler 1989). BED has been used as a radiobiological index in brachytherapy treatment evaluation since the 1980s. Several models have been proposed in the past for estimating cell survival after a given irradiation. Among them, the LQ model has been the most popular due to its simple mathematical form and presumed plausible mechanistic interpretation.

As discussed in Section 19.1.3, the LQ model postulates that the average fraction of cells that survive an irradiation in absence of cell proliferation is governed by Equation 19.7. When cell proliferation is present during irradiation, Equation 19.7 is modified to

$$S(D) = \exp[-\alpha D - \beta q(t)D^2 + \gamma \text{Max}\{0, (T - T_k)\}], \quad (19.28)$$

where $\gamma \equiv \ln(2)/T_d$ models the rate of cell proliferation, T_d is the effective cellular doubling time, T is total treatment duration, and T_k is the onset or lag time to the resumption of cell proliferation. Equation 19.28 assumes that cell repopulation is continuous and maintains a uniform proliferation rate throughout the treatment. The Lea–Catcheside dose protraction factor, $q(t)$, accounts for the effects of DSB repair during dose delivery. The explicit forms of $q(t)$ for several temporal patterns of radiation delivery encountered in brachytherapy are listed in Section 19.1.4 assuming DSB repair is monoexponential.

The mechanistic basis for the LQ survival model has been extensively reviewed in the literature (Sachs et al. 1997; Brenner et al. 1998). In the limit of small doses and dose rates, the LQ can be derived from the lethal and potentially lethal (LPL) model (Curtis 1986), the repair–misrepair (RMR) model (Tobias 1985), and the repair–misrepair–fixation (RMF) model (Carlson et al. 2008) using perturbation theory and other methods. The LPL, RMR, and RMF models (and, hence, the LQ model) are broadly consistent with the breakage and reunion theory of chromosome aberration formation, reviewed by Sachs et al. (1997) and Hlatky et al. (2002). α can be interpreted as the quantity per unit dose of lethally misrepaired DSB and lethal intratrack exchange-type chromosome aberrations, and β can be interpreted as the quantity per unit dose of lethal intertrack exchange-type chromosome aberrations formed through binary misrepair of two separate DSBs (Carlson et al. 2008). The third term in the exponent of Equation 19.28 characterizes the effect of clonogen proliferation on the number of cells surviving treatment. In using this form, it is assumed that cell repopulation is continuous and maintains a uniform proliferation rate throughout the treatment. The inclusion of cellular proliferation and dose protraction effects, for example, $q(t)$, enables explicit consideration of the effects

of treatment time on cell survival and the clinical application of the LQ formalism for analysis of the relative effectiveness of different treatment schemes used in EBRT and brachytherapy. The monoexponential repair model and the uniform cell proliferation used in Equation 19.28 are simplifications of the actual biological processes that occur in tumors and normal tissues (see Section 19.4.1 for further discussion). Deviations from these assumptions are expected, and their potential implications will be discussed in later sections.

Fowler (1989) introduced BED, as defined by Equation 19.25 in Section 19.1.7, to replace the nominal standard dose (NSD) and time–dose factor (TDF) tables that were traditionally used in comparing different fractionation schemes in EBRT. When cellular proliferation can be ignored, different radiation schemes that have the same BED would produce the same amount of cell kill in the same tissue. It is worth noting that BED is not a deliverable physical quantity, although it shares the same unit of Gy as physical dose. Indeed, BED is often confused in the literature with "biologically *equivalent* dose" such as the "the equivalent total dose which, when delivered in 2 Gy fractions of photon irradiation, produces the same biological effect (EQD2)" (Dale 2010). The confusion between "biologically effective" and "biologically equivalent" doses is easy to understand, given the similarity between the names and the fact that both are expressed in physical dose units (Gy). Fowler (2010) has recently proposed a change of the unit for BED to something other than the Gy to alleviate the confusion. The most commonly applied form of BED is obtained by substituting Equation 19.28 into Equation 19.25 such that

$$\text{BED} = D\left[1 + \frac{q(t)D}{\alpha/\beta}\right] - \frac{\gamma(T - T_k)}{\alpha}. \quad (19.29)$$

Equation 19.29 is also often written as

$$\text{BED} = D \cdot RE - \frac{\gamma(T - T_k)}{\alpha}, \quad (19.30)$$

where $RE \equiv [1 + q(t)D/(\alpha/\beta)]$ represents the relative effectiveness of the radiation treatment on the target tissue in the absence of cell proliferation. In subsequent sections, we review the historical application of BED in analyzing the radiobiological interplay of brachytherapy dose delivery with the biological characteristics of the irradiated cells.

19.2.3 Protracted Irradiation with Constant Dose Rate—Single Fraction LDR or HDR Brachytherapy

In classical LDR brachytherapy with temporary applications of long-lived radionuclide such as ^{137}Cs or ^{266}Ra, the duration of a procedure may last several days. Cell repopulation and sublethal damage repair during the therapy may be significant depending on the type of tissues being treated. However, because of the long decay half-life associated with these sources, the reduction in dose rate over the treatment course is relatively small, so that the dose rate in these applications may be considered approximately constant.

For constant dose rate, the dose-rate function has a simple mathematical form as given in Equation 19.11. The BED for such a treatment is then given by

$$\text{BED} = D\left[1 + \frac{D}{(\alpha/\beta)}\frac{2}{(\mu T)^2}\left(e^{-\mu T} + \mu T - 1\right)\right] - \gamma\frac{\left(T - T_k\right)}{\alpha}.$$

(19.31)

For a fixed dose, the treatment delivery time T is inversely proportional to dose rate, that is, $D/\dot{D}$. Changing the dose rate will affect the dose-delivery time and, hence, BED. It has been known for decades that lowering the dose rate generally leads to reduced radiobiological damage (Hall and Bedford 1964). Equation 19.31 indicates that the effect of dose rate on biological response (dose-rate effect) arises primarily from the presence of sublethal damage repair and cellular proliferation during the dose delivery.

In the absence of cell proliferation (i.e., $\gamma = 0$) and sublethal damage repair (i.e., $\mu \to 0$), the dose-rate function is unity, and there will be no dose-rate effect. In this special case, BED = $D[1 + D/(\alpha/\beta)]$. In the absence of cell proliferation (i.e., $\gamma = 0$) and with instantaneous sublethal damage repair (i.e., $\mu \to \infty$), the dose protraction factor goes to zero, and there is also no dose-rate effect. Under this condition, BED = D. Hence, the effect of changing the dose rate on BED (biological response) will be governed by the BED values between D and $D[1 + D/(\alpha/\beta)]$. In the following, we will use BED to illustrate the relative dependence of the dose-rate effect on dose rate and on the radiobiological characteristics of the irradiated tissues. These discussions can help us better understand the long-established clinical observations that protracting a radiotherapy exposure (i.e., reducing the dose rate) can yield a therapeutic gain between tumor control and normal tissue complications (Coutard 1932).

For clarity of discussion, we first examine the dose-rate effect on tissues that do not proliferate during treatment (or tissues with small cell proliferation rates where the effect can be neglected, e.g., muscle, central nervous system). Figure 19.1 shows BED as a function of dose rate for two types of tissue: those with a high α/β ratio (e.g., early-responding tissues) and those with a low α/β ratio (e.g., late-responding tissues). In this example, the prescribed dose = 60 Gy, sublethal damage repair half-life = 0.27 h, and cell proliferation is assumed negligible. Figure 19.1 illustrates that BED increases with increasing dose rate for both types of tissue. This behavior is consistent with the general observation that higher dose rates generally produce greater biological damages.

Note, however, that the rate at which BED increases is much greater for late-responding tissues than for early-responding tissues (e.g., typical tumors) as the dose rate is increased. For a more quantitative assessment of the relative changes of BED between these two types of tissue, Figure 19.2 shows the BED

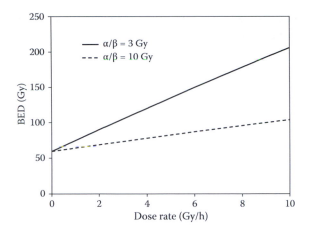

FIGURE 19.1 BED as a function of dose rate. (Courtesy of Z. Chen.)

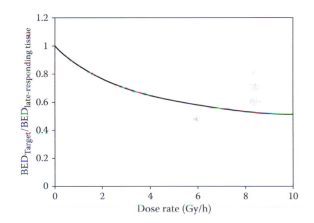

FIGURE 19.2 Ratio of BED of target to normal tissue as a function of dose rate. (Courtesy of Z. Chen.)

ratio between early- and late-responding tissues as a function of dose rate. It demonstrates that reducing the dose rate would lead to increased BED to early-responding tumor (e.g., α/β = 10 Gy) with respect to the BED delivered to late-responding normal tissues (e.g., α/β = 3 Gy) if the normal tissue receives the same dose as the tumor. Hence, relative to a treatment protocol established at a fixed dose rate (e.g., 3 Gy/h), protracting the dose delivery (i.e., reducing dose rate) would lead to an increased therapeutic advantage between tumor control and normal tissue complications. Figures 19.1 and 19.2, consistent with past clinical observations, confirm the currently accepted radiobiological philosophy that generally favors dose protraction while cautioning against the use of small numbers of high-dose fractions with respect to optimizing the tumor control against normal tissue sequelae.

The above discussions are based on three key assumptions: (1) the α/β ratio of the tumor under consideration is greater than that of irradiated normal tissues, (2) the normal tissues receive the same dose as the tumor, and (3) there is no cell proliferation. Although tumors usually have larger α/β ratios than late-responding normal tissues, assumption 1 may not hold for all types of tumors. For example, recent studies indicate that prostate cancer seems to have a low α/β ratio of ~3 Gy or less (see Section 19.4.2). When the α/β ratio of an intended target is comparable to or less than the surrounding normal tissues, the therapeutic advantage of protracting dose delivery discussed above would disappear. Indeed, there are a number of clinical trials being conducted to test hypofractionated dose delivery for prostate cancer.

Even when the α/β ratio of the intended target is greater than that of surrounding normal tissues, assumption 2 may be invalidated by designing more conformal dose distributions so that the normal tissue receives less dose than the tumor tissue. For example, intensity-modulated radiation therapy (IMRT) increases dose conformality substantially. Brachytherapy is also particularly capable of creating more conformal dose distributions with sharp dose falloffs for normal tissue dose sparing. When normal tissues receive doses less than the prescribed tumor dose, the BED calculated for the late-responding normal tissue will be reduced. This would have an immediate impact not only on normal tissue response but also on the relative efficacy of a particular treatment under consideration. For example, the additional dose sparing afforded to the organs at risk by HDR over LDR techniques due to better source positioning and fixation for the short period of dose delivery can alleviate the dose-rate disadvantage of HDR techniques (see Section 19.2.4 for additional details). However, if this dose reduction to normal tissue can be achieved by techniques from LDR to HDR, the general trend favoring LDR as exhibited in Figure 19.2 will still hold.

The effects of sublethal damage repair half-life and cell proliferation can also be examined in detail using Equation 19.31. The general trend exhibited in Figure 19.2 remains for cells with different repair rates. As the cell becomes less capable of repairing sublethal damage (e.g., increasing repair half-life from 0.27 to 3 h), the BED ratio between early- and late-responding tissues changes more dramatically with dose rate in the low dose-rate region. Similarly, cell proliferation has a larger effect in the low dose-rate range as longer dose-delivery times allow for a greater increase in the cell population due to proliferation.

The careful reader may challenge some of the statements made above on two levels. First is the question of whether a population representing a heterogeneous mixture of radiosensitivities and cell proliferation rates (e.g., a tumor) may be meaningfully characterized by a unique set of parameters (α, β, and γ). Nevertheless, as shown by Zaider and Minerbo (2000), TCP is determined to a great extent by the most radioresistant (as well as the fastest proliferating) moiety in the volume of interest, and this could be used to rationalize ignoring this sort of heterogeneity. The second question is this: what could possibly justify (and thus make credible) the notion of an inverse correlation between the average duration of the generation cycle and the ratio α/β? In a paper describing mathematically the effect of sublethal damage repair, cellular proliferation, and progression through the

mitotic cycle on cellular survival (Zaider et al. 1996), this effect is explained as follows: most of the slowly dividing cells (late-responding tissues) spend the irradiation time in G_0. In contrast, rapidly dividing cells (early-responding tissues) will cycle many times during the treatment; the duration of G_1 phase is much shorter, and for the other three phases of the cycle (S, G_2, and M), the ratio α/β can be in the range of 4–100 larger than in G_0 (Zaider 1996). This means that it is *the kinetics of cell proliferation* that explains the difference in radiosensitivity between early- and late-responding tissues. It also means that a discussion of the therapeutic ratio in terms of BED must necessarily account for cellular proliferation.

19.2.4 Multiple Acute Irradiations with HDR Brachytherapy

For single-fraction acute irradiation such as a single application of an HDR treatment, where cell repopulation and sublethal damage repair are negligible during irradiation, the dose protraction factor $q(t) \cong 1$. Equation 19.29 reduces to the more familiar form with BED given by

$$\text{BED} = D[1 + D/(\alpha/\beta)]. \tag{19.32}$$

As discussed in Section 19.2.3, single-fraction acute HDR dose delivery is biologically more potent than LDR dose delivery for both tumor and normal tissues. Moreover, the increase in BED is much greater to late-responding tissues than to early-responding tissues (e.g., tumor) when changing from LDR to HDR. To keep the complication rate of late-response normal tissues clinically acceptable, HDR brachytherapy is typically delivered with multiple fractions with relatively long separations between successive fractions. For HDR delivered with n total fractions and dose d per fraction,

$$\text{BED} = D[1 + d/(\alpha/\beta)] - \gamma \frac{T - T_k}{\alpha} \tag{19.33}$$

where $D = nd$, which is the total treatment dose. Intrafraction repair of sublethal damage is neglected in Equation 19.33 as the time needed to deliver a typical fraction is short compared to the separation between fractions. Equation 19.33 also assumes that the interfraction repair of sublethal damage is complete for typical interfraction separations used in HDR brachytherapy.

Radiobiological analysis of HDR brachytherapy should always be performed with full inclusion of the involved normal tissues. Proper inclusion of dose sparing (if any) to normal tissue achieved by the treatment technique under consideration would enable a more relevant analysis. For example, a study performed by Dale (1990) found that when geometrical sparing of critical tissues is available (by means of spacing or shielding), HDR treatment using a small number of fractions may be used in place of an LDR regime without loss of therapeutic ratio. In that study, he compared the BED of an LDR treatment of the cervix, which delivers 60 Gy in 72 h to the tumor, with that of HDR delivery using a different number of fractions and degrees of geometrical dose sparing. The calculation indicated that when there is a good degree of geometrical sparing in the LDR regime, very little further improvement in implant geometry is required in the corresponding HDR regime to enable the therapeutic ratio to be maintained with small fraction numbers. When there is no LDR geometrical dose sparing of the critical tissue, only a modest improvement in normal tissue dose sparing is required (around 10%–15%) to enable small fraction HDR treatments to be used without loss of the therapeutic ratio. Although this analysis is subject to inherent model limitations, Dale (1990) found that HDR treatment delivered in a small number of fractions may be more radiobiologically acceptable than previously thought. Indeed, an extensive analysis performed by Brenner and Hall (1991) comparing fractionated HDR and LDR regimens for intracavitary brachytherapy of the cervix also reached a similar conclusion. For other HDR applications, one should perform careful analysis of the dose distribution produced by a particular HDR applicator and treatment technique. Any additional dose sparing that can be achieved with the HDR technique to nearby critical normal structures would help to counter the reduction in the therapeutic ratio associated with HDR irradiation as discussed in Section 19.2.3.

19.2.5 Protracted Irradiation with Exponentially Decaying Dose Rates— Permanent Interstitial Brachytherapy

Permanent interstitial brachytherapy (PIB) typically uses radionuclides that have relatively short decay half-lives and emit low-energy photons of 20 to 40 keV (e.g., ^{125}I, [103]Pd, or [131]Cs). The use of low-energy photon sources ensures that a potentially curative radiation dose can be delivered to the target volume with minimum radiation exposure to the surrounding normal tissues. The use of these radionuclides subjects the target cells to continuous LDR irradiation with exponentially decreasing dose rates. As a result of different decay constants (or, equivalently, half-lives), the initial dose rate and the effective treatment time are significantly different for PIBs using different sources. Using prostate cancer as an example, the typical initial dose rates vary from ~7 cGy/h for ^{125}I to ~21 cGy/h for [103]Pd and up to ~30 cGy/h for [131]Cs. The time needed to deliver, for example, 90% of the total dose varies from approximately 1 month for the [131]Cs source to 2 and 6 months for the [103]Pd and ^{125}I sources, respectively. How do these different dose-delivery patterns affect the treatment response? How do we evaluate and compare the clinical effectiveness of PIBs with different sources? How do we translate the clinical experience gained from using one type of source to a new source with a different decay half-life? Led by the work of Dale (1985, 1989), the concept of BED has been used by many investigators to explore these questions and to examine the radiobiological interplays between the PIB dose-delivery patterns and the underlying tissue properties.

19.2.5.1 PIB Using Sources That Are Made of the Same Radionuclide

PIB is traditionally performed using the same radionuclide. In such applications, if the implanted source position and the involved tissue volume remain stationary, the dose rate at any given tissue subvolume (e.g., at location $\vec{r}$) is governed by a simple exponential function of time

$$\dot{D}(\vec{r},t) = \dot{D}_0(\vec{r})e^{-\lambda t}, \qquad (19.34)$$

where $\dot{D}_0(\vec{r})$ is the initial dose rate produced at the tissue subvolume $\vec{r}$ by all implanted sources, and λ is the decay constant of the radionuclide. For this dose-delivery pattern, Dale (1985, 1989) has derived an analytic form of BED with

$$\text{BED} = D(T)\text{RE}(T) - \gamma\frac{T - T_k}{\alpha}, \qquad (19.35)$$

where

$$\text{RE}(T) = 1 + \left(\frac{\beta}{\alpha}\right)\frac{\dot{D}_0}{(\mu - \lambda)} \times \frac{1}{1 - e^{-\lambda T}} \qquad (19.36)$$

$$\times \left\{ 1 - e^{-2\lambda T} - \frac{2\lambda}{\mu + \lambda}\left(1 - e^{-(\mu + \lambda)T}\right) \right\}.$$

In Equations 19.35 and 19.36, T denotes the length of elapsed treatment time and $D(T)$ the cumulative dose delivered up to T. To simplify the notation, we have omitted the explicit denotation of position $\vec{r}$ in Equation 19.36 and similarly in later sections. Note that Equations 19.35 and 19.36 are valid for a point dose or uniform irradiation. The impact of spatial dose variation with the volume of interest will be discussed in Section 19.3.3. In Equations 19.35 and 19.36, the radiobiological properties of the irradiated tissue, as discussed in Section 19.2.3, are characterized by a set of five parameters: α, β, μ, γ, and T_k. As expected, the BED depends not only on the physical characteristics of dose delivery ($\dot{D}_0$, λ, T) but also on the biological characteristics of the irradiated tissue (α, β, μ, γ, and T_k). It has been used to examine the biologic effect of mono-radionuclide PIB in terms of the prescribed dose, the radionuclide used, and their interplay with biological properties of the irradiated tissue as characterized by α, β, μ, γ, and T_k.

As pointed out in Section 19.1.7, the application of BED for permanent implants also requires the introduction of an "effective" treatment time, t_{eff}, to avoid the conceptual problem caused by cell proliferation in large T limit. Figure 19.3 plots the BED of an ^{125}I implant as a function of treatment time for different cell repopulation rates (for this illustration, the dose to full decay is assumed to be 145 Gy with $\alpha = 0.15$ Gy^{-1}, $\alpha/\beta = 3$ Gy, sublethal damage repair half-life = 0.27 h, and $T_k = 0$). In the absence of cell proliferation (the dashed curve on the top), BED increases

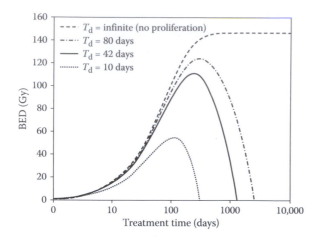

FIGURE 19.3 BED as a function of elapsed treatment time. (Courtesy of Z. Chen.)

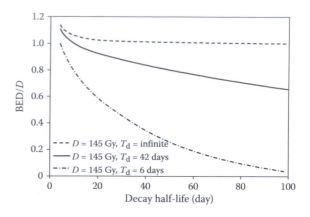

FIGURE 19.4 Ratio of BED to physical dose as a function of decay half-life. (Courtesy of Z. Chen.)

gradually with time at the start of the therapy, followed by a rapid increase between ~20 and ~270 days. After a gradual transition, the BED eventually levels off after ~450 days. The maximum BED value in this case was ~146.3 Gy. When cell repopulation is present, however, BED exhibits a characteristically different behavior at larger T values. Following a similar initial increase with time, the BED would reach a maximum value and begin to decrease monotonically (and turning into negative values eventually) as T further increases. The location of t_{eff} varies with tumor doubling time T_d, and the BED calculated at t_{eff} has been shown to underestimate the isoeffective dose by Zaider and Hanin (2007) (see Section 19.1.7 for a specific example). While the definition of t_{eff} is physically intuitive, the need to use t_{eff} as *the* time point for BED calculation may introduce additional uncertainties in the application of the BED model to tumors that continuously repopulate in PIB. The following discussions should be viewed in this context.

For nonproliferating tissues (e.g., most late-responding normal tissues), Equation 19.35 reduces to (by setting $T = \infty$)

$$\text{BED} = D \left\{ 1 + \frac{\lambda}{\mu + \lambda} \frac{D}{\alpha/\beta} \right\}. \quad (19.37)$$

The BED for these tissues is therefore dependent mainly on the choice of radionuclide (i.e., the decay half-life), the α/β ratio, and the sublethal damage repair half-life of the irradiated tissue in addition to the prescribed total dose.

The influence of total dose on BED is as expected, that is, BED increases with increasing total dose for a given radionuclide type and tissue characteristics. The influence of radioactive decay on BED is shown in Figure 19.4, where BED is plotted as a function of decay half-life for a nonproliferating tissue ($T_d = \infty$) and for tissues with two different repopulation rates ($T_d = 6$ and 42 days). In this example, total dose = 145 Gy and $\alpha = 0.15$ Gy^{-1}, $\alpha/\beta = 3$ Gy, sublethal damage repair half-life = 0.27 h, and $T_k = 0$. Sources with different decay half-lives may emit photons

of different energies, which may have different relative biological effectiveness (RBE). This radiation quality effect is not considered in this example. For a fixed total dose, using sources with shorter decay half-lives would require a higher initial dose rate (as $\dot{D}_0 = \lambda D$) and shorter effective treatment time. As a result, the BED increases with decreasing decay half-life.

For slowly proliferating tissues (e.g., late-responding normal tissues represented by the dashed curve in Figure 19.4), the rate of increase is modest at large decay half-lives but becomes rapid when the decay half-life is less than 20 days. The dose delivered by a radionuclide with a half-life of 5 days is biologically more effective (by ~10%) than that delivered by a radionuclide with a half-life of 60 days. The qualitative trend would remain for other values of α/β and μ. Because late-responding normal tissues usually have lower α/β ratio (e.g., 3 Gy) than that of early-responding normal tissues (e.g., 10 Gy), the biological effect (i.e., BED) of a given total dose is generally larger on late-responding tissues. Using BED, one can perform detailed analysis of the effects of a prescribed dose on early- and late-responding normal tissue and assess potential complications associated with a particular PIB treatment.

For proliferating tumor cells, a strong interplay between radioactive decay and tumor repopulation rate exists. While the general behavior remains that sources with shorter half-life are more effective in cell kill than sources with longer half-life, the difference is more dramatic for tissues with fast cell repopulation rates. For tissues with a doubling time of 42 days, a source with a half-life of 10 days (e.g., [131]Cs) is approximately 30% more effective than a source with a half-life of 60 days (e.g., ^{125}I) when the same total dose is given. This difference increases to 130% for tissues with a faster doubling time of 6 days. Therefore, for tumor eradication, sources with a shorter half-life are more effective than those with a longer half-life. However, one needs to balance the effects on normal tissues. Among the sources used in current clinical practice, [131]Cs would produce a greater biological effect on normal tissue than [103]Pd and ^{125}I if they were prescribed with the same total dose assuming the three sources produce similar normal tissue sparing. In practice, the dose prescribed to

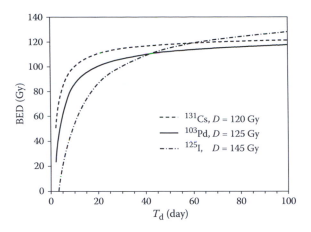

FIGURE 19.5 BED as a function of tumor cell doubling time. (Courtesy of Z. Chen.)

a radionuclide with a shorter half-life is generally smaller than that prescribed for a longer-half-life radionuclide. For example, in PIB for prostate cancer, the prescribed dose for ^{131}Cs, ^{103}Pd, and ^{125}I is approximately 120, 125, and 145 Gy, respectively, aimed to achieve similar effects on a tumor while maintaining acceptable normal tissue reactions. In fact, the lower photon energy emitted by ^{103}Pd often results in greater sparing to surrounding normal tissues, which would provide additional protection of the normal tissue. It should be noted that the nominal prescription dose currently used for each radionuclide does not provide a one-size-for-all solution to individual patients.

Figure 19.5 illustrates the theoretical BED achieved by each radionuclide for tumors with different doubling times. For a tumor with T_d of 42 days, all three radionuclides produce similar BED, with the largest value achieved by ^{131}Cs. For slow-growing tumors ($T_d > 60$ days), ^{125}I is slightly more effective. However, for fast-growing tumors ($T_d < 30$ days), ^{125}I is least effective compared to ^{103}Pd and ^{131}Cs. While ^{131}Cs with a prescribed dose of 120 Gy is more effective than ^{103}Pd and ^{125}I over a wide range of tumor doubling time, this advantage may be offset to some degree by edema-induced source variations as PIB using ^{131}Cs is more sensitive than using ^{125}I. The influence of α/β ratio and sublethal damage repair on biological response of PIB can also be explored in detail using Equation 19.35.

19.3 Clinical Applications

19.3.1 Introduction

Equation 19.35 has been used by many investigators in the examination of various issues related to prostate implants (Dale 1985; Fowler 1989; Dale and Jones 1998). For example, Ling et al. (1994) and others (Ling 1992; Lazarescu and Battista 1997; Dicker et al. 2000; Armpilia et al. 2003; Yue et al. 2005) have used it to assess the effects of dose heterogeneity associated with prostate implants and the RBE of different radionuclides. Other investigators have used such a model to compare the relative

effectiveness of LDR and HDR irradiations, the radiobiological effects of mixing sources with different decay half-lives, the impact of tumor shrinkage during the implant, the effects of prostate edema, and the biological effect of combining prostate brachytherapy with EBRT (Roy et al. 1993; Dale et al. 1994; Peschel et al. 1999; Yaes 2001; King 2002; Van Gellekom et al. 2002; Yue et al. 2002; Chen and Nath 2003; Li et al. 2003; Jani et al. 2004; Chen et al. 2008, 2011; Baltas et al. 2010). Recently, Stock et al. (2006) have also performed a dose response study for ^{125}I implants using BED as the implant quality index. The model is good for comparing different treatment techniques and other issues for which the absolute values of radiobiological parameters or model assumptions may not be critical. Note, however, that many radiobiological complexities are excluded by necessity, some of which are discussed in Section 19.1.4. In addition, as discussed in Section 19.1.7, BED does not take into account correctly the effects of cell proliferation, and concerns on the use of T_{eff} in BED calculation have been raised (Zaider and Hanin 2007). With these caveats in mind, we present a summary of a few applications below with the main purpose of highlighting the use of BED in permanent brachytherapy. The readers are referred to the original articles or related references for more detailed discussion and other applications.

19.3.1.1 Tumor Shrinkage during Therapy

Tumor regression during radiation therapy can introduce significant changes in PIB dosimetry as tissue subvolumes and radioactive sources contract. Normal tissues surrounding the target volume may also receive larger doses than originally planned. For example, Kreth et al. (1997) have reported clinical evidence of the potential hazard to normal tissues of the tumor shrinkage that occurs during permanent or temporary ^{125}I implants to low-grade gliomas. They have observed that, in some cases, the implanted glioma volume may shrink by 50% or more in 6 months.

To examine the potential implications of tumor shrinkage in PIB, Dale et al. (1994) have introduced a simple tumor shrinkage model and incorporated it into the BED formalism; for example, see Equation 19.35. The model assumes that the tumor shrinks isotropically during radiation delivery. In addition, the linear dimensions of an irradiated tumor shrink exponentially with time with a constant shrinkage rate. Under these conditions, Equations 19.35 and 19.36 can be rewritten by replacing -λ with ($2z - \lambda$) for PIBs with concurrent tumor shrinkage, where z is the linear shrinkage coefficient.

Using this model, Dale et al. (1994) have analyzed the implications of tumor shrinkage on both tumor and late-responding normal tissues [assuming the critical normal tissue receives the prescribed tumor dose, does not repopulate (i.e.. $\gamma = 0$), and $\alpha/\beta = 3$ Gy]. For tumors, their calculations indicated that tumor shrinkage would generally lead to delivered doses being greater than originally prescribed. The magnitude of increase depends on the tumor shrinkage rate, the radionuclide used, and the tumor doubling time. For ^{125}I PIB prescribed with conventional doses, very high biological doses to tumor and normal tissues can be expected when tumor repopulation is relatively modest and when there is even a modest degree of shrinkage. PIB using [103]Pd is generally better across a wide range of likely T_d values and has the added advantage that, even with general shrinkage of the treated volume, it will deliver lower doses to normal tissue than would PIBs using conventionally prescribed [125]I.

For normal tissue, Dale and Jones (1999) found that the dose enhancement depends on the shrinkage rate and the decay half-life of the radionuclide. For the shrinkage rate of 1.3×10^{-3} day^{-1}, the dose enhancement factor was 1.07 and 1.29 for [103]Pd and ^{125}I, respectively. The magnitude of the dose enhancement factor increases with increasing shrinkage rate, and the percentage increase is greater for radionuclides with longer half-lives. Using a radionuclide with shorter half-life could be one way to limit the shrinkage-induced dose enhancement to normal tissues. In addition, if the onset of shrinkage lags behind the start of treatment, the dose enhancement factor will be reduced further for shorter-lived radionuclides.

These analyses suggest that the criteria for successful radiotherapy with PIB may sometimes be quite different from those for EBRT, that is, effective brachytherapy may favor the deliberate choice of radionuclides that protract the effective treatment time in order to take advantage of the shrinkage mechanism (Dale and Jones 1999). However, improvements to the model may be needed when applying it to individual patients. For example, the BED equations apply only when centripetal tumor shrinkage occurs, tumor shrinkage may be asymmetrical, and shrinkage rates are unlikely to be truly constant.

19.3.1.2 BED for PIB Using Sources Made of Different Types of Radionuclide

Traditionally, multiple sources containing the same type of radionuclide are used in a given target volume. Nonetheless, clinical implants using a mixture of ^{125}I and [103]Pd sources have been reported (Chen and Nath 2003). Nuttens and Lucas (2008) have recently examined the possibility of using bi-radionuclide sources to exploit the potential advantages of the different decay half-lives.

When radionuclides of different decay half-lives are mixed in the same implant, the temporal dose-delivery pattern to a point within the implant is now defined by the decay characteristics of all radionuclide types that contribute dose. The resulting temporal pattern becomes biexponential, and its shape varies throughout the implant volume (as a function of seed locations). To properly evaluate such an implant, an index, such as BED, that can capture the interplay between this temporal dose-delivery pattern and the underlying tissue radiobiology is needed. Chen and Nath (2003) have generalized the Dale equation for single radionuclide, that is, Equation 19.35, and obtained a general formalism of BED for multiradionuclide PIB. Using this formalism, they have investigated the effects of mixing ^{125}I and [103]Pd sources in a clinical head-and-neck PIB and in animal model studies. Nuttens and Lucas (2008) have investigated the dose prescription issues for bi-radionuclide implants by numerical integration of the dose-rate function.

For head-and-neck implant using ^{125}I and [103]Pd sources, it was found that the resulting BED depends heavily on the relative dose contributions by the different types of radioactive sources used (Chen and Nath 2003). Clinically, if the dose prescription was established from existing implant experience that uses ^{125}I seeds only, mixing [103]Pd seeds with ^{125}I seeds would increase the effective cell kill for the same prescribed dose. On the other hand, if the dose prescription was established from implant experience of using [103]Pd seeds alone, mixing ^{125}I seeds with [103]Pd seeds in the same implant would create radiobiological "cold" spots (an increase in cell survival from the existing clinical experience) at locations where a major portion of the dose is contributed by the ^{125}I seeds. These "cold" spots can become significant for fast-growing tumors. The magnitude and spatial location of the "cold" spots can vary throughout the implant volume.

19.3.2 Multiple Protracted Irradiations with Constant Dose Rate—Pulsed-Dose-Rate Brachytherapy

Pulsed-dose-rate (PDR) brachytherapy (Brenner 1997; Dale and Jones 1998) has been developed to simulate a continuous LDR treatment using a series of multiple short HDR irradiation exposures, thereby limiting the potential reduction in the therapeutic ratio associated with HDR brachytherapy. The objective is to deliver a BED that is similar to continuous LDR. A single high-activity source is used to deliver a series of dose pulses that are typically on the order of minutes in duration. These pulses are repeated at regular intervals, typically hourly, over a course of several hours for total treatment times that can last up to several days. This prolonged temporal pattern of delivery results in a reduction of the effective dose rate. The dose rate of each pulse, separation between pulses, dose per pulse, and the total number of pulses all have an impact on the biological effectiveness of the treatment as a result of inter- and intra-pulse DNA damage repair and tumor cell proliferation. PDR brachytherapy therefore requires careful radiobiological analysis in advance in order to optimize the many physical treatment variables for the desired treatment effectiveness.

For pulsed brachytherapy with N pulses each of duration t and separated by radiation-free interval X (Dale and Jones 1998), the dose protraction factor can be formulated as

$$G(N,\mu,t,X) = \frac{2}{N\mu t}\left[1 - \frac{NY - SY^2}{N\mu t}\right], \qquad (19.38)$$

where

$$S = \frac{NW - W - NW^2Z + W^{N+1}Z^N}{(1-WZ)^2}, \qquad (19.39)$$

where $K = e^{-\mu X}$, $Z = e^{-\mu t}$, and $Y = 1 - Z$. The BED for PDR brachytherapy can now be calculated using Equation 19.31, where the total dose $D = \dot{D}Nt$. PDR offers a unique opportunity to optimize the temporal pattern of radiation delivery to achieve a similar or better therapeutic ratio as continuous LDR brachytherapy, while maintaining many of the logistical advantages of HDR brachytherapy with a remote afterloader (Brenner 1997).

19.3.3 Impact of Spatial Dose Heterogeneity Associated with Brachytherapy

The BED formulae discussed in Sections 19.2.3 to 19.3.2 are relevant for point doses or uniform dose irradiation. Brachytherapy dose distributions are inherently nonuniform due to the rapid reduction of photon fluence with increasing distance from a radioactive source. In PIB, the spatial variation of the dose rate is caused by both the sharp dose falloff around individual low-energy sources and the spatial arrangement of the implanted sources. Dose rates can even differ by more than a factor of 2 within the target volume. The radiobiological impact of this intrinsic dose heterogeneity can also be assessed using the BED concept.

Given a dose (or initial dose rate) distribution, the impact of dose heterogeneity can be evaluated using two methods: through analysis of the full 3D BED distribution on a voxel-by-voxel basis or by examining the effective BED of a volume of interest. In both cases, the volume of interest is first divided into small subvolumes such that the dose distribution within each subvolume can be considered uniform. The BED_i for each subvolume i with an initial dose rate of $\dot{D}_{0i}$ can then be calculated using the formulae discussed in previous sections. When the biological properties of each subvolume are known accurately, the calculated BED_i could explicitly account for spatial biological variation over the volume of interest. In practice, BED_i is usually calculated with nominal biological parameters assigned to the entire volume due to the lack of accurate information on spatial variations in the biological properties.

19.3.3.1 BED Distribution for Analyzing Radiobiological "Hot" and "Cold" Spots

A 3D iso-BED distribution can be constructed from individual BED_i for all subvolumes. BED distributions provide detailed information on the 3D variation of BED over the entire volume of clinical interest similar to the 3D dose distribution to which we are accustomed. The biological significance of "hot" or "cold" dose regions can be evaluated against the underlying anatomy, and radiobiologically significant "cold" spots can then be dealt with proactively in the planning of brachytherapy treatments. Similarly, BED-volume histograms can also be computed and analyzed.

19.3.3.2 Effective BED

With the knowledge of BED_i for all tissue subvolumes, an effective BED can also be computed for a given volume of interest. It provides a single index for the combined biologic effect of a brachytherapy dose distribution in the given volume. Mathematically, the BED for a given volume can be calculated as

$$BED = -\frac{1}{\alpha}\ln\left(\sum_i v_i e^{-\alpha \cdot BED_i}\right) \qquad (19.40)$$

where v_i is the fractional volume receiving the initial dose rate $\dot{D}_{0i}$ with $\sum_i v_i = 1$. v_i is directly related to the differential dose (or the initial dose rate) histogram of a brachytherapy treatment. The BED calculated with Equation 19.40 takes into account not only the time-dependent dose-rate variation, cell repopulation, and sublethal damage repair during the dose delivery but also the spatial heterogeneity of the dose-rate distribution in brachytherapy. Ling et al. (1994) have used Equation 19.40 in studying the effects of dose heterogeneity in permanent interstitial implants. Note that the BED calculated according to Equation 19.40 is preferentially weighted by subvolumes receiving low dose rates.

The use of Equation 19.40 implicitly assumes that the initial tumor burden and radiosensitivity are spatially uniform. Further, the RBE is assumed to be unity in the aforementioned analysis. Several radiobiological studies using in vitro cell lines have shown that the RBE of continuous LDR irradiation using low-energy photons can be significantly different from that of high-energy photons typically used in EBRT. The actual value of RBE is also affected by many factors such as LET, dose rate, and the biological properties of the irradiated tissue. The reported LET-induced RBE for ^{125}I (^{103}Pd) ranged from 1.2 to 1.5 (1.6 to 1.9) compared to photons of ^{60}Co (Ling et al. 1995; Nath et al. 2005). To determine the absolute equivalency between an LDR implant and an EBRT treatment, one needs to know the RBE specific to each irradiation condition and the tumor cells being irradiated.

19.4 Model Limitations and Areas for Improvement

19.4.1 Influence of Other Biological Processes

Many additional biological factors that potentially influence the effectiveness of brachytherapy treatments have been studied extensively. The simplest and most commonly used LQ model fails to capture several biological processes that are known to modulate the radiation response of cells, such as tumor hypoxia (Dasu and Denekamp 1998; Nahum et al. 2003; Carlson et al. 2006, 2011), low dose hyper-radiosensitivity (Marples and Collis 2008), bystander effects (Mothersill and Seymour 2001), and the possibility of additional biological targets other than the cellular DNA of tumor cells.

Tumor hypoxia has been observed in many human cancers and has been shown to correlate with local treatment failure in brachytherapy (Hockel et al. 1998; Knocke et al. 1999; Sundfor et al. 2000). Many groups (Dasu and Denekamp 1998; Nahum et al. 2003; Carlson et al. 2006, 2011) have quantified the radioprotective effect of hypoxia on tumor cell killing and tumor control. Extensions to the commonly used radiobiological models that incorporate the effects of cell cycle redistribution and reoxygenation have also been proposed (Zaider and Minerbo 1993; Brenner et al. 1995; Zaider 1996; Zaider et al. 1996). There are also data (Steel et al. 1987; Nelson et al. 1990; van Rongen et al. 1993; Stewart 2001) that suggest that DSBs are often rejoined with biphasic kinetics rather than through a simpler monoexponential decay process. The fast rate of DSB rejoining is typically on the order of 0.1 to 0.5 h, whereas the slow DSB rejoining rate is typically greater than 4 to 6 h. Understanding and quantifying repair rates are critical to quantifying the BED of continuous and fractionated treatments. An exhaustive discussion of these factors is beyond the scope of this chapter.

The applicability of the LQ model at high doses per fraction has been critically examined by many investigators (Dutreix et al. 1990; Guerrero and Li 2004; Park et al. 2008; Hanin and Zaider 2010). It has long been known that the LQ model is only an approximation to more sophisticated kinetic reaction rate models, and LQ predictions begin to deviate, for example, from RMR model predictions above ~5 Gy (Sachs et al. 1997). The LQ model has been shown to fit experimental survival data well up to ~10 Gy (Guerrero and Li 2004) and may even be appropriate for doses of up to ~15–18 Gy (Brenner 2008). However, when extrapolating to doses above 15 Gy, the LQ can exhibit an order-of-magnitude difference with the LPL model, overpredicting the level of cell killing. In addition, at higher doses, tumor cell killing might be enhanced by rapid endothelial cell apoptosis in the tumor vessels as vascular endothelial cell apoptosis has been reported to occur at doses greater than ~8–10 Gy (Garcia-Barros et al. 2003). While this is an interesting possibility, there is a need for more data before it can be an accepted mechanism for the effect of high dose fractions (Brown and Koong 2008). While not explicitly included in the models presented here, the α radiosensitivity parameter of the LQ model conceptually includes any type of molecular or cellular insult that is proportional to dose and triggers apoptosis in tumor cells.

19.4.2 Uncertainty in Radiosensitivity Parameter Estimates

The assumed parameter values reported in this chapter are not meant to be interpreted as the only biologically plausible parameters. It is widely known that intrinsic radiosensitivity is tumor- and tissue-specific and that there is a lack of sufficient data for many sites. Even for tumor sites such as prostate cancer where there is an abundance of laboratory and clinical data, there is much debate over the values that should be used in isoeffect and dose response modeling studies. For example, reported point estimates of α/β range from 0.5 to 8.3 and 1.1 to 8.4 for in vivo and in vitro data, respectively (Carlson et al. 2004; Dasu 2007). It is therefore crucial in any modeling study to determine the sensitivity of the results to the assumed radiosensitivity parameters by using estimates sampled over their observed or expected uncertainty range. In the absence of patient-specific radiobiological parameters, a self-consistent set of parameters may be used for relative comparison of radiobiological impacts of different treatment strategies. For example, American Association of Physicists in Medicine (AAPM) TG-137 (Nath et al. 2009) has recommended a set of radiosensitivity parameters for prostate cancer ($\alpha = 0.15$ Gy^{-1}, $\alpha/\beta = 3.0$ Gy, $T_d = 42$ days, $\tau = 0.27$ h) to increase the comparability of radiobiological indices reported by different institutions.

The challenge of obtaining accurate site-specific radiosensitivity parameters is confounded by another problem intrinsic to a radiobiological modeling approach. Interpatient variability in radiosensitivity exists among patient populations. This means that any individual patient may respond differently as compared with the average patient of a population. For example, Keall and Webb (2007) found that lognormal and normal distributions of interpatient tumor radiosensitivity heterogeneity more closely describe tumor control data than a single radiosensitivity value.

Intratumor variability also exists among individual patients because tumors and normal tissues are composed of cells with heterogeneous radiosensitivity. In addition to physical dose and dose-rate distributions, biological distributions of radiosensitivity are essential for a more realistic determination of radiobiological indices. These can include, but are not limited to, variations in the intrinsic radiosensitivity of cells across a tumor (e.g., α and α/β), the tissue architecture of the irradiated organs, clonogenic density, the proliferative capacity of the cell populations, and the rate at which sublethal damage is repaired. Other variable factors that affect radiobiological response such as the presence of hypoxic cells, cell cycle effects, and radiation-induced apoptosis should be considered as well for a complete radiobiological characterization. While considerable progress has been made in the last decade to incorporate these factors in the mathematical modeling of radiobiological responses, at present, not all factors are well understood or quantitatively characterized.

Radiobiological modeling tools should therefore not be used as absolute indices for predicting the treatment outcome of individual patients. When a given set of radiobiological parameters is used consistently, it can be a useful tool for assessing the relative effectiveness of different brachytherapy treatments and intercomparison of brachytherapy techniques with other treatment modalities such as EBRT. However, for any individual tumor, there will be a range of biological factors specific to that tumor cell population that will define its radiation response characteristics. Thus, any modeled parameters can only be an estimate based on a population having a range within which that individual will fall.

19.4.3 Clinical Interpretation

Brachytherapy, more than any other type of radiation therapy, presents the clinician with a bewildering array of variables that come together to constitute the final biological effect. This chapter has shown how these variables may be used in mathematical models to explore their interaction. Inevitably, such models are imperfect because the parameters they use are in some instances less precise than they might appear. Thus, additional factors should be taken into account, including the following.

Individual anatomy will vary from patient to patient, defining the clinical target volume but often more critically the normal tissue configuration. In the example of prostate implants, the urethral position will vary from patient to patient as will its caliber; it may also change from the size when a catheter is in situ to its natural uncatheterized shape, which, for a permanent seed implant, may reflect the majority of its irradiated period. The bowel and rectum vary considerably from hour to hour according to the passage of flatus and solid material, with consequent variations in dose and region irradiated when adjacent to an implant.

Implant design and source placement will be influenced by departmental convention, some preferring relatively few higher-energy sources and others more lower-energy sources. Placement may be influenced by many factors on the day of implant; a clear learning curve exists in the case of prostate brachytherapy, and poor setup, technical failure, or bad luck may all result in a suboptimal implant.

Clinical prescription isodose is often taken as a constant when describing brachytherapy regimes. In most instances, this will reflect the minimal peripheral isodose, but historically, it reflected the 85% isodose in the Paris dosimetry system and the 90% isodose in the Manchester rules, while others will use the 95% isodose as their prescription point. Furthermore, within a clinical target volume, there is enormous dose variation. However, this will depend upon dosimetric conventions within each department defining where hot spots may be and what dose levels or source dwell times are permitted to achieve defined dose constraints represented by D_{90}, V_{100}, and V_{150}.

Variations in tumor and normal tissue biological parameters during radiation delivery, while acknowledged, are largely ignored in the models. They will of course be individual to a given tumor or normal tissue and enormously difficult to predict or measure. However, we know that the main predictors of biological response, hypoxia, repopulation, repair, and reassortment change during and after radiation delivery, perhaps from minute to minute and certainly from hour to hour. Thus, assumptions using parameters for repair half-times and repopulation rates are, at best, an estimate, hopefully reflecting an average over the relevant period of radiation response.

19.4.4 Practice of Evidence-Based Medicine

Ultimately the brachytherapist is charged to prescribe a radiation dose balancing the imperative to provide the very best chance of tumor cure with the lowest possible risk of normal tissue damage. In some instances, the convention is clear; for example, 145-Gy peripheral dose using TG-43 formalism is well established for I^{125} prostate implants. However, this is far from the case for an HDR prostate implant that has been undertaken with many schedules described in the literature after 45-Gy external beam ranging from 15 Gy in three fractions to 22 Gy in two fractions to 15 Gy as a single dose, with many other options in between, and before variations in external beam prescription have been considered. A similar scenario exists in gynecological brachytherapy. It is in these settings that the models described here have their greatest utility for the clinician.

Clinical data from prospective randomized trials are the gold standard in medicine. The value of any model, such as a quantitative biological dose response model, is highest in assessing relative changes in conventional approaches in the absence of an abundance of good clinical data. The models presented in this work can be used to guide the selection of new or alternate treatment regimens in a careful and methodical way for treatment sites that do not already have successful conventional treatments. It is important to remember that uncertainties in radiosensitivity parameter estimates will have a larger impact on calculated dose prescriptions when the alternate treatment differs greatly from the conventional therapy. It is therefore wise to make small incremental changes away from conventional regimens instead of drastic changes to accepted clinical standard of care.

Acknowledgments

One of the authors (ZC) acknowledges partial support from NIH R01-CA134627.

References

Amols, H. I., Zaider, M., Hayes, M. K. and Schiff, P. B. 1997. Physician/patient-driven risk assignment in radiation oncology: reality or fancy? *Int J Radiat Oncol Biol Phys* 38:455–61.

Armpilia, C. I., Dale, R. G., Coles, I. P., Jones, B. and Antipas, V. 2003. The determination of radiobiologically optimized half-lives for radionuclides used in permanent brachytherapy implants. *Int J Radiat Oncol Biol Phys* 55:378–85.

Baltas, D., Lymperopoulou, G., Loffler, E. and Mavroidis, P. 2010. A radiobiological investigation on dose and dose rate for permanent implant brachytherapy of breast using 125I or 103Pd sources. *Med Phys* 37:2572–86.

Brenner, D. J. 1997. Radiation biology in brachytherapy. *J Surg Oncol* 65:66–70.

Brenner, D. J. 2008. The linear-quadratic model is an appropriate methodology for determining isoeffective doses at large doses per fraction. *Semin Radiat Oncol* 18:234–9.

Brenner, D. J. and Hall, E. J. 1991. Fractionated high dose rate versus low dose rate regimens for intracavitary brachytherapy of the cervix. I. General considerations based on radiobiology. *Br J Radiol* 64:133–41.

Brenner, D. J., Hlatky, L. R., Hahnfeldt, P. J., Hall, E. J. and Sachs, R. K. 1995. A convenient extension of the linear-quadratic model to include redistribution and reoxygenation. *Int J Radiat Oncol Biol Phys* 32:379–90.

Brenner, D. J., Hlatky, L. R., Hahnfeldt, P. J., Huang, Y. and Sachs, R. K. 1998. The linear-quadratic model and most other common radiobiological models result in similar predictions of time-dose relationships. *Radiat Res* 150:83–91.

Brown, J. M. and Koong, A. C. 2008. High-dose single-fraction radiotherapy: exploiting a new biology? *Int J Radiat Oncol Biol Phys* 71:324–5.

Carlson, D. J., Stewart, R. D., Li, X. A., Jennings, K., Wang, J. Z. and Guerrero, M. 2004. Comparison of in vitro and in vivo alpha/beta ratios for prostate cancer. *Phys Med Biol* 49:4477–91.

Carlson, D. J., Stewart, R. D. and Semenenko, V. A. 2006. Effects of oxygen on intrinsic radiation sensitivity: a test of the relationship between aerobic and hypoxic linear-quadratic (LQ) model parameters. *Med Phys* 33:3105–15.

Carlson, D. J., Stewart, R. D., Semenenko, V. A. and Sandison, G. A. 2008. Combined use of Monte Carlo DNA damage simulations and deterministic repair models to examine putative mechanisms of cell killing. *Radiat Res* 169:447–59.

Carlson, D. J., Keall, P. J., Loo, B. W., Jr., Chen, Z. J. and Brown, J. M. 2011. Hypofractionation results in reduced tumor cell kill compared to conventional fractionation for tumors with regions of hypoxia. *Int J Radiat Oncol Biol Phys* 79:1188–95.

Chen, Z. and Nath, R. 2003. Biologically effective dose (BED) for interstitial seed implants containing a mixture of radionuclides with different half-lives. *Int J Radiat Oncol Biol Phys* 55:825–34.

Chen, Z. J., Deng, J., Roberts, K. and Nath, R. 2008. On the need to compensate for edema-induced dose reductions in pre-planned (131)Cs prostate brachytherapy. *Int J Radiat Oncol Biol Phys* 70:303–10.

Chen, Z. J., Roberts, K., Decker, R., Pathare, P., Rockwell, S. and Nath, R. 2011. The impact of prostate edema on cell survival and tumor control after permanent interstitial brachytherapy for early stage prostate cancers. *Phys Med Biol* 56:4895–912.

Coutard, H. 1932. Roentgentherapy of epitheliomas of the tonsillar region, hypopharynx, and larynx, from 1920 to 1926. *AJR* 28:313–331, 343–348.

Curtis, S. B. 1986. Lethal and potentially lethal lesions induced by radiation—a unified repair model. *Radiat Res* 106:252–70.

Dale, R. G. 1985. The application of the linear-quadratic dose-effect equation to fractionated and protracted radiotherapy. *Br J Radiol* 58:515–28.

Dale, R. G. 1989. Radiobiological assessment of permanent implants using tumour repopulation factors in the linear-quadratic model. *Br J Radiol* 62:241–4.

Dale, R. G. 1990. The use of small fraction numbers in high dose-rate gynaecological afterloading: some radiobiological considerations. *Br J Radiol* 63:290–4.

Dale, R. G. 2010. The BJR and progress in radiobiological modelling. *Br J Radiol* 83:544–5.

Dale, R. G. and Jones, B. 1998. The clinical radiobiology of brachytherapy. *Br J Radiol* 71:465–83.

Dale, R. G. and Jones, B. 1999. Enhanced normal tissue doses caused by tumour shrinkage during brachytherapy. *Br J Radiol* 72:499–501.

Dale, R. G., Jones, B. and Coles, I. P. 1994. Effect of tumour shrinkage on the biological effectiveness of permanent brachytherapy implants. *Br J Radiol* 67:639–45.

Dasu, A. 2007. Is the alpha/beta value for prostate tumours low enough to be safely used in clinical trials? *Clin Oncol (R Coll Radiol)* 19:289–301.

Dasu, A. and Denekamp, J. 1998. New insights into factors influencing the clinically relevant oxygen enhancement ratio. *Radiother Oncol* 46:269–77.

Dicker, A. P., Lin, C. C., Leeper, D. B. and Waterman, F. M. 2000. Isotope selection for permanent prostate implants? An evaluation of 103Pd versus 125I based on radiobiological effectiveness and dosimetry. *Semin Urol Oncol* 18:152–9.

Dutreix, J., Cosset, J. M. and Girinsky, T. 1990. Biological equivalency of high single doses used in intraoperative irradiation. *Bull Cancer Radiother* 77:125–34.

Fowler, J. F. 1989. The linear-quadratic formula and progress in fractionated radiotherapy. *Br J Radiol* 62:679–94.

Fowler, J. F. 2010. 21 years of biologically effective dose. *Br J Radiol* 83:554–68.

Garcia-Barros, M., Paris, F., Cordon-Cardo, C. et al. 2003. Tumor response to radiotherapy regulated by endothelial cell apoptosis. *Science* 300:1155–9.

Guerrero, M. and Li, X. A. 2004. Extending the linear-quadratic model for large fraction doses pertinent to stereotactic radiotherapy. *Phys Med Biol* 49:4825–35.

Hall, E. J. and Bedford, J. S. 1964. Dose rate: its effect on the survival of Hela cells irradiated with gamma rays. *Radiat Res* 22:305–15.

Hanin, L. G. and Zaider, M. 2010. Cell-survival probability at large doses: an alternative to the linear-quadratic model. *Phys Med Biol* 55:4687–702.

Hlatky, L., Sachs, R. K., Vazquez, M. and Cornforth, M. N. 2002. Radiation-induced chromosome aberrations: insights gained from biophysical modeling. *Bioessays* 24:714–23.

Hockel, M., Schlenger, K., Hockel, S., Aral, B., Schaffer, U. and Vaupel, P. 1998. Tumor hypoxia in pelvic recurrences of cervical cancer. *Int J Cancer* 79:365–9.

Jani, A. B., Hand, C. M., Lujan, A. E. et al. 2004. Biological effective dose for comparison and combination of external beam and low-dose rate interstitial brachytherapy prostate cancer treatment plans. *Med Dosim* 29:42–8.

Keall, P. J. and Webb, S. 2007. Optimum parameters in a model for tumour control probability, including interpatient heterogeneity: evaluation of the log-normal distribution. *Phys Med Biol* 52:291–302.

Kellerer, A. M. and Rossi, H. H. 1978. A generalized formulation of dual radiation action. *Radiat Res* 75:471–488.

King, C. R. 2002. LDR vs. HDR brachytherapy for localized prostate cancer: the view from radiobiological models. *Brachytherapy* 1:219–26.

Knocke, T. H., Weitmann, H. D., Feldmann, H. J., Selzer, E. and Potter, R. 1999. Intratumoral pO2-measurements as predictive assay in the treatment of carcinoma of the uterine cervix. *Radiother Oncol* 53:99–104.

Kreth, F. W., Faist, M., Rossner, R., Birg, W., Volk, B. and Ostertag, C. B. 1997. The risk of interstitial radiotherapy of low-grade gliomas. *Radiother Oncol* 43:253–60.

Lazarescu, G. R. and Battista, J. J. 1997. Analysis of the radiobiology of ytterbium-169 and iodine-125 permanent brachytherapy implants. *Phys Med Biol* 42:1727–36.

Lea, D. E. and Catcheside, D. G. 1942. The mechanism of the induction by radiation of chromosome aberrations in Tradescantia. *J Genet* 44:216–45.

Li, X. A., Wang, J. Z., Stewart, R. D. and Dibiase, S. J. 2003. Dose escalation in permanent brachytherapy for prostate cancer: dosimetric and biological considerations. *Phys Med Biol* 48:2753–65.

Ling, C. C. 1992. Permanent implants using Au-198, Pd-103 and I-125: radiobiological considerations based on the linear quadratic model. *Int J Radiat Oncol Biol Phys* 23:81–7.

Ling, C. C., Roy, J., Sahoo, N., Wallner, K. and Anderson, L. 1994. Quantifying the effect of dose inhomogeneity in brachytherapy: application to permanent prostatic implant with 125I seeds. *Int J Radiat Oncol Biol Phys* 28:971–8.

Ling, C. C., Li, W. X. and Anderson, L. L. 1995. The relative biological effectiveness of I-125 and Pd-103. *Int J Radiat Oncol Biol Phys* 32:373–8.

Mackay, R. I., Hendry, J. H., Moore, C. J., Williams, P. C. and Read, G. 1997. Predicting late rectal complications following prostate conformal radiotherapy using biologically effective doses and normalized dose-surface histograms. *Br J Radiol* 70:517–26.

Marples, B. and Collis, S. J. 2008. Low-dose hyper-radiosensitivity: past, present, and future. *Int J Radiat Oncol Biol Phys* 70:1310–8.

Mothersill, C. and Seymour, C. 2001. Radiation-induced bystander effects: past history and future directions. *Radiat Res* 155:759–67.

Nahum, A. E., Movsas, B., Horwitz, E. M., Stobbe, C. C. and Chapman, J. D. 2003. Incorporating clinical measurements of hypoxia into tumor local control modeling of prostate cancer: implications for the alpha/beta ratio. *Int J Radiat Oncol Biol Phys* 57:391–401.

Nath, R., Bongiorni, P., Chen, Z., Gragnano, J. and Rockwell, S. 2005. Relative biological effectiveness of 103Pd and 125I photons for continuous low-dose-rate irradiation of Chinese hamster cells. *Radiat Res* 163:501–9.

Nath, R., Bice, W. S., Butler, W. M. et al. 2009. AAPM recommendations on dose prescription and reporting methods for permanent interstitial brachytherapy for prostate cancer: report of Task Group 137. *Med Phys* 36:5310–22.

Nelson, J. M., Braby, L. A., Metting, N. F. and Roesch, W. C. 1990. Multiple components of split-dose repair in plateau-phase mammalian cells: a new challenge for phenomenological modelers. *Radiat Res* 121:154–60.

Nuttens, V. E. and Lucas, S. 2008. Determination of the prescription dose for biradionuclide permanent prostate brachytherapy. *Med Phys* 35:5451–62.

Park, C., Papiez, L., Zhang, S., Story, M. and Timmerman, R. D. 2008. Universal survival curve and single fraction equivalent dose: useful tools in understanding potency of ablative radiotherapy. *Int J Radiat Oncol Biol Phys* 70:847–52.

Peschel, R. E., Chen, Z., Roberts, K. and Nath, R. 1999. Long-term complications with prostate implants: iodine-125 vs. palladium-103. *Radiat Oncol Investig* 7:278–88.

Roy, J. N., Anderson, L. L., Wallner, K. E., Fuks, Z. and Ling, C. C. 1993. Tumor control probability for permanent implants in prostate. *Radiother Oncol* 28:72–5.

Sachs, R. K. and Brenner, D. J. 1998. The mechanistic basis of the linear-quadratic formalism. *Med Phys* 25:2071–3.

Sachs, R. K., Hahnfeld, P. and Brenner, D. J. 1997. The link between low-LET dose-response relations and the underlying kinetics of damage production/repair/misrepair. *Int J Radiat Biol* 72:351–74.

Steel, G. G., Deacon, J. M., Duchesne, G. M., Horwich, A., Kelland, L. R. and Peacock, J. H. 1987. The dose-rate effect in human tumour cells. *Radiother Oncol* 9:299–310.

Stewart, R. D. 2001. Two-lesion kinetic model of double-strand break rejoining and cell killing. *Radiat Res* 156:365–78.

Stock, R. G., Stone, N. N., Cesaretti, J. A. and Rosenstein, B. S. 2006. Biologically effective dose values for prostate brachytherapy: effects on PSA failure and posttreatment biopsy results. *Int J Radiat Oncol Biol Phys* 64:527–33.

Sundfor, K., Lyng, H., Trope, C. G. and Rofstad, E. K. 2000. Treatment outcome in advanced squamous cell carcinoma of the uterine cervix: relationships to pretreatment tumor oxygenation and vascularization. *Radiother Oncol* 54:101–7.

Thames, H. D. and Hendry, J. H. 1987. *Fractionation in Radiotherapy.* Taylor & Francis, New York.

Tobias, C. A. 1985. The repair-misrepair model in radiobiology: comparison to other models. *Radiat Res Suppl* 8:S77–95.

Van Gellekom, M. P., Moerland, M. A., Kal, H. B. and Battermann, J. J. 2002. Biologically effective dose for permanent prostate brachytherapy taking into account postimplant edema. *Int J Radiat Oncol Biol Phys* 53:422–33.

Van Rongen, E., Thames, H. D., Jr. and Travis, E. L. 1993. Recovery from radiation damage in mouse lung: interpretation in terms of two rates of repair. *Radiat Res* 133:225–33.

Wang, J. Z., Li, X. A., D'souza, W. D. and Stewart, R. D. 2003. Impact of prolonged fraction delivery times on tumor control: a note of caution for intensity-modulated radiation therapy (IMRT). *Int J Radiat Oncol Biol Phys* 57:543–52.

Yaes, R. J. 2001. Late normal tissue injury from permanent interstitial implants. *Int J Radiat Oncol Biol Phys* 49:1163–9.

Yue, N., Chen, Z. and Nath, R. 2002. Edema-induced increase in tumour cell survival for 125I and 103Pd prostate permanent seed implants—a bio-mathematical model. *Phys Med Biol* 47:1185–204.

Yue, N., Heron, D. E., Komanduri, K. and Huq, M. S. 2005. Prescription dose in permanent (131)Cs seed prostate implants. *Med Phys* 32:2496–502.

Zaider, M. 1996. The combined effects of sublethal damage repair, cellular repopulation and redistribution in the mitotic cycle. II. The dependency of radiosensitivity parameters alpha, beta and t(0) on biological age for Chinese hamster V79 cells. *Radiat Res* 145:467–73.

Zaider, M. 1998a. Sequel to the discussion concerning the mechanistic basis of the linear quadratic formalism. *Med Phys* 25:2074–5.

Zaider, M. 1998b. There is no mechanistic basis for the use of the linear-quadratic expression in cellular survival analysis. *Med Phys* 25:791–2.

Zaider, M. and Hanin, L. 2007. Biologically-equivalent dose and long-term survival time in radiation treatments. *Phys Med Biol* 52:6355–62.

Zaider, M. and Hanin, L. 2011. Tumor control probability in radiation treatment. *Med Phys* 38:574–83.

Zaider, M. and Minerbo, G. N. 1993. A mathematical model for cell cycle progression under continuous low-dose-rate irradiation. *Radiat Res* 133:20–6.

Zaider, M. and Minerbo, G. N. 2000. Tumour control probability: a formulation applicable to any temporal protocol of dose delivery. *Phys Med Biol* 45:279–93.

Zaider, M., Wuu, C. S. and Minerbo, G. N. 1996. The combined effects of sublethal damage repair, cellular repopulation and redistribution in the mitotic cycle. I. Survival probabilities after exposure to radiation. *Radiat Res* 145:457–66.

20

Brachytherapy for Prostate Cancer

Alfredo Polo
Ramon y Cajal University Hospital

Michel Ghilezan
*Michigan Health Professionals/
21st Century Oncology*

20.1 Introduction

Interstitial brachytherapy with permanent seeds or temporary high-dose-rate (HDR) implants has received a renewed interest in the last 20 years. Several causes have contributed to this situation of brachytherapy in the treatment of prostate cancer. First of all, thanks to the routine use of the transrectal ultrasound and prostate-specific antigen (PSA) in screening for prostate cancer, there has been a phenomenon of increased incidence and another one of stage migration. Secondly, the technological developments have made interstitial brachytherapy a highly sophisticated precision technique. The appearance of new isotopes other than ^{125}I (^{103}Pd in 1986 and more recently ^{131}Cs) for permanent brachytherapy, developments in transrectal ultrasound (biplanar probe) and fundamentally the improvements introduced in dose planners, and the appearance of 3D ultrasound and real-time navigation systems have afforded brachytherapy a definitive push, which has brought it up to the same level as other local therapies for prostate cancer. Finally, the historical perspective has also put brachytherapy in its place. Encouraging clinical results for permanent seed implantation and temporary HDR brachytherapy confirm the efficiency and durability of brachytherapy in the treatment of prostate cancer.

In this chapter, we will review the role of permanent seed implantation and HDR brachytherapy [alone or in combination with external beam radiation therapy (EBRT)] to treat organ-confined prostate cancer. We will put clinical results in the context of the technological framework.

20.1.1 Epidemiology of Prostate Cancer

Prostate cancer is a major public health problem. In the last 20 years, an exponential rise in the number of patients diagnosed with prostate cancer has occurred. Increased longevity in older people and the introduction of transurethral resection of the prostate (TURP) and, more recently, of PSA serum screening programs resulted in an increase in prostate cancer incidence and cancer-related death rate (Futterer 2007).

Worldwide, an estimated 1 million men were diagnosed with prostate cancer in 2008, and more than two-thirds of cases are diagnosed in developed countries (Ferlay et al. 2008). The highest rates are in Australia/New Zealand, Western and Northern Europe, and Northern America. Prostate cancer risk is strongly related to age, with around 75% of cases occurring in men over 65 years and the largest number diagnosed in those aged 70–74. It is estimated from postmortem data that around half of all men in their 50s have histological evidence of cancer in the prostate, which rises to 80% by age 80, but less than 5% will die from this disease. This is a very important fact when considering population screening of asymptomatic men and treatment algorithms (Selley et al. 1997; Frankel et al. 2003).

With the advent of PSA screening and more attention to this disease by the medical community, the proportion of patients who present with early-stage disease has increased. Consequently, there has been a downward trend in the stage of prostate cancer determined at the time of diagnosis. In 2003, 85% of prostate cancers were local or regional at the time of diagnosis compared with 72% in 1993 (Futterer 2007). The evidence is insufficient to determine whether screening for prostate cancer with PSA reduces mortality from prostate cancer. Screening tests are able to detect prostate cancer at an early stage, but it is not clear whether this earlier detection and consequent earlier treatment lead to any change in the natural history and outcome of the disease. The observed trends may be due to screening or to other factors such as improved treatment. The benefit of screening on the mortality from this disease remains undetermined, and the results from two randomized trials show no effect on mortality through 7 years but are inconsistent beyond 7 to 10 years (Vis et al. 2008; De Vries et al. 2007).

20.1.2 Brief History of Prostate Brachytherapy

The first prostate implants were reported by Pasteau and Degrais in Paris in 1913 (Pasteau and Degrais 1913). On this basis, Hugh Hampton Young, head of the Urology Department of the Johns Hopkins (Baltimore, MD), developed a method of insertion of ^{226}Ra in the urethra, bladder, and rectum during the 1920s (Aronowitz 2002). Parallel to this, Benjamin Stockwell Barringer, at the New York Memorial Hospital, developed two prostate brachytherapy techniques using encapsulated radon (Barringer 1916). The first one consisted of the temporary insertion of vector needles containing ^{222}Rn, and the second was a forerunner of the permanent implants: Barringer had glass capillaries containing ^{222}Rn prepared, which were implanted transperineally through a trochar. These glass capillaries were subsequently placed in seed-form gold capsules so that dose distribution around the source would be improved considerably.

The initial outcomes of these groundbreaking techniques were very promising and were thus adopted by other centers as of the 1930s. A total of 80,000 patients were given radium therapy in 1931 alone. However, the limitations of the technique soon became apparent: in many cases, seed distribution was deficient. Open-sky implant modifications were more complex and within reach of a few specialists. Finally, the proportion of disease-free patients was very low. For these reasons, the use of brachytherapy in prostate cancer gradually diminished until it disappeared completely (Aronowitz 2002).

Prostate brachytherapy reemerged during the 1970s, when the open-sky retropubic implant technique was described and developed at the New York Memorial Hospital. The technique consisted of exposing the prostate directly to the operator through a medial laparotomy. An extraperitoneal lymphadectomy was performed first, followed by the ^{125}I seed implant. Dosimetric planning (total activity) was limited to the use of nomograms from the prostate size, whereas post-implant dosimetry was evaluated only in terms of "matched peripheral dose," a value that corresponded to the isodose that covered the same volume as that of the prostate obtained from an ellipsoid approach. This technique was applied for 15 years with very promising preliminary results (Zelefsky and Whitmore 1997). However, just like the techniques described above, a series of limitations on this modality of brachytherapy soon became apparent: first of all,

major surgery was required for the implant; moreover, the insertion of needles and seed tank was based on blind palpation of the prostate via the transrectal route, giving rise, on many occasions, to clearly deficient seed distribution, with overdosed or underdosed areas. Finally, clinical results were not consolidated, PSA was not available for patient follow-up, and a post-implant needle biopsy was not performed routinely. For all these reasons [linked to the improvements in the surgical technique and external radiotherapy (ERT)], the technique was abandoned, and radical prostatectomy was advocated as the gold standard in organ-confined disease.

During the 1980s, prostate brachytherapy became the focus of attraction, leading to the birth of the modern brachytherapy technique, which, with different modifications, is the one we now use. This new technique is the result of the convergence of technological developments (new isotopes, new imaging modalities, 3D planning systems), screening methods (routine use of PSA) ,and diagnosis methods (establishment of risk groups and other predictive methods), the study of the long-term outcomes of the classic series (identification of subgroups of patients presenting a more favorable response to brachytherapy), and the description of a new route for viewing the prostate by transrectal ultrasound (Fornage et al. 1983).

This all provided the foundations for "modern" prostate brachytherapy. Holm (1983) described the transperineal prostate implant technique guided by transrectal ultrasonography. This technique permitted a hitherto unknown precision and sowed the seeds for subsequent developments in this field. Paradoxically, Holm's group only treated 32 patients with this technique. However, they made an enormous contribution to its development with the use of the coordinate system, instruments, and accessories for needle insertion and contribution to the development of the multiplanar transrectal probe.

Before Holm's group abandoned the technique, Haakon Radge visited the Ultrasound and Urology Department of the University of Copenhagen (Denmark) and transferred the permanent implant with a transrectal guide technique to the United States, where they performed their first treatment in the Northwest Hospital of Seattle in 1985. Since then, the prostate brachytherapy has undergone a major boom, thanks to promotion, the technique has been fine-tuned, and long series of patients have been published with good long-term clinical outcomes.

HDR brachytherapy has been developed over the last 30 years thanks to major software and hardware developments in treatment planning and afterloading machines. This has improved dose delivery to the patient, reducing exposure of the staff to ionizing radiation. Nowadays, HDR brachytherapy has emerged as the ultimate conformal radiation therapy, allowing precise dose deposition on small volumes under direct image visualization.

HDR brachytherapy makes use of a single source, usually a 370-GBq (10Ci) ^{192}Ir source, although new isotopes are now being used or in the pipeline, secured at the end of a cable-driven wire. The source system (source and driven wire), plus additional systems for source position verification, radiation detection, and emergency removal of the source, is placed in a single afterloading machine. The trajectory of the source through the implanted catheters or needles is precisely programmed by a dedicated treatment planning system. The resulting isodose distribution can be optimized by modulating the dwell time of the source as a function of its trajectory within the implanted volume. This allows individualization of dose distributions, while essentially eliminating radiation exposure to the medical staff. Dose is delivered through large doses per fraction (usually >5 Gy), and some studies hypothesize that there could be a radiobiological advantage over low-dose-rate (LDR) brachytherapy depending on the alpha/beta value for tumors and organs at risk.

The HDR afterloading system enables highly conformal treatments to be given with all the advantages of a stepping source: only one source to replace once every 3 months (no source inventory required), no source preparation, programmable dose rate, optimization of dose distribution, radiation protection, and intracavitary, interstitial, intraoperative, and intraluminal treatments feasible using a single afterloader.

20.1.3 Role of Prostatic Brachytherapy for Prostate Cancer

The diagnosis of prostate cancer is increasing worldwide, and the majority of the cases are diagnosed in organ-confined stage, defined by a low PSA and low Gleason score. Many of the patients are asymptomatic and identified on PSA screening. Optimal treatment selection for these patients continues to be debated and comprises a range from active surveillance to radical radiotherapy and prostatectomy. The comparative effectiveness and harms of localized prostate cancer treatments are largely unknown, and no recommendations can be drawn from evidence-based medicine (Wilt et al. 2008). In the absence of these evidence-based recommendations, other decision methods have been described based on patient and provider interactions (Jani and Hellman 2008).

Today, treatment outcome from modern techniques can be expected to be better than the outcome for patients who were diagnosed in the 1980s and early 1990s. These improved outcomes are the result of stage migration, new technologies such as robotics for surgery and radiotherapy, functional imaging for treatment planning in radiotherapy, more conformal radiotherapy techniques, better implant techniques, and optimum use of hormone therapy (Peschel and Colberg 2003).

For most of these patients, brachytherapy (either by permanent radioactive seeds or temporary HDR implantation) will be an option providing effective high dose radiation therapy with a favorable side-effect profile, particularly when compared with radical external beam treatment or radical prostatectomy. Of importance in this group of relatively young sexually active men, it has the best chance among the radical treatment options of retaining erectile function (Sahgal and Roach 2007; Bossi 2010; Martinez et al. 2010).

20.2 Basic Aspects in Prostatic Brachytherapy

20.2.1 Patient Selection for Prostatic Brachytherapy

20.2.1.1 Patient Selection for LDR Brachytherapy

The surveys performed by the American Brachytherapy Society (ABS) or by the prostate brachytherapy quality assurance group (PBQAG) during the 1990s showed that the indications, techniques, treatment regimens, and dosimetry used varied widely. The most important areas of controversy were the selection criteria for single therapy or combined therapy, the role of hormone therapy, the role of preimplant dosimetry, the choice of isotope, and post-implant evaluation.

For this reason, the ABS through the PBQAG established recommendations for the use of permanent brachytherapy on the basis of the points of consensus and dissent. The results of these deliberations were published in 1999 and constitute the foundations on which most centers have supported their permanent seed prostate brachytherapy programs (Nag et al. 1999).

In Europe, an additional consensus effort was made, and a joint work group composed of urologists, oncologists, radiotherapists, and physicists was set up to issue recommendations in the name of the ESTRO, EAU, and EORTC. The results were published in 2000 (Ash et al. 2000).

The ACR, through the "ACR Appropriateness Criteria" program, also establishes recommendations for doctors and health providers on the most suitable therapeutic decisions in different areas of oncology. The recommendations are made by a panel of experts in diagnostic, interventionist radiology and radiotherapy oncology. The recommendations for the use of permanent brachytherapy in the treatment of prostate cancer were first approved in 1999 and revised in 2002 and 2006. This document can be obtained from the ACR Web site (http://www.acr.org).

All the recommendations referred to above establish criteria for different aspects of the technique: general criteria, biological criteria, functional criteria, and ballistic criteria.

- *General criteria:* Localized disease (no metastases), life expectancy above 5 years, absence of concomitant disease that gives rise to unacceptable surgical risks. Prior pelvic radiation is a factor to be considered, although it is not regarded as an absolute contraindication. Severe diabetes is also a relative contraindication, as is the existence of previous abdominal surgery. Patient age is not regarded as a criterion to be taken into account in the indication of a permanent seed treatment.
- *Biological criteria (of the tumor):* Permanent seed brachytherapy may be indicated as exclusive treatment in patients with prostate adenocarcinoma at clinical stage T1c–T2a with initial PSA < 10 ng/mL and Gleason grade 7. According to the ABS, combined treatment with EBRT and brachytherapy is recommended in patients with clinical stage T2b–T2c prostate adenocarcinoma or with

an initial PSA > 20 ng/mL or Gleason grade 8–10 (the ESTRO–EAU–EORTC recommendations do not consider combined treatment). Finally, for the group of patients with initial PSA between 10 and 20 ng/mL or with a combined Gleason grade 7, there is no agreement as to whether exclusive brachytherapy suffices or whether the addition of EBRT would be necessary.

A survey on the standards of treatment of prostate cancer conducted by the group of Frank et al. (2007) observed that a subgroup of patients, which may be framed within the classic definition of intermediate risk, could be treated suitably with exclusive brachytherapy. This subgroup would be composed of patients with tumors in stage T1c, without perineural invasion, with the presence of the tumor in less than 30% of the needle biopsy samples, and a Gleason index ≤ 7 or a PSA level on diagnosis between 10 and 20 ng/mL (Frank et al. 2007).

The RTOG 0232 trial seeks to prospectively analyze the differences between the use of brachytherapy as a single therapy versus the combination of external radiation with IMRT and brachytherapy as boost in the treatment of patients with intermediate risk tumors. Until the results of this study are mature and available, it does not seem possible to establish a firm recommendation on the use of brachytherapy as single therapy in prostate adenocarcinoma with intermediate risk.

- *Functional criteria (of the patient):* There is no suitable preimplant criterion to predict chronic urinary symptoms. The role of the International Prostate Symptom Score (IPSS) has been studied broadly with contradictory results, since it seems to be predictive of acute toxicity in the form of urethritis or acute urinary retention, but has not demonstrated an association with chronic symptoms. Patients with IPSS ≥ 20 are at risk of prolonged urinary retention or urethritis of 30%–40%. The joint ESTRO–EAU–EORTC recommendations advise brachytherapy for patients with IPSS ≤ 8.

  The usefulness of urine flow measurement has also been explored to predict the incidence of complications following brachytherapy. Qmax pressure figures are related to the incidence of acute urinary retention (Martens et al. 2006; Henderson et al. 2002). The joint recommendations of ESTRO–EAU–EORTC recommend reserving brachytherapy for patients with Qmax 15 mL/s.

  Prostate volume > 50 cc has sometimes been related to the incidence of urinary toxicity. However, series from reputable institutions have shown that large volume prostate implants are not necessarily associated with increased toxicity. The joint recommendations of ESTRO–EAU–EORTC recommend reserving brachytherapy for patients with prostate volume ≤ 40 cc.

  The presence of prior transurethral resection (TUR) has been considered to be a risk factor for the development of incontinence. Some series have shown incontinence rates of 11%–24% in patients undergoing prior TUR and 3% in patients without TUR (Grimm et al. 1996).

- *Ballistics criteria (of the technique):* The brachytherapy results may be conditioned by a series of factors that may affect dose distribution. They include prostate volume, the interference of the pubic arch, and the presence of previous TUR.

  Prostate volume has already been mentioned in relation to urinary toxicity under *Functional criteria.* Volume prostates > 50 cc present greater difficulty in implant planning and execution.

  The interference of the pubic arch occurs when the bone window that permits passage of the implant needle is too small or narrow, giving rise to deficient dose distribution.

  The presence of previous TUR has been related to the appearance of toxicity, as we have already seen, but it has also been regarded as a technical contraindication, since suitable dose distribution cannot be guaranteed due to the residual prostate tissue defect. However, feasibility depends fundamentally on the amount of resected tissue and defect healing (see Figure 20.1).

20.2.1.2 Patient Selection for HDR Brachytherapy

The same LDR brachytherapy selection criteria apply for HDR patients, as endorsed by GEC-ESTRO in their recommendations of 2005 (Kovacs et al. 2005). The ABS is in the process of issuing recommendations specifically designed for HDR candidates; nonetheless, there are very little differences compared with LDR patients. Overall, because of more dose optimization gland size over 60 cc, IPSS scores of up to 20 and a moderate TURP defect are not considered contraindications, and they have to be addressed on a case-by-case basis by the treating physician.

20.2.2 Target Definition in Prostatic Brachytherapy

Definitions for target volumes according to ICRU reports and organs at risk have been summarized recently for prostatic brachytherapy (Salembier et al. 2007).

(1) *Gross tumor volume (GTV):* the GTV corresponds to the gross palpable, visible, or clinically demonstrable location and extent of the malignant growth. Given the TNM definition for prostate cancer, GTV can only be defined for tumor stages larger than T1c. For these tumor stages (>T1c), the GTV definition can be useful, certainly in cases where the tumor area can be identified not only by digital rectal examination but also by radiological examinations including transrectal ultrasound, magnetic resonance imaging, and spectroscopy. In T3 disease extension through the capsule or into the seminal vesicles may be seen on imaging and included in the GTV.

(2) *Clinical target volume (CTV):* the CTV is the volume that contains the GTV and includes subclinical malignant disease at a certain probability level. Delineation of the CTV is based on the probability of subclinical malignant cells present outside the GTV. Prostate cancer has recognized

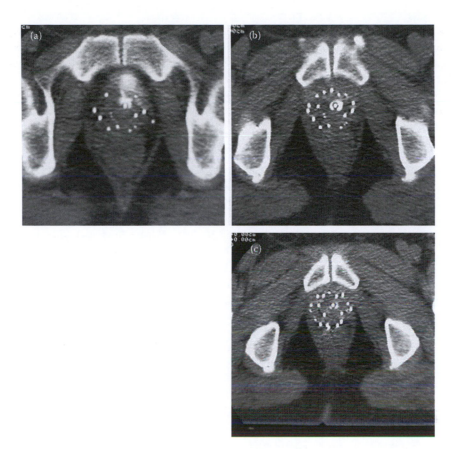

FIGURE 20.1 Appearance of a permanent implant in a patient who had previously received a TURP. (a) shows the interface with the bladder. (b) depicts the mid-gland lower right panel, apical portion. It can be seen that the tissue defect is filled with contrast. If the amount of tissue removed is small and the interval is more than 6 months since the intervention, then brachytherapy can be offered depending on the experience of the team. (Courtesy of A. Polo.)

paths of microscopic spread through the capsule and into the seminal vesicles, which may occur even in very early stage prostate cancer.

- *Probability of microscopic extension at different distances around the GTV:* it is well documented in surgical literature that prostate cancer is in the majority of cases a "whole gland" disease. Even in a very early stage, prostate cancer presents as a multifocal disease—both lobes can contain microscopic disease. Given this specific behavior, at least the whole prostate gland has to be considered as the "target" and included in the CTV.

- *Probability of subclinical invasion of the periprostatic tissues:* when available, a magnetic resonance scan of the prostate, ideally using an endorectal coil, should be performed for radiological staging and, in particular, to identify those patients with T3 disease. For very early stage tumors (T1c–T2), the probability of capsular penetration is related to the tumor stage (T1c < T2), iPSA, and Gleason score as demonstrated in the Partin tables. These tables show that even tumors with stage T1c and T2, independent of

Gleason score or iPSA, have a probability of established capsular penetration of at least 10%.

- *Extent of subclinical extraprostatic extension (EPE) of early prostate cancer:* studies on specimens obtained by radical prostatectomy show a tendency for clinical understaging of capsular penetration, with rates ranging from 40% to 60%. Only a few authors have focused on the geometrical extent of extraprostatic disease. The largest study included 376 specimens from patients undergoing radical retropubic prostatectomy using whole organ mount examination of the EPE (Davis et al. 1999). This identified EPE in 28% of examined cases. The radial EPE distance in these specimens had a mean of 0.8 mm (range 0.04–4.4 mm) and a median of 0.5 mm. Ninety-six percent of all specimens with EPE had a radial EPE distance $\leq$ 2.5 mm. All patients classified in the good prognostic risk group (PSA < 10, Gleason < 7) had a radial EPE distance < 3 mm.

The CTV should be the prostate gland with a margin: for T1–T2 prostate cancer, the CTV corresponds to the visible contour of the prostate expanded with a 3D volume expansion of 3 mm. This 3D expansion can be constrained to the anterior rectal wall (posterior direction) and the bladder neck (cranial direction). For T3 tumors, the CTV corresponds to the visible contour of the prostate including visible extension due to extracapsular growth, which is then expanded with a 3D expansion of 3 mm in each direction, with rectal and bladder constraints as above.

(3) *Planning target volume (PTV)*: the PTV surrounds the CTV with a margin to compensate for the uncertainties in treatment delivery. The PTV is a geometric concept, introduced for treatment planning. A margin must be added to the CTV either to compensate for expected physiological movements and variations in size, shape, and position of the CTV during therapy (internal margin) or for uncertainties (inaccuracies and lack of reproducibility) in patient setup during irradiation, which may be random or systematic. The CTV to PTV margin can be minimized in brachytherapy because there are no significant opportunities for setup error.

(4) *Different organs at risk can be defined in the preimplantation setting* (Salembier et al. 2007).

- *Prostatic urethra*: common practice to obtain visualization of the urethra is to use a urinary catheter. This should be a small gauge catheter, French gauge 10, to avoid distension of the urethra. The surface of the catheter is used to define the urethral surface from the prostatic base to apex. However, in practice, the urethra is not a circular structure, and an alternative that may give a more accurate anatomical picture is to instill aerated gel into the urethra prior to obtaining the ultrasound images.

- *Rectum*: using transrectal ultrasound, visualization of the anterior rectal wall is no problem but may introduce artifacts due to displacement and distension. Many simply outline the outer wall, and this should be regarded as the minimum requirement; others define outer and inner walls to define a doughnut. In terms of the critical cells in the rectum for late damage, the latter is probably more correct. For defining small volumes up to 5 cc, outlining the outer wall alone is therefore sufficient.

- *Penile bulb and/or neurovascular bundles (NVBs)*: currently this still requires investigation. However, William Beaumont Hospital (WBH) investigators found 16% of HDR patients whom develop erectile dysfunction (ED) within 2 years from their brachytherapy do so immediately, within 3 months after the implant, suggesting that trauma to NVBs rather than radiation-induced vascular and nerve damage is the cause of the early onset of ED (unpublished data). Therefore, an ultrasound (US) unit with Color Doppler functionality was introduced in routine clinical practice, enabling visualization of prostatic NVBs during needle implant. Although the ED rates improved since introduction of this method, it is difficult to ascertain the impact of Color Doppler imaging, as there was a solid trend over the last years toward recruiting younger, fitter patients for HDR brachytherapy, with better erectile function overall. Nonetheless, the 0–3 months ED rates dropped to 7% since integrating Color Doppler information in the intraoperative HDR planning.

20.2.3 Technical Aspects in Prostatic Brachytherapy

20.2.3.1 Technical Aspects for LDR Brachytherapy

20.2.3.1.1 *Implant Technique*

Modern seed implantation is performed under template guidance via a transperineal approach in a percutaneous procedure typically performed in an outpatient surgical setting (Yu et al. 1999). Holm et al. (1983) first described the use of transrectal ultrasound (TRUS) for precise guidance of transperineal seed insertion in 1983. The technique was further popularized by Blasko et al. (1993), Grimm et al. (2001), and Ragde et al. (2000a) and has evolved into the most popular modality to treat low-risk organ-confined prostate cancer during the past decade.

The ability to produce results at least equivalent to the pioneering centers of excellence is dependent upon achieving optimal dosimetric implants for every patient based on individual volume definition and dosimetry (Lee et al. 2003; Prestidge et al. 1998). The most common procedure uses a preplan before the implant (Kaplan et al. 2000). The patient is placed in the lithotomy position, and TRUS is used to acquire images through the prostate. From these ultrasound images, a treatment plan is generated. Subsequently, during the actual implantation in the operating room, the patient is repositioned, and, through manipulation of the transrectal probe and repositioning of the patient, a second series of ultrasound images is obtained to match the ultrasound images taken during the preplanning TRUS.

Intraoperative planned techniques have emerged in recent years in an attempt to overcome some of the troublesome characteristics of the preplanned technique (matching patient positioning is sometimes difficult, and prostate shape and volume can change from that in the preplan). The specific aims are to design the optimal treatment plan using intraoperative 3D anatomical information, to implement the treatment plan with precision, and to analyze the dosimetric outcome while the procedure is still in progress (Yu et al. 1999). Recent advances in technology allow treatment planning and dose calculations during the implantation procedure. This offers the opportunity to improve the quality of implants by appropriate modification in the seed implants and replanning during the procedure itself. The technology of intraoperative treatment planning raises unique challenges and opportunities for dose reporting in prostate implants. Inverse-planning optimization engines for optimal seed placement, 3D ultrasound (3D US) technology with real-time needle guidance (Prestidge et al. 1998), real-time seed detection on fluoroscopy, and the recent availability of flat-panel cone-beam tomography units are some examples of technological improvements already applied to seed implantation (Beaulieu et al. 2007).

According to the ABS, intraoperative planned techniques include intraoperative preplanning, interactive planning, and dynamic dose calculation (Nag et al. 2001).

- Intraoperative preplanning refers to the creation of a plan in the OR just before the implant procedure, with immediate execution of this plan (Nag et al. 2001). Intraoperative preplanning eliminates the conventional preplanning patient visit (a few days or weeks prior to the implant) by bringing the planning system into the OR. TRUS is performed in the OR, and the images are imported in real time into the treatment planning system. The target volume, rectum, and urethra are contoured on the treatment planning system either manually or automatically, and a treatment plan is generated. The prostate is implanted according to the plan. There are several reports on the use of intraoperative preplanning (Messing et al. 1999;

Beyer et al. 2000; Wilkinson et al. 2000; Gewanter et al. 2000; Matzkin et al. 2003; Woolsey et al. 2004; Shah et al. 2006). Intraoperative preplanning presents some advantages over the conventional two-step preplanned method: first, it avoids the need for two separate TRUS procedures. Second, patient repositioning and setup are no longer needed. Finally, dosimetric comparisons favor intraoperative preplanning, and probably this dosimetric advantage could translate into more favorable figures of biochemical control. However, intraoperative preplanning does not account for intraoperative changes in prostate geometry or deviations of needle position from the preplan.

- Interactive planning refers to an intraoperative stepwise refinement of the treatment plan using computerized dose calculations derived from image-based needle position feedback (Nag et al. 2001). An optimized treatment plan is performed in the operating room, the dose-volume histogram (DVH) is generated, and the plan is examined and validated. The needles are placed into the prostate, and the needle position is registered by the computer planning system. Based on the needle positions, the treatment plan is then regenerated and the implant proceeds. Individual needles can be adjusted, and a new plan is generated while in the operating room (see Figure 20.2).

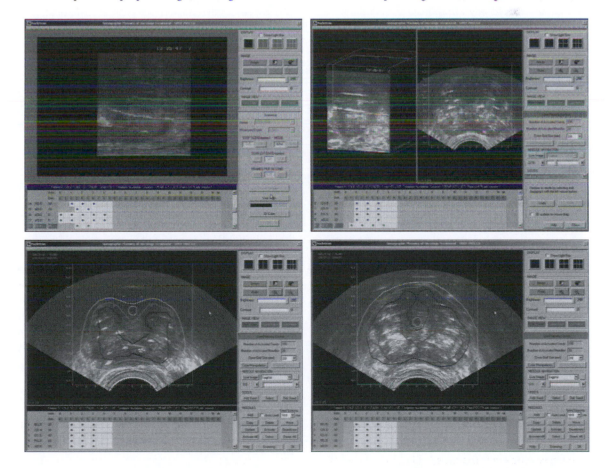

FIGURE 20.2   (See color insert.) Ultrasound appearance of an implant in real time. Shown here is the technique of interactive planning. Individual needles can be adjusted and a new plan generated while in the operating room. (Courtesy of A. Polo.)

The planned dose distribution is based on implanted needle position, and hence, interactive planning may not account for either seed movement after deposition or for prostate gland intraoperative movements and volume changes. There are several reports on the use of interactive planning for permanent seed implantation (Grado et al. 1998; Shanahan et al. 2002; Zelefsky et al. 2000a; Raben et al. 2004; Potters et al. 2006; Stock et al. 2000; Beaulieu et al. 2007).

- Dynamic dose calculation represents a paradigm shift in dose prescription and specification and source delivery for permanent seed implantation. It will mirror the image guided radiotherapy (IGRT) paradigm in EBRT in that an intended prescription dose is adaptively "painted" to a changing 3D target volume. This process of dose painting may result in alteration of a previously accepted isodose distribution and total implanted activity at any time until the end of the procedure when a satisfactory dose delivery is achieved. In comparison to interactive planning, dynamic dose calculation relies on full 3D dose planning based on the real position of the delivered seeds. Instead of having iterative dose calculations based on a surrogate for the seed, dynamic dose planning will constantly update dose distribution using continuous deposited seed position feedback. It will account for seed migration, needle motion and misalignment, and prostate gland changes in shape and volume (Nag et al. 2001). Specific technology is needed, with the potential to constantly update calculations of dose distribution (dynamic dose calculation) as the implant proceeds. It is essential that the exact seed position is known in three dimensions. Improvements in imaging are thus required, as TRUS image degradation with time is a major issue that impairs seed visualization. Several workflows have been outlined for dynamic dose calculation (Todor et al. 2003; Lee and Zaider 2003). Dynamic dose calculation overcomes some of the limitations of permanent seed implantation. However, the clinical value has to be proven. Clinical studies are needed to validate the importance of this approach.

20.2.3.1.2. *Post-Implant Dosimetry*

Post-implant dosimetry is the standard tool for assessing implant quality. Currently, both the ABS and the American Association of Physicists in Medicine (AAPM) recommend the execution of post-implant dosimetric analysis for all patients undergoing permanent seed implantation (Nag et al. 2000; Yu et al. 1999). The post-implant dosimetry permits the documentation of the actual dose delivered to the prostate and the OAR. Secondly, the calculated dosimetric parameters create a "learning curve" and help to refine the technique. Data obtained from post-implant dosimetry can be used for future clinical outcome analysis. Finally, post-implant dosimetry allows the comparison of clinical data sets from various institutions and, consequently, a quality control tool for multicentric clinical trials.

At present, CT-based post-implant analysis is the most commonly used method for carrying out quantitative dosimetric evaluations. Stock et al. (1998) found a dose–response relationship for ^{125}I prostate implants using CT-based dosimetry, identifying a cutoff value of 140 Gy predicting PSA relapse-free survival. In a recent comprehensive review of CT-based dosimetry parameters and biochemical control including 790 patients treated with permanent seed implantation, Potters et al. (2001) found a cutoff value of 90% of the prescribed dose predicting PSA relapse-free survival. Wallner et al. (1995) also showed that CT-based dosimetry can predict which patients are at higher risk of radiation-related morbidity. However, other authors did not find a correlation between CT-derived dosimetric parameters and biochemical outcome (Ash et al. 2006; Merrick et al. 2001a).

Those discrepancies can be related to uncertainties in organ delineation using CT images. The determination of the dose delivered to the target and to the OAR is highly dependent on how accurately the target volume is defined in the CT image set (Yu et al. 1999), making accurate and reproducible delineation a key point when evaluating seed implant quality by means of DVH analysis. Several studies have observed discrepancies in prostate volume, as outlined by TRUS, MRI, and CT, showing difficulties in differentiating the prostate gland from periprostatic tissues when using CT as well as significant differences in interobserver and intraobserver accuracy in outlining prostate on CT images (Polo et al. 2004). Due to the very small difference in density, the prostate gland is not well resolved from other adjacent soft-tissue structures, resulting usually in a prostate volume overestimation with respect to the TRUS-based volume used to plan the implant.

MRI has a better soft-tissue resolution, and MR image-based contours correlate better with TRUS-based evaluations and pathologic findings, and there is a better interobserver reproducibility, making it an attractive image modality for brachytherapy dosimetry (Polo 2010). CT–MRI image fusion (Figure 20.3) has been advocated by some authors as a sophisticated way of assessing the quality of a prostate seed implant reducing uncertainties in target volume outlining and offering more reliable dose–response relationships (Crook et al. 2004; Taussky et al. 2005; McLaughlin et al. 2004; Polo et al. 2004).

In conclusion, post-implant dosimetry is a tool for assessing implant quality after permanent seed implantation and to predict clinical outcome. It has many limitations, and some authors question the utility of this approach in the context of intraoperatively planned techniques (Potters et al. 2006). However, based on the current recommendations, CT-based post-implant dosimetry is an essential step in the clinical process of permanent seed implantation.

20.2.3.2 Technical Aspects for HDR Brachytherapy

HDR brachytherapy implants are template-based and performed under general or spinal anesthesia with the patient in lithotomy position, similar to LDR. Comparatively, HDR has a number of patient- and target-specific advantages over LDR therapy. They are summarized below:

(1) The overall treatment time reduction with HDR eliminates the uncertainties related to prostate volume changes

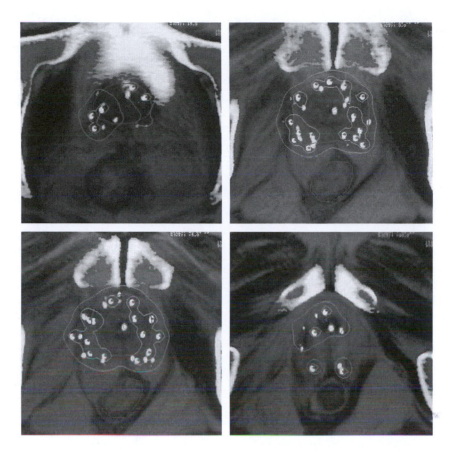

FIGURE 20.3 **(See color insert.)** Post-implant dosimetry images based on CT–MRI fusion. Prostate capsule definition is performed on MRI and the pelvic bones on the CT. The margin effect is shown and cold spots are viewed at the base and the apex of the prostate. (Courtesy of A. Polo.)

that occur during the weeks following a procedure (typical with LDR) due to trauma and swelling, or subsequent shrinkage due to postradiation fibrosis.

(2) HDR allows for improved accuracy of needle placement and radiation dose distribution through the use of intraoperative optimization software. New developments in HDR planning software by Nucletron (Nucletron Corp, Veenendaal, The Netherlands) enables inverse planning feedback for needle template position coordinates coupled with real-time dose modulation. HDR brachytherapy planning algorithms allow for double dose optimization, as it modulates the dwell times as well as the dwell positions for precise and accurate 3D spatial dose coverage. This "dose sculpting" capability allows using HDR inherent dose inhomogeneities to be positioned in areas of disease burden for potentially increased tumor control while limiting organ-at-risk doses, and thus treatment-related toxicities.

(3) From a radiobiology perspective, HDR may be favored as treatment delivery over a period of minutes, instead of weeks to months, does not allow malignant cells to repopulate, advance through the cell cycle, or recover from sublethal damage. Likewise, studies showed that if prostate cancer alpha/beta is low, in the range of 1.2–4, HDR biological effectiveness is higher than with LDR or EBRT.

(4) A single radioactive source may deliver treatment to large numbers of patients with HDR, whereas LDR requires radioactive sources to be purchased on a per case basis, leading to increased cost of treatment as compared to HDR.

(5) From a radiation safety perspective, HDR is ideal because patients are not radioactive when they return home. As such, patients do not need to follow special precautions such as limiting distance or duration of contact with another individual, children, or pregnant women as is necessary with LDR patients.

20.3 Role of LDR Brachytherapy in Treatment of Prostate Cancer

20.3.1 Disease Control

The analysis of brachytherapy outcomes is complicated due to the difficulty in comparing series to each other. For this purpose, classification criteria based on the combination of significant variables for biochemical disease-free survival (BDFS) were established. In this way, it is possible to establish homogeneous subgroups that can be compared to each other. Stratification by risk groups is a very powerful tool for reducing the dimensionality of the prognostic variables on combining three variables (two ordinals and one continuous) in a single ordinal. The validity of the groups or prognostic strata has already been well demonstrated in series with patients treated with ERT, brachytherapy, or a combination of both. The outcomes of permanent seed prostate brachytherapy for the most representative and expert centers in this technique are presented below.

- The Seattle group (initially composed of John Blasko, Peter Grimm, and Haakon Ragde) defines two risk groups: patients with clinical stage < T2b and GS < 7 were regarded as low-risk patients, whereas those with clinical stage ≥ T2b or GS ≥ 7 were regarded as high risk (Ragde et al. 2000b). The analysis of this group's outcomes is even more complicated since the original group split into two groups that continued to work in the Seattle area (Seattle Prostate Institute and Seattle Northwest Hospital) and continued to publish on the same series of patients treated initially, albeit without obtaining the same results due to the application of different definitions of relapse and different statistical methods for the projection of results.

  Ragde et al. (2000a) presented the series with the longest follow-up using modern brachytherapy techniques. Using the same cohort of patients as for the preceding communications, the authors found that for 229 patients with follow-up to 12 years, single therapy with seed implant achieved disease-free survival rates of 66%; furthermore, 79% of the patients with high-grade disease who were treated with brachytherapy and ERT also experienced long-term disease-free survival.

  In an update after 13 years (Ragde et al. 2001), the authors reported 769 consecutive patients with prostate cancer treated with brachytherapy alone. The patients were divided into two risk groups (low and high risk of extraprostatic disease) based mainly on clinical stage and Gleason index. Group 1 was composed of 542 patients regarded as low risk and who were treated with ^{125}I; group 2 was composed of 227 patients regarded as high risk and who were treated with [103]Pd. No patient was pathologically staged, and none received ablative therapy with androgens. Treatment failure was based on a modification of the ASTRO recommendations. A critical component in this modification is that the value of the third elevation of PSA was higher than 0.5 ng/mL. One hundred and thirty-seven patients were lost to follow-up. Thirteen patients died for non-cancer-related reasons within 18 months following the implant. This left 619 patients for evaluation, 441 in group 1 and 178 in group 2. The BDFS rates of the 619 patients after 3, 5, 10, and 13 years were 85%, 80%, 77%, and 77%, respectively. The BDFS rates of the 441 "low-risk" patients treated with ^{125}I after 3, 5, 10, and 13 years were 84%, 79%, 76%, and 76%, respectively. The BDFS rates of the 178 "high-risk" patients treated with [103]Pd after 3, 5, 10, and 13 years were 87%, 82%, 80%, and 80%, respectively.

- The Mount Sinai Hospital New York group of Richard Stock and Nelson Stone present very solid results on their own definition of risk groups (Stock and Stone 1997). Three groups are defined: low-risk group (T ≤ T2a, GS < 7, and PSA ≤ 10 ng/mL), intermediate-risk group (T2b–T2c or GS 7 or [10 < PSA ≤ 20]), and high-risk group (two or more intermediate risk criteria, or else [GS 8–10], or else [PSA > 20 ng/mL]).

  With these criteria, results are reported on 243 patients with a minimum follow-up of 5 years (median follow-up of 75 months with a range between 61 and 135 months). BDFS 8 years post-implant for low-, intermediate-, and high-risk patients was 88%, 81%, and 65%, respectively. Regarding this study, it should be mentioned that an event only takes place as of the sixth-year post-implant in the low-risk group, but that the three groups reach a "plateau" between the sixth- and eighth-year post-implant (Kollmeier et al. 2003).

  In 2009, investigators from this center published the results obtained in intermediate-risk patients with adenocarcinoma treated with brachytherapy with [125]I. The authors reported, with a median follow-up of 60 months and a minimum follow-up of 24 months post-implant, actuarial biochemical failure-free survival (BFFS) rates (SLFB), established in accordance with the RTOG/Phoenix criteria, of 86%. The total dose of radiotherapy, calculated according to the concept of biologically effective dose (BED) < 150 Gy2 versus ≥150 Gy2, was the only factor related to the SLFB ($p < 0.001$) of all those analyzed (PSA, Gleason, hormone treatment, stage, or treatment with ERT) (Ho et al. 2009).

- The Memorial Sloan Kettering Cancer Center (MSKCC) group led by Michael Zelefsky defines risk groups on the basis of the dose-scaling studies with ERT. The low-risk group is defined by (T1-2 and PSA ≤ 10 ng/mL, and GS < 7). The intermediate-risk group is defined by the violation of some of the preceding assumptions and the high-risk group by more than one. By applying these same criteria, the results on 248 patients treated with exclusive brachytherapy are described. BDFS after 5 years for low-, intermediate-, and high-risk groups were 88%, 77%, and 38%, respectively (Zelefsky et al. 2000b).

- The Boston group, led by Anthony D'Amico, defines the following risk groups: low risk (T ≤ T2a, PSA ≤ 10 ng/mL, and GS < 7), intermediate risk (T2b or [10 < PSA ≤ 20] or GS 7), and high risk (T2c, or PSA > 20 ng/mL or GS [8–10]). Two hundred eighteen patients undergoing exclusive brachytherapy were studied by applying this system. BDFS after 5 years for low-, intermediate-, and high-risk groups were 85%, 3%, and 5%, respectively (D'Amico et al. 1998).

- Battermann (2000) in the University Medical Centre of Utrecht revised 249 patients without previous treatment, treated by means of perineal implant between December 1989 and December 1998. The stage and grade were as follows: T(1), 121; T(2), 126; T(3), 2; well differentiated, 136; moderately differentiated, 100; undifferentiated, 15; not established, 8. The mean initial PSA level was 16.1 ng/mL (range between [<1.0] and 165 ng/mL). Tumor progression was defined, according to the ASTRO consensus, as any demonstrated local or distance relapse, beginning of hormone therapy, PSA > 10 ng/mL, or increase in PSA in three consecutive determinations over 6 months. With a median follow-up of 29.2 months (range 6–94 months), a total of 195 showed no evidence of disease (18 died for intercurrent reasons) and 54 showed evidence of disease (13 died with prostate cancer). Toxicity was found in 22 patients (urinary in 18 patients, 9 following TUR; 4 patients had intestinal problems, 1 presenting a rectal ulcer) (Battermann 2000; Hinnen et al. 2010).

- Beyer and Priestley (1997) reviewed 489 patients treated with exclusive brachytherapy. They were all clinically staged as T1 or T2 prostate adenocarcinomas with negative nodes. With a median follow-up of 35 months (range: 3–70), local actuarial clinical control is 83%. Stage and grade both predict outcome. Actuarial BDFS also correlates with stage, degree, and initial PSA. BDFS after 5 years was 94% for T1 tumors, 70% for unilateral T2, and 34% for T2c. Grade is also a predictive factor, ranging from 85% in low-grade tumors to 30% in the high-grade tumors. In a multivariate analysis, initial PSA is the factor that correlates most ($p < 0.0001$) with BDFS, local control, and clinical disease-free survival. Patients with a normal initial PSA enjoyed a BDFS of 93%, whereas those presenting PSA > 10 ng/mL had a BDFS of 40%. There were a few complications, with serious urinary urgency or dysuria in 4% and incontinence with proctitis in 1%.

- Grado et al. (1998) reported the results of 490 patients treated with radioactive seeds of ^{125}I or ^{103}Pd alone or in combination with additional ERT. Actuarial disease-free survival after 5 years was 79% (95% CI, 71%–85%), and the actuarial local control rate at 5 years was 98% (95% CI, 94%–99%). The minimum value (nadir) of post-treatment PSA and the level of pretreatment PSA were significant predictive factors of disease-free survival. In the patients with the PSA nadir < 0.5 ng/mL, disease-free survival at 5 years was 93% (95% CI, 84%–97%) compared to 25% (95% CI, 5%–53%) in patients whose PSA nadir was 0.5–1.0 ng/mL and 15% (95% CI, 3%–38%) in patients with a PSA nadir > 1.0 ng/mL.

- Merrick et al. (2001b) presented a long series of 425 patients treated with ultrasound-guided transperineal prostate brachytherapy using ^{103}Pd or ^{125}I for clinical stage T1b–T3a NxM0 (1997 AJCC) prostate adenocarcinoma. One hundred and ninety patients were treated with single therapy with ^{103}Pd or ^{125}I; 235 patients were given ERT followed by a boost with prostate brachytherapy; 163 patients were given neoadjuvant hormone therapy jointly with single therapy with ^{103}Pd or ^{125}I (77 patients) or jointly with EBRT and a boost of prostate brachytherapy (86 patients). The median follow-up was 31 months

(range 11–69 months). BDFS was defined according to the ASTRO definition. For the whole cohort, the actuarial biochemical non-evidence of disease (bNED) rate at 5 years was 94%. For patients with low, intermediate, and high risk of disease, the bNED rates after 5 years were 97.1%, 97.5%, and 84.4%, respectively. For patients not treated with hormones, 95.7%, 96.4%, and 79.9% of patients with low, intermediate, and high risk, respectively, were free of biochemical failure. The clinical and treatment parameters predictive of biochemical evolution included clinical stage, PSA pretreatment, Gleason index, risk group, age > 65 years, and neoadjuvant hormone therapy.

- The group of Guedea et al. collects the results of a joint analysis performed between five European centers including 1175 patients with low-risk (64%), intermediate-risk (28%), or high-risk (6%) prostate adenocarcinoma. BFFS in accordance with the ASTRO definition observed after 3 years was 91% (93%, 88%, and 80% for low, intermediate, and high risk, respectively) (Guedea et al. 2006).

- The RTOG publishes the results for a phase II study (RTOG 98-05) of exclusive brachytherapy in low-risk patients (PSA < 10, Gleason index 2-6, T1-T2a) (Lawton et al. 2007). Ninety-five patients were analyzed with a median follow-up of 5.3 years (0.4–6.5). At 5 years, 5 patients had local failure, 1 had evidence of local failure, and 6 (6%) had biochemical failure defined as an elevation of PSA on two consecutive occasions (if it reaches a level > 1 ng/mL) after the nadir. Other definitions of biochemical failure were analyzed: RTOG-Phoenix (nadir + 2 ng/mL), ASTRO (three successive increases in the PSA level above the nadir or PSA > 4 mg/mL), and Blasko (two consecutive increases in PSA or PSA > 4 ng/mL in patients with pre-implant PSA above 4 ng/mL or PSA above preimplant PSA in patients with preimplant PSA < 4 ng/mL). The actuarial biochemical disease rate according to the different criteria was 6% (protocol 98-05), 7% (ASTRO), 1% (RTOG-Phoenix), and 10% (Blasko). No patient died from prostate cancer, and global survival at 5 years was 96.7%.

- The St. James's University Hospital of Leeds group, led by Dan Ash, has recently published its results in treatment with brachytherapy with ^{125}I as single therapy in 1298 patients with prostate adenocarcinoma, classified in accordance with the criteria proposed by Zelefsky et al. for low risk (44%), intermediate risk (33%), or high risk (14%) (Henry et al. 2010). With a median follow-up of 5 years (range 2–12 years), the actuarial rates at 10 years of overall survival and specific cause reached 85% and 95%, respectively. BFFS at 10 years, according to the criteria established by ASTRO or RTOG/Phoenix, was 79.9% and 72.1%, respectively. In accordance with stratification by risk group, BFFS according to any of the definitions was significantly lower in the group of high-risk patients (60%/57.6%) versus the intermediate- (76.7%/73.5%) or low-risk groups (86.4%/72.3%, $p < 0.01$). The investigators observed a direct relationship between biochemical control of the disease and D_{90}, so that patients with $D_{90} \geq$ 140 Gy obtained significantly greater BFFS rates (88% versus 78%, $p > 0.01$) (Henry et al. 2010). A subsequent analysis of the same group, focusing exclusively on patients regarded as intermediate risk (Gleason 7, PSA $\leq$ 10 ng/mL) observed, with a median follow-up of 5 years, a BFFS of 82.4%/78% (depending on the criterion chosen, ASTRO or RTOG/Phoenix) in 187 patients analyzed. In patients with Gleason 3+4, the results were somewhat better than those observed in patients with Gleason 4+3 (86.7%/87.9% versus 85.2%/96.8%) albeit without statistical significance. As was observed in the global series of patients, the BFFS rates were better in patients with a $D_{90} \geq$ 140 Gy (50% versus 23%) albeit without reaching statistical significance ($p = 0.08$) (Munro et al. 2010).

20.3.2 Toxicity

Postpermanent implant toxicity is probably one of the main issues in which all the alternative treatments for localized tumors seek to be most effective. The choice of therapeutic option is conditioned in many patients by the different perception of toxicity associated with each one of the therapies. However, following the review of the literature, it is clear that while it may be asserted that treatment with brachytherapy is probably associated with a lower incidence of serious complications, it is also certain that the toxicity profile associated with each one of the therapies is different and that therefore direct comparison of them is not possible. Thus, it transpires that overall patient perception in the form of measuring quality of life following the different treatments does not differ much regardless of whether surgery or ERT or brachytherapy has been applied. This is important when informing patients and selecting optimal treatment.

20.3.2.1 Urinary Incontinence

Perhaps the most-feared urinary complication following the treatment of localized prostate cancer is incontinence. Published studies have reported a risk of urinary incontinence after prostate brachytherapy of 0% to 85% of the patients. This broad range stems from the different definitions used for incontinence and the way that data are collected. The highest rates come from studies in which patients reported this problem.

Table 20.1 shows a summary of the published results for urinary incontinence. It should be stated, generally speaking, that patients that receive a TUR as part of the treatment in any of the phases of the disease tend to course with a greater incidence of incontinence. Moreover, the rate of 83% reported by Talcott et al. (2001) corresponds to patients of the Seattle series treated with a homogeneous load of seeds, with high doses in urethra and a high incidence of fibrosis.

20.3.2.2 Acute Urinary Retention

Acute urinary retention requiring the use of post-implant catheter is a well-known side effect of the procedure (Bucci et al. 2002). The factors that predict which patients have a greater

TABLE 20.1 Results of Urinary Incontinence in Published Series (See Section 20.3.2)

Author	Number of Patients	Procedure	Incontinence (%)
Blasko et al. 1995	184	Implant	0
Talcott et al. 2001	105	Implant	15
Nag et al. 1995	32	Implant	19
Gelblum et al. 1999	693	Implant	0.7
Wallner et al. 1997	92	Implant	6
Storey et al. 2000	206	Implant	10
Benoit et al. 2000	2124	Implant	6.6
Zeitlin et al. 1998	212	Implant	3.8
Kaye et al. 1995	57	Implant	11
Stone and Stock 1999	301	Implant	0
Beyer and Priestley 1997	499	Implant	1
Talcott et al. 2001	13	TUR + implant	83
Ragde and Korb 2000	48	TUR + implant	12.5
Stone and Stock 1999	43	TUR + implant	0
Kaye et al. 1995	19	TUR + implant	22
Terk et al. 1998	6	Implant + TUR	0
Gelblum et al. 1999	28	Implant + TUR	17

risk of developing urinary obstruction are not very well known. As confirmed by other studies (Wallner et al. 1995; Desai et al. 1998), it has been assumed that at least part of urinary morbidity is related to the dose received through the urethra, which is the target volume. However, most of the obstructive symptoms take place after the procedure when the dose deposited in the tissues is low, which suggests a traumatic effect on the prostate gland or a predisposal to procedure-related obstruction, more than to the dose supplied (Keyes et al. 2006).

Different studies have reported a risk of catheterization between 5% and 22% (Vijverberg et al. 1992; Wallner et al. 1996; Blasko et al. 1991; Dattoli et al. 1996; Kaye et al. 1995); however, few have evaluated the predictive factors of this complication. Lee et al. (2000) show that the number of needles, pretreatment ultrasound prostate volume, planned target volume, and prostate volume by post-treatment CT scan were significant predictive factors. Their study was small, with only 91 patients treated in a nonhomogeneous form with ^{125}I or ^{103}Pd as single therapy or as a boost after ERT. Terk et al. (1998) carried out a multivariate analysis in 251 patients treated with single therapy ^{125}I or ^{103}Pd, which revealed that pretreatment IPSS and the use of neoadjuvant hormone therapy with ^{103}Pd were significant predictive factors of post-implant urinary retention, which occurs in 5.6% of patients. Bucci et al. (2002) demonstrated in 282 patients that urinary retention in most cases is an early event after implantation and that it resolves in most patients in less than 2 months. Fifteen percent of patients required a catheter after implantation, but only 3% needed it after 90 days. A high IPSS ($\geq$15) was predictive not only of the need for catheterization but also of a more prolonged duration. For these authors, diabetes entails an approximately twofold increase in the risk of catheterization, although the mechanism of this relationship is not clear.

20.3.2.3 Urethral Stricture

There are scant data on the incidence of urethral stricture after prostate brachytherapy. The Seattle and Memorial Sloan-Kettering groups reported an incidence of 10%–12% urethral stricture after 5 years (Ragde et al. 1997; Zelefsky et al. 1999a, 2000b). Most of the urethral strictures affect the bulbomembranous urethra. Ragde et al. (1997) reported that the strictures are normally short-lasting and easily solved by simple expansion of the meatus. At present, there are no compelling data to predict whether any clinical, treatment, or dosimetric parameter is associated with the development of urethral stricture after prostate brachytherapy.

20.3.2.4 Rectal Complications

The most feared complication after prostate brachytherapy is prostatorectal fistula and the consequent need for a colostomy. Benoit et al. (2000) reported a need for colostomy after prostate brachytherapy of 0.3% in the Medicare population. Most of the studies have reported that the need for colostomy is around 0% (Blasko et al. 1991; Beyer and Priestley 1997; Kaye et al. 1995; Kleinberg et al. 1994), although Wallner et al. (1996) reported that 3.3% of men of their cohort had a colostomy due to the development of a rectal fistula.

Other rectal complications are rectal ulcer, radical proctitis, and radical colitis. Benoit et al. (2000) reported 5.4% of men diagnosed with a rectal disorder after prostate brachytherapy (1.9% rectal fistula, 2.3% radical colitis, and 1.2% others).

20.3.2.5 Erectile Dysfunction

Many patients treated by means of permanent seed brachytherapy undergo the treatment because they expect their sexual function to be preserved. In fact, although the development of ED is a risk that is present after the treatment of localized prostate cancer with brachytherapy, the permanent implant of seeds is regarded as presenting the least risk of impotence of all the treatments commonly accepted for localized prostate cancer. These data are confirmed by a revision of the published studies on prostate brachytherapy, which report that between 79% (Stock et al. 1996) and 94% (Kleinberg et al. 1994) of men that have normal erections before treatment will maintain these erections after prostate brachytherapy. This low risk of ED is also demonstrated in the Medicare population survey published by Benoit et al. (2000) in which only 14 men were detected out of 2124 (0.7%) who had a penis prosthesis, and 179 men (8.4%) were diagnosed with ED after prostate brachytherapy (this figure is probably underestimated due to the method used to obtain the toxicity codes).

Most of the patients with brachytherapy-induced ED responded favorably to sildenafil citrate. Merrick et al. (2002) describe a high response rate in their series. When the potent patients were grouped with the patients with ED that used sildenafil citrate, the actuarial rate of preservation of potency after

6 years was 92%. Zelefsky et al. (1999b) reported that 90% of the patients that managed to obtain a partial erection after EBRT responded favorably to sildenafil, but only 52% of patients whose erections were flaccid after radiotherapy responded to sildenafil.

20.4 Role of HDR Brachytherapy in Treatment of Prostate Cancer

HDR prostate brachytherapy began in 1988 at Kiel University in Germany and soon after in 1991 at WBH in Royal Oak, MI, and at Seattle Prostate Institute (Stromberg et al. 1994). HDR brachytherapy was initially used as a boost in conjunction with EBRT for intermediate/high-risk prostate cancers, as a vehicle for dose escalation. As experience with HDR prostate brachytherapy accumulated, the ultrasound and computer technology evolved, and specific procedure and treatment toxicities were identified, HDR monotherapy trials were initiated and offered to low/intermediate-risk prostate cancer patients. The two HDR prostate brachytherapy approaches will be hereafter presented separately.

20.4.1 HDR Brachytherapy Boost

In the early 1990s, in an attempt to improve external-beam radiation treatment accuracy and targeting in prostate cancer, various studies were designed to optimize treatment planning and enhance treatment delivery toward more conformal EBRT (CRT) enabling treatments with higher total radiation doses with the goal of increasing tumor control and minimizing toxicity. Three major drawbacks of external beam CRT are day-to-day variations in internal anatomy secondary to organ motion (interfraction motion), temporal variations in internal anatomy, mainly organ deformation during actual RT delivery (intrafraction motion), and daily setup inaccuracies (setup errors). To overcome these problems, HDR brachytherapy was identified as potentially the ideal dose-escalating vehicle since interfraction and intrafraction motion as well as setup errors were nonissues with HDR.

With ultrasound guidance and the interactive online dosimetry system, organ motion and setup inaccuracies (as compared with EBRT) are insignificant because they do not occur or can be corrected during the procedure without increasing target volume margins. Common pitfalls of brachytherapy, including operator dependence and difficulty with reproducibility, have been eliminated with the intraoperative online planning system (Edmundson et al. 1993, 1995) (Figure 20.4).

- In 1991, at WBH, the first prospective phase I/II dose-escalating clinical trial of HDR brachytherapy boost combined with fractionated EBRT was activated. From November 1991 through November 1995, 58 patients received 45.6-Gy pelvic EBRT and three HDR ^{192}Ir boost implants of 5.5 to 6.5 Gy each (Martinez et al. 1995). They were compared with 278 similarly staged patients treated from January 1987 through December 1991 with EBRT

(a)

(b)

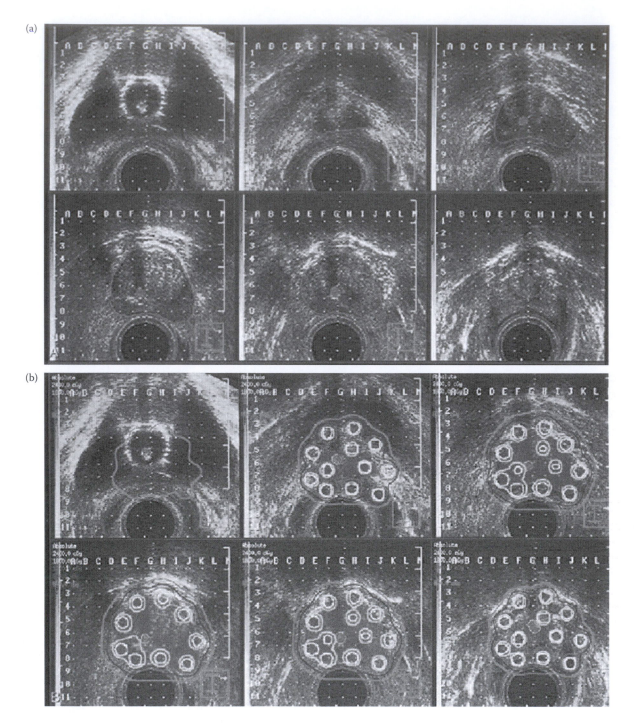

FIGURE 20.4   (See color insert.) Ultrasound-guided permanent implant planning. (a) Volume study with prostate and urethra identified. Note the eccentric position of the urethra. (b) Isodose distribution of the same patient. Red color wash is 100%; blue line indicates 75%. (Courtesy of M. Ghilezan.)

to prostate-only fields (median dose, 66.6 Gy) (Kestin et al. 2000). No patient received androgen deprivation. Patient outcome was analyzed for biochemical control. Biochemical failure was defined as a PSA level higher than 1.5 ng/mL and rising on two consecutive values. If serial post-treatment PSA levels were showing a continuous downward trend, failure was not scored. Median follow-up was 43 months for the conventionally treated group

and 26 months for the HDR boost group. The median pre-treatment PSA level was 14.3 ng/mL for the EBRT-alone group and 14.0 ng/mL for the HDR boost group. The median Gleason scores were 6 and 7, respectively, for the two groups. The biochemical control rate was significantly higher in the HDR boost treatment group. Three-year actuarial biochemical control rates were 85% and 52% for the conformal HDR boost and conventionally treated patients, respectively. In a multivariate analysis, the use of HDR boost and pretreatment PSA level were significant prognostic determinants of biochemical control. The 3-year actuarial rates of biochemical control for conformal HDR boost versus conventionally treated patients, respectively, were 83% versus 72% for a pretreatment PSA level of 4.1 to 10.0 ng/mL, 85% versus 47% for a PSA level of 10.1 to 20.0 ng/mL, and 89% versus 29% for a PSA level higher than 10 ng/mL. When the analysis was limited to patients in both groups with a minimum 12-month follow-up, the HDR boost group continued to show a higher biochemical control rate than the conventional radiation group (3-year actuarial rates of 86% versus 53%) (Kestin et al. 2000).

- The Beaumont group (Martinez et al. 2002) updated the series with an analysis of 207 patients treated on the dose-escalation HDR boost prostate brachytherapy trial. It demonstrated to be a precise and accurate dose-delivery system and a very effective treatment for patients with unfavorable prostate cancer. Using the same database, Brenner et al. (2002) reported a low α/β ratio of 1.2 showing high sensitivity to fractionation similar to the late responding tissues.

- With longer follow-up and a larger number of patients, Martinez et al. (2003) published the long-term results of the WBH prostate HDR dose-escalation trial. Data demonstrated that HDR boost improves biochemical control and cause-specific survival (CSS) in patients with prostate cancer and poor prognostic factors. At the 2005 ASTRO annual meeting, Vargas et al. (2006) from WBH reported the final analysis of the HDR boost dose-escalation trial. For the first time, an improvement on biochemical control led to a decrease in metastatic rate and improved OS.

- Galalae et al. (2004) reported on the collaborative trial between Kiel University in Germany, Seattle Tumor Institute, and WBH on long-term outcomes by risk factor using a conformal HDR brachytherapy boost for patients with localized prostate cancer during the PSA era. Similar results were found at the three institutions in 611 patients with prostate cancer harboring intermediate- and high-risk factors. With a mean follow-up of 5 years, the 5- and 10-year biochemical control was 77% and 73%, DFS was 67% and 49%, and CSS was 96% and 92%, respectively. The similarity in results at the three institutions gives credence to the reproducibility of the HDR brachytherapy boost treatment. Dose escalation greater than 95 Gy resulted in better 5-year biochemical control for conformal HDR boost (59% versus 85%; $p < 0.001$) for the entire cohort of hormonal-naïve men. Discriminating by risk factors, a striking dose-escalation effect was seen in the group of patients with two or three poor prognostic factors ($p = 0.02$ and $p < 0.001$). This unfavorable group has a remarkable 5-year biochemical control of 85%.

- These excellent results were confirmed by others showing that HDR prostate brachytherapy is a robust, safe, and reproducible treatment method (Bachand et al. 2009;

Pellizzon et al. 2008; Kalkner et al. 2007; Duchesne et al. 2007; Pistis et al. 2010).

- Two randomized trials demonstrated the superiority of adding HDR boost to EBRT compared with EBRT alone. In a study by Sathya et al. (2005), 104 patients with T2 and T3 prostate cancer with no evidence of metastatic disease were randomly assigned to EBRT of 66 Gy in 33 fractions during 6.5 weeks or to HDR boost of 35 Gy delivered to the prostate during 48 h plus EBRT of 40 Gy in 20 fractions during 4 weeks. The median follow-up was 8.2 years. In the HDR boost + EBRT arm, 17 patients (29%) experienced biochemical failure (BCF) compared with 33 patients (61%) in the EBRT arm ($p = 0.0024$). Eighty-seven patients (84%) had a postradiation biopsy; 10 (24%) of 42 in the HDR boost + EBRT arm had biopsy positivity compared with 23 (51%) of 45 in the EBRT arm (odds ratio, 0.30; $p = 0.015$). OS was over 90% for both treatment regimens.

  Hoskin et al. (2007) randomized 220 patients with prostate cancer and PSA < 50 ng/mL to receive either standard EBRT 55 Gy in 20 fractions over 4 weeks or a combined schedule comprising EBRT delivering 35.75 Gy in 13 fractions over 2.5 weeks followed by a temporary HDR afterloading implant delivering 17 Gy in two fractions over 24 h. With a median follow-up of 30 months, a significant improvement in actuarial biochemical relapse-free survival (bRFS) was seen in favor of the combined EBRT/brachytherapy schedule ($p = 0.03$). A lower incidence of acute rectal discharge was seen in the EBRT/brachytherapy group ($p = 0.025$), and other acute and late toxicities were equivalent. Patients randomized to brachytherapy had a significantly better FACT-P score at 12 weeks ($p = 0.02$) (Hoskin et al. 2007).

- The William Beaumont group addressed the question of long-term survival impact with a short course (≤ 6 months) of adjuvant androgen deprivation when a very high radiation dose was delivered to 934 patients treated with an HDR brachytherapy boost in a hypofractionated regime (Martinez et al. 2005). At 8 years, the addition of a course of 6 months or less of ADT to a very high hypofractionated radiation dose had not conferred a therapeutic advantage but added side effects and cost. Furthermore, for the most unfavorable group of patients harboring all three poor risk factors, there was a higher rate of distant metastasis and more prostate cancer–related deaths. This result questions the value of a short course of ADT and the impact on delaying curative treatment.

20.4.2 HDR Brachytherapy as Monotherapy

In 1997, the HDR brachytherapy as monotherapy was started at WBH. It consisted of a single implant followed by four fractions of 9.5 Gy delivered twice daily with a minimum of 6 h apart. The twice-per-day accelerated hypofractionated regime was selected based on HDR favorable radiobiological considerations and physical dose-delivery advantages of TRUS guidance (Martinez et al. 2001), with conformal intensity modulated real-time dosimetry of prostate HDR brachytherapy. This regimen has a BED of 266 Gy with an alpha/beta of 1.5, much higher than 81 Gy delivered in standard fractionation with EBRT, with a BED of only 178. For California Endocuritherapy (CET) Cancer Center, the HDR dose schedule is 42 Gy in six fractions (bid) in two separate implants 1 week apart. The biologic equivalence is 76 Gy in 2-Gy daily fractions of EBRT.

Patients with clinical stage II (T1c–T2a) disease, Gleason score less than 7 (unilobar, 3 + 4, no perineural invasion), and pretreatment PSA less than 12 ng/mL were treated with monotherapy. The majority of patients presented with what would be considered low-risk or favorable prostate cancer. Patients were offered either HDR or LDR brachytherapy as treatment options, and then the patient selected the brachytherapy modality. A short course of neoadjuvant androgen deprivation (<6 months) was used for downsizing the gland volume in 31% of WBH patients, in equal proportions between permanent seeds and HDR, and in 30% of the CET Cancer Center patients. All procedures were done under spinal anesthesia. Figure 20.5 depicts an HDR intraoperative implant using the Nucletron Swift guidance system.

Between January 1996 and December 2002, 378 consecutive patients with clinically localized prostate cancer were treated with accelerated hypofractionated brachytherapy as the sole treatment modality. Of the patients, 172 were treated with HDR brachytherapy alone using ^{192}Ir, and 206 patients were treated with LDR brachytherapy alone using [103]Pd.

For the implant procedure and for pain control during the entire treatment time, spinal anesthesia was administered following placement of an epidural catheter for analgesia. Dosimetry was continuously updated in real time based on the actual location of needles to compensate for organ distortion and motion and to ensure conformal coverage of the gland (Edmundson et al. 1993, 1995). Gold seed markers were then placed under TRUS guidance at the base and at the apex of the prostate to assess and measure possible interfraction needle displacement. Before delivery of the radiation, the entire prostate was imaged again, with final needle and urethral positions captured by TRUS, and a final treatment plan was created.

At CET, after recovery, the patient underwent a dual method of simulation radiography consisting of plain film localization for applicator adjustment and quality control, and a CT scan was performed. The images were downloaded to the "treatment-planning" computer, and a 3D reconstruction was carried out. A DVH and virtual images of the anatomy, CTV, and PTV were obtained.

20.4.2.1 Toxicity for HDR Monotherapy

The toxicity profile of HDR monotherapy was first described by Grills et al. (2004) from WBH, demonstrating less acute and

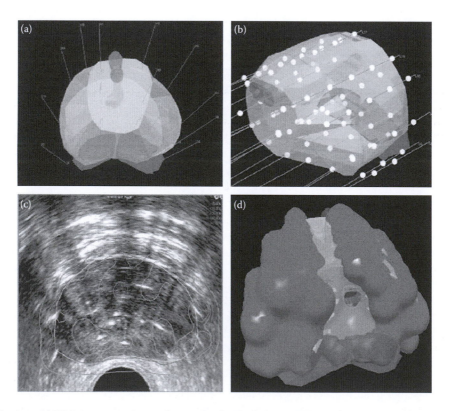

FIGURE 20.5   (See color insert.) HDR intraoperative implant using the Nucletron SWIFT guidance system. (a) 3D reconstruction of the prostate gland, urethral trajectory, and needle orientation. (b) Anatomic relationship of the prostate, the urethra, and the needles with the selected dwell positions on each needle. (c) Final TRUS-based intraoperative dosimetry and coverage of the PTV. (d) Dosimetric rendering of the prostate coverage by the 100% isodose cloud in red with urethral sparing from modulating the dwell times and dwell positions. (Courtesy of M. Ghilezan.)

chronic toxicity with HDR when compared with permanent seeds with ^{103}Pd. Also, the impotency rate was decreased in the HDR group of treated patients by half compared with permanent seeds. The following toxicity analysis is an updated report from the combined experience from WBH and CET (Martinez et al. 2010). The median follow-up for all patients was 4.1 years (range 8 to 12.3 years).

20.4.2.1.1 Acute Toxicity

HDR brachytherapy alone was associated with statistically significant reductions in the acute rates of dysuria (65% with ^{103}Pd seeds versus 38% with HDR monotherapy; $p <0.001$), as well as urinary frequency and/or urgency (^{103}Pd: 94% versus 53%, HDR; $p < 0.001$) and urinary retention (^{103}Pd: 43% versus 29%, HDR; $p = 0.012$). In addition to reduced acute genitourinary symptoms, HDR was also associated with lower rates of rectal pain (18% with LDR versus 7%, HDR; $p = 0.025$). The majority of acute toxicities in both groups were grade 1. Hormonal androgen ablation for gland downsizing was given to 31% of patients in both groups.

20.4.2.1.2 Chronic Toxicity

HDR brachytherapy alone was again associated with reduced urinary frequency and urgency (^{103}Pd: 54% versus 32%, HDR; $p < 0.001$). The majority of toxicities were grade 1. There were no differences in the remaining chronic toxicity rates of urinary incontinence or retention, hematuria, diarrhea, rectal pain, or rectal bleeding between the two treatment groups. The rate of urethral stricture requiring dilatation was 3% with HDR compared with 1% with ^{103}Pd ($p = 0.3$). The median time to development of urethral stricture was 17 months, with a range of 4 to 37 months. The cumulative proportion of chronic grade 3 toxicity by treatment modality did not differ between the two treatment types.

20.4.2.1.3 Erectile Dysfunction

Regardless of the use of adjuvant hormonal therapy, all cases were included for which complete pretreatment and posttreatment information was available. This included 169 patients, 61 of which were treated with HDR brachytherapy alone and 108 patients treated with permanent ^{103}Pd. The 5-year probability of ED was 33% for all patients with available preimplant data. Forty-one percent of LDR patients and 18% of HDR patients ($p = 0.03$) developed ED. The median times to ED for the HDR and LDR patient groups were 3.4 and 2.6 years, respectively.

20.4.2.2 Survival Outcomes

The 5-year actuarial outcomes for monotherapy showed no difference in terms of ASTRO definition for biochemical failure, cancer mortality, or OS between HDR alone versus permanent seeds. No difference in biochemical control using the ASTRO definition and the Phoenix definition can be seen by treatment modality.

20.4.2.2.1 Updated Analyses

An update of the largest prostate HDR monotherapy series of 248 patients from WBH and CET was published in 2009 with a median follow-up of 5 years (Martinez et al. 2010; Demanes et al. 2011). The 5-year Phoenix definition for biochemical control was 91%. In a comparable cohort of patients treated with permanent seed implants (^{103}Pd), the biochemical control was similar; nonetheless, the HDR patients experienced significantly less acute and chronic genitourinary and gastrointestinal toxicities.

The Beaumont group recently reported on a hypofractionated HDR monotherapy trial for low-risk prostate cancer that accrued 173 patients (Ghilezan et al. 2012). The total dose was 24 Gy for the first 50 patients, delivered in a single implant and two fractions, 6 h apart. The subsequent 50 patients received 27 Gy in two fractions in one day. There were no differences in acute or chronic toxicities between the two regimens with a median follow-up of 17 months. The maximum chronic GU and GI toxicities grade 2 or higher were 5% or lower with only urinary frequency/urgency being 16%. However, after 6 months, this issue resolved in almost all patients being 0% for the 24-Gy group and 4.8% for the 27-Gy group. Because of concerns that normal tissue repair may not be complete within the 6-h interfraction interval when using doses of 12–13.5 Gy, along with issues of patient discomfort and tolerance of epidural catheter-based controlled analgesia as well as logistics of work flow and time management, our 2 fractions/day protocol evolved to two separate fractions delivered 2–3 weeks apart. Despite a second implant and second anesthesia, the treatment tolerance improved, while patient satisfaction increased. Encouraged by this excellent tolerance and toxicity profile, we activated a new protocol in which HDR monotherapy is delivered in one single fraction of 19 Gy. Twenty-five patients were treated until October 2011, and with a median of 3 months of follow-up, there were no grade 3 or 4 acute GU or GI toxicities. For patients ineligible for single-fraction HDR, our current standard is two separate implants separated by 2–3 weeks. For patients with T1c disease, PSA < 10 ng/mL, Gleason score 6, and up to 3 cores positive out of 12, none of which with more than 50% tumor involvement, and patients of age 65 years or older are offered 12 Gy/2 fractions. All others are treated with 13.5 Gy/2 fractions.

Hoskin et al. (2012) recently published their experience with HDR monotherapy in a group of 197 patients treated with four regimens of 34 Gy/4 fractions, 36 Gy/4 fractions, 31.5 Gy/3 fractions, and ultimately 26 Gy/2 fractions, very similar to our 27 Gy/2 fractions schedule. The treatment regimens were designed to have similar biological effectiveness in terms of late morbidity and for the 13 Gy per fraction to be equitoxic to the 4×8.5 Gy regimen for early effects. The study predominantly recruited patients with intermediate- and high-risk disease (over 90%), as opposed to other HDR monotherapy series, and as such, 57/197 patients had androgen deprivation therapy for a median of 6 months. The late toxicity reported is 5% or less for grade 3 or higher GU, including urinary strictures requiring surgery; nonetheless, the authors used the four-point RTOG toxicity scale rather than the most commonly used five-point CTC scale, and therefore, the numbers may appear somewhat higher. The median time to stricture formation was 23 months (range, 5–40 months). In the 31.5-Gy group, there was no evidence that the urethral *D*30 dose was higher in patients with strictures, but mean PTV was significantly larger in patients who developed grade 3 late urinary obstruction (98.0 ± 8.1 cc versus 79.6 ± 3.0 cc; *p* = 0.04). No significant differences were detected between schedules, and there was no indication of a dose–response relationship. Overall, 95% of patients were free of biochemical relapse at 3 years, 99% for those at intermediate risk and 91% for high-risk disease (hazard ratio, 0.182; confidence interval, 0.026–0.848; *p* = 0.03), and 95% and 87%, respectively, at 4 years. The freedom from biochemical recurrence was significantly higher in patients whose time to nadir PSA was ≥15.7 months and whose concentration at nadir was ≤0.01 µg/L. With the caveat that a large number of these patients were on long-term ADT and that the number of patients and follow-up were relatively small, these results are remarkable and better than any other treatment methods or combinations, especially in a population of intermediate/high-risk patients.

Other centers around the world confirmed these promising results. In Europe, the group from Offenbach led by Zamboglou and Baltas obtained results similar to the Beaumont group, using the same dose and fractionation regimen of 38 Gy in 4 fractions and 2 days in favorable/intermediate-risk patients (Martin et al. 2004). This group was at the forefront of physics research and development of real-time ultrasound-based inverse planning and 3D dose optimization algorithms used in contemporary HDR treatment planning.

Australian researchers at Peter McCallum Cancer Center have published recently the results of a phase II study in which 79 favorable and intermediate-risk prostate cancer patients were included (Barkati et al. 2012). They administered HDR in a single implant and four fractionation schedules over 2 days (three fractions of 10, 10.5, 11, and 11.5 Gy) in a dose escalation study. The 3- and 5-year biochemical control rates (Phoenix definition) were 88% and 85%, respectively. There were no acute or chronic rectal toxicities. Grade 3 dysuria and urinary retention were under 10%. No grade 4 toxicities were recorded. They did not report any differences between the fractionation schedules.

In Japan, Yoshioka et al. (2012) pioneered a longer hypofractionated HDR schedule, that is, 54 Gy in nine fractions and 5 days. Their series of 112 patients have been reported recently with a median follow-up of 5.4 years. Although more than half of their patients were in the high-risk group with only 15 patients in the low-risk group, the 5-year results were remarkable (85%, 93%, and 79% for the low-, intermediate-, and high-risk groups, respectively). The toxicity profile was equally outstanding with no grade 4 acute or late toxicities and less than 4% grade 3 late toxicities.

However, a few questions arise with prostate HDR. First of all, there is no consensus on the most appropriate dose and fractionation schedule, as it is for LDR. Second, there are significant differences in dosimetry methods that are either CT-based or TRUS-based. As a consequence, DVH parameters and quality assurance guidelines are less standardized compared with LDR, and hence more difficult to define, as the HDR technique is generally institution-specific. Last but not least, HDR is more labor-intensive for the brachytherapy team and more time-consuming than LDR, for both patients and physicians, which limit somewhat its utilization. Consequently, the overall clinical experience, long-term results, and mature data from large phase II/III trials are sparse or lacking, a clear disadvantage compared with LDR prostate brachytherapy. Nonetheless, HDR reproducibility, short learning curve, consistency in producing high-quality and accurate treatment plans, short radiation delivery times overcoming the issues of prostate gland underdosing secondary to edema, seed migration or seed loss, and the emergence of promising results of regimens using only one to two treatments make HDR a very appealing treatment method for all prostate cancer risk groups. If, indeed, prostate cancer has a low alpha/beta ratio and thus it is very sensitive to dose per fraction, HDR would have a clear radiobiological advantage over LDR and EBRT having the potential to become the treatment of choice of prostate cancer among all radiation treatment options (Hoskin et al. 2012).

20.5 Conclusion

The technique of interstitial prostate brachytherapy has been significantly improved in the last 20 years. Both permanent seed implantation and temporary HDR brachytherapy can be used alone or in combination with external beam radiotherapy in the treatment of organ-confined prostate cancer. Prostate brachytherapy as an ambulatory, minimally invasive procedure with swift recovery, that lacks serious side effects or long-term sequelae, with the potential for better preservation of erectile function resulting in better quality of life, and ultimately that is financially less onerous compared to sophisticated photon or proton-based EBRT may become the primary, nonsurgical treatment approach for eligible prostate cancer patients.

References

Aronowitz, J.N. 2002. Dawn of prostate brachytherapy: 1915–1930. *Int J Radiat Oncol Biol Phys* 54:712–8.

Ash, D., Flynn, A., Battermann, J., de Reijke, T., Lavagnini, P., and Blank, L. 2000. ESTRO/EAU/EORTC recommendations on permanent seed implantation for localized prostate cancer. *Radiother Oncol* 57:315–21.

Ash, D., Al-Qaisieh, B., Bottomley, D., Carey, B., and Joseph, J. 2006. The correlation between D90 and outcome for I-125 seed implant monotherapy for localised prostate cancer. *Radiother Oncol* 79:185–9.

Bachand, F., Martin, A.G., Beaulieu, L., Harel, F., and Vigneault, E. 2009. An eight-year experience of HDR brachytherapy boost for localized prostate cancer: Biopsy and PSA outcome. *Int J Radiat Oncol Biol Phys* 73:679–84.

Barkati, M., Williams, S.G., Foroudi, F. et al. 2012. High-dose-rate brachytherapy as a monotherapy for favorable-risk prostate cancer: A phase II trial. *Int J Radiat Oncol Biol Phys* 82(5):1889–96.

Barringer, B.S. 1916. Radium in the treatment of carcinoma of the prostate and bladder. *JAMA* 67:1442–5.

Battermann, J.J. 2000. I-125 implantation for localized prostate cancer: the Utrecht University experience. *Radiother Oncol* 57:269–72.

Beaulieu, L., Evans, D.A., Aubin, S. et al. 2007. Bypassing the learning curve in permanent seed implants using state-of-the-art technology. *Int J Radiat Oncol Biol Phys* 67:71–7.

Benoit, R.M., Naslund, M.J., and Cohen, J.K. 2000. Complications after prostate brachytherapy in the Medicare population. *Urology* 55:91–6.

Beyer, D.C., Priestley, J.B. Jr. 1997. Biochemical disease-free survival following ^{125}I prostate implantation. *Int J Radiat Oncol Biol Phys* 37:559–63.

Beyer, D.C., Shapiro, R.H., and Puente, F. 2000. Real-time optimized intraoperative dosimetry for prostate brachytherapy: A pilot study. *Int J Radiat Oncol Biol Phys* 48:1583–9.

Blasko, J.C., Ragde, H., and Grimm, P.D. 1991. Transperineal ultrasound-guided implantation of the prostate: Morbidity and complications. *Scand J Urol Nephrol Suppl* 137:113–8.

Blasko, J.C., Grimm, P.D., and Ragde, H. 1993. Brachytherapy and organ preservation in the management of carcinoma of the prostate. *Semin Radiat Oncol* 3:240–9.

Blasko, J.C., Wallner, K., Grimm, P.D., and Ragde, H. 1995. Prostate specific antigen based disease control following ultrasound guided 125iodine implantation for stage T1/T2 prostatic carcinoma. *J Urol* 154:1096–9.

Bossi, A. 2010. Which modality for prostate brachytherapy? *Cancer Radiother* 14:488–92.

Brenner, D.J., Martinez, A.A., Edmundson, G.K., Mitchell, C., Thames, H.D., and Armour, E.P. 2002. Direct evidence that prostate tumors show high sensitivity to fractionation (low alpha/beta ratio), similar to late-responding normal tissue. *Int J Radiat Oncol Biol Phys* 52:6–13.

Bucci, J., Morris, W.J., Keyes, M., Spadinger, I., Sidhu, S., and Moravan, V. 2002. Predictive factors of urinary retention following prostate brachytherapy. *Int J Radiat Oncol Biol Phys* 53:91–8.

Crook, J., McLean, M., Yeung, I., Williams, T., and Lockwood, G. 2004. MRI-CT fusion to assess postbrachytherapy prostate volume and the effects of prolonged edema on dosimetry following transperineal interstitial permanent prostate brachytherapy. *Brachytherapy* 3:55–60.

D'Amico, A.V., Whittington, R., Malkowicz, S.B. et al. 1998. Biochemical outcome after radical prostatectomy, external beam radiation therapy, or interstitial radiation therapy for clinically localized prostate cancer. *JAMA* 280:969–74.

Dattoli, M., Wallner, K., Sorace, R. et al. 1996. ^{103}Pd brachytherapy and external beam irradiation for clinically localized, high-risk prostatic carcinoma. *Int J Radiat Oncol Biol Phys* 35:875–9.

Davis, B.J., Pisansky, T.M., Wilson, T.M. et al. 1999. The radial distance of extraprostatic extension of prostate carcinoma: Implications for prostate brachytherapy. *Cancer* 85:2630–7.

Demanes, D.J., Martinez, A.A., Ghilezan, M. et al. 2011. High-dose-rate monotherapy: Safe and effective brachytherapy for patients with localized prostate cancer. *Int J Radiat Oncol Biol Phys* 81(5):1286-92.

De Vries, S.H., Postma, R., Raaijmakers, R. et al. 2007. Overall and disease-specific survival of patients with screen-detected prostate cancer in the European randomized study of screening for prostate cancer, section Rotterdam. *Eur Urol* 51:366–74; discussion 374.

Desai, J., Stock, R.G., Stone, N.N., Iannuzzi, C., and DeWyngaert, J.K. 1998. Acute urinary morbidity following I-125 interstitial implantation of the prostate gland. *Radiat Oncol Investig* 6:135–41.

Duchesne, G.M., Williams, S.G., Das, R., and Tai, K.H. 2007. Patterns of toxicity following high-dose-rate brachytherapy boost for prostate cancer: Mature prospective phase I/II study results. *Radiother Oncol* 84:128–34.

Edmundson, G.K., Rizzo, N.R., Teahan, M., Brabbins, D., Vicini, F.A., and Martinez, A. 1993. Concurrent treatment planning for outpatient high dose rate prostate template implants. *Int J Radiat Oncol Biol Phys* 27 1215–23.

Edmundson, G.K., Yan, D., and Martinez, A.A. 1995. Intraoperative optimization of needle placement and dwell times for conformal prostate brachytherapy. *Int J Radiat Oncol Biol Phys* 33:1257–63.

Ferlay, J., Shin, H.R., Bray, F., Forman, D., and Mathers, C., DM P. 2008. GLOBOCAN 2008, Cancer Incidence and Mortality Worldwide: IARC CancerBase No. 10 [Internet].

Fornage, B.D., Touche, D.H., Deglaire, M., Faroux, M.J., and Simatos, A. 1983. Real-time ultrasound-guided prostatic biopsy using a new transrectal linear-array probe. *Radiology* 146:547–8.

Frank, S.J., Grimm, P.D., Sylvester, J.E. et al. 2007. Interstitial implant alone or in combination with external beam radiation therapy for intermediate-risk prostate cancer: A survey of practice patterns in the United States. *Brachytherapy* 6:2–8.

Frankel, S., Smith, G.D., Donovan, J., and Neal, D. 2003. Screening for prostate cancer. *Lancet* 361:1122–8.

Futterer, J.J. 2007. MR imaging in local staging of prostate cancer. *Eur J Radiol* 63:328–34.

Galalae, R.M., Martinez, A., Mate, T. et al. 2004. Long-term outcome by risk factors using conformal high-dose-rate brachytherapy (HDR-BT) boost with or without neoadjuvant androgen suppression for localized prostate cancer. *Int J Radiat Oncol Biol Phys* 58:1048–55.

Gelblum, D.Y., Potters, L., Ashley, R., Waldbaum, R., Wang, X.H., and Leibel, S. 1999. Urinary morbidity following ultrasound-guided transperineal prostate seed implantation. *Int J Radiat Oncol Biol Phys* 45:59–67.

Gewanter, R.M., Wuu, C., Laguna, J.L., Katz, A.E., and Ennis, R.D. 2000. Intraoperative preplanning for transperineal ultrasound-guided permanent prostate brachytherapy. *Int J Radiat Oncol Biol Phys* 48:377–80.

Ghilezan, M., Martinez, A., Gustason, G. et al. 2012. High-dose-rate brachytherapy as monotherapy delivered in two fractions within one day for favourable/intermediate-risk prostate cancer: Preliminary toxicity data. *Int J Radiat Oncol Biol Phys* 83(3):927–32.

Grado, G.L., Larson, T.R., Balch, C.S. et al. 1998. Actuarial disease-free survival after prostate cancer brachytherapy using interactive techniques with biplane ultrasound and fluoroscopic guidance. *Int J Radiat Oncol Biol Phys* 42:289–98.

Grills, I.S., Martinez, A.A., Hollander, M. et al. 2004. High dose rate brachytherapy as prostate cancer monotherapy reduces toxicity compared to low dose rate palladium seeds. *J Urol* 171:1098–104.

Grimm, P.D., Blasko, J.C., Ragde, H., Sylvester, J., and Clarke, D. 1996. Does brachytherapy have a role in the treatment of prostate cancer? *Hematol Oncol Clin North Am* 10:653–73.

Grimm, P.D., Blasko, J.C., Sylvester, J.E., Meier, R.M., and Cavanagh, W. 2001. 10-year biochemical (prostate-specific antigen) control of prostate cancer with (125)I brachytherapy. *Int J Radiat Oncol Biol Phys* 51:31–40.

Guedea, F., Aguilo, F., Polo, A. et al. 2006. Early biochemical outcomes following permanent interstitial brachytherapy as monotherapy in 1050 patients with clinical T1-T2 prostate cancer. *Radiother Oncol* 80:57–61.

Henderson, A., Cahill, D., Laing, R.W., and Langley, S.E. 2002. (125)Iodine prostate brachytherapy: Outcome from the first 100 consecutive patients and selection strategies incorporating urodynamics. *BJU Int* 90:567–72.

Henry, A.M., Al-Qaisieh, B., Gould, K. et al. 2010. Outcomes following iodine-125 monotherapy for localized prostate cancer: the results of Leeds 10-year single-center brachytherapy experience. *Int J Radiat Oncol Biol Phys* 76:50–6.

Hinnen, K.A., Battermann, J.J., van Roermund, J.G. et al. 2010. Long-term biochemical and survival outcome of 921 patients treated with I-125 permanent prostate brachytherapy. *Int J Radiat Oncol Biol Phys* 76:1433–8.

Ho, A.Y., Burri, R.J., Cesaretti, J.A., Stone, N.N., and Stock, R.G. 2009. Radiation dose predicts for biochemical control in intermediate-risk prostate cancer patients treated with low-dose-rate brachytherapy. *Int J Radiat Oncol Biol Phys* 75:16–22.

Holm, H.H., Juul, N., Pedersen, J.F., Hansen, H., and Stroyer, I. 1983. Transperineal 125iodine seed implantation in prostatic cancer guided by transrectal ultrasonography. *J Urol* 130:283–6.

Hoskin, P.J., Motohashi, K., Bownes, P., Bryant, L., and Ostler, P. 2007. High dose rate brachytherapy in combination with external beam radiotherapy in the radical treatment of prostate cancer: Initial results of a randomised phase three trial. *Radiother Oncol* 84:114–20.

Hoskin, P., Rojas, A., Lowe, G. et al. 2012. High-dose-rate brachytherapy alone for localized prostate cancer in patients at moderate or high risk of biochemical recurrence. *Int J Radiat Oncol Biol Phys* 82(4):1376–84.

Jani, A.B. and Hellman, S. 2008. Early prostate cancer: Hedonic prices model of provider-patient interactions and decisions. *Int J Radiat Oncol Biol* Phys 70:1158–68.

Kalkner, K.M., Wahlgren, T., Ryberg, M. et al. 2007. Clinical outcome in patients with prostate cancer treated with external beam radiotherapy and high dose-rate iridium 192 brachytherapy boost: A 6-year follow-up. *Acta Oncol* 46:909–17.

Kaplan, I.D., Holupka, E.J., Meskell, P. et al. 2000. Intraoperative treatment planning for radioactive seed implant therapy for prostate cancer. *Urology* 56:492–5.

Kaye, K.W., Olson, D.J., and Payne, J.T. 1995. Detailed preliminary analysis of 125iodine implantation for localized prostate cancer using percutaneous approach. *J Urol* 153:1020–5.

Kestin, L.L., Martinez, A.A., Stromberg, J.S. et al. 2000. Matched-pair analysis of conformal high-dose-rate brachytherapy boost versus external-beam radiation therapy alone for locally advanced prostate cancer. *J Clin Oncol* 18:2869–80.

Keyes, M., Schellenberg, D., Moravan, V. et al. 2006. Decline in urinary retention incidence in 805 patients after prostate brachytherapy: The effect of learning curve? *Int J Radiat Oncol Biol Phys* 64:825–34.

Kleinberg, L., Wallner, K., Roy, J. et al. 1994. Treatment-related symptoms during the first year following transperineal 125I prostate implantation. *Int J Radiat Oncol Biol Phys* 28:985–90.

Kollmeier, M.A., Stock, R.G., and Stone, N. 2003. Biochemical outcomes after prostate brachytherapy with 5-year minimal follow-up: Importance of patient selection and implant quality. *Int J Radiat Oncol Biol Phys* 57:645–53.

Kovacs, G., Pötter, R., Loch, T. et al. 2005. GEC/ESTRO-EAU recommendations on temporary brachytherapy using stepping sources for localised prostate cancer. *Radiother Oncol* 74:137–48.

Lawton, C.A., DeSilvio, M., Lee, W.R. et al. 2007. Results of a phase II trial of transrectal ultrasound-guided permanent radioactive implantation of the prostate for definitive management of localized adenocarcinoma of the prostate (radiation therapy oncology group 98-05). *Int J Radiat Oncol Biol Phys* 67:39–47.

Lee, E.K. and Zaider, M. 2003. Intraoperative dynamic dose optimization in permanent prostate implants. *Int J Radiat Oncol Biol Phys* 56:854–61.

Lee, N., Wuu, C.S., Brody, R. et al. 2000. Factors predicting for postimplantation urinary retention after permanent prostate brachytherapy. *Int J Radiat Oncol Biol Phys* 48:1457–60.

Lee, W.R., Moughan, J., Owen, J.B., and Zelefsky, M.J. 2003. The 1999 patterns of care study of radiotherapy in localized prostate carcinoma: A comprehensive survey of prostate brachytherapy in the United States. *Cancer* 98:1987–94.

Martens, C., Pond, G., Webster, D., McLean, M., Gillan, C., and Crook, J. 2006. Relationship of the International Prostate Symptom score with urinary flow studies, and catheterization rates following 125I prostate brachytherapy. *Brachytherapy* 5:9–13.

Martin, T., Baltas, D., Kurek, R. et al. 2004. 3-D conformal HDR brachytherapy as monotherapy for localized prostate cancer. A pilot study. *Strahlenther Onkol* 180:225–32.

Martinez, A., Gonzalez, J., Stromberg, J. et al. 1995. Conformal prostate brachytherapy: Initial experience of a phase I/II dose-escalating trial. *Int J Radiat Oncol Biol Phys* 33:1019–27.

Martinez, A.A., Pataki, I., Edmundson, G., Sebastian, E., Brabbins, D., and Gustafson, G. 2001. Phase II prospective study of the use of conformal high-dose-rate brachytherapy as mono-therapy for the treatment of favorable stage prostate cancer: A feasibility report. *Int J Radiat Oncol Biol Phys* 49:61–9.

Martinez, A.A., Gustafson, G., Gonzalez, J. et al. 2002. Dose escalation using conformal high-dose-rate brachytherapy improves outcome in unfavorable prostate cancer. *Int J Radiat Oncol Biol Phys* 53:316–27.

Martinez, A., Gonzalez, J., Spencer, W. et al. 2003. Conformal high dose rate brachytherapy improves biochemical control and cause specific survival in patients with prostate cancer and poor prognostic factors. *J Urol* 169:974–9; discussion 979–80.

Martinez, A.A., Demanes, D.J., Galalae, R. et al. 2005. Lack of benefit from a short course of androgen deprivation for unfavorable prostate cancer patients treated with an accelerated hypofractionated regime. *Int J Radiat Oncol Biol Phys* 62:1322–31.

Martinez, A.A., Demanes, J., Vargas, C., Schour, L., Ghilezan, M., and Gustafson, G.S. 2010. High-dose-rate prostate brachytherapy: an excellent accelerated-hypofractionated treatment for favorable prostate cancer. *Am J Clin Oncol* 33:481–8.

Matzkin, H., Kaver, I., Stenger, A., Agai, R., Esna, N., and Chen, J. 2003. Iodine-125 brachytherapy for localized prostate cancer and urinary morbidity: A prospective comparison of two seed implant methods-preplanning and intraoperative planning. *Urology* 62:497–502.

McLaughlin, P.W., Narayana, V., Kessler, M. et al. 2004. The use of mutual information in registration of CT and MRI datasets post permanent implant. *Brachytherapy* 3:61–70.

Merrick, G.S., Butler, W.M., Galbreath, R.W., and Lief, J.H. 2001a. Five-year biochemical outcome following permanent interstitial brachytherapy for clinical T1–T3 prostate cancer. *Int J Radiat Oncol Biol Phys* 51:41–8.

Merrick, G.S., Butler, W.M., Lief, J.H., and Galbreath, R.W. 2001b. Five-year biochemical outcome after prostate brachytherapy for hormone-naive men < or = 62 years of age. *Int J Radiat Oncol Biol Phys* 50:1253–7.

Merrick, G.S., Butler, W.M., Galbreath, R.W., Stipetich, R.L., Abel, L.J., and Lief, J.H. 2002. Erectile function after permanent prostate brachytherapy. *Int J Radiat Oncol Biol Phys* 52:893–902.

Messing, E.M., Zhang, J.B., Rubens, D.J. et al. 1999. Intraoperative optimized inverse planning for prostate brachytherapy: Early experience. *Int J Radiat Oncol Biol Phys* 44:801–8.

Munro, N.P., Al-Qaisieh, B., Bownes, P. et al. 2010. Outcomes from Gleason 7, intermediate risk, localized prostate cancer treated with Iodine-125 monotherapy over 10 years. *Radiother Oncol* 96:34–7.

Nag, S., Scaperoth, D.D., Badalament, R., Hall, S.A., and Burgers, J. 1995. Transperineal palladium 103 prostate brachytherapy: Analysis of morbidity and seed migration. *Urology* 45:87–92.

Nag, S., Beyer, D., Friedland, J., Grimm, P., and Nath, R. 1999. American Brachytherapy Society (ABS) recommendations for transperineal permanent brachytherapy of prostate cancer. *Int J Radiat Oncol Biol Phys* 44:789–99.

Nag, S., Bice, W., DeWyngaert, K., Prestidge, B., Stock, R., and Yu, Y. 2000. The American Brachytherapy Society recommendations for permanent prostate brachytherapy postimplant dosimetric analysis. *Int J Radiat Oncol Biol Phys* 46:221–30.

Nag, S., Ciezki, J.P., Cormack, R. et al. 2001. Intraoperative planning and evaluation of permanent prostate brachytherapy: Report of the American Brachytherapy Society. *Int J Radiat Oncol Biol Phys* 51:1422–30.

Pellizzon, A.C., Salvajoli, J., Novaes, P., Maia, M., and Fogaroli, R. 2008. Updated results of high-dose rate brachytherapy and external beam radiotherapy for locally and locally advanced prostate cancer using the RTOG-ASTRO Phoenix definition. *Int Braz J Urol* 34:293–301.

Pasteau, G. and Degrais, C. 1913. De l'emploi du radium dans le traitment des cancers de la prostate. *J Urol Med Chir* 4:341–66.

Peschel, R.E. and Colberg, J.W. 2003. Surgery, brachytherapy, and external-beam radiotherapy for early prostate cancer. *Lancet Oncol* 4:233–41.

Pistis, F., Guedea, F., Pera, J. et al. 2010. External beam radiotherapy plus high-dose-rate brachytherapy for treatment of locally advanced prostate cancer: The initial experience of the Catalan Institute of Oncology. *Brachytherapy* 9:15–22.

Polo, A. 2010. Image fusion techniques in permanent seed implantation. *J Contemp Brachyther* 3:98–106.

Polo, A., Cattani, F., Vavassori, A. et al. 2004. MR and CT image fusion for postimplant analysis in permanent prostate seed implants. *Int J Radiat Oncol Biol Phys* 60:1572–9.

Potters, L., Cao, Y., Calugaru, E., Torre, T., Fearn, P., and Wang, X.H. 2001. A comprehensive review of CT-based dosimetry parameters and biochemical control in patients treated with permanent prostate brachytherapy. *Int J Radiat Oncol Biol Phys* 50:605–14.

Potters, L., Calugaru, E., Jassal, A., and Presser, J. 2006. Is there a role for postimplant dosimetry after real-time dynamic permanent prostate brachytherapy? *Int J Radiat Oncol Biol Phys* 65:1014–9.

Prestidge, B.R., Prete, J.J., Buchholz, T.A. et al. 1998. A survey of current clinical practice of permanent prostate brachytherapy in the United States. *Int J Radiat Oncol Biol Phys* 40:461–5.

Raben, A., Chen, H., Grebler, A. et al. 2004. Prostate seed implantation using 3D-computer assisted intraoperative planning vs. a standard look-up nomogram: Improved target conformality with reduction in urethral and rectal wall dose. *Int J Radiat Oncol Biol Phys* 60:1631–8.

Ragde, H. and Korb, L. 2000. Brachytherapy for clinically localized prostate cancer. *Semin Surg Oncol* 18:45–51.

Ragde, H., Blasko, J.C., Grimm, P.D. et al. 1997. Interstitial iodine-125 radiation without adjuvant therapy in the treatment of clinically localized prostate carcinoma. *Cancer* 80:442–53.

Ragde, H., Korb, L.J., Elgamal, A.A., Grado, G.L., and Nadir, B.S. 2000a. Modern prostate brachytherapy. Prostate specific antigen results in 219 patients with up to 12 years of observed follow-up. *Cancer* 89:135–41.

Ragde, H., Grado, G.L., Nadir, B., Elgamal, A.A. 2000b. Modern prostate brachytherapy. *CA Cancer J Clin* 50:380–93.

Ragde, H., Grado, G.L., and Nadir, B.S. 2001. Brachytherapy for clinically localized prostate cancer: Thirteen-year disease-free survival of 769 consecutive prostate cancer patients treated with permanent implants alone. *Arch Esp Urol* 54:739–47.

Sahgal, A. and Roach, Mr. 2007. Permanent prostate seed brachytherapy: A current perspective on the evolution of the technique and its application. *Nat Clin Pract Urol* 4:658–70.

Salembier, C., Lavagnini, P., Nickers, P. et al. 2007. Tumour and target volumes in permanent prostate brachytherapy: A supplement to the ESTRO/EAU/EORTC recommendations on prostate brachytherapy. *Radiother Oncol* 83:3–10.

Sathya, J.R., Davis, I.R., Julian, J.A. et al. 2005. Randomized trial comparing iridium implant plus external-beam radiation therapy with external-beam radiation therapy alone in node-negative locally advanced cancer of the prostate. *J Clin Oncol* 23:1192–9.

Selley, S., Donovan, J., Faulkner, A., Coast, J., and Gillatt, D. 1997. Diagnosis, management and screening of early localised prostate cancer. *Health Technol Assess* 1:i, 1–96.

Shah, J.N., Wuu, C.S., Katz, A.E., Laguna, J.L., Benson, M.C., and Ennis, R.D. 2006. Improved biochemical control and clinical disease-free survival with intraoperative versus preoperative preplanning for transperineal interstitial permanent prostate brachytherapy. *Cancer J* 12:289–97.

Shanahan, T.G., Nanavati, P.J., Mueller, P.W., and Maxey, R.B. 2002. A comparison of permanent prostate brachytherapy techniques: Preplan vs. hybrid interactive planning with postimplant analysis. *Int J Radiat Oncol Biol Phys* 53:490–6.

Stock, R.G. and Stone, N.N. 1997. The effect of prognostic factors on therapeutic outcome following transperineal prostate brachytherapy. *Semin Surg Oncol* 13:454–60.

Stock, R.G., Stone, N.N., DeWyngaert, J.K., Lavagnini, P., and Unger, P.D. 1996. Prostate specific antigen findings and biopsy results following interactive ultrasound guided transperineal brachytherapy for early stage prostate carcinoma. *Cancer* 77:2386–92.

Stock, R.G., Stone, N.N., Tabert, A., Iannuzzi, C., and DeWyngaert, J.K. 1998. A dose-response study for I-125 prostate implants. *Int J Radiat Oncol Biol Phys* 41:101–8.

Stock, R.G., Stone, N.N., and Lo, Y.C. 2000. Intraoperative dosimetric representation of the real-time ultrasound-guided prostate implant. *Tech Urol* 6:95–8.

Stone, N.N. and Stock, R.G. 1999. Prostate brachytherapy: Treatment strategies. *J Urol* 162:421–6.

Storey, M.R., Pollack, A., Zagars, G., Smith, L., Antolak, J., and Rosen, I. 2000. Complications from radiotherapy dose escalation in prostate cancer: Preliminary results of a randomized trial. *Int J Radiat Oncol Biol Phys* 48:635–42.

Stromberg, J., Martinez, A., Benson, R. et al. 1994. Improved local control and survival for surgically staged patients with locally advanced prostate cancer treated with up-front low dose rate iridium-192 prostate implantation and external beam irradiation. *Int J Radiat Oncol Biol Phys* 28:67–75.

Talcott, J.A., Clark, J.A., Stark, P.C., and Mitchell, S.P. 2001. Long-term treatment related complications of brachytherapy for early prostate cancer: A survey of patients previously treated. *J Urol* 166:494–9.

Taussky, D., Austen, L., Toi, A. et al. 2005. Sequential evaluation of prostate edema after permanent seed prostate brachytherapy using CT-MRI fusion. *Int J Radiat Oncol Biol Phys* 62:974–80.

Terk, M.D., Stock, R.G., and Stone, N.N. 1998. Identification of patients at increased risk for prolonged urinary retention following radioactive seed implantation of the prostate. *J Urol* 160:1379–82.

Todor, D.A., Zaider, M., Cohen, G.N., Worman, M.F., and Zelefsky, M.J. 2003. Intraoperative dynamic dosimetry for prostate implants. *Phys Med Biol* 48:1153–71.

Vargas, C.E., Martinez, A.A., Boike, T.P. et al. 2006. High-dose irradiation for prostate cancer via a high-dose-rate brachytherapy boost: Results of a phase I to II study. *Int J Radiat Oncol Biol Phys* 66:416–23.

Vijverberg, P.L., Kurth, K.H., Blank, L.E., Dabhoiwala, N.F., de Reijke, T., and Koedooder, K. 1992. Treatment of localized prostatic carcinoma using the transrectal ultrasound guided transperineal implantation technique. *Eur Urol* 21:35–41.

Vis, A.N., Roemeling, S., Reedijk, A.M., Otto, S.J., and Schroder, F.H. 2008. Overall survival in the intervention arm of a randomized controlled screening trial for prostate cancer compared with a clinically diagnosed cohort. *Eur Urol* 53:91–8.

Wallner, K., Roy, J., and Harrison, L. 1995. Dosimetry guidelines to minimize urethral and rectal morbidity following transperineal I-125 prostate brachytherapy. *Int J Radiat Oncol Biol Phys* 32:465–71.

Wallner, K., Roy, J., and Harrison, L. 1996. Tumor control and morbidity following transperineal iodine 125 implantation for stage T1/T2 prostatic carcinoma. *J Clin Oncol* 14:449–53.

Wallner, K., Lee, H., Wasserman, S., and Dattoli, M. 1997. Low risk of urinary incontinence following prostate brachytherapy in patients with a prior transurethral prostate resection. *Int J Radiat Oncol Biol Phys* 37:565–9.

Wilkinson, D.A., Lee, E.J., Ciezki, J.P. et al. 2000. Dosimetric comparison of pre-planned and or-planned prostate seed brachytherapy. *Int J Radiat Oncol Biol Phys* 48:1241–4.

Wilt, T.J., MacDonald, R., Rutks, I., Shamliyan, T.A., Taylor, B.C., and Kane, R.L. 2008. Systematic review: Comparative effectiveness and harms of treatments for clinically localized prostate cancer. *Ann Intern Med* 148: 435–48.

Woolsey, J., Bissonette, E., Schneider, B.F., and Theodorescu, D. 2004. Prospective outcomes associated with migration from preoperative to intraoperative planned brachytherapy: A single center report. *J Urol* 172:2528–31.

Yoshioka, Y., Konishi, K., Sumida, I. et al. 2011. Monotherapeutic high-dose-rate brachytherapy for prostate cancer: Five-year results of an extreme hypofractionation regimen with 54 Gy in nine fractions. *Int J Radiat Oncol Biol Phys* 80:469–75.

Yu, Y., Anderson, L.L., Li, Z. et al. 1999. Permanent prostate seed implant brachytherapy: Report of the American Association of Physicists in Medicine Task Group No. 64. *Med Phys* 26:2054–76.

Zeitlin, S.I., Sherman, J., Raboy, A., Lederman, G., and Albert, P. 1998. High dose combination radiotherapy for the treatment of localized prostate cancer. *J Urol* 160:91–5; discussion 95–6.

Zelefsky, M.J. and Whitmore, W.F.J. 1997. Long-term results of retropubic permanent 125iodine implantation of the prostate for clinically localized prostatic cancer. *J Urol* 158:23–9; discussion 29–30.

Zelefsky, M.J., Wallner, K.E., Ling, C.C. et al. 1999a. Comparison of the 5-year outcome and morbidity of three-dimensional conformal radiotherapy versus transperineal permanent iodine-125 implantation for early-stage prostatic cancer. *J Clin Oncol* 17:517–22.

Zelefsky, M.J., McKee, A.B., Lee, H., and Leibel, S.A. 1999b. Efficacy of oral sildenafil in patients with erectile dysfunction after radiotherapy for carcinoma of the prostate. *Urology* 53:775–8.

Zelefsky, M.J., Yamada, Y., Cohen, G. et al. 2000a. Postimplantation dosimetric analysis of permanent transperineal prostate implantation: Improved dose distributions with an intraoperative computer-optimized conformal planning technique. *Int J Radiat Oncol Biol Phys* 48:601–8.

Zelefsky, M.J., Hollister, T., Raben, A., Matthews, S., and Wallner, K.E. 2000b. Five-year biochemical outcome and toxicity with transperineal CT-planned permanent I-125 prostate implantation for patients with localized prostate cancer. *Int J Radiat Oncol Biol Phys* 47:1261–6.

21

Brachytherapy for Gynecologic Cancers

Beth Erickson
Medical College of Wisconsin

Mahesh Kudrimoti
University of Kentucky College of Medicine

Christine Haie-Meder
Institut Gustave-Roussy

21.1 Incidence of Gynecologic Cancer

Gynecological cancers refer to cancers of the ovary, fallopian tubes, body of the uterus, cervix, vagina, and vulva. They comprise a heterogeneous group of cancers treated with differing strategies. Depending on the site of origin, brachytherapy (BT) has a diverse role in the management of these cancers. Cancer of the endometrium and cervix are the most commonly treated gynecological cancers worldwide. The incidence of cervical cancers is about 12,000/year in the United States, although the incidence is higher in the developing countries and parts of Asia and Africa (Jemal et al. 2008). It is a major problem accounting for almost 450,000 cases worldwide in 2002 and resulting in about 270,000 deaths. It is the third most common cancer in women worldwide. Nearly 80% of the cervical cancers are seen in the developing countries where it is the second most common cause of death in the female population. Increased screening and accessibility to health care are major reasons for the discordance in these rates. There is a strong association between the prevalence of human papilloma virus (HPV) infection and cervical cancer incidence. The other associations include smoking, use of oral contraceptives, multiplicity of sexual partners, age at first sexual intercourse, chronic immunosuppression, and history of sexually transmitted diseases. Adenocarcinoma of the endometrium

is the most common gynecologic cancer in the United States, accounting for an estimated 40,000 new patients in 2008 with nearly 7500 deaths (Ueda et al. 2008). Uterine sarcomas are relatively uncommon, representing less than 7% of uterine cancers.

Ovarian malignancies are a diverse group of tumors that can arise from the epithelial, stromal, or sex cord cells. Epithelial cancers are the most common ovarian cancers and account for 80% of the cases. These are highly fatal cancers with about 21,500 cases diagnosed in 2009 and about 14,000 deaths (Jemal et al. 2010). Over two-thirds of the cases are advanced at presentation in contrast to the endometrial cases. Other cancers, such as vulva and vaginal cancers, are seen in the elderly population. Vulva cancers are rare tumors, with an estimated 3500 patients per year in the United States (Jemal et al. 2010). They account for 3%–4% of gynecologic malignancies. There is a higher incidence in the elderly population (>70 years). Two-thirds of the cases arise from the labia majora/minora, and up to a fifth are locally advanced with involved inguinal lymph nodes at diagnosis. These may be associated with synchronous or metachronous gynecological tumors.

The most common vaginal cancers are metastatic with direct extension from the cervix above or the vulva below, or from lymphatic or hematogenous spread of nonvaginal primaries. According to International Federation of Gynecology and

Obstetrics (FIGO) staging, if the tumor involves the vulva or the cervical os, it is classified as arising from that structure, even if it is centered in the vagina. Only 10%–20% are primary vaginal tumors and account for ~2% of gynecologic malignancies (Jemal et al. 2010). Most of these tumors (60%) occur on the posterior wall, and 51% occur in the upper one-third of the vagina. Approximately 60% have had prior hysterectomy for a variety of reasons.

21.2 Strategies for Management

21.2.1 Cervical Cancers

Most cervical cancers are treated with radiation therapy. Young patients with Stages IA and IB1 are treated with surgery that usually involves a radical hysterectomy and a pelvic lymph node dissection. There is no difference in outcomes between radiation and surgery in these patients, but surgery has some quality-of-life advantages such as ovarian preservation in select cases, shorter treatment time, and better sexual satisfaction scores (Landoni et al. 1997). For patients with IB2 and higher, the recommended treatment is chemoradiation using either cis-platinum or 5FU combined with cis-platinum during the course of their external beam irradiation (Sherman et al. 2005). BT is a very important component of treatment in these stages of cervical cancers and will be discussed extensively in the following sections.

21.2.2 Endometrial Cancers

There is a general consensus that endometrial cancer is a surgical disease as almost 70% of these cancers present in early stages. The disease tends to follow a more indolent course compared to other gynecologic malignancies. Hysterectomy is often curative and helps select out patients needing additional therapy. The recommendations for adjuvant therapy are based on the grade of differentiation and histologic subtype; tumor size and location; the extent of myometrial invasion; cervical, parametrial, or adnexal spread; and extrauterine involvement (Creutzberg et al. 2000). A small percentage of these patients are inoperable secondary to medical comorbidities such as chronic obstructive pulmonary disease (COPD), congestive heart failure (CHF), obesity, uncontrolled diabetes, and others, which make these high-surgical-risk patients. These cases are best treated with either a combination of external beam irradiation and BT or BT alone, dependent on both clinical and social issues (Fishman et al. 1996). For those cases presenting in a more advanced stage, with cervical, parametrial, or nodal involvement, the option of external beam irradiation with or without BT, properly sequenced with surgical resection and chemotherapy, should be strongly considered.

21.2.3 Vaginal Cancers

For vaginal cancers, treatment is based on the extent of disease and involvement of adjacent structures and the performance status of the patient. Treatment recommendations include external

beam irradiation with or without chemotherapy and BT [either intracavitary or interstitial implantation (Frank et al. 2005; de Crevoisier et al. 2007)]. Some early cases with disease limited to the upper one-third of the vagina may be treated with surgical excision/resection with or without BT. In general, it is recommended to treat with definitive external beam irradiation and BT for most tumors. Particular attention should be paid to the extent of vaginal involvement as it affects both the external radiation fields, techniques and doses, and also the type of BT boost planned.

21.2.4 Vulva Cancers

Vulva cancers are usually treated with wide local resection followed by consideration of adjuvant radiation or chemoradiation based on operative findings and surgical pathology reports (Thomas et al. 1991). The extent of groin dissection is based on the characteristics of the primary tumor, location, and the proximity to the midline structures. Those tumors that present with involvement of the urethra, vagina, bladder, or the anal verge are considered to have the potential for significant morbidity if initial surgery is planned due to the extensive resection needed. Consideration for initial chemoradiation with assessment of response may enable less deforming surgery. In cases where there is no significant reduction of the tumor, additional external beam irradiation or BT may be indicated.

21.2.5 Other Gynecological Cancers

The other gynecologic tumors, including ovarian and fallopian tube, are mainly treated with surgical resection followed by chemotherapy (Chan et al. 2006; Fader and Rose 2007). In select cases, whole abdominal irradiation has been used historically as an alternative to chemotherapy (Du Bois et al. 2005). Most ovarian tumors respond to adequate doses of radiation, but it is difficult to deliver tumoricidal doses of radiation for gross residual disease in the upper abdomen due to the surrounding dose-limiting normal organs. The trials using whole abdominal radiation could not show a benefit over chemotherapy due to limitations of doses to the liver, kidneys, and bowels, so it has fallen out of favor as an initial therapeutic option. However, in our experience, ovarian cancers with relapse at the vaginal cuff can be treated with BT, analogous to recurrent endometrial cancers, with effective local control. Selective irradiation of nodal recurrences following chemotherapy can also be helpful in some patients.

21.3 BT Techniques

BT can be given using low-dose-rate (LDR), high-dose-rate (HDR), or pulsed-dose-rate (PDR) techniques. LDR is defined as a dose of 0.4–2 Gy per hour and HDR as a dose of >12 Gy per hour (ICRU 38 1985). For endometrial cancer, in the 2002 American Brachytherapy Society (ABS) survey, 69% of respondents used postoperative HDR vaginal BT for endometrial cancer, and the remaining, LDR (Small et al. 2005). In the most recent Quality Research in Radiation Oncology (QRRO) study, 55% of patients treated with definitive irradiation for cervical cancer between 2005 and 2007 were treated with HDR and 44% with LDR/PDR (Eifel et al. 2010).

21.3.1 Applicators

21.3.1.1 Intracavitary HDR Vaginal BT

Applicator selection is important in BT to customize the therapy to the individual case. As there is a significant variation in clinical presentation, anatomy, and patient comfort, a single applicator cannot treat all clinical scenarios. Practice pattern surveys by the ABS indicate use of a variety of applicators and dose prescriptions depending on the preference of the radiation oncologist and the institutional approach (Small et al. 2005). Commercial vendors have a range of applicators. Cylinders and ovoids are most commonly used for BT boosts to the vaginal cuff. Vaginal ovoids treat only the vaginal cuff, whereas vaginal cylinders can treat the entire vaginal canal to the introitus. The anatomical configuration of the vaginal cuff will determine applicator selection. A cylindrical vagina can be treated adequately by a vaginal cylinder. A range of diameters from 1.5 to 4.0 cm or more are available. In a minority of the patients with an irregular apex or deep lateral fornices, ovoids may be advantageous due to better proximity of the applicator to the vaginal mucosa and more homogeneous dose delivery. The ABS recommends vaginal cylinders in various lengths and diameters (ranging from 1.5 to 4 cm) to treat a variety of patients (Nag et al. 2000b). Cylinders of predetermined lengths, or segmented cylinders that can be assembled to the required length, should be used as clinically indicated. The ABS guidelines indicate that when only the upper half of the vagina is treated, the distal cylinder edge should stop short of the introitus as it may be more comfortable for the patient and avoid mucosal reactions. Condoms (check for latex allergy prior to placement) can be wrapped over the cylinders when they are inserted into the vagina for applicator hygiene. Most vaginal cylinders have a single, central channel and resultantly give a significant dose to the bladder and rectum as displacement packing cannot be used (Figure 21.1a). There are multichannel applicators available that facilitate better optimization of the dose at the apex of the vault or reduction of the dose to the bladder and/or rectum (Figure 21.1b). The commercially available multichannel applicator contains a central channel and six peripheral channels along the surface of the cylinder (Nag et al. 2000b). The cylinders are available with an initial diameter of 3.0 cm. The kit comes with additional sleeves, which can increase the diameter to 3.5 and 4.0 cm for better depth dose characteristics while decreasing the mucosal surface dose. Shielded cylinders are also available, and these selectively decrease absorbed dose to adjacent normal structures.

Vaginal ovoids are another way to treat the vaginal cuff. Fletcher-like shielded and unshielded vaginal ovoids in sizes of 2.0-, 2.5-, and 3.0-cm diameter are available (Figure 21.1c). For

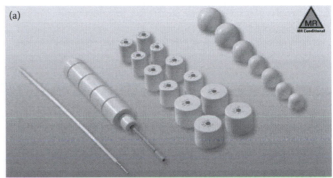

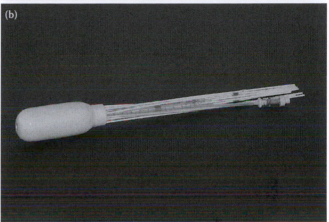

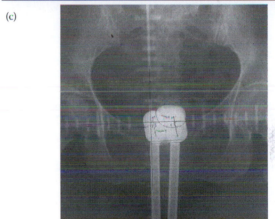

FIGURE 21.1 Vaginal BT applicators. (a) Domed vaginal cylinders. (Courtesy of Varian.) (b) Multichannel Capri vaginal applicator. (Courtesy of Varian.) (c) Radiograph of LDR vaginal ovoids. (Courtesy of B. Erickson.)

irradiation of narrow vaginas, Henschke-like hemispherical colpostats, 2.5 and 3.0 cm in diameter, with or without tungsten shielding, are also available. Shielded and unshielded retractors that can displace the rectum and bladder are available, but displacement can also be achieved with simple vaginal packing (Perez et al. 1985).

Another type of vaginal applicator is the intravaginal balloon. These applicators are similar to the MammoSite applicators, and their use has been modified for gynecological cancers. The theoretical advantage of this method is the conformation of the vaginal balloon to the shape of the vagina and avoidance of any potential drop in coverage because of the attenuation of the source along the long axis of the channel. Initial experience reports adequate coverage of the vault, acceptable toxicity, and no recurrences with a short follow-up period (Miller et al. 2010).

This is similar to the use of vaginal molds that customize the radiation according to individual patient anatomy (Magné et al. 2010). Electronic BT consists of a modified HDR BT technique using a kilovolt source, rather than ^{192}Ir. The purported advantage is that less shielding is required. Long-term results are not available with this mode of therapy, but initial experiences have not reported any grade III or IV reactions of the vaginal mucosa or pelvic organs (Dicker et al. 2010).

21.3.1.2 Intracavitary Cervical BT

21.3.1.2.1 *Importance of BT in Cervical Cancer*

When curative treatment is planned, patients with cervical carcinoma treated with definitive irradiation should receive a combination of external beam irradiation and BT. As revealed in the Patterns of Care Studies and several major retrospective series, recurrences and complications are decreased when BT is used in addition to external beam (Logsdon and Eifel 1999; Perez et al. 1983; Coia et al. 1990; Lanciano et al. 1991b). Use of an intracavitary implant was the single most important treatment factor in multivariate analysis for stage IIIB cervix cancer with respect to survival and pelvic control in the 1973 and 1978 PCS studies (Lanciano et al. 1991a). Retrospective series with external beam alone have proven marginal outcomes with this approach. The efficacy of BT is attributable to the ability of radioactive implants to deliver a higher concentrated radiation dose more precisely to tissues than external beam alone, which contributes to improved local control and survival. At the same time, surrounding healthy tissues such as the bladder and rectosigmoid are relatively spared due to the rapid falloff of dose around the applicators with distance. The external beam brings about tumor regression in intact cervical and vaginal cancers such that the residual tissue is brought within the range of the pear-shaped or cylindrical-shaped radiation dose distribution around standard applicators when there is optimal disease regression (Erickson and Wilson 1997).

21.3.1.2.2 *Applicators for BT HDR/PDR Tandem and Ring*

The applicators used to deliver PDR BT are identical to those used for HDR. The ring applicator is an adaptation of the Stockholm technique (Erickson et al. 2000). The commercially available tandem and ring applicators are available in metal and MR/CT compatible styles (Figure 21.2a). The plastic caps, which come with the non-MR/CT-compatible metal ring applicator, place the vaginal mucosa 0.5 cm from the source path, compared to the caps for the ovoids, which distance the vaginal mucosa from the source path by 1–1.5 cm. The short distance from the ring to vaginal mucosa can result in very high surface doses if fixed-weighting nonoptimized techniques are used (Noyes et al. 1995; Erickson et al. 2000). The bladder and rectum may also

(a)

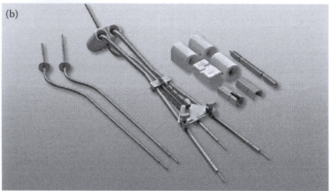

(b)

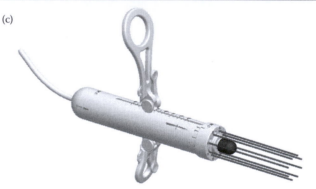

(c)

FIGURE 21.2 Intracavitary cervical applicators. (a) Tandem and ring. (Courtesy of Nucletron.) (b) Tandem and ovoids. (Courtesy of Varian.) (c) Tandem and multichannel cylinder. (Courtesy of Nucletron.)

receive higher doses with fixed-weighting nonoptimized dosimetry (Noyes et al. 1995). The tandems are available in lengths of 2–8 cm. Four ring tandem angles are available including 30°, 45°, 60°, and 90°. The 30° ring tandem angle applicator can bring the tandem very close to the bladder and is usually not advised. The shape of the isodose curves comparing the ring with tandem and ovoids will also have a different shape, and the volume of tissue irradiated will also differ (Erickson et al. 2000). The ring applicator is ideal for patients without deep lateral vaginal fornices. Its ease of insertion and predictable geometry make it a popular alternative to tandem and ovoids. Cervical seeds are used to confirm that the ring is up against the cervix on post-insertion imaging. Proper interlocking of the ring and tandem is necessary to have the tandem centered in the ring with its intended position.

21.3.1.2.3 LDR/HDR/PDR Tandem and Ovoids

21.3.1.2.3.1 Low Dose Rate
There are varieties of LDR applicators available for intracavitary BT. The best known are the Fletcher–Suit and Henschke tandem and ovoid (colpostat) applicators. The Fletcher applicator was modified in the 1960s for afterloading (Fletcher–Suit applicator) (Delclos et al. 1978; Haas et al. 1985; Perez et al. 1985) and in the 1970s to accommodate ^{137}Cs sources (Haas et al. 1985). In the 1970s, the Delclos mini-ovoid was developed for use in narrow vaginal vaults (Delclos et al. 1978, 1980; Haas et al. 1983, 1985). The ovoids are 2.0, 2.5, and 3.0 cm in diameter with and without shielding. The mini-colpostats have a diameter of 1.6 cm and a flat inner surface. The mini-ovoids do not have shielding added inside the colpostat, and this together with their smaller diameter produces a higher surface dose than regular ovoids with resultant higher doses to the rectum and bladder. Appropriate source strength and treatment duration adjustment are important considerations to prevent complications. Fletcher tandems are available in four curvatures, with the greatest curvature used in cavities measuring >6 cm and lesser curvatures used for smaller cavities. A flange with keel is added to the tandem once the uterine canal is sounded, which approximates the exocervix and defines the length of source train needed. The keel prevents rotation of the tandem after packing. The distal end of the tandem near the cap is marked so that rotation of the tandem after insertion can be assessed.

The Henschke tandem and ovoid applicator was initially unshielded (Perez et al. 1985; Henschke 1960) but later modified with rectal and bladder shielding (Hilaris et al. 1988; Mohan et al. 1985). It consists of hemispheroidal ovoids with the ovoids and tandem fixed together. Sources in the ovoids are parallel to the sources in the uterine tandem (Hilaris et al. 1988). The Henschke applicator may be easier to insert into shallow vaginal fornices in comparison to Fletcher ovoids.

21.3.1.2.3.2 HDR/PDR
The tandems and ovoids used with HDR and PDR are variations of the traditional Fletcher and Henschke applicators but are lighter, narrower, and smaller (Figure 21.2b). The ovoids are 2.0, 2.5, and 3.0 cm in diameter with and without shielding. The relationship of the colpostat to the handle is different between HDR/PDR and LDR colpostats so that the cable-driven HDR/PDR source can negotiate the angle between the handle and the colpostat. Standard Fletcher–Suit LDR colpostats are angled most often at 15° and sometimes at 30° with respect to the colpostat handles. This can lead to a different relationship between the tandem and the colpostats and between the colpostats and the cervix.

21.3.1.2.4 LDR and HDR/PDR Tandem and Cylinder

21.3.1.2.4.1 LDR and HDR/PDR
The Fletcher–Suit–Delclos tandem and cylinder applicator was designed to accommodate narrow vaginas and to treat varying lengths of the vagina when mandated by vaginal spread of disease. The LDR cylinders vary in size from 2.0 to 5.0 cm to accommodate varying vaginal sizes and the HDR/PDR from 2.0 to 4.0 cm (Delclos et al. 1980; Haas et al. 1983). MR- and CT-compatible versions of these applicators are also available to allow for image-based BT. Shielded and multichannel HDR/PDR tandem and cylinder applicators are also used at some institutions (Figure 21.2c).

A narrow vagina poses a therapeutic challenge. Use of vaginal cylinders may lead to a higher rate of local failure as the dose to the lateral cervix and pelvic sidewall is reduced in the absence of ovoids, which produce the optimum pear-shaped distribution. These patients also tend to receive lower total doses due to the proximity of the rectum and bladder. Packing cannot be used with cylinders to decrease the rectal and bladder doses (Crook et al. 1987; Cunningham et al. 1981). Additionally, vaginal cylinders increase the length of vagina and rectum and bladder treated, with an associated increase in complications (Esche et al. 1987a,b). Vaginal fistulae, rectal ulcers, and strictures are reported with increased frequency when vaginal cylinders are used. Alternatively, interstitial implantation should also be a consideration for patients with a narrow vagina or with distal vaginal disease.

21.3.1.2.4.2 Moulage LDR/HDR/PDR
At the Institut Gustave-Roussy in Villejuif, France, there has been a long tradition of use of a personalized applicator adapted to each patient, fabricated from individual vaginal impressions (see Figure 21.3; Albano et al. 2008; Magné et al. 2010). This mold technique was used previously for LDR BT and is used currently for PDR and HDR BT. The first step is the vaginal impression, with injection of a liquid paste into the vagina after the placement of two strips of gauze into each lateral vaginal fornix. After solidification, the impression is extracted from the vagina. This procedure does not require any anesthesia, apart from when performed in children. The tumor topography, the size of the exocervical tumor, and the extension within the vagina are perfectly identified on the vaginal impression. The second step consists of acrylic molded applicator fabrication, performed by technicians. The vaginal impression is plunged into liquid plaster. When dry, the plaster is cut into two pieces, the vaginal impression is removed, and after reconstitution of the plaster, a synthetic autopolymerized resin is poured in a thin layer. When this resin (Palapress) is dry, the two plaster pieces are removed, and the applicator is ready. The position

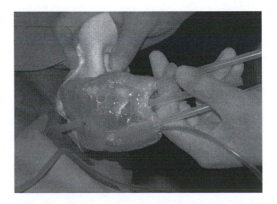

FIGURE 21.3 Customized moulage applicator with plastic guide tubes for tandem and vaginal sources. (Courtesy of C. Haie-Meder.)

of the vaginal catheters, which is determined by the radiation oncologist, is drawn according to the tumor limits. Vaginal catheters are basically located on each side of the cervical limits for cervical cancers. Their length depends on the tumor size and the vaginal extension. In postoperative prophylactic vaginal BT for endometrial cancer, three vaginal catheters are located within the mold: two anteriorly and one posteriorly. These catheters are fixed in place in the applicator. One hole is made at the level of the cervical os for the uterine tube insertion, and different holes are made in the applicator, allowing daily vaginal irrigation and vaginal mucosal herniation, which prevent mold displacement. This technique is also of interest in vaginal cancers, as the mold is perfectly adapted to the anatomy. The source position is adapted to the tumor topography. With this technique, no packing is necessary, as the mold by itself expands the vaginal walls.

21.3.1.2.4.3 Applicator Geometry It is important that applicators be inserted with care and precision. Geometrically favorable intracavitary implants improve outcome over suboptimal implants. In the 1978 and 1983 Patterns of Care study reported by Corn et al. (1994), a technically good implant correlated significantly with improved local control, with a trend toward improved survival. This has been confirmed in several large retrospective reviews, both with respect to local control and complications (Perez et al. 1983, 1984; Katz and Eifel 2000; Viswanathan et al. 2009b). The use of ultrasound to guide tandem insertion can be extremely helpful in negotiating a narrowed or obliterated endocervical canal and preventing perforation.

21.3.1.3 Interstitial Implantation

The limitations of intracavitary techniques contrast with the strengths of interstitial implantation. Interstitial implantation is helpful in patients with bulky disease, anatomical distortion such as narrow fornices, or an obliterated endocervical canal, vaginal spread of disease, or recurrent disease. Tumor volume and patient anatomy are key in this decision. One must select the appropriate applicator to "match" the disease and shape of the associated isodose distribution to encompass the disease (Erickson and Gillin 1997; Viswanathan et al. 2010).

The development of prefabricated perineal templates, through which stainless-steel needles were inserted and afterloaded, was pivotal in advancing interstitial techniques for the treatment of cervical and vaginal cancers. The template concept allows for a predictable distribution of needles inserted across the entire perineum through a perforated template according to an optimum pattern. Commercially available and institution-specific templates can be used to accommodate varying disease presentations. Stainless steel and plastic needles are used that are afterloaded with low- or high-activity ^{192}Ir sources. The Martinez Universal Perineal Interstitial Template (MUPIT, Beaumont Hospital, Royal Oak, MI) accommodates implantation of multiple pelvic-perineal malignancies and is available for both LDR and HDR applications (Martinez et al. 1984). The Syed–Neblett (Best Industries, Springfield, VA) is the other well-known template system that is commercially available (Syed et al. 1986).

Currently, there are three LDR Syed–Neblett templates of varying size and shape for use in implantation of gynecologic malignancies (Best Industries; GYN 1-36 needles; GYN 2-44 needles; GYN 3-53 needles). There are also templates that accommodate HDR needles. Free-hand interstitial implantation is also used selectively for small-volume vaginal and parametrial disease and is especially helpful in treating periurethral disease (Frank et al. 2005). The Vienna applicator is a modified ring applicator with needles and is used for smaller-volume disease than template-based approaches (see Figure 21.4a; Dimopoulos et al. 2006a; Kirisits et al. 2006). It is particularly suited to 3D image-based BT, as the tumor residuum can be accurately assessed with magnetic resonance imaging (MRI) and lateral dose provided to those parts of the tumor not covered by the intracavitary pear with the addition of the needles. There are also modified ovoid applicators (Utrecht applicator) that accommodate needles, enabling better coverage of disease in the parametrium with use of these interstitial techniques (see Figure 21.4b; Jurgenliemk-Schulz et al. 2009). These applicators can be used to improve lateral and superior coverage of the target volume by approximately 10 mm. Interstitial PDR implants should be performed according to the guidelines of a traditional LDR-BT system (i.e., Paris, Manchester, Quimby). The majority of published interstitial PDR-BT results to date have been achieved by investigators following the Paris system. Dose optimization is often beneficial but cannot substitute for poor-quality catheter placement.

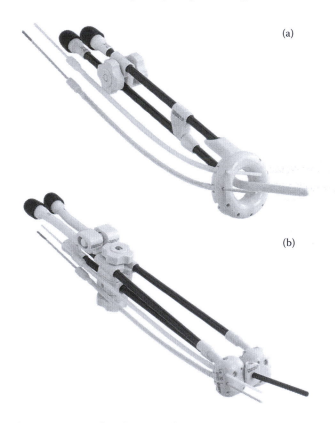

(a)

(b)

FIGURE 21.4 Combined interstitial/intracavitary CT/MR-compatible applicators including (a) Vienna ring and plastic needles and (b) Utrecht ovoids and plastic needles. (Courtesy of Nucletron.)

The Syed–Neblett system and MUPIT are particularly suited for treatment of vaginal disease as it is a combined intracavitary and interstitial system. The vaginal obturators that accompany the template are used to treat the vaginal surface, and the vaginal obturator needles can be strategically loaded to encompass disease from the fornices to the introitus. Along with the intracavitary uterine tandem, the obturator needles can also be advanced directly into the cervix as an interstitial application and may be essential in delivering tumoricidal radiation doses to the cervix by preventing a central "cold spot," especially in those circumstances when an intrauterine tandem cannot be inserted. Whenever possible, it is important to use a tandem along with the needles when there is an intact uterus. The tandem may extend dose superiorly throughout the uterine cavity, provide additional dose to the parametria, and increase the dose centrally in the implant where it is most needed (Viswanathan et al. 2009a). The more peripheral needles are used for implantation of the parametria, which is often underdosed in intracavitary approaches. Modifications of these standard templates have evolved and other innovative templates developed for vulvar, vaginal, and cervical carcinomas (Erickson and Gillin 1997). Interstitial PDR implants should be performed according to the guidelines of a traditional LDR-BT system (i.e., Paris, Manchester, Quimby). The majority of published interstitial PDR-BT results to date have been achieved by investigators following the Paris system. Dose optimization is often beneficial but cannot substitute for poor-quality catheter placement.

21.4 Treatment Planning: Dose Specification, Dosimetry Generation; Dose and Dose Fractionation Schemes

21.4.1 HDR Intracavitary Vaginal BT

There is controversy as to whether imaging of vaginal cylinders prior to treatment delivery is necessary. Some institutions use a catalog of plans based on applicator diameter and assume that the dose at 0.5 cm from the applicator is representative of the dose to the bladder and rectum. Imaging can, however, provide valuable information. It may be especially important when vaginal BT is combined with external beam irradiation as the dose limits to the organ at risk (OAR) can be exceeded in this setting. Simulation enables verification and documentation of the size of the applicator, applicator placement, and the relationship of the applicator and associated dose to the OAR (Holloway and Viswanathan 2010). Imaging has traditionally been with 2D film-based techniques with contrast in the rectosigmoid and bladder (Small et al. 2005; Holloway and Viswanathan 2010). If clips are placed at the vaginal apex, the proximity of the tip of the applicator to the top of the vagina can be accessed on the simulation films. Additionally, the relationship of the applicator to the bony landmarks can be assessed and used for confirmation films in subsequent fractions. The lateral films can be used to assess the angle of the applicator relative to the rectum and adjusted to a more horizontal position if needed to decrease dose to the rectum. Point doses for the bladder and rectosigmoid can be assessed and documented rather than a point dose estimate at 5 mm from the applicator. 3D imaging can provide more information and is increasingly performed due to the prevalence of CT simulators in most radiation oncology departments; (see Figure 21.5; Holloway and Viswanathan 2010). CT imaging allows assessment of the placement of the applicator, including the relationship to the vaginal apex, the conformity of the vaginal surface to the applicator, the depth of insertion, and the angulation of the cylinder and the relationship to the OAR. Air gaps are easily identified at the apex of the applicator or along its length and may lead to replacing the applicator with a larger diameter if feasible. Adjustment of the treatment length can be done to spare the vaginal introitus, and the dose-volume relationships of the OAR can be assessed at the time of the first fraction. This does not need to be done for every fraction, and subsequent 2D films can be acquired for the remaining.

21.4.1.1 Dose Specification

Typically, for HDR intracavitary vaginal BT, it is recommended that the proximal 3–5 cm of the vagina be treated. This was the

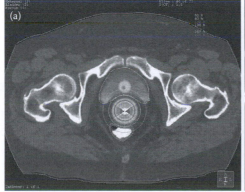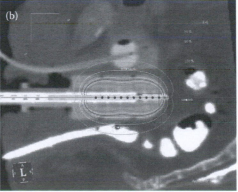

FIGURE 21.5 **(See color insert.)** Axial (a) and sagittal (b) CT images with domed vaginal applicator in good contact with the vaginal walls. Opacification of the bladder and rectum is shown for critical organ dose assessment. (Courtesy of B. Erickson.)

most common practice pattern noted in a survey of radiation oncologists in the United States (Small et al. 2005). The dose distribution should be optimized to deliver the prescribed dose either at the vaginal surface or at 0.5-cm depth, depending on the institutional policy. Doses can be specified in LDR or in an HDR/PDR equivalent. There has been a practice migration from using LDR vaginal BT to HDR, which affords better patient compliance and optimization. PDR BT allows the same optimization process as in HDR BT. The HDR dose fractionation scheme is dependent on the dose specification point and whether external beam radiation therapy (EBRT) is given. The dose specification point has to be clearly documented and used for reporting data. The usual EBRT dose (when given) is 45 Gy to the whole pelvis. Alternatively, 50.4 Gy has also been used in several trials. Suggested doses of HDR alone for vaginal cylinders are within a range of practice and institutional philosophy (Small et al. 2005). The Gynecologic Oncology Group (GOG) trials currently recommend using 7 Gy X 3 to a 5-mm depth, treating the upper one-third to two-thirds of the vagina, when no external irradiation is planned. Other fractionation schemes used include 5.5 Gy X 3 or 4.7 Gy X 4 specified at a depth of 5 mm from the surface of the applicator. When the dose specification is at the applicator surface, then 16.2 Gy X 2, 10.5 Gy X 3, 8.8 Gy X 4, or 7.5 Gy X 5 as well as other dose fractionation schemes have been used. The dose fractionation schemes used in a number of clinical trials and guideline papers are listed in Table 21.1. To allow for comparisons between HDR and LDR doses, the fraction size for HDR has to be calculated, using the linear–quadratic model, to give the EQD2 equivalent dose for tumor effects (Orton 1991b). A dose-modifying factor (CMF) of 0.7 was used by the ABS for recommendations, but any range between 0.55 and 0.7 seems to be clinically acceptable based on the worldwide experience (Nag et al. 2000a).

Optimization is needed to ensure adequate and uniform dose delivery. If vaginal cylinders are used, placing optimization points only on the lateral surface of the vagina without having optimization points at the apex can produce unacceptably high doses at the vaginal apex and possibly the overlying sigmoid or small bowel. The ABS recommends placing optimization points at the apex and along the curved portion of the cylinder dome,

in addition to the lateral surfaces, to make the dose more uniform across the straight and domed portions of the applicator. In evaluating the dose distribution, the source anisotropy can produce a lower dose at the vaginal apex, and isotropic dose calculations can result in as much as a 30% underdosing at the vaginal apex (Nag et al. 2000b). However, its clinical significance is not known. Multichannel cylinders tend to minimize the effect of anisotropy.

For vaginal ovoids, optimization points are placed at the lateral surface of the ovoids or at 0.5-cm depth into vaginal mucosa, according to the institutional dose specification policy. If the ovoids are separated, additional optimization points must be placed at the apex, midway between the ovoids, to avoid a cold region between them.

Higher doses of vaginal BT are needed in the setting of recurrent endometrial cancer or primary vaginal cancers. An LDR dose of at least 80 Gy was correlated with improved outcome in women treated for isolated vaginal recurrences with radiation alone in the series of Jhingran et al. (2003); likewise, higher doses of vaginal BT were correlated with improved outcome in the definitive management of vaginal cancer in the series of Frank et al. (2005).

21.4.2 Intracavitary Cervical BT

21.4.2.1 Dose Specification for Cervical Cancer: LDR

The Manchester system was developed in 1932 by Tod and Meredith (1938) and later modified in 1953 at the Holt Radium Institute (Tod and Meredith 1953). It standardized treatment with predetermined doses and dose rates directed to fixed points in the pelvis. The paracervical triangle was described as a pyramidal-shaped area with its base resting on the lateral vaginal fornices and its apex curving around with the anteverted uterus. "Point A" was defined as 2 cm lateral to the central canal of the uterus and 2 cm from the mucous membrane of the lateral fornix in the axis of the uterus. The definition of point A was modified in 1953 to be "2 cm up from the lower end of the intrauterine source and 2 cm laterally in the plane of the uterus, as the

TABLE 21.1 Dose Fractionation Schemes Used in Clinical Trials and Guideline Papers for Postoperative BT for Endometrial Cancer

Trial	External Pelvic RT Dose	Fractionation	Dose Rate	Depth of Prescription	Treated Length
PORTEC2	None	30 Gy X 1	LDR	5-mm depth	Half to upper third
PORTEC2	None	7 Gy X 3	HDR	5-mm depth	As above
RTOG 99-05	50.4 Gy	20 Gy X 1	LDR	Surface of applicator	As above
RTOG 99-05	50.4 Gy	6 Gy X 2	HDR	Surface of applicator	As above
RTOG 97-08	45 Gy	20 Gy X 1	LDR	Surface of applicator	As above
RTOG 97-08	45 Gy	6 Gy X 3	HDR	Surface of applicator	As above
GOG 184	50.4 Gy	7 Gy X 1	HDR	5 mm	As above
GOG 184	50.4 Gy	10 Gy X 1	LDR	Surface of applicator	As above
American Brachytherapy Society	45 Gy	5.5 Gy X 2	HDR	5-mm depth	As above
	45 Gy	4 Gy X 3	HDR	5-mm depth	As above
	45 Gy	8 Gy X 2	HDR	Surface of applicator	As above
	45 Gy	6 Gy X 3	HDR	Surface of applicator	As above

external os was assumed to be at the level of the vaginal fornices" (Tod and Meredith 1953).

Though dose specification definitions have varied from one institution to another, most have traditionally attempted to quantify doses in the paracervical region (point A) and at either point B or the pelvic wall (C or E), as well as the rectum and bladder (Potish and Gerbi 1986; Cunningham et al. 1981; Nag et al. 2002). Point dose calculations have not been consistently recorded for the sigmoid, vaginal mucosa, or cervix, but dose evaluation at these points may also be helpful. Maruyama et al. (1976) have defined a point T (tumor dose) located 1 cm above the cervical marker and 1 cm lateral to the tandem, which is usually two to three times the dose at point A. Vaginal surface dose rates, defined at the lateral ovoid surface, will vary based on the applicator diameter and available source strengths and should be in the range of 1.4–2.0 × the point A dose (Katz and Eifel 2000; Decker et al. 2001).

The advent of afterloading applicators and computerized dosimetry has brought about progressive change in dosimetric practice. The basic principle of LDR intracavitary prescription was to use a specific loading of sources for a defined time, determined by empirical experience, prescription rules, and computerized dosimetry used to generate point doses and isodose distributions (Batley and Constable 1967, Krishnan et al. 1990). A negotiation between the dose at point A and the normal tissue point doses on a case-by-case basis finally determines implant duration. Further evolution of dose specification using image-based techniques and volume-based dose specification has taken this process further. Now CT or MR simulation is possible with the advent of MR- and CT-compatible applicators as well as the computer software to use these images for treatment planning. Direct visualization of the tumor volume as well as the OAR is possible and allows for volume-based dose specification rather than point dose specification. This has been especially prevalent when using PDR and HDR treatment planning systems and will be discussed in depth later in this chapter.

Though frequently confused, the ICRU 38 system (Dose and Volume Specification for Reporting Intracavitary Therapy in Gynecology) is a dose reporting system, not a dose specification system (ICRU 38 1985). This was developed so that comparisons could be made between centers using different BT systems. It does provide definitions for determining dose to the bladder and rectum in addition to other characteristics of the implant, which can be used for reporting and is described in detail in Chapter 24.

21.4.2.2 Dose Specification for Cervical Cancer: HDR/PDR

Previously, most HDR/PDR regimens use a paracervical dose specification point (A) and other point doses just as in LDR (Nag et al. 2002). Rectal, bladder, sigmoid, and vaginal surface doses should always be specified or documented, and some assessment of dose to the pelvic lymph nodes and lateral parametria may also be of interest. With HDR and PDR, dose/fractionation schemes are required, and dose is specified at various points or volumes. Calculations using the linear–quadratic equation are used to convert LDR doses into the appropriate HDR or PDR dose/fractionation schemes (Orton 1991b). If 3D image-based dose specification is used, then use of the HR clinical target volume (CTV) as defined in the GEC ESTRO guidelines, in addition to or in place of point A, is an emerging trend. This allows for manipulation of the dose distribution according to the volume of the remaining disease and the proximity of the OAR. This will be described in Section 21.6 (Haie-Meder et al. 2005; Pötter et al. 2006).

The HDR system utilizes special vocabulary to describe certain functions and applications. Again, this vocabulary is applicable to PDR BT as well. A "dwell position" is a position at which the source is driven to stop or dwell. Dwell positions can be 2.5 and 5 mm apart. The active length is converted into a number of dwell positions. "Patient points" are points of interest at which the dose is calculated and are defined on the imaging studies. Examples include bladder, rectum, and sigmoid. "Applicator points" are points of interest at which the dose is calculated and are defined manually, inputting the coordinates. Typically, applicator points include point A and points on the surface of a ring, ovoid, or cylinder. "Dose points" are points at which the dose is optimized. In general, doses are specified or tracked at many of the same points used in LDR. For example, point T, which is 1 cm above the cervical os and 1 cm lateral to the tandem, defines dose within the cervix and should be 2–2.5 times the point A dose (Maruyama et al. 1976). It is used to detect underdosing or overdosing of the cervix when using optimization. At the Medical College of Wisconsin, we also place dose points laterally from four of the five ovoid dwell positions, which are assigned a weight of 140%–160% of the point A dose. This gives a dose distribution similar to a traditional LDR tandem and ovoids (Decker et al. 2001). To avoid underdosing endometrial extension of tumor, treating at least 4 cm above the exocervix so that point A is not in a region of dose constriction may be wise. Tandem lengths of 6–7 cm are typical. If there is endometrial extension, a longer tandem may be needed.

For the tandem and ring, a similar system is used, but dose points are placed on each side of the ring surface at a distance of 6 mm from each dwell position, at the surface of the plastic cap, and assigned a weight of 140%–160% of the point A dose. Typically, four dwells are activated on each side of the smallest ring (36 mm), five on each side of the medium ring (40 mm), and six on each side of the large ring (44 mm). The entire ring should never be activated, as this will cause high rectal, bladder, and vaginal doses. For the tandem and cylinder, a similar system is used with the exception that dose points are entered laterally from the dwell positions at the appropriate distances representing the cylinder surface. Due to the close proximity of the bladder and rectum in women with narrow vaginas requiring vaginal cylinders, it may be wise to decrease the vaginal surface dose to 100%–120% of the point A dose rather than 140%–160%. The optimization program then attempts to give the prescription dose at each of these points.

Optimization is a term used in HDR/PDR dosimetry planning: dose points are defined relative to the applicators and are used by the planning system to determine the dwell times necessary to deliver the specified dose to the points. The optimization process can be performed manually by manipulating the source dwell positions and dwell weights. Alternatively, some treatment planning systems also offer partially automatic optimization tools (i.e., dragging and dropping isodose lines through graphical optimization) or more automatic tools such as inverse-planning software. Caution must be applied when using such tools, as the resulting loading pattern and dose distribution can deviate significantly from standard techniques. During the optimization process, attention should also be paid to the more traditional parameters, which may influence toxicity, including the implant volume, the total reference air kerma (TRAK), and the spatial dose distribution of the implant, in addition to the point doses and dose-volume histogram (DVH) parameters to avoid unfavorable optimization. Hot and cold spots need to be carefully scrutinized as well.

With nonoptimized plans, fixed relative dwell weights are inputted into the planning system and are used to calculate the dose distribution. By altering the dwell times and dwell weights, it is possible to alter the dose distribution and thereby optimize tumor volume coverage and normal tissue exclusion. Excessive optimization can alter the pear shape to a less desirable configuration with the same point A dose (Kim et al. 1997; Cetingoz et al. 2001). Dose specification at point A alone can result in underdosing of the target tissues and overdosing of the dose-limiting tissues (Mai et al. 2001). It is important, therefore, to also monitor multiple points or volumes apart from point A when manipulating the dose distributions.

21.4.2.3 Dose Specification for Cervical Cancer: PDR

PDR planning allows dwell position activation at 2.5- or 5.0-mm intervals. At the Institut Gustave Roussy, an initial PDR-BT plan is prescribed in 0.5-Gy hourly pulses to standard reference points for intracavitary implants. The standard plan may then be optimized to improve tumor coverage and to reduce dose to the normal tissues (Chargari et al. 2009; De Brabandere et al. 2008; Lindegaard et al. 2008). The best correlation between tumor control/morbidity and dosimetric parameters has not been established yet. In some institutions, the prescribed dose per pulse is reduced (and the total number of pulses correspondingly increased) if the optimized plan demonstrates a dose rate greater than 0.60 Gy/h in the 2-cc OARs. This enables "radiobiological optimization" with PDR BT.

21.4.2.4 Dosimetry Generation: HDR/PDR

It is important to perform dosimetry for each fraction of an HDR/PDR regimen and each LDR insertion, even if the same applicator is used, as there may be quite a bit of variation in applicator position with each fraction as demonstrated in several series (Grigsby et al. 1993; Kim et al. 1996; Jones et al. 2004), resulting in a variable relationship of the applicator to the pelvic organs as well as applicator deformation of the pelvic organs (Christensen et al. 2001). Variables that may impact on applicator position are the relationship of the applicator to the uterus

and surrounding critical structures, vaginal packing and the type of sedation/anesthesia used, as well as use of the dorsal lithotomy versus legs-down position. The bladder and rectosigmoid may also change configuration due to changes in filling, and doses may vary between fractions. Uterine and sigmoid mobility may also impact on the dose distribution. Additionally, disease regression and vaginal narrowing will vary from fraction to fraction and can also result in changes in dose distribution. A change in the applicator can also result in changes in dose distribution as can changing the ovoid or ring size, ovoid separation, and tandem curvature. The ovoids may also change in separation and their relative position to the tandem over time, as they are often not fixed to the tandem (Kim et al. 1996). When treatment planning was not performed for each fraction and only the initial dosimetry was used, there was increased dose to at-risk structures (Jones et al. 2004). When using a tandem and ring, which has fixed geometry, several institutions perform dosimetry for the first fraction only (Ahmad et al. 1991; Gerszten et al. 1998; Han et al. 1996). This practice should be questioned because of the other factors described. The applicator position relative to the pelvic organs is the important factor impacting dose, rather than the relationship of the tandem to the ring.

21.4.2.5 Dose: LDR Intracavitary

Tod and Meredith found a dose range of 7000–9000 R within which a dose of 8000 R to point A seemed to result in the best survival with the lowest rate of "high dose effects" (Paterson 1952). Most subsequent series have found no clear-cut dose-response. Tumor size has a profound effect on both local control and survival, and this alone makes the relationship between dose and outcome problematic as patients with larger tumors generally receive higher doses of radiation (Perez et al. 1998). Some attempts have been made to correlate dose with tumor control despite these limitations. In patients with bulky endocervical carcinomas treated with definitive irradiation, Eifel et al. (1993) found that patients who received <6000 mg-h had a higher rate of pelvic disease recurrence at 5 years as well as a decreased actuarial survival at 5 years than those patients who received >6000 mg-h (>85–90 Gy). In the Patterns of Care analysis, Montana et al. (1995) concluded that the highest local control rate and disease-free survival was achieved with a total average paracentral dose of 75–85 Gy. A dose-response relationship for in-field pelvic control, however, could be documented only in Stage III disease, with the highest rate of pelvic control with paracentral doses >85 Gy (Lanciano et al. 1991a). Perez et al. (1983) also reported that in all stages except IB, higher doses of radiation delivered to the medial and lateral parametria with a combination of external beam and implant were correlated with a lower incidence of parametrial failures (<60 Gy versus 60–90 Gy versus >90 Gy to the medial parametria and <40 Gy versus 40–50 Gy versus 52.5–65 Gy versus >67.5 Gy to the lateral parametria). The subsequent 1988 publication documented a significant impact on the pelvic recurrence rate of increasing medial parametrial doses for stages IIB and III, in the dose ranges of <60, 60–75, and 75–90 Gy (Perez et al. 1988). At a minimum, it would appear that there is a dose range of perhaps 75–90 Gy to the medial parametria, which will enable local control without excessive complications. Doses below 70 Gy could result in a significant incidence of local recurrence, and doses above 90 Gy are difficult to deliver without exceeding normal tissue tolerance.

21.4.2.6 Dose-Fractionation Schemes: HDR

A review of the randomized and nonrandomized studies of HDR using multiple dose-fractionation schemes has been undertaken. Most centers used point A as a reference point, although the definition of point A varied. The dose per fraction at point A varied from 3 to 10.5 Gy, the number of fractions from 2 to 13, and the number of fractions per week from 1 to 3. Most centers used a schedule of 7–8 Gy/fx for three to six fractions. The external beam dose was variable (Fu and Phillips 1990). A survey of 56 international institutions treating a total of over 17,000 cervix patients with HDR found that the most common fractionation scheme was five fractions of 7.5 Gy to point A. Fractionation of the HDR treatments significantly influenced toxicity as morbidity rates were significantly lower for point A doses per fraction of less than or equal to 7 Gy compared with >7 Gy (Orton et al. 1991; Orton 1998). Fraction sizes <7.5 Gy have been recommended by the ABS (Nag et al. 2000a). The use of at least 5 HDR fractions and as many as 12 fractions to produce equal early effects with LDR, when the dose to the dose-limiting critical normal structures is less than 75% of the prescribed dose, has been advocated. This will result in late effects comparable to LDR (Brenner et al. 1991; Brenner 1992). A survey of 13 Japanese centers indicated that a dose per fraction of 6.0 Gy ± 1.1 was used most frequently to a total HDR dose of 27.9 Gy ± 4.2 given in 4.8 fractions (3–10) (Okawa et al. 1992). In a literature analysis by Petereit and Pearcy (1999) of the HDR fractionation schedules, a dose response relationship could not be identified for tumor control or complications. There is no consensus as to the optimal number of fractions and dose per fraction except that the choice will depend on the external beam dose and on whether central shielding is used, normal tissue doses, and the stage of disease. The linear–quadratic model has been suggested

as a guide to formulate the regimens chosen at each institution (Orton 1991b; Nag et al. 2000a). In order to convert HDR fractionation into equivalent dose in 2-Gy fraction sizes, a special interactive worksheet is available on both the ESTRO and ABS Web sites. Previously, the GOG protocols used 600 cGy X 5 to Point A, but this has now been reduced to 550 cGy per fraction due to an excessive rate of late complications, most of which were lower GI reported in a retrospective review (Forrest et al. 2010). The RTOG protocols allow 530–740 cGy/Fx, and four to seven fractions, depending on the external beam dose. Tables for combining various external beam doses with varying HDR fractions using the linear–quadratic formula and normal tissue–modifying factor have been provided with these protocols (Nag et al. 2000). A review of multiple HDR series reveals the lack of consensus on dose-fractionation schemes (Kuipers et al. 2001). At the Medical College of Wisconsin, we deliver 45 Gy to the whole pelvis dose followed by five fractions of 5.0–6.0 Gy for an LDR equivalent of 80–90 Gy. In an international survey, the mean cumulative external beam and BT EQD2 dose was 79 Gy for stages IB–IIA and 83 Gy for stages IIB–IVA (Viswanathan et al. 2012).

21.4.3 Interstitial BT: Dosimetry Generation, Dose Specification, Dose, and Dose Fractionation Schemes

Interstitial techniques should be based on 3D imaging using CT to identify the target volume and critical normal structures, in relation to the needle positions. Individualized computer-generated dosimetry is an integral part of interstitial dose delivery based on the CT-defined CTV (see Figure 21.6; Erickson and Gillin 1997; Erickson et al. 1996). Adequate postprocedure analgesia, often best achieved through an epidural approach, is essential and allows the needles and tandem to be manipulated outside the operating room if necessary. Modification of the planned source placement based upon the location of specific needles and critical structures can then be made before or after source loading.

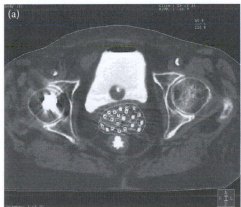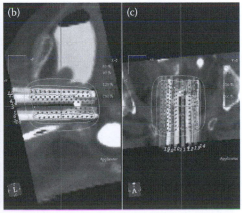

FIGURE 21.6 (See color insert.) CT-based transperineal BT dosimetric plan used in the treatment of a vaginal cancer in (a) axial, (b) sagittal, and (c) coronal views with needles and associated isodose curves. Note the relative sparing of the opacified rectum and bladder. (Courtesy of B. Erickson.)

With LDR techniques, an optimal isodose distribution was achieved through differential loading (core sources not more than half the activity of peripheral sources) of low-activity sources (Erickson et al. 1996). "Reference" dose rates of 60–80 cGy/h are optimal. The implant dose rates as well as the dose homogeneity and distribution could be manipulated by selectively changing the activity associated with a particular needle or needles or by selectively unloading, either immediately or during the implant, strategic needles in the pattern. Dosimetry can now be optimized by differential loading using the afterloader simulating LDR delivery with PDR or moving to fractionated HDR delivery. There may be some efficacy in making the dose heterogeneous in the center of the implant so as not to underdose the cervix. At MCW, the obturator surface and associated needles are given 120% of the peripheral dose. Attention to hot spots within the implanted volume is important, and doses greater than 120% of the peripheral dose should be limited to single needles or clusters of convergent needles. Typically, total LDR doses to the tumor volume or reference isodose from the implant range from 23 to 40 Gy over 2–4 days (Erickson et al. 1996). The total HDR dose will be approximately 60%–70% of the total LDR dose and will be given in divided fractions. With either approach, careful attention to significant hot spots within the implant and doses to the bladder, rectosigmoid, and vaginal surface is requisite to the best outcome. At the Medical College of Wisconsin, the standard prescription is 45-Gy external beam to the pelvis followed by five fractions of 4.5–5.5 Gy HDR to the periphery of the implant and 120% of this to the vaginal obturator surface.

For either LDR or HDR, one or two template and needle applications can be done 1–2 weeks following external beam. With HDR, one to two fractions can be delivered per day over a period of 2–5 days, separated by at least 6 h. The total LDR dose to the reference volume from the combined implant and external beam approximates 70–85 Gy over 8 weeks. With the modified Vienna ring applicator, a dose fractionation of 7 Gy X 4 is used (Dimopoulos et al. 2006a; Kirisits et al. 2006).

21.5 Clinical Results

21.5.1 Endometrial

BT has been consistently shown to decrease vaginal recurrences in endometrial cancers. There is controversy over the optimal management of early- and intermediate-stage disease, but the agreement in all cases seems to be that vaginal BT is the most efficient method to decrease the chance of a vaginal recurrence. Without adjuvant therapy (external beam RT or BT), patterns of failure for surgically staged I endometrioid uterine cancers are predominantly in the vagina and at distant sites. In GOG, 99 of 202 patients who received no adjuvant therapy or isolated vaginal or pelvic recurrences at 4 years were 7% overall and 13% in the high-intermediate risk (H-IR) subgroup (Keys et al. 2004). In the RT arm, at 4 years, there were 2% overall and 5% in the H-IR group. Most of the local recurrences were vaginal. For distant recurrences at 4 years, in the control arm, there were 8% overall

and 19% in the H-IR subgroup; in the RT arm at 4 years, there were 5% overall and 10% in the H-IR group. Pelvic radiation as used in GOG-99 and PORTEC was successful in reducing vaginal/pelvic recurrences (Keys et al. 2004; Creutzberg et al. 2000; Scholten et al. 2005). Even for patients presumed to be at high risk for failure, vaginal cuff failures predominate. In a retrospective series of 220 surgically staged IC patients, of the 121 (55%) who did not receive radiation therapy, 14 recurrences were observed (Straughn et al. 2003). Six of the 14 recurrences were vaginal, one was at another pelvic site, and seven recurrences occurred outside of the pelvis. In the 99 other IC patients who did receive postoperative radiation, there were 6 recurrences (5/6 at distant sites). The most recent update of the PORTEC 2 trial indicates a recurrence rate of 1.8% in intermediate-risk endometrial cancers in patients treated with vaginal BT alone (Nout et al. 2010). The patients selected for this trial included those with age >60 years with IC grade 1 or 2, stage IB grade 3, or stage IIa of any age with more than 50% myometrial invasion (this staging has been revised in the current staging manual). BT has been shown to decrease the rate of vaginal recurrences as well in higher stages of endometrial cancer. Prior randomized trials in early-stage disease showed that radiation reduces the risk of locoregional failures by approximately two-thirds, from 6%–14% without to 1.5%–5% with pelvic irradiation (Aalders et al. 1980; Creutzberg et al. 2000; Keys et al. 2004; ASTEC/EN 2009; Scholten et al. 2005), although distant failures persist (5%–10%) even in the favorable group of stage I and II patients. The PORTEC I trial suggested a higher risk of recurrence for older patients (>60 years old), and for those with stage IC, grade 3 disease or lymphovascular invasion. In the PORTEC I trial, patients with stage IC grade 3 disease received adjuvant pelvic external beam radiation alone (Creutzberg et al. 2004); the 5-year rate of pelvic relapse was 14%, distant relapse was 31%, and overall survival at 5 years was 58%. Patients with cervical involvement, clear-cell or papillary serous adenocarcinoma, or extra-uterine pelvic-confined disease, including pelvic nodal spread, have an increased risk of both local and distant metastases and require pelvic radiation to reduce the risk of a pelvic recurrence (Keys et al. 2004; Mundt et al. 2001). Patients with lymph node involvement have a 46% 3-year actuarial risk of vaginal and a 41% risk of pelvic relapse. The GOG 0184 randomized patients treated with surgery and pelvic and/or para-aortic radiation to cisplatin and doxorubicin with or without paclitaxel; 10% had a local regional recurrence, and 30% had a distant recurrence (Homesley et al. 2009). Because distant sites of failure are not infrequent, particularly in the H-IR groups, further trials are addressing the role of combining local and systemic therapies. Both the PORTEC and GOG-99 trials suggested that the primary benefit of pelvic radiation is in its ability to reduce vaginal cuff failures. The role of vaginal cuff BT (VCB) might be particularly important if it provides comparable control of vaginal failures and is associated with less toxicity. The toxicities of pelvic radiation reported in the two randomized studies included 3% acute grade 3–4 GI toxicity (GOG-99), 25% late complications (any grade) (PORTEC), and 3%–6% chronic GU/GI toxicity > grade 2 (Keys et al. 2004;

Creutzberg et al. 2000). These rates of toxicity appear to be greater than reported in many studies evaluating VCB. Patient tolerance and acceptance of VCB (one to five treatments, treatment course 2–3 weeks) compared to pelvic radiation (28 fractions over 5 weeks) is also greater. PORTEC II has shown that for intermediate-risk patients, vaginal BT achieves equivalent rates of vaginal relapse and survival compared with external beam with less pelvic toxicity and improved quality of life (Nout et al. 2010).

21.5.2 Cervical

There has been much debate about the outcome with HDR techniques versus the more traditional LDR techniques (Stewart and Viswanathan 2006; Petereit et al. 1999). There have been five randomized trials comparing HDR to LDR, detailed in Table 21.2. Shigematsu et al. (1983) published a Japanese study from September 1974 to December 1979, which showed higher 1-year rates of control in stages IIB and III (90% versus 77%) but a higher incidence of rectal complications in the HDR group (36% versus 25%). The absence of rectal retractors in some of these patients and the use of large fraction sizes (10 Gy) were implicated. The fraction size was subsequently reduced to 7 Gy (Inoue et al. 1978; Shigematsu et al. 1983). A follow-up by Teshima et al. (1993) from January 1975 to August 1983 noted no significant difference in survival at 5 years with HDR versus LDR but a slightly higher risk of complications with HDR (10%) versus LDR (4%). The Indian trial of Gupta et al. revealed a similar local control rate for the HDR and LDR groups of 80% and 85%, respectively. Rectal complications were lower and bladder and vaginal complications higher in the HDR versus LDR groups (Sharma et al. 1991). This was later updated by Patel et al., confirming that local control and 5-year survival were comparable with HDR and LDR. The incidence of rectosigmoid complications was 19.9% for LDR and 6.4% for HDR, which was statistically significant. Bladder morbidity was similar in both groups (Patel et al. 1994). A 2005 publication claims that the 9-Gy fraction size used is both safe and effective

with an actuarial grade 3 risk of 3.31% (Patel et al. 2005). Similarly, the randomized trials published by Lertsanguansinchai et al. (2004), Hareyama et al. (2002), and Rotte (1980) showed no difference in survival or complications for HDR versus LDR. Rotte found that complications decreased over time in the HDR arm as a result of intense treatment planning and optimization (Rotte 1990). A meta-analysis of the published literature on LDR versus HDR has also shown no difference in overall survival, local recurrence, and late complications for patients with stage I–III cervical cancer treated with HDR versus LDR (Viani et al. 2009).

21.5.3 Interstitial

The initial experience with LDR interstitial implantation was associated with a high rate of local recurrence as well as complications (Erickson and Gillin 1997). Fistula formation was much higher than in intracavitary series as was cervical and vaginal necrosis. This was due to excessive dose rates and doses within the implanted volume due to the loading patterns of the needles, obturator, and tandem, and the strength of the ^{192}Ir sources used. Modifications were suggested once these complications were published, and subsequent series showed improvement in outcome (Erickson and Gillin 1997; see Table 21.3). Use of CT-based planning has led to further understanding and manipulation of the complex dose distribution and the associated normal organ doses with resultant improvement in outcome (Erickson et al. 1996). It is clear that interstitial implantation is challenged by the bulk of disease present and the inherent poor prognosis of the patients who require such an intervention. It should be performed by brachytherapists who are experienced with this modality and requires a team of experienced physicists and dosimetrists to master the requisite complex treatment planning. Results from experienced institutions have improved over time with modifications in treatment planning and use of image-based techniques (Gupta et al. 1999; Syed et al. 2002). HDR techniques are becoming more prevalent

TABLE 21.2 Randomized Trials Comparing LDR with HDR BT in Cervical Cancer

Author	Number	Stage	LDR Dose	HDR Dose	External Beam	5-Year Survival LDR	5-Year Survival HDR	Complications LDR	Complications HDR
Patel et al.	482	I <3 cm	75 Gy/2f	35 Gy/1f	35–40 Gy	73%	78%	2.4%	0.4%
		I–III	38 Gy/4f	18 Gy/2f		62%	64% (stage II)		
						50%	43% (stage III)		
Teshima et al.	430	I	56 Gy/2f	28 Gy/4f	16–20 Gy	93%	85%	4%	10%
		II	57 Gy/2f	30 Gy/4f		78%	73%		
		III	58 Gy/2f	29 Gy/3f		47%	53%		
Hareyama et al.	132	IIA	50 Gy/4f	29.5 Gy/4f	30–40 Gy	Disease	specific	13%	10%
		IIB	40 Gy/3f	23.3 Gy/3 or 4f		87%	69% (stage II)		
		III	30 Gy/3f	17.3 Gy/2 or 3f		60%	51% (stage III)		
Lertsanguansinchai et al.	237	IB–IIIB	25–35 Gy/2f	15–16.6 Gy/2f	40–50 Gy	71%	68%	2.8%	7.1%
Shrivastava et al.	800	I–II	60 Gy/2f	35 Gy/5f	40 Gy	No difference in local control or toxicity			
		III	30 Gy/1f	21 Gy/3f		Abstract only with 18-month follow-up			

TABLE 21.3 Literature Series for LDR Interstitial Radiation for Gynecologic Cancers

LDR Series	Published Year	FIGO Stage	n	CR (%)	LC (%)	LRC (%)	DFS (%)	Serious Complications (n)	Follow-up (years)
Feder et al. (16)	1978	III–IVA	38	–	–	–	60[a]	–	1.3–3.3
Tak (40)	1978	IB–IVA	24	79	–	–	67	4	1–7
Gaddis et al. (41)	1983	IB–IVA	75	100	–	70	59	21	<1–5
Ampuero et al. (42)	1983	IB–IVA	21	85	59	–	43	24	2–3.4
Prempree et al. (37)	1983	IIB	49	84	–	–	65	8	5
Aristizabal et al. (43)	1983	IA–IVA	26	–	85	–	–	33	1–4
Martinez et al. (21)	1984	IIB–IIIB	37	84	–	83	57	5	1–7.5
Aristizabal et al. (19)	1986	IIB	45	100	–	–	77[b]	9	5
Syed et al. (17)	1986	IB–IVA	60	78	–	78	58(50)[a]	3	5
Erickson et al. (44)	1989	I–IV	–	–	74	–	–	–	–
Bloss et al. (45)	1992	IIB–IIIB	25	72	–	–	64	4	2
Rush et al. (46)	1992	IIB–IIIB	5	100	–	–	40	20	1.7
Fontanesi et al. (47)	1993	I–IV	–	–	87	–	60	20	<1–7
Hughes-Davies et al. (22)	1995	IIB–IIIB	49	68	25	–	32	12–20	3
Monk et al. (39)	1997	IIA–IVA	70	61	31	–	24(29)[a]	21	5
Nag et al. (48)	1998	IB–IVA	31	–	48	–	25	3	1–5.5
Gupta et al. (49)	1999	–	–	78	60	–	55	3–24	4.7
Present series	2001	IB–IVB	185	87	82	73	57(49)[a]	9	<1–18.6

Source: Syed et al. 2002. With permission. The reader is referred to the original article for the citations of the first column.

Note: Abbreviations: LDR = low dose rate; FIGO = International Federation of Gynecology and Obstetrics; CR = complete response; LC = local control; LRC = locoregional control; DFS = disease-free survival. Modifications to dose rates and doses within the implanted volume due to loading patterns of needles, obturator, and tandem were suggested once these complications were published, and subsequent series showed improvement in outcome.

[a] Stage IIIB.

[b] Actuarial.

due to the improved dose distributions produced with optimization and the decrease in radiation exposure afforded by remote afterloading. Initial results are very encouraging (Demanes et al. 1999; Beriwal et al. 2006; Viswanathan et al. 2010). Demanes et al. have reported an overall local tumor control of 94% with a mean follow-up of 40 months but an actuarial 5-year disease-free survival of 48% for the 62 stage IB–IVA patients treated. Distant metastases developed in 32%. Grade 3–4 complications occurred in 6.5% (Demanes et al. 1999). Beriwal et al. treated 16 patients with vaginal and cervical cancers using CT-guided techniques. The 5-year actuarial local control rate and cause-specific survival were 75% and 64%, respectively. There was a 7% actuarial 5-year rate of grade 3–4 complications (Beriwal et al. 2006).

21.6 Organs at Risk

21.6.1 Doses to Bladder and Rectosigmoid

The bladder and rectosigmoid are the organs of concern in the setting of combined external beam irradiation and BT. Attempts have been made to determine the maximum tolerated normal tissue doses. There is no consensus as to what these values should be. Point doses may or may not coincide with complication risk, as they do not account for the volume of the organ irradiated. They are also not defined consistently. Increasingly, dose and volume relationships are thought to be key in the development of

normal tissue complications, and new image-based dosimetry, including DVH analysis, will lead to new insights into the relation between dose volume and toxicity (Roeske et al. 1997). Use of tandem and cylinder applicators in a single line source rather than multisource applicators such as tandem and ovoids or ring can increase the volume of bladder and rectum treated, as can interstitial implants (Crook et al. 1987; Stryker et al. 1988). Stage, patient age, and medical comorbidities such as hypertension, diabetes, diverticulitis, or inflammatory bowel disease are also related to the risk of complications notwithstanding individual radiosensitivity.

21.6.1.1 Low Dose Rate

Maximum point bladder doses of 75 Gy and rectal doses of 70 Gy have traditionally been quoted as tolerance doses. The ICRU 38 report defines these points on AP and lateral radiographs at the time of each LDR BT insertion (ICRU 38 1985; Fletcher 1980), as discussed in Chapter 24. Contrast points have also been used to track the normal tissue doses in addition to the ICRU points. Retrospective series have demonstrated that with doses below 75–80 Gy to limited volumes of the bladder, the incidence of grade 3 or 4 complications is 5% or less, whereas with higher doses, a greater incidence of sequelae is noted (Eifel et al. 1995; Perez et al. 1999); however, in general, fixed point doses to the bladder are poorly correlated to late effects due to the variable nature of the bladder shape and volume. Retrospective analyses have shown that limited volumes of the rectum can tolerate approximately 75 Gy (external beam and BT) with acceptable morbidity (Eifel et al. 1995; Perez et al. 1999). Additionally, the ratio of dose to the rectum or bladder to the dose at point A is also important, with a low incidence of rectal (0.3% versus 5%) and bladder (2% versus 2%–5%) complications when this ratio was less than 80% (Perez et al. 1999). Bladder complications can appear much later than rectal and sigmoid complications (Eifel et al. 1995).

21.6.1.2 High Dose Rate

Acceptable normal tissue doses are even more uncertain in HDR. Various recommendations concerning normal tissue fraction size and total dose exist in the literature. Recommended maximums for the rectum included summed physical doses of external beam and HDR BT of <62–65 Gy and BED doses <125 (Sakata et al. 2002; Cheng et al. 2003; Takeshi et al. 1998; Chun et al. 2004; Chen et al. 2000; Ferrigno et al. 2001; Toita et al. 2003). Ferrigno et al. (2005) found that patients treated with a cumulative BED at bladder points above 125 Gy_3 had a higher risk of complications; however, again, correlation of bladder BED with radiation complications has been unreliable (Chen et al. 2004). In addition, using equivalent dose calculations, it would appear that doses of <70–75 Gy to 2 cc of rectum and sigmoid and <90 to 2 cc of bladder may be tolerable when using CT- or MR-guided DVH analysis (Georg et al. 2011).

Available clinical data suggest also that in addition to the total HDR dose, one of the most important factors in late complication development is the dose per fraction and the number of fractions (Stitt and Fowler 1992; Kapp et al. 1997; Wang et al. 1997). The organ most at risk of complications is the rectosigmoid, whereas the bladder complication risk is comparatively low (Ito et al. 1994; Ogino et al. 1995). Rectal and sigmoid complications occur earlier than bladder complications (Kapp et al. 1997; Wang et al. 1997). Bleeding is the most frequent rectal morbidity, occurring in approximately 30% of patients (Chun et al. 2004; Ito et al. 1994; Ogino et al. 1995; Chen et al. 1991). To avoid excessive morbidity, better physical dose distributions must be achieved with HDR to reduce doses to critical normal structures. Fowler has speculated that if only 80% of the tumor dose is received by the critical normal tissues, then 4-6 HDR fractions can be used, safely whereas 12-16 fractions would be required if the normal structures receive 90% of the Point A dose, and 30 fractions if the normal structures received 100% of the HDR dose (Stitt et al. 1992; Fowler 1990). This implies the need for rectal and bladder displacement, which can be achieved more reliably with the short treatment duration of HDR using rectal displacement and strict bladder filling protocols (Stitt 1992; Thomadsen et al. 1992). Retractors or gauze packing is used effectively to displace the bladder and rectum, but unfortunately, the sigmoid colon cannot be displaced due to its location above the packing and/or retractors, and hence, this is more vulnerable to complications (Orton 1991a). Rarely, mucosal damage leading to ulceration and strictures may form, which lead to progressively painful defecation and obstruction requiring fecal diversion in <5% of patients.

If using 2D planning, it is very important to choose points related to critical structures very carefully on the orthogonal films. Rectum above the level of the vaginal applicators and rectal retractor should be identified, and sigmoid in addition to rectal points should be evaluated, as should bladder points. Various methods for determining normal tissue doses have been described. Use of approximately 30–50 cc of contrast in the bladder and rectosigmoid can be helpful in identifying those portions of the organs in close proximity to the applicator. When possible, the doses to the normal critical structures should be less than the dose at point A, perhaps in the range of 50%–80%. The portions of the rectum and sigmoid that are above the range of the rectal retractor are most often the hot spots, and every effort must be made to decrease the dose to the rectosigmoid relative to the point A dose. Consideration to decreasing the dwell times or turning off dwell positions in the tandem should be given. Use of a "tapered tandem" will decrease sigmoid, bladder, and small bowel doses (Mai et al. 2001).

3D image-based BT has clearly revealed that dose to the rectum and sigmoid can be underestimated when using the ICRU 38 rectal point alone (Cheng et al. 2003; Holloway et al. 2009). It is interesting, however, that the dose to the ICRU rectal point and the D_{2cc} to the rectum correlate quite closely. It has also been demonstrated that the ICRU bladder point underestimates the bladder dose, and a point located 1.5 to 2 cm above the ICRU point is probably more relevant and especially helpful when 2D BT is performed (Barillot et al. 1994). Landmark studies using CT indicate that we significantly underestimate normal tissue doses when using 2D dosimetry, still in use at many institutions today (Stuecklschweiger et al. 1991; Kapp et al. 1992, 1997; Pötter et al. 2000). CT or MR imaging after applicator placement is much more accurate in defining the course of the OARs relative to the applicator (Figure 21.7). Appropriate image-based manipulation of the dose distribution can then follow to carefully reduce these doses while assuring an acceptable dose distribution around the tumor volume.

A recent analysis has been published by Georg et al. (2011) on DVH parameters and late effects in image-guided (MRI) adaptive BT on a total of 141 patients. With a median follow-up of 51 months, the overall 5-year actuarial side-effect rates were 12% for the rectum, 3% for the sigmoid, and 23% for the bladder. The population was divided into four groups—group 1: no side effects (grade 0); group 2: side effects (grades 1–4); group 3: minor side effects (grades 0–1); and group 4: major side effects (grades 2–4). The authors concluded that the D_{2cc} and the D_{1cc} were very relevant predictors of rectal toxicity, with statistically significant differences between groups 1 and 2 in all DVH parameters. A D_{2cc} of 75 Gy (EQD2) was considered as a clinical cutoff level for rectal morbidity prediction. For sigmoid side effects, there was no correlation between the DVH parameters and toxicity, as the number of events was too low. For bladder toxicity, the D_{2cc} was the strongest predictor of toxicity, with a cutoff at 100 Gy (EQD2).

Given the prevalence of CT simulators and MR scanners in many radiation oncology departments, this should

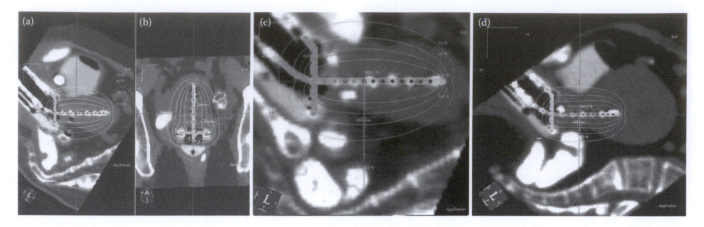

FIGURE 21.7   (See color insert.) CT-based dose distribution system in sagittal (a) and coronal (b) views showing the pear-shaped dose distribution of the tandem and ring. Sagittal views (c) of one implant with the sigmoid far from the associated dose distribution and in close proximity to the high dose region of the pear (d). Such loops may go unrecognized with film-based dosimetry. (Courtesy of B. Erickson.)

be considered the new standard for critical organ definition. Though CT is excellent at identifying the OAR, MR has better soft tissue resolution of the primary cervical tumor volume and its extensions and is ideal for defining a target for dose specification. MR-based BT has been formalized by the GEC ESTRO Gynecologic Working Group and is currently an integral part of the EMBRACE (an international study on MRI-guided BT in locally advanced cervical cancer) study, which will be discussed in detail below (Dimopoulos et al. 2006b; Haie-Meder et al. 2005; Pötter et al. 2006). Continued data collection using 3D image-based BT should give additional insights into the reasons for local failures and complications and allow for improvement in local control and late toxicities.

21.6.2 Doses to Vagina

The distal vagina is less tolerant of radiation than the proximal vagina, and the tolerance doses are in the range of 80–150 Gy (Hintz et al. 1980; Au and Grigsby 2003). Vaginal narrowing and shortening are late sequelae of radiation, which can alter and impede sexual function. Combined BT and external beam irradiation will cause more late effects than either modality alone. Use of a vaginal dilator and sexual intercourse can help in keeping the vagina patent (Grigsby et al. 1995; Au and Grigsby 2003). Rarely, with excessive doses of radiation, patients can develop vaginal necrosis. This is more of a risk at the introitus than at the vaginal apex (Hintz et al. 1980). Interstitial implants are more likely to cause necrosis than intracavitary implants.

Vaginal toxicity has also been evaluated with the use of image-guided (MRI) adaptive BT. DVH parameters have a high degree of uncertainty for vaginal dose estimation. They are influenced by the resolution of the sectional imaging, contouring accuracy, and applicator reconstruction. Vaginal morbidity was not significantly correlated to the D_{2cc} (Berger et al. 2007).

The uterus is very resistant to high doses of radiation as is evident in patients treated with external beam and BT for cervical cancer. Rare cervical necrosis can occur but is more commonly caused by recurrent tumor. Distinguishing recurrent disease from necrosis can be very difficult and will sometimes mandate surgical intervention.

Close follow-up of patients treated with definitive irradiation for cervical cancer is important, including clinical examinations and imaging. A normal PET scan at 3 months postradiation is prognostically important (Grigsby et al. 2003; Nakamoto et al. 2002). Early detection of a recurrence may lead to better salvage (Singh et al. 2005).

21.7 Summary of Recommendations

Up until recently, target volume assessment was exclusively based on clinical examination with documentation in three dimensions (height, width, and thickness). While anterior–posterior and lateral dimensions have always been accessible by digital clinical evaluation, the height of the tumor was clinically inaccessible, especially when the tumor extended to the endocervix and/or the endometrium. Through MRI, tumor size and configuration have been proven to be more appropriately assessed compared to clinical examination or with CT images. Accurate delineation of the gross tumor volume (GTV), definition and delineation of the CTV and planning target volume (PTV), as well as delineation of the critical organs have a direct impact on the BT procedure and dose specification, especially if it is possible to adapt the pear-shape isodose distribution to the tumor volume, while minimizing dose to the normal tissues. DVH analysis can be used to assess tumor coverage and normal organ sparing. With the data coming from MRI, it became clear that this should be applied to uterovaginal BT, with the use of a common language (Dimopoulos et al. 2006b).

In 2000, a working group was founded by GEC ESTRO [Gynaecological (GYN) GEC ESTRO WG], including physicians and physicists from different centers actively involved in this field at that time. The task was to describe basic concepts and terms for this approach and to work out a terminology that would enable other groups to use a common language for appropriately communicating their results (Haie-Meder et al. 2005).

From the start of the GYN GEC ESTRO WG, it became clear that major differences in tradition also existed within this group, which would have a major impact on developing a common 3D image-based approach: After a series of workshops using examples of cervix cancer at different stages with different responses after concomitant chemoradiotherapy, recommendations for contouring volumes and dose specification were proposed based on clinical experience and the dosimetric concepts of different institutions (Paris, Leuven, Vienna). Two main volumes of interest were identified. One derived from using point A as a reference point. This was based mainly on the residual tumor (GTV) at the time of BT, following external beam and chemotherapy, but also took into account the associated microscopic residual disease in the cervix and parametria in addition to the remaining GTV. From this, a CTV could be defined at the time of BT. The intent was to give a defined dose to this CTV appropriate to the stage, disease volume, and potential microscopic residual disease, aiming for 80 to 90 Gy for advanced-stage disease. This dose was comparable with the more traditional dose to point A.

The second, using ICRU 38 recommendations for the 60-Gy volume, started mainly from knowledge of the GTV at diagnosis (where the tumor had been prior to external beam) for defining the CTV at the time of BT. With this, a CTV was defined that included anatomically targeted safety margins in addition to the dimensions of the GTV at diagnosis. The total dose specified to this "larger" CTV was 60 Gy (ICRU 38 1985).

In order to take into account these major concepts that were founded on years of clinical experience, two CTVs were proposed:

(1) A "high-risk" CTV (HR CTV) with a major risk of local recurrence because of residual macroscopic disease following external beam and chemotherapy. The intent was to deliver a total dose as high as possible around 80–90 Gy to eradicate *all residual macroscopic tumors*.

(2) An "intermediate-risk" CTV (IR CTV) with a major risk of local recurrence in areas that corresponded to initial macroscopic extent of disease, with residual microscopic disease at the time of BT. The intent was to deliver a total radiation dose appropriate to cure significant microscopic disease in cervix cancer, which corresponded to a dose of at least 60 Gy.

Target definition had to respect natural anatomical borders defined by the anterior cervical wall/bladder wall, posterior cervix wall/rectum/sigmoid, and parametrial borders (anterior, posterior, lateral, inferior, superior). Contouring of the normal OARs, including the bladder, rectum, and sigmoid, is an integral part of this system. Dose constraints for the normal tissues for the first time are based on both dose and volume parameters rather than point doses alone. A D_{2cc} of <70–75 Gy for the rectum and sigmoid and 90 Gy for the bladder has been proposed as a starting point for restricting dose to the closely positioned pelvic organs. A D_{90} of 80–90 Gy for the HR CTV, depending on residual disease volume and critical organ sparing, is proposed as a standard for volume-based dose specification when using the MR-defined volumes inherent to this system. MR-based dosimetry with the associated DVH analysis can be pivotal in determining which patients may have underdosed tumors, requiring the addition of interstitial techniques, or over-dosed OAR requiring a change in the dose specification (Figure 21.8).

21.8 Future Considerations

As a first step, this comprehensive 3D image-based treatment approach was to be spread from dedicated specialized centers to a larger part of the radiation oncology community. The aim was not to modify clinical and dosimetric practices but to assess the feasibility, the value, and the reproducibility of such volumes of interest. Several publications have now validated the feasibility and the reproducibility of the definitions (Dimopoulos et al. 2009; Lang et al. 2006; Petric et al. 2008). DVH parameters could be analyzed retrospectively even if the optimization process was of limited complexity (e.g., with the use of cesium sources, optimization was achieved by adapting the time duration and the length of each radioactive source). Thirty-nine patients with early-stage cervical cancer treated with preoperative MRI-based BT were

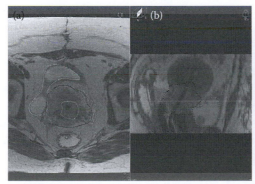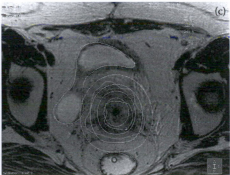

FIGURE 21.8 (See color insert.) Axial (a) and sagittal (b) MR images with contours of the GTV, HR CTV, and IR CTV following insertion of a tandem and ring applicator. The HR CTV is not adequately encompassed by the 100% isodose curve due to the volume of the remaining disease (c). Such a patient might benefit from the addition of interstitial techniques. (Courtesy of B. Erickson.)

reported. With a median follow-up of 4.4 years, only one latero-pelvic relapse occurred with a very low rate of long-term toxicities (Haie-Meder et al. 2009). In locally advanced cervical cancer, the same BT technique was applied in 84 patients with advanced cervical cancer (Haie-Meder et al. 2010). Of 31 observed failures, 10 patients presented with a local recurrence. The median D_{90} to the HR CTV was 79-Gy EQD2 $\alpha/\beta10$ (range, 53–122 Gy EQD2 $\alpha/\beta10$).

The second step was to explore the potential for local control improvement using these definitions and optimizing the dose distributions. At least two publications have shown an improvement in local control (Pötter et al. 2000; Chargari et al. 2009; Sturdza and Pötter 2011). A series of 145 patients was reported by Pötter at al. This series was divided into two periods: between 1998 and 2000, 71 patients were treated with MRI-based BT with no systematic protocol, which corresponded to a "learning period." After 2001, a systematic prospective protocol was performed using the HR CTV concept and delineation of the GTV, HR CTV, and OARs. Biological modeling was applied (using the linear–quadratic model) with dose-volume constraints of D_{2cc} of 75 Gy to the rectum and sigmoid and 90 Gy to the bladder (EQD2, $\alpha/\beta3$). The prescription dose was 85+ Gy (EQD2, $\alpha/\beta10$) to the HR CTV (D_{90}). If the D_{90} of the HR CTV was less than 85 Gy, then interstitial BT was combined with endocavitary BT. With a median follow-up of 50 months, 18 true pelvic recurrences were observed. A cutoff level was found at 87 Gy for the D_{90} HR CTV, with 3 recurrences above 87 Gy and 15 below 87 Gy ($p = 0/006$). For tumors larger than 5 cm, the improvement in local control translated into a gain in pelvic-free survival (43% for the 1998–2000 period versus 55% for the 2001–2003 period) and overall survival (28% for the 1998–2000 period versus 55% for the 2001–2003 period) (Pötter et al. 2000). Chargari et al. reported a series of 45 patients with advanced cervical cancer treated with 3D MRI-PDR intracavitary BT after concomitant chemoradiation. A dose of at least 15 Gy EQ D2$\alpha/\beta10$ was prescribed to 90% of the IR CTV, with a total dose limit of 80 Gy $\alpha/\beta3$ for the bladder and 70–75 Gy $\alpha/\beta3$ for the rectum and sigmoid (D_{2cc}). With a median follow-up of 26 months, no patient presented with a local recurrence, and only one grade 3 bladder complication was observed (Chargari et al. 2009).

Several articles have demonstrated a dosimetric improvement in DVH parameters (De Brabandere et al. 2008; Jurgenliemk-Schulz et al. 2009; Lindegaard et al. 2008). De Brabandère et al. analyzed the dosimetric data of 16 patients treated with PDR BT after external beam therapy, comparing the traditional 2D x-ray–based plans and 3D MRI-based plans. The aim was to assess the potential gain of MRI-based dose optimization, with the goal being to increase the D_{90} HR CTV dose above 85 Gy $\alpha/\beta10$. After optimization, an average dose increase of 3 Gy $\alpha/\beta10$ to the D_{90} HR CTV was achieved, while the D_{2cc} dose reduction was 7 ± 6 Gy $\alpha/\beta3$ in the bladder and 7 ± 4 Gy $\alpha/\beta3$ in the sigmoid colon (De Brabandere et al. 2008). The same type of comparison was reported in 21 patients treated with PDR BT by Lindergaard et al. DVH parameters of standard point A and MRI-based 3D optimized dose plans were compared. The results showed that optimization increased the minimal target dose (D_{100}) of the

high-risk CTV ($p < 0.007$) and decreased the minimal dose to 2 cc for the sigmoid significantly ($p = 0.03$) (Lindegaard et al. 2008). The same type of analysis was performed by Jürgenliemk-Schulz et al., who studied the impact of MRI-guided treatment planning on dose/volume parameters for PDR BT with an additional investigation of the potential benefit of an intra-cavitary/interstitial modification of the classical tandem and ovoid applicator. Twenty-four patients were studied: The total prescribed dose to Manchester point A or to D_{90} HR CTV was 80 Gy EQD2 $\alpha/\beta10$ in 17 patients (period I) and 84 Gy EQD2 $\alpha/\beta10$ in 7 patients (period II). The constraints to 2 cc of the OAR were 90 Gy EQD2 $\alpha/\beta3$ for the bladder and 75 Gy EQD2 $\alpha/\beta10$ for the rectum, sigmoid, and bowel. Most cases were treated with a traditional intracavitary tandem-ovoid applicator. In six patients, a newly designed combined IC/IS modification for the second PDR fraction was used, and the benefit of the interstitial part was investigated. The results showed that the average gain of MRI-guided optimization expressed in D_{90} HR CTV was 4 ± 9 Gy EQD2 $\alpha/\beta10$ ($p < 0.001$) and 10 ± 7 Gy EQD2 $\alpha/\beta10$ ($p = 0.003$) in the two periods. In all cases, the dose to 2 cc of the OAR met the constraints. In the group treated with the combined intracavitary/interstitial approach, the D_{90} HR CTV for the second PDR fraction was increased with 5.4 ± 4.2 Gy EQD2 $\alpha/\beta10$ ($p = 0.005$) and the D_{100} with 4.8 ± 3.1 Gy EQD2 $\alpha/\beta10$ ($p = 0.07$) (Jurgenliemk-Schulz et al. 2009). In order to further explore the potential of these definitions and their impact on local control, a prospective registry of 3D MRI-based BT data was initiated and is still ongoing as previously described through the EMBRACE study. Results from this study will be pivotal in further shaping the recommendations for image-based BT.

References

Aalders, J., Abder, V., Kolstad, P. et al. 1980. Post operative external irradiation and prognostic parameters in stage I endometrial cancer. Clinical histopathologic study of 540 patients. *Obst Gynecol* 56:419–27.

Ahmad, K., Kim, Y., Orton, C. et al. 1991. Fractionated high dose rate brachytherapy and concomitant teletherapy in the treatment of carcinoma of the cervix: Technique and early results. *Endocuriether Hypertherm Oncol* 7:117–24.

Albano, M., Dumas, I., and Haie-Meder, C. 2008. Brachytherapy at the Institut Gustave-Roussy: Personalized vaginal mould applicator: technical modification and improvement. *Cancer Radiother* 12:822–6.

ASTEC/EN 5 Study Group, Blake, P., Swart, A.M. et al. 2009. Adjuvant external beam radiotherapy in treatment of endometrial cancer: Pooled trail results, systematic review and Meta analysis. *Lancet* 373:137-46. Epub December 16, 2008.

Au, S. and Grigsby, P. 2003. The irradiation tolerance dose of the proximal vagina. *Radiother Oncol* 67:77–85.

Barillot, I., Horiot, J.C., Maingon, P. et al. 1994. Maximum and mean bladder dose defined from ultrasonography. Comparison with the ICRU reference in gynaecological brachytherapy. *Radiother Oncol* 30:231–8.

Batley, F. and Constable, W.C. 1967. The use of the Manchester system for treatment of cancer of the uterine cervix with modern after-loading radium applicators. *J Can Assoc Radiol* 18:396–400.

Berger, D., Dimopoulos, J., Georg, P. et al. 2007. Uncertainties in assessment of the vaginal dose for intracavitary brachytherapy of cervical cancer using a tandem-ring applicator. *Int J Radiat Oncol Biol Phys* 67:1451–9.

Beriwal, S., Bhatnagar, A., Heron, D. et al. 2006. High dose rate interstitial brachytherapy for gynecologic malignancies. *Brachytherapy* 5:218–22.

Brenner, D. 1992. HDR brachytherapy for carcinoma of the cervix: Fractionation considerations. *Int J Radiat Oncol Biol Phys* 22:221–2.

Brenner, D.J., Huang, Y., and Hall, E.J. 1991. Fractionated high dose-rate versus low dose-rate regimens for intracavitary brachytherapy of the cervix: Equivalent regimens for combined brachytherapy and external irradiation. *Int J Radiat Oncol Biol Phys* 21:1415–23.

Cetingoz, R., Ataman, O., Tuncel, N. et al. 2001. Optimization in high dose rate brachytherapy for utero-vaginal applications. *Radiother Oncol* 58:31–6.

Chan, J.K., Cheung, M.K., Husain, A. et al. 2006. Patterns and progress in ovarian cancer over 14 years. *Obstet Gynecol* 108:521–8.

Chargari, C., Magné, N., Dumas, I. et al. 2009. Physics contributions and clinical outcome with 3D-MRI-based pulsed-dose-rate intracavitary brachytherapy in cervical cancer patients. *Int J Radiat Oncol Biol Phys* 74:133–9.

Chen, M.S., Lin, F.J., Hong, C.H. et al. 1991. High-dose-rate afterloading technique in the radiation treatment of uterine cervical cancer: 399 cases and 9 years experience in Taiwan. *Int J Radiat Oncol Biol Phys* 20:915–9.

Chen, S., Liang, J., Yang, S. et al. 2000. The prediction of late rectal complications following the treatment of uterine cervical cancer by high dose rate brachytherapy. *Int J Radiat Oncol Biol Phys* 47:955–61.

Chen, S.W., Liang, J.A., Yeh, L.S. et al. 2004. Comparative study of reference points by dosimetric analyses for late complications after uniform external radiotherapy and high-dose-rate brachytherapy for cervical cancer. *Int J Radiat Oncol Biol Phys* 60:663–71.

Cheng, J.C.H., Peng, L.C., Chen, Y.H. et al. 2003. Unique role of proximal rectal dose in late rectal complications for patients with cervical cancer undergoing high-dose-rate intracavitary brachytherapy. *Int J Radiat Oncol Biol Phys* 57:1010–8.

Christensen, G., Carlson, B., Chao, C. et al. 2001. Image-based dose planning of intracavitary brachytherapy: Registration of serial-imaging studies using deformable anatomic templates. *Int J Radiat Oncol Biol Phys* 51:227–43.

Chun, M., Kang, S., Kil, H.J. et al. 2004. Rectal bleeding and its management after irradiation for uterine cervical cancer. *Int J Radiat Oncol Biol Phys* 58:98–105.

Coia, L., Won, M., Lanciano, R. et al. 1990. The patterns of care outcome study for cancer of the uterine cervix. *Cancer* 66:2451–6.

Corn, B.W., Hanlon, A.L., Pajak, T.F. et al. 1994. Technically accurate intracavitary insertions improve pelvic control and survival among patients with locally advanced carcinoma of the uterine cervix. *Gynecol Oncol* 53:294–300.

Creutzberg, C.L., van Putten, W.L., Koper, P.C. et al. 2000. Surgery and post operative radiotherapy versus surgery alone for patients with stage 1 endometrial carcinoma: Multi centre randomized trial. PORTEC study group. Post Operative radiation Therapy in endometrial carcinoma. *Lancet* 355:1404–11.

Creutzberg, C.L., van Putten, W.L., Warlam-Rodenhuis, C.C. et al. 2004. Outcomes of high risk IC, grade 3, compared with Stage I endometrial carcinoma patients: The post operative radiation therapy in endometrial carcinoma trail. *J Clin Oncol* 22:1234–41.

Crook, J.M., Esche, B.A., Chaplain, G. et al. 1987. Dose-volume analysis and the prevention of radiation sequelae in cervical cancer. *Radiother Oncol* 8:321–32.

Cunningham, D.E., Stryker, J.A., Velkley, D.E. et al. 1981. Routine clinical estimation of rectal, rectosigmoidal, and bladder doses from intracavitary brachytherapy in the treatment of carcinoma of the cervix. *Int J Radiat Oncol Biol Phys* 7:653–60.

De Brabandere, M., Mousa, A.G., Nulens, A. et al. 2008. Potential of dose optimization in MRI-based PDR brachytherapy of cervix carcinoma. *Radiother Oncol.* 88:217–26.

De Crevoisier, R., Sanfilippo, N., Gerbaulet, A. et al. 2007. Exclusive radiotherapy for primary squamous cell carcinoma of the vagina. *Radiother Oncol.* 85:362–70.

Decker, W., Erickson, B., Albano, K. et al. 2001. Comparison of traditional low dose rate to optimized and nonoptimized high dose rate tandem and ovoid dosimetry. *Int J Radiat Oncol Biol Phys* 50:561–7.

Delclos, L., Fletcher, G.H., Sampiere, V. et al. 1978. Can the Fletcher gamma ray colpostat system be extrapolated to other systems. *Cancer* 41:970–9.

Delclos, L., Fletcher, G.H., Moore, E.B. et al. 1980. Minicolpostats, dome cylinders, other additions and improvements of the Fletcher–Suit afterloadable system: Indications and limitations of their use. *Int J Radiat Oncol Biol Phys* 6:1195–206.

Demanes, D., Rodriguez, R., Bendre, D. et al. 1999. High dose rate transperineal interstitial brachytherapy for cervical cancer: High pelvic control and low complication rates. *Int J Radiat Oncol Biol Phys* 45:105–12.

Dicker, A., Puthawala, M., Thropay, J. et al. 2010. Prospective multi-center trial utilizing electronic brachytherapy for the treatment of endometrial cancer. *Radiation Oncol* 5:67.

Dimopoulos, J., Kirisits, C., Petric, P. et al. 2006a. The Vienna applicator for combined intracavitary and interstitial brachytherapy of cervical cancer: Clinical feasibility and preliminary results. *Int J Radiat Oncol Biol Phys* 66:83–90.

Dimopoulos, J., Schard, G., Berger, D. et al. 2006b. Systematic evaluation of MRI findings in different stages of treatment of cervical cancer: Potential of MRI on delineation of target, pathoanatomic structures, and organs at risk. *Int J Radiat Oncol Biol Phys* 64:1380–8.

Dimopoulos, J.C., De Vos, V., Berger, D. et al. 2009. Inter-observer comparison of target delineation for MRI-assisted cervical cancer brachytherapy: Application of the GYN GEC-ESTRO recommendations. *Radiother Oncol* 91:166–72.

Du Bois, A., Quinn, M., Thigpen, T. et al. 2005. 2004 consensus statement on the management of ovarian cancer: Final document of the 3rd International Gynecologic Cancer Intergroup Ovarian Cancer Consensus Conference. *Ann Oncol* 16(Suppl 8):viii 7–viii 12.

Eifel, P.J., Thoms, W.W., Smith, T.L. et al. 1993. The relationship between brachytherapy dose and outcome in patients with bulky endocervical tumors treated with radiation alone. *Int J Radiat Oncol Biol Phys* 28:113–8.

Eifel, P.J., Levenback, C., Wharton, J.T. et al. 1995. Time course and incidence of late complications in patients treated with radiation therapy for FIGO stage IB carcinoma of the uterine cervix. *Int J Radiat Oncol Biol Phys* 32:1289–300.

Eifel, P.J., Khalid, N., Erickson, B. et al. 2010. Patterns of radiotherapy practice for patients treated for intact cervical center in 2005-2–7: A QRRO study. *Int J Radiat Oncol Biol Phys* 78:S119–20.

Erickson, B. and Gillin, M. 1997. Interstitial implantation of gynecologic malignancies. *J Surg Oncol* 66:285–95.

Erickson, B. and Wilson, J.F. 1997. Clinical indications for brachytherapy. *J Surg Oncol* 65:218–27.

Erickson, B., Albano, K., and Gillin, M. 1996. CT-guided interstitial implantation of gynecologic malignancies. *Int J Radiat Oncol Biol Phys* 36:699–709.

Erickson, B., Jones, R., Rownd, J. et al. 2000. Is the tandem and ring applicator a suitable alternative to the high dose rate Selectron tandem and ovoid applicator? *J Brachytherapy Int* 16:31–144.

Esche, B.A., Crook, J.M., and Horiot, J.C. 1987a. Dosimetric methods in the optimization of radiotherapy for carcinoma of the uterine cervix. *Int J Radiat Oncol Biol Phys* 13:1183–92.

Esche, B.A., Crook, J.M., Isturiz, J. et al. 1987b. Reference volume, milligram-hours and external irradiation for the Fletcher applicator. *Radiother Oncol* 9:255–61.

Fader, A.N. and Rose, P.G. 2007. Role of surgery in ovarian carcinoma. *J Clin Oncol* 25:2873–33.

Ferrigno, R., Novaes, P., Pellizzon, A. et al. 2001. High dose rate brachytherapy in the treatment of uterine cervix cancer. Analysis of dose effectiveness and late complications. *Int J Radiat Oncol Biol Phys* 50:1123–35.

Ferrigno, R., Nishimoto, I.N., Novaes, P.E. et al. 2005. Comparison of low and high dose rate brachytherapy in the treatment of uterine cervix cancer. Retrospective analysis of two sequential series. *Int J Radiat Oncol Biol Phys* 62:1108–16.

Fishman, D.A., Roberts, K.B., Chambers, J.T. et al. 1996. Radiation therapy as exclusive treatment for medically inoperable patients with stage I and II endometroid carcinoma of the endometrium. *Gynecol Oncol* 61:189–96.

Fletcher, G.H. (ed.) 1980. *Textbook of Radiotherapy*, 3rd ed. Lea & Febiger, Philadelphia, PA, pp. 720–72.

Forrest, J.L., Ackerman, I., Barbera, L. et al. 2010. Patient outcome study of concurrent chemoradiation, external beam radiotherapy, and high-dose rate brachytherapy in locally advanced carcinoma of the cervix. *Int J Gynecol Cancer* 20:1074–8.

Fowler, J.F. 1990. The radiobiology of brachytherapy. In: *Brachytherapy HDR and LDR*. A.A. Martinez, C.G. Orton, and R.F. Mould (eds.). Nucletron, Columbia, pp. 121–37.

Frank, S.J., Jhingran, A., Levenback, C. et al. 2005. Definitive radiation therapy for squamous cell carcinoma of the vagina. *Int J Radiat Oncol Biol Phys* 62:138–47.

Fu, K.K. and Phillips, T.L. 1990. High-dose-rate versus low-dose-rate intracavitary brachytherapy for carcinoma of the cervix. *Int J Radiat Oncol Biol Phys* 19:791–6.

Georg, P., Lang, S., Dimopoulos, J. et al. 2011. Dose-volume histogram parameters and late side effects in magnetic resonance image-guided adaptive cervical cancer brachytherapy. *Int J Radiat Oncol Biol Phys* 79:356–62.

Gerszten, K., Faul, C., King, G. et al. 1998. High dose rate tandem and ring applicator movement with patient transfer from simulation to treatment room. *J Brachytherapy Int* 14:15–20.

Grigsby, P.W., Georgiou, A., Williamson, J.F. et al. 1993. Anatomic variation of gynecologic brachytherapy prescription points. *Int J Radiat Oncol Biol Phys* 27:725–9.

Grigsby, P., Russell, A., Bruner, D. et al. 1995. Late injury of cancer therapy on the female reproductive tract. *Int J Radiat Oncol Biol Phys* 31:1289–99.

Grigsby, P., Siegel, B., Dehdashti, F. et al. 2003. Post therapy surveillance monitoring of cervical cancer by FDG-PET. *Int J Radiat Oncol Biol Phys* 55:907–13.

Gupta, A., Vicini, F., Frazier, A. et al. 1999. Iridium-192 transperineal interstitial brachytherapy for locally advanced or recurrent gynecologic malignancies. *Int J Radiat Oncol Biol Phys* 43:1055–60.

Haas, J.S., Dean, R.D., and Mansfield, C.M. 1983. Fletcher–Suit–Delclos gynecologic applicator: evaluation of a new instrument. *Int J Radiat Oncol Biol Phys* 9:763–8.

Haas, J.S., Dean, R.D., and Mansfield, C.M. 1985. Dosimetric comparison of the Fletcher family of gynecologic colpostats 1950-1980. *Int J Radiat Oncol Biol Phys* 11:1317–21.

Haie-Meder, C., Pötter, R., Van Limbergen, E. et al. 2005. Recommendations for Gynecological (GYN) GEC-ESTRO Working Group (I): Concepts and terms in 3D image-based 3D treatment planning in cervix cancer brachytherapy with emphasis on MRI assessment of GTV and CTV. *Radiother Oncol* 74:235–45.

Haie-Meder, C., Chargari, C., Rey, A. et al. 2009. DVH parameters and outcome for patients with early-stage cervical cancer treated with preoperative MRI-based low dose rate brachytherapy followed by surgery. *Radiother Oncol* 93:316–21.

Haie-Meder, C., Chargari, C., Rey, A. et al. 2010. MRI-based low dose-rate brachytherapy experience in locally advanced cervical cancer patients treated by concomitant chemoradiotherapy. *Radiother Oncol* 96:161–5.

Han, I., Malviya, V., Chuba, P. et al. 1996. Multifractionated high dose rate brachytherapy with concomitant daily teletherapy for cervical cancer. *Gynecol Oncol* 63:71–7.

Hareyama, M., Sakata, K., Oouchi, A. et al. 2002. High dose rate vs. low dose rate intracavitary therapy for carcinoma of the uterine cervix. A randomized trial. *Cancer* 94:117–24.

Henschke, U.K. 1960. Afterloading applicator for radiation therapy of carcinoma of the uterus. *Radiology* 74:834.

Hilaris, B.S., Nori, D., and Anderson, L.L. 1988. Brachytherapy in cancer of the cervix. In: *An Atlas of Brachytherapy*. B.S. Hilaris, D. Nori, L.L. Anderson. (eds.). Macmillan Publishing Co., New York, NY, pp. 244–56.

Hintz, B.L., Kagan, A.R., Chan, P. et al. 1980. Radiation tolerance of the vaginal mucosa. *Int J Radiat Oncol Biol Phys* 6:711–6.

Holloway, C. and Viswanathan, A.N. 2010. Postoperative vaginal cylinder brachytherapy in an era of 3D imaging. In: *Gynecologic Radiation Therapy. Novel Approaches to Image-Guidance and Management*. A.N. Viswanathan, C. Kirisits, B.E. Erickson et al. (eds.). Springer-Verlag, Berlin, pp. 239–45.

Holloway, C., Racine, M.L., Cormack, R. et al. 2009. Sigmoid dose using 3D imaging in cervical-cancer brachytherapy. *Radiother Oncol* 93:307–10.

Homesley, H.D., Fillaci, V., Gibbons, S.K. et al. 2009. A randomized Phase III trial in advanced endometrial cancer of surgery and volume directed radiation followed by cisplatin and doxorubicin with and without paclitaxel. A gynecology oncology study. *Gynecol Oncol* 112:543–52.

International Commission on Radiation Units and Measurements (ICRU). 1985. Dose and volume specification for reporting and recording intracavitary therapy in gynecology. Report 38 of ICRU, ICRU Publications, Bethesda, MD.

Inoue, T., Hori, S., Miyata, Y. et al. 1978. High versus low dose rate intracavitary irradiation of carcinoma of the uterine cervix. A preliminary report. *Acta Radiol Oncol Radiat Phys Biol* 17:277–82.

Ito, H., Kutuki, S., Nishiguchi, I. et al. 1994. Radiotherapy for cervical cancer with high-dose rate brachytherapy correlation between tumor size, dose and failure. *Radiother Oncol* 31:240–7.

Jemal, A., Siegal, R., Ward, E. et al. 2008. Cancer Statistics 2008. *CA Cancer J Clin* 58:71–96.

Jemal, A., Siegel, R., Xu, J. et al. 2010. Cancer Statistics 2010. *CA Cancer J Clin* 60:277–300.

Jhingran, A., Burke, T., and Eifel, P.J. 2003. Definitive radiotherapy for patients with isolated vaginal recurrence of endometrial carcinoma after hysterectomy. *Int J Radiat Oncol Biol Phys* 56:1366–72.

Jones, N., Rankin, J., and Gaffney, D. 2004. Is simulation necessary for each high-dose-rate tandem and ovoid insertion in carcinoma of the cervix? *Brachytherapy* 3:120–4.

Jurgenliemk-Schulz, I.M., Tersteeg, R.J, Roesink, J.M. et al. 2009. MRI-guided treatment-planning optimization in intracavitary or combined intracavitary/interstitial PDR brachytherapy using tandem ovoid applicators in locally advanced cervical cancer. *Radiother Oncol* 93:322–30.

Kapp, K.S., Stuecklschweiger, G.F., Kapp, D.S. et al. 1992. Dosimetry of intracavitary placements for uterine and cervical carcinoma: Results of orthogonal film, TLD, and CT-assisted techniques. *Radiother Oncol* 24:137–46.

Kapp, K., Stueckschweiger, G., Kapp, D. et al. 1997. Carcinoma of the cervix: Analysis of complications after primary external beam radiation and Ir-192 HDR brachytherapy. *Radiother Oncol* 42:143–53.

Katz, A. and Eifel, P. 2000. Quantification of intracavitary brachytherapy parameters and correlation with outcome in patients with carcinoma of the cervix. *Int J Radiat Oncol Biol Phys* 48:1417–25.

Keys, H.M., Roberts, J.A., Brunetto, V.L. et al. 2004. A phase III trial of surgery with or without adjunctive external pelvic radiation therapy in intermediate risk endometrial adenocarcinoma: A gynecological oncology group study. *Gynecol Oncol* 92:744–51.

Kim, R., Meyer, J., Spencer, S. et al. 1996. Major geometric variation between intracavitary applications in carcinoma of the cervix: High dose rate vs. low dose rate. *Int J Radiat Oncol Biol Phys* 35:1035–8.

Kim, R., Caranto, J., Pareek, P. et al. 1997. Dynamics of pear-shaped dimensions and volume of intracavitary brachytherapy in cancer of the cervix: A desirable pear shape in the era of three-dimensional treatment planning. *Int J Radiat Oncol Biol Phys* 37:1193–9.

Kirisits, C., Lang, S., Dimopoulos, J. et al. 2006. The Vienna applicator for combined intracavitary and interstitial brachytherapy of cervical cancer: Design, application, treatment planning, and dosimetric results. *Int J Radiat Oncol Biol Phys.* 65:624–30.

Krishnan, L., Cytacki, E.P., Wolf, C.D. et al. 1990. Dosimetric analysis in brachytherapy of carcinoma of the cervix. *Int J Radiat Oncol Biol Phys* 18:965–70.

Kuipers, T., Mak, A., Van't Riet, A. et al. 2001. High dose rate brachytherapy in the treatment of cervical carcinoma: Review and current developments. *J Brachytherapy Int* 17:1–36.

Lanciano, R., Martz, K., Coia, L. et al. 1991a. Tumor and treatment factors improving outcome in stage IIIb cervix cancer. *Int J Radiat Oncol Biol Phys* 20:95–100.

Lanciano, R.M., Won, M., Coia, L.R., and Hanks, G.E. 1991b. Pretreatment and treatment factors associated with improved outcome in squamous cell carcinoma of the uterine cervix: A final report of the 1973 and 1978 patterns of care studies. *Int J Radiat Oncol Biol Phys* 20:667–76.

Landoni, F., Maneo, A., Colombo, A. et al. 1997. Randomized study of radical surgery versus radiotherapy for stage Ib–IIa cervical cancer. *Lancet* 350:535–40.

Lang, S., Nulens, A., Briot, E. et al. 2006. Intercomparison of treatment concepts for MR image assisted brachytherapy of cervical carcinoma based on GYNGEC-ESTRO recommendations. *Radiother Oncol* 78:185–93.

Lertsanguansinchai, P., Lertbutsayanukul, C., Shotelersuk, K. et al. 2004. Phase III randomized trial comparing LDR and HDR brachytherapy in treatment of cervical carcinoma. *Int J Radiat Oncol Biol Phys* 59:1424–31.

Lindegaard, J.C., Tanderup, K., Nielsen, S.K. et al. 2008. MRI-guided 3D optimization significantly improves DVH parameters of pulsed-dose-rate brachytherapy in locally advanced cervical cancer. *Int J Radiat Oncol Biol Phys* 71:756–64.

Logsdon, M. and Eifel, P. 1999. FIGO IIIB Squamous cell carcinoma of the cervix: An analysis of prognostic factors emphasizing the balance between external beam and intracavitary radiation therapy. *Int J Radiat Oncol Biol Phys* 43:763–75.

Magné, N., Chargari, C., SanFilippo, N. et al. 2010. Technical aspects and perspectives of the vaginal mold applicator for brachytherapy of gynecologic malignancies. *Brachytherapy* 9:274–7.

Mai, J., Erickson, B., Rownd, J. et al. 2001. Comparison of four different dose specification methods for high dose rate intracavitary radiation for treatment of cervical cancer. *Int J Radiat Oncol Biol Phys* 51:1131–41.

Martinez, A., Cox, R.S., and Edmundson, G.K. 1984. A multiple site perineal applicator (MUPIT) for treatment of prostatic, anorectal, and gynecologic malignancies. *Int J Radiat Oncol Biol Phys* 10:297–305.

Maruyama, Y., Van Nagell, J.R., Wrede, D.E. et al. 1976. Approaches to optimization of dose in radiation therapy of cervix carcinoma. *Radiology* 120:389–98.

Miller, D.A., Richardson, S., and Grigsby, P. 2010. A new method of anatomically conformal vaginal cuff HDR brachytherapy. *Gynecol Oncol* 116:413–8.

Mohan, R., Ding, I.Y., Toraskar, J. et al. 1985. Computation of radiation dose distributions for shielded cervical applicators. *Int J Radiat Oncol Biol Phys* 11:823–30.

Montana, G.S., Hanlon, A.L., Brickner, T.J. et al. 1995. Carcinoma of the cervix: patterns of care studies: Review of 1978, 1983, and 1988-1989 surveys. *Int J Radiat Oncol Biol Phys* 32:1481–6.

Mundt, A., Murpky, K.T., Rotmensh, J. et al. 2001. Surgery and Postoperative radiation therapy in FIGO stage IIIC endometrial cancer. *Int J Radiat Oncol Biol Phys* 50:1154–60.

Nag, S. and Gupta, N. 2000. A simple method of obtaining equivalent doses for use in HDR brachytherapy. *Int J Radiat Oncol Biol Phys* 46:507–13.

Nag, S., Erickson, B., Thomadsen, B. et al. 2000a. The American Brachytherapy Society recommendations for high-dose-rate brachytherapy for carcinoma of the cervix. *Int J Radiat Oncol Biol Phys* 48:201–11.

Nag, S., Erickson, B., Parikh, S. et al. 2000b. The American Brachytherapy Society recommendations for high-dose-rate brachytherapy carcinoma of the endometrium. *Int J Radiat Oncol Biol Phys* 48:779–90.

Nag, S., Chao, C., Erickson, B. et al. 2002. The American Brachytherapy Society recommendations for low-dose-rate brachytherapy for carcinoma of the cervix. *Int J Radiat Oncol Biol Phys* 52:33–48.

Nakamoto, Y., Eishburch, A., Achtyes, E. et al. 2002. Prognostic value of positron emission tomography using F-18 fluorodeoxyglucose in patients with cervical cancer undergoing radiotherapy. *Gynecol Oncol* 84:289–95.

Nout, R.A., Smit, V.T.H.B.M., Putter, H. et al. 2010. Vaginal brachytherapy versus pelvic external beam radiotherapy for patients with endometrial cancer of high-intermediate risk (PORTEC-2): An open-label, non-inferiority, randomized trial. *Lancet* 375:816–23.

Noyes, W., Peters, N., Thomadsen, B. et al. 1995. Impact of "optimized" treatment planning for tandem and ring, and tandem and ovoids, using high dose rate brachytherapy for cervical cancer. *Int J Radiat Oncol Biol Phys* 31:79–86.

Ogino, I., Kitamura, T., Okamoto, N. et al. 1995. Late rectal complication following high dose rate intracavitary brachytherapy in cancer of the cervix. *Int J Radiat Oncol Biol Phys* 31:725–34.

Okawa, T., Sakata, S., Kita-Okawa, M. et al. 1992. Comparison of HDR versus LDR regimens for intracavitary brachytherapy of cervical cancer: Japanese experience. In: *International Brachytherapy*. R.F. Mould (ed.). Nucletron International B.V., The Netherlands, pp. 13–7.

Orton, C.G. 1991a. HDR: forget not "time" and "distance". *Int J Radiat Oncol Biol Phys* 20:1131–2.

Orton, C.G. 1991b. Application of the linear quadratic model to radiotherapy for gynaecological cancers. *Selectron Brachytherapy J* Suppl 2:15–8.

Orton, C. 1998. High and low dose rate brachytherapy for cervical carcinoma. *Acta Oncol* 37:117–25.

Orton, C.G., Seyedsadr, M., and Somnay, A. 1991. Comparison of high and low dose rate remote afterloading for cervix cancer and the importance of fractionation. *Int J Radiat Oncol Biol Phys* 21:1425–34.

Patel, F., Rai, B., Mallick, I. et al. 2005. High-dose-rate brachytherapy in uterine cervical carcinoma. *Int J Radiat Oncol Biol Phys* 62:125–30.

Patel, F.D., Sharma, S.C., Negi, P.S. et al. 1994. Low dose rate vs. high dose rate brachytherapy in the treatment of carcinoma of the uterine cervix: A clinical trial. *Int J Radiat Oncol Biol Phys* 28:335–41.

Paterson, R. 1952. Studies in optimum dosage. The Mackenzie Davidson Memorial lecture. *Br J Radiol* 25:505–16.

Perez, C.A., Breaux, S., Madoc-Jones, H. et al. 1983. Radiation therapy alone in the treatment of carcinoma of the uterine cervix I. Analysis of tumor recurrence. *Cancer* 51:1393–402.

Perez, C.A., Breaux, S., Bedwinek, J.M. et al. 1984. Radiation therapy alone in the treatment of carcinoma of the uterine cervix II. Analysis of complications. *Cancer* 54:235–46.

Perez, C.A., Kuske, R., and Glasgow, G.P. 1985. Review of brachytherapy techniques for gynecologic tumors. *Encocuriether Hypertherm Oncol* 1:153–75.

Perez, C.A., Kuske, R.R., Camel, H.M. et al. 1988. Analysis of pelvic tumor control and impact on survival in carcinoma of the uterine cervix treated with radiation therapy alone. *Int J Radiat Oncol Biol Phys* 14:613–21.

Perez, C., Grigsby, P., Chao, C. et al. 1998. Tumor size, irradiation dose, and long-term outcome of carcinoma of uterine cervix. *Int J Radiat Oncol Biol Phys* 41:307–17.

Perez, C., Grigsby, P., Lockett, M. et al. 1999. Radiation therapy morbidity in carcinoma of the uterine cervix: Dosimetric and clinical correlation. *Int J Radiat Oncol Biol Phys* 44:855–66.

Petereit, D. and Pearcey, R. 1999. Literature analysis of high dose rate brachytherapy fractionation schedules in the treatment of cervical cancer: Is there an optimal fractionation schedule? *Int J Radiat Oncol Biol Phys* 43:359–66.

Petereit, D., Sakaria, J., Potter, D. et al. 1999. High dose rate vs. low dose rate brachytherapy in the treatment of cervical cancer: Analysis of tumor recurrence—The University of Wisconsin experience. *Int J Radiat Oncol Biol Phys* 45:1267–74.

Petric, P., Dimopoulos, J., Kirisits, C. et al. 2008. Inter- and intraobserver variation in HR-CTV contouring: Intercomparison of transverse and paratransverse image orientation in 3D-MRI assisted cervix cancer brachytherapy. *Radiother Oncol* 89:164–71.

Potish, R.A. and Gerbi, B.J. 1986. Cervical cancer: Intracavitary dose specification and prescription. *Radiology* 165:555–60.

Pötter, R., Knocke, T., Fellner, C. et al. 2000. Definitive radiotherapy based on HDR Brachytherapy with iridium 192 in uterine cervix carcinoma: Report on the Vienna University Hospital findings (1993–1997) compared to the preceding period in the context of ICRU 38 recommendations. *Cancer Radiother* 4:159–72.

Pötter, R., Haie-Meder, C., Van Limbergen, E. et al. 2006. Recommendations for Gynecological (GYN) GEC ESTRO Working Group (II): Concepts and terms in 3D image-based treatment planning in cervix cancer brachytherapy—3D volume parameters and aspects of 3D image-based anatomy, radiation physics radiobiology. *Radiother Oncol* 78:67–77.

Roeske, J., Mundt, A., Halpern, H. et al. 1997. Late rectal sequelae following definitive radiation therapy for carcinoma of the uterine cervix: A dosimetric analysis. *Int J Radiat Oncol Biol Phys* 37:351–8.

Rotte, K. 1980. A randomized clinical trial comparing a high dose-rate with a conventional dose-rate technique. *Br J Radiol Report* 17:75–9.

Rotte, K. 1990. Comparison of high dose rate afterloading and classical radium techniques for radiation therapy of cancer of the cervix uteri. In: *Brachytherapy HDR and LDR*. A.A. Martinez, C.G. Orton, and R.F. Mould (eds.). Nucletron, Columbia, pp. 110–8.

Sakata, K.I., Nagakura, H., Oouchi, A. et al. 2002. High-dose-rate intracavitary brachytherapy: Results of analyses of late rectal complications. *Int J Radiat Oncol Biol Phys* 54:1369–76.

Scholten, A.N., van Putten, W.L., Beerman, H. et al. 2005. PORTEC study group. Post operative radiotherapy for stage I endometrial carcinoma: Long term outcome of randomized PORTEC trial with central pathology review. *Int J Radiat Oncol Biol Phys* 63:834–8.

Sharma, S.C., Patel, F.D., Gupta, B.D. et al. 1991. Clinical trial of LDR versus HDR brachytherapy in carcinoma of the cervix. *Selectron Brachytherapy J* 5:75–9.

Sherman, M.E., Wang, S.S., Carreon, J. et al. 2005. Mortality trends for cervical squamous and adenocarcinoma in the United States relation to incidence and survival. *Cancer* 103:1258–64.

Shigematsu, Y., Nishiyama, K., Masaki, N. et al. 1983. Treatment of carcinoma of the uterine cervix by remotely controlled afterloading intracavitary radiotherapy with high-dose rate: a comparative study with a low-dose rate system. *Int J Radiat Oncol Biol Phys* 9:351–6.

Singh, A., Grigsby, P., Rader, J. et al. 2005. Cervix carcinoma, concurrent chemoradiotherapy, and salvage of isolated para-aortic lymph node recurrence. *Int J Radiat Oncol Biol Phys* 61:450–5.

Small, W. Jr, Erickson, B., and Kwakwa, F. 2005. American Brachytherapy Society survey regarding practice patterns of postoperative irradiation for endometrial cancer: Current status of vaginal brachytherapy. *Int J Radiat Oncol Biol Phys* 63:1502–7.

Stewart, A.J. and Viswanathan, A.N. 2006. Current controversies in high-dose-rate versus low-dose-rate brachytherapy for cervical cancer. *Cancer* 107:908–15.

Stitt, J.A. 1992. High-dose-rate intracavitary brachytherapy in gynecologic malignancies. *Oncology* 6:59–70.

Stitt, J.A., Fowler, J.F., Thomadsen, B.R. et al. 1992. High dose rate intracavitary brachytherapy for carcinoma of the cervix: the Madison system: I. Clinical and radiobiological considerations. *Int J Radiat Oncol Biol Phys* 24:335–48.

Straughn, J.M. Jr, Huh, W.K., Orr, J.M. et al. 2003. Stage IC adenocarcinoma of the endometrium. Survival comparisons of surgically staged patients with and without adjuvant irradiation therapy. *Gynecol Oncol* 89: 295–300.

Stryker, J., Bartholomew, M., Velkley, D. et al. 1988. Bladder and rectal complications following radiotherapy for cervix cancer. *Gynecol Oncol* 29:1–11.

Sturdza, A. and Pötter, R. 2011. Outcomes related to the disease and the use of 3D-based external beam radiation and image-guided brachytherapy. In: *Gynecologic Radiation Therapy. Novel Approaches to Image-Guidance and Management.* A.N. Viswanathan, C. Kirisits, B.E. Erickson et al. (eds.). Springer-Verlag, Berlin, pp. 263–82.

Stuecklschweiger, G.F., Arian-Schad, K.S., Poier, E. et al. 1991. Bladder and rectal dose of gynecologic high-dose-rate implants: comparison of orthogonal radiographic measurements with in vivo and CT-assisted measurements. *Radiology* 181:889–94.

Syed, A.M.N., Puthawala, A.A., Neblett, D. et al. 1986. Transperineal interstitial–intracavitary "Syed–Neblett" applicator in the treatment of carcinoma of the uterine cervix. *Endocuriether Hypertherm Oncol* 2:1–13.

Syed, A., Puthawala, A., Abdelaziz, N. et al. 2002. Long-term results of low-dose-rate interstitial–intracavitary brachytherapy in the treatment of carcinoma of the cervix. *Int J Radiat Oncol Biol Phys* 54:67–78.

Takeshi, K., Katsuyuki, K., Yoshiaki, T. et al. 1998. Definitive radiotherapy combined with high dose rate brachytherapy for stage III carcinoma of the uterine cervix: Retrospective analysis of prognostic factors concerning patient characteristics and treatment parameters. *Int J Radiat Oncol Biol Phys* 41:319–27.

Teshima, T., Inoue, T., Ikeda, H. et al. 1993. High-dose rate and low-dose rate intracavitary therapy for carcinoma of the uterine cervix. *Cancer* 72:2409–14.

Thomadsen, B.R., Shahabi, S., Stitt, J.A. et al. 1992. High dose rate intracavitary brachytherapy for carcinoma of the cervix: the Madison system: II. Procedural and physical considerations. *Int J Radiat Oncol Biol Phys* 24:349–57.

Thomas, G.M., Dembo, A.J., Bryson, S.C. et al. 1991. Changing concepts in the management of vulvar cancer. *Gynecol Oncol* 42:9–21.

Tod, M.C. and Meredith, W.J. 1938. A dosage system for use in the treatment of cancer of the uterine cervix. *Br J Radiol* 11:809–24.

Tod, M., and Meredith, W.J. 1953. Treatment of cancer of the cervix uteri—A revised "Manchester method". *Br J Radiol* 26:252–7.

Toita, T., Kakinohana, Y., Ogawa, K. et al. 2003. Combination external beam radiotherapy and high-dose-rate intracavitary brachytherapy for uterine cervical cancer: Analysis of dose and fractionation schedule. *Int J Radiat Oncol Biol Phys* 56:1344–53.

Ueda, S.M., Kapp, D.S., Cheung, M.K. et al. 2008. Trends in demographic and clinical characteristics in women diagnosed with corpus cancer and their potential impact on the increasing number of deaths. *Am J Obstet Gynecol* 198:218.e1–6.

Viani, G., Manta, G., Stefano, E. et al. 2009. Brachytherapy for cervix cancer: low-dose rate or high-dose rate brachytherapy—A meta-analysis of clinical trials. *J Exp Clin Cancer Res* 5;28:47.

Viswanathan, A.N., Cormack, R., Rawal, B. et al. 2009a. Increasing brachytherapy dose predicts survival for interstitial and tandem-based radiation for stage IIIB cervical cancer. *Int J Gynecol Cancer* 19:1402–6.

Viswanathan, A.N., Moughan, J., Small, W. et al. 2009b. Quality of cervical cancer brachytherapy implantation in RTOG prospective trials. *Int J Radiat Oncol Biol Phys* 75:S86.

Viswanathan, A.N., Erickson, B.E., and Rownd, J. 2010. Image-based approaches to interstitial brachytherapy. In: *Gynecologic Radiation Therapy. Novel Approaches to Image-Guidance and Management*. A.N. Viswanathan, C. Kirisits, B.E. Erickson et al. (eds.). Springer-Verlag, Berlin, pp. 247–59.

Viswanathan, A.N., Creutzberg, C.L., Craighead, P. et al. 2012. International brachytherapy practice patterns: A survey of the Gynecologic Cancer Intergroup (GCIG). *Int J Radiat Oncol Biol Phys.* 82(1):250–5.

Wang, C., Leung, S., Chen, H. et al. 1997. High-dose rate intracavitary brachytherapy (HDR-IC) in treatment of cervical carcinoma: 5-year results and implication of increased low-grade rectal complication on initiation of an HDR-IC fractionation scheme. *Int J Radiat Oncol Biol Phys* 38: 391–8.

Brachytherapy for Breast Cancer

Kara Lynne Leonard
Tufts University School of Medicine

Erik Van Limbergen
University Hospital Gasthuisberg

22.1 Introduction

Historical accounts of the use of brachytherapy in the treatment of breast cancer begin with the palliation of inoperable breast cancer by Dr. Gocht in 1897 (Gocht 1897). Later, Dr. Geoffrey Keynes pioneered the use of breast brachytherapy for breast conservation with operable breast cancer in the 1920s (Fletcher 1985; Aronowitz 2011). Now, 90 years later, with nearly 1.4 million new cases of breast cancer diagnosed in the world annually (Jemal et al. 2011), brachytherapy plays an important role in the treatment and cure of this disease.

22.2 Indications for Adjuvant Whole Breast Radiation Therapy

A series of clinical trials have demonstrated that breast conservation therapy, comprising a breast conserving surgery and adjuvant radiation therapy, provides equivalent survival as compared to mastectomy (Blichert-Toft et al. 1992; Jacobson et al. 1995; Arriagada et al. 1996; Veronesi et al. 2001; Fisher et al. 2002). These same studies showed that adjuvant radiation therapy is critical to in-breast tumor control. In fact, the omission of adjuvant radiation therapy resulted in-breast tumor recurrence rates roughly two-thirds higher. Subsequently, a meta-analysis has shown that improvements in local regional control related to whole breast irradiation (WBI) lead to long-term improvements

in breast cancer–specific survival (Early Breast Cancer Trialists' Collaborative Group (EBCTCG) 2005, 2011).

Adjuvant radiation therapy is now considered standard in breast conservation therapy for the treatment of women with early-stage breast cancer. Adjuvant radiotherapy is also integral to breast conservation therapy for ductal carcinoma in situ (DCIS) (Fisher et al. 1998, 1999; EORTC Breast Cancer Cooperative Group 2006; Cuzick et al. 2011; Holmberg et al. 2008; EBCTCG 2010). For both invasive and in situ disease, WBI to a dose of 42.5–50.4 Gy over a course of 4–5.5 weeks is most commonly used. This is typically followed by a tumor bed–directed boost. Two randomized Phase III clinical trials have demonstrated an in-breast tumor control benefit from the addition of a 10–16 Gy boost for patients with early-stage invasive disease (Romestaing et al. 1997; Bartelink et al. 1997). In the EORTC 22881-10882 trial, boost was delivered with either external beam (90%) or brachytherapy (8%) (Bartelink et al. 1997).

Analysis of local recurrences in patients treated without adjuvant WBI on the breast conservation trials demonstrated that the majority of in-breast tumor recurrences were within the tumor bed (Clark et al. 1992; Fisher et al. 1992; Liljegren et al. 1999; Veronesi et al. 2001). Similarly, observational studies have demonstrated that more than 90% of recurrences occur in the index quadrant (Van Limbergen et al. 1987; Vaidya et al. 2004; Clark et al. 1982). Based on these data, partial breast irradiation (PBI), targeting the postoperative tumor bed and a surrounding clinical target volume (CTV), has become popular in Europe and the United States.

22.2.1 Accelerated PBI (APBI)

Accelerated PBI (APBI) is more convenient than extended courses of external beam treatment to the whole breast. Women with restricted access to radiation therapy facilities often forgo a breast conservation option in favor of mastectomy for convenience sake (Athas et al. 2000; Farrow et al. 1992). The full APBI treatment can easily be delivered within 1 week, eliminating many of the complexities of coordinating daily care over a period of 4 to 6 weeks.

APBI was first investigated in clinical trials in the 1990s in the United Kingdom. A Phase I/II APBI trial was carried out at London's Guy's Hospital, and a Phase III trial was carried out at the Christie Hospital in Manchester. The Phase I/II trial resulted in a 37% in-breast tumor failure rate at 6 years median follow-up (Fentiman et al. 1996). The Phase III trial randomized patients to WBI or to APBI using a single electron-field approach. WBI therapy was associated with superior in-breast tumor control as compared to APBI (13% local failure versus 25% local failure, respectively) (Magee et al. 1996).

Subsequently, multiple institutions have demonstrated excellent local control and acceptable toxicity with APBI, most commonly using a dose of 34–38.5 Gy in 10 fractions, often delivered twice daily over 5 to 7 days (i.e., King et al. 2000; Ott et al. 2007; Shah et al. 2011; Polgar et al. 2008; Johansson et al. 2009; Arthur et al. 2008; Patel et al. 2008). The longest experience is with interstitial brachytherapy. The excellent long-term results of multiple Phase II trials and one randomized Phase III trial using multicatheter brachytherapy have renewed interest in APBI and resulted in the development of new radiation techniques such as intracavitary devices, seeds, electronic brachytherapy, external beam-based APBI, and intraoperative treatments with photons or electrons.

APBI trials have been initiated in the United States by the National Surgical Adjuvant Breast and Bowel Project/ Radiation Therapy Oncology Group (NSABP/RTOG 2011) and in Europe by the Groupe Européen de Curietherapie–European Society for Radiotherapy and Oncology (GEC-ESTRO 2011). Currently, four professional medical societies have published consensus criteria to identify patients suitable for APBI outside of clinical trials: (1) American Society for Radiation Oncology (ASTRO) (Smith et al. 2009); (2) GEC-ESTRO (Polgar et al. 2010); (3) American Brachytherapy Society (ABS 2012); and

TABLE 22.1 Recommendations for APBI from Professional Medical Societies

	ASTRO "Suitable" Patients (Smith et al. 2009)	ASTRO "Cautionary" Patients (Smith et al. 2009)	GEC-ESTRO "Low Risk" (Polgar et al. 2010)	GEC-ESTRO "Intermediate Risk" (Polgar et al. 2010)	ABS	ASBS
Age (years)	≥60	50–59	>50	40–50	≥50	≥45
Histology	IDC or other favorable subtypes (pure DCIS not allowed)	ILC; pure DCIS ≤ 3 cm	IDC without associated DCIS	ILC; IDC with associated DCIS	IDC	IDC or DCIS
Tumor size	≤ 2.0 cm; pT1	2.1–3.0 cm; pT0 or T2	<3.0 cm	T1–T2	≤3.0 cm	≤3.0 cm
Focality	Unicentric and unifocal	Clinically unifocal with total size 2.1–3.0 cm		Unifocal		
Grade	Any grade					
LVSI	No LVSI	LVSI: limited or focal	No LVSI			
EIC	No EIC	EIC ≤ 3.0 cm	No EIC			
ER status	ER+	ER–				
Margins	Negative by ≥ 2.0 mm	Close; <2.0 mm	Negative resection margins		No tumor at inked margin	Negative by ≥2.0 mm
Lymph node status	pN0 (i–, i+) by SLN biopsy or ALND		Lymph node negative	Limited nodal involvement (pN1a)	Lymph node negative by SLN biopsy or ALND (pN0)	Lymph node negative by SLN biopsy or ALND (pN0)
Genetic mutation status	BRCA1/2 mutation not present					
Treatment factors	Neoadjuvant chemotherapy not allowed					

Note: IDC = invasive ductal carcinoma; DCIS = ductal carcinoma in situ; LVSI = lymphovascular space invasion; EIC = extensive intraductal component; ER = estrogen receptor; pN0 = pathologically node negative; SLN = sentinel lymph node; ALND = axillary lymph node dissection; ILC = invasive lobular carcinoma.

(4) American Society of Breast Surgeons (ASBS 2011). A summary of these recommendations is displayed in Table 22.1. Generally, patients considered to be suitable for APBI are older and have smaller, biologically lower-risk tumors. Additionally, ASTRO identifies a group of patients who are unsuitable for treatment with APBI outside of clinical trial. These patients are younger than 50 years of age, have a known BRCA 1 or BRCA 2 mutation, have pure DICS, or have a tumor larger than 3 cm (or T3–T4 tumors). Patients with multicentric disease, microscopically multifocal disease measuring greater than 3 cm in total size, extensive intraductal component (EIC) measuring greater than 3 cm, extensive lymphovascular invasion (LVSI), positive margins, or those treated after the use of neoadjuvant chemotherapy, with no axillary evaluation, or with lymph-node positive disease are also grouped in this "unsuitable" category (Smith et al. 2009). Similarly, GEC-ESTRO identifies a group of patients in whom APBI (off of clinical trial) is contraindicated: patients younger than 40 years old with multifocal tumors or tumors larger than 3 cm, with positive or close resection margins, with extensive intraductal disease, LVSI, or lymph node involvement (Polgar et al. 2010).

22.3 Brachytherapy in Breast Cancer

The two main indications for the use of brachytherapy in the treatment of early-stage breast cancer are for adjuvant APBI as part of breast conservation therapy and for the breast boost. In the United States, brachytherapy is used in approximately 30% of APBI treatments. As previously mentioned, 8% of patients treated with boost in the EORTC boost trial were treated with ^{192}Ir brachytherapy implants (Bartelink et al. 1997). More rarely, breast brachytherapy is among the methods used as part of the adjuvant treatment for local recurrences in previously irradiated breasts (Hannoun-Levi et al. 2004; Resch et al. 2002a; Polgar et al. 2009) or for salvage treatment of chest wall recurrences (Fritz et al. 1997; Harms et al. 2001). In such situations, when clinical options are limited, brachytherapy may allow for delivery of definitive doses in previously irradiated tissues or tissues in close contact with radiosensitive normal structures.

22.3.1 Brachytherapy Boost Technique

Brachytherapy is an often-used technique for the delivery of the breast boost after adjuvant WBI as part of a breast conservation therapy. A freehand interstitial technique was first used to deliver the breast boost as a component of WBI. In current clinical settings, brachytherapy [high dose rate (HDR), pulsed dose rate (PDR), or low dose rate (LDR)] implants are not used as commonly as external beam electron, photon, or combined techniques. Brachytherapy techniques result in better local control rates (Mansfield et al. 1995; Hammer et al. 1994, 1998; Fourquet et al. 1995; Poortmans et al. 2004), and when applied properly, cosmetic results are comparable to or even better than those seen with linear accelerator-based boosts (Van Limbergen 2003).

22.3.2 Brachytherapy for APBI

APBI using brachytherapy provides an attractive alternative to 3D conformal radiotherapy (3D-CRT) APBI. With brachytherapy techniques, there is no requirement for planning target volume (PTV) expansions to account for setup error or patient motion. As such, the treated and irradiated volumes are much smaller with brachytherapy as compared to 3D-CRT treatment (Hammer et al. 1994, 1998; Patel et al. 2007a). The significance of the differences in treated volume is as of yet unclear. Initials reports of the 3D-CRT APBI technique developed at William Beaumont Hospital have shown acceptable toxicity rates with this method (Baglan et al. 2003; Vicini et al. 2003, 2007; Chen et al. 2010). Subsequent reports from Tufts/Brown using the same technique and from the University of Michigan using a similar, comparable technique have shown higher-than-expected rates of subcutaneous fibrosis and/or unacceptable cosmesis (Hepel et al. 2009; Jagsi et al. 2010). These two groups of investigators showed statistically significant associations between late toxicity and volume of treated breast. When larger volumes of normal breast tissue were treated to low or high doses, the risk of late toxicity was increased. These single-institution reports of higher-than-expected rates of late toxicity led the investigators of the ongoing NSABP-B39/RTOG-0413 to perform an interim analysis of the 1367 patients treated with APBI using the 3D-CRT technique. With a median time on study of 37 months, the rate of Grade 3–4 late toxicity in these patients was less than 3% (Julian et al. 2011). Although the current data remain conflicting, reducing the volume of breast receiving high and low doses may reduce the risk of late toxicity. Brachytherapy techniques have this advantage over 3D-CRT APBI treatments.

Interstitial brachytherapy was the first technique to be used for APBI and, as such, has the longest and most mature follow-up of all APBI techniques. The results of multiple series with follow-up of at least 4 years including a total of nearly 1000 patients are discussed in greater detail in Section 22.7. The annual recurrence rate of only 0.63% should be regarded as the standard to which newer techniques must be compared.

Using multicatheter interstitial brachytherapy techniques, an implant may be easily individualized to any patient and to any shape or size tumor bed, making it a universally available and versatile technique.

Newer techniques include MammoSite balloons with a single central catheter inside a balloon, Contura and MammoSite multichannel balloons, SAVI multichannel cage-like catheters, ClearPath hybrid catheters, and low-dose intraoperative photon (INTRABEAM) or electron (ELIOT) beams. Detailed descriptions of each of these techniques are provided below.

The biggest concern regarding brachytherapy APBI techniques relates to the placement of foreign bodies such as interstitial needles, plastic catheters, or intracavitary balloon or cage-like devices. The invasiveness of these treatments theoretically increases the risk for skin toxicity, wound complications, infections, seromas, or catheter failures. Moreover, breast brachytherapy, like most brachytherapy, is characterized by heterogeneous dose distributions within the target volume, placing the patient at risk for high dose–related complications.

22.4 Breast Brachytherapy Techniques

22.4.1 Interstitial Multicatheter Brachytherapy

22.4.1.1 Target Definition

Interstitial multicatheter brachytherapy (IMBT) is used to deliver both a breast boost and APBI. Accurate definition of the CTV and PTV are critical to the success of the interstitial brachytherapy. In some cases, an open procedure is performed, with catheters placed at the time of breast conservation surgery based on visualization and palpation of the surgical cavity under the guidance of the surgeon (Chen and Edmunson 2009; Kuske 2009). In most cases, however, a closed cavity technique is performed with the target volumes defined based on a postsurgical pretreatment CT scan. It is preferable for the surgeon to mark the tumor bed with radio-opaque clips that can be used to identify the radiographic boundaries of the tumor bed. The tumor bed is then expanded in three dimensions by 1–2 cm depending on the tumor type and on the tumor-free margin taken by the surgeon to create a CTV with careful attention to exclude the chest wall and 5 mm deep to the skin. Usually no PTV expansion margins have to be taken for IMBT, unless there is concern that the CTV can move along the axis of the implanted catheters.

For an IMBT breast boost following WBI, adequate CTV includes the tumor bed plus a 1.5–2.0 cm expansion, depending on the tumor type (EIC or no EIC), for tumors removed with negative margins. The CTV thus encompasses the tissue containing 80% of the microscopic tumor extensions around the primary tumor (Holland et al. 1985). In cases of surgically positive margins, expansions of up to 3.0 cm may be more appropriate.

22.4.1.2 Catheter/Needle Placement

The goal of the interstitial brachytherapy technique is appropriate placement of brachytherapy catheters to allow for sufficient dosimetric target coverage for delivery of APBI or the breast boost. Before proceeding with APBI using interstitial brachytherapy, patients must be confirmed as appropriate candidates for partial breast treatment. Following a breast conservation surgery, pathology should be carefully reviewed with respect to the appropriateness criteria detailed in Table 22.1. Each patient should undergo a postoperative planning CT scan to delineate the tumor bed and determine the technical feasibility of the interstitial implant. The target volume typically encompasses the tumor bed with a 1–2 cm expansion (the expansion radius depends on the length of the surgically tumor-free margins). The GEC-ESTRO breast cancer working group recommends a total margin of 2 cm including the surgical tumor-free margin and an additional margin up to 2 cm covered by the CTV. Similarly, the CT simulation scan can be used to determine catheter number, position, and placement relative to the tissue and to the other catheters. The details of techniques used at various institutions have been published, for example, Virginia Commonwealth

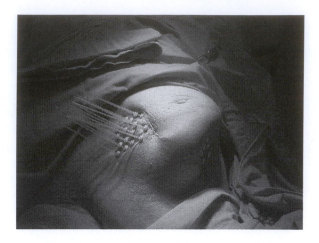

FIGURE 22.1 Interstitial catheter placement for APBI delivered with IMBT. (Courtesy of C. Polgar Limbergen.)

University, William Beaumont Hospital, and the University of Wisconsin/University of Arizona (Wazer et al. 2009; Strnad and Ott 2006).

At the time of the procedure, the patient should be positioned comfortably allowing for exposure of the breast to be treated. The patient should be placed in either the prone or supine position, as needles/catheters may be placed in either position, depending on the size of the breast (Patel et al. 2007a) and technique of the institution. Catheters may be placed under imaging (ultrasound or CT scan) guidance using an isocentric positioning technique or free hand. For placement, patients should undergo conscious sedation with prescription medications or general anesthesia. The skin should be cleaned with an antiseptic, and the procedure should be clean, with sterile draping. Based upon the planning imaging, the position of the catheters should be determined and the location of the entry and exit points marked on the patient's skin with a sterile pen. Many institutions use dual or triple plane templates to guide the needles. The areas of skin that will be traversed by needles should be numbed with 1%–2% lidocaine.

Needles/catheters are typically placed in one to three planes based on the thickness of the CTV. Catheters are typically placed in parallel planes 1.5–2.0 cm apart with 1.5–2.0 cm between catheters in the same plan to allow for coverage of the entire target volume. Typically, the deep plane is implanted first, followed by the superficial plane(s) (Figure 22.1). After the implant is completed, catheters to hold LDR sources or remotely afterloaded PDR or HDR sources may replace needles.

22.4.1.3 Dose Prescription

HDR is most typically used to deliver an APBI prescription dose of 30.3–34 Gy in 7–10 fractions. The multicatheter technique results in multiple small hotspots at overlap areas between catheters and planes, as displayed in Figure 22.2. Dosimetric coverage of 95% of the target volume receiving 95% of the dose is considered desirable. The volume of breast tissue receiving 150% of the prescribed dose (V_{150}) should be minimized; $V_{150} \geq 45$ cm^3 has been associated with increased risk of late toxicity (Wazer

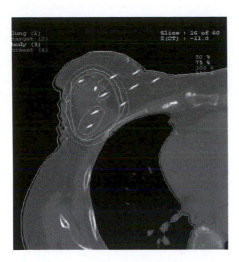

FIGURE 22.2 Dose distribution for APBI delivered with IMBT. (Courtesy of C. Polgar.)

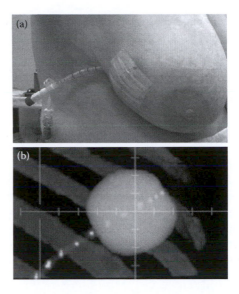

FIGURE 22.3 (a) Single-lumen MammoSite catheter placed in the tumor bed for APBI treatment and (b) corresponding digitally reconstructed radiograph (DRR). (Courtesy of K.L. Leonard.)

et al. 2006; Patel et al. 2007b). Carefully developed metrics can be used to ensure that the high dose region does not overlap the 5 mm of tissue immediately deep to the skin (Van Limbergen et al. 1990; Georg et al. 2005).

Breast boost doses delivered with IMBT typically are about 15–20 Gy with LDR or PDR and 8.5–10 Gy with HDR (Van Limbergen 2003; Georg et al. 2005; Resch et al. 2002b). This minimum target dose is typically prescribed to the periphery of the predetermined CTV/PTV and in accordance with the recommendations of ICRU 58 (ICRU 1997) is related as a percentage of the mean central dose (MCD) to allow for a description of dose inhomogeneity within the treated volume.

22.4.2 Intracavitary Balloons for HDR Brachytherapy

22.4.2.1 Single-Lumen Intracavitary Balloon Brachytherapy

The MammoSite is a single-lumen intracavitary balloon brachytherapy (single-lumen IBBT) device introduced to simplify the technical aspects of brachytherapy-based APBI (Figure 22.3). Similar to IMBT, single-lumen IBBT devices may be used for APBI or for the breast boost. MammoSite is most commonly used for ABPI with only a relatively few reports addressing the use of this device for delivery of breast boost (i.e., Shah et al. 2010). A central single-lumen catheter, which can accommodate remote afterloading of an HDR source, is surrounded by a dual-lumen silicone balloon catheter that expands 4–5 cm to fill the surgical tumor bed cavity. The device may be placed at the time of surgery with input from the surgeon or postoperatively in the outpatient setting. The former has been shown to be related to increased risk of seroma formation and, ultimately, higher rates of unacceptable cosmesis (Watkins et al. 2008; Evans et al. 2006). For the latter, percutaneous placement occurs under ultrasound and/or CT guidance by the radiation oncologist, surgeon, or

both. After placement of the catheter, the balloon is inflated to fill the cavity. A post placement CT scan is then obtained to verify appropriate location of the catheter and for treatment planning purposes.

22.4.2.2 Dose Prescription

When the HDR source is loaded into the central catheter, the dose most commonly prescribed for APBI is 34 Gy in 10 fractions, prescribed to 1 cm from the balloon surface that then falls off rapidly within the tissue (Figure 22.4). Thus, the dose is highest at the balloon surface. It is assumed that a 1.0-cm margin of compressed breast tissue around the inflated balloon corresponds to 1.6 cm of noncompressed breast tissue (Edmundson et al. 2002). Care must be taken to ensure that the balloon surface is not within 1.5 cm of the skin surface, as this has been shown to result in increased risk of skin toxicity because of

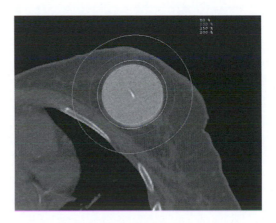

FIGURE 22.4 Dose distribution for APBI, 34 Gy in 10 fractions, delivered with MammoSite IBBT. (Courtesy of C. Polgar.)

overdosing of the skin (Belkacémi et al. 2009; Niehoff et al. 2006; Van Limbergen 2009).

22.4.2.3 Multilumen IBBT and Multilumen Cage-Like Intracavitary Brachytherapy

Multilumen IBBT catheters combine the simplified placement of balloon catheters with the more complex and flexible dosimetry of interstitial multicatheter techniques. There are currently at least two commercially available multilumen balloon catheters: the SenoRx Contura (Figure 22.5) and the MammoSite Multilumen (Figure 22.6), and at least two currently available multilumen cage-like catheters: the Strut-Adjusted Volume Implant (SAVI; Figure 22.7) and the ClearPath. These catheters are placed, just as the single-lumen catheters are, at the time of surgery or postoperatively. Because the SAVI is a nonballoon-based system, deployment of the cage-like peripheral lumens of the SAVI catheter results in a nonsymmetric lining of the

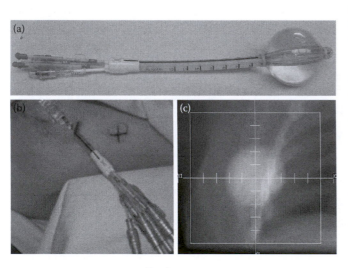

FIGURE 22.5 (a) SenoRx Contura catheter. (b) Image of Contura catheter placed postoperative for APBI. (c) DRR of patient treated with Contura-based APBI. (Courtesy of K.L. Leonard.)

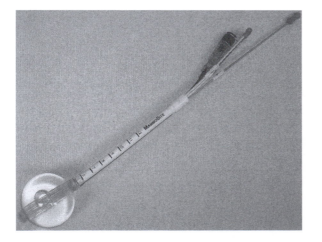

FIGURE 22.6 MammoSite multilumen catheter. (Courtesy of K.L. Leonard.)

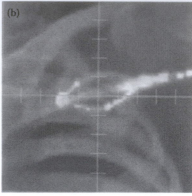

FIGURE 22.7 (a) Strut-Adjusted Volume Implant (SAVI) breast brachytherapy catheter. (b) DRR of SAVI catheter placed for APBI. (Courtesy of K.L. Leonard.)

lumpectomy cavity, allowing for unique dosimetric characteristics outlined below.

22.4.2.4 Dose Prescription

Similar to IMBT and single-lumen IBBT, 34 Gy in 10 fractions is the most commonly prescribed dose for treatment with multilumen IBBT and cage-like intracavitary devices. Each of the catheters/struts can be preferentially loaded to sculpt the dose, allowing for excellent PTV coverage, while limiting dose to organs at risk. Because these devices were developed to improve upon the dosimetry of single-lumen IBBT, there exist multiple studies examining the dosimetric characteristics of single-lumen IBBT and multilumen IBBT/multicatheter hybrid breast brachytherapy devices. Comparative dose distributions are presented in Figure 22.8.

In general, the skin dose is restricted to 125% of the prescribed dose, and the chest wall dose is limited to 145% of the prescribed dose while maintaining 95% of the PTV receiving 95% of the prescribed dose. However, it should be kept in mind that 125% of the prescribed dose to the skin will result in an equivalent dose of 61.6 Gy for late effects in 2-Gy fractions, which exceeds significantly the tolerance dose for skin telangiectasia that might result in more than 60% of grade 1, 30% of grade 2, and 5% of grade 3 telangiectasia (Turesson and Notter 1984). When the skin distance is small (<7 mm), the radial and posterior catheters may be preferentially loaded to allow for lower dose to skin. Decreasing the dose delivered from anterior dwell positions allows the skin dose to be kept as low as possible. For those patients on the Contura registry trial with skin distances ≥5 to <7 mm, the median skin dose was below 120% of the PD, and when skin thickness was <5 mm, the median skin dose was below 124% (Arthur et al. 2011). Similarly, in 11 patients with a minimum skin distance of 5 mm (deemed inadequate for treatment

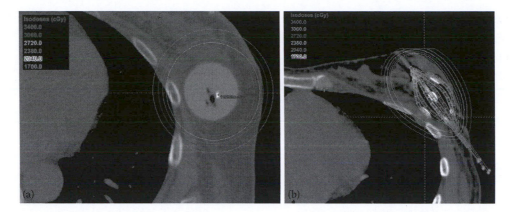

FIGURE 22.8 **(See color insert.)** (a) Comparison of dose distributions for APBI, 34 Gy in 10 fractions, delivered with Contura catheter, and (b) SAVI catheter. (Courtesy of K.L. Leonard.)

with MammoSite), treatment plans using the ClearPath catheter improved the median maximum skin dose from 5.5 Gy with MammoSite to 3.9 Gy with ClearPath (Beriwal et al. 2008).

Although all of these multisource devices help to sculpt the dose around the target and to significantly lower the dose delivered to the skin surface, the clinical benefit will only be achieved if the skin dose is reduced to less than 45 Gy in 2-Gy equivalents (EQD2) and to less than 90% of the prescription isodose.

The Contura device has been shown to limit skin dose without compromising PTV coverage (Scanderberg et al. 2008; Wilder et al. 2009; Brown et al. 2011; Cuttino et al. 2011; Arthur et al. 2011). The design of the Contura catheter, specifically the vacuum port, which allows for removal of seroma fluid or air, permits improved conformity of the balloon to the lumpectomy cavity. Among the 144 patients reviewed by Arthur et al. (2011) as part of a Phase IV trial using the Contura, 92% and 89% of cases met dose restriction to the skin (125%) and chest wall (145%), respectively, while maintaining that 95% of the PTV received 95% of the prescribed dose.

The SAVI, with its cage-like struts, also provides excellent tumor bed conformity, measured at greater than 99% (Bloom et al. 2011). As mentioned previously, the asymmetric deployment of this device results in unique dosimetry with dose hot spots slightly higher than those seen with IMBT. Dosimetric analyses of the SAVI have shown lower skin dose with lower dose homogeneity as compared to the MammoSite (Manoharan et al. 2010; Gurdalli et al. 2011; Bloom et al. 2011). Gurdalli et al. (2011) demonstrated dose homogeneity deemed acceptable by NSABP B-39/RTOG 0413 trial standards using the SAVI.

23.4.3 Electronic 50-kV Brachytherapy Balloon

The Axxent, an electronically generated 50-kV photon source embedded within a balloon delivery system, was introduced by Xoft (Figure 22.9). This electronic brachytherapy balloon (EBBT) delivery system possesses the depth-dose characteristics of a low-energy photon source resulting in a roughly sphere-like dose distribution with increased attenuation and more rapid fall-off. Similar to balloons for HDR ^{192}Ir brachytherapy, the balloon is implanted either at the time of surgery or postoperatively. Treatment with this device, however, can be delivered under convenient circumstances. For the reasons mentioned above, treatment may be delivered without shielding and therefore is significantly less expensive.

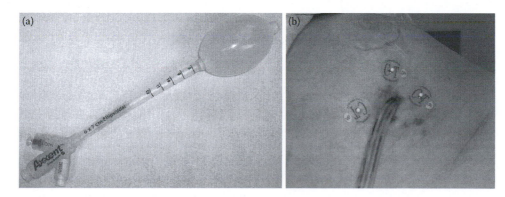

FIGURE 22.9 (a) Xoft Axxent brachytherapy catheter, (b) placed postoperatively in the lumpectomy cavity of a patient for APBI treatment. (Courtesy of K.L. Leonard.)

22.4.3.1 Dose Prescription

For APBI with EBBT devices, 34 Gy in 10 fractions is prescribed to 1 cm beyond the balloon surface. In a dosimetric comparison between MammoSite HDR breast brachytherapy and Xoft Axxent 50-kV breast brachytherapy, the electronic brachytherapy was associated with a statistically significantly lower values for ipsilateral breast $V_{150\%}$ (41.8% versus 59.4%; $p < 0.05\%$), ipsilateral breast $V_{50\%}$ (19.8% versus 13%; $p < 0.05$), ipsilateral lung $V_{30\%}$ (3.7% versus 1.1%; $p < 0.05$), and heart $V_{5\%}$ (59.2% versus 9.4%; $p < 0.05$) (Dickler et al. 2007).

22.4.4 Intraoperative Radiotherapy Using INTRABEAM or ELIOT

Targeted intraoperative radiotherapy (IORT) has been used both for tumor bed boost and for APBI (Vaidya et al. 2009, 2010, 2011). Treatment with these devices is considered a form of brachytherapy governed by AAPM Task Group-182. The INTRABEAM device, produced by the Zeiss company, has a comparable 50-kV source. This device comprises a portable x-ray tube that delivers 50-kV x-rays via a conical applicator paired with a spherical tip measuring 2–5 cm in diameter with eight sizes available. The spherical cone has to be inserted operatively, and the dose is delivered in a single fraction. The applicator is positioned in the center of the tumor bed, and radiation is delivered for 20–35 min to the target tissue surrounding the lumpectomy cavity. Careful postoperative pathologic analysis will identify any patients with adverse features considered to be at too high of risk for local recurrence to be treated solely with this technique. Women with any of these features should be treated with WBI to standard doses, minus a boost, which is considered to have already been delivered intraoperatively.

The ELIOT technique uses 6–12 MeV electrons via a mobile linear accelerator to deliver APBI or boost perioperatively in the operating theatre (Veronesi et al. 2010).

22.4.4.1 Dose Prescription

Using the INTRABEAM, the surface of the tumor bed typically receives 20 Gy in one intraoperative fraction, which is attenuated to 5–7 Gy at 1 cm depth. Similarly, the boost dose is 18–20 Gy at the surface of the applicator and 5–7 Gy at 1 cm into the surrounding tissues. Using the ELIOT intraoperative electron technique, a single fraction delivers 21 Gy directly to the tumor bed (Veronesi et al. 2010).

22.4.5 Permanent Breast Seed Implant

A newer, permanent breast seed brachytherapy implant (PBSI) technique has been pioneered at the University of Toronto (Pignol et al. 2006, 2009). Following breast conservation surgery, women return to the department of radiation oncology within a few weeks for treatment. After assessment and delineation of a postoperative tumor bed and seroma using CT scan and ultrasound, planning for placement of permanent ^{103}Pd

seeds commences. A tumor bed is considered appropriate if it contained an invasive cancer ≤ 3 cm, with surgical margins ≥ 2 mm, a clearly defined surgical bed > 5 mm from the skin surface, fluid cavity ≤ 2.5 cm, and PTV (lumpectomy cavity plus 1.5-cm margin less 5 mm deep to the skin surface) 70 cm^3 or smaller.

For the seed placement procedure, the PTV projection (determined from previous imaging studies) is outlined on the breast skin. Under ultrasound guidance, a stabilizing fiducial needle is placed in the tumor cavity and a template is attached. Stranded seeds are then placed into the surgical cavity via the template with preloaded needles as previously planned.

The procedure has a limited indication since a small percentage of referrals were eligible for the study and approximately half were technically implantable; many had to be cancelled after CT- or ultrasound-based simulation because of medial localization (which seems to be not suitable for the seed implant technique), a seroma 2.5 cm^3 or larger (risk of floating of the stranded seeds), an implant area larger than 4.5×5.5 cm^2 (implantation grid restrictions), or volumes larger than 70 cm^3 (radioprotection constraints).

Other concerns about permanent seeds were formulated by Godinez and Gombos (2006). Apart from difficulties of tumor bed localization, and correct positioning of the seeds at time of implantation, the dosimetric uncertainties related to tumor bed shrinking and a geometry, which is continuously changing with elapsing time after surgery. The tumor bed evaluation remains a best estimate, which may be temporarily good enough and acceptable for external beam or temporary implants but insufficient for permanent seed implantation. Seed migration may also be a more prominent problem in the loose fibro-fatty breast tissue and the changing seroma than it is in the firm compact prostate. Seeds may migrate to another quadrant of the breast, toward the skin or to remote areas of the body, immediately or at an unpredictable time later in the follow-up.

22.4.5.1 Dose Prescription

Using the pretreatment planning CT scan, 3D planning is used to prescribed 90 Gy to the minimum peripheral isodose line that covers the PTV. Immediately and then 2 months post-implant, patients undergo a second CT scan to calculate $V_{100\%}$ and $V_{200\%}$ as well as maximum point dose to the skin.

22.4.6 Noninvasive Image-Guided Breast Brachytherapy

AccuBoost, a mammography-based noninvasive image-guided breast brachytherapy (NIBB) system, has been developed to deliver the breast boost using HDR ^{192}Ir breast brachytherapy. Details of this delivery system, which comprises mammography paddles, a mammography image console, and surface applicators that attach to the mammography paddles to deliver treatment, are described in detail elsewhere (Figure 22.10) (Sioshansi et al. 2011; Hepel and Wazer 2011).

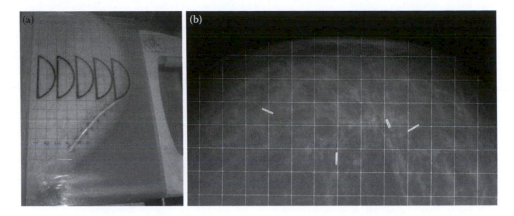

FIGURE 22.10 (a) Breast compression using mammography paddles, and (b) associated mammography image using the AccuBoost system for image-guided breast boost. (Courtesy of K.L. Leonard.)

22.4.6.1 Dose Prescription

AccuBoost is most typically used to deliver a boost dose of 10–14 Gy in five to seven fractions (Hamid et al. 2011). Monte Carlo dosimetric simulations and measured doses have been used to develop a treatment nomogram to easily deliver the prescription dose using parallel-opposed applicators along a single axis or orthogonal axes (Rivard et al. 2009a,b; Yang et al. 2011). Since dose is delivered in two directions with perpendicular breast compression, actual 3D dosimetry is not possible with this technique.

22.5 Comparison of Techniques

A careful comparison of each of the breast brachytherapy techniques is available in a recent review of breast brachytherapy written by Hepel and Wazer (2011) and outlined in Table 22.2.

22.6 Clinical Results of Breast Brachytherapy Boost

In general, interstitial brachytherapy boost results in superior local control compared to external beam boost, despite a smaller treated volume. The better local control rates seen with brachytherapy may be due to the comparatively higher nominal and biological effective dose of the interstitial implants. Among the studies comparing the local control rates for the breast boost delivered with breast brachytherapy to those delivered with external beam electrons or photons, retrospective analyses (i.e., Mansfield et al. 1995; Hammer et al. 1994, 1998) as well as a randomized trial (Fourquet et al. 1995), show better local control in patients treated with a brachytherapy boost. Moreover, a subgroup analysis of the EORTC 22881-10882 trial comparing local control in those treated with a brachytherapy boost to local control in those treated with an external beam boost showed a nonsignificant trend toward improved control in implanted patients (2.5% versus 4.5%, $p = 0.09$) (Poortmans et al. 2004). It

is important to remember that patients were not randomized for boost modality in this large study.

Touboul et al. (1995) showed nonsignificant differences in local failure rates between those treated with interstitial brachytherapy boost and those treated with external beam electron boosts (8.1% and 13.5%, respectively, at 10 years; $p = 0.32$). However, both groups were not strictly comparable; implanted patients were younger and less frequently treated with quadrantectomy. Similarly, the data from Wazer et al. (1997) suggested a nonsignificantly higher rate of local failure with interstitial brachytherapy boost than with linear-accelerated based boost. Importantly, the implanted patients in this series were substantially younger than the nonimplanted patients (51 versus 62 years).

The dose rate likely also impacts local control of the breast boost. Mazeron et al. (1991) reported that 5-year local control was 84% with dose rates higher than 50 cGy/h compared to 74% for lower dose rates. Thus, these authors recommended a dose rate of 60 cGy/h to maximize local control. Similarly, Deore et al. (1993) reported on 289 cases and found a significant increase in local failure rate with LDRs < 30 cGy/h as compared to higher LDRs up to 160 cGy/h (24% versus 5%–9%, respectively; $p < 0.05$).

Generally, external-beam photon or electron boost techniques are used more commonly than brachytherapy techniques for the breast boost, perhaps because of anecdotal assumptions that brachytherapy boost results in inferior cosmetic outcomes. Given the relatively small treated volumes and skin sparing with brachytherapy boost, good cosmetic outcomes are expected. Historically, the first two experiences of interstitial brachytherapy boost reported poor outcomes, demonstrating that technical performance is critical (Ray and Fish 1983; Fowble et al. 1986). Indeed, at least two randomized trials have evaluated cosmetic outcomes in patients randomized to receive either external beam photon/electron boost or HDR brachytherapy boost. The first trial compared cosmesis in patients treated with definitive radiation therapy alone including either a 20-Gy boost delivered with ^{60}Co external beam or a 20-Gy boost delivered with [192]Ir

TABLE 22.2   Comparison of Breast Boost and APBI Brachytherapy Techniques

Technique	Device	Advantages	Disadvantages
IMBT		• Most mature clinical follow-up. • Flexible to conform to complex tumor bed geometry.	• Invasive. • Multiple catheter placement may cause discomfort. • Catheter placement requires specialized expertise.
Single-lumen IBBT	MammoSite	• Simple insertion technique. • Simple spherical dosimetric geometry. • Large clinical experience, just beginning to mature.	• Invasive. • Fixed dosimetric geometry, not flexible to shape.
Multilumen IBBT	Contura; MammoSite Multilumen	• Simple insertion technique. • Simple spherical dosimetric geometry. • Improved flexibility to shape dose but limited.	• Invasive. • Improved flexibility to shape dose but limited. • Limited clinical experience.
Multilumen cage-like intracavitary brachytherapy	SAVI; Clear Path	• Simple insertion technique. • Flexibility to shape dose.	• Invasive. • Multiple hotspots at catheter–tissue interface. • Limited clinical experience.
EBBT	Xoft Axxent	• Simple insertion technique. • Simple spherical dosimetric geometry. • No vault shielding required. • Reduced heart, lung, and nontarget breast dose.	• Invasive. • Fixed dosimetric geometry. • Increased surface dose. • Limited clinical experience.
PBSI		• Convenient, single-day procedure with increased access to remote areas. • Flexible to conform to complex tumor bed geometry. • LDR may improve therapeutic ratio.	• Invasive. • Permanent seeds may be unacceptable to some patients. • Not appropriate for large CTVs or seroma cavities.
IORT	Intrabeam; Eliot	• Single-day procedure. • Not invasive beyond surgery.	• Relatively low treatment dose. • Short clinical follow-up. • Expensive.
NIBB	AccuBoost	• Noninvasive. • Breast immobilization and image guidance. • Sparing of non-target breast tissue compared with external beam techniques.	• Skin dose may be increased if there is significant skin overlap between orthogonal axes.

Source: Adapted from Hepel, J.T. and Wazer, D.E., *Brachytherapy*. August 3. Epub ahead of print, 2011.

Note: APBI = accelerated partial breast irradiation; IMBT = interstitial multicatheter brachytherapy; IBBT = intracavitary balloon brachytherapy; EBBT = electronic balloon brachytherapy; PBSI = permanent breast seed implant; LDR = low dose rate; CTV = clinical target volume; IORT = intraoperative radiotherapy; NIBB = noninvasive image-guided breast brachytherapy.

implant (Fourquet et al. 1995). After adjusting for prognostic and treatment factors, the brachytherapy boost was associated with lower in-breast tumor recurrence. Cosmetic outcome was comparable between the two groups. A second trial in women treated with breast-conserving surgery followed by a boost randomized women to a 15-Gy boost delivered with ^{192}Ir implant or with linear-accelerated-based combination of photon and electrons (Vass and Bairati 2005). Local control was excellent and comparable between groups, and there was no statistically significant difference in overall cosmetic outcome between the two groups. When measured by a digitized scoring system, objective breast retraction was greater among those treated with brachytherapy boost.

More recently, a retrospective review of patients at the University of Pennsylvania treated with ^{192}Ir boost has shown that these patients were less likely to have good or excellent cosmesis 1 year post-treatment than matched controls treated with electron boost (Hill-Kayser et al. 2011). The trend persisted throughout follow-up, but at 5 years, there was no significant difference in overall cosmesis between the two groups. There was also no difference between rates of local recurrence, freedom

from distant metastases, overall survival, or patterns of failure between groups.

Table 22.3 outlines the results of studies comparing the local control, cosmesis, or both, seen with external beam breast boost versus interstitial brachytherapy breast boost. In general, interstitial brachytherapy boosts resulted in superior local control and, if applied properly, equivalent or superior rates of good/excellent cosmesis.

The early results of the INTRABEAM technique as a boost (Vaidya et al. 2010, 2011) demonstrated excellent clinical results. At 5 years, this 20-Gy boost delivered to an unselected T1–T2 breast cancer population (including high-grade tumors in 29% of patients and pN+ disease in 29% of patients) resulted in a 1.74% local failure.

The newer AccuBoost surface brachytherapy technique has been shown to improve localization of the breast boost and to reduce the larger treated volumes inherent with external beam breast boost (Sioshansi et al. 2011). Results of a recently published registry study demonstrate the clinical ease and safety of this technique (Hamid et al. 2011).

TABLE 22.3 Comparison of External Beam and Interstitial Brachytherapy Techniques for Delivery of Breast Boost

Study	5-year Local Control Rates			Rates of Good/Excellent Cosmesis		
	External Beam	Brachytherapy	Comparison	External Beam	Brachytherapy	Comparison
Ray and Fish 1983				97%	74%	Not reported
Fowble et al. 1986				96%	88%	Not reported
Olivotto et al. 1989				85%	58%	$p < 0.03$
Sarin et al. 1993				40%	87%	$p < 0004$
De la Rochefordière et al. 1992	7%	8%	$p = ns$	97%	95%	$p = ns$
Mansfield et al. 1995	18% (10 years)	12% (10y)	$p < 0.05$	95%	91%	$p = ns$
Touboul et al. 1995[a]	8% (5 years); 15.5% (10 years)	5.5% (5 years); 8.1% (10 years)	$p = 0.32$	83%	62%	$p < 0.001$[a]
Perez et al. 1996	6%	7%	$p = ns$	81%		
Wazer et al. 1997	3.2% (5 years); 3.2% (7 years)	3.9% (5 years); 9% (7 years)	$p = ns$[b]	68% (20 Gy); 84% (10–14 Gy)	90%	$p = 0.001; p = 0.01$
Hammer et al. 1994	8.2%	4.3%	$p < 0.04$	70%	88%	$p < 0.0001$
Poortmans et al. 2004	4.5%	2.5%	$p = 0.09$			
Polgar et al. 2001	5.8% (4.5 years)	7.7% (4.5 years)	$p = 0.69$			

[a] Groups not strictly comparable: Implanted patients were younger (91% below 50 years, versus 70% below 50 years for electrons) and less frequently treated with quadrantectomy.

[b] All patients had close margins < 2 mm, but implanted patients were younger with a median age of 51 years versus 62 years ($p = 0.0001$).

22.7 Clinical Results of APBI

Of all the breast brachytherapy techniques described, IMBT has the most mature clinical follow-up data. Multiple Phase I-II trails both in Europe and the United States have demonstrated the safety and efficacy of LDR, PDR, and HDR brachytherapy for APBI. Table 22.4 outlines the clinical results of trials using IMBT for APBI with at least 4 years of follow-up. Among these trials, the in-breast tumor recurrence rate ranged from 0% to 9.3% (0.63% per year) with a good/excellent cosmesis rate of 75%–100%.

In a Phase III trial comparing WBI to PBI delivered with either 50-Gy external beam irradiation (31.2%) or with 36.4 Gy in seven fractions, HDR IMBT (68.8%) local recurrence was

TABLE 22.4 Local Failure Rates after Multicatheter Brachytherapy as APBI Modality after Breast-Conserving Surgery in Selected Patients (Review of Literature)

Institute	Selection	BT Type	N	Median Follow-up	Local failure	LF Per Year
Ochsner Clinic USA (King et al. 2000)	<4 cm, N0; N1bi; SM clear	LDR/HDR; 45/32–34 Gy	150	6.25	2%	0.3%
German-Austrian Phase II; (Ott et al. 2007)	≤3 cm; SM clear >2 mm pNo, ER +; ≥ 35 years	HDR 32 Gy; PDR 50 Gy	274	4 years	2.2%	0.55%
William Beaumont Hospital (Shah et al. 2011)	<3 cm, N0 N1bi; SM clear >2 mm	LDR/HDR; 50/32–34 Gy	199	9.6 years	4%	0.4%
NIO Budapest Phase II; (Polgar et al. 2008)[ac]	<2 cm, N0-1a; > 40 y	HDR; 30.3–36.4	45	12 years	9.3%	0.8%
Örebro Medical Center (Johansson et al. 2009)	<5 cm N0-1; SM clear	PDR; 50 Gy	43	7 years	6%	0.8%
RTOG 95-17 Phase II; (Arthur et al. 2008)	≤3 cm non ILA; SM clear pNo, 1a	LDR 45 Gy; HDR 34 Gy	99	6.7 years	6.1%	0.9%
Univ. Wisconsin (Patel et al. 2008)	≤3 mm non ILA; SM clear, pNo-1a	HDR 32–34 Gy	90 HR[d]; 183 LR[e]	4 years	2.9%	0.7%
NIO Budapest Phase III (Polgar et al. 2007)[bc]	<2 cm No-1a	HDR 36.4 Gy	88	6.8 years	4.5%	0.7%
		Total:	979	6.6 years	4.1%	0.63%

Note: SM = surgical margins.

[a] Results similar to matched pair controls.

[b] Randomized Phase III trial.

[c] Also patients < 40 years old included.

[d] HR = high risk < 50 years, ER −, pn1a.

[e] +LR = low risk: the rest.

equivalent in the WBI and PBI arms (3.1% versus 4.7% with 66 months median follow-up; $p = 0.05$) (Polgar et al. 2007). Cosmetic outcome was good or excellent in 77.6% of patients treated with PBI and in 62.9% of patients treated with WBI ($p = 0.009$). Among those treated with PBI, there was no significant difference in the rates of good/excellent cosmesis between those treated with IMBT and those treated with external beam PBI (81.2% versus 70%).

Apart from IMBT, all APBI techniques have short follow-up, inadequate for full assessment of local control and cosmetic outcome results. It is important to remember that median time to in-breast tumor failure is 7 years. Poor cosmesis typically begins to develop after 12–24 months, and telangiectasia and fibrosis may take even longer to develop (up to 5 years in some cases).

Keisch et al. (2005) published the results of the first Phase I/II trial designed to test the safety and efficacy of the MammoSite device for delivery of PBI or breast boost. The purpose of this trial was to obtain FDA clearance for the device. Initially, 70 patients were enrolled on the trial, with 54 able to have the device implanted and 43 qualifying for treatment on protocol. All 43 patients treated on the trial were treated with MammoSite-based APBI. In order to be eligible, patients had to be 45 years of age or older with tumors 2 cm or less in size. Histology had to be invasive ductal, and margins and nodes had to be negative. The MammoSite device was placed within 10 weeks of surgery if the cavity was considered suitable to accommodate the device. All patients were treated with 34 Gy in 10 fractions prescribed to a point 1 cm from the balloon surface, and treatment was delivered twice daily over 5 to 7 days. No severe side effects were reported with this trial, and the most commonly reported effects were erythema, desquamation, infection, and seroma formation. At 1 month, 88% of patients had good/excellent cosmesis. With medium-term follow-up, in-breast tumor control was 100% and 82%–83% of patients had good/excellent cosmesis. Infection developed in 9.3% of patients, telangiectasias were seen in 39.5% of patients, asymptomatic fat necrosis developed in 9.3%, and seroma formation occurred in 32.6% (symptomatic seroma formation occurred in 12% of patients). There was a statistically significant relationship between cosmesis and skin spacing with smaller skin-to-balloon distance associated with fair/poor cosmesis (Keisch et al. 2005; Benitez et al. 2004). A skin spacing of >7 mm was associated with better cosmesis. Thirty-five percent of patients developed localized fibrosis, and 7.5% of patients developed asymptomatic fat necrosis. Patient satisfaction was 100%.

The American Society of Breast Surgeon registry trial and five additional single- or multi-institutional trials have demonstrated results similar to those reported by the FDA trial (Vicini et al. 2011; Soran et al. 2007; Chen et al. 2007; Cuttino et al. 2008; Dragun et al. 2007; Chao et al. 2007). Among these studies, the local recurrence rate was low, ranging from 0% to 5% at 5 years. The rates of good/excellent cosmesis were 88%–91% at 5 years. Chao et al. (2007) replicated the findings from the FDA trial showing that skin spacing ≥ 7 mm was associated with better cosmesis, whereas the data from Cuttino et al. (2008) showed

that skin spacing ≤ 6 mm was statistically significantly associated with severe skin reaction and telangiectasia.

Unexpected complexities involved with MammoSite balloon brachytherapy have been discussed extensively by Prosnitz and Marks (2006) and by Van Limbergen (2009). The exact timing of perioperative or postoperative insertion in open or closed cavities, the conformance of the balloon to the cavity and to the asymmetrical target, and the balloon-to-skin distance have been associated with higher rates of toxicity. Intraoperative placement of single-lumen IBBT has been shown to increase the risk of seroma, and balloon-to-skin distance < 15 mm has been shown to increase the risk of skin toxicity. Authors also discussed the high costs of single-lumen IBBT HDR treatments and the lack of long-term follow-up available from Phase II studies performed on a limited number of patients (Belkacémi et al. 2009; Niehoff et al. 2006; Van Limbergen 2009).

Direct comparisons between the Contura and MammoSite have shown improved chest wall and skin dose with the multicatheter IBBT device (Wilder et al. 2009; Brown et al. 2011; Cuttino et al. 2011;, Arthur et al. 2011). For example, the group from Virginia Commonwealth University showed that even with similar skin distances, the mean skin dose was significantly lower with Contura than with MammoSite (67% versus 82% of the prescribed dose, respectively; $p < 0.05$) (Cuttino et al. 2011). Similarly, the mean maximum rib dose was 82% of prescription dose with Contura and 105% of the prescription dose with MammoSite ($p < 0.05$).

Preliminary clinical results of APBI delivered with the SAVI appear promising (Yashar et al. 2009, 2011). Among the first 100 patients treated with this device at two institutions, local control was 99%, with a median follow-up of 21 months. With this very short follow-up, the rate of telangiectasia was 1.9% and hyperpigmentation occurred in 9.8% of the patients. Grade 2 fibrosis, symptomatic seroma, and asymptomatic fat necrosis each developed in 1.9% of the patients.

A very early publication from a clinical trial of treatment with EBBT reported toxicity results from 11 patients treated with APBI using the Xoft Axxent (Ivanov et al. 2011). Grade 2 or less breast fibrosis developed in two patients. There were no in-breast tumor recurrences with 1-year follow up. With this very short follow-up, cosmesis was good or excellent in all patients.

The results from the inaugural Phase I/II trial of 67 patients treated with PBSI demonstrated 100% local control at a median follow-up of 32 months (Pignol et al. 2009). As previously described, only a small fraction of patients (4.6%) initially enrolled met criteria for implantation. The procedure was well tolerated, and the acute skin reaction developed in 42% of the patients; 10.4% of the patients developed desquamation. Skin reaction was less common when the skin received less than 85% of the prescribed dose. One patient developed an infectious abscess.

The TARGIT-A trial has demonstrated excellent clinical results and statistical noninferiority of TARGIT as compared to WBI with respect to in-breast tumor control (Vaidya et al. 2010). The four-year recurrence rate was only 1.2% in the TARGIT arm and 0.95% in the WBI arm ($p = 0.45$). Moreover, rates of major toxicity were similar in the TARGIT and WBI arms (3.3% versus 3.9%, respectively). Skin breakdown, infection, and seroma formation were the most common clinically significant adverse events, and each occurred in less than 3% of the patients in both arms. However, the mean follow-up was only 25 months. Thus, these early results should be carefully interpreted.

The Milan Group reported on a large group of 1822 patients with unifocal tumors up to 2.5 cm treated with the ELIOT system between 2000 and 2008 (Veronesi et al. 2010). After a median follow-up of 36.1 months, 3.6% in-breast recurrences were reported. Local side effects were mainly liponecrosis (4.2%) and fibrosis (1.8%).

22.8 Overview of Ongoing Phase III Clinical Trials

There are currently two ongoing randomized trials, one in the United States (NSABP B-39/RTOG 0413) and one in Europe (GEC-ESTRO), comparing APBI to WBI. When the NSABP B-39/RTOG 0413 trial initially opened, eligible patients included those with Stage 0, I, or II breast cancer resected by lumpectomy with a tumor size ≤ 3 cm and ≤ 3 histologically positive lymph nodes. Patients are randomized to WBI (50 Gy in 25 fractions or 50.4 Gy in 28 fractions) followed by an optional boost or APBI according to one of the following schemas to be chosen at the discretion of the treating physician: (1) 34 Gy in 3.4-Gy fractions using multicatheter brachytherapy, or (2) 34 Gy in 3.4-Gy fractions using MammoSite balloon catheter or other intracavitary device, or (3) 38.5 Gy in 3.85-Gy fractions using 3D-CRT. Initially, interstitial brachytherapy and MammoSite catheters were the only two brachytherapy modalities allowed on the trial. Subsequently, however, the trial was modified to include treatment with MammoSite Multilumen, Contura, and SAVI. Because the trial rapidly accrued low-risk patients, the eligibility is currently limited to women younger than 50 years of age with DCIS or invasive cancer and women of any age with ER negative tumors or positive lymph nodes. As of the last report of data from the ongoing NSABP B-39/RTOG 0413 trial, 3738 patients are enrolled (86.9% of target accrual); 1367 of these have been randomized to APBI (Julian et al. 2011). The majority of patients enrolled on the trial thus far have been treated with 3D-CRT; however, brachytherapy techniques have been frequently employed.

The European GEC-ESTRO breast cancer working group APBI trial randomized women ≥40 years of age with Stage 0, I, or II breast cancer with tumor size < 3 cm with pN0/Nmic nodal status to IMBT-based APBI or external beam WBI. APBI was delivered with HDR (32 Gy in eight 4.0-Gy fractions or 30.3 Gy in seven 4.3-Gy fractions) or PDR 50 Gy in 0.6–0.8 Gy pulses hourly. The WBI dose was either 50.4 Gy in 28 fractions followed by a tumor bed boost or 50 Gy in 25 fractions followed by a tumor bed boost.

References

ABS 2012. American Brachytherapy Society Breast Brachytherapy Task Group. Available at: http://www.american brachytherapy.org/guidelines/abs_breast_brachytherapy_taskgroup.pdf. Last accessed on August 13, 2012.

Aronowitz, J.N. 2011. Partial breast irradiation by brachytherapy. *Brachytherapy* 10:427–31.

Arriagada, R., Le, M.G., Rochard, F., and Contesso, G. 1996. Conservative treatment versus mastectomy in early breast cancer: patterns of failure with 15 years of follow up data. Institut Gustave-Roussy breast cancer group. *J Clin Oncol* 14: 1558–64.

Arthur, D.W., Winter, K., Kuske, R.R. et al. 2008. A Phase II trial of brachytherapy alone after lumpectomy for select breast cancer: Tumor control and survival outcomes of RTOG 95-17. *Int J Radiat Oncol Biol Phys* 72:467–73.

Arthur, D.W., Vicini, F.A., Todor, D.A., Julian, T.B., and Lyden, M.R. 2011. Improvements in critical dosimetric endpoints using the Contura multilumen balloon breast brachytherapy catheter to deliver accelerated partial breast irradiation: Preliminary dosimetric findings of a Phase IV trial. *Int J Radiat Oncol Biol Phys* 79:26–33.

ASBS 2011. American Society of Breast Surgeons. Available at: http://www.breastsurgeons.org/statements/PDF_Statements/APBI_statement_revised_100708.pdf. Last accessed on October 17, 2011.

Athas, W.F., Adams-Cameron, M., Hunt, W.C., Amir-Fazli, A., and Key, C.R. 2000. Travel distance to radiation therapy and receipt of radiotherapy following breast-conserving surgery. *J Natl Cancer Inst* 92:269–71.

Baglan, K.L., Sharpe, M.B., Jaffray, D. et al. 2003. Accelerated partial breast irradiation using 3D conformal radiation therapy (3D-CRT). *Int J Radiat Oncol Biol Phys* 55:302–11.

Bartelink, H., Horiot, J.-C., Poortmans, P.M. et al. 1997. Impact of a higher radiation dose on local control and survival in breast-conserving therapy of early breast cancer: 10-yr results of the randomized boost versus no boost EORTC 22881-10882. *J Clin Oncol* 25:3259–65.

Belkacémi, Y., Cauvet, M.P., Giard, S. et al. 2009. Partial breast irradiation as sole therapy for low risk breast carcinoma: Early toxicity, cosmesis and quality of life results of a MammoSite brachytherapy Phase II study. *Radiother Oncol* 90:23–9.

Benitez, P.R., Chen, P.Y., Vicini, F.A. et al. 2004. Partial breast irradiation in breast conserving therapy by way of interstitial brachytherapy. *Am J Surg* 188:355–64.

Beriwal, S., Coon, D., Kim, H., Haley, M., Patel, R., and Das, R. 2008. Multicatheter hybrid breast brachytherapy: A potential alternative for patients with inadequate skin distance. *Brachytherapy* 7:301–4.

Blichert-Toft, M., Rose, C., Anderson, J.A. et al. 1992. Danish randomized trial comparing breast conservation therapy with mastectomy: Six years of life-table analysis. Danish breast cancer cooperative group. *J Natl Cancer Inst Monogr* 11:19–25.

Bloom, E.S., Kirsner, S., Mason, B.E. et al. 2011. Accelerated partial breast irradiation using the strut-adjusted volume implant single-entry hybrid catheter in brachytherapy for breast cancer in the setting of breast augmentation. *Brachytherapy* 10:178–83.

Brown, S., McLaughlin, M., Pope, D.K. et al. 2011. A dosimetric comparison of the Contura multilumen balloon breast brachytherapy catheter vs. the single-lumen MammoSite balloon device in patients treated with accelerated partial breast irradiation at a single institution. *Brachytherapy* 10:68–73.

Chao, K.K., Vicini, F.A., Wallace, M. et al. 2007. Analysis of treatment efficacy, cosmesis, and toxicity using the MammoSite breast brachytherapy catheter to deliver accelerated partial-breast irradiation: The William Beaumont hospital experience. *Int J Radiat Oncol Biol Phys* 69:32–40.

Chen, P. and Edmundson, G. 2009. The William Beaumont Hospital technique of interstitial brachytherapy. In: *Accelerated Partial Breast Irradiation*, 2nd ed. D.E. Wazer, D.W. Arthur, and F.A. Vicini (eds.), Springer-Verlag, New York, pp. 203–18.

Chen, P.Y., Wallace, M., Mitchell, C. et al. 2010. Four-year efficacy, cosmesis, and toxicity using three-dimensional conformal external beam radiation therapy to deliver accelerated partial breast irradiation. *Int J Radiat Oncol Biol Phys* 76:991–7.

Chen, S., Dickler, A., Kirk, M. et al. 2007. Patterns of failure after MammoSite brachytherapy partial breast irradiation: A detailed analysis. *Int J Radiat Oncol Biol Phys* 69:250–31.

Clark, R.M., Wilkinson, R.H., Mahoney, L.J., Reid, J.G., and MacDonald, W.D. 1982. Breast cancer: A 21-year experience with conservative surgery and radiation. *Int J Radiat Oncol Biol Phys* 8:967–79.

Clark, R.M., McCulloch, P.B., Levine, M.N. et al. 1992. Randomized clinical trial to assess the effectiveness of breast irradiation following lumpectomy and axillary dissection for node-negative breast cancer. *J Natl Cancer Inst* 84:683–9.

Cuttino, L.W., Keisch, M., Jenrette, J.M. et al. 2008. Multi-institutional experience using the MammoSite radiation therapy system in the treatment of early-stage breast cancer: 2-year results. *Int J Radiat Oncol Biol Phys* 71:107–14.

Cuttino, L.W., Todor, D., Rosu, M., and Arthur, D.W. 2011. A comparison of skin and chest wall dose delivered with multicatheter, Contura multilumen balloon, and MammoSite breast brachytherapy. *Int J Radiat Oncol Biol Phys* 79:34–8.

Cuzick, J., Sestak, I., Pinder, S. et al. 2011. Effect of tamoxifen and radiotherapy in woman with locally excised ductal carcinoma in situ: Long-term results from the UK/ANZ DCIS trial. *Lancet Oncol* 12:21–9.

De la Rochefordière, A., Abner, A.L., Silver, B., Vicini, F., Recht, A., and Harris, J.R. 1992. Are cosmetic results following conservative surgery and radiation therapy for early breast cancer dependent on technique? *Int J Radiat Oncol Biol Phys* 23:925–31.

Deore, S.M., Sarin, R., Dingshaw, K.A., and Shrivastava, S.K. 1993. Influence of dose rate and dose per fraction on clinical outcome of breast cancer treated by external beam irradiation plus iridium-192 implants. Analysis of 289 cases. *Int J Radiat Oncol Biol Phys* 26:601–6.

Dickler, A., Kirk, M.C., Seif, N. et al. 2007. A dosimetric comparison of MammoSite high-dose-rate brachytherapy and Xoft Axxent electronic brachytherapy. *Brachytherapy* 6:164–8.

Dragun, A.E., Harper, J.L., Jenrette, J.M., Sinha, D., and Cole D.J. 2007. Predictors of cosmetic outcome following MammoSite breast brachytherapy: A single- institution experience of 100 patients with two years of follow up. *Int J Radiat Oncol Biol Phys* 68:354–8.

Early Breast Cancer Trialists' Collaborative Group (EBCTCG), Clarke, M., Collins, R., Darby, S. et al. 2005. Effects of radiotherapy and of differences in the extent of surgery for early breast cancer on local recurrence and 15-year survival: An overview of the randomised trails. *Lancet* 366:2087–106.

Early Breast Cancer Trialists' Collaborative Group (EBCTCG), Correa, C., McGale, P., Taylor, C. et al. 2010. Overview of the randomized trials of radiotherapy in ductal carcinoma in situ of the breast. *J Natl Cancer Inst Monogr* 2010:162–77.

Early Breast Cancer Trialists' Collaborative Group (EBCTCG), Darby, S., McGale, P., Correa, C. et al. 2011. Effect of radiotherapy after breast-conserving surgery on 10-year recurrence and 15-year breast cancer death: Meta-analysis of individual patient data for 10,801 women in 17 randomised trials. *Lancet* 378:1707–16.

Edmundson, G.K., Vicini, F.A., Chen, P.Y., Mitchell C., and Martinez, A.A. 2002. Dosimetric characteristics of the MammoSite RTS, a new breast brachytherapy applicator. *Int J Radiat Oncol Biol Phys* 52:1132–9.

EORTC Breast Cancer Cooperative Group; EORTC Radiotherapy Group, Bijker, N., Meijnen, P., Peters, J.L. et al. 2006. Breast-conserving treatment with or without radiotherapy in ductal carcinoma-in-situ: ten-year results of European Organisation for Research and Treatment of Cancer randomized Phase III trial 10853—A study by the EORTC Breast Cancer Cooperative Group and EORTC Radiotherapy Group. *J Clin Oncol* 24:3381–7.

Evans, S.B., Kaufman, S.A., Price, L.L., Cardarelli, G., DiPetrillo, T.A., and Wazer, D.E. 2006. Persistent seroma after intraoperative placement of MammoSite for accelerated partial breast irradiation: Incidence, pathologic anatomy, and contributing factors. *Int J Radiat Oncol Biol Phys* 65:333–9.

Farrow, D.C., Hunt, W.C., and Samet, J.M. 1992. Geographic variation in the treatment of localized breast cancer *N Eng J Med* 326:1097–101.

Fentiman, I.S., Poole, C., Tong, D. et al. 1996. Inadequacy of iridium implant as sole radiation treatment for operative breast cancer. *Eur J Cancer* 32A:608–11.

Fisher, B., Dignam, J., Wolmark, N. et al. 1998. Lumpectomy and radiation therapy for the treatment of intraductal breast cancer: Findings from National Surgical Adjuvant Breast and Bowel Project B-17. *J Clin Oncol* 16:441–52.

Fisher, B., Dignam, J., Wolmark, N. et al. 1999. Tamoxifen in treatment of intra- ductal breast cancer: National Surgical Adjuvant Breast and Bowel Project B-24 randomised controlled trial. *Lancet* 353:1993–2000.

Fisher, B., Anderson, S., Bryant, J. et al. 2002. Twenty-year follow up of a randomized trial comparing total mastectomy, lumpectomy, and lumpectomy plus irradiation for the treatment of invasive breast cancer. *N Engl J Med* 347:1233–41.

Fisher, E.R., Anderson, S., Redmond, C., and Fisher, B. 1992. Ipsilateral breast tumor recurrence and survival following lumpectomy and irradiation: Pathological findings from NSABP protocol B-06. *Semin Surg Oncol* 8:161–6.

Fletcher, G.H. 1985. History of irradiation in the primary management of apparently regionally confined breast cancer. *Int J Radiat Oncol Biol Phys* 11:2132–42.

Fourquet, A., Campana, F., Mosseri, V. et al. 1995. Iridium-192 versus cobalt-60 boost in 3–7 cm breast cancer treated by irradiation alone: final results of a randomized trial. *Radiother Oncol* 34:114–20.

Fowble, B., Solin, L.J., Martz, K.L., Pajak, T.F., and Goodman, R.L. 1986. The influence of the type of boost (electrons vs. implant) on local control and cosmesis in patients with stages I and II breast cancer undergoing conservative surgery and radiation. *Int J Radiat Oncol Biol Phys* 12:150 [abstract].

Fritz, P., Berns, C., Anton, H.W. et al. 1997. PDR brachytherapy with flexible implants for interstitial boost after breast-conserving surgery and external beam radiation therapy. *Radiother Oncol* 45:23–32.

Georg, P., Georg, D., and Van Limbergen, E. 2005. The use of the source-skin distance measuring bridge indeed reduces skin telangiectasia after interstitial boost in breast conserving therapy. *Radiother Oncol* 74:323–30.

Gocht, H. 1897. Therapeutische verwendung der Röntgenstrahlen (Therapeutic use of x-rays). *Fortschr Geb Röntgenstr* 1:14–28.

Godinez, J. and Gombos, E.C. 2006. A permanent breast seed implant as partial-breast radiation therapy for early-stage patients: In regards to Keller et al. *Int J Radiat Oncol Biol Phys* 64:1611.

GEC-ESTRO 2011. Groupe Européen de Curiethérapie-European Society for Radiotherapy and Oncology (GEC-ESTRO). Available at: http://www.apbi.uni-erlangen.de/. Last accessed on October 25, 2011.

Gurdalli, S., Kuske, R.R. Jr., Quiet, C.A., and Ozer, M. 2011. Dosimetric performance of Strut-Adjusted Volume Implant: A new single-entry multicatheter breast brachytherapy applicator. *Brachytherapy* 10:128–35.

Hamid, S., Arthur, D., Shah, S. Jr. et al. 2011. A multi-institutional study of implementation, tolerance, and early clinical results with noninvasive image-guided breast brachytherapy for tumor bed boost. *Brachytherapy* 10:S37 [abstract].

Hammer, J., Seewald, D.H., Track, C., Zoidl, J.P., and Labeck, W. 1994. Breast cancer: Primary treatment with external-beam radiation therapy and high-dose rate iridium implantation. *Radiology* 193:573–7.

Hammer, J., Track, C., Seewald, D.H. et al. 1998. Breast Cancer: External beam radiotherapy and interstitial iridium implantation—10-year clinical results. *Eur J Cancer* 34S:32 [abstract].

Hannoun-Levi, J.M., Houvenaeghel, G., Ellis, S. et al. 2004. Partial breast irradiation as second conservative treatment of local breast cancer recurrence. *Int J Radiat Oncol Biol Phys* 60:1385–92.

Harms, W., Krempien, R., Hensley, F.W., Berns, C., Wannenmacher, M., and Fritz, P. 2001. Results of chest wall reirradiation using pulsed-dose-rate (PDR) brachytherapy molds for breast cancer local recurrences. *Int J Radiat Oncol Biol Phys* 49:205–10.

Hepel, J.T. and Wazer, D.E. 2011. A comparison of brachytherapy techniques for partial breast irradiation. *Brachytherapy*. August 3. Epub ahead of print.

Hepel, J.T., Tokita, M., MacAusland, S.G. et al. 2009. Toxicity of three-dimensional conformal radiotherapy accelerated partial breast irradiation. *Int J Radiat Oncol Biol Phys* 75:1290–6.

Hill-Kayser, C.E., Chacko, D., Hwang, W.T., Vapiwala, N., and Solin, L.J. 2011. Long-term clinical and cosmetic outcomes after breast conservation treatment for women with early-stage breast carcinoma according to the type of breast boost. *Int J Radiat Oncol Biol Phys* 79:1048–54.

Holland, R., Veling, S.H., Mravunac, M., and Hendriks, J.H. 1985. J. Histologic multifocality of Tis, T1 T2 breast carcinomas. Implications for clinical trials of breast conserving therapy. *Cancer* 56:979–90.

Holmberg, L., Garmo, H., Granstrand, B. et al. 2008. Absolute risk reductions for local recurrence after postoperative radiotherapy after sector resection for ductal carcinoma in situ of the breast. *J Clin Oncol* 26:1247–52.

International Commission on Radiation Units and Measurements (ICRU). 1997. Dose and Volume Specification for Reporting Interstitial Therapy. Report 58 of ICRU, ICRU Publications, Bethesda, MD.

Ivanov, O., Dickler, A., Lum, B.Y., Pellicane, J.V., and Francescatti, D.S. 2011. Twelve-month follow up results of a trial utilizing Axxent electronic brachytherapy to deliver intraoperative radiation therapy for early-stage breast cancer. *Ann Surg Oncol* 18:453–58.

Jacobson, J.A., Danforth, N.D., Cowan, K.H. et al. 1995. Ten-year results of a comparison of conservation with mastectomy in the treatment of stage I and II breast cancer. *N Engl J Med* 332:907–11.

Jagsi, R., Ben-David, M.A., Moran, J.M. et al. 2010. Unacceptable cosmesis in a protocol investigating intensity-modulated radiotherapy with active breathing control for accelerated partial-breast irradiation. *Int J Radiat Oncol Biol Phys* 76:71–8.

Jemal, A., Bray, F., Center, M.M., Ferlay, J., Ward, E., and Forman, D. 2011. Global cancer statistics. *CA Cancer J Clin* 61:69–90.

Johansson, B., Karlsson, L., Liljegren, G., Hardell, L., and Persliden, J. 2009. Pulsed dose rate brachytherapy as the sole adjuvant clinical study. *Radiother Oncol* 90:30–5.

Julian, T.B., Constantino, J.P., Vicini, F.A. et al. 2011. Early Toxicity Results with 3-D Conformal External Beam Therapy (CEBT) from the NSABP B-39/RTOG 0413 Accelerated Partial Breast Irradiation (APBI) Trial. *Int J Radiat Oncol Biol Phys* 81S:S7 [abstract].

Keisch, M., Vicini, F., Scroggins, T. et al. 2005. Thirty-nine months results with the MammoSite breast brachytherapy applicator: details Regarding cosmesis, toxicity, and local control in partial breast irradiation. *Int J Radiat Oncol Biol Phys* 63S:56 [abstract].

King, T.A., Bolton, J.S., Kuske, R.R., Fuhrman, G.M., Scroggins, T.G., and Jiang, X.Z. 2000. Long-term results of wide-field brachytherapy as the sole method of radiation therapy after segmental mastectomy for T(is, 1, 2) breast cancer. *Am J Surg* 180:299–304.

Kuske, R.R. 2009. Brachytherapy techniques: The University of Wisconsin/Arizona approach. In: *Accelerated Partial Breast Irradiation*, 2nd ed. D.E. Wazer, D.W. Arthur, and F.A. Vicini (eds.), Springer-Verlag, New York, pp. 219–46.

Liljegren, G., Holmberg, L., Bergh, J. et al. 1999. 10-yr results after sector resection with or without postoperative radiotherapy for stage I breast cancer: A randomized trial. *J Clin Oncol* 17:2326–33.

Magee, B., Swindell, R., Harris, M., and Banerjee, S.S. 1996. Prognostic factors for breast recurrence after conservative breast surgery and radiotherapy: Results from a randomized trial. *Radiother Oncol* 39:223–7.

Manoharan, S.R., Rodriguez, R.R., Bobba, V.S., and Chandrashekar, M. 2010. Dosimetry evaluation of SAVI-based HDR brachytherapy for partial breast irradiation. *J Med Phys* 35:131–6.

Mansfield, C.M., Komarnicky, L.T., Schwartz, G.F. et al. 1995. Ten-year results in 1070 patients with stages I and II breast cancer treated by conservative surgery and radiation therapy. *Cancer* 75:2328–36.

Mazeron, J.-J., Simon, J.M., Crook, J. et al. 1991. Influence of dose rates on local control of breast carcinoma treated by external beam irradiation plus iridium implant. *Int J Radiat Oncol Biol Phys* 21:1173–7.

Niehoff, P., Polgar, C., Ostertag, H. et al. 2006. Clinical experience with the MammoSite® Radiation Therapy System for intracavitary brachytherapy of breast cancer—Results from an international Phase II trial. *Radiother Oncol* 79:316–20.

Olivotto, I.A., Rose, M.A., Osteen, R.T. et al. 1989. Late cosmetic outcome after conservative surgery and radiotherapy: Analysis of causes of cosmetic failure. *Int J Radiat Oncol Biol Phys* 17:747–53.

Ott, O.J., Hildebrandt, G., Pötter, R. et al. 2007. Accelerated partial breast irradiation with multi-catheter brachytherapy: Local control, side effects and cosmetic outcome for 274 patients. Results of the German-Austrian multi-centre trial. *Radiother Oncol* 82:281–6.

Patel, R.R., Becker, S.J., Das, R.K., and Mackie, T.R. 2007a. A dosimetric comparison of accelerated partial breast irradiation techniques: Multicatheter interstitial brachytherapy, three-dimensional conformal radiotherapy, and supine versus prone helical tomotherapy. *Int J Radiat Oncol Biol Phys* 68:935–42.

Patel, R., Ringwala, S., Forouzannia, A. et al. 2007b. Clinical fat necrosis in patients treated with multi-catheter APBI: A 3D CT-based clinical correlation. *Int J Radiat Oncol Biol Phys* 69S:S217 [abstract].

Patel, R.R., Christensen, M.E., Hodge, C., Adkison, J.B., and Das, R.K. 2008. Clinical outcome analysis in "high-risk" versus "low-risk" patients eligible for National Surgical Adjuvant Breast and Bowel B-39/Radiation Therapy Oncology Group 0413 trial: Five-year results. *Int J Radiat Oncol Biol Phys* 70:970–3.

Perez, C.A., Taylor, M.E., Halverson, K., Garcia, D., Kuske, R.R., and Lockett, M.A. 1996. Brachytherapy or electron beam boost in conservation therapy of carcinoma of the breast; A non-randomized comparison. *Int J Radiat Oncol Biol Phys* 34:995–1007.

Pignol, J.P., Keller, B., Rakovitch, E., Sankreacha, R., Easton, H., and Que, W. 2006. First report of a permanent breast ^{103}Pd seed implant as adjuvant radiation treatment for early- stage breast cancer. *Int J Radiat Oncol Biol Phys* 64:176–81.

Pignol, J.P., Rakovitch, E., Keller, B.M., Sankreacha, R., and Chartier C. 2009. Tolerance and acceptance results of a palladium-103 permanent breast seed implant Phase I/II study. *Int J Radiat Oncol Biol Phys* 73:1482–8.

Polgar, C., Orosz, Z., Fodor, J. et al. 2001. The effect of high-dose rate brachytherapy and electron boost on local control and side effects after breast conserving surgery: first results of the randomized Budapest breast boost trial. *Radiother Oncol* 60:10–27.

Polgar, C., Fodor, J., Major, T. et al. 2007. Breast-conserving treatment with partial or whole breast irradiation for low-risk invasive breast carcinoma—5 yr results of a randomized trial. *Int J Radiat Oncol Biol Phys* 69:694–702.

Polgar, C., Major, T., Lövey, K., Frohlich, G., Takacsi-Nagy, Z., and Fodor, J. 2008. Hungarian experience on partial breast irradiation versus whole breast irradiation: 12-year results of a Phase II trial and updated results of a randomized study. *Brachytherapy* 7:91–2 [abstract].

Polgar, C., Sulyak, Z., Tibor, M., Fröhlich, G., Takacsi-Nagy, Z., and Fodor, J. 2009. Reexcision and perioperative high dose rate brachytherapy in the treatment of local relapse after breast conservation: an alternative to salvage mastectomy. *J Contemp Brachytherapy* 1:131–6.

Polgar, C., Van Limbergen, E., Pötter, R. et al. 2010. Patient selection for accelerated partial breast irradiation (APBI) after breast conserving surgery. Recommendations of the Groupe Européen de Curiethérapie–European Society for Therapeutic Radiology and Oncology Breast cancer working group, based on clinical evidence. *Radiother Oncol* 94:264–73.

Poortmans, P.H., Bartelink, H., Horiot, J.C. et al. 2004. The influence of the boost technique on local control in breast conserving treatment in the EORTC boost versus no boost randomized trial. *Radiother Oncol* 72:25–33.

Prosnitz, L. and Marks, L. 2006. Partial breast irradiation: A cautionary note. *Int J Radiat Oncol Biol Phys* 65: 319–21.

RTOG/NSABP 2011. Radiation Therapy Oncology Group (RTOG) and National Surgical Adjuvant Breast and Bowel Project (NSABP). Available at: http://www.rtog.org/LinkClick.aspx?fileticket=olCDavtP2jk%3D&tabid=40. Last accessed on October 30, 2011.

Ray, G.R. and Fish, V.J. 1983. Biopsy and definitive radiation therapy in stage I and II adenocarcinoma of the female breast: Analysis of cosmesis and the role of electron beam supplementation. *Int J Radiat Oncol Biol Phys* 9:813–8.

Resch, A., Fellner, C., Mock, U. et al. 2002a. Locally recurrent breast cancer: pulsed dose rate brachytherapy for repeat irradiation following lumpectomy—A second chance to preserve the breast. *Radiology* 225:713–8.

Resch, A., Pötter, R., Van Limbergen, E. et al. 2002b. Long term results (10 years) of intensive breast conserving therapy including a high-dose and large-volume interstitial brachytherapy boost (LDR/HDR) for T1/T2 breast cancer. *Radiother Oncol* 63:47–58.

Rivard, M.J., Melhus, C.S., Wazer, D.E., and Bricault, R.J. Jr. 2009a. Dosimetric characterization of round HDR ^{192}Ir AccuBoost applicators for breast brachytherapy. *Med Phys* 36:5027–32.

Rivard, M.J., Melhus, C.S., Granero, D., Perez-Calatayud J., and Ballester, F. 2009b. An approach to using conventional brachytherapy software for clinical treatment planning of complex, Monte Carlo-based brachytherapy dose distributions. *Med Phys* 36:1968–75.

Romestaing, P., Lehingue, Y., Carrie, C. et al. 1997. Role of a 10 Gy boost in the conservative treatment of early breast cancer. Results of a randomized clinical trial in Lyon, France. *J Clin Oncol* 15:963–8.

Sarin, R., Dinshaw, K.A., Shrivastava, S.K. et al. 1993. Therapeutic factors influencing the cosmetic outcome and late complications in the conservative management of early breast cancer. *Int J Radiat Oncol Biol Phys* 27:285–92.

Scanderberg, D.J., Yashar, C., Rice, R., and Pawlicki, T. 2008. Clinical implementation of a new HDR brachytherapy device for partial breast irradiation. *Radiother Oncol* 90:36–42.

Shah, A.P., Strauss, J.B., Kirk, M.C., Chen, S.S., and Dickler, A. 2010. A dosimetric analysis comparing electron beam with the MammoSite brachytherapy applicator for intact breast boost. *Phys Med* 26:80–7.

Shah, C., Antonucci, J.V., Wilkinson, J.B. et al. 2011. Twelve-year clinical outcomes and patterns of failure with accelerated partial breast irradiation versus whole-breast irradiation: Results of a matched-pair analysis. *Radiother Oncol* 100:210–4.

Sioshansi, S., Rivard, M.J., Hiatt, J.R., Hurley, A.A., Lee, Y., and Wazer, D.E. 2011. Dose modeling of noninvasive image-guided breast brachytherapy in comparison to electron beam boost and three-dimensional conformal accelerated partial breast irradiation. *Int J Radiat Oncol Biol Phys* 80:410–6.

Smith, B.D., Arthur, D.W., Buchholz, T.A. et al. 2009. Accelerated partial breast irradiation consensus statement from the American Society for Radiation Oncology (ASTRO). *Int J Radiat Oncol Biol Phys* 74:987–1001.

Soran, A., Evrensel, T., Beriwal, S. et al. 2007. Placement technique and the early complications of balloon breast brachytherapy: Magee-Womens hospital experience. *Am J Clin Oncol* 30:152–5.

Strnad, V. and Ott, O.J. 2006. *Partial Breast Irradiation Using Multicatheter Brachytherapy.* Zuckschwerdt Verlag München, Wien.

Touboul, E., Belkacemi, Y., Lefranc, J.P. et al. 1995. Early breast cancer: influence of type of boost (electrons vs iridium-192 implant) on local control and cosmesis after conservative surgery and radiation therapy. *Radiother Oncol* 34:105–13.

Turesson, I. and Notter, G. 1984. The influence of fraction size in radiotherapy on the late normal tissue reaction (I + II). *Int J Radiat Oncol Biol Phys* 10:593–606.

Vaidya, J.S., Tobias, J.S., Baum, M. et al. 2004. Intraoperative radiotherapy for breast cancer. *Lancet Oncol* 5:165–73.

Vaidya, J., Baum, M., Tobias, J. et al. 2009. Targeted Intraoperative Radiotherapy (TARGIT) boost after breast conserving surgery results in a remarkably low recurrence rate in a standard risk population: 5 year results. *Cancer Res* 69S:4104 [abstract].

Vaidya, J.S., Joseph, D.J., Tobias, J.S. et al. 2010. Targeted intraoperative radiotherapy versus whole breast radiotherapy for breast cancer (TARGIT-A trail): an international, prospective, randomized, non-inferiority, Phase 3 trial. *Lancet* 376:91–102.

Vaidya, J.S., Baum, M., Tobias, J.S. et al. 2011. Long-term results of targeted intraoperative radiotherapy (TARGIT) boost during breast-conserving surgery. *Int J Radiat Oncol Biol Phys* 81:1091–7.

Van Limbergen, E. 2003. Indications and technical aspects of brachytherapy in breast conserving treatment of breast cancer. *Cancer Radiothérapie* 7:107–20.

Van Limbergen, E. 2009. Accelerated partial breast irradiation with intracavitary balloon brachytherapy may be not as simple as it was supposed to be. *Radiother Oncol* 91:147–9.

Van Limbergen, E., Van den Bogaert, W., Van der Schueren, E., and Rijnders, A. 1987. Tumor excision and radiotherapy as primary treatment of breast cancer. Analysis of patient and treatment parameters and local control. *Radiother Oncol* 8:1–9.

Van Limbergen, E., Briot, E., and Drijkoningen, M. 1990. The source-skin distance measuring bridge: A method to avoid radiation telangiectasia in the skin after interstitial therapy for breast cancer. *Int J Radiat Oncol Biol Phys* 18:1239–44.

Vass, S. and Bairati, I. 2005. A cosmetic evaluation of breast cancer treatment: A randomized study of radiotherapy boost technique. *Int J Radiat Oncol Biol Phys* 62:1274–82.

Veronesi, U., Marubini, E., Mariani, L. et al. 2001. Radiotherapy after breast-conserving surgery in small breast carcinoma: Long-term results of a randomized trial. *Ann Oncol* 12:997–1003.

Veronesi, U., Orecchia, R., Luini, A.P. et al. 2010. Intraoperative radiotherapy during breast conserving surgery: A study on 1,822 cases treated with electrons. *Breast Cancer Res Treat* 124:141–51.

Vicini F.A., Remouchamps, V., Wallace, M. et al. 2003. Ongoing clinical experience utilizing 3D conformal external beam radiotherapy to deliver partial-breast irradiation in patients with early-stage breast cancer treated with breast-conserving therapy. *Int J Radiat Oncol Biol Phys* 57:1247–53.

Vicini, F.A., Chen, P., Wallace, M. et al. 2007. Interim cosmetic results and toxicity using 3D conformal external beam radiotherapy to deliver accelerated partial breast irradiation in patients with early-stage breast cancer treated with breast-conserving therapy. *Int J Radiat Oncol Biol Phys* 69:1124–30.

Vicini, F., Beitsch, P., Quiet, C. et al. 2011. Five-year analysis of treatment efficacy and cosmesis by the American Society of breast surgeons MammoSite breast brachytherapy registry trial in patients treated with accelerated partial breast irradiation. *Int J Radiat Oncol Biol Phys* 79:808–17.

Watkins, J.M., Harper, J.L., Dragun, A.E. et al. 2008. Incidence and prognostic factors for seroma development after MammoSite breast brachytherapy. *Brachytherapy* 7:305–9.

Wazer, D.E., Kramer, B., Schmid, C., Ruthazer, R., Ulin, K., and Schmidt-Ullrich, R. 1997. Factors determining outcome in patients treated with interstitial implantation as a radiation boost for breast conservation therapy. *Int J Radiat Oncol Biol Phys* 39:381–93.

Wazer, D.E., Kaufman, S., Cuttino, L., DiPetrillo, T.A., and Arthur, D.W. 2006. Accelerated partial breast irradiation: An analysis of variables associated with late toxicity and long-term cosmetic outcome after high-dose-rate interstitial brachytherapy. *Int J Radiat Oncol Biol Phys* 64:489–95.

Wazer, D.E., Arthur, D.W., and Vicini, F.A. 2009. *Accelerated Partial Breast Irradiation*, 2nd ed. Springer-Verlag, New York.

Wilder, R.B., Curcio, L.D., Khanijou, R.K. et al. 2009. A Contura catheter offers dosimetric advantages over a MammoSite catheter that increase the applicability of accelerated partial breast irradiation. *Brachytherapy* 8:373–8.

Yang, Y., Melhus, C.S., Sioshansi, S., and Rivard, M.J. 2011. Treatment planning of a skin-sparing conical breast brachytherapy applicator using conventional brachytherapy software. *Med Phys* 38:1519–25.

Yashar, C.M., Blair, S., Wallace, A., and Scanderberg, D. 2009. Initial clinical experience with the strut-adjusted volume implant brachytherapy applicator for accelerated partial breast irradiation. *Brachytherapy* 8:367–72.

Yashar, C.M., Scanderbeg, D., Kuske, R. et al. 2011. Initial clinical experience with the Strut-Adjusted Volume Implant (SAVI) breast brachytherapy device for Accelerated Partial-Breast Irradiation (APBI): First 100 patients with more than 1 year of follow up. *Int J Radiat Oncol Biol Phys* 80:765–70.

23

Brachytherapy for Other Treatment Sites

Vratislav Strnad
University Hospital Erlangen

Beth Erickson
Medical College of Wisconsin

Mahesh Kudrimoti
*University of Kentucky
College of Medicine*

23.1 Introduction

Historically, brachytherapy was used long before the development of external beam radiation kilovoltage, orthovoltage, and megavoltage machines. Radium was the most commonly used radioactive isotope and used to treat a variety of tumors and nonmalignant conditions. Brachytherapy is highly successful in the eradication of cervical cancers and other gynecological malignancies. It is, however, also used to treat several nongynecological cancers and some nonmalignant conditions. Over the years, the techniques and treatment methods have been better defined and continue to evolve from low-dose-rate (LDR) to high-dose-rate (HDR) to pulsed-dose-rate (PDR) methods. Use of radium has been abandoned in favor of safer radioactive sources. Some areas where implants are clinically used include the head/neck area, where there is a plethora of experience in organ preservation and quality-of-life issues; for lung cancers, implants are mainly used to palliate symptoms; for esophageal and skin cancers, implants offer a chance of a durable clinical response; for

sarcomas, anal and rectal cancers, and uveal melanomas, there is emphasis on functional outcome. The use of brachytherapy to treat some nonmalignant conditions such as the pterygium and keloids is also discussed in the following chapter.

23.2 Head-and-Neck Cancer Tumors

23.2.1 Introduction

Brachytherapy represents an important part of oncological care for a substantial part of patients with head-and-neck cancer. In principle, the patients should meet the following basic requirements for brachytherapy:

- Good cooperation and motivation of the patient.
- General anesthesia is available (with the exceptions of the lips and the nasopharyngeal region).
- A clinical examination with a precise diagram has been performed. Panendoscopy is also preferable.
- CT and/or MRI examination results are available.

Due to the complexity as well as the differences in both the techniques and results of brachytherapy for different head-and-neck tumors, due to their dependence on their specific anatomical location, we prefer to subdivide brachytherapy options on the basis of their individual anatomical location (lips, tongue, oral base, oropharyngeal, nasopharyngeal, and cheek).

23.2.2 Lip Carcinoma

23.2.2.1 Indications, Patient Selection Criteria

Small tumors (<4 cm) represent a classic indication for the brachytherapy alone. Beyond this, interstitial brachytherapy is also indicated postoperatively where the following risk factors exist (Strnad and Kovács 2010):

- Close resection margins
- R1 status
- G3 differentiation
- L1 lymphatic invasion
- V1 vascular invasion

For tumors > 4 cm, interstitial brachytherapy is indicated as a boost only once an adequate level of tumor reduction has been achieved subsequent to external beam irradiation. Significant bone infiltration represents a contraindication.

23.2.2.2 Recommendations for Target Definition, Dose Prescription, Dose, and Dose Rate

23.2.2.2.1 Target Definition

The target area (PTV = CTV) comprises the palpable and visible tumor or tumor bed (GTV), surrounded by a 10-mm-wide safety margin in all directions (Figure 23.1), which respects the natural, anatomical borders. Beyond the oral angle, the rule of defining PTV does not apply. However, depending on the individual anatomy, the PTV generally includes the entire thickness

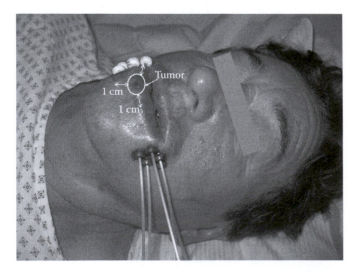

FIGURE 23.1 Example of an implant of a lip carcinoma with two localizations—left and right. (Courtesy of V. Strnad, University Hospital Erlangen.)

of the affected lip. When contouring, the following rules should be observed:

(1) Contouring the CTV (= PTV) alone is adequate.
(2) The entire lip thickness should be involved.
(3) Safety margins in the lateral and caudal (cranial) directions are GTV + 10 mm in all directions—respecting the natural, anatomical limits.

23.2.2.2.2 Dose Prescription

Dose specification should be preferably done analogous to ICRU-58. For assessment of quality assurance, at least the parameter such as the dose nonuniformity ratio (DNR) and V_{100} should be documented.

23.2.2.2.3 Dose and Dose Rate

We recommend the following dose concepts:

- Brachytherapy only
  - LDR/PDR procedure: 0.45–0.55 Gy/h; up to 60–75 Gy
  - HDR procedure: 5 × 3–4 Gy/week; up to 35–40 Gy
- Brachytherapy as boost
  - LDR/PDR procedure: 0.45–0.55 Gy/h; up to 12–24 Gy
  - HDR procedure: 5 × 2 Gy/week; up to 12–24 Gy

23.2.2.2.4 Techniques

Lip carcinoma implantation is preferably performed under local anesthesia (2% procaine solution) using sterile catheters. Once the necessary. Once the necessary number of needles has been placed (their parallel position must be permanently observed), single-leader plastic catheters are threaded and secured in place with plastic knobs. The implantation should be made, taking the Paris system into account, so that the PTV is completely enclosed by the plastic catheters (Figure 23.1). The plastic catheters are easily adjusted to the individual tumor shape and are very well tolerated by patients.

23.2.2.2.5 Results

Local tumor control can be achieved in 90%–98% of patients (Beauvois et al. 1994; Rovirosa-Casino et al. 2006). The most frequently occurring later side effects include a mild depigmentation among 2.5%–17% of all patients, telangiectasia among 15%, and various degrees of fibroses among 8% of all patients. The most serious subsequent effect is the development of a tissue necrosis in up to 5% of all patients. The necrosis risk depends directly on the total dose and the dose rate.

23.2.3 Oral Cavity: Floor of Mouth Cancer, Oral Tongue Carcinoma

23.2.3.1 Indications, Patient Selection Criteria

Interstitial brachytherapy can be carried out both as brachytherapy alone or postoperatively. Generally, tumors < 30 mm with no broad contact with the mandible are suitable for sole brachytherapy (Figure 23.2).

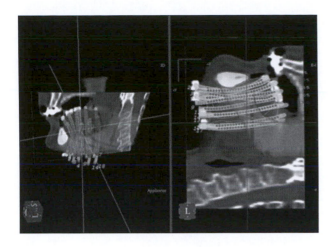

FIGURE 23.2 Example of an implantation of tumor of floor of mouth. (Courtesy of V. Strnad, University Hospital Erlangen.)

Postoperatively, interstitial brachytherapy alone is indicated if one or more of the following risk factors are present: resection margins < 5 mm, tumor infiltration depth > 5 mm, L1-status, or G3 differentiation. Naturally, brachytherapy alone is justified only if the lymph node status was negative. For tumors < 40–50 mm and a positive lymph node status, interstitial brachytherapy as a boost in combination with external beam irradiation is indicated, particularly when postoperative resection margins are <5 mm, an R1-situation is presented, or other risk factors such as L1, V1, or G3 are present (Strnad and Kovács 2010).

We see a significant infiltration of the mandible by the tumor as an absolute contraindication for interstitial brachytherapy. An infiltration of the gingival area merely represents a relative contraindication. Interstitial brachytherapy is possible in cases of gingival infiltration if no more than approximately 1 cm of the mucosal membrane is affected, and no more than two plastic catheters are to be placed in the infiltration.

Interstitial brachytherapy can also represent an effective treatment method for previously irradiated patients. For these patients, we indicate brachytherapy both as a primary treatment, as well as postoperatively when the tumor diameter does not exceed 4 cm, and when there is no bone infiltration and close-meshed control conditions are assured.

23.2.3.2 Recommendations for Target Definition, Dose Prescription, Dose, and Dose Rate

23.2.3.2.1 Target Definition

The target area comprises the tumor/tumor bed and a 5- to 15 mm safety area in all directions, with natural borders to the mandible and the skin. The target area is defined by the implantation area. Subsequent to the implantation, the following rules (Strnad and Kovács 2010) for contouring of GTV and PTV should be followed (Figure 23.3):

(1) Contour the GTV based on the CT scan and on *clinical experiences* related to tumor size and location.
(2) Contour PTV = CTV: at a lateral distance of GTC +5 (–10) mm in all directions (safety margins)—natural, anatomical borders must be respected (lower jaw, lingual edge, skin).

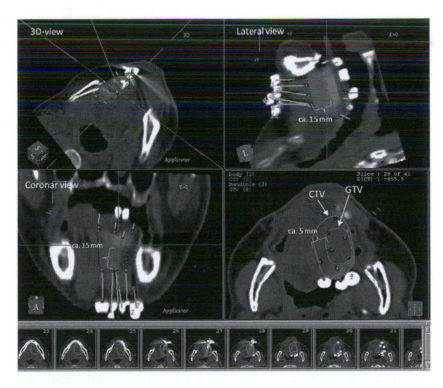

FIGURE 23.3 **(See color insert.)** Examples of the contouring rules for cancer of the mobile tongue (refer to the text for more details). (Courtesy of V. Strnad, University Hospital Erlangen.)

(3) Contour PTV = CTV at a depth of GTC +15 mm here too, the natural, anatomical borders must be respected (skin).

23.2.3.2.2 Dose Prescription

The dose can be specified either in accordance with the recommendation of ICRU-58 using the "mean central dose" (MCD), Dref = 85% × MCD, or in accordance with the rules of intensity-modulated brachytherapy planning (reference dose should be prescribed no more than 1 cm from the lateral catheter, while the maximum catheter surface dosage should not exceed 400% of the reference dosage). Both the dose distribution and the anatomy must be displayed in 2D as well as 3D images, while the quality of the implant must be examined not only subjectively and visually but also through the employment of valid quality indices. Additional optimization may be carried out as required. Care must, however, be taken, particularly in postoperative situations (no more tumor tissue present), to ensure that optimization is not too strong or that maximum isodose values arise, as this will significantly increase the risk of soft tissue necrosis. This could occur particularly in the case of intensity-modulated brachytherapy. In general, we recommend that, for an average implant size of 30–40 cm³ (typically corresponding to 85% of the MCD), the area of the 150% isodose does not exceed approximately 10 cm³ by any great extent and that all important quality parameters are documented (Strnad and Kovács 2010). Deviating values should be accepted only in individually justified cases. Also using intensity-modulated brachytherapy, the DNR should not exceed 0.42.

The following represent the most important quality assurance factors that, as far as possible, should be maintained to maximize the treatment effect and minimize side effects (Melzner et al. 2007; Simon et al. 1993):

(1) V_{100} < 28–30 cm³
(2) V_{150} < 10 cm³
(3) V_{100} > 95%–90%
(4) V_{90} ~ 100%
(5) D_{90} > 100%
(6) Dose rate/pulse dose 0.4–0.6 Gy/h
(7) MCD < 0.64 Gy/h
(8) Distance between plastic tubes: 7–14 mm
(9) DNR < 0.36 (LDR/PDR), DNR < 0.42 (HDR-intensity-modulated brachytherapy)
(10) Total dose 60–65 Gy (postoperative or combined with chemotherapy), 65–70 Gy (for advanced tumors that have not been surgically treated and without chemotherapy)
(11) Time interval between external beam irradiation and brachytherapy < 20 days

23.2.3.2.3 Dose and Dose Rate

The following doses are recommended:

- Brachytherapy only, without surgery
  - LDR/PDR procedure: 0.4–0.6 Gy/h; up to 60–70 Gy
- Postoperative brachytherapy only

  - LDR/PDR procedure: 0.4–0.6 Gy/h; up to 60–65 Gy
- Brachytherapy as a boost
  - LDR/PDR procedure: 0.4–0.6 Gy/h; up to 10–24 Gy
  - HDR procedure: 1–2× per day, 2.0–3.5 Gy (with at least a 6-h break between administrations), 5 days/week; up to 35–60 Gy (depending on the selected fraction dose)

We recommend using an LDR or a PDR procedure for curative purposes. Currently, only controversial results are available for the HDR procedure, which still require verification from studies. If HDR treatment is selected, fractionated, interstitial brachytherapy with a fractionation dose of ≤3.5 Gy should be performed outside the studies.

23.2.3.2.4 Techniques

Plastic catheter implantation has shown itself to be the most effective technique for treating oral tongue and floor of mouth carcinomas. Other techniques are used less frequently. The catheters are implanted using hypodermic needles. The needles are inserted in sequence and as parallel as possible into the tumor/tumor bed from the submental region. With their assistance, the plastic catheters can then be threaded.

Care should be taken to ensure that a spacing of 7–14 mm (ideally, 10–12 mm) (Melzner et al. 2007; Simon et al. 1993; Lapeyre et al. 2008) is maintained between the needles and plastic catheters and that all lie parallel. Either buttons can be used to secure the plastic catheters to the back of the tongue, or the loop can be continued. During implantation, care must be taken to ensure that the catheters along the periphery of the tumor bed/tumor are placed in healthy tissue, as this is the only way to reliably ensure an adequate safety gap of 5–10 mm.

23.2.3.2.5 Results

Results of interstitial brachytherapy for oral cavity cancer depend particularly on the tumor size and the used total dose. It is apparent that the best results are achieved with smaller tumors (T1, T2) with tumor control rates of 80%–94% (Decroix and Labib 1983; Mazeron and Richaud 1984; Mazeron et al. 1991; Strnad et al. 2000; Strnad 2004). Other factors such as bone infiltration, tumor grade, lymph node involvement, dose rate, and the total dose are also decisive with regard to the prognosis. As analysis carried out by Mazeron et al. (1991) has shown, the highest level of tumor control is achieved at a dose rate of ≥0.5 Gy/h and a total dose of ≥62.5 Gy.

Brachytherapy also represents an important option as a highly effective treatment method for reirradiation of patients with recurrent tumor, as well as for patients with secondary tumors. Brachytherapy allows tumor control in approximately 70% of such patients (Narayana et al. 2007; Strnad et al. 2003).

Generally, as late side effects, osteoradionecroses and soft tissue necrosis may occur. In the LDR brachytherapy, soft tissue necrosis was reported in 20%–30% of the patients, while osteoradionecrosis was observed in 4%–10% of patients (Decroix and

Labib 1983; Pernot et al. 1996). The necrosis rate increases as the volume of the reference total dose increases, as it nears the lower jaw, and as the dose rate rises above 0.55 Gy/h (Mazeron et al. 1991). When PDR and 3D CT-based planning are used, the incidence of both osteoradionecrosis as well as soft tissue necrosis is significantly lower. Surgical treatment of this serious side effect is required only in 2%–4% of patients (Lapeyre et al. 2008; Strnad et al. 2000, 2005; Strnad 2004).

23.2.4 Oropharynx

23.2.4.1 Indications, Patient Selection Criteria

For oropharyngeal carcinoma, interstitial brachytherapy may be indicated depending on the tumor size and anatomy. Implants can be done in every location of the oropharynx (base of tongue, including the vallecula, tonsils, soft palate), but no neighboring regions (larynx/hypopharynx/nasopharynx) should be affected. The tumor should not exceed a diameter of 4–5 cm. Tumor infiltration of the bone and tumor manifestations in the retromolar triangle generally represent contraindications for interstitial brachytherapy.

In the case of oropharyngeal carcinoma, brachytherapy is generally indicated only as a boost. Interstitial brachytherapy alone is justified only for tumors < 1 cm or for recurrent tumors. Postoperatively, interstitial brachytherapy is indicated, particularly where the resection margins are less than 5 mm or where an R1 resection or other local recurrence risk factors such as lymphatic vascular invasion or grade 3 are present.

Similar to the oral cavity carcinomas, interstitial brachytherapy can also represent an effective treatment for patients with base-of-tongue cancers who had been previously irradiated (Narayana et al. 2007; Strnad et al. 2003).

23.2.4.2 Recommendations for Target Definition, Dose Prescription, Dose, and Dose Rate

23.2.4.2.1 Target Definition

The target area comprises the tumor/tumor bed and 5–10 mm safety margins in all directions. In most instances, this requires implants throughout the entire base of tongue (or the entire tonsillar area or the entire soft palate). Precise knowledge regarding the panendoscopy findings, results of the CT/MRI examinations, and the surgical report together with the histology are essential. Additional procedures—contouring the GTV and CTV—are identical to those for the determination of the target area for oral cavity carcinomas.

23.2.4.2.2 Dose Specifications

Dose specifications can be made either analogous to ICRU-58 or to other established systems. For base-of-tongue implants, the procedure in no way differs from the dose specification for oral tongue (see above). As a rule and so as to avoid any possible loop underdosages, the central plane should be defined along the center of the soft palate/tonsils and, respectively, in the tonsil base or along lateral soft palate borders for implants with one to two

tonsillar and soft palate loops. Additional details with regard to dose specification and quality assurance are the same as for oral cavity carcinoma brachytherapy.

23.2.4.2.3 Dose and Dose Rate

The following doses are recommended:

- Brachytherapy only, without surgery
  - LDR/PDR procedure: 0.4–0.6 Gy/h; up to 65–70 Gy
- Brachytherapy alone, postoperative
  - LDR/PDR procedure: 0.4–0.6 Gy/h; up to 60–65 Gy
  - HDR procedure: dose fraction < 3 Gy, total dose 40–60 Gy
- Brachytherapy as boost
  - LDR/PDR procedure: 0.4–0.6 Gy/h; up to 10–30 Gy
  - HDR procedure: 2–6× 3–4 Gy, 2× per day (with at least a 6-h break), 5 days/week; up to 12–24 Gy

23.2.4.2.4 Techniques

As a rule, the implants are performed under general anesthesia with nasal intubation. Tracheotomy is not recommended as a routine procedure. The technique for tongue base implantation is identical to that for mobile tongue implantation (see above).

We recommend the procedure in accordance with Pernot et al. (1996) for performing the implantation in the tonsillar fossa or the soft palate (Figure 23.4).

23.2.4.2.5 Results

In contrast to mobile tongue and floor-of-mouth carcinomas, the best results can be achieved by a combination of brachytherapy and external beam radiotherapy with local control rates up to 94% (Lusinchi et al. 1989; Pernot et al. 1996).

All patients develop temporary, acute grade II–III° radio mucositis. The risk of necrosis depends on the implant quality (refer to the details above), dose rate, and the total dose and can occur in the form of severe complications in 2%–9% of patients (Mazeron and Richaud 1984; Pernot et al. 1996; Strnad 2004; Strnad et al. 2005).

23.2.5 Buccal Mucosa

23.2.5.1 Indications, Patient Selection Criteria

Interstitial brachytherapy alone, both with or without a surgical procedure, is indicated particularly for tumors measuring < 4 cm and located in the front two-thirds of the buccal mucosa. For larger tumors or tumors located in the posterior third of the buccal mucosa, brachytherapy is indicated only as a boost. Contraindications include an infiltration of the bone or the gingiva and large, diffusely infiltrated tumors in the posterior third of the buccal mucosa.

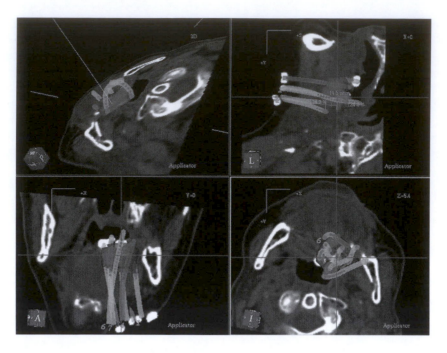

FIGURE 23.4 **(See color insert.)** Typical example of an implant of a tumor in the region of the base of tongue and tonsillar fossa. (Courtesy of V. Strnad, University Hospital Erlangen.)

23.2.5.2 Recommendations for Target Definition, Dose Prescription, Dose, and Dose Rate

23.2.5.2.1 Target Definition

The tumor/tumor bed with a safety margin of 0.5–1 cm in all directions represents the target. The entire thickness of the buccal mucosa should always be included. Due to the proximity to the upper and lower jaws and in order to avoid osteoradionecrosis, limitations may arise in the cranial and caudal directions.

For target delineation, the following rules should be followed (Strnad and Kovács 2010):

(1) Contouring CTV (= PTV) alone is adequate.
(2) The entire cheek thickness should be included if possible.
(3) Safety margins in the lateral and in the caudal and cranial directions are GTV + 5–10 mm in all directions—respecting the natural, anatomical limits.

23.2.5.2.2 Dose Specifications

The dose specification can conform to either ICRU-58 or to another established system.

23.2.5.2.3 Dose and Dose Rate

The following dose concepts are recommended:

- Brachytherapy alone:
  - LDR/PDR procedure: 0.4–0.6 Gy/h; up to 60–75 Gy
- Brachytherapy as a boost
  - LDR/PDR procedure: 0.4–0.6 Gy/h; up to 10–24 Gy

- HDR procedure: fractionated dose of <3 Gy, for two daily fractions; total dose 30–40 Gy

23.2.5.2.4 Techniques

We recommend performing the implantation using plastic catheters under general anesthesia. Regardless of the tumor size, several plastic catheters are implanted in one or two planes in the buccal mucosa in such a way that the catheters located along the periphery are implanted in healthy tissue (Figure 23.5). In this regard, an intraoperative ultrasound examination can prove helpful. As a rule, with the patient's mouth closed, the catheters lie parallel to the lower jaw. Some authors recommend the

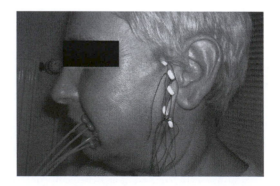

FIGURE 23.5 Single-plane implant of a cheek tumor using single leader plastic tubes. (Courtesy of V. Strnad, University Hospital Erlangen.)

implantation of an additional loop at the implantation border, as required (Lapeyre et al. 1995).

23.2.5.2.5 *Results*

Treatment results depend, in particular, on the tumor status and size—tumor control of about 80%–90% has been reported (Lapeyre et al. 1995, 2008; Strnad 2004). As a late side effect, soft tissue necrosis can develop in up to 12% of all patients (Lapeyre et al. 1995).

23.2.6 Nasopharynx

23.2.6.1 Indications, Patient Selection Criteria

Brachytherapy is usually indicated for a boost of nasopharyngeal surface tumors smaller than 1 cm. As a sole treatment, the indication is possible only for a recurrence in previously irradiated patients. We view tumors larger than 1 cm, bone infiltration, and tumors growing outside the nasopharyngeal region as contraindications to brachytherapy.

23.2.6.2 Recommendations for Target Definition, Dose Prescription, Dose, and Dose Rate

23.2.6.2.1 *Target Definition*

The target area comprises the tumor itself and as large a surrounding safety margin as possible to permit the entire nasopharyngeal roof and lateral walls to be included. Selecting only the nasopharyngeal area is seldom adequate when defining the target area. The reference depth should not exceed 10 mm. The target area should always be examined endoscopically and should be verified with CT or MRI.

For target delineation, the following rules should be followed (Strnad and Kovács 2010):

(1) Contouring CTV (= PTV) alone is adequate.
(2) The entire thickness of the nasopharyngeal mucous membrane should, if possible, be included.
(3) Safety margins in the lateral directions are GTV + at least 10 mm in all directions.

23.2.6.2.2 *Dose Prescription*

As a rule, the dose is specified at the border between the soft tissue and the adjacent skull base. In most instances, this bone border lies at a depth of approximately 5–10 mm. A CT-based dosage specification of D_{90}–D_{95} ($V_{100} > 95\%$–90%) should be assured. No other quality parameters need to be documented.

23.2.6.2.3 *Dose and Dose Rate*

The following dose concepts are recommended:

- Brachytherapy as a boost
  - HDR procedure: 4–6× 2–4 Gy
- Brachytherapy alone in the case of a recurrent tumor
  - LDR/PDR procedure: 0.4–0.6 Gy/h; up to 60 Gy
  - HDR procedure: 2–5 Gy with at least a 6-h break between fractions; total dose 20–40 Gy

23.2.6.2.4 *Techniques*

A variety of treatment options are available for brachytherapy of nasopharyngeal tumors—both interstitial as well as intracavitary. We prefer the intracavitary techniques. Even on ambulatory patients, the intracavitary brachytherapy can be carried out without problems. The disadvantage of intracavitary brachytherapy is that the source path travels down the center of the balloon. Due to the high surface doses, this can result in ulcers/necroses of the soft palate.

23.2.6.2.5 *Results*

Intracavitary brachytherapy as a boost after EBRT allows local control rates up to 80%–94% (Chang et al. 1996; Thiagarajan et al. 2006). Compared with external beam irradiation, a boost using brachytherapy appears to significantly reduce the probability of a recurrence (Wang 1997). The side effects of intracavitary brachytherapy for nasopharyngeal carcinomas are, as a rule, limited to grade III mucositis. Severe subsequent effects such as necroses can be observed in approximately 7% of patients (Wang 1997).

23.3 Esophageal Cancer

23.3.1 Introduction

Survival rates of more than 30% at 3 years are currently achievable with a combination of chemotherapy and external beam irradiation in locally advanced esophageal cancer; however, with these regimens, the locoregional recurrence/local progression rates of approximately 40% are still high. Though surgical resection has been advocated to improve local control, most patients are not candidates for aggressive surgical resection as a part of the trimodality management of esophageal cancers. Brachytherapy is one of the most important radiotherapeutic techniques available with the potential for improving durable local control and maintaining the patency of the esophageal lumen permitting adequate swallowing (Gaspar et al. 1997). There is a wide variation of application of this technique across the world, with the most aggressive use reported from centers in Asia and Europe (Sai et al. 2005; Okawa et al. 1999; Sur et al. 2002).

23.3.2 Patient Selection

Important selection criteria are the tumor size and location, extra-esophageal extension of tumor, regional or other lymph nodal disease. Histology as a predictor has not been established, although most clinical experience is described with respect to squamous cell carcinomas. The most suitable cases are those with tumors less than 10 cm and with midthoracic location without involvement of peri-esophageal structures such as bronchus, aorta, pericardium, roots of great vessels, and tracheoesophageal fistulas. Critical clinical judgment should be used in treating tumors approaching the gastroesophageal junction, as there may be extension of the tumor into the cardia of the stomach, and those tumors involving the proximal cervical esophagus, as there are reports of treatment sequelae resulting in development of fistulas (Gaspar et al. 1997; Gaspar 1999).

23.3.2.1 Recommendations for Target Definition, Dose Prescription, Dose, and Dose Rate

There is wide variation in the literature regarding dose prescription, and consideration should be given to the goals of therapy. Also taken into account should be the consideration of external beam dose, length of the esophagus to be treated, concurrent chemotherapy regimens, use of alternative therapy, and strategies such as stents, laser resections, and photodynamic therapy. In general, as more modalities such as full dose chemotherapy are clinically incorporated into patient management, the practice is to reduce the dose and number of fractions delivered by brachytherapy (Coia et al. 1991; Vuong et al. 2005).

The current recommendations are to treat the entire length of the mucosal esophageal lesion as visualized endoscopically with a 1–2 cm margin. Dose can be prescribed either to a point 1 cm from the source axis or to the distance related to the applicator surface—usually at 5 mm. Dose prescription to a reference point 10 mm from the source axis is most often chosen if small diameter applicators are used. We recommend to record and report, in any case, the dose at 5-mm depth because this reporting is independent of the prescription mode and the size and type of applicator (Gaspar et al. 1997).

Various dose rates have been reported, including LDR and HDR. For those patients receiving external beam irradiation along with brachytherapy, and with good expected life expectancy and good Karnofsky performance status/Eastern Cooperative Oncology Group (KPS/ECOG) performance scores, it is recommended that the external dose is limited to 45–50 Gy at 1.8 to 2 Gy per fraction.

If HDR is being used, an additional 10–12 Gy is given at 5–6 Gy/fraction, one fraction per week, 2–3 weeks after completion of external beam irradiation. If LDR is being used, then the prescribed dose is 10–20 Gy (Gaspar et al. 1997).

If the goal is palliation, then the treatment can be individualized based on the clinical situation. The treatment approach can include the use of 10–25 Gy delivered by HDR in two to four fractions or 20–40 Gy with LDR (40–100 cGy/h) in a single fraction for recurrent tumors, or it can combine a short course of external irradiation of 30–40 Gy in 2–3 Gy/fraction with an HDR (10–14 Gy) or LDR boost (20–25 Gy) for those with a poor KPS (less than 70) or ECOG scores and poor life expectancy. Fraction sizes larger than 7 Gy/fraction should be avoided on account of the long-term sequelae experienced in the clinical trials (Gaspar et al. 1997; Sur et al. 2002).

23.3.2.2 Techniques

To avoid serious side effects, we recommend using an applicator with an external diameter no less than 0.6 cm, preferably 12–15 mm. One should keep in mind that only applicators with a diameter ≥ 12 mm ensure that the significant part of the steep dose gradient (overdoses) is inside the applicator and not inside of the esophageal mucosa. The insertion of the applicator is usually done in conjunction with the endoscopist. The patient is sedated prior to the procedure. The preliminary endoscopy is performed, and then the length of the lesion is determined. Thereafter, the applicator is placed, and the tip of the catheter is advanced past the tumor by a 2–3 cm margin. Thereafter, CT simulation is performed, and the target length is determined from the simulation. The source is inserted based on the modality used, and the radiation protection sequence is initiated based on local guidelines (Gaspar et al. 1997).

23.3.2.3 Results

Various trials have reported results based on measuring several different end points including local control, overall survival, maintenance of swallowing function, and need for salvage therapy. The largest clinical experience from an Asian series reports no change in survival but an increase in local control with the addition of brachytherapy. The treatment experience from Japan for early-stage esophageal cancer treated with external radiation with or without brachytherapy, with surgery for salvage, was 60% at 5 years. Around 20% required salvage esophagectomy. The 2-year local control with and without brachytherapy was 79% and 54%, respectively, favoring the addition of brachytherapy. A Canadian experience with brachytherapy before XRT or chemoradiation reports local control rates of 75% at 2 years and a 5-year survival of 28%. An International Atomic Energy Agency multinational study for palliation of dysphagia in advanced esophageal cancer reported an overall dysphagia-free survival of 7.1 months with a median survival of 7.9 months (Gaspar et al. 1997; Sai et al. 2005; Sur et al. 2002; Okawa et al. 1999).

23.3.3 Summary

Esophageal brachytherapy is a very good tool for management of esophageal cancer. It is helpful in palliation of dysphagia and improvement of local control, although it is difficult to demonstrate a survival advantage. Dose fractionation is an important component when considering combining brachytherapy with other modalities, especially when using HDR. Various fractionation schemes have been used worldwide, but attention has to be given to the prescription of dose, and clinical judgment is needed in balancing the dose of external beam and brachytherapy when used in conjunction. Care should be taken to avoid performing brachytherapy on the same week as the chemotherapy delivery is planned as excessive toxicity has been reported in the clinical trials that combined chemotherapy and brachytherapy the same week.

23.4 Anal/Rectal Cancers

23.4.1 Introduction

The optimum dose needed to eradicate anal cancers is debated. It is unclear if external beam or brachytherapy boosting to escalate dose is more optimum. Brachytherapy requires a skill set to deliver dose in an area prone to ulceration and necrosis with respect to normal tissue dose and volume constraints. For rectal cancers, preoperative irradiation is the standard approach. Intraluminal brachytherapy can be used to deliver such preoperative therapy. Brachytherapy applications can be used to treat transanal resection beds, and rarely, but fortuitously, definitive brachytherapy can also be used for frail patients.

23.4.2 Indications, Patient Selection Criteria

Interstitial brachytherapy for anal canal/verge cancers can be used as a boost for bulky tumors following external beam irradiation (EBRT), with or without chemotherapy, with the goal of increasing local control while preserving anal function. Interstitial brachytherapy as a boost can be recommended 1 to 2 months following external beam irradiation for patients with residual tumor after EBRT or anal tumors larger than 4 cm (T2–T3) and for deeply infiltrating tumors (Papillon et al. 1989).

Furthermore, for interstitial brachytherapy for anal cancer, the following criteria should be met:

(1) Tumor size does not exceed half the circumference of the anal canal.
(2) Tumor size does not exceed 5–10 mm (exceptionally 15 mm) in thickness and 4–5 cm in length.
(3) Tumor does not widely infiltrate the surface of the perianal skin.
(4) Tumor does not infiltrate the external anal sphincter.
(5) No rectovaginal fistula is present at diagnosis.

Brachytherapy has also been used in conjunction with external beam irradiation in select patients with rectal cancers. Typically, this is reserved for patients who are medically unfit for surgery or refuse surgery. Interstitial implantation of residual disease following endocavitary irradiation of early rectal tumors (T1–T2, 4–10 cm from the anal verge) was described by Papillon using the fork applicator for tumors that did not exceed 2 cm in width and 3 cm in length, 6 weeks following contact x-ray therapy (Papillon et al. 1989; Gerard et al. 2002). For more advanced (T2–T3) rectal tumors that were moderately differentiated and within 8 cm of the anal verge, a single-plane template-guided implant, 8 weeks following external beam and contact therapy, was a relative indication for brachytherapy (Papillon et al. 1989; Gerard et al. 2002). According to Gerard et al. (2002), patients best suited for this approach include those with T2–T3, N0–N1, M0 adenocarcinoma of the middle or lower rectum involving less than two-thirds of the circumference and <6 cm from the anal verge.

Coatmeur et al. (2004) reported on the use of endocavitary irradiation with or without interstitial brachytherapy in 124 T1–T2 rectal patients. Interstitial implantation was used as a boost following endocavitary irradiation in patients who had an infiltrated tumor base. These tumors were ideally <3 cm or slightly larger and involved less than half the lumen. The infiltrative component was evaluated after the second fraction of

endocavitary contact therapy had caused regression of the exophytic component. This combination was indicated for tumors in the lower part of the rectal wall that were not amenable to low anterior resection (Coatmeur et al. 2004). Grimard et al. (2009) reported the use of template-guided interstitial brachytherapy following full-thickness transanal excision of T1 and T2 rectal cancers at risk of recurrence because of residual subclinical disease.

Frail patients with rectal cancers often cannot undergo surgical removal of their tumors. Intraluminal brachytherapy has been used to increase the dose above what can be safely given with external beam irradiation. Similar to contact therapy, intraluminal brachytherapy can be used to escalate dose. Intraluminal radiation has also been used for recurrent tumors and to palliate bleeding, discharge, and pain, either with or without external beam (Corner et al. 2010; Kaufman et al. 1989; Marijnen 2007).

23.4.2.1 Recommendations for Target Definition, Dose Prescription, Dose, and Dose Rate

23.4.2.1.1 Target Definition

The clinical target volume is defined as the palpable and visible tumor (MR, transanal ultrasound) before EBRT with a safety margin of 5 mm. The volume of the PTV should not be more than 20–30 cm³.

23.4.2.1.2 Dose Prescription

23.4.2.1.2.1 Anal Cancer
Dose specification should be preferably done according to ICRU-58 (ICRU 1997). Typically, boost doses of 15–20 Gy are given with low-dose-rate (LDR) interstitial brachytherapy over 15–28 h using ^{192}Ir (Papillon and Gerard 1987; Papillon et al. 1989; Gerard et al. 2002). Pulsed-dose-rate (PDR) (Gerard et al. 1999; Roed et al. 1996; Bruna et al. 2006) and high-dose-rate (HDR) (Doniec et al. 2006; Saarilahti et al. 2008; Oehler-Jänne et al. 2007) techniques have recently been reported for treatment of anal tumors in addition to the more traditional LDR approach. For PDR, doses of 10–25 Gy with a median of 15 Gy in 30 h have been recommended, depending on the external beam dose (Gerard et al. 1999). HDR brachytherapy has been used by Doniec et al. for anal cancers. Two fractions of 4–6 Gy are given with intracavitary and/or interstitial techniques following 45 Gy of external beam irradiation and chemotherapy (Doniec et al. 2006). Saarilahti et al. (2008) used interstitial HDR in one or two 5–6 Gy fractions specified to the Paris system reference isodose using CT-based planning following 48 Gy of intensity modulated radiation therapy (IMRT) and chemotherapy (Saarilahti et al. 2008).

23.4.2.1.2.2 Rectal Cancer

Interstitial Brachytherapy
Dose specification should be preferably done according to ICRU-58. Coatmeur et al. (2004) reported that T2 ultrasound-staged rectal patients were given a boost of 20–30 Gy over 2–3 days, following contact therapy. Grimard et al. (2009) delivered 45–50 Gy to the reference isodose following transanal excision of rectal tumors.

Endoluminal Brachytherapy

Definitive treatment consisted of either chemoradiation (45 Gy) followed by intraluminal brachytherapy of 12 Gy at 1 cm in two fractions or intraluminal brachytherapy alone delivering up to 36 Gy at 1 cm in six fractions, two to three times per week. Palliative HDR brachytherapy schedules were typically 10 Gy specified at 1 cm in a single dose (Corner et al. 2010).

Preoperative intraluminal therapy for rectal cancer was described by Yanagi et al. (1997). Moderate-dose radiation of 16–40 Gy and high-dose radiation of 40–80 Gy specified at 0.5 cm was given using the RAL remote afterloading device. Jakobsen et al. (2006) used external beam and intraluminal brachytherapy to escalate the preoperative rectal tumor dose to 65 Gy. The intraluminal boost consisted of 5 Gy in one fraction specified at 1 cm from the applicator surface. Using image guidance, Vuong et al. specified a preoperative dose of 26 Gy in four daily HDR fractions of 6.5 Gy over 1 week. The dose was specified at the tumor radial margins and the intramesorectal deposits on pretreatment MR. Adequate radial dose coverage was delivered up to 2.5 cm from the rectal lumen. This was used in an attempt to downstage tumors prior to surgery due to the high luminal doses delivered while sparing the adjacent small bowel in comparison to external beam irradiation (Vuong et al. 2002, 2005).

23.4.2.1.3 Techniques

23.4.2.1.3.1 Interstitial Brachytherapy for Anal Cancer
Traditionally, these have been small-volume implants using the Papillon–Gerard guide needle technique (Papillon et al. 1989). The implant is intended to cover the same quadrants of the anal canal circumference as were initially involved by the tumor using trocar pointed metal guide needles inserted through Papillon's template (Figure 23.6) or similar other templates (Figure 23.7). A typical implant consists of three to five needles spaced 1 cm apart, and for a tumor with thickness of more than 10 mm, more than five needles in two planes (volume implant) are needed (Figure 23.8).

FIGURE 23.7 Template for volume implants for anal carcinoma. (Courtesy of V. Strnad, University Hospital Erlangen.)

As described above, no more than 1/2 of the circumference of the lumen should be implanted to prevent strictures. Stainless-steel needles are inserted through the template holes with a guiding finger in the rectum, to the appropriate depth, and later afterloaded with ^{192}Ir. A large rubber catheter is placed in the anorectal lumen to displace the needles from the contralateral wall (Papillon et al. 1989; Gerard et al. 2002).

23.4.2.1.3.2 Interstitial Brachytherapy for Rectal Cancer
Grimard et al. described use of both single- and double-plane implants with customized templates following full-thickness transanal excision of rectal cancers to treat the remaining tumor bed. The single-plane template was 15 mm thick and had 12-mm spacing between predrilled holes for the 17 gauge needles. These needles were 10 or 15 cm in length. A rectal tube was inserted after needle insertion to keep the anal canal open and to push the mucosa opposite the needles, away from the treatment volume. The double-plane template had an internal ring with 10-mm spacing between holes and a second outer ring 10 mm beyond. This was used for anorectal tumors and tumors within 5 cm of the anal verge and in older patients. Insertion of the outer ring needles was done after the inner ring and rectal tube (Grimard et al. 2009).

With rectal tumors more than 5 cm from the anal verge, an interstitial approach without a template is needed, unlike that described for lower rectal tumors by Papillon and Gerard. In this setting, an ^{192}Ir preloaded steel fork applicator with two straight hollow prongs, 4 cm long and 16 mm apart (Arplay, Inc), are inserted for small rectal tumors. The implant is performed under local anesthesia with the patient in the knee-chest position. The fork is inserted 1 cm below the tumor and pushed superiorly, parallel to the rectal ampulla, and kept in place by a rubber drain sutured to the anal margin skin (Papillon et al. 1989; Gerard et al. 2002).

23.4.2.1.3.3 Endoluminal Brachytherapy for Rectal Cancer
Corner et al. (2010) use HDR intraluminal techniques as a boost with external beam or as palliation for rectal cancers. The applicators were inserted with a local anesthetic gel and inserted to a depth to treat the length of the tumor with a 2-cm margin. Tumors were localized by palpation if low in the rectum or by

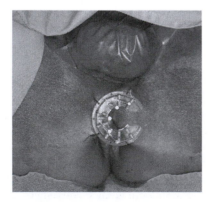

FIGURE 23.6 Typical clinical example of an implant for T1-anal cancer using five steel needles and Papillon's template. (Courtesy of V. Strnad, University Hospital Erlangen.)

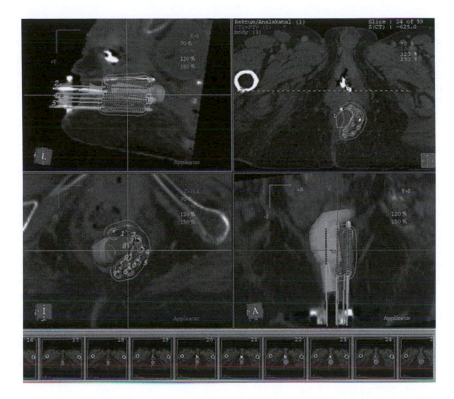

FIGURE 23.8 **(See color insert.)** Example of 2D and 3D imaging of a volume implant for anal carcinoma. (Courtesy of V. Strnad, University Hospital Erlangen.)

clips placed at the superior and inferior aspects of higher lesions, visible on treatment planning radiographs. For most patients, a single line source 2 cm in diameter applicator was used. When the tumor was not circumferential, use of segmental shielding in commercially available applicators was used to shield 25%–50% of the rectal circumference. In patients in whom the tumor is beyond the reach of the 15-cm applicator, a flexible catheter within a nasogastric tube has been used (Corner et al. 2010).

Jakobsen et al. (2006) used a 15-cm-long, 2-cm-diameter cylindrical applicator to deliver preoperative intraluminal radiation for rectal cancers. It was recommended that lead shielding could be added to reduce dose to the uninvolved parts of the lumen (Jakobsen et al. 2006).

Vuong et al. prefer the eight-channel flexible endorectal applicator (Nucletron BV, Veenendaal, Netherlands) rather than the traditional 2-cm single-channel HDR rectal applicators. Only the circumferential channels in direct contact with the tumor are loaded to spare normal rectum (Vuong et al. 2002, 2005). At the time of endoscopy, clips are placed to mark the proximal and distal margins of the tumor. The applicator is inserted in the CT scanner to the level optimum for coverage of the radio-opaque clips. The applicator is made with a balloon-type device that is optimally inflated to immobilize the applicator and brings the circumferential catheters in close contact with the tumor. For subsequent fractions, a radiograph is taken after applicator insertion and compared with the reference DRR with subsequent adjustments as needed for either depth of applicator insertion or applicator rotation (Vuong et al. 2005).

23.4.2.1.4 Results

23.4.2.1.4.1 Anal Canal Papillon treated a series of 221 patients with anal cancer at the Centre Leon Berard with external beam of 35 Gy in 15 fractions with 5FU and Mitomycin C, followed 2 months later by an LDR interstitial implant of 15–20 Gy. At 3 years, 63% were alive with anal preservation. Among the 118 patients alive and well at 5 years, the rate of anal preservation was 93%. Tumor size predicted outcome (Papillon et al. 1989). Use of Papillon's technique was reported at Centre Hospitalier Lyon Sud, with 95 of the 108 patients treated with an interstitial implant boost. Fifty-one also received chemotherapy. A complete response was observed in 104/108 (96%) of the patients, 2 months after treatment was finished. Anal preservation was possible in 85% (92/108) of the patients. Seven APRs needed to be performed because of severe necrosis, and three diverting colostomies. Necrosis typically occurred 10–18 months posttreatment. Complications were increased in larger tumors (T3–T4), and larger numbers of iridium-192 sources, the total brachytherapy treatment time, and the total brachytherapy dose were statistically correlated with the risk of necrosis. The mean dose in the group with complications was 23 Gy versus 19 Gy in the group without complications. Wagner et al. (1994) suggested that dose rates of <80 cGy/h may be helpful in limiting anorectal complications.

In a series of 95 patients from the same institution reported by Gerard et al., all received 5FU and cisplatin with external beam, followed 8 weeks later by a brachytherapy boost according to the Papillon–Gerard technique. Cancer-specific survival at 5 and 8 years was 90% and 86%, respectively, and the colostomy-free survival was 71% and 67%. Eighty-two percent (78/95) preserved their anus, and in these 78 patients, anal sphincter function was excellent or good in 72 (92%). The response of the tumor at 8 weeks after pelvic irradiation and before the brachytherapy was the only significant variable in multivariate analysis. Fourteen patients developed painful necrosis of the anus at a median time of 11 months. Hyperbaric oxygen therapy was helpful in some (Gerard et al. 1998).

Peiffert et al. (1997) reported the Nancy experience using the Papillon–Gerard technique. In this series, cancer-specific survival was 75% at 5 years, and sphincter preservation occurred in 75%. A reduction in brachytherapy dose and volume as well as a reduction in the delay between the completion of external beam and brachytherapy reduced the risk of grade 3 complications and increased sphincter preservation to 84% (Peiffert et al. 1997). Similar results are reported in several other series (Ng et al. 1988; Sandhu et al. 1998).

For T3 and T4 anal canal presentations, Hwang et al. (2004) reported local-regional control in 84% of the 31 patients treated, 7 of whom required APR. Seventy-four percent remained colostomy-free. Seven patients developed post-implant rectal ulceration, and eight developed radiation proctitis with an HDR interstitial boost (Hwang et al. 2004). Oehler-Jänne et al. (2007) also used an HDR interstitial boost of 14 Gy/7 fractions and compared this to an external beam boost of 14 Gy following external beam irradiation of 45 Gy with chemotherapy, with no improvement in outcome with the addition of brachytherapy. In the Kiel series using HDR intracavitary or interstitial brachytherapy following external beam irradiation and chemotherapy for anal cancer, Doniec et al. (2006) reported local failure in 5 of the 50 patients treated (10%), with a 5-year disease-specific survival of 82%. In 80% of the patients, sphincter function was preserved. Fraction size was decreased from 6 to 4 Gy after one patient developed sphincter necrosis following intracavitary HDR (Doniec et al. 2006).

A French study reported treatment of 71 patients with anal canal cancer with chemoradiation and a PDR interstitial boost of 17.8 Gy (10–25 Gy) to the 85% reference isodose. The 2-year actuarial local control rate was 90%. Ten patients developed grade III complications and two developed grade IV. Two patients required APR because of necrosis. The 2-year colostomy-free survival rate was 89%. Single-plane implants were recommended over double plane to reduce toxicity. Additionally, brachytherapy was recommended to be performed 2–4 weeks after chemoradiation rather than 6–8 weeks (Bruna et al. 2006).

23.4.2.1.4.2 Rectum For the 90 patients with early rectal tumors (T1–T2) treated with endocavitary irradiation and the iridium fork, Papillon reported that 74.4% were alive with anal preservation. For 62 patients with advanced disease (T2–T3), 58% were alive at 4 years with anal preservation. Radiation-induced necrosis was rare with the techniques described (Papillon et al. 1989).

Gerard et al. (2002) treated 63 patients with T2–T3 ultrasound-defined rectal tumors with a combination of external beam, contact therapy, and ^{192}Ir interstitial implant with curative intent. With a median follow-up of 54 months, the local tumor control rate was 63% and with salvage surgery was 73% (46/63). The 5-year overall survival rate was 64.4%: 84% for T2 tumors and 53% for T3 tumors. No severe grade 3–4 toxicity was seen. Late rectal bleeding occurred in 24 patients and lasted for 2–3 years. In 12 patients (19%), grade 2 rectal necrosis was seen and healed within 3–6 months. Good anorectal function was maintained in 92% of surviving patients (Gerard et al. 2002).

Grimard et al. (2009) reported 5-year local control rates of 76% with a median time to recurrence of 8 months. Ten had positive margins and 22 negative, although margin status and T stage did not correlate with recurrence. Patients with rectal tumors were more likely to recur than those with anorectal tumors. Five of the eight recurrences were salvaged surgically. Eighty-four percent retained an intact sphincter. There were four cases of grade 2 necrosis and one case of stool incontinence. The 5-year disease-specific survival was 85%, and the overall survival was 78% (Grimard et al. 2009).

Necrosis occurred in 13 of 17 patients treated with PDR, with colostomy required in 8, in the series of Roed et al. (1996), but with local control achieved in 83%. Single-plane implants were less likely to produce necrosis than larger multiplane implants (Roed et al. 1996). Gerard et al. (1999) speculated that the larger volume treated with a volume rather than single-plane implants as well as an increased number in needles and spacing between needles of 1.2–2.0 cm may have been contributory.

Using intraluminal HDR, Corner et al. (2010) treated 79 patients with medical contraindications to surgery or with advanced or recurrent disease in need of palliation of bleeding. Objective local tumor response was observed in 41/48 patients, with 28 (58%) showing a complete response and 13 (27%) a partial response. Rectal bleeding was controlled in 63%, with a median duration of symptom response of 3 months. The median survival of patients treated with palliative intent was 6 months, and it was 18.5 months for those treated with curative intent. Six patients had late toxicity (rectal ulcer—3, stricture—2, fistula—1) (Corner et al. 2010). Yanagi et al. (1997) noted improved rates of sphincter preservation in 115 patients who received preoperative intraluminal brachytherapy in comparison to those who did not.

Jakobsen et al. (2006) noted a pathologic complete response in 27% of their patients treated with 65-Gy preoperative external beam and brachytherapy, and they found microscopic disease only in another 27%. Vuong et al. (2007) treated 285 patients with rectal cancer from 1998 to 2007 as previously delineated. The median follow-up was 54 months, and the actuarial local recurrence rate was 5%, the disease-free survival rate 65%, and the overall survival rate 68% at 5 years (Vuong et al. 2007). Proctitis was observed 7–10 days after brachytherapy in all patients treated. A postoperative leak rate of 10% and a perineal wound infection rate of 12% were observed, which was not an increase over traditional rates. The most common surgical difficulty was localizing the tumor bed since in two-thirds of the patients, there was no palpable residual tumor. There was a 27% complete pathological response rate with an additional 37% having a few foci of microscopic residual cells. A local recurrence in the tumor bed was reported in four patients (Vuong et al. 2007).

23.4.3 Summary

Brachytherapy applications can be invaluable in the treatment of anorectal tumors. Using brachytherapy cure is possible with preservation of this important organ with intact function. Attention to radiation dose, dose rate, implant volume, and sequencing with external beam or surgery is important.

23.5 Lung Cancer

23.5.1 Introduction

Lung cancer is a significant epidemiological problem worldwide. Despite improvements in medical care, most cancers present in advanced stages. External beam irradiation with concurrent chemotherapy is the most commonly used modality for stage II through IIIB lung cancer. Brachytherapy has evolved as an important tool in symptom management for tumors with an incomplete response to external beam irradiation or with patients having local recurrence with a significant endobronchial component. Uncontrolled or recurrent disease results in a myriad of troublesome symptoms such as chronic cough refractory to medications, shortness of breath, hemoptysis, atelectasis, and obstructive pneumonias. Most of these symptoms are attributed to uncontrolled endobronchial disease causing airway irritability (Kelly et al. 2000). Although the use of endobronchial brachytherapy is largely palliative, there are reports of use of this modality in early-stage lung cancers from centers in France (Marsiglia et al. 2000).

23.5.2 Patient Selection

The main indication is palliative therapy of significant, symptomatic endotracheal, or endobronchial symptoms as listed earlier. Beyond that, in select cases, brachytherapy can be used as sole modality of therapy in patients unable to tolerate external radiation therapy and surgery in T1–T2 lung cancers.

Generally, patients should be able to tolerate bronchoscopy, have normal coagulation factors, and be off of anticoagulants and antiplatelet agents. Patients need to have prior bronchoscopy and biopsy documenting the presence of an endobronchial component. Patients without an endobronchial component do not benefit from this modality of therapy. Tumor characteristics, such as location and degree of obstruction, should be documented endoscopically and with chest radiographs. Photodocumentation is helpful to assess tumor response in follow-up (Kelly et al. 2000; Marsiglia et al. 2000; Senan et al. 2000; Hennequin et al. 2007).

23.5.2.1 Recommendations for Target Definition, Dose Prescription, Dose, and Dose Rate

There are a range of schedules described with this treatment modality based on local institutional practice. In general, the active length is defined by the length of the disease seen at bronchoscopy and a 2-cm proximal and distal margin. The distance to the prescription point is based on the tumor location, but usually, 1 cm from the catheter is satisfactory.

23.5.2.1.1 Technique

Patients are usually given IV sedation for the bronchoscopy. Continuous pulse oximetry with monitoring of oxygen saturation in blood and periodic blood-pressure monitoring must be performed. Bronchoscopy should be performed nasally rather than orally as it stabilizes the catheter and prevents displacement from cough and discomfort. After documenting the tumor, a 5 or 6 French nylon catheter is placed in the airway lumen. An attempt should be made to push it 4 cm or so past the distal extent of the tumor as visualized during endoscopy. Two catheters may be used if necessary to cover a disease site, and these should be individually identified with different dummy markers placed at the time of simulation. Once the catheters are placed, these are secured to the patient's nose, and the patient is moved to the simulation/planning room. Orthogonal x-rays are usually obtained for planning, but the patient may undergo CT simulation if facilities for CT-based planning are available. CT-based dosimetry can be advantageous in reducing the mucosal doses in homogeneity, but is of course also more time consuming, and faster algorithms will be needed in the future in view of the advantage of CT-based dosimetry (Kelly et al. 2000; Senan et al. 2000; Hauswald et al. 2010).

23.5.2.1.2 Results

High-dose brachytherapy is an effective modality to treat recurrent tumors. Improvement in symptoms is seen in almost two-thirds of the cases after the procedure. Repeat bronchoscopy shows almost an 80% local response in these advanced cases. The 1-year complication rate is about 13%, with a fatal hemoptysis rate of around 5%. As a sole modality of treatment in advanced lung cancers, a 17% CR and 71% PR was noted. One-year survival was 35%, with a 20% recurrence rate in those surviving 1 year. For those early-stage cancers treated with brachytherapy alone, local failure was seen in 15%. Another series from France reports CR in 80% of cases and a 5-year survival of 22%, with only 29% of patients having cancer-related mortality (Kelly et al. 2000; Hennequin et al. 2007; Hauswald et al. 2010; Lo et al. 1995).

23.5.3 Summary

Endobronchial brachytherapy offers excellent palliation of symptoms in recurrent and locally advanced lung cancer with a significant endobronchial component of disease. It has been used with curative intent in a small number of cases with good results.

23.6 Penis Carcinoma

23.6.1 Introduction

Penis carcinomas are rare disease, representing a mere 1% of tumors that attack males (Gerbaulet 1991; Solsona et al. 2004). Their frequency increases in less developed countries. Chronic infections caused by smegma or by the human papilloma virus (HPV) and due to poor hygiene represent risk factors in the development of penis carcinomas (Gregoire et al. 1995). In Europe and the United States, the primary treatment is effective surgery. The European Association of Urology recommends organ-preserving treatment only for stages Ta and T1 G1-2. A partial or complete penis amputation is preferred in stages T1 G3 and stage T2. Here, brachytherapy is mentioned as an option only for tumors that have attacked less than 50% of the penile glands (Solsona et al. 2004). Apparently, the possibilities offered by brachytherapy are somewhat underestimated, and often, too little is known about the "brachytherapy" treatment option. The advantages of brachytherapy include organ maintenance and reduced psychosexual patient stress (Gregoire et al. 1995).

23.6.2 Indications, Patient Selection Criteria

Indications for brachytherapy alone include a histologically confirmed carcinoma up to 4 cm in size and restricted to the penile glands, which has not infiltrated the penile body.

Surgical treatment is preferred for tumors > 4 cm. Should the patient refuse this option, a combined external beam radiation therapy and brachytherapy approach can be carried out (McLean et al. 1993).

23.6.2.1 Recommendation for Target Definition, Dose Prescription, Dose, and Dose Rate

23.6.2.1.1 *Target Definition*

The target area for brachytherapy (CTV) comprises the macroscopic extent of the tumor (GTV), plus a 5–10 mm safety area in both the lateral direction as well as downward into the penile glans. The standard for GTV definition is considered to be the examination of the penis using ultrasound. Optionally, an MRI examination can be helpful. These examinations allow the physician to define the likeliest tumor infiltration depth and to consequently decide whether the entire penile glans requires treatment (generally the case).

23.6.2.1.2 *Dose Prescription*

The dose prescription should be done in accordance with the ICRU-58 rules in such a manner that the tumor and its microscopic proliferation are covered by the reference isodose (ICRU 1997). The urethra should be viewed as the critical organ and should, as far as possible, be protected.

23.6.2.1.3 *Dose and Dose Rate*

Depending on the tumor size, a total dose of 60–70 Gy at a dose rate of 0.4–0.5 Gy/h represents the aim for LDR/PDR treatment.

A dose rate (pulse dose) of 0.6 Gy/h must not be exceeded and, as much as possible, the CTV should be maintained at less than 22 cm³. Exceeding either of these limits will result in a significant increase in the complication rate (de Crevoisier et al. 2009). No published data for larger series involving fractionated HDR treatment are available. Individual case studies report 54 Gy in 27 fractions over the course of 42 days for an individual bolus.

23.6.2.1.4 *Technique*

Templates, preferably with a 10–18 mm spacing between the needles, should be used for the implantation (Figures 23.9 and 23.10). The guidelines of the Paris system should be followed in order to achieve as homogeneous a radiation dose distribution as possible. Simultaneously, all rules of image-guided brachytherapy, image-guided implantation, and dose optimization should be realized (Kovács and Strnad 2010). In addition, the implant area should be <30 cm³ (Rozan et al. 1995).

23.6.2.1.5 *Results*

Data in the literature indicate that the survival rate among T1 and T2 tumors treated with interstitial LDR brachytherapy is up to 90% after 10 years and that 80%–85% local control can be achieved (Crook et al. 2005; de Crevoisier et al. 2009; Gerbaulet 2002; Naeve et al. 1993; Pow-Sang et al. 2002; Rozan et al. 1995). While 20%–30% of patients developed side effects, the penis remained intact in 72%–80% of all patients. Compared with surgical methods, this represents the greatest advantage of brachytherapy.

The complication rates are closely related to the applied dose and the treated area. Most frequently, telangiectasias, dyspigmentation, fibroses, or scleroses that do not affect the patient's quality of life occur. Less frequently, urethral stenoses (9%–35%) and necroses (6%–22%) are reported (Crook et al. 2005; de Crevoisier et al. 2009; Gerbaulet 2002; Pow-Sang et al. 2002; Rozan et al. 1995).

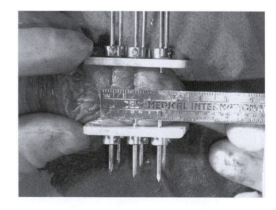

FIGURE 23.9 Example of interstitial brachytherapy of cancer of the glans penis—implant of whole glans penis using rigid needles and template. (Courtesy of V. Strnad, University Hospital Erlangen.)

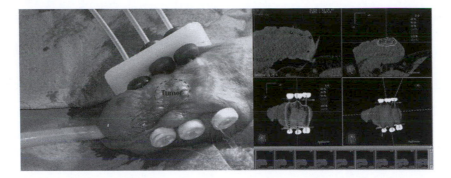

FIGURE 23.10 **(See color insert.)** Example of interstitial brachytherapy of small cancer of the penis using plastic catheters and resulting dose distribution. (Courtesy of V. Strnad, University Hospital Erlangen.)

23.7 Skin Cancer

23.7.1 Introduction

Skin cancer is a very common cancer and is the most frequently detected cancer. In comparison to its incidence, the nonmelanomatous cancers of the skin have very low mortality rates. They are, however, a source of anxiety, disfigurement, and morbidity. Basal cell cancers rarely metastasize to lymph nodes, but squamous cell cancers can spread to lymph nodes and distant organs and can be fatal if they are not treated. The other skin cancers that arise from the skin appendages tend to have an aggressive natural history with rapid spread systemically.

23.7.2 Patient Selection

The main indications for brachytherapy are T1–T2 skin cancers typically located at areas such as the hands, scalp, ears, and facial areas where cosmesis is of paramount importance. The patients selected for these treatments are typically patients with primary or recurrent cancers on the face (any area including the nose, eyelids, cheek, upper lip, lower lip, ear, scalp) and are either squamous or basal cell cancers measuring 0.5 to 8 cm in size.

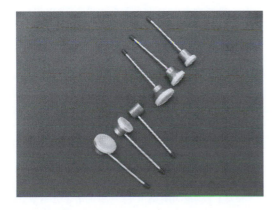

FIGURE 23.11 Leipzig applicators. (Courtesy of V. Strnad, University Hospital Erlangen.)

Patients with surgical resection and positive or close margins are also candidates for postoperative brachytherapy. Brachytherapy can also be used as a boost for T2–T3 skin cancers after external beam radiation therapy.

Most of these patients are treated with surface contact applicators (Figure 23.11) or surface molds rather than interstitial implantation (Rio et al. 2005; Guix et al. 2000).

23.7.2.1 Recommendations for Target Definition, Dose Prescription, Dose, and Dose Rate

The target for squamous cell or basal cell carcinomas is the visible tumor lesion plus 2–3 mm (Morphea-like basal cell carcinomas 5–7 mm) and the maximum depth of the infiltration of the skin due to the tumor.

For LDR and PDR brachytherapy using molds, the fractionation scheme is similar to that used in external beam therapy delivering doses between 60 and 66 Gy to tumors less than 4 cm and 75–80 Gy for those more than 4 cm. In cases where nonskin cancers such as lymphomas are treated, the dose is limited to 36 Gy.

For HDR brachytherapy using surface contact applicators, fractions of 5.0 Gy (2× weekly) up to a total dose of 30.0–40.0 Gy (six to eight fractions) are recommended.

The dose is always prescribed to the maximum depth of the tumor infiltration (up to 0.5–0.8 cm from the skin surface) (Rio et al. 2005; Guix et al. 2000).

23.7.2.1.1 Technique

Use of the surface mold and of surface applicators in general will be discussed.

For surface molds, first the number of catheters to be placed and the shape of the mold depend upon the patient anatomy. Initially, the impression of the treated area is obtained using commercially available putty material. A plaster mold of the patient is then obtained. Over this, a cast of 5 mm of polymethyl methacrylate is placed. The catheters are laid on this based on the CTV coverage needed and secured with adhesive. Usually, three to seven catheters spaced 0.5 to 1 cm are placed. The potency of the catheters is checked by having a dummy run.

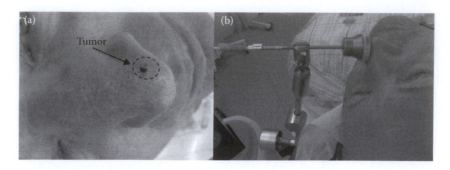

FIGURE 23.12 Example of treatment of a basal cell carcinoma of nose (a) using Leipzig applicator with a 20-mm diameter (b). (Courtesy of V. Strnad, University Hospital Erlangen.)

An additional layer of methacrylate is added once the catheter patency is verified, and the ends of the molds are secured to the patient by adding adhesive tape. Treatment planning is performed on commercially available software to deliver a homogeneous dose to the reference depth depending on the situation. CT-based planning should be performed particularly by use of large molds and in case of curved surfaces of the treated skin area. The dose calculation can be also verified with TLDs prior to treating the patient. The variation in dose at sites between the catheters should be less than 5%.

The surface applicators (e.g., Leipzig applicator; see Figure 23.11) have different sizes—typically between 10 and 25 mm. The size of the applicator is chosen depending on the tumor size, and the reference isodose for dose prescription is chosen depending on the depth of infiltration. For the treatment itself, it is necessary to fix the applicator in the most suitable position (Figure 23.12) and to instruct the patient not to move for some minutes.

23.7.2.1.2 Results

The 5-year local control with these types of treatment ranges from 85% to 99% depending on the condition treated (Guix et al. 2000; Rio et al. 2005). Recurrent cancers have a lower control rate than primary tumors. Recurrence is seen within the first 24 months. For cases treated on the dorsum of the hands and fingers, there was no change in the range of movements of the fingers, grip strength, sensation, and an excellent cosmesis has been reported (Rio et al. 2005; Guix et al. 2000; Somanchi et al. 2008).

23.7.3 Summary

Even though most skin cancers are treated with several other modalities, HDR brachytherapy affords creativity, customizability, and excellent cosmetic results with comparable local control. Dose fractionation and prescription vary on individual experience, but the overall local control reported is more than 80%–90%.

23.8 Sarcomas

23.8.1 Introduction

Radiation has an important therapeutic role in the management of soft tissue sarcomas. Over the years, several trials including

radiation as a treatment arm have established the role for radiation therapy to improve local control and limb salvage. These experiences have used radiation therapy preoperatively, postoperatively, or intraoperatively using interstitial implants or intraoperative electrons, depending on institutional practices. The focus of this section will be the role of interstitial implants using LDR or HDR brachytherapy for improving local control and limb salvage in soft tissue sarcomas. The use of this modality has varied depending upon the institution, and so the American Brachytherapy Society has established guidelines for standardizing the practice of this art of therapy (Nag et al. 2001).

23.8.2 Patient Selection

This procedure should be undertaken as a team effort involving the surgeon, radiation oncologist, medical oncologist, pathologist, radiologist, and the other support staff including physical therapy and rehabilitation. The pathologist should be experienced in grading sarcomas as there is a wide variety in grading these lesions. The patient selection for this procedure includes patients with sarcomas of the extremity or superficial trunk of more than 5 cm with an intermediate- to high-grade histology in whom the surgical margins are negative. Brachytherapy alone is not sufficient when the CTV cannot be adequately covered by an implant, normal tissue tolerance is exceeded beyond therapeutic benefit, margins are positive, the overlying skin is infiltrated, or there is clinical or histological proof of skin lymphatic involvement. Brachytherapy may be considered as a boost following a course of external beam irradiation in intermediate- to high-grade sarcomas if the intraoperative findings indicate a need for additional therapy postoperatively. There is no clear established role for brachytherapy in low-grade sarcomas. In tumors less than 5 cm, certain situations may warrant a boost to the tumor bed using brachytherapy. These include positive margins, uncertain margins, surgical field contamination, or deep lesions (Nag et al. 2001; Muhic et al. 2008; Delaney et al. 2007).

23.8.2.1 Recommendations for Target Definition, Dose Prescription, Dose, and Dose Rate

The CTV is defined as the volume of tissue with risk of microscopic spread and includes the preoperative imaged tumor and intraoperatively defined tumor bed (GTV) with 2-cm safety margins. Clips should be placed to identify the tumor bed. The CTV should be completely encompassed by the catheters.

There is a range of doses recommended based on the clinical and patient situation. The treatment has to be individualized based on the plan of therapy. LDR or PDR brachytherapy may be used as a monotherapy or as a boost after external irradiation. Intraoperative or fractionated HDR brachytherapy might also be used, but clinical experience is limited to a few institutions, and these have not been extensively tested in randomized trials.

LDR/PDR brachytherapy doses as monotherapy are recommended at 45–50 Gy delivered over 4–6 days. The recommended dose rate (pulse dose) is 35–60 cGy/h; as a boost, the LDR/PDR doses in the range of 15–20 Gy at 45 cGy/h are prescribed.

There is no consensus on HDR dosing, and it has been used in ranges of 3–9 Gy delivered daily or twice daily (Nag et al. 2001).

Treatment planning requires simulation after recovery from anesthesia. Localization radiographs with dummy sources placed in catheters are obtained. The catheters are individually identified with unique dummies, and an orthogonal radiograph is obtained to identify the sources. The CTV is outlined by the radiation oncologist on each plane usually about 2 cm apart and perpendicular to the catheters. CT-based dose calculations are strongly recommended. Dose prescription should preferably be done according to ICRU-58—the minimal target dose (Dmin) corresponds usually to 80%–90% of the mean central dose. If other prescription rules are used, the Dmin and MCD should be reported always. DVHS including the D_{90}, D_{100}, V_{100}, V_{150}, and V_{200} should also be documented, although clinical correlation with outcomes is lacking (Nag et al. 2001; Delaney 2004).

23.8.2.1.1 Technique

Most cases require a single-plane implant. Attention should be paid to the entrance of the catheters in the skin. They should be placed at least 1 cm or more past the incision. It should also be taken into account that the entry should permit loading and unloading the sources in the catheters. Catheter size depends upon the source to be used. The catheters may be placed transversely or parallel to the incision. The spacing between the catheters should be 1 to 1.5 cm. Gel foam or other material or intact muscle should be placed to separate the catheters from neurovascular bundles if they are not microscopically involved by the tumor. The catheters can be tacked to the muscle or fascia to secure them in place. There are no concrete recommendations about the margins to be chosen, and decision should be made intraoperatively. In general, the catheters should extend at least 1–2 cm past the CTV transversely and 2 cm past the CTV longitudinally. Consideration should be given to the anatomical constraints, adjacent normal tissue exposure, use of external beam therapy, and grade of the tumor when planning the margins. Care should be taken to seal the ends of the catheters to prevent infection. Catheters embedded in a Vicryl mesh are often used to implant surgical beds in the chest and abdomen, and often, tissue expanders or omental grafts may be placed to displace the catheters from sensitive structures such as bowels or nerves. The surgical incision should be closed without tension. The catheters should have dummy sources at all times except when loaded with the sources to keep them patent. Source loading should be performed no sooner than 5–7 days after wound closure (Pisters et al. 1996; Andrews et al. 2004; Laskar et al. 2007).

Intraoperative HDR (IOHDR) brachytherapy is another alternative to LDR or PDR implants. There are several commercially available IOHDR applicators that can be placed on the tumor bed to perform intraoperative HDR. Care must be taken to move away critical structures during this procedure as the tolerance of nervous tissue to HDRs is poor.

23.8.2.1.2 Results

LDR implants have been shown to be beneficial in improving local control in randomized trials. The control rates with and without brachytherapy at 76-month follow-up is reported at 89% with LDR brachytherapy versus 66% without (Llácer et al. 2006). The wound complication was related to the timing of loading the catheters with those loaded within 5 days, showing almost three times more complications than those loaded after 5 days (14%) (Pisters et al. 1996). Other nonrandomized series report control rates ranging from 65% to 89% with varying degrees of follow-up. LDR brachytherapy as a boost has control rates ranging from 70% to 95% with varying ranges of follow-up (Llácer et al. 2006; Davis 1999).

Experience with HDR brachytherapy is more limited than with LDR brachytherapy. However, small series report control rates of 50%–100% with shorter range of follow-up.

23.8.3 Summary

Soft tissue sarcomas can be adequately treated with LDR, PDR, or HDR brachytherapy or intraoperative HDR implants. Patient selection and excellent teamwork and coordination between multiple departments are needed to successfully implement this kind of a program.

23.9 Pediatric Malignancies

23.9.1 Introduction

Pediatric cancers are a heterogeneous group of tumors that are predominantly treated with chemointensive regimens. The role for radiation therapy is mainly seen in tumors involving the CNS, soft tissue sarcomas, or orbital tumors. There are very little randomized data on the role of implants in these patients as most are treated on protocols, and often, brachytherapy is not allowed on clinical trials on account of quality control issues. Most experience has been institutional.

23.9.2 Patient Selection

The most commonly treated childhood malignancies are soft tissue sarcomas. All treatment decisions are individualized based on the age and histology. Brachytherapy is a good alternative to external beam therapy as the integral dose to the patient can be reduced. The treatment and techniques of implanting the soft tissue sarcomas are very similar to those of adults, but attention should be paid to the developing structures such as joints and long ends of bones and nerves. The number of catheters used may be significantly lower than those used for adults. The procedure has to be integrated into the treatment regimen as chemotherapy is often used in high doses, and it is preferable to avoid performing the implant when the neutrophil count nadir. Brachytherapy alone is useful when dealing with low-volume disease, and the combination of external beam and brachytherapy may be useful when there is extensive residual disease after surgery or high-volume disease. Attention should be paid to the dose received by the caregivers during the implant as children need additional nursing care and constant supervision is often required. ^{192}Ir is the preferred radionuclide of choice, but ^{125}I permanent implants may be better to reduce the dose to the gonads (Merchant et al. 2000; Jabbari et al. 2009; Blank et al. 2010). The use of brachytherapy is also site- and size-dependent. Tumors larger than 4 cm after a course of chemotherapy are not candidates for brachytherapy alone (Gerbaulet et al. 1992). For such patients, surgery or external radiotherapy (or both) should be considered in addition to brachytherapy. For those patients with a complete or near-complete response and small-volume disease after chemotherapy, brachytherapy can be considered as the sole radiation modality, more so in vaginal tumors. The advantage of brachytherapy over EBRT lies in the ability to increase local control with a decrease in the probability of late complications, especially skeletal growth and sexual development. Brachytherapy not only conserves the organ but also retains the potential for future fertility and successful pregnancy. In a study (Haie-Meder et al. 2004) assessing the effect of dose on myometrial contractility, a dose of more than 30 Gy to the uterus was seen to greatly impair myometrial contractility. Thus, brachytherapy should be considered actively when a dose higher than 30 Gy to the uterine myometrium is to be delivered. Brachytherapy restricted to the vagina or vulva has a fewer consequences on fertility.

23.9.2.1 Recommendations for Target Definition, Dose Prescription, Dose, and Dose Rate

23.9.2.1.1 Technique and Results

Recommendations are similar to soft tissue tumors in adults (see Section 23.8), but for children, the target volume must be as small as possible—consequently, the safety margins are usually smaller than 2 cm as a compromise between target coverage and protection of surrounding tissues, minimizing the impact of radiation dose on tissue growth and functions (Merchant et al. 2000; Blank et al. 2010).

23.9.2.1.1.1 Central Nervous Tumors in Children   The most extensive experience in implanting a variety of CNS malignancies in the United States is with the use of permanent and temporary ^{125}I implants in children. Permanent ^{125}I implants or temporary ^{192}Ir implants were placed after surgical resection of the tumor. Thirteen of 17 patients had local control, and there

was one treatment-related death. The overall morbidity reported was very low (Healey et al. 1995).

Orbital rhabdomyosarcomas treated with brachytherapy using tailor-made molds have been reported from the Netherlands. Individual molds made from silicone were created to fit tumor beds. Holes were drilled through the molds for placement of the iridium catheters. The CTV was defined 5 mm from the surface of the mold. LDR or PDR was used. The dose to the PTV was between 40 and 50 Gy. The control rate was 71% in group I and 85% in group II, and the 5-year survival was reported at 92% (Blank et al. 2010). Both series show a low level of long-term toxicity with the use of brachytherapy, despite which general use of brachytherapy is low.

23.9.2.1.1.2 Urogenital Clear Cell Adenocarcinoma   Most of the experience of using brachytherapy for clear cell adenocarcinoma is from the Institute Gustave-Roussy, Villejuif, Paris (IGR). The target volume is defined similarly as for gynecological tumors in adults, respecting the anatomy and age of the child. Safety margins between 10 and 20 mm are recommended. The technique of brachytherapy is the same as for adult patients with carcinoma of the cervix or vagina. The total dose varies from 32 to 60 Gy depending on risk factors and aim of the treatment. The brachytherapy makes it possible to realize an organ-sparing therapy for 70% of patients and to achieve a 2-year survival rate between 57% (stage III–IV) and 83%–95% (stages I and II).

23.9.2.1.1.3 Genital Tract Rhabdomyosarcomas   The technique of gynecological brachytherapy as developed at IGR is anatomy-adapted with the use of the mold applicator, as has been discussed in Chapter 21 on gynaecological cancers. The IGR technique is based on the use of molded applicators for both cervical and vaginal rhabdomyosarcomas (Gerbaulet et al. 1985). In short, the first step consists of a vaginal impression obtained under anesthesia that accurately shows the topography and extension of the tumor as well as the anatomy of the vagina. Since this vaginal impression is very accurate in providing a description of the tumor, it represents an essential step in tumor assessment. A vaginal mold is then made from the vaginal impression. This molded applicator is patient specific, even for very irregular target volumes. Patients are treated with low-dose-rate (LDR) brachytherapy or pulsed-dose-rate brachytherapy. For highly infiltrating tumors, a treatment combining interstitial implants and intracavitary brachytherapy is used.

Reported experiences focusing on brachytherapy for rhabdomyosarcoma of the female genital tract are rare and are highly institution dependent. A large series of genital tract rhabdomyosarcomas in girls has been reported from IGR, France (Magné and Haie-Meder 2007).

Spunt et al. (2005) reported the late toxic effects in 26 children with pelvic rhabdomyosarcomas (seven vaginal) treated by a multimodality combination, with a median follow-up of more than 20 years. Nineteen patients received EBRT and three received brachytherapy. The late effects (genitourinary and gastrointestinal) were substantial. EBRT and brachytherapy resulted in long-term ovarian and uterine dysfunction. Since the radiation dose needed to destroy 50% of immature oocytes is less than 2 Gy, ovarian transposition is required to protect developing ovaries. Brachytherapy has a major advantage over EBRT in the preservation of healthy hormonal activity on account of the lower ovarian doses delivered during therapy. Using brachytherapy not only conserves the uterus but also retains the potential for future pregnancy. In a study (Haie-Meder et al. 2004) assessing the effect of dose on myometrial contractility, a dose of more than 30 Gy to the uterus was seen to greatly impair myometrial contractility. Thus, brachytherapy should be considered actively when a dose higher than 30 Gy to the uterine myometrium is to be delivered. Brachytherapy restricted to the vagina or vulva also has fewer consequences on fertility.

In 1989, the first report by Gerbaulet emphasized the possibility of gynecological function preservation and showed the need for ovarian transposition to protect fertility (Gerbaulet et al. 1989). Flamant et al. (1979, 1990, 1998) have reported on the sequelae in 17 girls with vulval and vaginal rhabdomyosarcoma treated with LDR brachytherapy, including the first series of 14 patients reported by Gerbaulet et al. Before brachytherapy, eight patients underwent oophoropexy. Twelve pubescent or postpubescent girls were studied for long-term sequelae with a follow-up of longer than 10 years: 11 had normal puberty, and 2 had a total of three healthy children delivered by caesarean section. Only one patient had ovarian insufficiency (oophoropexy was not done), with an estimated dose of 10–15 Gy to both ovaries. Only one patient underwent hysterectomy after LDR brachytherapy. Eleven patients had normal menses, and ten had normal menarche (one after hormonal replacement). Five girls had only minor telangiectasias; three were sexually active. Three needed minor surgical correction to allow adequate sexual activity. Four girls had severe sequelae (digestive, genital, and urinary tract). Fifteen girls were free of disease.

An update in 1999 by Martelli reported the overall results of 38 girls treated in two different International Society of Paediatric Oncology (SIOP) protocols with genital-tract rhabdomyosarcoma (at vulval, vaginal, and uterine locations) (Martelli et al. 1999). Twenty-seven patients had a primary tumor in the vulva or vagina, whereas 11 had a uterine tumor. Thirteen were given chemotherapy only, and 17 needed local treatment to achieve local control. No difference in outcome was seen between the different locations in the genital tract. Among girls with residual disease after chemotherapy, brachytherapy was an adequate alternative to radical surgery or EBRT. Magné et al. (2006) reported an update and reappraisal of the brachytherapy experience at IGR in the management of vulval and vaginal rhabdomyosarcoma, focusing on long-term outcomes. From 1971 to 2005, data for 39 patients (including the group reported by Flamant et al.) given brachytherapy as a part of treatment were retrospectively analyzed. Of these patients, 20 were treated before 1990, when the initial tumor extension was included in the brachytherapy volume; however, since then, the dose prescription was changed to address only residual disease after chemotherapy. Acute and late side effects were classified with the Common Terminology

Criteria for Adverse Events (version 3.0, CTCAE). The median follow-up was 8.4 years (range 10 months to 30 years), and median age was 16.3 months at diagnosis. Tumors were located in the vagina of 26 girls and vulva of 6 girls. Botryoid histology was present in more than 75% of cases. A multidisciplinary approach of surgery, chemotherapy, and brachytherapy was used for all patients. Five-year overall survival was 91%. Six of the 39 patients presented with local or metastatic relapse, and acute side effects were manageable. Of the 20 patients treated before 1990, 6 had grade 1–2 renal or genitourinary function symptoms, and 15 patients (75%) presented sequelae of vaginal or urethral sclerosis and stenosis. Four patients also had follow-up treatment for psychological disturbance. After 1990, two of 19 patients had acute side-effects with maximum grade 1–2 symptoms of renal or genitourinary function impairment, and four patients (20%) had vaginal or urethral sclerosis and stenosis. None of the complications needed surgery. Between the two groups (before and after 1990), a substantial difference was seen in rates of acute and long-term sequelae. Twenty patients were older than 12 years at the last follow-up. Seventeen patients had begun or had had a normal puberty, 3 had a total of four healthy children (of whom two had not undergone previous ovarian transposition), and 14 had normal menses. Three of the 20 patients older than 12 years had some ovarian insufficiency (oophoropexy was not done), with an estimated total dose of 10–15 Gy to both ovaries, and 1 had undergone hysterectomy after brachytherapy.

It should be noted that the brachytherapy as reported by IGR has never been directly compared with other therapeutic strategies, surgery or ERBT (or both), in a randomized fashion and perhaps never will be. In a series of 151 female patients with a genital-tract tumor treated on the Intergroup Rhabdomyosarcoma Study Group protocols I–IV in the United States, Arndt et al. (2001) reported that 31% received external radiation, combined with surgery in 19%. Results showed an overall 5-year outcome of 87%. This is somewhat similar to the experience at IGR using brachytherapy. The German experience from the CWS German Cooperative Soft Tissue Sarcoma study on rhabdomyosarcoma of the urinary bladder and vagina has been reported by Leuschner et al. (2001). At 10 years of follow-up, patients with a botryoid rhabdomyosarcoma in the urinary bladder or vagina had 91% survival compared with 73% for patients with classic rhabdomyosarcoma. Patients with vaginal rhabdomyosarcoma had a more favorable prognosis than those with tumors in urinary sites.

These results show that brachytherapy compares favorably with other techniques.

23.9.3 Summary

The brachytherapy for pediatric tumors is a highly effective treatment with great potential to reduce long-term serious side effects and preservation of function of organs.

For a successful pediatric brachytherapy, a closest cooperation between radiation oncologist, pediatric oncologist, surgeon, radiologist, and pathologist is "conditio sine qua non."

23.10 Uveal Melanoma

23.10.1 Introduction

Preservation of sight and esthetic appearance are critical goals in treating ocular and orbital malignancies. Brachytherapy allows delivery of a high radiation dose, while sparing the lens, cornea, lacrimal gland, retina, and optic nerve. Though rare, melanoma is the most common primary intraocular malignancy and arises in the uveal tract, which is the pigmented layer of the eye, including the iris, ciliary body, and choroid. Choroidal melanomas are the most common uveal tract tumors.

23.10.2 Indications, Patient Selection Criteria

The size and location of ocular melanomas will determine the optimum therapeutic intervention. In addition to the expert exam of the referring ophthalmologists, standard A and B ocular ultrasound is used to help stage uveal melanomas. Ultrasound reveals the apical height of the tumor, and ophthalmoscopic exam reveals the basal dimensions. Patients with small uveal melanomas (<2.5 mm height and <10 mm in largest basal dimension) are typically observed and treated only if there is interval growth of the tumor. Patients with medium-sized lesions (>2.5 to 10 mm height and <12–16 mm largest basal diameter) have the option of enucleation or plaquing (Figure 23.13). Those patients with large melanomas (>12–16 mm largest basal dimension and >10 mm in height) are usually enucleated. However, plaquing is recommended for some large melanomas if more than 3 mm from the optic nerve. Visual outcomes, however, may be compromised in these patients with large lesions because of the location of the lesions in close proximity to the optic nerve and macula and the inherent high radiation dose gradients within the base of the tumor while specifying the treatment dose at the apex. Extrascleral extension, diffuse ring or multiple lesions, peripapillary lesions, lesions involving the iris, or lesions at an angle of > ½ or more of the ciliary body are contraindications to plaque

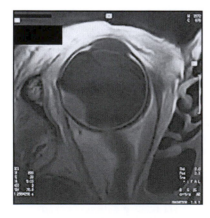

FIGURE 23.13 Example of a typical uveal melanoma 5 mm high and 8 mm in diameter indicated for brachytherapy. (Courtesy of V. Strnad, University Hospital Erlangen.)

therapy. Visual outcome improves as the distance increases from the plaque or tumor to the macula (Nag et al. 2003).

23.10.2.1 Recommendations for Target Definition, Dose Prescription, Dose, and Dose Rate

Stallard (1966), using custom cobalt-60 (^{60}Co) plaques, demonstrated the radiosensitivity of choroidal melanomas following 7000 to 14,000 Gy delivered over 7 to 14 days. Equivalent 5-year survival rates to those of enucleation were achieved in selected patients. Approximately half the patients, however, experienced ocular morbidity, decreased visual acuity, and even enucleation. [60]Co is a high-energy gamma emitter and caused damage to the retina, optic nerves, and lenses. To retain vision, alternate sources were developed, including ^{192}Ir, [106]Ru, and ^{125}I, whose emission energies are less than ^{60}Co and more readily shielded (Nag et al. 2003). Though to a lesser degree than ^{60}Co, ^{192}Ir also has high-energy gamma emissions and was associated with increased dose to personnel and surrounding ocular and extraocular structures.

[125]I is a gamma emitter but at a much lower energy, with better preservation of vision due to this inherent difference. Because of its radiation safety issues and satisfactory tissue penetration, ^{125}I was chosen for treating choroidal melanoma in the Collaborative Ocular Melanoma Study (COMS) and is the most common isotope used in the United States. [103]Pd has also been of recent interest (Margo 2004). The low gamma emission of ^{125}I and [103]Pd is easily absorbed by the gold plaque, resulting in less radiation exposure of personnel and the surrounding ocular and extraocular structures. Compared with ^{125}I, [103]Pd has lower energy and more rapid dose falloff with the possibility of decreasing ocular complications (Finger et al. 2002).

[106]Ru, which has gained popularity in Europe, is a beta emitter and has an even more rapid dose falloff, allowing dose concentration to the base while minimizing dose to the contralateral ocular structures. [106]Ru, however, may result in a higher scleral dose and a lower apex dose for tumors > 3 mm in apical height and should be preferably used for treatment of small melanomas ≤ 5 mm in apical height (Nag et al. 2003).

For plaque fabrication, the ophthalmologist provides the apical height of the tumor as well as the basal dimensions. A fundus diagram with orientation of the tumor borders relative to the surrounding structures should include the optic nerve, foveola, equator, ora serrata, and center of the lens. The basal dimensions at the center of the tumor in the direction of the macula and optic disc and the minimum distance from the tumor edge to the macula and the optic disc should be documented (Nag et al. 2003). The location and dimensions must be transferred to the treatment planning system to allow an accurate calculation of tumor and critical structure radiation doses (Nag et al. 2003). The COMS dose prescription system is used as a guideline by many for dose specification (COMS 2001). The initial COMS prescribed dose, which was administered to the apex of the tumor or to within 5 mm of the interior surface of the sclera, was 100 Gy. Dose was specified at the apex of the tumor for tumors > 5 mm in height and at 5 mm from the inner scleral surface for those < 5 mm in height (Jampol et al. 2002). In 1996, this was changed to 85 Gy when the dosimetry formalism of the American Association of Physicists in Medicine (AAPM) Task Group 43 was applied (Nath et al. 1995).

The American Brachytherapy Society (ABS) recommends 85 Gy to the apex of the tumor, with the 85-Gy isodose passing through the prescription point and encompassing the entire tumor, using the AAPM TG-43 formalism for the calculation of dose (Nag et al. 2003).

The COMS allowed dose rates to the dose specification points between 0.43 and 1.05 Gy/h with treatment duration of 3–7 days. Reports of dose rates < 0.6 Gy/h by Quivey et al. (1996) have revealed inferior control rates. The ABS suggests dose rates for ^{125}I of 0.6–1.05 Gy/h (Nag et al. 2003). The dose rates for ^{106}Ru are typically 2–12 Gy/h, and sometimes also < 2 Gy/h. The optimum dose and dose rate for control of choroidal melanomas, however, are not defined due to the significant gradients of dose and dose rate over the treatment volume, largely dependent on tumor height (Quivey et al. 1996). Knowledge of the dose rate and total dose at the apex and the base of the index lesion is pivotal. The dose rate is adjusted by varying the number and activity of the seeds (total activity) and is also dependent on the apical height of the tumor. The scleral dose may be three to four times the apical dose. It is speculated by some that the high dose to the vascular supply in the base may be more important in tumor control than the dose at the apex.

A preplan is created before the plaque is placed. With knowledge of the plaque size and number of seeds, seeds of a determined strength are ordered to achieve a preselected dose rate, which will determine the implant duration. Uniform and nonuniform source strengths can be used, the latter requiring special attention to seed location and plaque orientation. Once the seeds arrive, a final preplan is run with the exact seed strengths to determine the total implant time (Nag et al. 2003).

23.10.2.1.1 Technique and Applicators

23.10.2.1.1.1 Applicators There are rimmed and unrimmed plaques as well as customized and prefabricated plaques. The radioactive sources are glued into the plaque either in a silastic insert or directly onto the undersurface of the plaque, and the entire applicator is sterilized prior to placement. Rimmed plaques prevent side scatter of dose, thus minimizing dose to adjacent extraocular structures. The COMS study brought with it standardized, prefabricated rimmed gold plaques with silastic seed carriers. The standard COMS plaques are circular, with inner diameters of 12, 14, 16, 18, and 20 mm carrying 8, 13, 13, 21, and 24 ^{125}I seeds, respectively (Figure 23.14a). For select larger lesions, 22-mm plaques are also available but were not used in the COMS study. The seeds are embedded in a predesigned pattern in the associated seed carriers and glued in place. With these plaques, >99% of the photons generated from ^{125}I or ^{103}Pd seeds are absorbed by the gold plaque with inherent direction of the prescribed radiation into the eye. Notched plaques can be used for peripapillary tumors and those close to the optic nerve. A margin of at least 2–3 mm on all sides of the lesion is optimal. Suture holes are placed at the periphery of the plaque for fixation to the scleral surface.

The ^{106}Ru applicators are as well in different sizes (12–20 mm) and forms (Figure 23.15) with a nominal activity of 13–26 MBq (0.35–0.7 mCi).

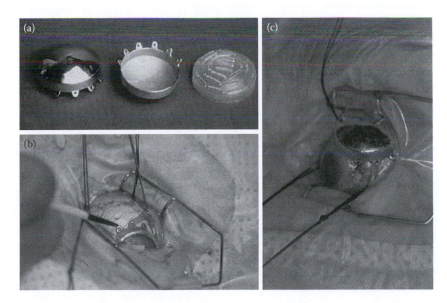

FIGURE 23.14 Typical sequence of a positioning of the COMS plaque: (a) COMS plaque with seeds placed in the plastic insert; (b) transparent "dummy" plaque placed to optimize plaque location; (c) gold plaque sutured to sclera surface over the choroidal melanoma. (Courtesy of B. Erickson, Medical College of Wisconsin.)

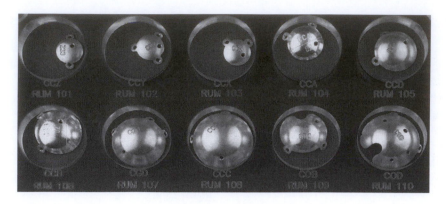

FIGURE 23.15   Example of the most common forms of ruthenium-106 applicators. (Courtesy of V. Strnad, University Hospital Erlangen.)

23.10.2.1.1.2 Technique Plaquing entails superimposing a radioactive plaque on the sclera, over the tumor, with a 2–3 mm margins on all sides. The conjunctiva is excised and retracted, and if necessary, ocular muscles are temporarily detached. A retraction suture is used to rotate the globe for visualization and plaque placement. The plaque is placed by the ophthalmologist under local or general anesthesia with indirect ophthalmoscopy and transillumination through the pupil to localize and mark the lesion to the scleral surface (Figure 23.16). A dummy plaque, which is transparent and has the same size as the active plaque, can be used to mark and position the sutures on the sclera and to verify the position and orientation of the plaque in relation to the tumor (Figure 23.14). A 2–3 mm margin is desired around the edges of the tumor. Once accurate placement of the dummy plaque has been realized, it is removed, and the active plaque is then slipped over these sutures and stitched in place. Verification of the plaque's position relative to the tumor is performed after the plaque is stitched in place and repositioned if suboptimal. The conjunctiva is then reapproximated, and the eye rotated into normal position. The plaque is left in place to deliver the intended dose over 3–7 days. After the determined treatment duration is reached, the plaque is removed in the operating room. Inpatient and outpatient options will depend on the radiation safety rules of the designated state or country.

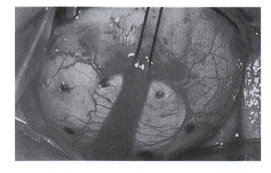

FIGURE 23.16   Example of marking of the borders of uveal melanoma on the sclera surface. (Courtesy of V. Strnad, University Hospital Erlangen.)

23.10.2.1.2 Results

Tumor size correlates directly with survival and impacts treatment decisions. In addition to disease-free survival, preservation of useful vision, cosmetic appearance, and quality of life are important treatment considerations. Enucleation had been considered the standard treatment for ocular melanomas, but in an effort to preserve vision, episcleral plaque radiotherapy was developed as an innovative alternative.

To resolve some of the controversies comparing enucleation to radioactive plaquing, the COMS was developed (Margo 2004). This was a multicenter, national, prospective randomized trial funded by the National Eye Institute of the National Institutes of Health, initiated in 1987 and completed in 1998. Patients with small, medium, and large tumors were accrued to the study. Those with medium-sized tumors were randomized between enucleation and ^{125}I plaque irradiation. The study closed in 1998 after accrual of 1317 patients with medium-sized tumor. Survival curves demonstrated no difference in survival between the enucleation and plaque arms (Jampol et al. 2002). Local treatment failure was 10.3% at 5 years, with 88% retaining their eyes at 5 years (Jampol et al. 2002). Visual loss, however, was noted to be progressive over time in the irradiated patients, with 43% of treated eyes having a visual acuity of 20/200 by 3 years posttreatment (COMS 2001). Other than dose, factors found to have visual prognostic significance were pretreatment vision, diabetes, tumor height and shape, proximity to the macula, and retinal detachment. Visual acuity declined at a rate of approximately two lines per year on average (COMS 2001). Complications after ocular brachytherapy are related to total dose, dose rate, and dose/volume relationships in addition to tumor size and location (Jones et al. 2002). The most common late complication is radiation retinopathy for posterior presentations and glaucoma and cataracts for anterior presentations. Of course, enucleation results in immediate loss of vision, whereas changes in visual function following radiation are more gradual.

The Mayo Clinic experience with ^{125}I was reported by Jensen et al. (2005). The 5-year disease-specific survival was 91%. Initial local control at 5 years was 92%, with all recurrences subsequently salvaged. The median time to maximum tumor response was 33 months. Vision stayed the same or improved in 25% of patients, and 44% of patients maintained visual acuity better than 20/200. Thirteen percent experienced chronic pain. Dose rates to the tumor apex greater than 90–100 cGy/h were associated with increased systemic control but worse radiation toxicity. The authors concluded that low dose rates (<75–90 cGy/h) to the apex should be avoided as this reduces systemic control (Jensen et al. 2005).

Correa et al. (2009) reported treatment of 120 patients with small, medium, and select large melanomas treated with I-125 plaques for ocular melanoma. The 5- and 8-year cause-specific survival was 85.7%, and local failure was 11.8% at 5 years and 27.3% at 8 years (Correa et al. 2009).

Finger et al. (2002) reported the outcome of 100 patients treated with Pd-103 plaque radiotherapy followed for a mean of 4.6 years. Local control was achieved in 96% with six secondary enucleations. Thirty-five percent lost six or more lines of vision, and 73% had vision of 20/200 or better, concluding comparable local control with I-125 but better visual function with Pd-103, due to the emission of lower-energy photons less likely to affect the posterior orbit. Pd-103 can be used to treat small, medium, and large melanomas (Finger et al. 2002).

[106]Ru has been used to treat small-to-medium choroidal melanomas. Several recent reports described a local control in 93%–98% of the patients after 5 years (Damato et al. 2005; Heindl et al. 2007). The most recent study reported by Verschueren et al. (2010) confirms these results. The 5-year actuarial local control by 425 patients was 96%, and the 5-year metastases-free survival was 76.5%. Two- and 5-year rates free of radiation complications were 60% and 35%. The 5-year enucleation rate was 4.4%. Cosmetic and functional eye preservations rates were 96% and 52% at 5 years (Verschueren et al. 2010).

23.10.3 Summary

Ocular brachytherapy is an innovative and effective alternative to enucleation in properly selected patients. Useful vision can be retained in most patients with equivalent survival to those treated with enucleation. Attention to the optimum isotope, dose, and dose rate relative to the size and location of the index lesions will enhance outcome.

23.11 Benign Diseases (Keloids and Pterygium)

23.11.1 Introduction

Keloids are benign skin tumors that can form either as the result of a skin injury or spontaneously. A differentiation must be made between keloids and hypertropic skin scarring, which exhibits a significantly better prognosis.

Pterygium (wing skin) is a complex, benign illness of the conjunctiva—a proliferation of the connective tissue that attacks the cornea. Pterygium is operable and, as a rule, benign, although there is a tendency toward recurrence.

23.11.2 Indications, Patient Selection Criteria

Keloids: An already existing keloid or known tendencies toward keloid formation serve as the primary indications.
Pterygium: An existing pterygium represents an indication.

23.11.2.1 Recommendations for Target Definition, Dose Prescription, Dose, and Dose Rate

23.11.2.1.1 Target Definition

Keloids: The target area for interstitial brachytherapy subsequent to keloid excision comprises the entire surgical scar, together with a 1-cm safety area in all directions.
Pterygium: The target area for surface brachytherapy employing ^{90}Sr comprises the scar region subsequent to surgical pterygium excision, together with a 2- to 3-mm-wide safety margin.

23.11.2.1.2 Dose Prescription

Keloids: The dose should be specified to a 5–10 mm space from the catheter axis. The best results were found for dose specification at a distance of 10 mm (Guix et al. 2001).
Pterygium: The dose should be specified on the surface of the cornea.

23.11.2.1.3 Dose, Dose Rate

Keloids: The following dose concept is recommended:

- HDR brachytherapy: 4×3 Gy/10 mm.

The first fraction should be administered within 60–90 min subsequent to the scar excision.
Pterygium: The following dose concept is recommended:

- HDR brachytherapy with [90]Sr
- $2-3 \times 10$ Gy/applicator surface
- $1 \times 20-25$ Gy/applicator surface

The first fraction should be administered within hours (no more than 24 hours) after the excision.

For recurring pterygium, it has also been shown that a biological equivalent dosage (BED) between 30 and 40 Gy with ^{90}Sr is totally adequate to reduce recurrence to less than 10% (Kal et al. 2009).

23.11.2.1.4 Techniques

Keloids: Catheter implantations can be carried out under either local or general anesthetic. Generally, after the keloid excision, a plastic catheter is positioned in the center of the scar (wherever possible, no deeper than 1 cm below the skin surface). The scar is then closed using a nonreabsorbing material. The plastic catheters must cover the entire length of the scar and must be accessible from both scar ends (Figure 23.17).
Pterygium: Standard treatment takes place under a local anesthetic using eye drops containing procaine. After application of the anesthetic, a lid spreader is carefully inserted around the exterior region of the pterygium to permit complete access. After this, the physician brings the surface of the ^{90}Sr applicator

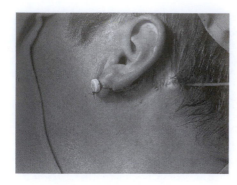

FIGURE 23.17 Implant of the single leader plastic tube in the scar after excision of a keloid in the region of the earlobe. (Courtesy of V. Strnad, University Hospital Erlangen.)

into contact with the pterygium conjunctiva for a precise time which is necessary for the prescribed dose.

23.11.2.1.5 Results

Keloids: The recurrence rate after interstitial brachytherapy after surgery for keloid varies between 5% and 20% (Arnault et al. 2009; Bertiere et al. 1990; Garg et al. 2004; Guix et al. 2001; Clavere et al. 1997; Escarmant et al. 1993; Maalej et al. 2000; Narkwong and Thirakpuht 2006; Veen and Kal 2007). Conflicting reports comparing brachytherapy and external beam radiation therapy (EBRT) exist. Some authors report significantly higher recurrence rates when employing EBRT, with rates ranging from 63% to 100% of all cases (Borok et al. 1988). Other authors indicate comparable results (Kovalic and Perez 1989; Kutzner et al. 2003). Undeniably, the major difference between brachytherapy and EBRT lies particularly in the fact that, with the use of brachytherapy, significantly smaller amounts of healthy tissue remain within the irradiated area. Due to the fact that very small tissue volumes are treated, the probability of subsequent effects such as fibroses, pigmentation, and skin atrophies is extremely low. The incidence of these side effects can be seen as being equal to zero (Guix et al. 2001). Neither are any acute side effects to be anticipated.

Pterygium: Recurrence rate after surface brachytherapy for pterygium conjunctiva is only 5%–10% (Alaniz-Camino 1982; Beyer 1991; Fukushima et al. 1999; Isohashi et al. 2006; Jürgenliemk-Schulz et al. 2004; MacKenzie et al. 1991; Nishimura et al. 2000; Parayani et al. 1994; Viani et al. 2008; Wesberry and Wesberry 1993; Wilder et al. 1992).

The majority of the authors used very high doses here. More recently published analyses (Kal et al. 2009) confirm that the total BED should correspond to a value between 30 and 40 Gy.

Only mild, acute side effects such as eye irritation, photophobia, thinner sclera, or postoperative granulomas have been noted as side effects subsequent to surface treatment of a pterygium. In rare cases, scleral or corneal necrosis has been observed, primarily the result of bacterial, fungal, or viral infections, and not appearing until several years after treatment.

References

23.2 Head-and-Neck Cancer Tumors

Beauvois, S., Hoffstetter, S., Peiffert, D. et al. 1994. Brachytherapy for lower lip epidermoid cancer: Tumoral and treatment factors influencing recurrences and complications. *Radiother Oncol* 33:195–203.

Chang, J.T., See, L.C., Tang, S.G. et al. 1996. The role of brachytherapy in early-stage nasopharyngeal carcinoma. *Int J Radiat Oncol Biol Phys* 36:1019–24.

Decroix, Y. and Labib, A. 1983. Traitement de la lesion primaire dans les cancers de la langue mobile. Experience de l'institut Curie (1959–1972). *Ann Otolaryngol Chir Cervicofac* 100:549–55.

Lapeyre, M., Peiffert, D., Malissard, L., Hoffstetter, S., and Pernot, M. 1995. An original technique of brachytherapy in the treatment of epidermoid carcinomas of the buccal mucosa. *Int J Radiat Oncol Biol Phys* 33:447–54.

Lapeyre, M., Bellière, A., Hoffstetter, S., and Peiffert, D. 2008. Brachytherapy for head and neck cancers (nasopharynx excluded). *Cancer Radiother* 12:515–21.

Lusinchi, A., Eskandari, J., Son, Y. et al. 1989. External irradiation plus curietherapy boost in 108 base of tongue carcinomas. *Int J Radiat Oncol Biol Phys* 17:1191–7.

Mazeron, J.J., and Richaud, P. 1984. Compte Rendu de la XVIIIe réunion de Groupe Européen de Curietherapie. *J Eur Radiother* 5:50–6.

Mazeron, J.J. Simon, J.M., Le Pechoux, C. et al. 1991. Effect of dose rate on local control and complications in definitive irradiation of T1-2 squamous cell carcinomas of mobile tongue and floor of mouth with interstitial iridium-192. *Radiother Oncol* 21:39–47.

Melzner, W.J., Lotter, M., Sauer, R., and Strnad, V. 2007. Quality of interstitial PDR-brachytherapy-implants of head-and-neck-cancers: Predictive factors for local control and late toxicity? *Radiother Oncol* 82:167–73.

Narayana, A., Cohen, G.N., Zaider, M. et al. 2007. High-dose-rate interstitial brachytherapy in recurrent and previously irradiated head and neck cancers—Preliminary results. *Brachytherapy* 6:157–63.

Pernot, M., Hoffstetter, S., Malissard, L. et al. 1996. Interet de l'association radiotherapie externe-curietherapie dans les carcinomes de la region veloamygdalienne. Etude statistique d'une serie de 361 patients. *Bull Cancer Radiother* 83:40–6.

Rovirosa-Casino, A., Planas-Toledano, I., Ferre-Jorge, J. et al. 2006. Brachytherapy in lip cancer. *Med Oral Patol Oral Cir Bucal* 11:223–9.

Simon, J.M., Mazeron, J.J., Pohar, S. et al. 1993. Effect of inter-source spacing on local control and complications in brachytherapy of mobile tongue and floor of mouth. *Radiother Oncol* 26:19–25.

Strnad, V. 2004. Treatment of oral cavity and oropharyngeal cancer. Indications, technical aspects, and results of interstitial brachytherapy. *Strahlenther Onkol* 180:710–7.

Strnad, V. and Kovács, G. 2010. Head and neck cancer. In: *Practical Manual of Brachytherapy* [German]. V. Strnad, R. Pötter, and G. Kovács (eds.). UniMed Verlag AG, Bremen, pp. 170–87.

Strnad, V., Lotter, M., Grabenbauer, G., and Sauer, R. 2000. Early results of pulsed-dose-rate interstitial brachytherapy for head and neck malignancies after limited surgery. *Int J Radiat Oncol Biol Phys* 46:27–30.

Strnad, V., Geiger, M., Lotter, M., and Sauer, R. 2003. Role of brachytherapy for head and neck cancer: Retreatment in previously irradiated area. *Brachytherapy* 2:158–63.

Strnad, V., Melzner, W., Geiger, M. et al. 2005. Role of interstitial PDR-brachytherapy in the treatment of oral and oropharyngeal cancer: A single-institute experience of 236 patients. *Strahlenther Onkol* 181:762–7.

Thiagarajan, A., Lin, K., Tiong, C.E. et al. 2006. Sequential external beam radiotherapy and high-dose-rate intracavitary brachytherapy in T1 and T2 nasopharyngeal carcinoma: An evaluation of long-term outcome. *Laryngoscope* 116:938–43.

Wang, C.C. 1997. Carcinoma of the nasopharynx. In: *Radiation Therapy for Head and Neck Neoplasms*, 3rd ed. Wiley-Liss, New York, pp. 257–60.

23.3 Esophageal Cancer

Coia, L.R., Engstrom, P.F., and Paul, A.R. 1991. Long term results of infusional 5FUm mitomycin C, and radiation as primary management of esophageal carcinoma. *Int J Radiat Oncol Biol Phys* 20:29–36.

Gaspar, L.E. 1999. Esophageal brachytherapy: A phantom menace. *Int J Radiat Oncol Biol Phys* 45:549–50.

Gaspar, L.E., Nag, S., Herskovic, A. et al. 1997. American Brachytherapy Society (ABS) consensus guidelines for brachytherapy of esophageal cancer. *Int J Radiat Oncol Biol Phys* 38:127–32.

Okawa, T., Dokiya, T., and Nishio, M. 1999. Multi-institutional randomized trial of external radiotherapy with and without intraluminal brachytherapy for esophageal cancer in Japan. *Int J Radiat Oncol Biol Phys* 45:623–8.

Sai, H., Mitsumori, M., Araki, N. et al. 2005. Long term results of definitive radiotherapy for stage I esophageal cancer. *Int J Radiat Oncol Biol Phys* 62:1339–44.

Sur R., Levin, V., Donde, B. et al. 2002. Prospective randomized trial of HDR brachytherapy as a sole modality in palliation of advanced esophageal carcinoma—An International Atomic Energy Agency study. *Int J Radiat Oncol Biol Phys* 53:127–33.

Vuong, T., Szego, P., David, M. et al. 2005. The safety and usefulness of high dose rate endoluminal brachytherapy as a boost in the treatment of patients with esophageal cancer with external beam radiation with and without chemotherapy. *Int J Radiat Oncol Biol Phys* 63:758–64.

23.4 Anal/Rectal Cancers

Bruna, A., Gastelblum, P., Thomas, L. et al. 2006. Treatment of squamous cell anal canal carcinoma (SCACC) with pulsed dose rate brachytherapy: A retrospective study. *Radiother Oncol* 79:75–9.

Coatmeur, O., Truc, G., Barillot, I. et al. 2004. Treatment of T1-T2 rectal tumors by contact therapy and interstitial brachytherapy. *Radiother Oncol* 70:177–82.

Corner, C., Bryant, L., Chapman, C. et al. 2010. High-dose-rate afterloading intraluminal brachytherapy for advanced inoperable rectal carcinoma. *Brachytherapy* 9:66–70.

Doniec, J.M., Schniewind, B., Kovacs, G. et al. 2006. Multimodal therapy of anal cancer added by new endosonographic-guided brachytherapy. *Surg Endosc* 20:673–8.

Gerard, J.P., Ayzac, L., Hun, D. et al. 1998. Treatment of anal canal carcinoma with high dose radiation therapy and concomitant fluorouracil-cisplatinum. Long-term results in 95 patients. *Radiother Oncol* 46:249–56.

Gerard, J.P., Mauro, F., Thomas, L. et al. 1999. Treatment of squamous cell anal canal carcinoma with pulsed dose rate brachytherapy. Feasibility study of a French cooperative group. *Radiother Oncol* 51:129–31.

Gerard, J.P., Chapet, O., Ramaioli, A. et al. 2002. Long-term control of T2-T3 rectal adenocarcinoma with radiotherapy alone. *Int J Radiat Oncol Biol Phys* 54:142–9.

Grimard, L., Stern, H., and Spaans, J. 2009. Brachytherapy and local excision for sphincter preservation in T1 and T2 rectal cancer. *Int J Radiat Oncol Biol Phys* 74:803–9.

Hwang, J.M., Rao, A.R., Cosmatos, H.A. et al. 2004. Treatment of T3 and T4 anal carcinoma with combined chemoradiation and interstitial ^{192}Ir implantation: A 10-year experience. *Brachytherapy* 3:95–100.

ICRU 1997. Report 58: Dose and volume specification for reporting interstitial Therapy. International Commission on Radiation Units and Measurements. Bethesda, MD.

Jakobsen, A., Mortensen, J.P., Bisgaard, C. et al. 2006. Preoperative chemoradiation of locally advanced T3 rectal cancer combined with an endorectal boost. *Int J Radiat Oncol Biol Phys* 64:461–5.

Kaufman, N., Nori, D., Shank, B. et al. 1989. Remote afterloading intraluminal brachytherapy in the treatment of rectal, rectosigmoid, and anal cancer: A feasibility study. *Int J Radiat Oncol Biol Phys* 17:663–8.

Marijnen, C.A.M. 2007. External beam radiotherapy and high dose rate brachytherapy for medically unfit and elderly patients. *Clin Oncol* 19:706–10.

Ng Ying Kin, N.Y.K., Pigneux, J., Auvray, H. et al. 1988. Our experience of conservative treatment of anal canal carcinoma combining external irradiation and interstitial implant: 32 cases treated between 1973 and 1982. *Int J Radiat Oncol Biol Phys* 14:253–9.

Oehler-Jänne, C., Seifert, B., Lütolf, U.M. et al. 2007. Clinical outcome after treatment with a brachytherapy boost versus external beam boost for anal carcinoma. *Brachytherapy* 6:218–26.

Papillon, J., and Gerard, J.P. 1987. Anorectal Region. In: *Modern Brachytherapy*. B. Pierquin, J.F. Wilson, and D. Chassagne (eds.). MASSON Publishing USA, Inc., New York, pp. 249–61.

Papillon, J., Montbarbon, J.F., Gerard, J.P. et al. 1989. Interstitial curietherapy in the conservative treatment of anal and rectal cancers. *Int J Radiat Oncol Biol Phys* 17:1161–9.

Peiffert, D., Bey, P., Pernot, M. et al. 1997. Conservative treatment by irradiation of epidermoid cancers of the anal canal: Prognostic factors of tumoral control and complications. *Int J Radiat Oncol Biol Phys* 37:313–24.

Roed, H., Engelholm, S., Svendsen, L. et al. 1996. Pulsed dose rate (PDR) brachytherapy of anal carcinoma. *Radiother Oncol* 41:131–4.

Saarilahti, K., Arponen, P., Vaalavirta, L. et al. 2008. The effect of intensity-modulated radiotherapy and high dose rate brachytherapy on acute and late radiotherapy-related adverse events following chemoradiotherapy of anal cancer. *Radiother Oncol* 87:383–90.

Sandhu, A.P.S., Symonds, R.P., Robertson, A.G. et al. 1998. Interstitial iridium-192 implantation combined with external radiotherapy in anal cancer: Ten years experience. *Int J Radiat Oncol Biol Phys* 40:575–81.

Vuong, T., Belliveau, P.J., Michel, R.P. et al. 2002. Conformal preoperative endorectal brachytherapy treatment for locally advanced rectal cancer. *Dis Colon Rectum* 45:1486–95.

Vuong, T., Devic, S., Moftah, B. et al. 2005. High-dose-rate endorectal brachytherapy in the treatment of locally advanced rectal carcinoma: Technical aspects. *Brachytherapy* 4:230–5.

Vuong, T., Devic, S., and Podgorsak, E. 2007. Long-term data for high dose rate endorectal brachytherapy as a neoadjuvant therapy for patients with operable rectal cancer. In: *Choices in Advanced Radiotherapy*. R. F. Mould (ed.). Nucletron BV, Veenendaal, The Netherlands, pp. 65–72.

Wagner, J.P., Mahe, M.A., Romestaing, P. et al. 1994. Radiation therapy in the conservative treatment of carcinoma of the anal canal. *Int J Radiat Oncol Biol Phys* 29:17–23.

Yanagi, H., Kusunoki, M., Kamikonya, N. et al. 1997. Results of preoperative intraluminal brachytherapy combined with radical surgery for middle and lower rectal carcinomas. *J Surg Oncol* 65:76–81.

23.5 Lung Cancer

Hauswald, H., Soiber, E., Rochet, N. et al. 2010. Treatment of recurrent bronchial carcinoma: Role of high dose rate endoluminal brachytherapy. *Int J Radiat Oncol Biol Phys* 77:373–7.

Hennequin, C., Bleichner, O., Tredaniel, J. et al. 2007. Long-term results of endobronchial brachytherapy: A curative treatment? *Int J Radiat Oncol Biol Phys* 67:425–30.

Kelly, J.F., Delclos, M.E., Morice, R.C. et al. 2000. High Dose Rate endobronchial brachytherapy effectively palliates symptoms due to airway tumors: The 10 year M.D. Anderson Cancer center experience. *Int J Radiat Oncol Biol Phys* 48:697–702.

Lo, T.C.M., Girschvich, L., Healey, G. et al. 1995. Low dose rate versus high dose rate intraluminal brachytherapy for malignant endo bronchial tumors. *Radiother Oncol* 35:193–7.

Marsiglia, H., Baldeyrou, P., Lartigau, E. et al. 2000. High dose rate brachytherapy as sole modality for early stage endobronchial carcinoma. *Int J Radiat Oncol Biol Phys* 47:665–72.

Senan, S., Lagerwaard, F., de Pan, C.A. 2000. CT assisted method of dosimetry in brachytherapy of lung cancer. *Radiother Oncol* 55:75–80.

23.6 Penis Carcinoma

Crook, J.M., Jezioranski, J., Grimard L. et al. 2005. Penile brachytherapy: Results for 49 patients. *Int J Radiat Oncol Biol Phys* 62:460–7.

de Crevoisier, R., Slimane, K., Sanfilippo, N. et al. 2009. Long-term results of brachytherapy for carcinoma of the penis confined to the glans (N- or NX). *Int J Radiat Oncol Biol Phys* 21:1–7.

Gerbaulet, A. 1991. Cancers de la verge. In: *Cancers uro-genitaux*. A. Steg and F. Eschwege (eds.). Flammarion Medicine Science, Paris, pp. 406–24.

Gerbaulet, A. 2002. Tumors of the penis. In: *Oxford Textbook of Oncology*, 2nd ed. R.L. Souhami, I. Tannock, P. Hohenberger, and J.C. Horiot (eds.). Oxford University Press, Oxford, pp. 2047–56.

Gregoire, L., Cubilla, A.L., Reuter, V.E. et al. 1995. Preferential association of human papilloma virus with high-grade histologic variants of penile-invasive squamous-cell carcinoma. *J Natl Cancer Inst* 87:1705–9.

ICRU 1997. Report 58: Dose and volume specification for reporting interstitial therapy. International Commission on Radiation Units and Measurements. Bethesda, MD.

Kovács, G. and Strnad, V. 2010. Penis cancer. In: *Practical Manual of Brachytherapy* [German]. V. Strnad, R. Pötter, G. Kovács (eds.). UniMed Verlag AG, Bremen, pp. 165–9.

McLean, M., Akl, A.M., Warde, P. et al. 1993. The results of primary radiation therapy in the management of squamous cell carcinoma of the penis. *Int J Radiat Oncol Biol Phys* 25:623–8.

Naeve, F., Neal, A.J., Hoskin, P.J., and Hope-Stone, H.F. 1993. Carcinoma of the penis: A retrospective review of treatment with iridium mould and external beam radiation. *Clin Oncol* 5:207–10.

Pow-Sang, M.R., Benavente, V., Pow-Sang, J.E. et al. 2002. Cancer of the penis. *Cancer Control* 9:305.

Rozan, R., Albuisson, E., Giraud, B. et al. 1995. Interstitial brachytherapy for penile carcinoma: Multicentric survey (259 patients). *Radiother Oncol* 36:83–93.

Solsona, E., Algaba, F., Horenblas, S. et al. 2004. EAU guidelines on penile cancer. *Eur Urol* 46:1–8.

23.7 Skin Cancer

Guix, B., Finestress, F., Tello, J.I. et al. 2000. Treatment of skin carcinomas of the face by high dose arte brachytherapy and custom made surface molds. *Int J Radiat Oncol Biol Phys* 47:95–102.

Rio, E., Bardet, E., Ferron, C. et al. 2005. Interstitial brachytherapy of periorificial skin carcinomas of the face: A retrospective study of 97 cases. *Int J Radiat Oncol Biol Phys* 63:753–7.

Somanchi, B.V., Stabton, A., Webb, M. et al. 2008. Hand function after high dose rate brachytherapy for squamous cell carcinoma of the skin of the hand. *Clin Oncol* 9:691–7.

23.8 Sarcomas

Andrews, S.F., Anderson, P.R., Eisenberg, B.L. et al. 2004. Soft tissue sarcomas treated with post operative external beam radiation therapy with and without low dose rate brachytherapy. *Int J Radiat Oncol Biol Phys* 59:475–80.

Davis, A.M. 1999. Functional outcome in extremity soft tissue sarcoma. *Semin Radiat Oncol* 9:360–8.

Delaney, T.F. 2004. Optimizing radiation therapy and post treatment function in management of extremity soft tissue sarcomas. *Curr Treat Opt Oncol* 5:463–76.

Delaney, T.F., Kepka, L., Goldberg, S.I. et al. 2007. Radiation therapy for local control of soft tissue sarcomas resected with positive margins. *Int J Radiat Oncol Biol Phys* 67:1460–9.

Laskar, S., Bahl, G., Puri, A. et al. 2007. Pulse dose rate perioperative interstitial brachytherapy for soft tissue sarcomas-prognostic factors and long term results of 155 patients. *Ann Surg Oncol* 14:560–7.

Llácer, C., Delannes, M., Minsat, M. et. al. 2006. Low dose intraoperative brachytherapy in soft tissue sarcomas involving neuro vascular structure. *Radiother Oncol* 78:10–6.

Muhic, A., Hovgaard, D., Peterson, M.M. et al. 2008. Local control and survival in patients with soft tissue sarcomas treated with limb sparing surgery in combination with interstitial brachytherapy and external radiation. *Radiother Oncol* 88:382–7.

Nag, S., Shasha, D., Janjan, N. et al. 2001. The American Brachytherapy Society recommendation for brachytherapy of soft tissue sarcomas. *Int J Radiat Oncol Biol Phys* 49:1033–43.

Pisters, P.W., Harrison, L.B., Leung, D.H.Y. et al. 1996. Long term results of prospective randomized trial of adjuvant brachytherapy in soft tissue sarcoma. *J Clin Oncol* 14:859–68.

23.9 Pediatric Malignancies

Arndt, C.A.S., Donaldson, S.S., Anderson, J.R. et al. 2001. What constitutes optimal therapy for patients with rhabdomyosarcoma of the female genital tract? *Cancer* 91:2454–68.

Blank, L., Koedooder, K., Van Der Grient, H.N.S. et al. 2010. Brachytherapy as a part of the multidisciplinary treatment of childhood rhabdomyosarcomas of the orbit. *Int J Radiat Oncol Biol Phys* 77:1463–9.

Flamant, F., Chassagne, D., Cosset, J.M. et al. 1979. Embryonal rhabdomyosarcoma of the vagina in children. Conservative treatment with curietherapy and chemotherapy. *Eur J Cancer* 15:527–32.

Flamant, F., Gerbaulet, A., Nihoul-Fekete, C. et al. 1990. Long-term sequelae of conservative treatment by surgery, brachytherapy, and chemotherapy for vulval and vaginal rhabdomyosarcoma in children. *J Clin Oncol* 8:1847–53.

Flamant, F., Rodary, C., Rey, A. et al. 1998. Treatment of non-metastatic rhabdomyosarcoma in childhood and adolescence. Results of the second study of the International Society of Paediatric Oncology: MMT84. *Eur J Cancer* 34:1050–62.

Gerbaulet, A., Panis, X., Flamant, F., and Chassagne, D. 1985. Iridium afterloading curietherapy in the treatment of pediatric malignancies. The Institut Gustave Roussy experience. *Cancer* 56:1274–9.

Gerbaulet, A., Esche, B.A., Haie-Meder, C. et al. 1989. Conservative treatment for lower gynecological tract malignancies in children and adolescents: the Institut Gustave-Roussy experience. *Int J Radiat Oncol Biol Phys* 17:655–8.

Gerbaulet, A., Haie, C., Michel, G. et al. 1992. Combined radio-surgical treatment in early invasive cervix carcinoma according to prognostic factors. Experience of the Gustave-Roussy Institute. *Eur J Gynaecol Oncol* 13:256–61.

Haie-Meder, C., Morice, P., Paris, B. et al. 2004. Consequences of brachytherapy (BT) on uterine myocontractility. *Radiother Oncol* 71:54.

Healey, E., Shamberger, R., Grier, H. et al. 1995. A 10-year experience of pediatric brachytherapy. *Int J Radiat Oncol Biol Phys* 32:451–5.

Jabbari, S., Andolino, D., Weinberg, V. et al. 2009. Successful treatment of high risk and recurrent pediatric desmoids using radiation as a component of multimodality management. *Int J Radiat Oncol Biol Phys* 75:177–82.

Leuschner, I., Harms, D., Mattke, A. et al. 2001. Rhabdomyosarcoma of the urinary bladder and vagina. A clinicopathologic study with emphasis on recurrent disease: A report from the Kiel pediatric tumor registry and the German CWS study. *Am J Surg Pathol* 25:856–64.

Magné, N., and Haie-Meder, C. 2007. Brachytherapy for genital tract rhabdomyosarcomas in girls: Technical aspects, reports and perspectives. *Lancet* 8:725–9.

Magné, N., Oberlin, O., Martelli, H. et al. 2006. Vulval and vaginal rhabdomyosarcoma in children: the Institut Gustave Roussy brachytherapy experience with a particular attention on long term outcome. *Int J Gynecol Cancer* 16(Suppl 3):610.

Martelli, H., Oberlin, O., Rey, A. et al. 1999. Conservative treatment for girls with nonmetastatic rhabdomyosarcoma of the genital tract: A report from the study committee of the International Society of Pediatric. *J Clin Oncol* 17:2117–22.

Merchant, T.E., Parsh, N., Del Valle, P.L. et al. 2000. Brachytherapy for pediatric soft tissue sarcoma. *Int J Radiat Oncol Biol Phys* 46:427–32.

Spunt, S.L., Sweeney, T.A., Hudson, M.M. et al. 2005. Late effects of pelvic rhabdomyosarcoma and its treatment in female survivors. *J Clin Oncol* 23:7143–51.

23.10 Uveal Melanoma

COMS 2001. Collaborative Ocular Melanoma Study (COMS) randomized trial of I-125 brachytherapy for medium choroidal melanoma. I. Visual acuity after 3 years. COMS Report No. 16. *Ophthalmology* 108:348–66.

Correa, R., Pera, J., Gómez, J. et al. 2009. 125 I episcleral plaque brachytherapy in the treatment of choroidal melanoma: A single-institution experience in Spain. *Brachytherapy* 8:290–6.

Damato, B., Patel, I., Campbell, I.R. et al. 2005. Local tumor control after 106Ru brachytherapy of choroidal melanoma. *Int J Radiat Oncol Biol Phys* 63:385–91.

Finger, P.T., Berson, A., Ng, T. et al. 2002. Palladium-103 plaque radiotherapy for choroidal melanoma: An 11-year study. *Int J Radiat Oncol Biol Phys* 54:1438–45.

Heindl, L.M., Lotter, M., Strnad, V. et al. 2007. High dose brachytherapy for uveal and ciliar melanoma using 106 Ruthenium. *Ophthalmology* 104:149–57.

Jampol, L.M., Moy, C.S., Murray, T.G. et al. 2002. The COMS randomized trial of iodine 125 brachytherapy for choroidal melanoma: IV. Local treatment failure and enucleation in the first 5 years after brachytherapy. COMS Report No. 19. *Ophthalmology* 109:2197–206.

Jensen. A.W., Petersen. I.A., Kline. R.W. et al. 2005. Radiation complications and tumor control after 125 I plaque brachytherapy for ocular melanoma. *Int J Radiat Oncol Biol Phys* 63:101–8.

Jones, R., Gore, E., Mieler, W. et al. 2002. Post treatment visual acuity in patients treated with episcleral plaque therapy for choroidal melanomas: Dose and dose rate effects. *Int J Radiat Oncol Biol Phys* 52:989–95.

Margo, C.E. 2004. The collaborative ocular melanoma study: An overview. *Cancer Control* 11:304–9.

Nag, S., Quivey, J.M., Earle, J.D. et al. 2003. The American Brachytherapy Society recommendations for brachytherapy of uveal melanomas. *Int J Radiat Oncol Biol Phys* 56:544–55.

Nath, R., Anderson, L.L., Luxton, G. et al. 1995. Dosimetry of interstitial brachytherapy sources: Recommendations of the AAPM Radiation Therapy Committee Task Group No. 43. *Med Phys* 22:209–34.

Quivey, J.M., Augsburger, J., Snelling, L. et al. 1996. 125 I Plaque therapy for uveal melanoma. Analysis of the impact of time and dose factors on local control. *Cancer* 77:2356–62.

Stallard, H.B. 1966. Radiotherapy for malignant melanoma of the choroid. *Br J Ophthal* 50:147–55.

Verschueren, K.M.S., Creutzberg, C.L., Schalij-Delfos, N.E. et al. 2010. Long-term outcomes of eye-conserving treatment with Ruthenium106 brachytherapy for choroidal melanoma. *Radiother Oncol* 95:332–8.

23.11 Benign Diseases (Keloids and Pterygium)

Alaniz-Camino, F. 1982. The use of postoperative beta radiation in the treatment of pterygia. *Ophthalmic Surg* 3:1022–5.

Arnault, J.P., Peiffert, D., Latarche, C. et al. 2009. Keloids treated with postoperative Iridium192 brachytherapy: a retrospective study. *J Eur Acad Dermatol Venereol* 17:807–13.

Bertiere, M.N., Jousset, C., Marin, J.L., and Baux, S. 1990. Intérêt de l'irradiation interstitielle des cicatrices cheloides par Iridium 192. A propos de 46 cas. *Ann Chir Plast Esthet* 35:27–30.

Beyer, D.C. 1991. Pterygia: Single-fraction post-operative beta irradiation. *Radiology* 178:569–71.

Borok, T., Bray, M., Sinclair, I. et al. 1988. Role of ionizing irradiation for 393 keloids. *Int J Radiat Oncol Biol Phys* 15:865–70.

Clavere, P., Bedane, C., Bonnetblanc, J.M. et al. 1997. Postoperative interstitial radiotherapy of keloids by iridium 192: A retrospective study of 46 treated scars. *Dermatology* 195:349–52.

Escarmant, P., Zimmermann, S., Amar, A. et al. 1993. The treatment of 783 keloid scars by iridium 192 interstitial irradiation after surgical excision. *Int J Radiat Oncol Biol Phys* 26:245–51.

Fukushima, S., Onoue, T., and Onoue, T. et al. 1999. Postoperative irradiation of pterygium with 90Sr eye applicator. *Int J Radiat Oncol Biol Phys* 43:597–600.

Garg, M., Weiss, P., Sharma, A. et al. 2004. Adjuvant high dose rate brachytherapy (192Ir) in the management of keloids which have recurred after surgical excision and external radiation. *Radiother Oncol* 73:233–6.

Guix, B., Henríquez, I., Andrés, A. et al. 2001. Treatment of keloids by high-dose-rate brachytherapy: A seven-year study. *Int J Radiat Oncol Biol Phys* 50:167–72.

Isohashi, F., Inoue, T., and Xing, S. et al. 2006. Postoperative irradiation for pterygium: Retrospective analysis of 1,253 patients from the Osaka University Hospital. *Strahlenther Onkol* 182:437–42.

Jürgenliemk-Schulz, I.M., Hartman, L.J.C., and Roesink, J.M. et al. 2004. Prevention of pterygium recurrence by postoperative single-dose beta irradiation: A prospective randomized clinical double-blind trial. *Int J Radiat Oncol Biol Phys* 59:1138–47.

Kal, H.B., Veen, R.E., and Jürgenliemk-Schulz, I.M. 2009. Dose-effect relationships for recurrence of keloid and pterygium after surgery and radiotherapy. *Int J Radiat Oncol Biol Phys* 74:245–51.

Kovalic, J.J. and Perez, C.A. 1989. Radiation therapy following keloidectomy: A 20 year experience. *Int J Radiat Oncol Biol Phys* 17:77–80.

Kutzner, J., Schneider, L., and Seegenschmiedt, M.H. 2003. Strahlentherapie des Keloids in Deutschland Patterns-of-Care-Studie - Ergebnisse einer Umfrage. *Strahlenther Onkol* 179:54–8.

Maalej, M., Frikha, H., Bouaouina, N., and Ben Abdallah, M. 2000. Place de la curiethérapie dans le traitement des chéloïdes. A propos de 114 cas. *Cancer Radiother* 4:274–8.

MacKenzie, F.S., Hirst, L.W., Kynaston, B. et al. 1991. Recurrence rate and complications after beta irradiation for pterygia. *Ophthalmology* 98:1776–80.

Narkwong, L. and Thirakhupt, P. 2006. Postoperative radiotherapy with high dose rate iridium 192 mould for prevention of earlobe keloids. *J Med Assoc Thai* 89:428–33.

Nishimura, Y., Nakai, A., Yoshimasu, T. et al. 2000. Long-term results of fractionated strontium-90 therapy for pterygia. *Int J Radiat Oncol Biol Phys* 46:137–41.

Parayani, S.B., Scott, W.P., Wells Jr. J.W. et al. 1994. Management of pterygium with surgery and radiation therapy. The North Florida Pterygium Study Group. *Int J Radiat Oncol Biol Phys* 28:101–3.

Veen, R. and Kal, H. 2007. Postoperative high-dose-rate brachytherapy in the prevention of keloids. *Int J Radiat Oncol Biol Phys* 69:1205–8.

Viani, G.A., Stefano, E.J., De Fendi, L.I., and Fonseca, E.C. 2008. Long-term results and prognostic factors of fractionated strontium-90 eye applicator for pterygium. *Int J Radiat Oncol Biol Phys* 72:1174–9.

Wesberry Jr., J.M., and Wesberry Sr., J.M. 1993. Optimal use of beta irradiation in the treatment of pterygia. *South Med* 86:633–7.

Wilder, R.B., Buatti, J.M., Kittelson, J.M. et al. 1992. Pterygium treated with excision and postoperative beta irradiation. *Int J Radiat Oncol Biol Phys* 23:533–7.

VI

Developments in Clinical Brachytherapy

24

Developments in Gynecologic Brachytherapy Dose Recording and Reporting: From ICRU Report 38 to Image-Guided Brachytherapy

John E. Mignano
Tufts University School of Medicine

24.1 Preface

It has been more than 25 years since ICRU Report 38, "Dose and Volume Specification for Reporting Intracavitary Therapy in Gynecology" was first published (ICRU 1985). This report provided a basis for source specification and dose reporting with reference to anatomic landmarks during gynecologic brachytherapy (BT) procedures. The recommended reference volume concept should provide means to compare treatment outcomes among various clinical facilities. The use of computerized dose calculation facilitated treatment in patients with "abnormal" anatomy and for whom the calculations provided by the previously employed Stockholm, Paris, and Manchester systems were not optimal.

ICRU Report 38 provided guidelines for recording and reporting dose points in adjacent normal tissues (bladder and rectum) and representative for the lymphatic system during the use of either low-dose-rate (LDR) or high-dose-rate (HDR) intracavitary BT (ICBT) for the treatment of cervical cancer. Along with information contained in ICRU Report 29 (ICRU 1978), there was a rational basis for dose calculation, recording, and reporting during both fractionated pelvic external beam radiotherapy (EBRT) and ICBT for the treatment of cervical cancer and other gynecologic malignancies.

The salient points of ICRU Report 38 are as follows:

(1) Gamma-emitting sources replace radium.
(2) Old written directive treatment parameters likely not appropriate for use with new source types.
(3) SI units accepted as standard terminology.
(4) Computer-generated treatment calculations allow dose distributions in multiple planes.

Taken together, the recommendations provided within ICRU Report 38 would promote uniformity in treatment prescriptive parameters and facilitate comparison of treatment outcomes among facilities. To help standardize reporting of gynecologic ICBT, ICRU Report 38 provided a list of minimum data to be recorded for each procedure. These included

A. A description of technique used, that is, HDR versus pulsed dose rate (PDR) versus LDR, type of applicator employed
B. The total reference air *kerma*
C. The reference volume
D. The absorbed dose at standard reference points (i.e., bladder and rectal points)
E. Time dose or fractionation pattern

Point A historically has had near universal use (except for the French school) for at least two reasons: (1) it may (or may not) delineate the lateral extent of the parametria, and (2) it can be readily defined even in the most rudimentary of clinical practices and thereby provides some universal standard for comparison of clinical outcomes in disparate facilities. The use of point A (and B) gained wide acceptance in the early era of gynecologic BT given the technical limitations of the day. Shortcomings primarily related to the imaging and computational equipment available at that time demonstrated the inadequacy of dose plans providing dose distributions in a limited number of planes or point doses within adjacent normal structures in predicting outcomes (Pötter et al. 2001). As such, a move was made to integrate 3D imaging modalities and better treatment algorithms. These new technologies have greatly extended the utility of ICBT in the treatment of gynecologic malignancies, and more detailed imaging modalities have facilitated individualized care for these patients. Given the evolution of techniques and technology available since ICRU Report 38 was published in 1985, it is appropriate to review and update this report's recommendations at this time.

24.2 Introduction

From a historical perspective, BT has been established as an essential part of definitive radiation treatment of cervical cancer, and for locally advanced disease, conformal EBRT and concurrent platinum-based chemotherapy are often employed (Pötter et al. 2008). ICRU Report 38 provided a framework for BT (primarily LDR) in which the dose prescription was based on reference points delineated on 2D orthogonal plain films (ICRU 1985). The authors of ICRU Report 38 realized that ICBT was often used in conjunction with EBRT, particularly when treating locally advanced disease. They considered that the concepts of target volume, treatment volume, and irradiated volume be used during ICBT just as was the case for EBRT practiced at that time. Though ICBT developed more extensively during the early days of radiation therapy probably as a result of the available technology and equipment for clinical practice (i.e., simple physical applicators and sources having a defined and predictable rate of decay), concepts subsequently more extensively developed during the clinical use of EBRT (target volume, treatment volume, and irradiated volume) could then carry over into clinical ICBT practice. The authors did, however, note that the steep dose gradient around the sources made the comparison of point dose within the target and treatment volumes not meaningful, except as maximal, minimal, and perhaps average point doses within the target and treatment volumes. A further difficulty lies in the inability to easily integrate physical (and biological) doses from the ICBT and EBRT treatment plans into a single unified treatment plan as the generation of dose-volume histograms (DVHs) was difficult and tedious until the advent of 3D imaging, better algorithms, and high-speed computation. Thus, while this report and the referenced technology of the day represented a major improvement over the previous era of LDR BT for cervical cancer employing Manchester, Paris, or Stockholm plans, there were still considerable deficiencies. While HDR ICBT was practiced in the decade (or longer) after the release of ICRU Report 38, treatment planning was most often done with 2D imaging (even when 3D imaging was available), and doses were most often prescribed and recorded at the points A and B and rectal and bladder points as had been used for decades with LDR ICBT and the Manchester system.

24.3 Concepts Used in ICBT

24.3.1 Treatment Techniques

24.3.1.1 Radium Substitutes

In the spring of 1985, when ICRU Report 38 was published, HDR BT and the use of individualized computerized BT treatment planning were just starting to become widely utilized. Thus, while ^{192}Ir HDR remote afterloading BT was available in a limited number of facilities, LDR BT remained the treatment standard of the day. ICRU Report 38 did tout the advantages of ^{192}Ir and ^{137}Cs over ^{226}Ra as an LDR BT source. The advantage was primarily in regard to improved radiologic safety and increased specific activity of the newer sources relative to ^{226}Ra sources. The report did note that the high specific activity of the newer radionuclides, ^{192}Ir in particular, allowed for miniaturization of the BT sources. The use of afterloading HDR ICBT applicators in conjunction with computerized BT treatment planning provided a means to generate dose distributions tailored to the local extent of disease. The authors of ICRU Report 38 did caution that the radiobiology of HDR treatment was not the same as for LDR treatment, and doses typically employed in LDR BT if used for HDR BT would result in significantly different clinical outcomes in both tumor and normal tissues.

24.3.1.2 Simulation of Linear Sources

LDR sources used in gynecologic ICBT are typically linear and ~2 cm in length and ~4 mm in diameter. Commercially supplied HDR sources are typically much smaller (5 mm in length and ≤1.1 mm in diameter) and can be considered point sources. As a linear source can be simulated by a series of point sources, the HDR source can be used to provide isodose lines that replicate those obtained with one or more linear sources typically employed during LDR BT. The ability to use short dwell times and small interval distances for source placement within the applicator provides greater flexibility in dose delivery than is obtainable for standard LDR sources, in addition to being able to use small applicators given the smaller physical source size.

24.3.1.3 Dose Rates

The HDR source provides dose rates in excess of 20 cGy per minute (>12 Gy per hour) at the prescription isodose line, whereas LDR sources deliver 3.3 cGy or less per minute (<2 Gy per hour) at the prescription isodose line. Medium dose rate treatment

(2 to 12 Gy per hour) is seldom employed clinically. (More will be said later about the effect of dose rate on biologic response.)

24.3.1.4 Afterloading Techniques

Whereas LDR ICBT applicators typically utilize manually afterloaded sources, modern HDR ICBT employs computer-controlled remotely driven sources. This eliminates the potential for exposure of clinical staff during normal ICBT practice, and this represents a significant advantage over manually afterloaded ICBT. A more important clinical advantage includes the ability to tailor radiation delivery so as to maximize dose within a target volume and minimize dose in adjacent normal structures. Remote afterloading HDR BT was developed and used in limited clinical roles in the mid-1970s and was commercially available in only limited numbers in the late 1970s. It was not until this technology became widely available in clinical practice in the mid- to late 1980s coupled with the broad array of high-speed computers with dedicated treatment planning algorithms that the ability to readily individualize treatment become utilized. The subsequent wide availability of high-resolution CT and MR imaging and the ability to generate 3D image reconstruction prompted the development of BT applicators compatible with those imaging modalities. Moreover, the radiobiology of HDR BT was not fully appreciated until even later. As such, ICRU Report 38 as published in 1985 could not be considered the definitive guide for BT in the modern era. To fully take advantage of HDR remote afterloading BT, a number of issues would have to be addressed. These would include better image definition of the "target volume" and adjacent normal structures, improved dosimetry, and improved understanding of time–dose relationships.

24.3.2 Absorbed-Dose Pattern and Volume Definitions

24.3.2.1 Absorbed-Dose Pattern

ICRU Report 38 did discuss the differences in soft tissue absorbed dose from EBRT employing photon beams and ICBT employing gamma emitters. In the time that has passed since the first publication of this report, there has been better modeling of the dose distributions within biologic tissues, and with the use of better immobilization and localization devices and image guidance during both EBRT and ICBT, practitioners have been able to achieve more accurate and reproducible dose delivery within target tissues. With the exception of somewhat broader application of proton EBRT and the demise of ^{252}Cf ICBT (at least in the Western Hemisphere), the changes in absorbed dose patterns have been evolutionary and not revolutionary, and will not be further discussed.

24.3.2.2 Volumes

Treatment planning in the early days of BT was based on clinician's physical evaluation, which included limited visualization of the tumor volume (initially limited to the clinical examination and "educated" fingertips) and 2D plain-film imaging. Additional treatment of "regions at risk" was based on the clinician's understanding of the usual natural progression of unchecked disease. Plain-film imaging also provided reference points for dose measurement within adjacent dose-limiting structures (i.e., bladder and rectum). The advent of modern imaging technology, first CT in the 1970s and then MR in the 1980s, more recently along with CT/MR compatible BT applicators represented a significant move forward. MR IGBT (image-guided BT) is a leading contender for the standard of care in the modern era, and a number of societies have recommended that this practice be widely accepted and employed.

24.3.2.2.1 Treatment Volume

High-speed dedicated BT treatment planning systems for HDR BT in conjunction with optimized external beam treatment plans represents the current standard of care in comprehensive radiotherapy treatment planning. These systems, when used with high-resolution CT and MR scanners, permit rapid generation of accurate dose distributions within tissue volumes. The DVH data can be displayed in 3D and 4D formats, which represents a tremendous step forward from the limitations of gynecologic BT based on 2D plain-film dosimetry with critical target dose reporting generally limited to a geometric reference point, which may or may not represent a critical dose within the presumptive target or critical structure in a particular patient. The DVH provides more relevant treatment information than that provided by 2D isodose lines superimposed on orthogonal simulation films.

24.3.2.2.2 Reference Volume

ICRU Report 38 recommended recording and reporting the reference volume for ICBT, which was the volume enclosed by a reference isodose surface. The use of defined reference volumes would facilitate comparison of treatment results among different facilities. The use of modern treatment planning systems and 3D (and 4D) imaging equipment allows rapid DVH information generation for reference volumes pertinent to a particular treatment outcome.

24.3.2.2.3 Irradiated Volume

In ICRU Report 38, the irradiated volume is the volume of tissue receiving a specified absorbed dose. Once again, modern planning systems and imaging equipment facilitate rapid DVH generation. In this way, V_5, V_{20}, V_{30}, etc. (the volume receiving a specified absorbed dose in Gy at standard fractionation) can be recorded and reported so as to accumulate data that may support a specific threshold for a certain event.

24.3.2.2.4 Organs at Risk

For treatment of gynecologic malignancies, the organs at risk (OARs) during ICBT would be primarily the rectum and bladder and possibly the ureters and sigmoid colon. DVH information for all of these structures can be recorded and reported for future efforts in lessening clinically significant treatment-related complications.

24.3.3 Specification of Radioactive Sources

ICRU Report 38 recommended that the reference air-*kerma* rate be recorded and reported, as appropriate (the quantity is expressed in µGy/h at 1 m). This task has been simplified since 1985 in that radium sources are no longer used in the Western world due to radiologic safety concerns regarding the possibility of these sources rupturing during clinical use. As a result of recent world events and concerns regarding diversion for terrorist activities, LDR cesium (^{137}Cs) and cobalt (^{60}Co) sources are no longer commercially available in the United States. They are still available in other parts of the world, although a political reserve may grow in those regions for the same reasons. Thus, by default, essentially all ICBT is carried out with computer-controlled ^{192}Ir remote afterloading equipment. As Varian and Nucletron are the two major manufacturers of this equipment, there has been a resultant uniformity of source specifications with source lengths of 5 mm and diameters of 1.1 mm or less being typically employed. Reference air *kerma* of ~3.5 µGy/h at 1 m is typical of ^{192}Ir HDR sources.

24.4 Recommendations for Reporting Absorbed Doses and Volumes in Intracavitary Therapy

24.4.1 Introduction

ICRU Report 38 noted that the soft tissue absorbed dose from ICBT was so heterogeneous throughout the target volume that concepts of maximum, mean, median, and modal target absorbed dose as defined in ICRU Report 29 for conventional external beam therapy were not relevant. It was suggested that the minimum absorbed target dose was the only useful concept.

ICRU Report 38 dealt with ICBT primarily for the treatment of cervical cancer. There was a presumption that the region of interest (ROI) anatomy of each patient was similar and could be defined with relatively similar geometric reference points and that the positioning of applicators and sources would be relatively constant from patient to patient.

24.4.2 Description of Technique

Per recommendations of ICRU Report 38, there would be recording and reporting of the source characteristics including radionuclide, reference air-*kerma* rates, and descriptions of source size, shape, and cladding filtration. Also to be recorded and reported was a description of simulation of sources as a linear source or point source and description of source movement, when applicable. Similarly there would be a description provided for the type of applicator used and whether it contained additional high-density shielding material.

The major impact of ICRU Report 38, at least from a clinician's viewpoint, was that it made relatively simple recommendations for applicator placement, source loading, and simulation in preparation for treatment of cervical cancer via ICBT that should ensure a relatively uniform degree of treatment practice from one facility to another. Treatment recommendations were made in regards to dose and dose rates at points A and B. Doses (and dose rates) at reference points within the bladder and rectum could also be recorded, although it was unclear whether a point dose would accurately reflect the potential for late toxicity. The downside was that treatment was relatively uniform patient to patient in spite of the local extent of disease. A skilled clinician could draw upon their experience to modify the prescription for total delivered dose, but the limited ability to visualize the local extent of disease made this a highly subjective practice. Overall, the quality of care was impaired by the limited imaging and simulation technology of the era.

24.5 The Modern Era (Post-1985)

The recommendations contained within ICRU Report 38 provided a framework for improved quality of ICBT and, by extension, improved outcomes for patients with cervical cancer. The potential gains were blunted by the limitations of technology available when ICRU Report 38 was being formulated. Starting in the late 1980s, LDR sources and applicators were largely abandoned in North American and Western European practices, as the potential for individually optimized dose distributions obtainable with HDR sources and remote afterloading equipment was realized.

24.5.1 Imaging and Applicators

The maximal utility of this technology was still limited by its reliance at that time upon plain-film dosimetry, as widespread availability of CT simulation was limited, and the limited computational power of the BT planning systems. In the mid-1990s, 3D external beam treatment planning based primarily on CT imaging was being installed at many facilities in developed countries. This provided relatively fast, simple, and accurate volume-based treatment plans. Subsequently, BT planning systems were developed with similar hardware and software improvements to permit volume-based BT planning. The development first of CT-compatible and then CT/MRI-compatible applicators allowed advanced imaging technology to be used, thus permitting rapid acquisition of accurate treatment and tissue volumes and generation of volume-based HDR BT plans.

CT imaging was initially used almost exclusively, and BT planning performed with these images did provide the potential for rapid 3D rendering and DVH generation, as tumor and adjacent normal structures appeared to be clearly visible. In this way, CT-based BT planning did provide more realistic calculation of dose within tumor and adjacent OARs as conventional orthogonal film plans tended to overestimate the dose within the target and underestimate the OAR dose (Onal et al. 2009). While CT imaging–based treatment planning did represent a substantial step forward relative to orthogonal film-based treatment planning, it still had limitations. For example, investigators noted that CT imaging overestimated tumor volumes resulting in increased dose delivered to adjacent normal tissue (Viswanathan et al. 2006).

There is common agreement on the superiority of MR, as compared to CT, for determining both the initial extent of cervical tumor (Pötter et al. 2006, 2008; Viswanathan et al. 2006; Dimopoulos et al. 2009a; Nag et al. 2004; Haie-Meder et al. 2005; Mayr et al. 2010) and response to treatment as seen at the time of BT and beyond (Pötter et al. 2008; Dimopoulos et al. 2009a; Haie Meder et al. 2005; Mayr et al. 2010). Despite the advances in imaging and BT technology, cervical cancer remains a clinically staged disease, and improvements in outcomes remain incremental. Platinum-based chemotherapy is often employed concurrently with EBRT in the treatment of locally advanced disease (Pötter et al. 2008). Advanced imaging modalities are routinely employed in treatment planning and are aiding in the delivery of 3D conformal EBRT and BT. MRI is capable of assessing tumor size with an accuracy of ± 0.5 cm. It is also capable of assessing parametrial extension correctly 77%–96% of the time (Pötter et al. 2008). In this regard, MRI is clearly superior to CT or other clinical or radiographic processes. Assessment of nodal metastases remains poor with either CT or MRI, and the degree of nodal enlargement remains a primary criterion (Pötter et al. 2008).

It was hypothesized that better imaging would lead to more accurate BT dosimetry and ultimately better treatment outcomes as a result of improved local control (resulting from dose escalation and/or improved dose distribution) and/or reduced toxicity (from lower maximal dose to OAR and/or smaller volumes of OAR receiving large doses). A report on 145 patients with locally advanced cervical cancer would suggest that dose escalation would be possible without exceeding normal tissue tolerance (Pötter et al. 2007). Despite this, MR-image-guided BT was not rapidly embraced. This was no doubt in large part due to the scarcity of MR simulators and probably also to reluctance to change institutional practice and accept the technical inferiority of CT-based BT.

24.5.2 Volumes

Though the value of diagnostic MRI in visualizing cervical cancer had been demonstrated, work was carried out at European centers in Vienna, Paris, and Leuven relatively early to provide evidence that MR-image-guided BT was practicable and clinically valuable (Haie-Meder et al. 2005; Pötter et al. 2006; Gregoire et al. 2004). Standard radiation therapy terminology as well as ICRU reference point and volume definitions were reviewed and modified or supplemented as needed to be meaningful and appropriate to image-guided therapy. Some volumes and terms such as the gross tumor volume (GTV), clinical target volume (CTV), and OARs are self-explanatory. Others such as planning treatment volume (PTV) are not applicable to BT as the applicator moves along with the treatment volume during any physiological movement, and there should be no treatment volume uncertainty within the patient as BT plan is generated with 3D image guidance while the applicator is in place. There

may be a small degree of uncertainty (+/– 1 mm) as to the source location within a standard HDR tandem applicator of minimal curvature and a slightly greater degree of uncertainty (+/– 3 to 4 mm) as to source location within the ring of HDR tandem and ring applicator (Tanderup et al. 2010). Though one may usually be able to extend the dose distribution superiorly along the tandem within the uterus without undue risk of increased risk to OAR, expansion of the radial margin is not recommended. For the case of BT, the applicator's spatial relationship to the GTV and CTV is essentially fixed, whereas the relationship of treatment portals in EBRT is subject to much greater degrees of uncertainty (even when IGRT and high-tech immobilization techniques are utilized). As such, a small PTV will always be present when EBRT is practiced (Tanderup et al. 2010). As a start toward standardized practice across many centers, the authors suggested that total dose and fractionation size and schedule of the EBRT and BT should be recorded and reported, as should the conversion factors (α/β_{tumor} and $\alpha/\beta_{normal\ tissue}$) used for calculation of biological dose from physical dose (Gregoire et al. 2004).

Subsequently, collaborative and cooperative efforts began in European and North American centers to develop new recommendations for BT in the treatment of localized cervical cancer to take advantage of new and improved imaging capabilities and planning systems. In North America, representatives from the Gynecology Oncology Group, Radiologic Physics Center, American Brachytherapy Society, American College of Radiology, American College of Radiology Imaging Network, American Association of Physicists in Medicine, Radiation Therapy Oncology Group, and American Society for Therapeutic Radiology and Oncology formed the North American Image-Guided Brachytherapy Working Group (Nag et al. 2004). As the name of this working group (WG) implies, their goal was to generate guidelines for image-guided cervical cancer BT. The group recommended the use of T2-weighted MRI with a pelvic surface coil and MRI-compatible BT applicators. They indicated that imaging must be performed with the applicator in place and the patient in the treatment position. Specific terminology for image-based BT was proposed but has been largely superseded by recommendations put forth by the European WGs.

Working in parallel with the North American Image-Guided Brachytherapy WG was the European gynecologic GEC-ESTRO WG, which produced two reports providing image-guided BT treatment recommendations (Haie-Meder et al. 2005; Pötter et al. 2006). This WG formed as an initiative of GEC-ESTRO in 2000 to support advanced imaging-based 3D treatment planning for localized cervical cancer BT. In contrast to the ABS group, it consisted of experts from centers who already had several years of experience with MRI-based treatment. The first report of the WG described basic concepts and terms so that multiple facilities with different available technologies and clinical practices would have a "common language" for prescriptive parameters and treatment result reporting. One of the observations first made was that MR imaging allowed more accurate assessment of tumor volume and extension than is achievable with CT imaging. Recommendations were made for observing and recording

GTV at diagnosis (GTV_D) and at start of BT (GTV_B). The CTV is similarly determined at diagnosis (CTV_D) and at completion of EBRT/start of BT (CTV_B). The CTV_D may be determined not only radiographically but also clinically, and takes into account parametrial, vaginal, and uterine extension observed on physical examination. Delineation of GTV and CTV is based on the clinical examination at diagnosis and at BT and series of sectional images (preferably fast spin-echo T2-weighted MRI) obtained at the same time points (Kirisits et al. 2005). The GTV_D includes gross tumor discernible on clinical evaluation (visual and manual examination) and also that which is visualized as high signal intensity on T2-weighted MRI (Haie-Meder et al. 2005). Also discernable on these images would be the normal gynecologic structures including the cervix, uterus, vagina, parametria, bladder, and rectum. The GTV_B would include macroscopic tumor at the time of BT as is detected clinically and on imaging (again preferentially with T2-weighted MRI). If a patient is treated with BT alone or "upfront" BT, then the GTV_B would be the GTV_D.

The CTV is divided into two categories: a high-risk CTV (HR CTV) and an intermediate-risk CTV (IR CTV). The HR CTV for BT (HR CTV_B) would include the GTV_B and paracervical disease (presumed or biopsy proven) at the time of BT. Treatment doses for this volume would be appropriately selected for macroscopic disease. The IR CTV_B is at significant risk for microscopic disease and would be dosed accordingly. This volume would typically extend 5 to 15 mm beyond the HR CTV_B and is tailored to take into account factors such as primary tumor size and location, degree of response to previous treatment, probable areas of tumor spread, and borders with adjacent structures (see Figure 24.1). For example, in cases of limited disease burden (tumor < 4 cm maximum diameter at diagnosis), the margin in the AP/PA direction is limited to ~5 mm by the posterior bladder and anterior rectal walls. Ten millimeters would be a typical margin in the cranial/caudal direction. The lateral margin is also typically 10 mm into the parametria (the inner one-third of the parametria), but is increased a further 5 mm in cases of suspected tumor

spread. When more extensive disease is present at diagnosis, the patients would complete a planned course of EBRT. At that point, the IR CTV_B is based on the difference in visualized tissue volume between GTV_D and the imaging obtained at the time of initial BT (Haie-Meder et al. 2005).

Different margins are employed depending on both the initial extent of disease and the degree of regression obtained by the completion of EBRT. For example, in cases of complete regression, the IR CTV includes the GTV_D, and the HR CTV is the GTV_B. In cases of partial response to EBRT, the IR CTV includes a 10-mm margin beyond the GTV_B and HR CTV_B. In the case of no meaningful response, the IR CTV would be a 10-mm margin beyond the GTV_D (see Figure 24.2). Lastly, if there was disease progression through EBRT, BT would probably not be feasible, and a GTV greater than GTV_D would likely be addressed via other interventions. In no case is an additional margin added to

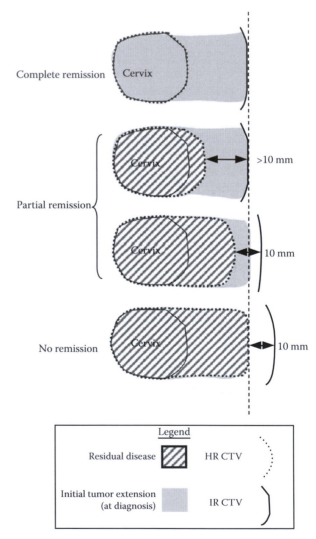

FIGURE 24.2 Schematic diagram for cervix cancer, extensive disease, poor remission after EBRT with GTV, gray zones on MRI, and HR CTV and IR CTV for definitive treatment: coronal and transverse view. (Reproduced from Haie-Meder, C. et al., *Radiother Oncol* 74, 235–45, 2005. With permission.)

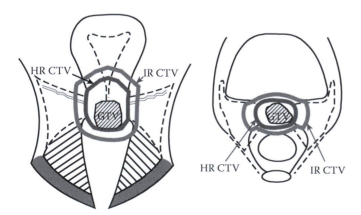

FIGURE 24.1 Schematic diagram for cervix cancer, limited disease, with GTV, HR CTV, and IR CTV for definitive treatment: coronal and transverse view. (Reproduced from Haie-Meder, C. et al., *Radiother Oncol* 74, 235–45, 2005. With permission.)

the BT treatment plan for organ motion or patient setup uncertainty, so the final CTV is the PTV (Haie-Meder et al. 2005).

24.5.3 HDR Radiobiology

Two European facilities contributed data on two patients each with stage IIB disease for the initial work by the WG (Haie-Meder et al. 2005). Among the data recorded were organ/target volumes within the 60-, 75-, and 90-Gy (physical dose converted to biological dose with α/β_{tumor} typically 10 and α/β_{OAR} typically 3 to provide biological equivalence to standard 2-Gy fractional dose, EQD2) isodoses. Point A doses and the volume within the isodose covering point A were not recorded at one of the facilities. Despite the institutional practice differences, the recorded dose-volume results were quite similar.

A systemic review of radiographic anatomy was undertaken to increase the uniformity of contouring GTV and CTV. The CTV had been previously divided into an HR CTV and an IR CTV. The HR CTV was defined as the volume difference in the GTV_D and GTV_B. This volume received 80–90 Gy (EQD2). The IR CTV was more difficult to define, although the concept of treating a volume of tissue to an intermediate dose (60 Gy_{EQD2}) was reasonable and similar to the "reduced fields" or "boost fields" commonly employed in EBRT. Initial thoughts were to extend the IR CTV some fixed distance (8–15 mm) beyond the HR CTV, but this was thought to be not always clinically appropriate (Haie-Meder et al. 2005). In the case of locally extensive paracervical disease in which effective isodose coverage would be difficult to achieve with either a tandem and ovoids or a tandem and ring applicator, an interstitial implant with Syed or Vienna applicator may be utilized (Dimopoulos et al. 2006; Kirisits et al. 2006).

24.5.4 Data Comparison

An accurate clinical examination is the result of a number of factors, not the least of which is accumulating experience. It appears that reviewing MRI and contouring the appropriate volume also improve as experience is accumulated. A report documented the variation in contouring among the member facilities and even of staff within a facility. In this case, members of the WG did a review and compared volumes for 19 BT patients. This work did show a high degree of uniformity for the GTV and HR CTV contours. There were, however, significant differences in the IR CTV (Dimopoulos et al. 2009b). Members of the group did demonstrate that the contouring could be performed more quickly in the sagittal or coronal planes than in the axial plane, but not less accurately (Petric et al. 2008).

As more centers contributed to the WG, a set of recommendations were put forward as a synthesis of the clinical experiences at these facilities (Haie-Meder et al. 2005; Pötter et al. 2006). These recommendations include a biologically equivalent dose of 80–90 Gy or more (EQD2) delivered to the peripheral margins of the GTV (GTV_B). This dose would be comparable to the point A dose used in other centers at the time. The HR CTV is a volume at high risk of local recurrence due to the high probability of clinically occult and residual microscopic disease. The HR CTV is to be treated with a dose sufficient to sterilize all microscopic and a substantial burden of macroscopic disease. An area of intermediate risk of local recurrence (IR CTV) would encompass the distal margin of the volume represented by the difference in the GTV_B and GTV_D and would have an intermediate risk of microscopic disease. It would be treated with a dose sufficient to sterilize all microscopic disease.

These concepts were further refined with the WG second report (Pötter et al. 2006). In this report, the recommendations of GEC-ESTRO WG (II) further refined the definition and anatomic delineation of the treatment volumes and ROIs in gynecologic BT. The recommendations were the synthesis of the work performed at three European centers as had been GEC-ESTRO WG (I). This multicenter collaboration provided recommendations for standardization of dose and dose-volume reporting. Once again the linear–quadratic 2-Gy radiobiologic dose equivalent (EQD2) was used for calculation of both EBRT and BT dose contribution to the total dose delivered. The recommendations strongly favored MRI as superior to CT for delineating the GTV and normal adjacent uterine tissue, although CT is very suitable (as is MRI) for contouring bladder, rectum, sigmoid, intestine, and vagina. Image acquisition parameters and QA are very important as inaccuracies of a few millimeters in the *X*-, *Y*-, or *Z*-planes during contouring can lead to significant variation in the calculated dose given the influence of the inverse square law during BT (Tanderup et al. 2008).

In addition to providing dose and dose-volume parameters for 3D image-based cervical cancer BT, the WG has made recommendations for the transition from orthogonal plain film-based BT planning (Pötter et al. 2006). It was recommended that the traditional ICRU 38 Report dose points (i.e., rectal and bladder reference point doses) be recorded and reported. Additionally, mean and maximum doses within the rectum, sigmoid, and bladder could be reported. This would allow some degree of comparison with OAR DVH information.

Once 3D image-based BT planning is implemented, then dose volume parameters of interest would include those providing some degree of the quality of treatment of the GTV and CTV. Such parameters would include the D_{100} and D_{90}, which are the minimum doses delivered to 100% and 90%, respectively, of the volume of interest. They are easily derived from DVH information and can be quickly converted to EQD_2 doses. It should be noted that D_{100} is extremely sensitive to perturbations in the apparent contours of the ROI, and as such, D_{90} is a more useful tool as it is not greatly influenced by small changes in contours of the ROI (Pötter et al. 2006). The volume of tissue receiving a specified dose is also of interest (i.e., V_X is the volume receiving dose X). This dose can be recorded and reported as physical dose, but reporting biological dose facilitates comparison of data among multiple facilities employing different treatment technologies. As biological dose is usually referenced to 2-Gy fractions, the volume of an ROI receiving a dose would typically be reported as $V(85\ Gy_{EQD2})$ or $V(60\ Gy_{EQD2})$ with an α/β specified to allow conversion of physical dose to biological dose.

It is recommended that V_{150} and V_{200} be reported as this would indicate the proportion of the GTV and CTV receiving doses of 150%–200% of the reference dose. These values would indicate the volume of the target receiving supratherapeutic doses (assuming the prescriptive dose is therapeutic), but biologic doses in these regions become difficult to calculate as the fractional dose and dose rates are more than 1.5 or 2 times the prescribed dose and dose rate. 3D image-based BT planning will allow the replacement of the ICRU 38 Report bladder and rectal reference point doses with the minimal dose within a small volume of OAR. The dose delivered to 0.1, 1.0, and 2.0 cm³ of an adjacent organ is thought to be predictive of late toxicity. These volumetric reference doses are likely found to be more valuable than the OAR reference point doses recommended in the ICRU 38 Report. D_{50} and D_{30} may also be of potential values as in predicting toxicity from high dose volumes within the target and should likely be recorded and reported (Pötter et al. 2006). One additional point to consider is that as the total dose within a high or low dose region changes, the fractional dose also changes, and this may have a significant radiobiological effect. One group studied the plans of five cervical cancer patients being treated with intensity-modulated radiotherapy (IMRT)–EBRT (45 Gy in 25 fractions) and four fractions of HDR BT (7 Gy each) or an equivalent dose delivered via PDR BT. The physical doses were converted to 2-Gy biologically equivalent doses (EQD2). The authors then reviewed plans that assumed uniform doses within the volume treated by EBRT and also by BT, and then summed these doses, versus a plan in which the DVH for each of these volumes are summed. GEC-ESTRO has explored this topic and has termed the former case as "parameter adding," while the latter is called "distributions adding." It is still unclear whether either of these methods would be most appropriate in the majority of cases, and members of GEC-ESTRO continue to study this topic (Van de Kamer et al. 2010).

The use of $V(85 \text{ Gy}_{EQD2})$ or $V(60 \text{ Gy}_{EQD2})$ provides biological dose guidelines in the modern era of 3D image-guided BT from the vast published reference base of BT dose-outcome data (largely LDR BT). As a starting point, $V(85 \text{ Gy}_{EQD2})$ or $V(60 \text{ Gy}_{EQD2})$ appears to be reasonable minimal doses for treatment of the HR CTV and IR CTV, respectively, with a still higher biologic dose delivered to the GTV (Pötter et al. 2006). (Though better treatment plans may allow for substantial dose escalation over time.)

It should be noted that despite similar goals of furthering advanced image-based BT, the North American Image-Guided Brachytherapy WG and the gynecologic GEC-ESTRO WG did not put forth a unified set of recommendations until the middle of the first decade of the twenty-first century (Nag et al. 2004; Pötter et al. 2005). This was probably reflective of different treatment philosophies expressed in different countries, as well as regional differences in utilization of technologies (EBRT, chemotherapy, and/or surgery). The different treatment philosophies were at least partially due to regional availability of and reliance upon different technologies.

At present, BT planning is carried out most frequently with dedicated CT simulators due to the widespread availability of these devices in most radiation oncology facilities (in contrast to the relative scarcity of MR simulators) (Viswanathan and Erickson 2010). Moreover, some would argue in favor of CT-based planning because of MRI's absence of electron density information and cortical bone signal as well as the presence of intrinsic spatial distortion and magnetic susceptibility artifact in most MR-acquired images (Barillot and Reynaud-Bougnoux 2006). Proponents of MR-based BT planning would argue that MR gives a vastly superior definition of the tumor volume, and most modern BT is carried out with ¹⁹²Ir sources with which tissue density variation has a much reduced influence on dose calculations. Similarly, MRI distortion becomes much less of an issue when the ROI is located within the central field of view (as is the case in gynecologic BT). The ability to directly acquire images in sagittal and coronal planes rather than relying on the reconstruction of axially acquired images is an additional benefit of MRI relative to CT.

One of the first prospective clinical trials in the United States on real-time MR image-guided BT in the treatment of gynecologic cancer was published in 2006 (Viswanathan et al. 2006). Despite clear superiority in image quality and tumor volume delineation in MRI-based BT planning, this group noted that MR-based planning is seldom utilized (Viswanathan and Erickson 2010). This report on 141 radiation oncologists indicated that while the majority of BT practitioners (55%) do use CT-based planning (43% still use orthogonal film planning), only 2% use MR-based planning. Moreover, while both CT- and MR-based treatments facilitate DVH generation and BT dose prescription to be based on tumor volume, the majority (76%) of radiation oncologists continue to prescribe to point A and record dose per ICRU Report 38 dose point prescriptives alone. Moreover, 20% of respondents did not use any sort of image guidance during applicator insertion or for applicator placement verification intraoperatively. Approximately 50% of the surveyed practitioners used ultrasound during insertion, while 37% used fluoroscopy. The lead author commented that, "While 70% of ABS survey respondents used CT imaging after applicator insertion, the available images are not necessarily used optimally for patient care. With additional training and awareness regarding the patient benefits of 3D images, the utilization of available imaging technology should significantly increase in the near future, resulting in substantially improved outcomes for patients" (Viswanathan and Erickson 2010).

24.5.5 Ongoing and Future Work

The potential advantages of high-tech, high-quality, and high-cost imaging modalities for improving patient outcomes are currently under investigation in a large, multi-institution, European randomized clinical trial. The EMBRACE (European Study on MRI-guided Brachytherapy in Locally Advanced Cervical Cancer) trial is enrolling patients with locally advanced cervical cancer for treatment of curative intent (EMBRACE 2011). Pelvic MRI is performed at diagnosis and for treatment planning at least for the first BT session. (3D conformal pelvic EBRT ± chemotherapy is also utilized.) The impetuses for this trial are the recently published recommendations by societies of European (GEC-ESTRO) and American (ABS) practitioners of BT regarding 3D and adaptive (4D) ICBT in the treatment of gynecologic malignancies (Dimopoulos et al. 2009a; Nag et al. 2004; Haie-Meder et al. 2005). GEC-ESTRO recommendations are supported by data in published reports of the systematic clinical use of MRI (and to a lesser extent CT) based planning of ICBT (Tanderup et al. 2010; Viswanathan and Erickson 2010; Wachter-Gerstner et al. 2003). Both societies have agreed to use GEC-ESTRO recommendations I and II as the basis of future trials (Haie-Meder et al. 2005; Pötter et al. 2006). The EMBRACE trial has, as its goal, a broader acceptance and implementation of MR-based BT planning. Its primary aim is to "…introduce MRI based BT in a multicenter setting within the frame of a prospective observational study and to correlate image based DVH parameters for the clinical target volume and organs at risk with outcome" (EMBRACE 2011). The study seeks to enroll 600 patients with locally advanced cervical cancer (stage IB1–IVA). The majority of patients will likely be stage IIA (37%) and stage IIIB (32%). It is hoped that the results of this trial will help "…develop prognostic and predictive statistical models for clinical outcomes including volumetric, dosimetric, clinical, and biological risk factors as well as radiobiological parameter estimates that will allow precise risk assessment in individual patients and aid in the development of new treatment protocols" (EMBRACE 2011).

24.5.6 Clinical Practice Recommendations

While the formal update of ICRU Report 38 is awaited and there is hope that the results of large cooperative trials such as the EMBRACE study will provide clarity and guidance for the practitioner of gynecologic BT, there are several recommendations that can be made in regard to information recording and reporting for ICBT. T2-weighted imaging should be performed at the time of diagnosis and at each simulation for BT. The degree of regression of tumor volume at the completion of EBRT and each session of BT can be recorded. At a minimum, the D_{90} for GTV, HR CTV, and IR CTV should be calculated, recorded, and reported, as should be D_{2cc} and $D_{0.1cc}$ for the adjacent normal structures. Other parameters may be found to have prognostic significance as more data from increasing numbers of facilities are accumulated. [At the time of publication of this chapter, the ICRU Report 38 revision committee has been formed and has been working for more than 1 year on the formal update of ICRU Report 38. The ABS is currently writing new recommendations for IGBT that include prescriptive guidelines, in addition to contouring, recording, and reporting guidelines (and, in this way, differs from GEC-ESTRO WG recommendations; Haie-Meder et al. 2005; Pötter et al. 2006).]

References

Barillot, I., Reynaud-Bougnoux, A. 2006. The use of MRI in planning radiotherapy for gynaecologic tumours. *Cancer Imaging* 6:100–6.

Dimopoulos, J.C.A., Kirisits, C., Petric, P. et al. 2006. The Vienna applicator for combined intracavitary and interstitial brachytherapy of cervical cancer: Clinical feasibility and preliminary results. *Int J Radiat Oncol Biol Phys* 66:83–90.

Dimopoulos, J.C., Schirl, G., Baldinger, A., Helbich, T.H., Potter, R. 2009a. MRI assessment of cervical cancer for adaptive radiotherapy. *Strahlenther Onkol* 185:282–7.

Dimopoulos, J.C., De Vos, V., Berger, D. et al. 2009b. Inter-observer comparison of target delineation for MRI-assisted cervical cancer brachytherapy: Application of the GYN GEC-ESTRO recommendations. *Radiother Oncol* 91:166–72.

EMBRACE. 2011. An International Study on MRI-Guided Brachytherapy in Locally Advanced Cervical Cancer. Web site homepage, https://www.embracestudy.dk/. Last accessed February 2011.

Gregoire, V., Pötter, R., Wambersie, A. 2004. General principles for prescribing, recording, and reporting therapeutic irradiation. *Radiother Oncol* 73 (Suppl. 2):57–61.

Haie-Meder, C., Pötter, R., Van Limbergen, E. et al. 2005. Recommendations from gynecologic (GYN) GEC-ESTRO working group (I): Concepts and terms in 3D image-based treatment planning in cervix cancer brachytherapy with emphasis on MRI assessment of GTV and CTV. *Radiother Oncol* 74:235–45.

International Commission on Radiation Units and Measurements (ICRU). 1978. *Dose Specification for Reporting External Beam Therapy with Photons and Electrons*. Report 29 of ICRU, ICRU Publications, Washington, DC.

International Commission on Radiation Units and Measurements (ICRU). 1985. *Dose and Volume Specification for Reporting and Recording Intracavitary Therapy in Gynecology*. Report 38 of ICRU, ICRU Publications, Bethesda, MD.

Kirisits, C., Pötter, R., Lang, S. et al. 2005. Dose and volume parameters for MRI-based treatment planning in intracavitary brachytherapy for cervical cancer. *Int J Radiat Oncol Biol Phys* 62:901–11.

Kirisits, C., Lang, S., Dimopoulos, J., Berger, D., Georg, D., Pötter, R. 2006. The Vienna applicator for combined intracavitary and interstitial brachytherapy of cervical cancer: Design, application, treatment planning, and dosimetric results. *Int J Radiat Oncol Biol Phys* 65:624–30.

Mayr, N.A., Wang, J.Z., Lo, S.S. et al. 2010. Translating response during therapy into ultimate treatment outcome: A personalized 4-dimensional MRI tumor volumetric regression approach in cervical cancer. *Int J Radiat Oncol Biol Phys* 76:719–27.

Nag, S., Cardenes, H., Chang, S. et al. 2004. Proposed guidelines for image-based intracavitary brachytherapy for cervical carcinoma: Report from the image-guided brachytherapy working group. *Int J Radiat Oncol Biol Phys* 60:1160–72.

Onal, C., Arslan, G., Topkan, E. et al. 2009. Comparison of conventional and CT-based planning for intracavitary brachytherapy for cervical cancer: Target volume coverage and organs at risk doses. *J Exp Clin Cancer Res* 28:95–104.

Petric, P., Dimopoulos, J., Kirisits, C., Berger, D., Hudej, R., Pötter, R. 2008. Inter- and intraobserver variation in HR-CTV contouring: Intercomparison of transverse and paratransverse image orientation in 3D-MRI assisted cervix cancer brachytherapy. *Radiother Oncol* 89:164–71.

Pötter, R., Van Limbergen, E., Gerstner, N., Wanbersie, A. 2001. Survey of the use of ICRU 38 in recording and reporting cervical cancer brachytherapy. *Radiother Oncol* 58:11–8.

Pötter, R., Dimopoulos, J., Kirisits, C. et al. 2005. Recommendations for image-based intracavitary brachytherapy of cervix cancer: The GEC-ESTRO Working Group point of view: In regards to Nag *et al.* (*Int J Radiat Oncol Biol Phys* 2004; 60:1160–72.) *Int J Radiat Oncol Biol Phys* 62:293–5; author reply 295–6.

Pötter, R., Haie-Meder, C., Van Limbergen, E. et al. 2006. Recommendations from gynecologic (GYN) GEC-ESTRO working group (II): Concepts and terms in 3D image-based treatment planning in cervix cancer brachytherapy—3D dose volume parameters and aspects of 3D image-based anatomy, radiation physics, radiobiology. *Radiother Oncol* 78:67–77.

Pötter, R., Dimopoulos, J., Georg, P. et al. 2007. Clinical impact of MRI assisted dose volume adaptation and dose escalation in brachytherapy of locally advanced cervical cancer. *Radiother Oncol* 83:148–55.

Pötter, R., Fidarova, E., Kirisits, C., Dimopoulos, J. 2008. Image-guided adaptive brachytherapy for cervix carcinoma. *Clin Oncol* 20:426–32.

Tanderup, K., Hellebust, T.P., Lang, S. et al. 2008. Consequences of random and systemic reconstruction uncertainties in 3D-based brachytherapy in cervical cancer. *Radiother Oncol* 89:156–63.

Tanderup, K., Pötter, R., Lindegaard, J.C., Berger, D., Wambersie, A., Kirisits, C. 2010. PTV margins should not be used to compensate for uncertainties in 3D image guided intracavitary brachytherapy. *Radiother Oncol* 97:495–500.

Van de Kamer, J.B., de Leeuw, A.A.C., Moerland, M.A., Jurgenliemk-Schulz, I.-M. 2010. Determining DVH parameters for combined external beam and brachytherapy treatment: 3D biological dose adding for patients with cervical cancer. *Radiother Oncol* 94:248–53.

Viswanathan, A.N., Erickson, B.A. 2010. Three-dimensional imaging in gynecologic brachytherapy. *Int J Radiat Oncol Biol Phys* 76:104–9.

Viswanathan, A.N., Cormack, R., Holloway, C.L. et al. 2006. Magnetic resonance-guided interstitial therapy for vaginal recurrence of endometrial cancer. *Int J Radiat Oncol Biol Phys* 61:91–9.

Wachter-Gerstner, N., Wachter, S., Reinstadler, E., Fellner, C., Knocke, T.H., Pötter, R. 2003. The impact of sectional imaging on dose escalation in endocavitary HDR-brachytherapy of cervical cancer: Results of a prospective comparative trial. *Radiother Oncol* 68:51–9.

25

In Vivo Dosimetry in Brachytherapy

Joanna E. Cygler
The Ottawa Hospital Cancer Center and University of Ottawa

Kari Tanderup
Aarhus University Hospital

Sam Beddar
M.D. Anderson Cancer Center

José Pérez-Calatayud
Hospital La Fe

25.1 Introduction

As discussed in the following sections, in vivo dosimetry has been utilized for different purposes during the radiation therapy of cancer patients. Initially, this technique was used to verify dose to the organs adjacent to the implanted volume. However, verifications of the accuracy of the treatment delivery for complex treatment techniques have been added to the list of applications of in vivo dosimetry. This chapter provides an overview of various in vivo dosimetry systems and discusses their advantages and limitations.

25.1.1 Assessment of Organ Dose

In vivo dosimetry has been used in brachytherapy for decades. It can be used to assess dose to organs at risk (OAR) in order to avoid overexposure of such organs. Typical sites for OARs in case of gynecological and prostate cancers are the rectum (Joslin et al. 1972), bladder, and urethra (Brezovich et al. 2000; Cygler et al. 2006; Bloemen-van Gurp et al. 2009a). Due to the large dose gradients present in brachytherapy, it is a major challenge to measure an organ dose that will be representative of a potentially critical dose received by that organ. The major difficulty for

bladder and rectum is to accurately position the dosimeter at the location that represents the most exposed part of the organ wall (Kapp et al. 1992). The anatomy of the urethra is more favorable for performing accurate in vivo dosimetry since this organ is smaller and less deformable.

Introduction of 3D image-guided brachytherapy treatment planning was a significant step forward to more accurate calculations of tumor and OAR doses as well as dose-volume histograms (DVHs) (Pötter et al. 2006; Major et al. 2011). OAR dose assessment based on 3D imaging is, in most cases, much more relevant and precise than in vivo dosimetry. However, in a number of situations, TG-43 calculations (Rivard et al. 2004) are associated with significant uncertainties due to lack of correction for heterogeneities and variable scatter conditions. This is true, in particular, for low-energy photon emitters (Landry et al. 2010; Afsharpour et al. 2010), in regions close to the skin (e.g., breast brachytherapy), or when shielded applicators (e.g., gynecological brachytherapy) are used. Improved dose calculation tools are underway (Ballester et al. 2009; Thomson and Rogers 2010; see also Chapter 11). During implementation of these new dose calculation approaches, in vivo dosimetry can be used to verify their accuracy and to minimize uncertainties in dose assessment and reporting (Mangold et al. 2001).

Even in regions where the TG-43 dose calculations are sufficiently accurate, there may be discrepancies between image-based calculated and delivered doses because the anatomy and position of applicators can change between imaging and dose-delivery times. In certain situations, reimaging and recalculation of the dose just before dose delivery may be done in order to improve the accuracy of dose assessment or to detect if applicator adjustment and a new dose optimization are needed. For example, if two fractions of high dose rate (HDR) are delivered on succeeding days, it may be appropriate to perform imaging before the second fraction to assess if the applicator position and/or topography of OARs have changed. However, in some cases, it may be impractical to perform reimaging because it requires extra resources and also an additional transfer of the patient, for example, if computed tomography (CT), magnetic resonance, or positron emission tomography (PET) imaging is used. In case of prostate, it is simply impossible to perform ultrasound (US) imaging corresponding to the delivery situation, since the rectal US probe is removed before HDR treatment delivery (or at the end of seed implantation). In vivo dosimetry can be a useful alternative to demonstrate whether there are deviations from planned organ dose caused by changes in anatomy (Cygler et al. 2006; Bloemen-van Gurp et al. 2009b).

25.1.2 Overall Verification of Treatment Plan

In addition to measurement of organ doses, in vivo dosimetry can be used as part of an overall verification of the brachytherapy procedure (Van Dyk et al. 1993). Underexposure and overexposure of the dosimeter can be an indication that the treatment is not progressing as planned. Deviations may be caused by different issues such as organ movement or deformation, errors related to imaging, dose planning, transfer of patient plan, and dose delivery. In such context, the purpose of in vivo dosimetry becomes broader than verification of organ doses, since it also aims for an overall check of whether the planned dose is actually delivered correctly to the patient. In this setting, the objective of in vivo dosimetry for quality assurance is to verify that a patient receives the prescribed dose value.

Verification of brachytherapy delivery by in vivo dosimetry can be performed either off-line (i.e., determination of the dose after treatment delivery) or in real time. Some dosimeters are feasible only for off-line verification because their reading is not possible during the brachytherapy treatment, for example, as in the case of thermoluminescent dosimeter (TLD) (Brezovich et al. 2000; Das et al. 2007) and alanine (Ciesielski et al. 2003; Schultka et al. 2006). Real-time measurements have been demonstrated with dosimeters that continuously measure dose rate, for example, diodes (Alecu and Alecu 1999; Tanderup et al. 2006), MOSFETs (Cygler et al. 2006; Bloemen-van Gurp et al. 2009a,b), scintillators (Lambert et al. 2006), and radioluminescence (RL) (Andersen et al. 2009a,b). Real-time dosimetry is particularly interesting for quality assurance of HDR and pulsed-dose-rate (PDR) stepping source treatments, since dose rate measurements are an obvious way to monitor that the dose is correctly building

up as the source steps through the implant. Basically, a dose rate measurement is an indication of the distance between the source and the dosimeter, and therefore, real-time dosimetry can reveal if the source is not in the expected position (Nose et al. 2008). Errors can be identified instantaneously, and the treatment can be terminated. Although measurement of instantaneous dose rate has great potential, real-time dosimetry has been mainly used until now for evaluation of time-integrated dose. Most studies are based on evaluation of the total dose per treatment session rather than the dose from individual dwell positions. However, a few reports introduced the possibility of using the instantaneous dose rate for quality assurance (Andersen et al. 2009b; Tanderup et al. 2006; Kertzscher et al. 2011). These studies demonstrate that time-resolved dosimetry is more sensitive for error detection than dose verification based on integral dose.

The ability of in vivo dosimetry to detect actual errors depends critically on where the detector probes can be placed and how well these positions can be localized. Positional uncertainties of the detectors will translate into less sensitivity toward detection of source positioning errors. Furthermore, probes placed far away from the source will be insensitive to detection of source positioning errors due to a low signal. Some papers on in vivo dosimetry measurements report frequent deviations between measured and calculated doses of more than 20% (Nose et al. 2008). The deviations between calculated and measured doses are often attributed to positional uncertainties of the detector caused by independent movements of the detector and the applicator. Such results indicate that positional uncertainty of the detectors may result in significant uncertainty of the measured dose. Therefore, in order to improve the sensitivity of in vivo dosimetry to detect errors, it is essential to address improved detector positioning and fixation.

25.1.3 Errors Encountered in Brachytherapy

International Commission on Radiological Protection (ICRP) reports 86 (ICRP 2000) and 97 (ICRP 2005) as well as International Atomic Energy Agency (IAEA) safety report series 17 (IAEA 2000) contain recorded errors occurring in brachytherapy. Brachytherapy errors and accidents are mainly related to human errors. In addition, some errors are caused by mechanical events. Mechanical HDR events have been related to control units, computers, source cables, catheters, and applicators. Human errors include incorrect medical indication, source strength, patient identification, diagnosis or area of treatment, prescription, data entry, catheter, or applicator. Some of these errors could have been detected by in vivo dosimetry, for example, a number of source positioning errors, which caused the radiation to be delivered outside the prescribed volume, resulting in underdosage of the target. Some specific source positioning errors have been associated with the HDR afterloading techniques and have not been seen with the low dose rate (LDR). Examples of such errors include applicator reconstruction errors, use of wrong applicator length or offset, wrong source step size, interchanged guide tubes, or afterloader malfunctions

(IAEA 2000). Using a built-in dummy (i.e., nonactive) check source, afterloader safety systems can detect certain dose-delivery errors (e.g., mechanical obstruction of the source or improperly connected guide tubes). Recently, Koedooder et al. (2008) have analyzed afterloader error log files for 1300 treatment sessions and concluded that PDR brachytherapy is a safe treatment modality. However, this conclusion may not be completely warranted, since current afterloader safety systems do not detect all possible errors that may occur, such as interchanged guide tubes or reconstruction mistakes. This means that an unknown number of brachytherapy errors remain undetected. One may discover errors leading to overdosage of critical organs only at a late stage or posttreatment, when a patient presents with clinical complications. In contrast, errors leading to underdosage of the target volume will reduce the chance of cure, and such errors may go unnoticed.

25.1.4 Need for In Vivo Dosimetry

The practice of using in vivo dosimetry varies considerably among countries, and systematic use of in vivo dosimetry is not often incorporated as a standard procedure in remote afterloading brachytherapy. The differences in utilization of in vivo dosimetry in various countries are due to insufficient precision of current in vivo dosimetry systems, manpower issues, and presence/absence of specific legislation. Recently, France introduced a law that requires performing some form of in vivo dosimetry on all radiotherapy patients. The future of in vivo dosimetry depends very much on the development of robust in vivo dosimetry concepts and detectors. Broad dissemination of in vivo dosimetry requires development of accurate and affordable ways to perform in vivo dosimetry. It has to be demonstrated that in vivo dosimetry is actually sensitive to errors and uncertainties that may have clinical effect. However, it still remains to be proven that in vivo dosimetry can eliminate treatment errors and compensate for uncertainties. Furthermore, incorporation of routine in vivo dosimetry into clinical practice requires a procedure that is both straightforward and practical. The practice will not be feasible if it requires an excessive amount of manpower for calibration, quality assurance of the detectors, and analysis of measurements.

Last and most important, there is currently no full overview of the type of errors happening in brachytherapy and the rate of occurrence of these errors. Recent AAPM Report 138 addresses only uncertainties pertaining to single-source dosimetry preceding clinical use (DeWerd et al. 2011). Dosimetric uncertainties during treatment delivery are not included in the detailed analysis. Current reports of brachytherapy accidents and errors are mainly based on retrospective collection of events. Prospective investigations have not yet been performed at a large multicenter scale. In order to make progress in brachytherapy safety in the future, it is essential to establish a more systematic description of errors. Currently, there are many brachytherapy errors that are not identified by standard safety systems. If better in vivo dosimetry systems are available, there is a possibility

to improve the detection of errors and their frequencies. Such progress has potential to facilitate significant improvements in quality assurance of brachytherapy by (1) making it possible to focus on the most relevant sources of uncertainties and errors, and (2) advancing the use of in vivo dosimetry systems in order to prevent and avoid significant uncertainties and brachytherapy errors.

25.2 Application of TLDs

Thermoluminescence (TL) refers, in general, to the emission of visible light from a heated solid that has been previously exposed to ionizing radiation. There has been extensive work done to study TL and its applications in the last four decades. The first extensive book on TL dosimetry was published in 1968 by Cameron et al. (1968). The general theory and background of the solid-state physics of TL have been well covered by Horowitz (1984) and McKeever (1985).

Excellent sources of the theory TL dosimetry and TL dosimeters (TLDs) as applied in radiological and radiation oncology sciences are provided by Attix (1986), Baltas et al. (2007), and DeWerd et al. (2009). Therefore, the details of TL dosimetry will not be covered in this chapter, and the reader should refer to Chapter 6 on experimental dosimetry and to the above references. In this chapter, we will present the practical aspects of using TLDs for the specific tasks of performing in vivo dosimetry in brachytherapy.

25.2.1 TLD Materials

The most common TL material used for dosimetry is lithium fluoride (LiF). This material is usually doped with magnesium (Mg) and titanium (Ti), often denoted as LiF:Mg,Ti, and is available in many different formulations (TLD-100, TLD-600, TLD-700) depending on their % content of ^{6}Li and ^{7}Li isotope (DeWerd et al. 2009). TLD materials other than the doped LiF do exist and are usually categorized as either low- or high-Z materials. Low-Z materials include LiF and beryllium oxide, which are approximately tissue equivalent. However, calcium sulfate ($CaSO_4$), calcium fluoride (CaF_2), and aluminum oxide (Al_2O_3) are considered high-Z materials (Kron 1999). The physical characteristics including the advantages and disadvantages of these materials have been well presented and discussed by Kron (1999).

25.2.2 Proper Handling and Use of TLDs

TLDs come in many different forms. The most commonly used forms are chips/rods (often called pellets) or ribbons of compressed powder, Teflon matrix (often termed Teflon discs), or loose powder (see Figure 25.1). Each form needs to be properly handled and has a different processing method that one needs to be very careful with in order to achieve acceptable reproducibility and, more importantly, good accuracy. The methods of calibrating and reading each type of TLD can be found elsewhere

(Kron 1999; Brezovich et al. 2000) and will not be dealt with in this chapter. It is important to mention that great care needs to be taken when handling TLDs, and associated instrumentation required reading their TL response. The two main disadvantages of these dosimeters are the time invested in them during the entire process (calibration, preparation of the samples, annealing the samples, packaging and getting them ready for use, readout process) and the waiting period between irradiation and readout to reduce the contribution of short-lived peaks to the TLD output. The solid form of TLDs is reusable, and after each irradiation and readout of their response, they can be cleaned (known as annealing) and be ready for the next use.

TLDs have been used extensively for the dosimetry of linear accelerator beams including verification of machine output (photon and electron beams). TLDs are used for mailed audit systems to check beams, like done by RPC, Equal-ESTRO, and IAEA. This application is discussed in more detail in Chapter 31. They are also used extensively for external beam dosimetry including patient treatment delivery verification and monitoring of the dose received by pacemakers and implantable cardioverter defibrillators. TLDs are also used for the dosimetry of LDR brachytherapy sources where the dose rates are low in general and the energies are below 380 keV. In addition to patient dosimetry, they have also proved very useful for ^{192}Ir HDR brachytherapy source characterization (Meigooni et al. 1988; Goetsch et al. 1991). In practice, LiF TLDs are preferred because they are almost energy-independent in the high-energy range used in radiation therapy (Co-60 up to 25 MV x-rays). However, in brachytherapy dosimetry, the response of each TLD type must be characterized for linearity and energy dependence.

25.2.3 In Vivo Dosimetry Performed on Patients

The most remarkable and significant studies using TLDs have been deployed in the prostatic urethra as a minimally invasive

FIGURE 25.1 Most commonly used forms of TLDS. http://www.tld .com.pl/tld/index.html. (Courtesy of RADCARD s.c.)

verification tool for HDR brachytherapy of the prostate. In two studies, a series of TLD rods were placed in a Foley catheter along the prostatic urethra during HDR brachytherapy delivery. Both studies used LiF:Mg,Ti rods (Harshaw TLD 100) of 1-mm diameter and 6-mm length. The TLD dose measurements were compared with the dose values calculated by the treatment planning system (TPS) to evaluate the agreement between the planned and delivered doses (Brezovich et al. 2000; Das et al. 2007). These studies indicated that urethral TLD measurements could provide a reliable measurement of the dose delivered in prostate HDR brachytherapy.

When considering seven treatment fractions given to four patients, Brezovich et al. (2000) found that the highest measured urethral dose was within 11.7% ± 6.2% of the planned dose for any single patient, with an average discrepancy of –1.7% ± 6.2%. The mean measured urethral dose for any single patient was within 10.4% ± 6.4% of the planned dose, with an average discrepancy of –1.5% ± 4.4%. In a more extensive study of 50 patients, Das et al. (2007) reported an average discrepancy of –1.6 ± 13.7% between the planned and measured mean doses.

The study by Das et al. (2007) included TLD measurements in a rectal catheter for 37 of the 50 patients. Rectal measurements proved to be more difficult to interpret, as there was more variability of TLD position between planning and treatment. Rectal measurements showed greater variability than the urethral ones, which was attributed to the rectal catheter movement.

In another study, Gonzalez et al. (2005) used the above same type of TLD rods to estimate the real dose that the rectum receives during HDR when treating prostate patients with external radiotherapy and two sessions of HDR (11.5 Gy/sessions) along the course of treatment. The study included 27 patients. It had a goal to confirm, that the dose calculated by the TPS (Plato BPSv14.2) was correct, keeping in mind that during the irradiation, the trans-rectal US probe was absent. Consequently, the conditions were different from those during the real-time planning and catheter implantation, and the dose distribution calculated in rectum was theoretically inferior. It was established that as the detector–source distance increased from 1 to 8 cm, the TLD response varied by 15%. This behavior was attributed to spectral changes in the radiation field and taken into account in the dose calculation.

The TLD readings were also corrected for the response nonlinearity in the dose range 0.5 to 16 Gy.

Gonzalez et al. (2005) and Prada et al. (2007) studied the effect of transperineal injection of hyaluronic acid (Restylane sub-Q) in HDR treatments with the purpose of separating the prostate gland from the rectum and reducing the dose to the latter. They used TLD rods for in vivo dosimetry in the rectum and urethra of 27 patients. The difference between the urethra mean dose calculated by the TPS and the one measured with the TLDs was 0.2% ± 6.5%. The measured median rectal dose, when normalized to the median urethral dose, demonstrated a decrease in dose from 47.1% to 39.2% ($p < 0.001$) with or without injection. For an HDR boost dose of 11.5 Gy, the rectum mean D_{max} reduction was from 7.08 to 5.07 Gy, $p < 0.001$, and the rectum mean D_{mean} drop was from 6.08 to 4.42 Gy, $p < 0.001$, post-HA injection.

25.3 Application of Diodes

25.3.1 Principles of Operation

Silicon diode detectors are widely used for the measurement of electron and photon beams. The operating principles of these detectors are well known (Grusell and Rikner 1984; Bomford et al. 2003). The relatively high silicon density results in a very high number of ionizations when the diode is exposed to radiation. There is no need for a large polarizing voltage because the contact potentials within the diode are sufficient to prevent ion recombination. The sensitive part of the diode is the junction between the p- and the n-type silicon. The p-type silicon is the silicon with boron or aluminum impurities, which absorb electrons from the surrounding silicon, leaving positive (p) "holes" in the material. On the other side of the junction, the phosphorus impurities donate negative (n) electrons to the silicon. This imbalance of composition results in a contact potential at the junction between the two dissimilar materials and a mopping up of the "free" ions, creating a few-micrometers-thick depletion layer. If the atoms in this depletion layer are ionized by radiation, then negative ions (electrons) will be attracted to the positively charged phosphorus impurities in the n-type silicon, and the positive "holes" will diffuse toward the boron impurities in the p-type silicon. This flow of ions constitutes an ionization current proportional to the incident dose.

Although all these detectors use both n-type and p-type material, some are described as p-type while others as n-type. The label identifies which material forms the larger part of the junction (Grussel and Rikner 1984; Huyskens et al. 2001; Bomford et al. 2003), being the p-type when the conduction is due to the movement of electrons rather and not the positive holes. Diodes can be operated with or without bias. In the photovoltaic mode (without bias), the generated current is proportional to the dose rate. More details about the diode properties for radiation dosimetry can be found in the works of Grusell and Rikner (1984), Rikner and Grusell (1985, 1987), Huyskens et al. (2001), Yorke et al. (2005), and Zhu and Saini (2009).

25.3.2 Brachytherapy Applications

Diode-based dosimetry systems are routinely used in some clinics for brachytherapy, especially in treatments of gynecological cancers. Currently, commercially available systems are from Isorad and PTW. A typical diode-based in vivo dosimetry system, electrometer, and different diode models are shown in Figure 25.2.

There are several applications of diodes for in vivo brachytherapy dosimetry described in the literature. Piermattei et al. (1995) studied the sensitivity of a cylindrical p-type silicon detector (EDD-5 from Scanditronix) in air and water using different photon beams (from 30-keV x-rays to ^{60}Co). The authors investigated the reliability of compensating the energy dependence of the

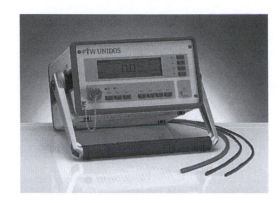

FIGURE 25.2 PTW Unidos system for in vivo dosimetry in brachytherapy. http://www.ptw.de/unidos_in-vivo_dosemeter.html. (Courtesy of PTW, Freiburg.)

diode response for ^{192}Ir brachytherapy sources. Lead filter caps around the diode ranging in thickness from 0.15 to 0.25 mm were used to minimize the dependence of the detector response as a function of the photon energy. Using such filters ensured that diode sensitivity was constant over a wide range of dose rate of clinical interest. The authors concluded that a single shielding filter around the diode of the 0.22-mm thickness is sufficient to obtain accurate dosimetry for ^{192}Ir, ^{137}Cs, and ^{60}Co brachytherapy sources.

Das et al. (1996) simulated p-type silicon diode (Scanditronix electron-field diode) absolute response as a function of detector geometry and photon energy with the Monte Carlo photon transport (MCPT) code. The purpose of their study was to assess the accuracy of that code. Monte Carlo simulation was used to calculate the absorbed dose to the active volume of the detector per unit air kerma. A diode was irradiated in air using an HDR source (^{192}Ir, MicroSelectron) and superficial x-ray beams. Diode calibration was done in a ^{60}Co teletherapy unit attenuated by a lead filter that reduced the dose rate to the value within the range of diode response linearity. A 4-mm-thick polystyrene build-up cap was added for both ^{60}Co and ^{192}Ir irradiations. They concluded that the absolute response of the diode was directly proportional to the dose absorbed by the active volume of the detector and independent of beam quality. MCPT simulation accurately accounts for energy-response and self-attenuation artifacts that cause the measured detector response to vary by nearly an order of magnitude with incident photon energy. The study clearly demonstrates that MCPT simulation is able to predict accurately (within 3%) absolute dose rates near brachytherapy sources (Das et al. 1996).

Alecu et al. (1997) proposed an in vivo dosimetry program with diodes. The goal of this program was to avoid misadministrations in HDR brachytherapy. The authors calibrated the Isorad (model 114200) diode in a $25 \times 25 \times 20$ cm^3 polystyrene phantom using an ^{192}Ir VariSource. They also determined the correction factors needed to account for the source-to-detector distance and the active length of the source. They performed 36 measurements for endobronchial and esophageal implants.

Diodes were placed on the patient's anterior surface. For treatment planning calculations, the diode position was determined using orthogonal x-ray films. The expected diode readings were calculated for the total treatment time per fraction and for the initial 20% of the treatment time. The authors established that if the readings for the initial 20% of treatment time were outside a tolerance range, the treatment should be stopped and rechecked. They found agreement between measured and calculated dose values within ±10%, with largest deviations up to 15% for multicatheter implants. They have concluded that in vivo dosimetry with diodes can be used as a quick and reliable method to help avoid misadministration and provides both a confidence check for the treatment and records of the actual delivered dose.

The previous method was extended to intracavitary gynecological implants. In this case, Alecu and Alecu (1999) measured the rectal dose using an energy-compensated diode Isorad (model 1141) positioned within a rectal marker (radiation opaque marker Shadow Form, I.Z.I.). The diodes were calibrated in a $25 \times 25 \times 20$ cm^3 polystyrene phantom with an ^{192}Ir source. They were mounted on a graduated plastic semirigid rod, so that they could be positioned accurately and the depth of insertion within the rectal tube could be verified. The rectal points were identified in the TPS (Cadplan BT, Varian Associates), and their coordinates relative to the distal end of the rectal marker were determined. Diode readings were monitored during the treatment, and if the readings for the initial 20% of the total dwell time were outside a tolerance range, the treatment was stopped and rechecked. The results of this study found an agreement between calculated and in vivo measured values within 5% for the in-phantom total dose measurements and 11% ($k = 2$) for the dose delivered during the initial 20% of the total dwell time. The maximum discrepancy between the measured and calculated values was 15%. The authors concluded again that their in vivo dosimetry program provided both a confidence check for the treatment and recorded verification of the delivered dose.

Sakata et al. (2002) analyzed measured rectal doses in intracavitary HDR brachytherapy to examine the incidence of radiation-induced late rectal complications. They placed a linear array of five detectors (distance between them: 10 mm) fixed to a wooden pole into the rectum close to the surface of the anterior rectal wall. The source was an HDR ^{60}Co. Orthogonal x-ray films were used for treatment planning to localize the tandem and ovoid applicator and the diodes. A total of 105 patients have been monitored with the five detector arrays. Good agreement was found between the TPS (Modulex, CMS) calculated and measured doses. The major source of the difference between the calculated and measured rectal doses was due to the applicators and dosimeter movement during the treatment delivery, as verified by x-ray fluoroscopy.

Waldhäusl et al. (2005) characterized specific diode probes: rectum probes (type 9112 from PTW) consisting of five semiconductors separated by 15 mm and single detector bladder probes (type 9113 from PTW). They compared doses measured and computed for the rectum and bladder ICRU reference points (ICRU 1985). For calibration, the diodes were placed in PMMA cubes of $12 \times 12 \times 12$ cm^3 immersed in a water tank of $30 \times 30 \times 20$ cm^3. The diodes were calibrated with an ^{192}Ir source (MicroSelectron) using cylindrical ionization chambers as a reference dosimeter. As calibration coefficients of diodes vary with time, weekly and daily calibrations were performed. Orthogonal radiographs (AP and lateral) were used for reconstruction (PLATO, BPS v13.2) of the applicator and points of interest, such as position of the in vivo probes and the ICRU rectum and bladder reference points. The reported rectal dose was the maximum dose measured with the probe. The bladder dose was measured in the center of the Foley balloon. If the differences between planned and measured doses exceeded 10%, the possible shift that might have occurred in the probe position was evaluated. In vivo dosimetry in patients showed differences between calculated and measured doses ranging from −31% to +90% (mean 11%) for the rectum and from −27% to +26% (mean 4%) for the bladder. Shifts in probe position of 2.5 mm for the rectal probe and 3.5 mm for the bladder probe caused dose differences exceeding 10%. The doses calculated at the ICRU reference points differed from the measured ones between −61% and 156% (mean 29%) in the rectum and between 12% and 162% (mean 58%) in the bladder. In summary, the authors estimated the overall uncertainty of the diode measurements to be 7%. They concluded that the diode accuracy and reproducibility were sufficient for clinical applications and that daily calibration did not improve the accuracy of the measurements. They recommended weekly calibration of the diode as sufficient for clinical use. Their study indicated that, in general, the rectal and bladder doses measured in vivo tended to underestimate the doses at the ICRU rectum and bladder reference points. The differences between the measured and calculated doses were attributed to the probe movement and to the possible change of the diode sensitivity at larger distances from the radioactive source. The results of this study underscore the importance of the accurate detector position during brachytherapy in vivo dosimetry.

Tanderup et al. (2006) developed a mathematical model that used rectal in vivo dosimetry data to determine the spatial relationship between applicator and rectal diodes during intracavitary PDR brachytherapy treatment. The aim of this method was to monitor the movement of the source and to assure that it was in the correct position. They employed five diodes spaced by 1.5 cm (PTW type 9112/K) and an ^{192}Ir source (GammaMed 12i). Orthogonal AP and lateral x-ray films were taken to evaluate the position of the applicator and the rectal dosimeter. The mathematical model is based on minimization of squared deviation between calculated and measured dose rates, and it quantifies the geometric stability by transforming measured rectal dose rates into relative positions of applicator and rectal diodes. The method is relevant for PDR treatments, where the geometry of source position during any pulse can be compared to that during the first pulse. Fourteen cervix cancer patients were studied. Relative movements of applicator and rectal diodes in lateral (x), longitudinal (y), and anterior–posterior (z) directions were of the order of 1–2 mm. The authors concluded that intrafraction variations of applicator and diode positions were small, compared to interfraction variations, which indicate that the geometry varies considerably more between fractions than during PDR. The spatial relation between rectal dosimeter and applicator was very stable during extended PDR treatments. This method can be also used for HDR treatments to evaluate the applicator–probe geometry, but since the geometry can be reconstructed only after the fraction has been given, this method cannot be used to interrupt the treatment in case of applicator displacements.

Ghahramani et al. (2008) investigated dependence of diode response on temperature, detector–source distance, and dose rate for MDR/LDR intracavitary brachytherapy. They also studied the angular dependence of these detectors with respect to the radiation source. They used flexible probes from PTW/Germany (rectal probe type 9112 and bladder probe 9113) and Cs-137 sources (Selectron LDR/MDR afterloading unit by Nucletron). Their study showed that there was no significant variation in the response of diodes with temperature (0.08%/°C), dose rate, and distance from the source for MDR brachytherapy (0.13%/cGy/h and −0.04%/mm, respectively). However, there was a significant decrease in diode response with respect to the diode-radioactive source angle; the diode responses decreased linearly with the diode-source angle, increasing from 0° to 53°. The authors recommended application of a correction factor as a function of the diode-radioactive source distance.

Palvolgyi (2009) studied the consistency of the Fletcher–Suit applicator geometry and the stability of position of a rectal probe for several HDR treatment fractions. In vivo dose measurements were performed with a five-channel rectal semiconductor probe (PTW, Germany). The study was focused on the differences between pretreatment and posttreatment geometry. In case of a difference bigger than 5% between the dose values measured versus calculated at points of the semiconductor detectors, the applicator and rectal probe position were verified with posttreatment reconstruction images. This work does not include details about diode dosimetric methodology: calibration, correction factors, uncertainty, etc.

The literature reviewed above illustrates that diode systems can be routinely used for in vivo dosimetry in brachytherapy. This is due to many advantages of such systems, which include overall robustness, high sensitivity, and the ability to measure both dose and dose rate. The main disadvantages of diodes are energy, temperature, and directional dependencies. To overcome the energy and temperature dependence, users are advised to calibrate these detectors at a temperature similar to the human body (37°C) and in the radiation field of the source used for patient treatment.

25.4 Application of MOSFET Detectors

25.4.1 Principles of Operation

Metal–oxide–semiconductor field-effect transistors (MOSFETs) belong to the semiconductor type of radiation detectors. These detectors have been used in radiation therapy for over 15 years

(Soubra et al. 1994; Ramani et al. 1997; Ramaseshan et al. 2002, 2004; Scalchi and Francescon 1998; Scarantino et al. 2004; Scalchi et al. 2005; Kinhikar et al. 2006a,b). Principles of MOSFET operation and dosimetry have been described in detail elsewhere (Sze 1981; Thomson et al. 1984; Ma 1989; Soubra et al. 1994; Sedra and Smith 2004; Scarantino et al. 2008; Cygler and Scalchi 2009); therefore, in this chapter, we will provide only a short description of these devices. Figure 25.3a shows the schematic cross section of a p-channel MOSFET showing the main functional elements of the detector device: source, drain, gate, silicon (Si) substrate, silicon oxide (SiO$_2$), and channel (inversion layer) (Soubra et al. 1994).

The measurement of dose is based on the gate threshold voltage shift, ΔV_{th}, which is required to maintain constant current flow between the source and drain (see Figure 25.3b). When ionizing radiation passes through the SiO$_2$ layer, electron–hole pairs are formed. Holes (+ charged) are trapped at the Si/SiO$_2$ interface. Trapped charge acts to screen the gate potential, and a higher value of V_{th} is required to switch the MOSFET "on." More details can be found in the work of Cygler and Scalchi (2009). Depending on their design, MOSFETs can operate in active or passive mode during the radiation exposure. Active mode means

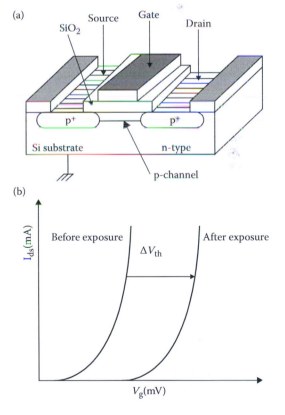

FIGURE 25.3 (a) Schematic cross section of a p-channel MOSFET showing the crucial elements of the device: source, drain, gate, silicon (Si) substrate, silicon oxide (SiO$_2$), and channel (inversion layer). (b) Source-to-drain current, I_{ds}, versus gate voltage, V_g. (Adapted from Soubra, M. et al., *Med Phys*, 21, 567–72, 1994. With permission.)

that during the radiation exposure, MOSFETs have a positive bias applied to the gate. Positive gate bias during irradiation reduces the recombination of electron–hole pairs and moves holes faster to the Si/SiO$_2$ interface. This results in a higher sensitivity of the MOSFET. For operation in passive mode, no bias is applied to the gate during the radiation exposure. In principle, dosimetric properties of MOSFET detectors cannot be generalized. They strongly depend on the detector size and construction. As with all solid-state dosimeters, MOSFET sensitivity depends on radiation energy, increasing for lower energies (Wang et al. 2005; Cygler et al. 2006; Cygler and Scalchi 2009; Panettieri et al. 2007; Bloemen-van Gurp et al. 2009a). For photon energies lower than 200 keV, sensitivity of MOSFET detector increases by a factor of 3–6, reaching a maximum of around 5–40 keV, depending on the kind of the detector (Wang et al. 2005). Therefore, it is recommended to calibrate these detectors in the radiation field energy they are to be used. MicroMOSFETs have practically isotropic response to radiation (Cygler et al. 2006; Bloemen-van Gurp et al. 2009a). All MOSFETs have a finite life depending on the accumulated dose and construction of the detector. Typically, MOSFET life is finished when the total threshold voltage reaches 10–20 V. The exact value depends again on the detector sensitivity and construction. More details about dosimetric characteristics of various MOSFET detectors can be found in a recent review (Cygler and Scalchi 2009).

25.4.2 Nonimplantable Detectors

There are two major commercial suppliers of MOSFET dosimetry systems: Best Medical Canada (previously Thomson-Nielsen) and Sicel Technologies Inc. Best Medical Canada provides single MOSFET detectors and MOSFET linear arrays consisting of five detectors. Individual MOSFETs come in two sizes: standard and microMOSFETs. The last one is the detector more suitable for brachytherapy because of its smaller size (external dimensions: $3.5 \times 1 \times 1$ mm) and isotropic response to radiation (Cygler and Scalchi 2009). Figure 25.4 shows a mobile MOSFET Patient Dose

Verification System and detectors available from Best Medical Canada.

OneDose/OneDosePlus MOSFET detectors from Sicel, shown in Figure 25.5, allow for a single-time-use dose measurement. These detectors are cable-free and capable of saving dose information along with the time and date of the measurement. OneDose detectors are precalibrated by the manufacturer. They were originally intended for use with high-energy photons and electrons, for doses from 1 to 500 cGy (Halvorsen 2005; Briere et al. 2006). Kinhikar et al. (2006a) characterized OneDose MOSFET in an HDR ^{192}Ir field. They observed dose linearity in the range of 0.3–5 Gy. The angular response of the detector was within $\pm 6\%$, and temperature dependence between 20°C and 37°C was within 1.5%.

In recent years, a new implantable MOSFET detector has been introduced for dose measurement inside the tumor (Scarantino et al. 2004; Beddar et al. 2005; Briere et al. 2005; Beyer et al. 2007). Implantable MOSFETs are also calibrated by the manufacturer (Sicel Technologies Inc.). They are gaining popularity, since they can directly measure the dose inside the tumor, which is particularly challenging when complex radiotherapy techniques are used.

In principle, as in the case of other solid-state detectors, the response of MOSFETs depends on temperature. The exception is the single MOSFET detector from Best Medical Canada, which due to the dual-MOSFET-dual-bias circuitry is temperature independent (Soubra et al. 1994). For other types of MOSFETs, including the linear MOSFET array, temperature correction factors should be applied (Cheung et al. 2004; Bloemen-van Gurp et al. 2009a).

In brachytherapy, MOSFETs have been used for both in vivo dosimetry (Cygler et al. 2006; Kinhikar et al. 2006a,b; Bloemen-van Gurp et al. 2009b) and source characterization measurements (Niu et al. 2004; Fagerstrom et al. 2008). MOSFET's small size, instant readout, and waterproof design make these detectors suitable for measurements in both LDR and HDR brachytherapy treatments. They have been used in LDR (Cygler et al.

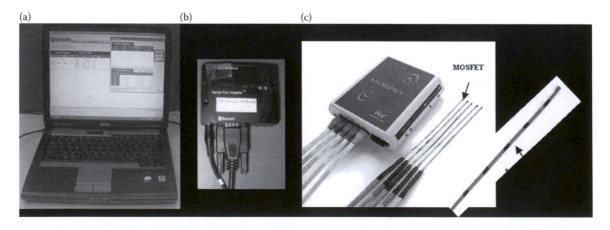

FIGURE 25.4 MobileMOSFET Patient Dose Verification System and MOSFET detectors: (a) computer, (b) Bluetooth wireless transceiver, and (c) Reader/Bias Box module and MOSFET detectors. (Courtesy of Best Medical, Canada.)

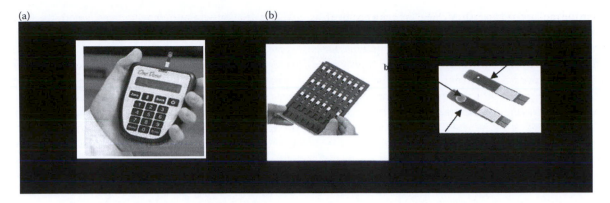

FIGURE 25.5 OneDose system and MOSFET detectors: (a) reader, and (b) MOSFETs. (Courtesy of G. Beyer.)

2006; Bloemen-van Gurp et al. 2009b) and HDR (Kinhikar et al. 2006b) treatments to verify the dose delivered to the patient. The high-sensitivity microMOSFET dosimeter (TN-1002RDM) with the high bias supply was used for in vivo dosimetry during LDR prostate brachytherapy (Cygler et al. 2006). For this application, a microMOSFET was inserted inside a urinary Foley catheter placed in the patient's urethra. The detector was moved along the urethra in 1-cm intervals. At each position, a reading was taken for 10 min, which could later be related to the initial dose rate at each measurement point. The resulting dose profile along the urethra can be used as a tool to evaluate the overall quality of the implant. The analysis of the urethral dose profile provided the information about the maximum dose received by the urethra, the prostate base and apex coverage, and the length of the prostatic urethra receiving at least the minimum peripheral dose (MPD). The study on 10 patients demonstrated that the maximum initial dose rate at the urethra varied between 10 and 16 cGy/h, corresponding to a total absorbed dose of 205–328 Gy. A similar study was performed later by Bloemen-van Gurp et al. (2009b) using a MOSFET array consisting of five detectors separated by 2 cm from each other. Using an array of MOSFETs allows for fast, simultaneous measurement of the dose profile along the entire urethra length.

Kinhikar et al. (2006b) used the OneDose MOSFET for skin dose measurements during internal mammary chain irradiation with HDR brachytherapy in the carcinoma of the breast.

Very thin Centre for Medical Radiation Physics MOSkin devices were used for ^{192}Ir-HDR dosimetry. Phantom measurements with MOSkins agreed within 5% with the treatment plan calculated doses (Qi et al. 2007).

25.4.3 Implantable Detectors

The first prototype detector manufactured by Sicel Technologies (Morrisville, NC) consisted of a single p-channel MOSFET, a data acquisition chip, a microprocessor, and a copper coil, all encapsulated in a glass tube 3.25 mm in diameter and 25 mm in length (Figure 25.6b). The circuit is powered by a current induced in the coil by an external handheld antenna connected to a radiofrequency reader. The dosimeter is passive during irradiation and powered only for the measurement of the threshold voltage. Beddar et al. (2005) have studied the dosimetric characteristics of these MOSFET detectors in vitro under irradiation from a ^{60}Co source. The detectors showed dose reproducibility generally within 5% or better, with the main sources of error being temperature fluctuations occurring between the

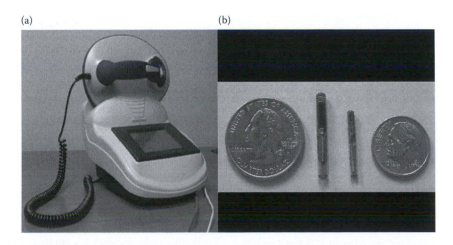

FIGURE 25.6 Dose verification system: (a) reader and wand and (b) implantable MOSFET detector. (Courtesy of G. Beyer.)

pre-irradiation and post-irradiation measurements as well as detector angular orientation. Their preliminary in vitro results showed that true in vivo dosimetry measurements are feasible and can be performed remotely using telemetric technology, and the agreement improved substantially when the dosimeters were placed in an environment with well-controlled temperature. A two-transistor dosimeter, in which the threshold voltages from the individual MOSFETs are averaged, was expected to provide a more accurate response than a single-transistor device.

To date, the original prototype has been successfully implanted into the breast, prostate, rectum, lung, and thigh for in vivo dosimetry during external beam radiation therapy, as reported by the manufacturer (Black et al. 2005; Scarantino et al. 2005). The device evolved into the commercial detector composed of a smaller capsule, 2.1 mm in diameter and 20 mm in length, and a dual MOSFET design. The signals from each detector are averaged, which results in better reproducibility and somewhat less sensitivity to temperature fluctuations (Figure 25.6a). Although the capsule is containing a millimeter-sized MOSFET, the detecting area (near the top of the capsule) is micrometer-sized and therefore can be considered a point detector. More than half the capsule contains a copper antenna used to communicate with the RF reader. The detectors are precalibrated by the manufacturer. In a study to characterize the performance of these new detectors, Briere et al. (2007) found the average dose response to be within 5.5% (2σ) for doses up to 20 Gy and 6.5% (2σ) for doses up to 74 Gy, and these specifications were adopted by the manufacturer. For doses significantly larger or smaller than the calibration dose of 200 cGy/fraction, the difference between the measured and delivered doses was found to be greater.

These MOSFETs, including the reader system (shown in Figure 25.6), are marketed under the name of *Dose Verification System* (DVS). The initial clinical studies performed to obtain FDA approval concluded that the DVS is safe for long-term implantation for the purpose of measuring radiation dose in breast and prostate cancer. Of the 126 implant procedures performed on breast and prostate subjects, 13 resulted in mild or moderately rated surgical complications, which were subsequently resolved. Out of the 126 implants (38 breast patients with a total of 72 detectors and 30 prostate patients with a total of 54 detectors), 4 detectors (3 in the breast and 1 in the prostate) migrated over time. Seven of the 126 dosimeters implanted (5.5%) were reported as device malfunctions due to the failure of the retention mechanism to deploy from the cannula on the first attempt. In summary, Scarantino et al. (2008) have noted only grade I/II adverse events of pain and bleeding. There were only four instances of dosimeter migration of >5 mm. A deviation of ≥7% in cumulative dose was noted in 7 of 36 (19%) for breast cancer patients. In prostate cancer patients, a ≥7% deviation was noted in 6 of 29 (21%) and 8 of 19 (42%) during initial and boost irradiations, respectively.

When considering application of this type of detector to sources of radiation aside from external beam, one must consider their energy response. If there is significant energy dependence

when used for brachytherapy sources, where the energy spectrum changes with depth, it may be challenging to assign a single correction factor that would be valid for all depths. Fagerstrom et al. (2008) studied the response of these detectors to ^{192}Ir HDR radiation. The authors found the detectors to be dose-rate-independent in the range of 22 to 84 cGy/min. However, the detectors did exhibit a higher sensitivity to ^{192}Ir radiation than to ^{60}Co. No simple correction factor could be determined, as the ratio of the readings varied with accumulated dose. Further, there was also an energy dependence within the ^{192}Ir measurements themselves, which was measured by placing the detectors at different depths in a specially designed phantom. The change in detector sensitivity, which differed by more than 10% for measurements performed at 3 and 5 cm, varied with accumulated dose. There was also a strong longitudinal dependence, whereby the detector response decreased by up to 16% when the detector's copper coil was aligned between the sensitive region and the incident radiation. The detector did show a weak rotational dependence, consistent with the results from external beam measurements.

Based on experiments to date, the application of these types of implantable detectors for brachytherapy in patients remains yet to be seen.

25.5 OSL and RL Dosimetry

The principle of optically stimulated luminescence (OSL) dosimetry is similar to that of TL. Electrons or holes are trapped in crystal defects when exposed to ionizing irradiation (Yukihara and McKeever 2008). The traps are stable at room temperature, and the crystal acts as a passive dosimeter. By optical stimulation, the electrons/holes are released to the conduction/valence bands, and afterwards, the recombination energy is emitted as light. The luminescence signal is a surrogate for accumulated absorbed dose in the crystal. Furthermore, prompt recombination of holes and electrons during irradiation gives rise to immediate RL. The RL signal depends on dose rate. In this way, two different luminescence signals can be used for measuring accumulated dose and dose rate, respectively. An optical fiber system can be used to connect the crystal to a reader containing a stimulation source and detection system. The fiber carries both the stimulation signal to the crystal and the luminescence signal from the crystal to the detector (Figure 25.7). Carbon-doped aluminum oxide (Al_2O_3:C) is currently the most commonly used OSL material due to its high sensitivity (Bøtter-Jensen et al. 2003; Akselrod et al. 1990) and negligible fading after irradiation (Bøtter-Jensen 1997).

OSL and RL dosimetry with Al_2O_3:C was initially developed and characterized in external beam radiotherapy for clinical dosimetric measurements and was later explored with ^{192}Ir sources. Subsequently, commercial OSL systems have been developed. However, exploitation of the Al_2O_3:C RL signal still requires in-house solutions for reading and processing the signal.

The Al_2O_3:C dosimetry system has several advantages with regard to brachytherapy in vivo dosimetry. The main advantage

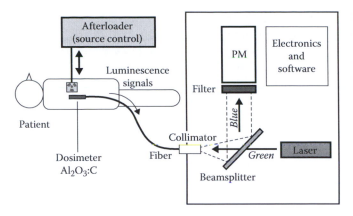

FIGURE 25.7 Fiber-coupled OSL/RL system with instrumentation containing laser stimulation source and photomultiplier (PM) detector. (Courtesy of Andersen et al. 2009a.)

is the high sensitivity, which makes it possible to use small dosimeters. The high spatial resolution is of clear benefit for measurements in the high-gradient brachytherapy dose distribution. Furthermore, for both RL and OSL systems, stability and reproducibility have been found to be excellent (below 2.5% SD). Further advantages of the Al_2O_3:C dosimetry systems are the temperature stability (better than $-0.2\%/°C$ to $0.6\%/°C$) (Edmund et al. 2006) and the limited angular dependence (Jursinic 2007; Andersen et al. 2009a).

The effective atomic number of Al_2O_3:C is 11.3 for photon energy range where photoelectric effect dominates, and this results in an overresponse for lower-energy photons (<300 keV). For brachytherapy, there is a considerable change of the energy spectrum with increasing distance from the source, and it means that the crystal-to-water dose ratio increases with distance to the source. Monte Carlo calculations have shown an overresponse of 4%, 6%, and 13% at depths of 35, 50, and 100 mm, respectively (Andersen et al. 2009a), when the dosimeter was calibrated at a distance of 10 mm. The sensitivity is 6% higher with ^{192}Ir than with megavoltage energies (Jursinic 2007), and therefore, it is most convenient to calibrate the dosimeters in the field of the relevant brachytherapy source.

The absorbed dose dependence of the Al_2O_3:C OSL is affected by the measurement protocol. It has been found that the dose response is linear for doses smaller than 10 Gy for real-time OSL (Polf et al. 2002) and up to at least 4 Gy for the commercial InLight microStar reader (Landauer, Inc) (Viamonte et al. 2008). Supralinearity has been found for larger doses (Akselrod et al. 2007). The Al_2O_3:C RL sensitivity is highly dependent on the dose received by the crystal due to filling of deep and dosimetry traps, and therefore, the RL signal has a highly nonlinear relationship to dose rate. However, the sensitivity changes are reproducible, and a correction can be incorporated into the measurement protocol (Andersen et al. 2006).

A particular challenge of the RL system is the stem effect. The stem effect arises from luminescence and Cerenkov irradiation

that is created in the fiber cable. The relative magnitude of the stem effect depends on (1) the geometric configuration of the fiber, crystal, and source, and (2) the sensitivity of the crystal, which in turn is dependent on the accumulated dose in the crystal (Andersen et al. 2009a). The stem effect can be as large as 20% in unfortunate configurations where the source is close to the fiber but far away from the crystal and where the crystal is exposed to a low dose (less than 0.25 Gy). However, for dose accumulations larger than 1 Gy, the stem effect will be less than 5% in most clinical situations. For OSL measurements, there is no stem effect, since the readout is performed after irradiation.

So far, the use of OSL and RL for in vivo dosimetry in brachytherapy is still limited, with only one study reporting dose measurements in patients (Andersen et al. 2009b). In this study, in vivo dosimetry was performed in five cervix cancer patients who received PDR brachytherapy. Al_2O_3:C crystals grown by Landauer were used with PMMA optical fibers connected to ME-03 RL/OSL readers (Risø DTU, Denmark). The crystals were 2 mm long and 0.5×0.5 mm^2 in the cross-sectional area, which meant they were sufficiently small to fit into commercial brachytherapy needles. The measured doses corresponded well to doses calculated in four out of five patients. For the last patient, a significant deviation between measured and calculated doses was detected. It was interpreted as due to displacement of the crystal. The RL measurements were found to provide good visualization of the progression and stability of the brachytherapy dose delivery. It was concluded that RL dosimetry could have potential for online detection of brachytherapy errors.

25.6 Application of Plastic Scintillation Detectors

25.6.1 Plastic Scintillation Detectors

A plastic scintillation detector (PSD) is composed of a scintillation component that emits light proportionally to the dose deposited in its sensitive volume. The light produced in the detector is then optically coupled to an optical fiber guide and transmitted toward a photodetector.

The properties of PSDs have been studied for high-energy external beams (Beddar et al. 1992a,b). High spatial resolution, linearity with dose, independence of response for megavoltage energies, temperature independence, and water equivalence are among the advantages that have been demonstrated for such detectors. Cerenkov light production has also been identified in the optical guide (Beddar et al. 1992a, 2004). This light component is produced in the fiber when struck by radiation over a certain energy threshold, which depends on the fiber material, and needs to be removed to perform accurate dosimetry. Different methods have been proposed to efficiently account for this spurious effect (Archambault et al. 2006). Using these detectors, the possibility of real-time in vivo dosimetry has been demonstrated under external beam radiation (Archambault et al. 2010), and accuracy of better than 1% has been achieved.

25.6.2 In Vivo Dosimetry for ^{192}Ir HDR Brachytherapy

PSDs could be very useful to perform an accurate online in vivo dosimetry during ^{192}Ir HDR brachytherapy (Lambert et al. 2006, 2007; Therriault-Proulx et al. 2009). Typical sizes of PSDs could be easily inserted in catheters used in this modality. Because the energy emission spectrum of an ^{192}Ir radiation source is mainly over the Cerenkov production threshold energy, the need for a removal technique has been stressed for clinically relevant situations (Lambert et al. 2006). Lambert et al. (2007) performed a comparative phantom study of PSDs and other commercially available detectors (MOSFETs, diamond detectors, and TLDs). Based on size, accuracy, and real-time possibilities, the authors claimed that PSDs showed the best combination of characteristics to perform dosimetry during ^{192}Ir HDR brachytherapy.

A study using a prostate phantom was performed by Therriault-Proulx et al. (2009). The authors used an RGB photodiode as the photodetection component, allowing for the removal of the Cerenkov component through the polychromatic approach. Thirteen catheters were implanted in the phantom (Figure 25.8a): 12 for dose delivery and 1 for the PSD. The phantom has been CT-scanned with a radio-opaque marker at the detector position (Figure 25.8b). Planning based on those images was performed with a PLATO workstation (Nucletron, Netherlands), and the treatment was delivered using an HDR microSelectron afterloader. In vivo dose monitoring was performed using the PSD dosimetry system. The dose rates were measured in real time during the treatment delivery. The average accumulated dose measured at the PSD location for five consecutive treatment deliveries was 275 ± 14 cGy and was found to be in good agreement with the expected dose of 275 cGy as calculated by the TPS.

Another study using an array of 16 PSDs in an insertable applicator that enables quality assurance of the treatment delivery and provides an alert to potential radiation accidents during HDR brachytherapy treatments has been performed by Cartwright et al. (2010). The system presented is capable of measuring doses for 1-s exposures with an uncertainty between 2% and 3% for most of the PSDs. The extent of such an in vivo dosimetry system would be to allow the clinicians to carry out

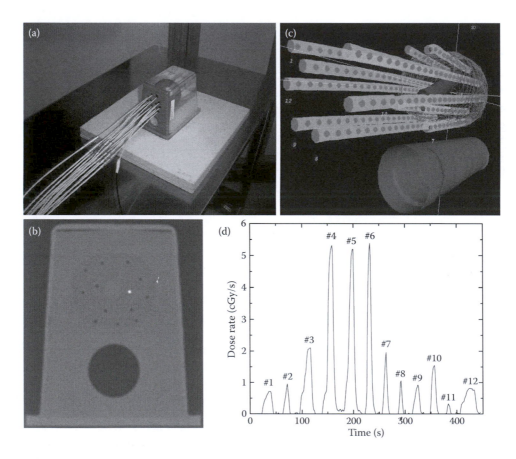

FIGURE 25.8 Overview of the procedure performed by Therriault-Proulx et al. (2009). (a) Thirteen catheters were implanted in the phantom: 12 for dose delivery and 1 for PSD insertion. (b) Prostate phantom was CT-scanned with a radio-opaque marker at the detector position. (c) Treatment planning was performed with the PLATO system (Nucletron, Netherlands). (d) Real-time dose rate measured using the PSD system during the entire treatment delivery.

dose escalation to the tumor volume while avoiding rectal side effects.

Based on the results shown so far by the different groups involved in the field, the use of PSD is promising as a quality-assurance approach when performing ^{192}Ir HDR brachytherapy treatments.

25.7 Electron Paramagnetic Resonance Dosimetry

Electron paramagnetic resonance (EPR) or electron spin resonance (ESR) spectroscopy is a technique for studying chemical species that have one or more unpaired electrons, for example, free radicals created during irradiation of some materials. In EPR dosimetry, the peak-to-peak amplitude of the first-derivative EPR spectrum of radiation-induced radicals in the dosimeter is used to monitor the absorbed dose. Solid-state EPR spectroscopy has been used for dosimetry of ionizing radiation for many years (Regulla 2000). The applications include retrospective dosimetry based on EPR signals obtained from tooth enamel or fingernails (Alekhin et al. 1982; Reyes et al. 2008, 2009), geological dating (Ikeya 1975), and dosimetry in radiation therapy (Regulla and Deffner 1982; de Angelis et al. 1999; Olsson et al. 2002; Ciesielski et al. 2003; Schultka et al. 2006; Anton et al. 2009; Al-Karmi 2010).

For EPR dosimetry in radiation therapy, L-α-alanine is the most commonly used material, although recently, some new materials such as lithium formate have been tried (Waldeland et al. 2010).

L-α-alanine is a nonessential amino acid that occurs in the form of a white and odorless crystal powder. In addition to powder, alanine can be formed in the shape of rods, pellets, films, etc. (Farahani et al. 1993). Irradiation of the amino acid L-α-alanine produces stable free radicals according to the reaction shown in Figure 25.9.

The concentration of these radicals is proportional to the absorbed dose and can be measured by EPR spectroscopy. Alanine, similar to Fricke solutions, belongs to the chemical type of dosimeters, since the determination of dose is based on a measurement of chemical changes induced by radiation.

Figure 25.10 shows a typical first derivative of the EPR signal of L-α-alanine irradiated to various doses (Ciesielski et al. 2003). One can see that alanine as a dosimeter is relatively insensitive at low doses. This is its main drawback.

Apart from the temporal stability of its signal, alanine as a detector of ionizing radiation has many attractive features. It has effective atomic number, $Z_{eff} = 6.79$, similar to that of water

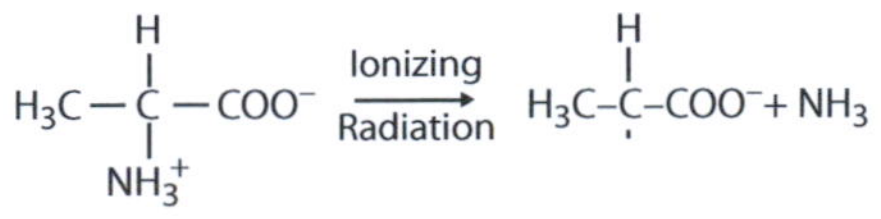

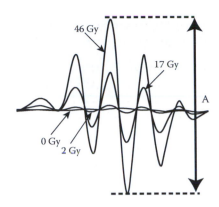

FIGURE 25.10 EPR spectra of alanine detectors irradiated with typical single-fraction doses (0.5, 1.0, and 2.0 Gy). The spectrum marked with 0 Gy is the background spectrum from nonirradiated detector. (Reprinted from *Int J Radiat Oncol Biol Phys*, 56, Ciesielski, B. et al., In vivo alanine/EPR dosimetry in daily clinical practice: A feasibility study, 899–905, Copyright (2003). with permission from Elsevier.)

(7.42). Therefore, the absorption of radiation energy by alanine is similar to that in water. The energy response of this dosimeter is relatively flat above 150 keV. The absorbed dose coefficient is constant up to about 10 kGy. For doses above 6 Gy, it has precision comparable to that of the TL and dosimetry with LiF powder (Mehta and Girzikowsky 1996).

Another useful feature of alanine is its independence of dose rate up to about 108 Gy/s (Regulla and Deffner 1982). Alanine response is almost independent of temperature (Deffner and Regulla 1980; Nagy et al. 2000) and humidity (Garcia et al. 2004), which makes it an attractive radiation dosimeter in tropical countries in situations of poor climate control. Similarly, as in the case of MOSFETs and OSL, the reading of the EPR signal is nondestructive, which allows for permanent dose storage and multiple analyses of the irradiated samples.

In spite of all these attractive features, alanine has not been commonly used in clinical dosimetry. This is mostly due to its relatively low sensitivity to low doses of irradiation and to the fact that it requires relatively expensive ESR equipment to read the signal.

In brachytherapy, alanine has been used by Schaeken and Scalliet (1996), Olsson et al. (2002), and Calcina et al. (2005) in HDR source dosimetry and characterization; by de Angelis et al. (1999) in ^{137}Cs dosimetry; by Soares et al. (2001) for beta dosimetry of ophthalmic applicators; and by Anton et al. (2009) for in-phantom dosimetry of the urethra in prostate HDR treatments.

Alanine has been successfully used for in vivo dosimetry of gynecological insertions by Schultka et al. (2004) and Schultka et al. (2006). The detectors had a form of small cellulose capsules (external diameter 5 mm; length 15 mm) filled with 0.5 g of alanine powder. The capsules were sealed in waterproof Parafilm pockets.

They used various methods to help them with the visualization of the detectors.

$$H_3C-\overset{\overset{\textstyle H}{|}}{\underset{\underset{\textstyle NH_3^+}{|}}{C}}-COO^- \xrightarrow[\text{Radiation}]{\text{Ionizing}} H_3C-\overset{\overset{\textstyle H}{|}}{\underset{}{C}}-COO^- + NH_3$$

FIGURE 25.9 Formation of free radicals upon irradiation of alanine. (Courtesy of J. Cygler.)

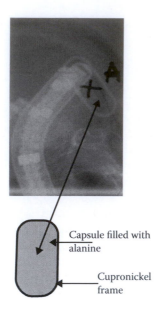

FIGURE 25.11 Radiograph of alanine detector placed in the vagina of a cervix patient. The cross marks the central point of the detector in which the dose was calculated by RTP. (From Schultka, K. et al., *Radiat Prot Dosimetry*, 120, 171–5, 2006. With permission.)

Figure 25.11 presents the radiograph of detector placed in the vagina of the cervix cancer patient. The diagram below the radiograph shows placement of a cupronickel frame, allowing localization of the detectors on the radiograph (Schultka et al. 2006).

Their study underscores the importance of the utmost accurate detector localization in order to obtain meaningful results from the dose measurements in high dose gradient fields such as encountered in brachytherapy.

The advantages and disadvantages of EPR alanine dosimetry are as follows:

- Advantages
  - Small dependence on dose rate
  - Almost independent of energy
  - Wide dose range
  - Small, portable
  - No bias required
  - Nondestructive type of measurement
  - Very little fading with time
- Disadvantages
  - Delay in reading (off-line dosimetry system)
  - Requires ESR equipment not easily available in the clinic
  - Relatively insensitive for lower doses
  - Some dependence on temperature and humidity

25.8 Summary

In vivo dosimetry in brachytherapy is not yet commonly utilized. Some errors in treatment delivery go undetected and therefore uncorrected, compromising the treatment outcomes. Performing systematic in vivo dosimetry in brachytherapy will allow for detecting and eliminating brachytherapy errors that cannot be identified by standard safety systems. Only by doing systematic in vivo dosimetry can these errors and their frequencies be detected, leading to improvement of quality assurance of brachytherapy by (1) making it possible to focus on the most relevant sources of errors, and (2) advancing the use of in vivo dosimetry systems in order to prevent brachytherapy errors.

There are currently a number of different in vivo dosimeters available for brachytherapy procedures. Choosing an optimal detector will always be a compromise between different requirements and constraints of the detector and anatomical site. The choice of the dosimeter will also depend on the purpose of doing in vivo dosimetry. There are a number of detector and system characteristics that are of particular importance for in vivo dosimetry in brachytherapy. Among them are (1) ability to measure accumulated dose and/or dose rate, (2) size of the detector, (3) sensitivity, (4) energy dependence, (5) dose rate dependence, (6) temperature dependence, (7) angular dependence, (8) sensitivity changes depending on previous irradiation, (9) patient comfort, and (10) the overall practicability of doing the measurements. Table 25.1 summarizes some features of various brachytherapy in vivo dosimetry systems.

TABLE 25.1 Characteristics of Various in Vivo Dosimetry Detectors

Detector	Cables	Bias Voltage	Dose	Dose Rate	Energy Dependence	Temperature Dependence	Angular Dependence
TLD	–	–	+	–	+	–	–
Diode	+	+/–	+	+	+	+	+
MOSFETs	+/–	+/–	+	–	+	+/–	+/–
OSL/RL	+/–	–	+/–	+/–	+	+/–	+/–
PSD	+/–	–	+	+	+	–	–
EPR	–	–	+	–	–	+	–

Note: "+" dependence present; "-" no dependence present; "+/–" dependence may be present, or somewhat, or for given detector types.

References

Afsharpour, H., Pignol, J.P., Keller, B. et al. 2010. Influence of breast composition and interseed attenuation in dose calculations for post-implant assessment of permanent breast 103Pd seed implant. *Phys Med Biol* 55:4547–61.

Akselrod, M.S., Bøtter-Jensen, L., and McKeever, S.W.S. 2007. Optically stimulated luminescence and its use in medical dosimetry. *Radiat Meas* 41(Sup. 1):S78–S99.

Akselrod, M.S., Kortov, V.S., Kravetsky, D.J., and Gotlib, V.I. 1990. Highly sensitive thermoluminescent anion-defective alpha-Al2O3:C single crystal detectors. *Radiat Prot Dosimetry* 32:15–20.

Al-Karmi, A.M. 2010. Dosimetric evaluation of alanine-in-glass dosimeters at clinical dose levels using high-energy X-rays from a linear accelerator. *Radiat Meas* 45:133–5.

Alecu, R. and Alecu, M. 1999. In-vivo rectal dose measurements with diodes to avoid misadministrations during intracavitary high dose rate brachytherapy for carcinoma of the cervix. *Med Phys* 26:768–70.

Alecu, R., Feldmeier, J.J., Court, W.S., Alecu, M., and Orton, C.G. 1997. A method to avoid misadministrations in high dose rate brachytherapy. *Med Phys* 24:259–61.

Alekhin, I.A., Babenko, S.P., Kraitor, S.N. et al. 1982. Experience from application of radiolyoluminescence and electron paramagnetic resonance to dosimetry of accidental irradiation. *At Energ* 53:546–50.

Andersen, C.E., Marckmann, C.J., Aznar, M.C. et al. 2006. An algorithm for real-time dosimetry in intensity-modulated radiation therapy using the radioluminescence signal from Al2O3:C. *Radiat Prot Dosimetry* 120:7–13.

Andersen, C.E., Nielsen, S.K., Greilich, S. et al. 2009a. Characterization of a fiber-coupled Al2O3:C luminescence dosimetry system for online in vivo dose verification during ^{192}Ir brachytherapy. *Med Phys* 36:708–18.

Andersen, C.E., Nielsen, S.K., Lindegaard, J.C., and Tanderup, K. 2009b. Time-resolved in vivo luminescence dosimetry for online error detection in pulsed dose-rate brachytherapy. *Med Phys* 36:5033–43.

Anton, M., Wagner, D., Selbach, H.J. et al. 2009. in vivo dosimetry in the urethra using alanine/ESR during ^{192}Ir HDR brachytherapy of prostate cancer—A phantom study. *Phys Med Biol* 54:2915–31.

Archambault, L., Beddar, A.S., Gingras, L., Roy, R., and Beaulieu, L. 2006. Measurement accuracy and cerenkov removal for high performance, high spatial resolution scintillation dosimetry. *Med Phys* 33:128–35.

Archambault, L., Briere, T.M., Pönisch, F. et al. 2010. Toward a real-time in vivo dosimetry system using plastic scintillation detectors. *Int J Radiat Oncol Biol Phys* 78:280–7.

Attix, F.H. 1986. *Introduction to Radiological Physics and Radiation Dosimetry*. John Wiley and Sons, Inc, New York.

Ballester, F., Granero, D., Pérez-Calatayud, J., Melhus, C.S., and Rivard, M.J. 2009. Evaluation of high-energy brachytherapy source electronic disequilibrium and dose from emitted electrons. *Med Phys* 36:4250–6.

Baltas, D., Sakelliou, L., and Zamboglou, N. 2007. *The Physics of Modern Brachytherapy for Oncology*. CRC Press, Taylor & Francis, Boca Raton, FL.

Beddar, A.S., Mackie, T.R., and Attix, F.H. 1992a. Water-equivalent plastic scintillation detectors for high-energy beam dosimetry: I. Physical characteristics and theoretical consideration. *Phys Med Biol* 37:1883–900.

Beddar, A.S., Mackie, T.R., and Attix, F.H. 1992b. Water-equivalent plastic scintillation detectors for high-energy beam dosimetry: II. Properties and measurements. *Phys Med Biol* 37:1901–13.

Beddar, A.S., Suchowerska, N., and Law, S.H. 2004. Plastic scintillation dosimetry for radiation therapy: Minimizing capture of Cerenkov radiation noise. *Phys Med Biol* 49:783–90.

Beddar, A.S., Salehpour, M., Briere, T.M., Hamidian, H., and Gillin, M.T. 2005. Preliminary evaluation of implantable MOSFET radiation dosimeters. *Phys Med Biol* 50:141–9.

Beyer, G.P., Scarantino, C.W., Prestidge, B.R. et al. 2007. Technical evaluation of radiation dose delivered in prostate cancer patients as measured by an implantable MOSFET dosimeter. *Int J Radiat Oncol Biol Phys* 69:925–35.

Black, R.D., Scarantino, C.W., Mann, G.G., Anscher, M.S., Ornitz, R.D., and Nelms, B.E. 2005. An analysis of an implantable dosimeter system for external beam therapy. *Int J Radiat Oncol Biol Phys* 63:290–300.

Bloemen-van Gurp, E.J., Haanstra, B.K., Murrer, L.H. et al. 2009a. in vivo dosimetry with a linear MOSFET array to evaluate the urethra dose during permanent implant brachytherapy using iodine-125. *Int J Radiat Oncol Biol Phys* 75:1266–72.

Bloemen-van Gurp, E.J., Murrer, L.H., Haanstra, B.K. et al. 2009b. In vivo dosimetry using a linear MOSFET-array dosimeter to determine the urethra dose in 125I permanent prostate implants. *Int J Radiat Oncol Biol Phys* 73:314–21.

Bomford, C.K., Kunkler, I.H. et al. 2003. *Walter and Miller Textbook of Radiotherapy. Radiation Physics, Therapy and Oncology*. Elsevier Churchill Livingstone, Oxford, UK.

Bøtter-Jensen, L. 1997. Luminescence techniques: Instrumentation and methods. *Radiat Meas* 17:749–68.

Bøtter-Jensen, L., McKeever, S.W.S., and Wintle, A.G. 2003. *Optically Stimulated Luminescence Dosimetry*. Elsevier, Amsterdam.

Brezovich, I.A., Duan, J., Pareek, P.N., Fiveash, J., and Ezekiel, M. 2000. In vivo urethral dose measurements: A method to verify high dose rate prostate treatments. *Med Phys* 27:2297–301.

Briere, T.M., Beddar, A.S., and Gillin, M.T. 2005. Evaluation of precalibrated implantable MOSFET radiation dosimeters for megavoltage photon beams. *Med Phys* 32:3346–9.

Briere, T.M., Lii, J., Prado, K., Gillin, M.T., and Beddar, A.S. 2006. Single-use MOSFET radiation dosimeters for the quality assurance of megavoltage photon beams. *Phys Med Biol* 51:1139–44.

Briere, T.M., Gillin, M.T., and Beddar, A.S. 2007. Implantable MOSFET detectors: Evaluation of a new design. *Med Phys* 34:4585–90.

Calcina, C.S., de Almeida, A., Rocha, J.R., Abrego, F.C., and Baffa, O. 2005. Ir-192 HDR transit dose and radial dose function determination using alanine/EPR dosimetry. *Phys Med Biol* 50:1109–17.

Cameron, J.R., Suntharalingam, N., and Kenney, G.N. 1968. *Thermoluminescent Dosimetry*. University of Wisconsin Press. Madison, WI.

Cartwright, L.E., Suchowerska, N., Yin, Y., Lambert, J., Haque, M., and McKenzie, D.R. 2010. Dose mapping of the rectal wall during brachytherapy with an array of scintillation dosimeters. *Med Phys* 37:2247–55.

Cheung, T., Butson, M.J., and Yu, P.K. 2004. Effects of temperature variation on MOSFET dosimetry. *Phys Med Biol* 49:N191–6.

Ciesielski, B., Schultka, K., Kobierska, A., Nowak, R., and Peimel-Stuglik, Z. 2003. In vivo alanine/EPR dosimetry in daily clinical practice: A feasibility study. *Int J Radiat Oncol Biol Phys* 56:899–905.

Cygler, J.E., and Scalchi, P. 2009. MOSFET dosimetry in radiotherapy. In: *Clinical Dosimetry Measurements in Radiotherapy*. D.W.O. Rogers and J.E. Cygler (eds.). Medical Physics Publishing, Madison, WI. Medical Physics Monograph No. 34, pp. 941–77.

Cygler, J.E., Saoudi, A., Perry, G., and Morash, C.E.C. 2006. Feasibility study of using MOSFET detectors for in vivo dosimetry during permanent low-dose-rate prostate implants. *Radiother Oncol* 80:296–301.

Das, R.K., Li, Z., Perera, H., and Williamson, J.F. 1996. Accuracy of Monte Carlo photon transport simulation in characterizing brachytherapy dosimeter energy-response artefacts. *Phys Med Biol* 41:995–1006.

Das, R., Toye, W., Kron, T., Williams, S., and Duchesne, G. 2007. Thermoluminescence dosimetry for in-vivo verification of high dose rate brachytherapy for prostate cancer. *Australas Phys Eng Sci Med* 30:178–84.

De Angelis, C., Onori, S., Petetti, E., Piermattei, A., and Azario, L. 1999. Alanine/EPR dosimetry in brachytherapy. *Phys Med Biol* 44:1181–91.

Deffner, U. and Regulla, D.F. 1980. Influences of physical parameters on high-level amino acid dosimetry. *Nucl Instrum Methods* 175:134–5.

DeWerd, L.A., Bartol, L.J., and Davis, S.D. 2009. Thermoluminescence dosimetry. In: *Clinical Dosimetry for Radiotherapy*. D.W.O. Rogers and J.E. Cygler (eds.). Medical Physics Publishing, Madison, WI, pp. 815–40.

DeWerd, L., Ibbott, G., Meigooni, A.S. et al. 2011. A dosimetric uncertainty analysis for photon-emitting brachytherapy source: Report of the AAPM Task Group No. 138 and GEC-ESTRO. *Med Phys* 38:782–801.

Edmund, J.M., Andersen, C.E., Marckmann, C.J., Aznar, M.C., Akselrod, M.S., and Bøtter-Jensen, L. 2006. CW-OSL measurement protocols using optical fibre Al2O3:C dosemeters. *Radiat Prot Dosimetry* 119:368–74.

Fagerstrom, J.M., Micka, J.A., and DeWerd, L.A. 2008. Response of an implantable MOSFET dosimeter to ^{192}Ir HDR radiation. *Med Phys* 35:5729–37.

Farahani, M., Eichmiller, F.C., and McLaughlin, W.L. 1993. New method for shielding electron beams used for head and neck cancer treatment. *Med Phys* 20:1237–41.

Garcia, R.M.D., Desrosiers, M.F., Attwood, J.G. et al. 2004. Characterization of a new alanine film dosimeter: Relative humidity and post-irradiation stability. *Radiat Phys Chem* 71:375–9.

Ghahramani, F., Allahverdi, M., and Jaberi, R. 2008. Dependency of semiconductor dosimeter responses, used in MDR/LDR brachytherapy, on factors which are important in clinical conditions. *Rep Prac Oncol Radiother* 13:29–33.

Goetsch, S.J., Attix, F.H., Pearson, D.W., and Thomadsen, B.R. 1991. Calibration of ^{192}Ir high-dose-rate afterloading systems. *Med Phys* 18:462–7.

Gonzalez, J., Rodriguez, A., Crelgo, D. et al. 2005. 289 Dose determination in rectum and urethra with TLD dosimeters in prostate brachytherapy with high dose rate (HDR). *Radiother Oncol* 76 (Supplement 2):S135.

Grusell, E. and Rikner, G. 1984. Radiation damage induced dose rate non-linearity in an N-type silicon detector. *Acta Radiol Oncol* 23:465–9.

Halvorsen, P.H. 2005. Dosimetric evaluation of a new design MOSFET in vivo dosimeter. *Med Phys* 32:110–7.

Horowitz, Y.T. 1984. *Thermoluminescence and Thermoluminescent Dosimetry*. CRC Press, Boca Raton, FL.

Huyskens, D.P., Bogaerts, R., Verstraete, J. et al. 2001. *Practical Guidelines for the Implementation of In Vivo Dosimetry with Diodes in External Radiotherapy with Photon Beams (Entrance Dose)*. ESTRO booklet #5. ESTRO, Brussels.

IAEA. 2000. IAEA Safety Report Series No. 17. 2000. Lessons learned from accidental exposuresin radiotherapy. International Atomic Energy Agency. IAEA, Vienna.

ICRP Publication 86. 2000. Prevention of accidental exposures to patients undergoing radiation therapy. International Commission on Radiological Protection. ICRP Publication 86, Pergamon Press, Oxford.

ICRP Publication 97. 2005. Prevention of high-dose-rate brachytherapy accidents. International Commission on Radiological Protection, 2005. Annals of the ICRP, Elsevier. ICRP Publication 97. In: *Ann ICRP* 35 (2).

Ikeya, M. 1975. Dating a stalactite by electron paramagnetic resonance. *Nature* 255 (5503):48–50.

International Commission on Radiation Units and Measurements (ICRU). 1985. Dose and volume specification for reporting and recording intracavitary therapy in gynecology. Report 38 of ICRU, ICRU Publications, Bethesda, MD.

Joslin, C.A., Smith, C.W., and Mallik, A. 1972. The treatment of cervix cancer using high activity 60 Co sources. *Br J Radiol* 45:257–70.

Jursinic, P.A. 2007. Characterization of optically stimulated luminescent dosimeters, OSLDs, for clinical dosimetric measurements. *Med Phys* 34:4594–604.

Kapp, K.S., Stuecklschweiger, G.F., Kapp, D.S., and Hackl, A.G. 1992. Dosimetry of intracavitary placements for uterine and cervical carcinoma: Results of orthogonal film, TLD, and CT-assisted techniques. *Radiother Oncol* 24:137–46.

Kertzscher, G., Andersen, C.E., Siebert, F.A., Nielsen, S.K., Lindegaard, J.C., and Tanderup, K. 2011. Identifying after-loading PDR and HDR brachytherapy errors using real-time fiber-coupled Al(2)O(3):C dosimetry and a novel statistical error decision criterion. *Radiother Oncol* 100:456–62.

Kinhikar, R.A., Sharma, P.K., Tambe, C.M., and Deshpande, D.D. 2006a. Dosimetric evaluation of a new OneDose MOSFET for Ir-192 energy. *Phys Med Biol* 51:1261–8.

Kinhikar, R.A., Sharma, P.K., Tambe, C.M. et al. 2006b. Clinical application of a OneDose MOSFET for skin dose measurements during internal mammary chain irradiation with high dose rate brachytherapy in carcinoma of the breast. *Phys Med Biol* 51:N263–8.

Koedooder, K., van Wieringen, N., van der Grient, H.N., van Herten, Y.R., Pieters, B.R., and Blank, L.E. 2008. Safety aspects of pulsed dose rate brachytherapy: analysis of errors in 1,300 treatment sessions. *Int J Radiat Oncol Biol Phys* 70:953–60.

Kron, T. 1999. Dose measuring tools. In: *The Modern Technology of Radiation Oncology*. J. Van Dyk (ed.). Medical Physics Publishing, Madison, WI, pp. 753–821.

Lambert, J., McKenzie, D.R., Law, S., Elsey, J., and Suchowerska, N. 2006. A plastic scintillation dosimeter for high dose rate brachytherapy. *Phys Med Biol* 51: 5505–16.

Lambert, J., Nakano, T., Law, S., Elsey, J., McKenzie, D.R., and Suchowerska, N. 2007. In vivo dosimeters for HDR brachytherapy: A comparison of a diamond detector, MOSFET, TLD, and scintillation detector. *Med Phys* 34:1759–65.

Landry, G., Reniers, B., Murrer, L. et al. 2010. Sensitivity of low energy brachytherapy Monte Carlo dose calculations to uncertainties in human tissue composition. *Med Phys* 37:5188–98.

Ma, T.P. 1989. Process-induced radiation effects. In: *Ionizing Radiation Effects in MOS Devices and Circuits*. T.P. Ma and P.V. Dressendorfer (eds.). John Wiley & Sons, New York, pp. 401–42.

Major, T., Polgar, C., Lövey, K., and Fröhlich, G. 2011. Dosimetric characteristics of accelerated partial breast irradiation with CT image-based multicatheter interstitial brachytherapy: A single institution's experience. *Brachytherapy* 10:421–6.

Mangold, C.A., Rijnders, A., Georg, D., Van Limbergen, E., Pötter, R., and Huyskens, D. 2001. Quality control in interstitial brachytherapy of the breast using pulsed dose rate: Treatment planning and dose delivery with an Ir-192 afterloading system. *Radiother Oncol* 58:43–51.

McKeever, S.W.S. 1985. *Thermoluminescence of Solids*. Cambridge University Press, Cambridge.

Mehta, K. and Girzikowsky, R. 1996. Alanine-ESR dosimetry for radiotherapy. IAEA experience. *Appl Radiat Isot* 47:1189–91.

Meigooni, A.S., Meli, J.A., and Nath, R. 1988. Influence of the variation of energy spectra with depth in the dosimetry of ^{192}Ir using LiF TLD. *Phys Med Biol* 33:1159–70.

Nagy, V., Puhl, J.M., and Desrosiers, M.F. 2000. Advancements in accuracy of the alanine dosimetry system: Part 2: The influence of the irradiation temperature. *Radiat Phys Chem* 57:1–9.

Niu, H., Hsi, W.C., Chu, J.C., Kirk, M.C., and Kouwenhoven, E. 2004. Dosimetric characteristics of the Leipzig surface applicators used in the high dose rate brachy radiotherapy. *Med.Phys* 31:3372–7.

Nose, T., Koizumi, M., Yoshida, K. et al. 2008. In vivo dosimetry of high-dose-rate interstitial brachytherapy in the pelvic region: Use of a radiophotoluminescence glass dosimeter for measurement of 1004 points in 66 patients with pelvic malignancy. *Int J Radiat Oncol Biol Phys* 70:626–33.

Olsson, S., Bergstrand, E.S., Carlsson, A.K., Hole, E.O., and Lund, E. 2002. Radiation dose measurements with alanine/agarose gel and thin alanine films around a ^{192}Ir brachytherapy source, using ESR spectroscopy. *Phys Med Biol* 47:1333–56.

Palvolgyi, J. 2009. The consistency of Fletcher-Suit applicator geometry and of the rectal probe's position in high dose rate brachytherapy treatment fraction of cervix carcinoma. *J Contemp Brachytherapy* 1:154–6.

Panettieri, V., Duch, M.A., Jornet, N. et al. 2007. Monte Carlo simulation of MOSFET detectors for high-energy photon beams using the PENELOPE code. *Phys Med Biol* 52:303–16.

Piermattei, A., Azario, L., Monaco, G., Soriani, A., and Arcovito, G. 1995. p-type silicon detector for brachytherapy dosimetry. *Med Phys* 22:835–9.

Polf, J.C., McKeever, S.W., Akselrod, M.S., and Holmstrom, S. 2002. A real-time, fibre optic dosimetry system using Al2O3 fibres. *Radiat Prot Dosimetry* 100:301–4.

Pötter, R., Haie-Meder, C., Van Limbergen, E. et al. 2006. Recommendations from gynaecological (GYN) GEC ESTRO working group (II): concepts and terms in 3D image-based treatment planning in cervix cancer brachytherapy-3D dose volume parameters and aspects of 3D image-based anatomy, radiation physics, radiobiology. *Radiother Oncol* 78:67–77.

Prada, P.J., Fernández, J., Martinez, A.A. et al. 2007. Transperineal injection of hyaluronic acid in anterior perirectal fat to decrease rectal toxicity from radiation delivered with intensity modulated brachytherapy or EBRT for prostate cancer patients. *Int J Radiat Oncol Biol Phys* 69:95–102.

Qi, Z.-Y., Deng, X.-W., Huang, S.-M. et al. 2007. Verification of the plan dosimetry for high dose rate brachytherapy using metal–oxide–semiconductor field effect transistor detectors. *Med Phys* 34(6):2007–13.

Ramani, R., Russell, S., and O'Brien, P. 1997. Clinical dosimetry using MOSFETs. *Int J Radiat Oncol Biol Phys* 37:959–64.

Ramaseshan, R., Lam, T., Perkin, G. et al. 2002. In-vivo dosimetry for IMRT using MOSFET dosimeter. *Med Phys* 29:1943.

Ramaseshan, R., Kohli, K.S., Zhang, T.J. et al. 2004. Performance characteristics of a microMOSFET as an in vivo dosimeter in radiation therapy. *Phys Med Biol* 49:4031–48.

Regulla, D. 2000. From dating to biophysics—20 years of progress in applied ESR spectroscopy. *Appl Radiat Isot* 52(5):1023–30.

Regulla, D.F. and Deffner, U. 1982. Dosimetry by ESR spectroscopy of alanine. *Int J Appl Radiat Isot* 33:1101–14.

Reyes, R.A., Romanyukha, A., Trompier, F. et al. 2008. Electron paramagnetic resonance in human fingernails: the sponge model implication. *Radiat Environ Biophys* 47(4):515–26.

Reyes, R.A., Romanyukha, A., Olsen, C., Trompier, F., and Benevides, L.A. 2009. Electron paramagnetic resonance in irradiated fingernails: variability of dose dependence and possibilities of initial dose assessment. *Radiat Environ Biophys* 48(3):295–310.

Rikner, G. and Grusell, E. 1985. Selective shielding of a p-Si detector for quality independence. *Acta Radiol Oncol* 24:65–9.

Rikner, G. and Grusell, E. 1987. General specifications for silicon semiconductors for use in radiation dosimetry. *Phys Med Biol* 32:1109–17.

Rivard, M.J., Coursey, B.M., DeWerd, L.A. et al. 2004. Update of AAPM Task Group No. 43 Report: A revised AAPM protocol for brachytherapy dose calculations. *Med Phys* 31:633–74.

Sakata, K., Nagakura, H., Oouchi, A. et al. 2002. High-doserate intracavitary brachytherapy: Results of analyses of late rectal complications. *Int J Radiol Oncol Biol Phys* 54:1369–76.

Scalchi, P. and Francescon, P. 1998. Calibration of a MOSFET detection system for 6-MV in vivo dosimetry. *Int J Radiat Oncol Biol Phys* 40:987–93.

Scalchi, P., Francescon, P., and Rajaguru, P. 2005. Characterization of a new MOSFET detector configuration for in vivo skin dosimetry. *Med Phys* 32:1571–8.

Scarantino, C.W., Ruslander, D.M., Rini, C.J., Mann, G.G., Nagle, H.T., and Black, R.D. 2004. An implantable radiation dosimeter for use in external beam radiation therapy. *Med Phys* 31:2658–71.

Scarantino, C.W., Rini, C.J., Aquino, M. et al. 2005. Initial clinical results of an in vivo dosimeter during external beam radiation therapy. *Int J Radiat Oncol Biol Phys* 62(2):606–13.

Scarantino, C.W., Prestidge, B.R., Anscher, M.S. et al. 2008. The observed variance between predicted and measured radiation dose in breast and prostate patients utilizing an in vivo dosimeter. *Int J Radiat Oncol Biol Phys* 72:597–604.

Schaeken, B. and Scalliet, P. 1996. One year of experience with alanine dosimetry in radiotherapy. *Appl Radiat Isot* 47:1177–82.

Schultka, K., Ciesielski, B., Serkies, K. et al. 2004. In vivo dosimetry using electron paramagnetic resonance in L-alanine in gynecological low dose rate brachytherapy. *Nowotwory Journal of oncology* 54: 560–3.

Schultka, K., Ciesielski, B., Serkies, K., Sawicki, T., Tarnawska, Z., and Jassem, J. 2006. EPR/alanine dosimetry in LDR brachytherapy—A feasibility study. *Radiat Prot Dosimetry* 120:171–5.

Sedra, A.S. and Smith, K.C. 2004. *Microelectronic Circuits*. Oxford University Press. Oxford.

Soares, C.G., Vynckier, S., Järvinen, H. et al. 2001. Dosimetry of beta-ray ophthalmic applicators: Comparison of different measurement methods. *Med Phys* 28:1373–84.

Soubra, M., Cygler, J.E., and Mackay, G. 1994. Evaluation of a dual bias dual metal oxide-silicon semiconductor field effect transistor detector as radiation dosimeter. *Med Phys* 21:567–72.

Sze, S.M. 1981. *Physics of Semiconductor Devices. MOSFET*. Wiley & Sons, New York.

Tanderup, K., Christensen, J.J., Granfeldt, J., and Lindegaard, J.C. 2006. Geometric stability of intracavitary pulsed dose rate brachytherapy monitored by in vivo rectal dosimetry. *Radiother Oncol* 79:87–93.

Therriault-Proulx, F., Guillot, M., Gingras, L., Beaulieu, L., and Beddar, S. 2009. Importance of contamination signal removal on HDR brachytherapy in vivo dosimetry when using a scintillating fiber dosimeter. *Med Phys* 36(6):2536–7.

Thomson, I., Thomas, R.E., and Berndt, L. P. 1984. Radiation dosimetry with MOS sensors. *Rad Prot Dos* 6:121–4.

Thomson, R.M. and Rogers, D.W. 2010. Monte Carlo dosimetry for 125I and 103Pd eye plaque brachytherapy with various seed models. *Med Phys* 37:368–76.

Van Dyk, J., Barnett, R.B., Cygler, J.E., and Shragge, P.C. 1993. Commissioning and quality assurance of treatment planning computers. *Int J Radiat Oncol Biol Phys* 26:261–73.

Viamonte, A., da Rosa, L.A., Buckley, L.A., Cherpak, A., and Cygler, J.E. 2008. Radiotherapy dosimetry using a commercial OSL system. *Med Phys* 35:1261–6.

Waldeland, E., Hörling, M., Hole, E.O., Sagstuen, E., and Malinen, E. 2010. Dosimetry of stereotactic radiosurgery using lithium formate EPR dosimeters. *Phys Med Biol* 55:2307–16.

Waldhäusl, C., Wambersie, A., Pötter, R., and Georg, D. 2005. In-vivo dosimetry for gynaecological brachytherapy: Physical and clinical considerations. *Radiother Oncol* 77:310–7.

Wang, B., Xu, X.G., and Kim, C.H. 2005. Monte Carlo study of MOSFET dosemeter characteristics: dose dependence on photon energy, direction and dosemeter composition. *Radiat Prot Dos* 113:40–6.

Yorke, E.D., Alecu R., Ding, L. et al. 2005. *Diode In-Vivo Dosimetry for Patients Receiving External Beam Radiation Therapy*. AAPM Report No. 87. Report of Task Group 62. Medical Physics Publishing, Madison, WI.

Yukihara, E.G. and McKeever, S.W. 2008. Optically stimulated luminescence (OSL) dosimetry in medicine. *Phys Med Biol* 53: R351–79.

Zhu, T.C. and Saini, S.A.S. 2009. Diode dosimetry for megavoltage electron and photon beams. In: *Clinical Dosimetry Measurements in Radiotherapy*. D.W.O. Rogers and J.E. Cygler (eds.). Medical Physics Publishing, Madison WI, Medical Physics Monograph No. 34, pp. 913–39.

26

Special Brachytherapy Modalities

Christopher S. Melhus
Tufts University School of Medicine

Mark J. Rivard
Tufts University School of Medicine

Ravinder Nath
Yale University School of Medicine

26.1 Introduction

This chapter addresses a number of special applications of brachytherapy that use a unique or unusual radiation source or treatment delivery system. Specifically, chapter sections cover electronic brachytherapy, microsphere brachytherapy, eye plaque brachytherapy, intravascular brachytherapy (IVBT), and neutron brachytherapy. The first three are accepted standards-of-care for selected patients; the fourth was a standard-of-care, but current utilization is very low due to competing therapies; and the last is still considered experimental. The objective of this chapter is to broadly describe these modalities and summarize the available clinical and professional guidance.

26.2 Electronic Brachytherapy

Unlike traditional radionuclide-based brachytherapy sources, electronic brachytherapy utilizes x-rays produced by miniature x-ray tubes using an externally applied electric potential to accelerate electrons and produce bremsstrahlung photons from a high-atomic number target. These radiation sources have the advantage and the disadvantage of not emitting ionizing radiation when de-energized, in contrast to radionuclide sources that continually emit radiation with exponential decay. It should be noted that regulatory requirements for electronic brachytherapy source handling and management are different than those for radionuclide-based brachytherapy sources. In the United States,

the use of electronic brachytherapy devices is regulated by individual states instead of the Nuclear Regulatory Commission.

26.2.1 Overview

Commercially available electronic brachytherapy sources operate at peak voltages below 100 kVp, with the most common devices operating at ≤50 kVp. Advantages of electronic brachytherapy include adjustable dose rate, capacity to adjust dwell times and dose rate dynamically to provide more flexibility in optimization of dose distributions (limited to systems with translatable source position), and often, variable photon spectra through variable kVp. Compared to high-dose-rate (HDR) ^{192}Ir brachytherapy, another potential advantage is the lessened need for shielding; however, some radionuclide-based systems may necessitate even less shielding than an electronic source, for example, high-activity beta-emitting sources. Current disadvantages for this modality compared to HDR ^{192}Ir include need to provide an external power source, lack of constancy of output compared to radionuclide decay, variations in source-to-source dose rate and dose distributions, increased source size, limited dose rate at depth in tissue, limited source lifespan, and the need to measure the output of a tube preceding each clinical use. In addition to these disadvantages, there is the potential for risk of electrical shock to the patient and for blood clotting due to heat production in nearby blood vessels, and there is limited evidence of treatment efficacy and healthy tissue toxicities.

Polyenergetic bremsstrahlung x-ray spectra for electronic brachytherapy sources are generally different than the photon and electron spectra from radionuclide sources. While some investigations have demonstrated increased relative biological effectiveness (RBE) for low-energy photons (see for example Shridhar et al. 2010), radiobiological studies directly comparing 50-kVp electronic brachytherapy photons to HDR ^{192}Ir (E_{avg} = 0.4 MeV) have not been performed. Prescription doses for electronic brachytherapy generally follow those for HDR ^{192}Ir brachytherapy. Similarly, disease sites approved by the U.S. Food and Drug Administration (FDA) for electronic brachytherapy mimic those for HDR ^{192}Ir brachytherapy.

There are requirements (Williamson et al. 1998) by the American Association of Physicists in Medicine (AAPM) to have source calibrations traceable to a primary standard at the National Institute of Standards and Technology (NIST). This is mainly to set uniform and high standards for posting on the joint AAPM/Radiological Physics Center (RPC) source registry for inclusion in multi-institutional U.S. clinical trials (Radiological Physics Center 2011). However, electronic brachytherapy devices are sufficiently different from low-energy photon-emitting brachytherapy seeds such that the AAPM may apply different dosimetry prerequisites. These AAPM guidelines are under development.

26.2.2 Devices

There are currently two commercial electronic brachytherapy devices approved for use by the FDA: the Zeiss INTRABEAM and Xoft Axxent. Both systems also have CE Mark for European distribution. While other electronic brachytherapy systems have been announced, they have not been commercialized, for example, the Advanced X-Ray Technologies, Inc., needle-like source (Gutman et al. 2004) or the Nucletron source (Nucletron 2008). As such, the INTRABEAM and Axxent platforms are discussed in more detail in the following sections.

26.2.2.1 INTRABEAM

The INTRABEAM system (see Figure 26.1) was initially developed by the Photoelectron Corporation (Dinsmore et al. 1996) to perform radiosurgery procedures. In 1997, the FDA approved the INTRABEAM system for use in intracranial irradiation. The 510(k) summary stated that the device was substantially equivalent to an HDR ^{192}Ir source or stereotactic radiosurgery with a Leksell Gamma Knife (Elekta, Inc.) (FDA 2011a). While the INTRABEAM and Gamma Knife were both designed for treatment of intracranial lesions, it is interesting to note that the FDA determined that brachytherapy using this system was substantially equivalent to stereotactic radiosurgery. The indications for use were modified in 1999 to allow for intracavitary treatment or irradiation of the tumor bed or tumor margin in intraoperative radiation therapy (IORT). Distribution of the INTRABEAM product line was later performed by Carl Zeiss Surgical GmbH. In 2009, the FDA approved the INTRABEAM for use in partial breast irradiation.

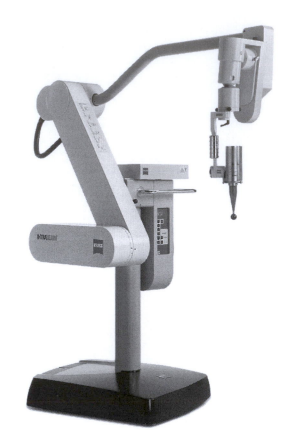

FIGURE 26.1 Carl Zeiss INTRABEAM x-ray brachytherapy system. (Courtesy of Carl Zeiss, Oberkochen, Germany.)

INTRABEAM uses a miniature electron gun to direct electrons onto a target located within a treatment probe that can be placed directly in tissue or at the center of an applicator. The electrons are generated outside of the treatment probe and directed into a 100-mm-long, 3.2-mm-diameter drift tube. At the distal end of the drift tube is a beryllium hemisphere coated with an approximately 1-µm-thick gold target (Beatty et al. 1996). The gold must not only be thick enough to stop a majority of the electrons but also be thin enough not to significantly attenuate clinically useful photons (>10 keV). The outside of the tube is coated with a nickel film to which a titanium nitride layer is adhered, with these layers providing a biocompatible surface as well as additional low-energy photon attenuation. The electron beam is spread across the surface of the gold hemispherical target to improve dose isotropy. For intracavitary breast treatments, the probe is placed within a solid plastic (polyetherimide) sphere of varying diameter (1.5 to 5.0 cm) (Eaton and Duck 2010).

The INTRABEAM operates at 30, 40, and 50 kVp at selected currents from 5 to 40 µA. Beatty et al. (1996) reported a dose rate of 2.5 cGy/s at 1-cm depth from a 3-mm-diameter treatment probe in water for 40 kVp and 40 µA. For intracavitary treatment following breast lumpectomy, a dose of 5 to 7 Gy at 1-cm depth correlated to a dose of 18 to 20 Gy at the spherical applicator surface (Vaidya et al. 2006). Approximate treatment times are 25 and 38 min for a 3.5- and 5.0-cm-diameter spherical applicator

(Vaidya et al. 2002). At 50 kVp, dose to breast tissue is 20% lower than dose to liquid water based on differences in the mass energy absorption coefficients (Melhus and Rivard 2006; Hubbell and Seltzer 1995). The INTRABEAM x-ray target has a lifespan of approximately 100 h before it needs to be replaced.

Dosimetry data published for the INTRABEAM system have typically not followed any brachytherapy dosimetry formalism. Initial dosimetry data published by Beatty et al. (1996) measured the beam half value layer (HVL), emitted photon spectra, 1D depth dose, and dose anisotropy about the target. Eaton and Duck (2010) revisited characterization of dose distributions for the INTRABEAM system by evaluating source HVL and depth dose characteristics for the 1.5-, 3.0-, 3.5-, and 5.0-cm-diameter applicators. They followed the Institution of Physics and Engineering in Medicine and Biology protocol for kilovoltage x-ray sources (Klevenhagen et al. 1996), which converts absorbed dose to air (measured in air) to absorbed dose to water (in water) using appropriate air-kerma calibration and backscatter factors. Percentage depth dose curves were in good agreement with the manufacturer's recommendations, showing <9% discrepancy in absolute dose for depths < 1 cm and <5% discrepancy in relative dose for depths > 1 cm (Eaton and Duck 2010). The largest differences were for the smallest applicator for which measurements were the most sensitive to errors in source-to-detector positioning.

26.2.2.2 Axxent

Xoft, Inc. developed a miniature x-ray source (see Figure 26.2) in the late 1990s with the intended purpose for IVBT. After the widespread introduction in the United States of drug-eluting stents in 2003, IVBT was essentially discontinued in most clinics. In 2005, the FDA approved the Axxent for balloon-based brachytherapy in support of partial breast irradiation. Since that time, the FDA has expanded approval to other disease sites. Specifically, endometrial therapy using vaginal cylinders was approved in 2008 (see also Dickler et al. 2010), and applicators for skin or IORT were approved in 2009. In 2010, Xoft, Inc. was acquired by iCAD, Inc. (Nashua, NH).

The Axxent source is a disposable, single-use miniature x-ray tube at the end of a flexible, water-cooled sheath attached to a control console. Cooling water is allowed in close contact with the target anode, permitting high output with heat dissipation. The tube can be operated between 20 and 50 kVp. At a standard operating voltage of 50 kV and tube current of 300 μA, the air-kerma strength is 1.4 kGy/h at 1 cm, which can be compared to 0.4 kGy/h at 1 cm for a 370-GBq (10 Ci) HDR ^{192}Ir source. An Axxent tube can continuously operate for dozens of hours, although sources generally do not last as long under noncontinuous operation. A separate controller unit containing a precision motor is available; this unit translates the source through a channel or applicator to provide a shaped dose distribution. With modest use factors and short irradiation times, it may be possible to treat patients in an unshielded room, for example, an operating room, or to use only mobile shields. For treatment planning, the well-known AAPM Task Group 43 (TG-43) report on dosimetry formalism (Nath et al. 1995; Rivard et al. 2004a)

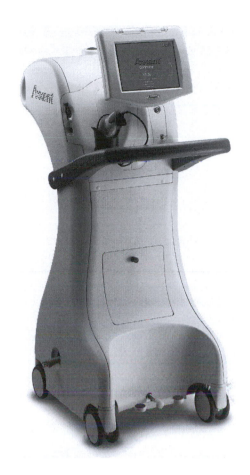

FIGURE 26.2 Xoft Axxent electronic brachytherapy system. (Courtesy of iCad, Nashua, NH.)

has been successfully applied to the Axxent electronic brachytherapy source from dose distributions obtained using Monte Carlo (MC) simulations and ionization chamber measurements in phantoms (Rivard et al. 2006). In 2011, NIST established a new calibration facility for the Axxent source to measure reference air-kerma rates and photon spectra. The process of associating well-ionization chamber response to air kerma is under investigation and will be disseminated to the AAPM Accredited Dosimetry Calibration Laboratories (ADCLs).

26.2.3 Guidance for Electronic Brachytherapy

Through task groups within the Emerging Technology Committee, the American Society for Radiation Oncology (ASTRO) issues guidance for early adopters of new technologies. In 2010, ASTRO published a report on electronic brachytherapy (Park et al. 2010) with the intent to provide an overview of delivery systems, a review of existing and potential clinical applications, comparison of electronic brachytherapy to competing modalities, and a summary of safety issues. ASTRO included the INTRABEAM and Axxent systems in their evaluation.

Further, the AAPM has formed TG-182 on recommendations for electronic brachytherapy quality management. This task group aims to cover all applicable anatomical sites and treatment systems in their report. The task group charge is to:

(1) Review manufacturer-suggested QA procedures
(2) Develop a risk-based set of QM procedures, both for the treatment systems and for patient treatment plans (including techniques, frequencies, and tolerances, along with statements on required training, connectivity with computer networks, and licensing/regulations)
(3) Suggest designs for needed QA and treatment delivery tools that do not yet exist
(4) Suggest quality improvement procedures

26.3 Microsphere Brachytherapy

Microsphere brachytherapy is a process in which the radioactivity is delivered within a carrier that is microscopic, for example, on the order of tens of micrometers. The average diameters range from 20 to 60 μm. To some extent, microsphere brachytherapy is closely related to nuclear medicine techniques that deliver therapeutic radiopharmaceuticals, such as iodine-based thyroid therapies for Graves' disease. In comparison to traditional brachytherapy methods employing readily identified, approximately centimeter-sized sealed sources, dose distributions cannot be calculated using the standard AAPM TG-43U1 report for microsphere brachytherapy. Further, it is not possible to confirm the total number or absolute activity of sources administered during a procedure, and most dosimetry methods in microsphere brachytherapy can only approximate the delivered dose.

26.3.1 Overview

Microsphere brachytherapy was first applied in animal and human systems in the early 1950s (Kennedy et al. 2004). Many different materials such as glass, polylactic acid, resin, micropolymer, plastic, ceramic, and starch have been employed in the construction of the microspheres. For current microsphere brachytherapy, the radioactivity is impregnated, doped, infused, or attached to the microsphere. The large majority of microsphere brachytherapy is performed in the liver, where the technique utilizes the vascular flow of blood. In a healthy liver, the hepatic artery supplies only 20% to 30% of the blood. However, in metastatic tumors and in hepatocellular carcinoma, 80% to 100% of the blood is delivered by the hepatic artery. Microspheres with impregnated radionuclides (generally beta emitters) can be delivered intravascularly through the hepatic artery as a bimodal therapy: local irradiation of the tumor and embolization of tumor vasculature to enhance treatment efficacy. If produced at an optimal size, microspheres are able to gain entry to the malignancy and become permanently lodged. Specifically, they should not be too large to prevent diffusion in the tumor but must be larger than about 10 μm to prevent entry into venous capillaries (Kennedy et al. 2004).

Microsphere brachytherapy treatment is always preceded by a preliminary angiographic procedure to evaluate whether the patient has favorable blood circulation for the procedure. Inappropriate blood circulation could allow for significant deposition of microspheres outside of the liver with a corresponding dosage to nontarget tissues. A catheter is percutaneously inserted into the femoral artery and directed to the hepatic artery by an interventional radiologist using fluoroscopy. Several arteriograms are performed to assess liver blood flow and verify that the patient is a candidate for the technique. Following these arteriograms, ^{99m}Tc-labeled macroaggregated albumin (MAA) is infused into the proposed treatment site as a surrogate for the ^{90}Y microspheres. Subsequent whole-body imaging with a gamma camera, which is generally performed in nuclear medicine, determines the fraction of injected radionuclide (^{99m}Tc-MAA) that was shunted to the lungs or other critical organs during the preliminary study. Due to the high risk of radiation pneumonitis, delivering 20% or more of the activity to the lungs is a contraindication for microsphere brachytherapy. This preliminary angiographic procedure may also be used to identify the ratio of target healthy liver compared to healthy liver or the proportion of activity that might shunt to other gastrointestinal organs at risk.

26.3.2 Yttrium-90 Microspheres

Since the 1980s, nearly all microsphere brachytherapy of the liver has been performed using ^{90}Y, although numerous radionuclides have been evaluated for this indication over the past 60 years, including ^{32}Cr, ^{46}Sc, ^{51}Cr, ^{57}Co, ^{63}Zn, ^{113}Tn, ^{125}I, ^{140}Ba, ^{153}Gd, ^{153}Sm, ^{166}Ho, and ^{198}Au (Kennedy et al. 2004). ^{90}Y is a pure beta emitter; the maximum beta energy is 2.3 MeV with a mean energy of 0.9 MeV. The half-life ($t_{1/2}$) is 64 h (NUDAT 2008).

Brachytherapy allows for excellent dose conformality and low healthy liver doses. External beam radiotherapy or stereotactic body radiosurgery present challenges as healthy liver tissue tolerance is approximately 30 to 35 Gy to the whole liver (Choi et al. 2008), and a therapeutic dose may require as much as 70 Gy. Unlike encapsulated sources that follow the AAPM TG-43 report dosimetry, ^{90}Y microsphere brachytherapy dosimetry follows the Medical Internal Radiation Dose (MIRD) methodology commonly employed in nuclear medicine (Loevinger et al. 1991). If one assumes a uniform distribution of radioactivity in the target tissue and local absorption of radiation dose (i.e., ignoring bremsstrahlung losses), the dose rate $\dot{D}$ can be calculated using Equation 26.1:

$$\dot{D} = k\frac{A}{m}\langle E \rangle. \tag{26.1}$$

In the equation, A is the radionuclide activity (Bq), m is the mass of target tissue (kg), $\langle E \rangle$ is the average energy released per decay [MeV/(Bq s)], and k represents a constant that yields preferred units (e.g., J/MeV to provide Gy/s). Doses to the whole liver, lungs, or other gastrointestinal organs are estimated using the result of the preliminary ^{99m}Tc-MAA imaging study. Some institutions have performed an MAA-SPECT study in combination with the lung-shunt study and convolved the SPECT signal distribution with Monte Carlo–calculated dose kernels for ^{90}Y.

These and other quantitative dosimetry methods for microsphere brachytherapy are an area of active research.

26.3.2.1 TheraSphere

Nordian, Inc. (Kanata, Canada) manufactures and distributes TheraSpheres, an insoluble glass microsphere (20 to 30 μm diameter) with imbedded ^{90}Y. Glass microspheres are embedded with stable ^{89}Y and subsequently activated to ^{90}Y in a nuclear reactor (Giammarile et al. 2011). The FDA has approved TheraSpheres for clinical use under the 1999 Humanitarian Device Exemption (HDE) Act. Specifically, the FDA approval clears use of TheraSpheres in the United States for "radiation treatment or as a neoadjuvant to surgery or transplantation in patients with unresectable hepatocellular carcinoma who can have placement of appropriately positioned hepatic arterial catheters" (FDA 2011b). HDE designation means that the device cannot be sold for profit, except under specific conditions, and use of the device requires facility-specific Institutional Review Board approval except for specific emergencies. The procedure currently qualifies for reimbursement through public and private payers. Individual glass TheraSpheres contains approximately 2.5 kBq per sphere, with total activities between 3 and 20 GBq available (Dezarn et al. 2011; Giammarile et al. 2011).

26.3.2.2 SIR-Spheres

SIRTex Medical, Inc. (Lane Cove, New South Wales, Australia) manufactures SIR-Spheres, a 20 to 60 μm resin micropolymer sphere tagged with ^{90}Y. SIR-Spheres were approved for use by the FDA in 2002 for therapeutic treatment of colorectal tumors that have metastasized to the liver. SIR-Spheres are administered with adjuvant intrahepatic artery chemotherapy and are contraindicated for patients that have had prior external beam radiotherapy to the liver. An individual resin SIR-Spheres is impregnated with an activity of approximately 40 to 70 Bq per sphere; spheres are available in 3-GBq batches. Because of the lower activity per sphere, more SIR-Spheres are delivered in comparison to a typical TheraSphere procedure, which can enhance the embolic effect of the microspheres (Giammarile et al. 2011).

26.3.3 Guidance for Microsphere Brachytherapy

The AAPM created TG-144 in 2006 to make recommendations on dosimetry and procedures for ^{90}Y microsphere brachytherapy for liver cancer. The charge of the task group was to:

(1) Review current microsphere products and the specific protocols associated with the use of each
(2) Develop a consensus on a minimum set of microsphere-specific quality assurance tasks
(3) Review and evaluate current (and potential) methods and models for radiation dose calculations to liver and other organs from ^{90}Y microspheres that are useful in clinical practice for guiding and evaluating individual patient dosimetry
(4) Recommend a traceable ^{90}Y standard and a method for relating it to local dose calibrators (a NIST calibration standard exists for free ^{90}Y, and some work has been completed toward transferring this standard to ^{90}Y contained within a given microsphere)
(5) Develop a consensus regarding multiple imaging modalities used in planning and monitoring ^{90}Y radioactive microspheres in liver brachytherapy
(6) Consider alternate calibration standards for ^{90}Y microspheres
(7) Discuss and quantify uncertainties associated with this treatment method

The TG-144 report (Dezarn et al. 2011) has been completed and approved by the AAPM and was published in 2011. The European Association of Nuclear Medicine published procedure guidelines (Giammarile et al. 2011) for the intra-arterial treatment of liver cancer and liver metastases using radioactive compounds. The procedure guidelines include ^{131}I-Lipiodol and ^{90}Y-based therapies of liver malignancies and provided a summary of patient selection criteria, modality characteristics, facility requirements, dosimetry methods, advantages, disadvantages, and areas of future research.

26.4 Eye Plaque Brachytherapy

26.4.1 Overview

The most common eye tumor in the United States is ocular melanoma, including choroidal and uveal melanomas, with approximately 2,570 new cases per year (American Cancer Society 2011). Early treatments for ocular melanoma involved implantation of ^{222}Rn sources within the tumor. Brachytherapy applied to the surface of the globe, plaque brachytherapy or plesio-brachytherapy, has been used with a variety of radionuclides, for example, ^{60}Co, ^{103}Pd, ^{106}Ru/^{106}Rh, ^{125}I, ^{192}Ir, and ^{198}Au (Nag et al. 2003). Currently, the most frequently used radionuclides in addition to ^{125}I are ^{106}Ru/^{106}Rh (Taccini et al. 1997), ^{103}Pd (Finger et al. 2009), and ^{131}Cs (Leonard et al. 2011).

26.4.2 Collaborative Ocular Melanoma Study Group

Historically, the standard-of-care for ocular melanoma had been enucleation. However, ophthalmologists desired a treatment technique to preserve the eye yet maintain overall survival rates. In 1987, a 10-year multi-institutional randomized clinical trial was started by the U.S. National Eye Institute. The purpose was to formally evaluate the role of radiation therapy in the treatment of ocular melanoma. This clinical trial was called the Collaborative Ocular Melanoma Study (COMS). There were two branches of this study:

(1) Evaluate the survival benefit for large-sized tumors (i.e., >10 mm in apical height) using external beam radiation therapy (EBRT) prior to eye removal.
(2) Evaluate survival rates for medium-sized tumors (i.e., between 2.5 and 10 mm in apical height) between eye plaque brachytherapy and enucleation.

COMS ultimately enrolled over 1300 patients who were followed for at least 5 years, with up to 15 years of follow-up for some patients. The two study branches ultimately demonstrated (1) no survival benefit for large-sized tumors receiving EBRT prior to enucleation (Hawkins 2004), and (2) equivalent 5-year survival rates for medium-sized tumors receiving either eye plaque brachytherapy or enucleation (COMS 2006). This latter branch study generated enthusiasm for eye plaque brachytherapy given the clear equivalence demonstrated in the trial and quality of the data.

Brachytherapy was administered in the COMS medium-sized tumor trial using low-dose-rate (LDR) [125]I sources that were loaded into an elastic polymer insert that was glued within a gold-alloy plaque. The gold-alloy plaque served to shield the healthy eye structures behind the plaque and to collimate the radiation toward the lesion. Upon measuring the source strength, the sources are inserted into the seed holder insert. Then, the insert is placed within the gold-alloy plaque concavity and glued in place. The entire assembly is sterilized for subsequent suturing to the eye by the ophthalmologist. COMS plaque diameters of 12 to 20 mm in 2-mm increments were produced to treat medium-sized tumors while providing a radial margin of 2 to 3 mm around the tumor basal extent. Though not included in the COMS trial, 10- and 22-mm plaque diameters were later developed to increase the range of treatable lesions. After COMS, additional radionuclides beyond ^{125}I were considered, with some still in clinical use.

26.4.3 Eye Plaque Dosimetry

Initially, the prescribed dose in the COMS protocol was 100 Gy delivered to the tumor apex. Upon accounting for Ti K-edge characteristic x-rays (about 4.5 keV) and their influence on brachytherapy source air-kerma strength measurements (Kubo et al. 1998a) with negligible contributions beyond the 1-mm elastic polymer insert, this prescription dose was revised to 85 Gy. The initial COMS protocol allowed dose rates between 0.42 and 1.05 Gy/h. However, the 2003 American Brachytherapy Society (ABS) report increased the lower dose rate limit to 0.60 Gy/h due to reports of improved control rates with shorter implant durations (Nag et al. 2003). Critical structures included the optic nerve, retina, macula, fovea (vision center of macula with a high density of cones to generate the central visual field), lens, and sclera (fibrous outer layer of the eye).

COMS utilized a brachytherapy dose calculation approach similar to the AAPM TG-43 dosimetry formalism (Nath et al. 1995; Rivard et al. 2004a). This dosimetry formalism was based on the following assumptions: the 1D point source formalism, limited dose anisotropy corrections, the shielding/scattering effects of the gold-alloy plaque being ignored, air/tissue interface effects being neglected, and the elastic polymer insert being assumed to be water-equivalent. The validity of these assumptions was subsequently tested using measurements and Monte Carlo–based radiation transport simulations by several research groups. These issues have been recently examined (Rivard et al. 2011) and indicate that dose calculation errors range from 10% to 20% for dose prescriptions and factors of 10 at the plaque periphery. Modern dose calculation methods using Monte Carlo–based radiation transport simulations (Melhus and Rivard 2008; Thomson et al. 2008) or TG-43 hybrid techniques (Rivard et al. 2009a; Yang et al. 2011) will improve the capabilities of brachytherapy treatment planning systems (Rivard et al. 2009b, 2011).

26.4.4 Guidance for Eye Plaque Brachytherapy

Because eye plaque brachytherapy was the focus of a multicenter National Institutes of Health clinical trial and has been practiced for decades, there is more guidance available for this modality in comparison to the other modalities in this chapter. The COMS Manual of Procedures includes a protocol for application of COMS plaques in accordance with the clinical trial (COMS 2011). The ABS published a comprehensive review of clinical aspects of eye plaque brachytherapy (Nag et al. 2003) and provided recommendations for clinical practice. The dosimetry calculation formalism employed by the COMS was revised by the AAPM in the TG-43 update (Rivard et al. 2004a) to account for changes in the recommended brachytherapy dosimetry formalism. In addition, the AAPM established TG-129 in 2006 to investigate eye plaque dosimetry. The charge of the task group is to:

(1) Review and assess the literature on eye plaque dosimetry
(2) Verify or update the dosimetry calculations in articles
(3) Recommend consensus dosimetry parameters including heterogeneity correction factors for the appliances and the effects of lack of backscatter as a function of the plaque position

The AAPM TG-129 has prepared a report issuing dosimetric standards and quality assurance guidelines (Chiu-Tsao et al. in press) and is soon expected to be published. Toward evaluating charge (3), a multi-institutional dosimetry study was performed for a 12-mm-diameter COMS plaque loaded with common ^{125}I and [103]Pd sources using three commercial treatment planning systems and two Monte Carlo codes (Rivard et al. 2011). This publication may be used by clinicians to benchmark independent eye plaque dose calculations.

26.5 Intravascular Brachytherapy

IVBT was an important application of radiotherapy to address a serious nonmalignant disease (restenosis following percutaneous angioplasty of cardiac arteries) for about one decade starting in the mid-1990s. IVBT involves temporary placement of brachytherapy sources within cardiac stents in which *in-stent restenosis* has occurred or to prevent restenosis following percutaneous stent placement in atherosclerotic patients. Radiation delivered during or following angioplasty improved restenosis occurrence rates (to less than 10%) in comparison to unirradiated lesions (35% to 40%) as reported by AAPM TG-60 (Nath et al. 1999). While many radionuclides have been considered for IVBT, only a few have been used extensively for clinical purposes.

26.5.1 Radionuclide Energies and Radiation Ranges

Typical lesion thickness for IVBT is a few millimeters, yet there are healthy tissues immediately peripheral to the lesion that require protection from radiation. Due to the short range of their beta emissions, ^{32}P and ^{90}Sr/^{90}Y have the greatest potential for avoiding dose to healthy tissues beyond the target distance of 0.2 cm. The maximum beta energy of ^{32}P is 1.7 MeV, while those of ^{90}Sr and ^{90}Y are 0.5 and 2.3 MeV, respectively. Higher beta energies provide deeper maximum range, with 0.5, 1.7, and 2.3 MeV corresponding to 0.2, 0.7, and 1.1 cm, respectively. Given the 0.2-cm target distance, it may appear that ^{32}P is more favorable than ^{90}Sr/^{90}Y for IVBT. However, for treatment planning calculations, ^{32}P has a $t_{1/2}$ = 14 days and is more sensitive to radioactive disintegration calculations than ^{90}Sr/^{90}Y. For example, ^{90}Sr has $t_{1/2}$ = 29 years and beta decays to ^{90}Y with $t_{1/2}$ = 64 h, which is also a beta emitter. Consequently, a ^{90}Sr/^{90}Y source decays about 0.2% in 1 month, while ^{32}P decays by 77% over the same duration. The other IVBT radionuclide used extensively in patients is ^{192}Ir. It has $t_{1/2}$ = 74 days and emits photons with an average energy of 0.4 MeV. Its dose rate diminishes approximately as $1/r^2$ and consequently does not terminate as quickly with increasing distance in comparison to beta-emitting radionuclides.

26.5.2 IVBT Source Dimensions

Due to the size of cardiac arteries, IVBT source dimensions need to be small; smaller sources have lower dose rates for a given radionuclide having a fixed specific activity. Another clinical constraint is to not occlude the vessel beyond several minutes to deliver radiotherapy. Therefore, radionuclides with high specific activity and consequently smaller dimensions are ideal for IVBT. Given that higher specific activities are associated with the shorter-lived radionuclides, smaller sources are, in general, produced using shorter-half-life radionuclides. For example, the outer diameters for the ^{32}P and ^{90}Sr/^{90}Y sources used for IVBT were 0.5 and 1.2 mm, respectively, compared with ^{192}Ir at 0.8 mm (Chiu-Tsao et al. 2007). Due to the intraluminal diameters typical in the treatment of coronary disease, all three sources are feasible, and devices for clinical use have been fabricated. IVBT sources are positioned within cardiac catheters that further increase the total diameter of the device. As a result, the ^{90}Sr/^{90}Y sources are limited to 3.5 F catheters that could not be positioned at all cardiac sites and for all patients. To facilitate a wide scope of use, the manufacturer supports clinical applications to treat endovascular disease such as in peripheral arteries using femoropopliteal balloon angioplasty (Nath et al. 1999).

26.5.3 Current Status of IVBT

With more than 0.5 million angioplasties performed annually in the United States, IVBT was an important brachytherapy option for the treatment of a nonmalignant proliferative disorder, with immense public health impact. The procedure was performed in about 100 centers in the United States at its peak (Devlin 2003; Tripuraneni 2003), delivering a therapeutic dose of 18.4 Gy at 2 mm in approximately 4 min. The logistics of setting up an IVBT program were complex, requiring input from interventional radiologists, cardiologists, radiation oncologists, medical physicists, and health physicists. Also, the FDA ruled that this radiotherapy modality presented a significant risk to the public and required further clinical trials and long-term follow-up studies. Meanwhile, pharmaceutical firms also recognized the market potential and had a breakthrough with development of drug-eluting stents using paclitaxel. The FDA approved drug-eluting stents for de novo lesions, not for in-stent restenosis. Off-label use for in-stent restenosis became prevalent, in large part to avoid use of radioactive materials and the requirement for a multidisciplinary team to treat what had largely been patients under the control of the cardiologist (Thomadsen et al. 2010). Currently, there are less than a dozen centers in the United States still using IVBT, most using IVBT for in-stent restenosis following a failed coronary stent. Long-term clinical results using drug-eluting stents for in-stent restenosis are favorable (Dangas et al. 2010); there will still be a need for IVBT for patients not eligible for coronary artery bypass surgery.

There is only one commercially available IVBT system, the Novoste Beta-Cath System (see Figure 26.3) from Best Vascular, Inc. (Norcross, GA). The Beta-Cath was CE-marked in 1999 and received FDA approval the following year. A coil containing small ^{90}Sr/^{90}Y source capsules is hydraulically propelled from a shielded, handheld treatment device into position within a

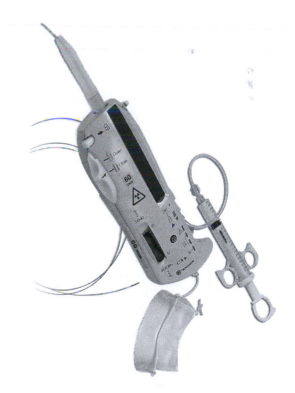

FIGURE 26.3 Best Novoste Beta-Cath IVBT system. (Courtesy of Best Vascular, Inc., Norcross, GA.)

single-use catheter. After the therapeutic dose has been delivered, the source is hydraulically returned to the treatment device.

26.5.4 Guidance for IVBT

AAPM TG-60 (Nath et al. 1999) was initiated in 1995 to evaluate and summarize the state of the art in IVBT physics. The report evaluated radionuclide sources employing ^{32}P, ^{90}Sr/^{90}Y, and ^{192}Ir and provided consensus dosimetry data for use with each system. The TG-60 report was complemented by TG-149 (Chiu-Tsao et al. 2007), which provided updated dosimetry data for three commercial IVBT systems employing the aforementioned radionuclides. The updated TG-43 dosimetry formalism (Rivard et al. 2004b) was adopted for TG-149, with a modification for beta particles from the ^{32}P-based delivery system.

In 1999, the ABS endorsed the AAPM TG-60 recommendations for IVBT dosimetry and addressed the lack of clinical guidance (Nag et al. 1999). This review summarized IVBT clinical trials performed or in process, examined patient selection criteria, listed indications and contraindications, and suggested areas of future research.

26.6 Neutron-Emitting Sources

Therapies using neutrons and high linear energy transfer (LET) radiations have shown certain desirable characteristics and advantages over photon-based treatments. These advantages include limited particle range, reduced oxygen effect, decreased cell-cycle dependence, and increased RBE. While radionuclide sources emitting photon and/or beta radiation are common, it is unusual for a radionuclide to emit high-LET radiation and also have sufficiently long half-life and specific activity to be clinically useful in brachytherapy. There are many long-lived alpha emitters; however, they are not amenable for therapeutic use in encapsulated form due to their short range and rapid attenuation by any encapsulation. Single-proton or double-proton radioactivity was confirmed by experiment in the early 1980s and 2002, respectively. Proton decay modes require extremely proton-rich nuclei that are generally formed during high-energy (GeV) spallation reactions, $t_{1/2} \sim 10^{-15}$ s, with longer-lived states approaching ~250 μs (Blank 2009). In any case, protons do not have the same radiobiological advantages as neutrons. There are a handful of neutron-emitting radionuclides with sufficient half-life and specific activity for use in brachytherapy. With $t_{1/2} =$ 2.645 years, ^{252}Cf balances high specific activity with a clinically useful half-life and is the most widely studied neutron-emitting radionuclide.

26.6.1 Overview

^{252}Cf is a manmade trans-plutonium radionuclide that undergoes alpha decay; however, a nonnegligible fraction of nuclear transitions (3.09%) undergo spontaneous fission (NUDAT 2008) where neutrons are emitted. ^{252}Cf neutrons are similar in energy to fission neutrons in a nuclear reactor, having a mean energy of

2.1 MeV, and are emitted at a rate of 2.314×10^9 s^{-1} mg^{-1} (Martin et al. 1997). The relatively long half-life of ^{252}Cf compared to ^{192}Ir, currently used in most commercial HDR brachytherapy systems, is desirable because it allows for relatively infrequent source exchanges and the potential for related cost savings (Rivard et al. 2004b).

Oak Ridge National Laboratory (ORNL) manufactures californium isotopes from highly enriched curium (Cm) targets (primarily ^{244}Cm with significant ^{246}Cm and americium isotopes) irradiated in the high-flux isotope reactor for up to 1 year (Martin et al. 2000). Combinations of neutron activation and beta decay in the 0.1-kg target leads to successively higher atomic numbers. For example, neutron activation of ^{248}Cm produces ^{249}Cm ($t_{1/2} = 64$ min), beta decays to ^{249}Bk ($t_{1/2} = 330$ days), and beta decays to ^{249}Cf ($t_{1/2} = 351$ years), and sequential neutron capture reactions produce californium isotopes with atomic numbers from 250 to 254 (NUDAT 2008). Due to the lengthy target irradiation time needed to produce ^{252}Cf and the extensive radiochemistry needed to separate californium isotopes from the curium target, ^{252}Cf is both extremely rare and expensive. As of this writing, the production of ^{252}Cf is approximately 3 mg every 3 years at about $0.1 million/mg. Rivard et al. (2004b) have estimated that hospital costs for an HDR ^{252}Cf brachytherapy program would not exceed 30% more than the costs for a standard-of-care HDR ^{192}Ir brachytherapy program.

Clinical use of the neutron-emitting radionuclide ^{252}Cf was first proposed by Schlea and Stoddard (1965). They noted that early radiobiological studies of external beam neutron therapy demonstrated effectiveness in anoxic tumors that were typically resistant to conventional photon therapies. They also described the advantage of intracavitary and interstitial brachytherapy treatment over external beam neutron therapy, which generally includes higher radiation doses to healthy tissue (Schlea and Stoddard 1965). The first medical source containing ^{252}Cf was modeled after a radium needle (Maruyama et al. 1997).

Brachytherapy dosimetry for ^{252}Cf sources was aggressively pursued in the early 1970s once medical sources were available. Calculations by Krishnaswamy (1972) used MC methods to derive a point source model for four different source types, including the applicator tube (AT) source with source length $L =$ 1.5 cm. Each point source was subsequently used in a numerical integration to calculate the dose to points up to 5 cm along and away from the source. Neutron kerma factors were not reported, and historical photon cross sections were employed. Colvett et al. (1972) followed with measurements using paired ion chambers (tissue-equivalent plastic and aluminum) in a large volume of tissue-equivalent liquid to simulate infinite scatter conditions. Measurements from 1 to 5 cm on the transverse plane agreed within 0.94 ± 0.02 with Krishnaswamy's calculations; however, the measurements did not agree favorably with other reported measurements. Differences between measurements were later attributed to differences in phantom size, which resulted in varying scatter conditions, and to differences in chamber or ^{252}Cf source calibration (Colvett et al. 1972; Windham et al. 1972). Windham et al. (1972) calculated dose distributions for [252]Cf sources using a 1D discrete ordinate code, 21 energy groups, and neutron kerma factors published by Ritts et al. (1969). Calculations by Windham et al. were in good agreement with the Krishnaswamy calculations (1.01 ± 0.03) and Colvett et al. measurements (0.97 ± 0.02) for $0.5 \leq r \leq 5.0$ cm on the transverse plane. Anderson (1973) examined these and six other publications (three calculations and three measurements) in a review article on [252]Cf dosimetry. The measurements of Colvett et al. were recommended for [252]Cf, although Colvett et al. provided photon data only for AT-type sources. The calculations of Krishnaswamy for non-AT-type needle sources were the recommended alternative. These data were reported as along-and-away tables up to 5 cm from the source center.

26.6.2 Early Work with [252]Cf Sources

Thousands of patients at various centers worldwide have been treated over the past 40 years with [252]Cf-based brachytherapy for diverse lesions from head to toe (Maruyama et al. 1997). The mixed neutron and photon radiation has been shown to be particularly effective for large tumors and traditionally radioresistant tumors, such as melanomas and late-stage cervical carcinomas. This increased effectiveness is attributed to a number of radiological characteristics of neutrons, such as increased RBE, decreased oxygen effect, and decreased cell-cycle dependence. Clinical observations include acute tissue effects comparable to photon-based brachytherapy, dose rate independence, local tumor control, and rapid tumor regression with an RBE of 6 for LDR brachytherapy and an RBE of 3 for HDR brachytherapy. In the United States, Maruyama treated many disease sites and published nearly 100 clinical papers on the subject. Most notably, cervical cancers exhibited positive results for advanced, late-stage (III–IV) bulky tumors, including a 54% 5-year survival for early (i.e., before photon therapy) [252]Cf implants compared to 12% 5-year survival with conventional therapy for stage IIIB tumors (Maruyama et al. 1985). Similarly, Tačev et al. (2003) noted an 18.9% increase in 5-year survival for stage IIB and IIIB cervical cancers compared to conventional photon therapy due to a significant decrease in local relapse. Similar results were obtained after 12-year follow-up (Tačev et al. 2004). Positive local control was observed for several diseases and sites, including traditionally radioresistant melanomas that achieved complete local control with external [252]Cf plaque irradiation (Vtyurin et al. 1994). The reader is advised to study the work of Maruyama et al. (1997) for a detailed overview and summary of clinical results with [252]Cf-based neutron brachytherapy for head-and-neck, skin, gynecological, rectal, esophageal, prostate, and brain tumors.

26.6.3 Recent Advances

The most recent determinations of [252]Cf neutron dose distribution near an LDR AT-type source were published by Yanch and Zamenhof (1992) and by Rivard (2000). Yanch and Zamenhof calculated along-and-away in water tables using MC methods

for AT-type sources. Absorbed-dose-rate (cGy/h) tables were provided for the neutron dose rate and the total dose, including neutron dose, primary photon dose, and secondary photon dose from neutron capture in water. Rivard (2000) adapted the TG-43 formalism for a generalized source of variable active length and provided brachytherapy dosimetry parameters for the neutron dose component. Toward converting physical dose to biologically equivalent dose, RBE factors for HDR [252]Cf neutrons were estimated from accelerator-produced neutron data by Rivard et al. (2005). The RBE for HDR [252]Cf neutrons was observed to vary as a function of fraction dose, with values of 3.2 and 2.5 for fraction doses of 0.5 and 2.0 Gy, respectively.

[252]Cf brachytherapy in the United States was applied using manually loaded LDR sources delivering below 2 Gy/h to the prescription point during patient treatment (Maruyama et al. 1997). Both manual loading and LDR treatment increase radiological risk to the brachytherapist and support personnel who must attend to patient needs during long treatment sessions. Considering these issues, there was interest to employ [252]Cf brachytherapy in the HDR regime. In addition, remote afterloading technology, which is well documented and successful for HDR ^{192}Ir brachytherapy, was being considered for [252]Cf (Martin et al. 1997). Measurement of the mixed-dose distribution about a clinical HDR [252]Cf source has not been attempted because sources comparable in size and geometry to conventional HDR ^{192}Ir sources have become possible only due to advances in radiochemistry (Martin et al. 1997). Furthermore, while radiochemistry now allows an HDR source containing >1 mg/mm[3] of [252]Cf, such a source has not yet been constructed.

As the main supplier of californium isotopes, ORNL is working to produce an HDR [252]Cf brachytherapy source (> 1.0 mg) that would be compatible with the HDR ^{192}Ir brachytherapy remote afterloading environment. As of this writing and to the best knowledge of the authors, there is no institution in the Western Hemisphere that utilizes [252]Cf clinically. Remotely afterloaded [252]Cf brachytherapy has been performed in China since 1999 using the Linden Neutron Knife remote afterloader (Tang et al. 2002); however, after 10 years of clinical use, there have been no publications in Western medical or scientific journals.

28.6.4 Guidance for Neutron Brachytherapy

In 2007, the AAPM and the European Society for Therapeutic Radiology and Oncology formed TG-167 on requirements for investigational sources for interstitial brachytherapy. In this report, there is a section specific to [252]Cf brachytherapy describing the applicable FDA regulations, primary calibration standards, treatment planning possibilities, and the level of clinical experience in comparison to other brachytherapy modalities. At the time of writing, this report is expected to be published in late 2012. However, there are no commercially available [252]Cf brachytherapy sources in use or in active development in North America or Europe. In North America, only NIST offers [252]Cf source-strength calibrations; the ADCLs offer no [252]Cf services.

Clinicians pursuing this treatment modality should conservatively apply recommendations for conventional brachytherapy, such as TG-56 for general brachytherapy (Nath et al. 1997) and/or TG-59 for HDR brachytherapy (Kubo et al. 1998b). Mixed-modality photon and neutron dosimetry can be performed following the dosimetry formalism of TG-43 (see, e.g., Rivard 2000).

References

American Cancer Society. 2011. *Cancer Facts & Figures 2011.* Atlanta, GA.

Anderson, L.L. 1973. Status of dosimetry for ^{252}Cf medical neutron sources. *Phys Med Biol* 18:779–99.

Beatty, J., Biggs, P.J., Gall, K. et al. 1996. A new miniature x-ray device for interstitial radiosurgery: Dosimetry. *Med Phys* 23:53–62.

Blank, B. 2009. One- and two-proton radioactivity. In: *The Euroschool Lectures on Physics with Exotic Beams*, Vol. III. J.S. Al-Khalili and E. Roeckl (eds.), Lect Notes Phys 764. Springer, Berlin, pp. 153–201.

Chiu-Tsao, S.-T., Schaart, D.R., Soares, C.G., and Nath, R. 2007. Dose calculation formalisms and consensus dosimetry parameters for intravascular brachytherapy dosimetry: Recommendations of the AAPM Therapy Physics Committee Task Group No. 149. *Med Phys* 34:4126–58.

Chiu-Tsao, S.-T., Astrahan, M.A., Finger, P.T. et al. in press. Dosimetry of ^{125}I and ^{103}Pd COMS eye plaques for intraocular tumors: Report of Task Group 129 by the AAPM and ABS. *Med Phys.*

Choi, B.O., Choi, I.B., Jang, H.S. et al. 2008. Stereotactic body radiation therapy with or without transarterial chemoembolization for patients with primary hepatocellular carcinoma: Preliminary analysis. *BMC Cancer* 8:351.

Colvett, R.D., Rossi, H.H., and Krishnaswamy, V. 1972. Dose distributions around a californium-252 needle. *Phys Med Biol* 17:356–64.

COMS. 2006. The COMS randomized trial of iodine 125 brachytherapy for choroidal melanoma: V. Twelve-year mortality rates and prognostic factors: COMS report No. 28. *Arch Ophthalmol* 124:1684–93.

COMS. 2011. Publications of the COMS. Available at http://www.jhu.edu/wctb/coms/general/publicat/pubs.htm, last accessed October 5, 2011.

Dangas, G.D., Claessen, B.E., Caixeta, A., Sanidas, E.A., Mintz, G.S., and Mehran, R. 2010. In-stent restenosis in the drug-eluting stent era. *J Am Coll Cardiol* 56:1897–907.

Devlin, P.M. 2003. The future of coronary artery radiation therapy—The CART before the (unbridled) horse. *Brachytherapy* 2:73–4.

Dezarn, W.A., Cessna, J.T., DeWerd, L.A. et al. 2011. Recommendations of the American Association of Physicists in Medicine on dosimetry, imaging, and quality assurance procedures for ^{90}Y microsphere brachytherapy in the treatment of hepatic malignancies. *Med Phys* 38:4824–45.

Dickler, A., Mohamed Y.P., Thropay J.P., Bhatnagar, A., and Schreiber, G. 2010. Prospective multi-center trial utilizing electronic brachytherapy for the treatment of endometrial cancer. *Radiat Oncol* 5:67.

Dinsmore, M., Harte, K.J., Sliski, A.P. et al. 1996. A new miniature x-ray source for interstitial surgery: Device description. *Med Phys* 23:45–52.

Eaton, D.J. and Duck, S. 2010. Dosimetry measurements with an intra-operative x-ray device. *Phys Med Biol* 55: N359–69.

Food and Drug Administration. 2011a. Photoelectron Corporation photon radiosurgery system—K964947. Available at http://www.accessdata.fda.gov/cdrh_docs/pdf/K964947.pdf, last accessed October 5, 2011.

Food and Drug Administration. 2011b. TheraSphere®—H980006. Available at http://www.accessdata.fda.gov/scripts/cdrh/cfdocs/cftopic/pma/pma.cfm?num=H980006, last accessed October 5, 2011.

Finger, P.T., Chin, K.J., and Duvall, G. 2009. Palladium-103 ophthalmic plaque radiation therapy for choroidal melanoma: 400 treated patients. *Ophthalmology* 116:790–6.

Giammarile, F., Bodei, L., Chiesa, C. et al. 2011. EANM procedure guideline for the treatment of liver cancer and liver metastases with intra-arterial radioactive compounds. *Eur J Nucl Med Mol Imaging* 38:1393–406.

Gutman, G., Sozontov, E., Strumban, E., Yin, F.F., Lee, S.W., and Kim, J.H. 2004. A novel needle-based miniature x-ray generating system. *Phys Med Biol* 49:4677–88.

Hawkins, B.S. 2004. The Collaborative Ocular Melanoma Study (COMS) randomized trial of pre-enucleation radiation of large choroidal melanoma: IV. Ten-year mortality findings and prognostic factors. COMS report number 24. *Am J Ophthalmol* 138:936–51.

Hubbell, J.H. and Seltzer, S.M. 1995. *Tables of X-ray Mass Attenuation Coefficients and Mass Energy-Absorption Coefficients. NISTIR 5632.* National Institute of Standards and Technology, Gaithersburg, MD.

Kennedy, A.S., Nutting, C., Coldwell, D., Gaiser, J., and Drachenberg, C. 2004. Pathologic response and microdosimetry of ^{90}Y microspheres in man: Review of four explanted whole livers. *Int J Radiat Oncol Biol Phys* 60:1552–63.

Klevenhagen, S.C., Auckett, R.J., Harrison, R.M., Moretti, C., Nahum, A.E., and Rosser, K.E. 1996. The IPEMB code for practice for the determination of absorbed dose for x-rays below 300 kV generating potential (0.035 mm Al–4 mm Cu HVL; 10–300 kV generating potential). *Phys Med Biol* 41:2605–25.

Krishnaswamy, V. 1972. Calculated depth dose tables for californium-252 sources in tissue. *Phys Med Biol* 17:56–63.

Kubo, H.D., Coursey, B.M., Hanson, W.F. et al. 1998a. Report of the ad hoc committee of the AAPM Radiation Therapy Committee on ^{125}I sealed source dosimetry. *Int J Radiat Oncol Biol Phys* 40:697–702.

Kubo, H.D., Glasgow, G.P., Pethel, T.D., Thomadsen, B.R., and Williamson, J.F. 1998b. High dose-rate brachytherapy treatment delivery: Report of the AAPM Radiation Therapy Committee Task Group No. 59. *Med Phys* 25:375–403.

Leonard, K.L., Gagne, N.L., Mignano, J.E., Duker, J.S., Bannon, E.A., and Rivard, M.J. 2011. A 17-year retrospective study of institutional results for eye plaque brachytherapy of uveal melanoma using ^{125}I, [103]Pd, and [131]Cs and historical perspective. *Brachytherapy* 10:331–9.

Loevinger, R., Budinger, T., and Watson, E. 1991. *MIRD Primer for Absorbed Dose Calculations, Revised Edition.* Society of Nuclear Medicine, New York.

Martin, R.C., Laxson, R.R., Miller, J.H., Wierzbicki, J.G., Rivard, M.J., and Marsh D.L. 1997. Development of high-activity [252]Cf sources for neutron brachytherapy. *Appl Radiat Isot* 48:1567–70.

Martin, R.C., Knauer, J.B., and Balo, P.A. 2000. Production, distribution and applications of californium-252 neutron sources. *Appl Radiat Isot* 53: 785–92.

Maruyama, Y., Kryscio, R., Van Nagell, J.R. et al. 1985. Clinical trial of [252]Cf neutron brachytherapy vs. conventional radiotherapy for advanced cervical cancer. *Int J Radiat Oncol Biol Phys* 11:1475–82.

Maruyama, Y., Wierzbicki, J.G., Vtyurin, B.M., and Kaneta, K. 1997. Californium-252 neutron brachytherapy. In: *Principles and Practice of Brachytherapy.* S. Nag (ed.). Futura Publishing Co., Inc., Armonk, NY, pp. 649–87.

Melhus, C.S. and Rivard, M.J. 2006. Approaches to calculating AAPM TG-43 brachytherapy dosimetry parameters for ^{137}Cs, ^{125}I, ^{192}Ir, [103]Pd, and [169]Yb sources. *Med Phys* 33:1729–37.

Melhus, C.S. and Rivard, M.J. 2008. COMS eye plaque brachytherapy dosimetry simulations for [103]Pd, ^{125}I, and [131]Cs. *Med Phys* 35:3364–71.

Nag, S., Cole, P.E., Crocker, I. et al. 1999. The American Brachytherapy Society perspective on intravascular brachytherapy. *Cardio Rad Med* 1:8–19.

Nag, S., Quivey, J. M., Earle, J.D., Followill, D., Fontanesi, J., and Finger, P.T. 2003. The American Brachytherapy Society recommendations for brachytherapy of uveal melanomas. *Int J Radiat Oncol Biol Phys* 56:544–55.

Nath, R., Anderson, L.L., Luxton, G., Weaver, K.A., Williamson, J.F., and Meigooni, A.S. 1995. Dosimetry of interstitial brachytherapy sources: Recommendations of the AAPM Radiation Therapy Committee Task Group No. 43. *Med Phys* 22:209–34.

Nath, R., Anderson, L.L., Meli, J.A., Olch, A.J., Stitt, J.A., and Williamson, J.F. 1997. Code of practice for brachytherapy physics: Report of the AAPM Radiation Therapy Committee Task Group No. 56. *Med Phys* 24:1557–98.

Nath, R., Amols, H., Coffey, C. et al. 1999. Intravascular brachytherapy physics: Report of the AAPM Radiation Therapy Committee Task Group No. 60. *Med Phys* 26:119–52.

Nucletron. 2008. Nucletron announces the development of an electronic brachytherapy solution. Available at http://www.nucletron.com/en/NewsAndEvents/News/Pages/NUCLETRONANNOUNCESTHEDEVELOPMENTOFANELECTRONICBRACHYTHERAPYSOLUTION.aspx, last accessed October 5, 2011.

NUDAT 2.4, 2008. National Nuclear Data Center, Brookhaven National Laboratory. Available at http://www.nndc.bnl.gov/nudat2/index.jsp, last accessed October 5, 2011.

Park, C.C., Yom, S.S., Podgorsak, M.B. et al. 2010. American Society for Therapeutic Radiology and Oncology (ASTRO) Emerging Technology Committee report on electronic brachytherapy. *Int J Radiat Oncol Biol Phys* 76:963–72.

Radiological Physics Center. 2011. Joint AAPM/RPC registry of brachytherapy sources meeting the AAPM dosimetric prerequisites. Available at http://rpc.mdanderson.org/RPC/BrachySeeds/Source_Registry.htm, last accessed October 5, 2011.

Ritts, J.J., Solomito, M., and Stevens, P.N. 1969. Calculation of neutron fluence-to-kerma factors for the human body. *Nucl App Tech* 7:89–99.

Rivard, M.J. 2000. Neutron dosimetry for a general [252]Cf brachytherapy source. *Med Phys* 27:2803–15.

Rivard, M.J., Coursey, B.M., DeWerd, L.A. et al. 2004a. Update of AAPM Task Group No. 43 Report: A revised AAPM protocol for brachytherapy dose calculations (AAPM Report No. 84). *Med Phys* 31:633–74.

Rivard, M.J., Kirk, B.L., Stapleford, L.J., and Wazer, D.E. 2004b. A comparison of the expected costs of high dose rate brachytherapy using [252]Cf versus [192]Ir. *Appl Radiat Isot* 61:1211–6.

Rivard, M.J., Melhus, C.S., Zinkin, H.D. et al. 2005. A radiobiological model for the relative biological effectiveness of high-dose-rate [252]Cf brachytherapy. *Rad Res* 164: 319–23.

Rivard, M.J., Davis, S.D., DeWerd, L.A., Rusch, T.W., and Axelrod, S. 2006. Calculated and measured brachytherapy dosimetry parameters in water for the Xoft Axxent X-Ray Source: An electronic brachytherapy source. *Med Phys* 33:4020–32.

Rivard, M.J., Melhus, C.S., Granero, D., Pérez-Calatayud, J., and Ballester, F. 2009a. An approach to using conventional brachytherapy software for clinical treatment planning of complex, Monte Carlo-based brachytherapy dose distributions. *Med Phys* 36:1968–75.

Rivard, M.J., Venselaar, J.L.M., and Beaulieu, L. 2009b. The evolution of brachytherapy treatment planning. *Med Phys* 36:2136–53.

Rivard, M.J., Chiu-Tsao, S.-T., Finger, P.T. et al. 2011. Comparison of dose calculation methods for brachytherapy of ocular tumors. *Med Phys* 38:306–16.

Schlea, C.S. and Stoddard, D.H. 1965. Californium isotopes proposed for intracavity and interstitial therapy with neutrons. *Nature* 206:1058–9.

Shridhar, R., Estabrook, W., Yudelev, M. et al. 2010. Characteristic 8 keV X rays possess radiobiological properties of higher-LET radiation. *Rad Res* 173:290–7.

Taccini, G., Cavagnetto, F., Coscia, G., Garelli, S., and Pilot, A. 1997. The determination of dose characteristics of ruthenium ophthalmic applicators using radiochromic film. *Med Phys* 24:2034–7.

Tačev, T., Ptáčková, B., and Strnad, V. 2003. Californium-252 ([252]Cf) versus conventional gamma radiation in the brachytherapy of advanced cervical carcinoma. *Strahlenther Onkol* 179:377–84.

Tačev, T., Ptáčková, B., and Strnad, V. 2004. Comparison of twelve-year treatment results of advanced cervical carcinoma with californium-252 vs. gamma radiation only in brachytherapy. *Radiother Oncol* 71(S2): (abstract) S8.

Tang, X., Cheng, C., and Xu, X. 2002. Cf-252 neutron brachytherapy in China. 5th International Topical Meeting on Industrial Radiation and Radioisotope Measurement Applications, Bologna, Italy. (abstract) 51.

Thomadsen, B.R., Heaton 2nd, T.H., Jani, S.K. et al. 2010. Off-label use of medical products in radiation therapy: Summary of the Report of AAPM Task Group No. 121. *Med Phys* 37:2300–11.

Thomson, R.M., Taylor, R.E., and Rogers, D.W.O. 2008. Monte Carlo dosimetry for ^{125}I and ^{103}Pd eye plaque brachytherapy. *Med Phys* 35:5530–43.

Tripuraneni, P. 2003. The future of CART in the era of drug eluting stents: 'It's not over until it's over'. *Brachytherapy* 2:74–6.

Vaidya, J.S., Baum, M., Tobias, J.S., Morgan, S., and D'Souza, D. 2002. The novel technique of delivering targeted intraoperative radiotherapy (Targit) for early breast cancer. *Eur J Surg Oncol* 28:447–54.

Vaidya, J.S., Baum, M., Tobias, J.S. et al. 2006. Targeted intraoperative radiotherapy (TARGIT) yields very low recurrence rates when given as a boost. *Int J Radiat Oncol Biol Phys* 66:1335–8.

Vtyurin, B.M., Medvedev, V.S., Anikin, V.A. et al. 1994. Neutron brachytherapy in the treatment of melanoma. *Int J Radiat Oncol Biol Phys* 28:703–9.

Williamson, J.F., Coursey, B.M., DeWerd, L.A., Hanson, W.F., and Nath, R. 1998. Dosimetric prerequisites for clinical use of new low energy photon interstitial brachytherapy sources. *Med Phys* 25:2269–70.

Windham, J.P., Shapiro, A., and Keriakes, J.G. 1972. Calculated neutron dose rates for implantable californium-252 sources. *Phys Med Biol* 17:493–502.

Yanch, J.C. and Zamenhof, R.G. 1992. Dosimetry or ^{252}Cf sources for neutron radiotherapy with and without augmentation by boron neutron capture therapy. *Rad Res* 131:249–56.

Yang, Y., Melhus, C.S., Sioshansi, S., and Rivard, M.J. 2011. Treatment planning of a skin-sparing conical breast brachytherapy applicator using conventional brachytherapy software. *Med Phys* 38:1519–25.

Advanced Brachytherapy Technologies: Encapsulation, Ultrasound, and Robotics

Mark J. Rivard
Tufts University School of Medicine

Luc Beaulieu
*Centre Hospitalier Universitaire
de Québec and Université Laval*

Marinus A. Moerland
University Medical Center Utrecht

27.1 Introduction

Beyond the special brachytherapy modalities mentioned in the previous chapter are cutting-edge advances in the field of brachytherapy. These may be categorized into three sections as new capsule designs for brachytherapy sources, developments in ultrasound (US) imaging for brachytherapy planning and treatment delivery, and robotic brachytherapy for implantation with US or magnetic resonance imaging (MRI) guidance.

27.2 New Brachytherapy Source Designs

Chapter 3 examined the radionuclides and radiological properties of sources that have been, are currently, or are being considered for clinical use. This section examines established radionuclides but considers potential variations in source encapsulation toward diversifying brachytherapy utilization and improving the patient experience.

27.2.1 Non-Titanium Encapsulation

Since 1965, brachytherapy seeds have been encapsulated in titanium based on the work of physicist Don Lawrence (Aronowitz 2010) for low-energy photon-emitting sources such as ^{125}I, ^{103}Pd, and ^{131}Cs. Due to the relatively high cross section of Ti ($Z = 22$) compared to tissue for the photoelectric effect at photon energies of 0.02 to 0.03 MeV, a significant number of Ti K-edge

characteristic x-rays are emitted. These characteristic x-rays (about 4.5 keV) significantly influence measurement of air-kerma strength yet do not deposit dose in tissue beyond a couple of millimeters (Kubo 1985). Four decades after Lawrence's invention, the International Brachytherapy S.A., now Eckert & Ziegler BEBIG S.A., considered replacing titanium with another biocompatible material for brachytherapy seeds.

The dosimetric properties of the OptiSeed103 (model 1032P ^{103}Pd source) were examined in the middle of the last decade (Figure 27.1). Bernard and Vynckier (2005) performed Monte Carlo (MC) simulations using the MCNP4C radiation transport code. The photoelectric cross-section libraries accompanying the release of this code were based on inaccurate results obtained from the 1960s. Consequently, their dose rate constant (Λ) value calculated in water was 0.712 ± 0.043 cGy/h/U. This is 7.5% higher than the 0.665 ± 0.014 cGy/h/U value obtained by Wang and Hertel (2005), who also used MC methods. However, Wang and Hertel used the MCNP5 radiation transport code that included modern cross-section libraries. To resolve this discrepancy, Khan et al. (2008) examined the model 1032P seed using a photon spectrometry technique. They obtained a Λ value of 0.664 ± 0.025 cGy/h/U, which was in 0.2% agreement with that measured by Wang and Hertel. For comparison with Ti-encapsulated seeds such as the model 200 ^{103}Pd source, Λ is 0.686 cGy/h/U (Rivard et al. 2004).

Khan et al. (2008) state that the purported advantages of replacing the Ti encapsulation with plastic are to reduce

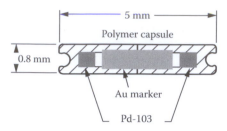

FIGURE 27.1 Model 1032P [103]Pd source with a polymer encapsulation. (Courtesy of M.J. Rivard.)

encapsulation-induced spectral hardening of transmitted [103]Pd photons and to improve uniformity of the dose distribution. Bernard and Vynckier (2005) state that "replacement of the titanium encapsulation by a biocompatible polymer modifies the shape of the radiation field around the source as well as reduces the amount of isotope that must be incorporated to deliver the same dose rate relative to a titanium encapsulated seed." The model 1032P [103]Pd source Λ is higher by only 3% than the model 200 [103]Pd source, and the radial dose function at 5 cm is also only about 3% higher than the model 200 [103]Pd source. At 2 and 5 cm, the 2D anisotropy function values along the source long axis are 0.76 and 0.72, respectively, with 0.53 and 0.51 for the model 200 [103]Pd source. Also, the OptiSeed103 seed was shown by Afsharpour et al. (2008) to be the source having dose distributions most similar to the TG-43–based formalism by Nath et al. (1995) in terms of interseed attenuation. It is no surprise that the model 1032P [103]Pd source is no longer available on the market, given that these differences with Ti-encapsulated sources do not seem worthwhile.

27.2.2 Thin Seeds

Since the introduction of permanent prostate implants using [222]Rn seeds, the practice of low-dose-rate (LDR) brachytherapy has used seeds 5 mm in length and 0.8 mm in outer diameter for over 90 years. The currently most popular vendor of LDR ^{125}I seeds, model 6711 by Oncura of GE Healthcare, has developed the ThinSEED (model 9011) to investigate the potential for reduced implantation trauma and subsequent edema following surgery (Moran 2012). The model 9011 source is significantly thinner than the model 6711 source (Rivard 2009; Kennedy et al.

FIGURE 27.2 Depiction of the model 9011 ^{125}I source (top) in comparison to the model 6711 ^{125}I source (bottom) with outer diameters of 0.77 and 0.51 mm, respectively; both sources have lengths of 4.56 mm. (Courtesy of M.J. Rivard.)

2010) and has 56% less cross-sectional area, while retaining the same active and encapsulated lengths, and similar dose distributions and Λ as the model 6711 for compatibility with existing brachytherapy infrastructure. Research is underway to assess the imaging properties and clinical outcomes associated with this diameter reduction (Figure 27.2).

27.2.3 Absorbable Brachytherapy Seeds

Another creative modification of existing LDR seed design is to make the seed disappear following deposition of the therapeutic dose. Use of an absorbable seed would facilitate clinical applications such as permanent breast brachytherapy (Pignol et al. 2006) where the presence of the capsule and imaging marker complicates patient follow-up and disease monitoring.

A search of the Internet on "absorbable brachytherapy device" reveals that a number of companies have protected intellectual property on this topic and received substantial financial awards from the U.S. government as research incentives. However, there do not appear yet to be any devices approved by the U.S. Food and Drug Administration (FDA) or that have received Conformité Européenne (CE) Marking, which means compliance with European patient-safety regulations.

In the absence of documented properties of absorbable sources, their characteristics may still be elucidated. The majority of the radionuclide would need to decay before being biologically absorbed by the patient. This would include breakdown of the encapsulation material, radioactive carrier, and any radio-opaque marker [assuming compatibility with conventional computed tomography (CT)-based localization tools]. For low-energy photon-emitting radionuclides such as [131]Cs, [103]Pd, and ^{125}I with disintegration half-lives of 9.7, 17.0, and 59.4 days, respectively, 99% of the radiation dose is delivered after 9, 16, and 56 weeks. Given the typical timeframe for disease progression and also the thyroid-seeking nature of iodine, it seems that ^{125}I is not a suitable candidate. Both cesium and palladium are relatively inert. However, issues not normally considered such as influence of typically shielded beta emissions or other radiological features require close examination. Other radionuclides may also be considered for this potential source type.

27.2.4 Elongated LDR Sources

High-dose-rate (HDR) brachytherapy sources can be translated within a catheter in dwell steps of 1 to 5 mm. Given that the active length of the source is typically less than 5 mm, this provides direct dosimetric overlap among dwell positions and simulates a continuous line segment. The resultant dose distribution minimizes dosimetric variations due to the presence of encapsulation and endwelds such as typically observed with cold spots in between LDR sources. An advantage of elongated LDR sources is to emulate the dose uniformity provided by having a uniform length of radioactivity without the dosimetric variation often associated with strands of LDR brachytherapy seeds. There are currently three elongated LDR sources of interest: [192]Ir wire, the [103]Pd RadioCoil, and the [103]Pd CivaString.

27.2.4.1 LDR [192]Ir Wire

LDR ^{192}Ir wires for temporary implant are currently more common in Europe than in North America and are suitable for a busy clinic requiring the inventory of these single-use devices (being cut to length for a specific patient implant). Wires (0.3-mm outer diameter) have a radioactive ^{192}Ir core of 0.1 mm in diameter surrounded by a 0.1-mm-thick nonradioactive Pt sheath to control radiocontamination (van der Laarse et al. 2008). Their dosimetric attributes have been researched (Ballester et al. 1997; Karaiskos et al. 2001; Pérez-Calatayud et al. 2003). Typically, 10 wires of 14-cm length are ordered.

CIS Bio and Bebig (now Eckert & Ziegler BEBIG S.A.) have also produced 0.5- and 0.6-mm radioactive ^{192}Ir wires. The same thickness of nonradioactive Pt sheath covered the active core. The material was used in different forms, like single and double pins ("hairpins") for direct implantation in floor-of-mouth tumors. TG-43 data have been derived by MC calculations, such as by Karaiskos et al. (2001), and are listed in the ESTRO Braphyqs TG-43 Web site (http://www.estro.org/estroactivities/Pages/TG43DOSIMETRICPARAMETERSIr-192LDRWIRESOURCES.aspx).

27.2.4.2 LDR [103]Pd RadioCoil

The RadioCoil source was researched extensively in the last decade by the group of Ali Meigooni (see, e.g., Meigooni et al. 2004). This source of [102]Rh-enriched wire was cyclotron-activated to produce [103]Pd. A curled helical ribbon of 0.8-mm outer diameter and 0.7-mm hollow inner diameter permits great flexibility for clinical implantation. However, the company (RadioMed Corp.) did not produce a product that was subsequently implanted in patients.

27.2.4.3 LDR [103]Pd CivaString

A product currently under development is the LDR [103]Pd CivaString by CivaTech Oncology. While no patients have been treated at the time of this writing, the U.S. FDA has provided a 501(k) clearance (K082159) in 2008, and a prototype was developed in 2011 (Figure 27.3). From the figure, discrete units of [103]Pd (black) are segmented with radio-opaque markers (yellow). The source comes in lengths of 1 to 6 cm and is contained in a bioabsorbable material.

27.2.4.4 Calibration Challenges

Using an appropriately long and calibrated reentrant well ionization chamber, the physicist will assay the cut length and compare the measured S_K/cm or RAKR/cm result to that provided by the vendor (Pérez-Calatayud et al. 2003). At this time, there

FIGURE 27.3 CivaString [103]Pd source. (Courtesy of M.J. Rivard.)

are no societal guidelines for accepted tolerances when comparing to the manufacturer's calibration certificate. However, the AAPM and ESTRO are working to set a guidance document on this topic for elongated brachytherapy sources, with a tolerance at approximately ±5%.

The principal challenge to calibrate elongated brachytherapy sources is based on their lengths. There are no calibration inter-comparisons among the primary standard calibration laboratories (PSDLs) currently performed for elongated high-energy sources such as LDR [192]Ir (Vollans and Wilkinson 2000). Low-energy LDR [103]Pd sources are measured at PSDLs using a collimated aperture that typically does not accommodate active lengths ≥ 10 mm. Calibrations are performed through cross-calibrations using shorter-length sources by assuming uniform radioactivity as a function of source length and with reentrant well ionization chambers having special holders and long collecting volumes (DeWerd et al. 2006; Meigooni et al. 2006; Paxton et al. 2008).

27.2.4.5 Treatment Planning Challenges

From observations of LDR ^{192}Ir wire sources, elongated sources take a curvilinear shape when implanted within the patient. Consequently, treatment planning cannot be performed with a rigid straight-line segment. Characterization of the summation dose of [103]Pd line segments for the RadioCoil has been examined by Meigooni's group (Meigooni et al. 2005; Awan et al. 2006, 2008; Dini et al. 2007) and for a variety of source types by Bannon et al. (2011).

The polar coordinate system of the AAPM TG-43 dose calculation formalism depicts uniform dose as a function of radial distance and polar angle. However, the dose distributions of elongated brachytherapy sources, especially low-energy photon emitters such as [103]Pd, tend to be better sampled uniformly with a cylindrical coordinate system. Bannon et al. (2011) have shown that both coordinate systems may provide accurate depictions of dose distributions, but high-resolution dosimetry parameters are needed to minimize parameter interpolation errors and subsequent dose calculation errors. High-energy sources such as ^{192}Ir are less subject to these interpolation errors than low-energy sources such as [103]Pd.

27.2.5 AAPM/ESTRO TG-167 Report

The AAPM and ESTRO have prepared a joint Task Group report (TG-167, Ravinder Nath is chair) on recommendations for new or innovative brachytherapy sources/devices or applications. These include the following categories:

(1) New HDR ^{192}Ir sources/afterloaders
(2) New LDR ^{125}I or [103]Pd sources
(3) LDR [131]Cs sources
(4) Intravascular brachytherapy (IVBT) sources
(5) Long half-life HDR sources (^{60}Co and ^{137}Cs)
(6) Elongated [103]Pd and ^{192}Ir wire sources
(7) Intermediate energy sources ([169]Yb, [241]Am, [170]Tm)

(8) Neutron-emitting sources (^{252}Cf)

(9) ^{90}Y microspheres

(10) Electronic brachytherapy sources

(11) Intracavitary balloons

(12) New eye plaques

Of these categories, there are separate Task Group reports in process or recently published on IVBT sources (TG-149, Chiu-Tsao et al. 2007), elongated ^{103}Pd and ^{192}Ir wire sources (TG-143, Ravinder Nath is chair), ^{90}Y microspheres (TG-144, Dezarn et al. 2011), electronic brachytherapy sources (TG-182, Bruce Thomadsen is chair), and new eye plaques (TG-221, Christopher Melhus and Rowan Thomson are cochairs). There is no coordinated societal investigation of the novel source designs (non-Ti encapsulated, thin diameters, absorbable materials) described in the prior subsections.

TG-167 was formed in November 2007 and charged with the following tasks:

(1) To review the current practice in the marketplace, current FDA requirements for new brachytherapy sources, and the difficulties encountered regarding the dosimetry of novel brachytherapy sources for permanent implantation using examples such as the historical introduction of TheraSeed by Theragenics and, more recently, Cs-131 by IsoRay Medical.

(2) To review critical physical, dosimetric, and radiobiological issues that arise when a novel source is introduced for permanent interstitial brachytherapy such as calibration traceability, accuracy of dosimetry parameters, and choice of prescribed dose for equivalent results. Also, the need to receive pertinent information about the source design from the manufacturer will be outlined so that a proper evaluation of a novel source can be accomplished. The primary focus of this task group will be on prostate implantation.

(3) To develop consensus guidelines on the methodology for these dosimetry issues that can minimize the potential risk to the human subjects and maximize the efficiency with which these novel sources can be widely adopted for the benefit of cancer patients. The overall objective is to highlight the critical issues and missing information that may affect the clinical response so that investigators address them as a deliberate part of their study design.

The TG-167 report covers specific issues that can arise regarding regulatory approvals, calibration requirements, reference dosimetry, and treatment planning requirements for the 12 aforementioned brachytherapy source categories. Guidance is provided on when to use the sources or devices under standard-of-care clinical use, as off-label use (as described by Thomadsen et al. 2010), and within a research setting under a clinical trial.

27.3 US Guidance in Brachytherapy

27.3.1 Introduction

US has been used for a variety of medical imaging procedures for decades. It is a relatively inexpensive modality compared to CT or MRI and involves nonionizing radiations with wavelength in the megahertz range. US is also intrinsically a real-time imaging device. It has a high repetition rate of 15 to 60 images per second. Thus, even when viewing a dimensional (2D) image, the time dimension is always compounded, and measurements are performed on an image capture at a given time point. A popular example is in vivo imaging of a fetus. While penetration of the sound waves is limited to a few centimeters, the tissue contrast within that range is usually very good and sometimes second only to MRI (Tewari et al. 1996).

The aforementioned characteristics make US a desirable imaging modality not only for diagnostic purposes but also for interventional procedures. Surgery is looking more and more to US, in particular, in the field of robotic surgery (Stoianovici 2000). Prostate biopsies and brachytherapy constitute the most widely used applications of real-time US guidance (Holm et al. 1983, 1990; Narayan et al. 1995). In retrospect, US is the main reason permanent seed implant has become such a popular procedure. The use of a convenient transrectal US (TRUS) probe allows for direct contact of the TRUS probe with the rectal wall and a direct access to the prostate. The TRUS probe is mounted on a support that ensures a stable, hands-free, and reproducible geometry. The support further includes a movable portion, a stepper, along the insertion axis, to permit 2D measurements at known distances (interslice or retraction distances). This manual motion allows for the accumulation of multiple 2D views of the prostate and, finally, the reconstruction of a 3D model that is used for treatment planning purposes. Treatment delivery is also validated with TRUS imaging. This is possible because a template guide is used for needle insertion; the template is also in a fixed geometry relative to the TRUS probe and the patient anatomy, all linked together by the planning system software. This is a key feature of precise image-guided procedures.

The success of TRUS-guided prostate brachytherapy has not yet been reproduced for other disease sites. A proposition was made for a US-guided breast brachytherapy system (De Jean et al. 2009), but no commercial systems exist. Various methods using handheld probes for breast brachytherapy (Chen and Vicini 2007) and GYN applicator insertion guidance (Davidson et al. 2008) have been reported but are used more as a visual aid than for quantitative purposes.

27.3.2 Advanced US Guidance and Augmented Reality

2D axial image acquisition and needle guidance has been at the heart of permanent seed implants since it was first proposed by Holm in the 1980s (Holm et al. 1983). However, the TRUS probe can have two sets of crystals for biplanar imaging, axial and sagittal (side-firing crystals). Using sagittal images for needle guidance has the advantage of following the needle insertion path throughout the prostate without any probe motion (assuming that the needle is inserted in the image plane). The other potential advantage is that the rotational motion of the TRUS probe needed to reach the insertion plane of a given needle is expected

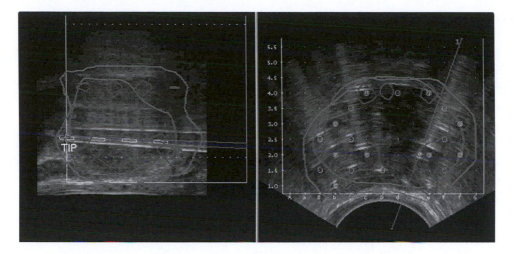

FIGURE 27.4 **(See color insert.)** Illustration of sagittal needle insertion guidance and augmented reality. An axial view from the planning system is shown on the right. The related live US image with augmented reality elements is displayed on the left. See text for details. The figure is a screen capture from the SPOT system (Nucletron BV—an Elekta company, Veenendaal, The Netherlands.) (Courtesy of L. Beaulieu.)

to induce less displacement of the prostate than the stepper motion, which alters the prostate shape as the probe moves in and out to cover the volume in axial mode.

Vendors have built upon this approach and provide end users with systems that generate more information than would normally be available with the original TRUS systems. Figure 27.4 is given as an example illustrating the possibilities available when US images are combined with planning system elements. The right panel shows the typical axial view from the planning system with isodose lines, needle positions (open green circles), and seeds (green dots). The left panel provides new information; it shows the live sagittal image upon which real-time information is superimposed. Namely, the expected or virtual needle track is represented in pink with the expected seed position as a green cylinder within the needle and the resultant isodose lines. It is further possible to change the needle track, or even seed positions, to match the actual position (the bright white reflection below the virtual needle) and visualize the change in dose coverage in real time (Polo et al. 2010). Combination of real-time US images with critical real-time procedure information in a single view is called *augmented reality*. For prostate brachytherapy, this real-time imaging provides the brachytherapy team supplemental information to achieve increased delivery precision (Beaulieu et al. 2007). This advanced imaging also provides constant feedback on any alternative delivery choice made by the physicians (i.e., change in needle position due to pubic arch interference) by displaying the dosimetric consequence in real time. However, while needles can clearly be seen on US, seeds cannot be visualized with confidence using current devices (further discussed in Section 27.3.3.2).

The process described above can be further automated by having a motorized US probe that is computer controlled (Tong et al. 1996). In such a case, both the image acquisition and needle insertion guidance steps can be fully integrated within the

treatment planning. Having a motorized US probe also allows moving beyond 2D imaging.

27.3.3 From 2D to 3D Imaging

Nowadays, 3D images are integrated into the radiotherapy planning process. Without it, advanced procedures such as intensity-modulated radiation therapy or radiosurgery would not be possible. Images are usually acquired in two ways, either a combination of back-to-back 2D images or volumetric acquisition (e.g., cone-beam CT or CBCT). In general, volumetric acquisition has the advantage of being free of partial volume effects and allows for representation in any planes with the same image quality. This usually provides more information for contouring, planning, and treatment delivery guidance. Figure 27.5

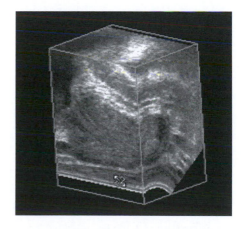

FIGURE 27.5 Example of a volumetric TRUS acquisition of a prostate using a motorized side-firing TRUS probe, same system as Figure 27.4. (Courtesy of L. Beaulieu.)

illustrates a 3D TRUS image "cube" of an intraoperative preimplant prostate. As with CBCT, the prostate volume (middle of the cube) can be displayed in any plane without artifacts.

There are various possibilities to acquire 3D US images. In the next subsection, the general methods are described briefly.

27.3.3.1 Implementation of 3DUS

There are a limited number of ways to generate 3D US data (Prager et al. 2010; Fenster et al. 2001). Starting with simple probes, Figure 27.6 provides an overview of the approaches that can be taken.

In prostate brachytherapy, methods a and b are both used for image acquisition. Method a has been used for three decades by manually moving the TRUS probe through the use of a stepper (as described in Section 27.3.1). Method b has been available commercially in brachytherapy systems for about a decade. Such motions can be accomplished manually or through mechanical means. In either case, the position and angle of the probe must be accurately known in order to prevent introduction of reconstruction artifacts. The TRUS probe, method b, is probably the easiest method since it provides a physical restraint. However, handheld probes must be tracked. This can be done through motorized apparatus, infrared (Fraser et al. 2010), or radiofrequency (RF) tracking devices mounted on the probes in order to know at all times the probe position and angle (Leotta et al. 1997; Cleary et al. 2005). These are the most common methods used for US devices found in radiation oncology centers. Other approaches are possible (Fenster et al. 2001).

Advances in US imaging probes are also very promising. It is now possible to find various probes for which mechanical motion of the crystals is included directly within the probes themselves. The principles are the same as depicted in Figure 27.6, but the probe is kept in static contact with the tissue with internal translation/rotation of the crystals (Prager et al. 2010).

Another promising approach consists of using 2D phased arrays. This implementation of 3DUS requires no mechanical probe motion. Instead, each element of a 2D array is controlled electronically to adjust scanning direction and focus depth. The

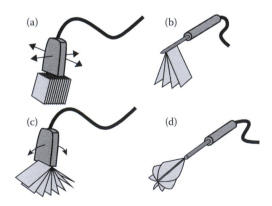

FIGURE 27.6 Generation of 3DUS data by motion of an US probe: (a) linear displacement, (b) rotation from a side-firing probe, (c) tilting, and (d) rotation for an end-firing probe. (Courtesy of L. Beaulieu, adapted from Fenster 2001.)

volume of interest is swept by diverging broad US beams, and volumetric information is obtained in arbitrary planes. While not yet used in brachytherapy, a TRUS probe equipped with such technology would not require any motion and thus eliminate motion influence on prostate shape, and would also make 3D US an interesting guidance modality for other disease sites.

27.3.3.2 Needle Tracking and Seed Detection under US Imaging

Having a 3D computer-controlled US device linked to a computerized planning system enables other advanced tasks. Real-time needle tracking using planning information as a starting point and small probe motions around the expected insertion plane has been demonstrated using clinically used TRUS probes (Draper et al. 2000; Ding and Fenster 2003; Wei et al. 2005). It was shown that a needle can be segmented and its progression tracked without slowing the insertion process. Automated needle tracking has potential such as providing real-time feedback without any user intervention (e.g., feedback on possible dosimetric consequences if no correction to needle path is made) and also for robotic implantation procedures (Mozer et al. 2009); see Section 27.4.

Individual seed detection remains an elusive task based solely on b-mode US imaging. While needle insertion and even the needle track after retraction can be clearly observed via US (in particular, on real-time sagittal images), individual seeds are difficult to distinguish from other structures and fluids present after needle retraction. Studies indicate success rates usually < 90% even after image processing to account for the image before and after needle insertion (and the needle path completely reconstructed going in and out based on the approach discussed above). This success rate is not high enough to provide confidence in the clinical value of the dosimetric parameters. At least one seed manufacturer (GE Oncura) tried to build a seed (model 6733) that had more echogenic properties (Blake et al. 2000; Meigooni et al. 2002; Sowards and Meigooni 2002; Han et al. 2003; Xue et al. 2005; Ding et al. 2006). However, the clinical detection rates were insufficient. Considering that treatment outcome is directly linked to the dose coverage, the real-time detection of seeds using US will be a major breakthrough for TRUS-guided prostate brachytherapy.

27.3.4 Beyond B-Mode Imaging for Brachytherapy

For the most part, US devices used for brachytherapy (and biopsy) purposes are displaying in grayscale mode a map of the envelope of the backscatter signals. This is the B-mode US imaging. The raw RF signal undergoes much processing before being displayed. As such, detailed information of the US interaction with the tissues is lost in the process. While high-resolution B-mode devices could reach resolution of 0.1 to 1 mm, information of even smaller structures could be contained in the raw signals. Nowadays, the signal processing via powerful commodity central processing unit/graphical processing unit might enable

US imaging to move beyond the traditional B-mode images and lead to significant improvement in US imaging and allow for real-time functional US (Polo et al. 2010). Moreover, it has recently been demonstrated that the processing of the RF signal could lead to significant progress in detection of individual brachytherapy seeds (Mamou and Feleppa 2007; Wen et al. 2010).

Imaging of blood flow via Doppler and Power Doppler is being investigated and part of the state-of-the-art US imaging toolkit for cancer detection. This is related to the fact that angiogenesis plays an important role in tumor growth, leading to increase in blood vessel pathways that could be detected via US (Kay et al. 1998). This US imaging technique has been applied to the prostate (Cho et al. 1998; Cheng and Rifkin 2001; Kuligowska et al. 2001) and breast (Surry et al. 2002; LeCarpentier et al. 2008), in particular, to guide the biopsy procedure. The combination of Power Doppler and US contrast agents (such as microbubbles) further leads to increased detection of the microvessel densities (Frauscher et al. 2002). Furthermore, blood flow in the neurovascular bundles could be visualized.

A good example of US raw RF signal analysis is that of elastography (Ophir et al. 1991). Elastography measures tissue density or stiffness under mechanical deformation. This can be accomplished by applying and relaxing pressure to the US transducer. It has been used successfully for in vivo identification of cancer masses in the breast, prostate, and cervix (Garra et al. 1997; Thomas et al. 2007; Aigner et al. 2010). In the prostate, elastography is being investigated for US-guided biopsy procedures (Garra 2011).

US RF signal analysis has further been shown to be able to image cell apoptosis at high frequencies (Czarnota et al. 1999). However, recently, the same team of researchers has demonstrated that even frequencies as low as 10 MHz might be used for such purpose (Vlad et al. 2011), much closer to what is currently used in our clinics. While still a long way from our clinic, this application constitutes an excellent illustration of the level information that can be extracted from advanced signal processing of the US signal.

Closer to our clinical practice, an US-based tissue characterization technique for the prostate has recently been developed by Advanced Medical Diagnostic (AMD, Waterloo, Belgium) and commercialized by BK Medical ApS (Herlev, Denmark) under the name Histoscanning (Braeckman et al. 2008). From a 3D US acquisition, the RF signal is processed, and an algorithm is used to classify/separate tissues based on expected classes of characteristics. A European multicentric study comparing this technology to histopathology found a sensitivity, positive predicted value and negative predictive value as high as DCE-MRI for lesion of 0.2 cm^3 or more (Simmons et al. 2011). This technology is further developed for breast and ovarian cancer (Lucidarme et al. 2010). MRI was seen as the only imaging avenue (DiBiase et al. 2002) for application to prostate brachytherapy, in particular, for boost (Pouliot et al. 2004; Gaudet et al. 2010) or focal brachytherapy.

27.3.5 US and Robotics

As described in previous sections, US possesses excellent qualities as a real-time guidance imaging device for robotic

brachytherapy. The AAPM Task Group 192 has identified 13 different brachytherapy robotic systems at the time of writing. The vast majority are designed to be used in conjunction with US imaging for prostate brachytherapy. Based on the task they achieve, they can be classified into three main categories: robotic template (Salcudean et al. 2008; Fichtinger et al. 2008; Bax et al. 2011), seed delivery and needle retraction robots (Van Gellekom et al. 2004a), and a device that performs needle insertion (Wei et al. 2004; Meltsner et al. 2007; Podder et al. 2007, 2011; Hungr et al. 2009). Finally, another US-guided robotic system is dedicated to lung brachytherapy (Trejos et al. 2008). Presently, a single device (seedSelectron from Nucletron BV) is available commercially for permanent prostate implant. With this system, the needles are inserted manually using virtual needle guidance under TRUS. The seedSelectron uses seeds and spacers from cartridges to build. A drive-wire delivers the seeds inside a transfer tube that is manually connected to a previously inserted needle in the patient. A robotic arm performs the needle retraction while maintaining the seed/spacer sequence in place with the drive-wire, which is retracted once the needle is outside the prostate capsule. Removing the needle from the patient is then performed manually. A robotic template is potentially the next step for both HDR and seed implants. Replacing the physical template with a robotic one would enable various needle/catheter spacings but also potentially the insertion of angulated needles. The latter could help in bringing prostate brachytherapy to patients, which would normally be excluded due to pubic arch interference, for example. Robotic template guidance is also minimally disruptive from the current clinical workflows. Full robotic needle insertion clearly needs good-quality imaging and real-time needle tracking. The operating parameters, limitations, and safety features for robotic needle insertion systems must be defined with the utmost care. At this time, no such system is commercially available.

27.4 Robotic Devices for MRI-Guided Brachytherapy

27.4.1 Introduction

MRI is valuable in soft-tissue interventions over other image modalities, such as US and CT, due to its superior soft tissue contrast (Tempany et al. 2008; Jürgenliemk-Schulz et al. 2009; Peters et al. 2009; Hambrock et al. 2010). It establishes better visualization, localization, and delineation of the target and surrounding critical structures at the moment of intervention (Kerkhof et al. 2008; Tempany et al. 2008; Jürgenliemk-Schulz et al. 2009; Peters et al. 2009; Hambrock et al. 2010; Kühn et al. 2010). However, accessibility to the patient is restricted to the internal geometry of generally available MRI systems.

At several institutes, robotic devices have been developed to overcome this limitation and to perform real-time MRI-guided needle interventions (DiMaio et al. 2007; Fischer et al. 2008; Melzer et al. 2008; Muntener et al. 2008; Morikawa et al. 2009; Schouten et al. 2010; van den Bosch et al. 2010b). Although the principles of these robotic devices are different, they have common challenges to face before they can be safely applied on patients. To ensure that such a device is safe for use in an MRI environment, several general test methods were formulated (Schaefers 2008; Yu et al. 2011). MRI-related challenges concern accessibility, magnetic compatibility, needle placement accuracy, and safety.

27.4.2 Patient Accessibility

For robot design, the available space inside an MR scanner is important. The closed bore and the open MR scanner have their own strengths and limitations. Due to the gap between the two poles of an open MR scanner, the patient is accessible from almost any angle. This gap can be in either the horizontal (Gossmann et al. 2008) or vertical directions (van den Bosch et al. 2006). Such an open scanner has two drawbacks.

(1) The first is the direction of the static magnetic field. This field is oriented vertically for a horizontal gap and horizontally for a vertical gap. Since the intervention needle will generally be positioned perpendicular to this field, the susceptibility artifact caused by the needle will be present along the entire needle in the MR image (Guermazi et al. 2003; Müller-Bierl et al. 2004; Lagerburg et al. 2008; Kühn et al. 2010).

(2) Another drawback is the lower magnetic field strengths that are available for open systems compared to closed-bore MR scanners, resulting in lower image quality (Machann et al. 2008).

In a closed-bore MR scanner, the direction of the static magnetic field is oriented along the longitudinal axis of the MR bore, making it more logical to align the intervention needle with this field. Consequently, the susceptibility artifact caused by the needle is confined to the tip of the needle in the MR image (Guermazi et al. 2003; Müller-Bierl et al. 2004; Lagerburg et al. 2008).

Higher magnetic field strength causes a better signal-to-noise ratio (SNR) and resultant image quality (Machann et al. 2008); however, there are also disadvantages of using high field strengths for robotic interventions. Static field distortions and signal intensity artifacts caused by the robot and the needle become more prominent at higher field strengths (Guermazi et al. 2003; Machann et al. 2008; Peters et al. 2009). Also, the wavelength of the electromagnetic RF waves that are needed to generate an MR image becomes shorter, which can increase the risk of serious tissue heating around the intervention needle tip due to resonating RF waves along the needle (Dempsey et al. 2001; Yeung et al. 2002; Machann et al. 2008). At field strengths exceeding 3 T, the shorter RF waves can lead to destructive interferences of superimposed RF waves inside the human body, resulting in local regions of signal losses (Machann et al. 2008; van den Bergen et al. 2009). The tissue RF power deposition increases with higher field strengths, which might result in heat sensations by the patient (Machann et al. 2008). Moreover,

magnetic forces on metallic devices generally correlate with the magnetic field (gradient) strength of the MR scanner (Schenck 2000; Dempsey et al. 2002; Shellock 2002). A 1.5-T closed-bore MR scanner provides sufficient SNR for adequate image quality, and its field orientation is beneficial to place the needle along the static magnetic field. Van Gellekom et al. (2004b) showed that with a divergent needle insertion method, the entire prostatic gland could be reached with a robotic device that fits between the legs of the patients. Other tumor sites will require other creative geometric approaches.

27.4.3 Robot MR Compatibility

A robot that performs well outside a magnetic field may cause risks when used within a magnetic field. Potential risks are ferromagnetism, induced currents, and signal intensity distortions in the MR image.

27.4.3.1 Ferromagnetism

Ferromagnetic structures may become dangerous projectiles when taken into the scanning room (Condon et al. 2001; Zimmer et al. 2004). They tend to move due to magnetic forces in regions where a spatial magnetic field gradient is present (Schenck 2000; Dempsey et al. 2002). The force increases with the magnetic susceptibility of the material and the gradient magnitude. In general, the spatial gradients are maximal near the magnet portal (Dempsey et al. 2002). When placed in a magnetic field, the structure can rotate as a result of a magnetic torque (Schenck 2000; Dempsey et al. 2002). In addition, ferromagnetic structures greatly distort the field uniformity inside the magnet, destroying image quality. The use of ferromagnetic materials is generally prohibited in all MR scanner types. All robotic components should therefore be tested for nonferromagnetism before entering the MR scanning room.

The robot developed at the University Medical Center Utrecht (UMCU) in the Netherlands is pneumatically and hydraulically driven and consists of polymers and nonferromagnetic materials such as brass, copper, titanium, and aluminum (Figure 27.7).

FIGURE 27.7   UMCU robot. (Courtesy of M.A. Moerland.)

FIGURE 27.8 Johns Hopkins robot. (Courtesy of M.A. Moerland, adapted from Muntener et al. 2008.)

All components of the UMCU robot were tested for nonferromagnetism with the aid of a low-field hand magnet. No forces were measured except for a small potentiometer. Nevertheless, this sensor is used to monitor the displacement of the buffer stop (van den Bosch et al. 2010b) that limits the maximal needle insertion depth per tap as a backup to the MRI. The small potentiometer was securely attached to the massive robot. No forces were observed when sliding the robot into a 1.5-T MR scanner.

The robot developed by Muntener et al. (2008) is built of nonmagnetic and dielectric materials, such as ceramics, fiberglass, plastics, and rubber (Figure 27.8). Pneumatic motors are used to provide the robot with five degrees of freedom to position the end effector. The end effector has one degree of freedom to set the depth of needle insertion.

27.4.3.2 Induced Currents

Eddy currents are evoked by a change in magnetic flux through conducting pathways and produce magnetic fields counteracting the original change in magnetic field (Robertson et al. 2000; Dempsey et al. 2002; Graf et al. 2006). The change in magnetic flux can be caused by time-varying magnetic fields, such as during gradient switching, resulting in image artifacts and vibrations of the conductive component (Graf et al. 2006). Moreover, eddy currents are induced when a conductive component moves

through a static magnetic field gradient or rotates in a static magnetic field, causing motion resistance (Robertson et al. 2000; Dempsey et al. 2002). This may hamper the performance of a robotic device. If a device contains conductive materials, precautions should be taken to prevent or minimize the area of loops and to electrically isolate critical structures.

27.4.3.3 Signal Intensity Distortions in the MR Image

A material with a susceptibility other than its surrounding medium (e.g., human tissue) locally distorts the static magnetic field, leading to intravoxel dephasing of the spins (Guermazi et al. 2003; Graf et al. 2005). The intravoxel dephasing results in signal intensity losses in the MR image. This artifact (susceptibility artifact) is less prominent in spin-echo (SE) images compared to gradient echo (GE) images due to the extra 180° refocusing pulse that can compensate for static magnetic field inhomogeneities (Port and Pomper 2000; Guermazi et al. 2003; Graf et al. 2005). In GE images, the artifact can be reduced by shortening the echo time (Port and Pomper 2000; Guermazi et al. 2003; Müller-Bierl et al. 2004), increasing the read-out gradient strength (Port and Pomper 2000), aligning the material with the static magnetic field (Port and Pomper 2000; Guermazi et al. 2003; Müller-Bierl et al. 2004; Lagerburg et al. 2008; Kühn et al. 2010), and using an MR scanner with lower field strength (Port and Pomper 2000; Guermazi et al. 2003; Müller-Bierl et al. 2004).

Eddy currents due to gradient switching may also cause signal intensity artifacts, the so-called conductivity-induced artifacts (Graf et al. 2005). These artifacts increase with the strength and slew rate of the applied gradients (Graf et al. 2005). In addition to signal intensity artifacts, both phenomena may induce geometric image distortions (Haacke et al. 1999; Guermazi et al. 2003).

As an example, Figure 27.9 shows signal intensity artifacts in balanced steady-state free precession (bSSFP) images caused by the UMCU robotic device. In Figure 27.9a and b, black bands can be distinguished. The artifact bands are circularly shaped around a single point. Closer investigation of the robot revealed a small ferromagnetic pin at this location. The scan was repeated after substitution with a nonferromagnetic component. Consequently, the density of the black bands was strongly reduced (Figure 27.9c and d).

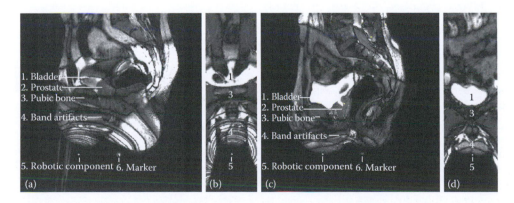

FIGURE 27.9 Examples of signal intensity losses in a bSSFP MR image caused by the UMCU robotic device (a, b) with ferromagnetic component (c, d) after substituting the ferromagnetic component with a nonferromagnetic component. (Courtesy of M.R. van den Bosch.)

27.4.4 Needle Tip Placement Accuracy

In MRI-guided interventions, needle placement accuracy may be hampered by geometric image distortions, the susceptibility artifact at the needle tip, tissue deformation during needle insertion, and needle deflection. These first three issues will be examined in the three subsequent sections, while the latter cause will be discussed in Section 27.4.5.

27.4.4.1 Geometric Image Distortions

Spins that are spatially encoded at the wrong position lead to geometric image distortions. These distortions hamper the needle placement accuracy, when the needle is shifted with respect to the target. In general, geometric image distortions (Bakker et al. 1992) arise from nonlinearity of the gradients (gradient errors) and inhomogeneity of the static magnetic field (static field errors). The displacement due to the gradient error is independent of the orientation and strength of the applied gradient (Bakker et al. 1992). The size of the shift caused by the static error is inversely proportional to the gradient strength, and its direction is affected by the direction and polarity of this gradient (Bakker et al. 1992; Jezzard and Clare 1999).

The geometric image distortions can be induced by both the scanner and the object. Once the scanner-induced geometric errors are known, a correction for these errors can be performed (Bakker et al. 1992). The object-induced static field errors are caused by the difference in susceptibility between the robotic device and surrounding media. Eddy currents in the robotic device, caused by gradient switching, may induce gradient errors (Haacke et al. 1999; Guermazi et al. 2003).

The geometric errors induced by a robotic device can be quantified using a grid phantom of equally spaced tubes as shown in Figure 27.10 (parameters for this particular phantom are as follows: intertube distance = 4.2 cm, tube length = 34 cm, and diameter = 4 mm, tubes filled with $MnCl_2$-doped water). The phantom was imaged in a 1.5-T Achieva MR scanner with and without the UMCU robot present. Two bSSFP scans were acquired with anterior as the readout gradient direction in the

first scan and posterior in the second scan. In both scans, the readout gradient was 4.5 mT/m. Both datasets were summed to create an overlay. Furthermore, the geometric image correction provided by the software of the MR scanner was applied to correct for the scanner-induced geometric image distortions.

In Figure 27.11, the white dots represent the cross sections of the phantom tubes. The robot was at the feet side of the phantom, where two shifted dots (belonging to the same tube) can be distinguished in the constructed overlays (Figure 27.11a and b). The dot shift arises in the direction of the readout gradient and is due to the static magnetic field error caused by the robot. Its magnitude equals half of the interdot distance and decreases with distance from the robot, for example, 5.3 mm at the outer tube position close to the robot to 1.5 mm in the scanner isocenter. This corresponds to a static field distortion of 16 and 5 ppm, respectively. In the scans acquired without the robot present, only single dots can be distinguished, implying that the static field error caused by the MR scanner is negligible.

The magnitude of the gradient error is defined as the distance between the mean dot position and the expected tube position based on the geometry of the phantom (cross point of the dashed lines in Figure 27.11). As illustrated in Figure 27.11b and d, the geometric image correction provided by the software of the MR scanner can strongly reduce the magnitude of the gradient error (average tube position moves toward expected one). The differences in the gradient error between the situation with and without the robot present are negligible, except in the feet head direction close to the robot. This implies that the robot slightly affects the magnitude of the gradient error in this direction, with a gradient error increase < 2 mm.

27.4.4.2 Susceptibility Artifact at Needle Tip

As described in Section 27.4.3.3, a material with susceptibility other than its surrounding medium induces a signal intensity artifact in the MR image. The appearance of this artifact depends on many geometric factors, for example, material, shape, diameter, and orientation (Port and Pomper 2000; Guermazi et al. 2003; Müller-Bierl et al. 2004; Lagerburg et al. 2008; Kühn et al.

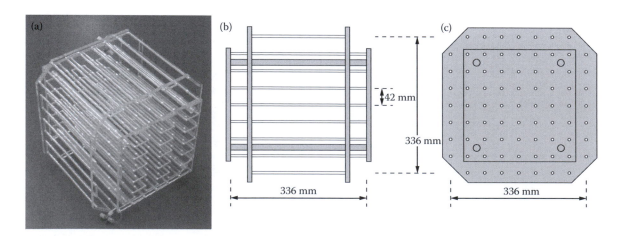

FIGURE 27.10 (a) Tube phantom and (b and c) details of phantom geometry. (Courtesy of M.A. Moerland.)

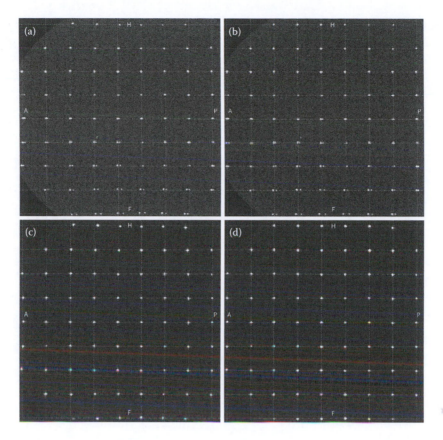

FIGURE 27.11　(a) Overlay of two sagittal bSSFP phantom scans with opposed directions of the readout gradient and the robot at the feet side; (b) overlay of both scans after applying the geometric image correction; (c) overlay of two bSSFP phantom scans with opposed directions of the readout gradient for a situation without the robot; (d) after applying the geometric image correction. (Courtesy of M.R. van den Bosch.)

2010), and acquisition parameters, for example, sequence type, echo time, and strength of readout gradient (Port and Pomper 2000; Guermazi et al. 2003; Müller-Bierl et al. 2004; Graf et al. 2005). In general, an intervention needle induces such an artifact, which may decrease the needle tip placement accuracy.

Figure 27.12 illustrates the artifact dependency on the readout direction and on the susceptibility difference between the intervention needle and the surrounding medium. A titanium needle ($L = 20$ cm, $d = 1.65$ mm) was inserted into two different phantoms to a depth of 10 cm in a 1.5-T Achieva scanner.

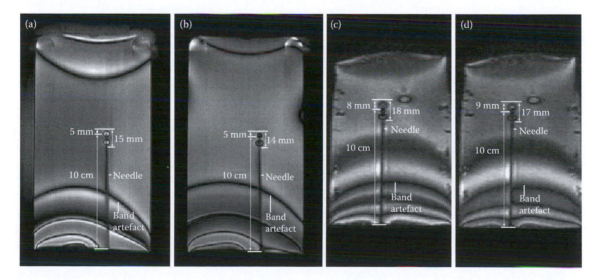

FIGURE 27.12　Needle artifact in gelatine phantom for sagittal image with (a) readout direction F and (b) readout direction H. Needle artifact in cheese phantom for coronal image with (c) readout direction F and (d) readout direction H. (Courtesy of M.R. van den Bosch.)

The first phantom was a plastic cylinder (L = 20 cm, d = 11 cm) filled with gelatine. The second phantom was a piece of Dutch cheese. Two bSSFP scans with opposed directions of the readout gradient were acquired. The readout gradient strength was high to minimize the geometric image distortions. The readout directions were F and H.

In the MR image of the gelatine phantom, the tip artifact is 15 mm in the scan with readout direction F and equals 14 mm in the scan with readout direction H. For the cheese phantom, the tip artifact is 18 and 17 mm in the image with readout directions F and H, respectively. The distance between the needle tip and the end of the artifact amounted to 5 mm in the gelatine phantom and 8–9 mm in the cheese phantom. This difference is caused by the susceptibility difference of the phantom material.

The uncertainty in the needle tip localization can be reduced by the use of needle materials closer to human tissue and the optimization of the image sequence, for example, larger readout gradient strength, smaller echo time, and addition of an extra 180° refocusing pulse to compensate for static magnetic field inhomogeneities (Port and Pomper 2000; Guermazi et al. 2003; Müller-Bierl et al. 2004; Graf et al. 2005; Lagerburg et al. 2008; Kühn et al. 2010). Furthermore, a marker placed on the robotic device, for example, on a needle guide close to the patient, can be used as a reference point to estimate the required needle insertion depth to reach the target (Muntener et al. 2008).

27.4.4.3 Tissue Deformation during Needle Insertion

The third source that hampers the needle placement accuracy is the tissue deformation during needle insertion, as observed in experiments on patients (Lagerburg et al. 2006b; van den Bosch et al. 2010b). These deformations are unpredictable (Lagerburg et al. 2005) but can be minimized using advanced insertion techniques, such as needle tapping (Lagerburg et al. 2006a), shooting (Muntener et al. 2008), and axial needle rotation (Abolhassani et al. 2006).

27.4.5 Safety

Regarding safety to the patient, there are two main challenges that MRI-guided robotic devices have to cope with, namely, needle deflection and RF-induced heating.

27.4.5.1 Needle Deflection

Needle deflection is the primary source of needle placement error, especially for needles with an asymmetrical bevel (Blumenfeld et al. 2007). Moreover, it can lead to undesired piercing of critical structures. By stepwise needle insertion and real-time MRI, needle trajectory can be controlled (van den Bosch et al. 2010b). If necessary, the needle can be retracted for reinsertion. Figure 27.13 demonstrates the feasibility of real-time needle tracking in a patient study. Fast bSSFP scans were acquired in a few seconds to monitor the needle trajectory in two orthogonal planes (coronal and sagittal) with the intersection line on the needle. At the tip of the needle, a susceptibility artifact is clearly visible.

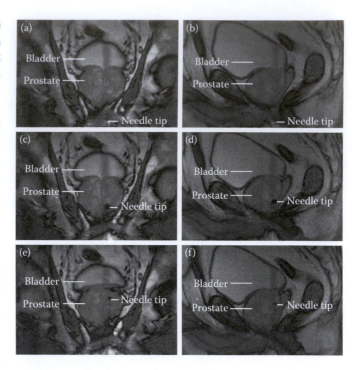

FIGURE 27.13 Coronal (a, c, e) and sagittal (b, d, f) bSSFP images during needle tapping at different time points. (Courtesy of M.A. Moerland.)

27.4.5.2 RF-Induced Heating

The elongated conductive needle can act as a dipole antenna, which interacts with the electromagnetic RF field applied to generate an MR image (Konings et al. 2000; Dempsey et al. 2001; Pictet et al. 2002; Yeung et al. 2002, 2007; Park et al. 2007; van den Bosch et al. 2010a). The altered electric field is strongly increased at the tip of the needle, resulting in tissue heating in this region (Park et al. 2007). In case of resonance, temperature increases exceeding 35°C were reported in phantom experiments with long conductive wires (Konings et al. 2000; Dempsey et al. 2001). The resonance properties and the amount of heat deposition depend on many factors such as electrical properties (Dempsey et al. 2001; Yeung et al. 2002) and volume (Pictet et al. 2002; Yeung et al. 2007) of the surrounding medium, the needle position within the MR bore (Yeung et al. 2007), the needle insertion depth or length (Konings et al. 2000; Dempsey et al. 2001; Yeung et al. 2002, 2007; van den Bosch et al. 2010a), and the sequence RF power (Konings et al. 2000; Yeung et al. 2002).

Due to these multiple factors, the amount of tissue heating is hard to predict. To investigate whether the intended intervention is safe with respect to the RF heating, several phantom experiments need to be performed that mimic worst-case scenarios for the patient. Mostly, fiber-optic temperature sensors are positioned at the needle tip to measure the temperature increase (Konings et al. 2000; Dempsey et al. 2001; Pictet et al. 2002; Yeung et al. 2002, 2007; van den Bosch et al. 2010a). Since the heating is deposited locally, the measured temperature is highly dependent on the location of the temperature sensor with respect to the needle tip (Park et al. 2007; Ehses et al. 2008). Another way to monitor the induced heating is by performing MR thermometry (Ehses et al. 2008). However, MR thermometry is not possible close to the needle tip due to the susceptibility artifact (Ehses et al. 2008). Van den Bosch et al. (2010a) developed a method based on RF-induced image artifacts to investigate the RF safety, and Overall et al. (2010) developed a qualitative technique to estimate the safety risk based on reversed RF polarization. The last two methods are effective for detecting potentially dangerous situations noninvasively, although they do not measure the exact heat deposition at the needle tip.

At 1.5 T, typical wavelengths in body tissue range from 43 cm in muscle to 105 cm in tissue with fat content (Mohsin et al. 2008). The wavelength in air is larger, namely, 470 cm (Dempsey et al. 2001). This implies that (in theory) the minimal in vivo resonance length (half of wavelength) is 21.5 cm when a conductive needle is completely surrounded by muscle. Nevertheless, the resonance peak can be broad and differ from the calculated one (Yeung et al. 2002). It is also reported that the tip heating increases quadratically with off-center distance (Yeung et al. 2007). The risk of thermal injury is small for interventions at 1.5 T with a titanium needle and insertion depth smaller than 15 cm close to the scanner isocenter.

The risk of tissue heating at the needle tip can be further reduced by using coated needles (Yeung et al. 2002) or can even be excluded by using nonconductive needle materials. In theory, the RF waves can also induce currents in conductive robotic components, resulting in heating in the surrounding tissue. Therefore, any contact of conductive robotic components (except the needle) to the patient should be avoided.

References

Abolhassani, N., Patel, R., and Moallem, M., 2006. Control of soft tissue deformation during robotic needle insertion. *Minim Invasive Ther Allied Technol* 15:165–76.

Afsharpour, H., D'Amours, M., Coté, B. et al. 2008. A Monte Carlo study on the effect of seed design on the interseed attenuation in permanent prostate implants. *Med Phys* 35:3671–9.

Aigner, F., Pallwein, L., Junker, D. et al. 2010. Value of real-time elastography targeted biopsy for prostate cancer detection in men with prostate specific antigen 1.25 ng/ml or greater and 4.00 ng/ml or less. *J Urol* 184:913–7.

Aronowitz, J.N. 2010. Don Lawrence and the "k-capture" revolution. *Brachytherapy* 9:373–81.

Awan, S.B., Meigooni, A.S., Mokhberiosgouei, R. et al. 2006. Evaluation of TG-43 recommended 2D-anisotropy function for elongated brachytherapy sources. *Med Phys* 33:4271–9.

Awan, S.B., Dini, S.A., Houssain, M. et al. 2008. Cylindrical coordinate-based TG-43U1 parameters for dose calculation around elongated brachytherapy sources. *J Appl Clin Med Phys* 9:123–42.

Bakker, C.J.G., Moerland, M.A., Bhagwandien, R., and Beersma, R. 1992. Analysis of machine-dependent and object-induced geometric distortion in 2DFT MR imaging. *Magn Reson Imaging* 10:597–608.

Ballester, F., Hernandez, C., Pérez-Calatayud, J. et al. 1997. Monte Carlo calculation of dose rate distributions around ^{192}Ir wires. *Med Phys* 24:1221–8.

Bannon, E.A., Yang, Y., and Rivard, M.J. 2011. Accuracy assessment of the superposition principle for evaluating dose distributions of elongated and curved [103]Pd and ^{192}Ir brachytherapy sources. *Med Phys* 38:2957–63.

Bax, J., Smith, D., Bartha, L. et al. 2011. A compact mechatronic system for 3D ultrasound guided prostate interventions. *Med Phys* 38:1055–69.

Beaulieu, L., Evans, D., Aubin, S. et al. 2007. By passing the learning curve in permanent seed implants using state-of-the-art technology. *Int J Radiat Oncol, Biol, Phys* 67:71–7.

Bernard, S. and Vynckier, S. 2005. Dosimetric study of a new polymer encapsulated palladium-103 seed. *Phys Med Biol* 50:1493–504.

Blake, C.C., Elliot, T.L., Slomka, P.J. et al. 2000. Variability and accuracy of measurements of prostate brachytherapy seed position in vitro using three-dimensional ultrasound: An intra- and inter-observer study. *Med Phys* 27:2788–95.

Blumenfeld, P., Hata, N., DiMaio, S. et al. 2007. Transperineal prostate biopsy under magnetic resonance image guidance: a needle placement accuracy study. *J Magn Reson Imaging* 26:688–94.

Braeckman, J., Autier, P., Garbar, C. et al. 2008. Computer-aided ultrasonography (HistoScanning): A novel technology for locating and characterizing prostate cancer. *BJU Int* 101:293–8. http://www.histoscanning.com/, last accessed January 25, 2012.

Chen, P.Y. and Vicini, F.A. 2007. Partial breast irradiation: Patient selection, guidelines for treatment, and current results. *Front Radiat Ther Oncol* 40:253–71.

Cheng, S., and Rifkin, M.D. 2001. Color Doppler imaging of the prostate: important adjunct to endorectal ultrasound of the prostate in the diagnosis of prostate cancer. *Ultrasound Q* 17:185–9.

Chiu-Tsao, S-T., Schaart, D.R., Soares, C.G., and Nath, R. 2007. Dose calculation formalisms and consensus dosimetry parameters for intravascular brachytherapy dosimetry: Recommendations of the AAPM Therapy Physics Committee Task Group No. 149. *Med Phys* 34:4126–58.

Cho, J.Y., Kim, S.H., and Lee, S.E. 1998. Diffuse prostatic lesions: role of color Doppler and power Doppler ultrasonography. *J Ultrasound Med* 17:283–7.

Cleary, K., Zhang, H., Glossop, N. et al. 2005. Electromagnetic tracking for image-guided abdominal procedures: overall system and technical issues. *IEEE EMBC* 2005, Shanghai, China, September 1–4, p. 268 (abstract).

Condon, B., Hadley, D.M., and Hodgson, R. 2001. The ferromagnetic pillow: a potential MR hazard not detectable by a hand-held magnet. *Br J Radiol* 74:847–51.

Czarnota, G.J., Kolios, M.C., Abraham, J. et al. 1999. Ultrasound imaging of apoptosis: high-resolution non-invasive monitoring of programmed cell death in vitro, in situ and in vivo. *Br J Cancer* 81:520–7.

Davidson, M.T.M., Yuen, J., D'Souza, D.P., and Batchelar, D.L. 2008. Image-guided cervix high-dose-rate brachytherapy treatment planning: Does custom computed tomography planning for each insertion provide better conformal avoidance of organs at risk? *Brachytherapy* 7:37–42.

De Jean, P., Beaulieu, L., and Fenster, A. 2009. Three-dimensional ultrasound system for guided breast brachytherapy. *Med Phys* 36:5099–106.

Dempsey, M.F., Condon, B., and Hadley, D.M. 2001. Investigation of the factors responsible for burns during MRI. *J Magn Reson Imaging* 13:627–31.

Dempsey, M.F., Condon, B., and Hadley, D.M. 2002. MRI safety review. *Semin Ultrasound CT MR* 23:392–401.

DeWerd, L.A., Micka, J.A., Holmes, S.M. et al. 2006. Calibration of multiple LDR brachytherapy sources. *Med Phys* 33:3804–13.

Dezarn, W.A., Cessna, J.T., DeWerd, L.A. et al. 2011. Recommendations of the American Association of Physicists in Medicine on dosimetry, imaging, and quality assurance procedures for ^{90}Y microsphere brachytherapy in the treatment of hepatic malignancies. *Med Phys* 38:4824–45.

DiBiase, S.J., Hosseinzadeh, K., Gullapalli, R.P. et al. 2002. Magnetic resonance spectroscopic imaging-guided brachytherapy for localized prostate cancer. *Int J Radiat Oncol, Biol, Phys* 52:429–38.

DiMaio, S.P., Pieper, S., Chinzei, K. et al. 2007. Robot-assisted needle placement in open MRI: System architecture, integration and validation. *Comput Aided Surg* 12:15–24.

Ding, M. and Fenster, A. 2003. A real-time biopsy needle segmentation technique using Hough transform. *Med Phys* 30:2222–33.

Ding, M., Wei, Z., Gardi, L. et al. 2006. Needle and seed segmentation in intra-operative 3D ultrasound-guided prostate brachytherapy. *Ultrasonics* 44S1:e331–6.

Dini, S.A., Awan, S.B., Dou, K. et al. 2007. TG-43U1 parameterization of elongated RadioCoil ^{103}Pd brachytherapy sources. *J Appl Clin Med Phys* 8:60–75.

Draper, K.J., Blake, C.C., Gowman, L. et al. 2000. An algorithm for automatic needle localization in ultrasound-guided breast biopsies. *Med Phys* 27:1971–9.

Ehses, P., Fidler, F., Nordbeck, P. et al. 2008. MRI thermometry: Fast mapping of RF-induced heating along conductive wires. *Magn Reson Med* 60:457–61.

Fenster, A., Downey, D. B., and Cardinal, H.N. 2001. Three-dimensional ultrasound imaging. *Phys Med Biol* 46:R67–99.

Fichtinger, G., Fiene, J.P., Kennedy, C.W. et al. 2008. Robotic assistance for ultrasound-guided prostate brachytherapy. *Med Image Anal* 12:535–45.

Fischer, G.S., Iordachita, I., Csoma, C. et al. 2008. MRI-compatible pneumatic robot for transperineal prostate needle placement. *IEEE/ASME Trans Mechatron* 13:295–305.

Fraser, D.J., Chen, Y., Poon, E. et al. 2010. Dosimetric consequences of misalignment and realignment in prostate 3DCRT using intramodality ultrasound image guidance. *Med Phys* 37:2787–95.

Frauscher, F., Klauser, A., Volgger, H. et al. 2002. Comparison of contrast enhanced color Doppler targeted biopsy with conventional systematic biopsy: Impact on prostate cancer detection. *J Urol* 167:1648–52.

Garra, B.S. 2011. Elastography: Current status, future prospects, and making it work for you. *Ultrasound Q* 27:177–86.

Garra, B.S., Cespedes, E.I., Ophir, J. et al. 1997. Elastography of breast lesions: initial clinical results. *Radiology* 202:79–86.

Gaudet, M., Vigneault, E., Aubin, S. et al. 2010. Dose escalation to the dominant intraprostatic lesion defined by sextant biopsy in a permanent prostate I-125 implant: A prospective comparative toxicity analysis. *Int J Radiat Oncol, Biol, Phys* 77:153–9.

Gossmann, A., Bangard, C., Warm, M. et al. 2008. Real-time MR-guided wire localization of breast lesions by using an open 1.0-T imager: Initial experience. *Radiology* 247:535–42.

Graf, H., Steidle, G., Martirosian, P. et al. 2005. Metal artifacts caused by gradient switching. *Magn Reson Med* 54:231–4.

Graf, H., Lauer, U.A., and Schick, F. 2006. Eddy-current induction in extended metallic parts as a source of considerable torsional moment. *J Magn Reson Imaging* 23:585–90.

Guermazi, A., Miaux, Y., Zaim, S. et al. 2003. Metallic artefacts in MR imaging: effects of main field orientation and strength. *Clin Radiol* 58:322–8.

Haacke, E.M., Brown, R.W., Thompson, M.R., and Venkatesan, R. 1999. *Magnetic Resonance Imaging—Physical Principles and Sequence Design*, 1st ed. John Wiley & Sons, New York, p. 847.

Hambrock, T., Somford, D.M., Hoeks, C. et al. 2010. Magnetic resonance imaging guided prostate biopsy in men with repeat negative biopsies and increased prostate specific antigen. *J Urol* 183:520–7.

Han, B.H., Wallner, K., Merrick, G. et al. 2003. Prostate brachytherapy seed identification on post-implant TRUS images. *Med Phys* 30:898–900.

Holm, H.H., Juul, N., Pedersen, J.F., Hansen, H., and Stroyer, I. 1983. Transperineal ^{125}iodine seed implantation in prostatic cancer guided by transrectal ultrasonography. *J Urol* 130:283–6.

Holm, H.H., Torp-Pedersen, S., and Myschetzky, P. 1990. Transperineal seed-implantation guided by biplanar transrectal ultrasound. *Urology* 36:249–52.

Hungr, N., Troccaz, J., Zemiti, N., and Tripodi, N. 2009. Design of an ultrasound-guided robotic brachytherapy needle-insertion system. *Conf Proc IEEE Eng Med Biol Soc* 1:250–3.

Jezzard, P. and Clare, S. 1999. Sources of distortion in functional MRI data. *Hum Brain Mapp* 8:80–5.

Jürgenliemk-Schulz, I.M., Tersteeg, R.J.H.A., Roesink, J.M. et al. 2009. MRI-guided treatment-planning optimisation in intracavitary or combined intracavitary/interstitial PDR brachytherapy using tandem ovoid applicators in locally advanced cervical cancer. *Radiother Oncol* 93:322–30.

Karaiskos, P., Papagiannis, P., Angelopoulos, A. et al. 2001. Dosimetry of ^{192}Ir wires for LDR interstitial brachytherapy following the AAPM TG-43 dosimetric formalism. *Med Phys* 28:156–66.

Kay, P.A., Robb, R.A., and Bostwick, D.G. 1998. Prostate cancer microvessels: A novel method for three-dimensional reconstruction and analysis. *Prostate* 37:270–7.

Kennedy, R.M., Davis, S.D., Micka, J.A., and DeWerd, L.A. 2010. Experimental and Monte Carlo determination of TG-43 dosimetric parameters for the model 9011 ThinSEED™ brachytherapy source. *Med Phys* 37:1681–8.

Kerkhof, E.M., Raaymakers, B.W., van der Heide, U.A. et al. 2008. Online MRI guidance for healthy tissue sparing in patients with cervical cancer: An IMRT planning study. *Radiother Oncol* 88:241–9.

Khan, S., Chen, Z.J. and Nath, R. 2008. Photon energy spectrum emitted by a novel polymer-encapsulated ^{103}Pd source and its effect on the dose rate constant. *Med Phys* 35:1403–6.

Kline, R.W., Gillin, M.T., Grimm, D.F. et al. 1985. Computer dosimetry of ^{192}Ir wire. *Med Phys* 12: 634–8.

Konings, M.K., Bartels, L.W., Smits, H.F.M., and Bakker, C.J.G. 2000. Heating around intravascular guidewires by resonating RF waves. *J Magn Reson Imaging* 12:79–85.

Kubo, H. 1985. Exposure contribution from Ti K x rays produced in the titanium capsule of the clinical ^{125}I seed. *Med Phys* 12:215–20.

Kühn, J.P., Langner, S., Hegenscheid, K. et al. 2010. Magnetic resonance-guided upper abdominal biopsies in a high-field wide-bore 3-T MRI system: Feasibility, handling, and needle artifacts. *Eur Radiol* 20:2414–21.

Kuligowska, E., Barish, M.A., Fenlon, H.M., and Blake, M. 2001. Predictors of prostate carcinoma: Accuracy of gray-scale and color Doppler US and serum markers. *Radiology* 220:757–64.

Lagerburg, V., Moerland, M.A., Lagendijk, J.J.W., and Battermann, J.J. 2005. Measurement of prostate rotation during insertion of needles for brachytherapy. *Radiother Oncol* 77:318–23.

Lagerburg, V., Moerland, M.A., Konings, M.K. et al. 2006a. Development of a tapping device: A new needle insertion method for prostate brachytherapy. *Phys Med Biol* 51:891–902.

Lagerburg, V., Moerland, M.A., van Vulpen, M., and Lagendijk, J.J.W. 2006b. A new robotic needle insertion method to minimise attendant prostate motion. *Radiother Oncol* 80:73–7.

Lagerburg, V., Moerland, M.A., Seppenwoolde, J.H., and Lagendijk, J.J.W. 2008. Simulation of the artefact of an iodine seed placed at the needle tip in MRI-guided prostate brachytherapy. *Phys Med Biol* 53:59–67.

LeCarpentier, G.L., Roubidoux, M.A., and Fowlkes, J.B. 2008. Suspicious breast lesions: Assessment of 3D Doppler US indexes for classification in a test population and fourfold cross-validation scheme. *Radiology* 249:463–70.

Leotta, D.F., Detmer, P.R., and Martin, R.W. 1997. Performance of a miniature magnetic position sensor for three-dimensional ultrasound imaging. *Ultrasound Med Biol* 23:597–609.

Lucidarme, O., Akakpo, J.P., Granberg, S. et al. 2010. A new computer-aided diagnostic tool for non-invasive characterisation of malignant ovarian masses: Results of a multicentre validation study. *Eur Radiol* 20:1822–30.

Machann, J., Schlemmer, H.P., and Schick, F. 2008. Technical challenges and opportunities of whole-body magnetic resonance imaging at 3T. *Phys Med* 24:63–70.

Mamou, J. and Feleppa, E.J. 2007. Singular spectrum analysis applied to ultrasonic detection and imaging of brachytherapy seeds. *J Acoust Soc Am* 121:1790–801.

Meigooni, A.S., Dini, S.A., Sowards, K. et al. 2002. Experimental determination of the TG-43 dosimetric characteristics of EchoSeed™ model 6733 ^{125}I brachytherapy source. *Med Phys* 29:939–42.

Meigooni, A.S., Zhang, H., Clark, J.R. et al. 2004. Dosimetric characteristics of the new RadioCoil ^{103}Pd wire line source for use in permanent brachytherapy implants. [Erratum appears in *Med Phys.* 2006 Aug;33(8):3077] *Med Phys* 31:3095–105.

Meigooni, A.S., Awan, S.B., Rachabatthula, V. et al. 2005. Treatment planning consideration for prostate implants with the new linear RadioCoil ^{103}Pd brachytherapy source. *J Appl Clin Med Phys* 6:23–36.

Meigooni, A.S., Awan, S.B., Dou, K. et al. 2006. Feasibility of calibrating elongated brachytherapy sources using a well-type ionization chamber. *Med Phys* 33:4184–9.

Meltsner, M.A., Ferrier, N.J., and Thomadsen, B.R. 2007. Observations on rotating needle insertions using a brachytherapy robot. *Phys Med Biol* 52:6027–37.

Melzer, A., Gutmann, B., Remmele, T. et al. 2008. INNOMOTION for percutaneous image-guided interventions: Principles and evaluation of this MR- and CT-compatible robotic system. *IEEE Eng Med Biol Mag* 27:66–73.

Mohsin, S.A., Sheikh, N.M., and Saeed, U. 2008. MRI-induced heating of deep brain stimulation leads. *Phys Med Biol* 53:5745–56.

Moran B.J. 2012. Comparison of Health Related Quality of Life and Other Clinical Parameters Between ThinSeed™ and OncoSeed™ for Permanent Low Dose Rate Implantation in Localized Prostate Cancer. Available at http://clinicaltrials .gov/ct2/show/study/NCT01379742, last accessed January 25, 2012.

Morikawa, S., Naka, S., Murakami, K. et al. 2009. Preliminary clinical experiences of a motorized manipulator for magnetic resonance image-guided microwave coagulation therapy of liver tumors. *Am J Surg* 198:340–7.

Mozer, P.C., Partin, A.W., and Stoianovici, D. 2009. Robotic image-guided needle interventions of the prostate. *Rev Urol* 11:7–15.

Müller-Bierl, B., Graf, H., Lauer, U. et al. 2004. Numerical modeling of needle tip artifacts in MR gradient echo imaging. *Med Phys* 31:579–87.

Muntener, M., Patriciu, A., Petrisor, D. et al. 2008. Transperineal prostate intervention: Robot for fully automated MR imaging-system description and proof of principle in a canine model. *Radiology* 247:543–9.

Narayan P., Gajendran, V., Taylor, S.P. et al. 1995. The role of transrectal ultrasound-guided biopsy-based staging, preoperative serum prostate-specific antigen, and biopsy Gleason score in prediction of final pathologic diagnosis in prostate cancer. *Urology* 46:205–12.

Nath, R., Anderson, L.L., Luxton, G. et al. 1995. Dosimetry of interstitial brachytherapy sources: Recommendations of the AAPM Radiation Therapy Committee Task Group No. 43. *Med Phys* 22:209–34.

Ophir, J., Céspedes, I., Ponnekanti, H. et al. 1991. Elastography: A quantitative method for imaging the elasticity of biological tissues. *Ultrason Imaging* 13:111–34.

Overall, W.R., Pauly, J.M., Stang, P.P., and Scott, G.C. 2010. Ensuring safety of implanted devices under MRI using reversed RF polarization. *Magn Reson Med* 64:823–33.

Park, S.M., Kamondetdacha, R., and Nyenhuis, J.A. 2007. Calculation of MRI-induced heating of an implanted medical lead wire with an electric field transfer function. *J Magn Reson Imaging* 26:1278–85.

Paxton, A.B., Culberson, W.S. DeWerd, L.A. et al. 2008. Primary calibration of coiled ^{103}Pd brachytherapy sources. *Med Phys* 35:32–8.

Pérez-Calatayud, J., Ballester, F., Granero, D. et al. 2003. Influence of the non-homogeneity of Ir-192 wires in calibration of well chambers. *Phys Med Biol* 48:3961–8.

Peters, N.H.G.M., Meeuwis, C., Bakker, C.J.G. et al. 2009. Feasibility of MRI-guided large-core-needle biopsy of suspicious breast lesions at 3 T. *Eur Radiol* 19:1639–44.

Pictet, J., Meuli, R., Wicky, S., and van der Klink, J.J. 2002. Radiofrequency heating effects around resonant lengths of wire in MRI. *Phys Med Biol* 47:2973–85.

Pignol, J.P., Keller, B., Rakovitch, E. et al. 2006. First report of a permanent breast seed implant PBSI as adjuvant radiation for early stage breast cancer. *Int J Radiat Oncol, Biol, Phys* 64:176–81.

Podder, T.K., Ng, W.-S., and Yu, Y. 2007. Multi-channel robotic system for prostate brachytherapy. *Conf Proc IEEE Eng Med Biol Soc* 2007:1233–6.

Podder, T.K., Buzurovic, I., Huang, K. et al. 2011. Reliability of EUCLIDIAN: An autonomous robotic system for image-guided prostate brachytherapy. *Med Phys* 38:96–106.

Polo, A., Salembier, C., Venselaar, J., Hoskin, P. et al. 2010. Review of intraoperative imaging and planning techniques in permanent seed prostate brachytherapy. *Radiother Oncol* 94:12–23.

Port, J.D. and Pomper, M.G. 2000. Quantification and minimization of magnetic susceptibility artifacts on GRE images. *J Comput Assist Tomogr* 24:958–64.

Pouliot, J., Kim, Y., Lessard, E. et al. 2004. Inverse planning for HDR prostate brachytherapy used to boost dominant intra-prostatic lesions defined by magnetic resonance spectroscopy imaging. *Int J Radiat Oncol Biol Phys* 59:1196–207.

Prager, R.W., Ijaz, U.Z., Gee, A.H., and Treece, G.M. 2010. Three-dimensional ultrasound imaging. *Proc Inst Mech Eng* 224:193–223.

Rivard, M.J. 2009. Monte Carlo radiation dose simulations and dosimetric comparison of the model 6711 and 9011 ^{125}I brachytherapy sources. *Med Phys* 36:486–91.

Rivard, M.J., Coursey, B.M., DeWerd, L.A. et al. 2004. Update of AAPM Task Group No. 43 Report: A revised AAPM protocol for brachytherapy dose calculations (AAPM Report No. 84). *Med Phys* 31:633–74.

Robertson, N.M., Diaz-Gomez, M., and Condon, B. 2000. Estimation of torque on mechanical heart valves due to magnetic resonance imaging including an estimation of the significance of the Lenz effect using a computational model. *Phys Med Biol* 45:3793–807.

Salcudean, S.E., Prananta, T.D., Morris, W.J., and Spadinger, I. 2008. A robotic needle guide for prostate brachytherapy. *IEEE Int Conf Robotics Automation. ICRA* 2008:2975–81.

Schaefers, G. 2008. Testing MR safety and compatibility: an overview of the methods and current standards. *IEEE Eng Med Biol Mag* 27:23–7.

Schenck, J.F. 2000. Safety of strong, static magnetic fields. *J Magn Reson Imaging* 12:2–19.

Schouten, M.G., Ansems, J., Renema, W.K. et al. 2010. The accuracy and safety aspects of a novel robotic needle guide manipulator to perform transrectal prostate biopsies. *Med Phys* 37:4744–50.

Shellock, F.G. 2002. Biomedical implants and devices: assessment of magnetic field interactions with a 3 0-Tesla MR system. *J Magn Reson Imaging* 16:721–32.

Simmons, L.A., Autier, P., Zát'ura, F. et al. 2011. Detection, localisation and characterisation of prostate cancer by Prostate HistoScanning(™). *BJU Int* 110:28–35.

Sowards, K.T. and Meigooni, A.S. 2002. A Monte Carlo evaluation of the dosimetric characteristics of the EchoSeed™ model 6733 ^{125}I brachytherapy source. *Brachytherapy* 1:227–32.

Stoianovici, D. 2000. Robotic surgery. *World J Urol* 18:289–95.

Surry, K.J., Smith, W.L., Campbell, L.J. et al. 2002. The development and evaluation of a three-dimensional ultrasound-guided breast biopsy apparatus. *Med Image Anal* 6:301–12.

Tempany, C., Straus, S., Hata, N., and Haker, S. 2008. MR-guided prostate interventions. *J Magn Reson Imaging* 27:356–67.

Tewari, A., Indudhara, R., Shinohara, K. et al. 1996. Comparison of transrectal ultrasound prostatic volume estimation with magnetic resonance imaging volume estimation and surgical specimen weight in patients with benign prostatic hyperplasia. *J Clin Ultrasound* 24:169–174.

Thomadsen, B.R., Heaton 2nd, T.H., Jani, S.K. et al. 2010. Off-label use of medical products in radiation therapy: Summary of the Report of AAPM Task Group No. 121. *Med Phys* 37:2300–11.

Thomas, A., Kümmel, S., Gemeinhardt, O., and Fischer, T. 2007. Real-time sonoelastography of the cervix: Tissue elasticity of the normal and abnormal cervix. *Acad Radiol* 14:193–200.

Tong, S., Downey, D. B., Cardinal, H.N., and Fenster, A. 1996. A three-dimensional ultrasound prostate imaging system. *Ultrasound Med Biol* 22:735–46.

Trejos, A.L., Lin, A.W., Mohan, S., Bassan, H. et al. 2008. MIRA V: An integrated system for minimally invasive robot-assisted lung brachytherapy. *IEEE Int Conf Robotics Automation. ICRA* 2008:2982–7.

van den Bergen, B., van den Berg, C.A.T., Klomp, D.W.J., and Lagendijk, J.J.W. 2009. SAR and power implications of different RF shimming strategies in the pelvis for 7T MRI. *J Magn Reson Imaging* 30:194–202.

van den Bosch, M.A.A.J., Daniel, B.L., Pal, S. et al. 2006. MRI-guided needle localization of suspicious breast lesions: Results of a freehand technique. *Eur Radiol* 16:1811–7.

van den Bosch, M.R., Moerland, M.A., Lagendijk, J.J.W. et al. 2010a. New method to monitor RF safety in MRI-guided interventions based on RF induced image artefacts. *Med Phys* 37:814–21.

van den Bosch, M.R., Moman, M.R., van Vulpen, M. et al. 2010b. MRI-guided robotic system for transperineal prostate interventions: Proof of principle. *Phys Med Biol* 55:133–40.

van der Laarse, R., Ganero, D., Pérez-Calatayud, J. et al. 2008. Dosimetric characterization of Ir-192 LDR elongated sources. *Med Phys* 35:1154–61.

Van Gellekom, M.P., Moerland, M.A., Wijrdeman, H.K., and Battermann, J.J. 2004a. Quality of permanent prostate implants using automated delivery with seedSelectron versus manual insertion of RAPID Strands. *Radiother Oncol* 73:49–56.

Van Gellekom, M.P.R., Moerland, M.A., Battermann, J.J., and Lagendijk, J.J.W. 2004b. MRI-guided prostate brachytherapy with single needle method-a planning study. *Radiother Oncol* 71:327–32.

Vlad, R.M., Kolios, M.C., and Czarnota, G.J. 2011. Ultrasound imaging of apoptosis: Spectroscopic detection of DNA-damage effects at high and low frequencies. *Methods Mol Biol* 682:165–87.

Vollans, S.E. and Wilkinson, J.M. 2000. Calibration of pre-cut iridium-192 wires for low dose rate interstitial brachytherapy using a Farmer-type ionization chamber. *Br J Radiol* 73:201–5.

Wang, Z. and Hertel, N.E. 2005. Determination of dosimetric characteristics of OptiSeed™ a plastic brachytherapy ^{103}Pd source. *Appl Radiat Isot* 63:311–21.

Wei, Z., Wan, G., Gardi, L. et al. 2004. Robot-assisted 3D-TRUS guided prostate brachytherapy: System integration and validation. *Med Phys* 31:539–48.

Wei, Z., Gardi, L., Downey, D.B., and Fenster, A. 2005. Oblique needle segmentation and tracking for 3D TRUS guided prostate brachytherapy. *Med Phys* 32:2928–41.

Wen, X., Salcudean, S.T., and Lawrence, P.D. 2010. Detection of brachytherapy seeds using 3-D transrectal ultrasound. *IEEE Trans Biomed Eng* 57:2467–77.

Xue, J., Waterman, F., Handler, J., and Gressen, E. 2005. Localization of linked ^{125}I seeds in postimplant TRUS images for prostate brachytherapy dosimetry. *Int J Radiat Oncol, Biol, Phys* 62:912–9.

Yeung, C.J., Susil, R.C., and Atalar, E. 2002. RF safety of wires in interventional MRI: Using a safety index. *Magn Reson Med* 47:187–93.

Yeung, C.J., Karmarkar, P., and McVeigh, E.R. 2007. Minimizing RF heating of conducting wires in MRI. *Magn Reson Med* 58:1028–34.

Yu, N., Gassert, R., and Riener, R. 2011. Mutual interferences and design principles for mechatronic devices in magnetic resonance imaging. *Int J Comput Assist Radiol Surg* 6:473–88.

Zimmer, C., Janssen, M.N., Treschan, T.A., and Peters J. 2004. Near-miss accident during magnetic resonance imaging by a "flying sevoflurane vaporizer" due to ferromagnetism undetectable by handheld magnet. *Anesthesiology* 100:1329–30.

Intraoperative Brachytherapy

Joost Jan Nuyttens
Erasmus MC–Daniel den
Hoed Cancer Centre

Subir Nag
Northern California Kaiser
Permanente and Stanford
School of Medicine

28.1 Introduction

Intraoperative (high-dose-rate) brachytherapy (HDR-IORT) delivers a high dose of radiation during surgery to the tumor bed or the tumor while the normal tissues are protected from radiation by moving the critical structures away from the radiation field or by shielding the anatomically fixed tissues. By using this technique, complications can be reduced and/or the tumor dose escalated.

The terminology and abbreviations used in intraoperative radiotherapy literature can be confusing. Intraoperative radiotherapy can be performed with brachytherapy (HDR-IORT) or electron beam radiotherapy (IOERT). Implantation of radiation seeds at the time of surgery for continuous delivery of radiation is, strictly speaking, not intraoperative brachytherapy but permanent perioperative brachytherapy. Similarly, inserting hollow catheters into tumor or tumor bed at the time of surgery and subsequently inserting radiation sources will be referred to as perioperative removable brachytherapy and not as intraoperative radiotherapy. In this chapter, only the intraoperative brachytherapy will be discussed.

Intraoperative brachytherapy can be used for all kinds of tumors but is mainly used as a boost in the treatment of sarcomas, rectal cancers, and breast cancers. The intraoperative technique of breast cancer is addressed in Chapter 22.

Some centers use the intraoperative boost technique for all patients with locally advanced, recurrent, or inoperable rectal cancer, whereas other centers use it only when a close or positive margin is found on frozen section during the resection of the rectal cancer. For the sarcomas, the technique is used as an upfront boost during the operation, and postoperative radiotherapy of 50 Gy is given later (Alektiar et al. 2000b).

28.2 Technical Aspects

28.2.1 Shielded Facility in OR versus Radiation Oncology

A shielded operating room (OR) is required for a dedicated fixed HDR or mobile HDR-IORT facility, either in the Radiation Oncology Department or in the hospital operating suite. The shielding can be accomplished by lining an existing OR with lead, using an existing shielded treatment room, or constructing a room with appropriately thick concrete walls.

Another option being used for HDR-IORT by some institutions, where completely shielded ORs are unavailable, is to treat the patient within a lead-lined box permanently placed in the OR (a room within a room) after resection has been accomplished and the applicator for HDR-IORT has been positioned. Such a solution has been created at Duke University in Durham, NC.

Institutions not having a dedicated shielded OR can perform IORT by moving the anesthetized patient from the OR to the radiation oncology department for HDR-IORT. A special transportation cart and strict procedural policy are required to facilitate the transfer of the patient. Another alternative is to build an unshielded OR adjacent to the shielded radiation treatment room. Hence, the patient will have to be moved only for a short distance for IORT treatment. The latter situation exists and has functioned well at Medical College of Ohio in Toledo and at Thomas Jefferson University in Philadelphia, PA, and some other centers. However, it requires the institution to provide or at least to consider the availability of OR services such as specimen transport, blood bank support, sterilization, pharmacy, etc., in a location remote from the routine ORs.

28.2.2 Applicators

Several applicators are used and usually consist of a 5- to
10-mm-thick pad, which is used as a template. The pad is usually made of flexible silicon with 1-cm-spaced parallel source
guide tubes running through the center of the template. In the
Memorial Sloan-Kettering Cancer Center, New York, an 8-mm-
thick Harrison–Anderson–Mick (HAM) applicator is used. In
the Netherlands, in the Rotterdam Erasmus MC–Daniel den
Hoed Cancer Center, a 5-mm-thick pad (Flexible Intraoperative
Template, FIT type) is used (see Figure 28.1). The Ohio State
University, Columbus, uses different types of applicators, which
can be made of foam, rigid Delrin, or silicon of 1 cm thickness.
Strassmann et al. (2000) presented a 10-mm-thick pad made of
10-mm-diameter spherical pellets. Huber et al. (1996) used a
commercially available vinyl plastic on a synthetic oil base with
a thickness of 1 cm. Figure 28.2 shows a commercially available
flab material by Nucletron, Veenendaal, the Netherlands.

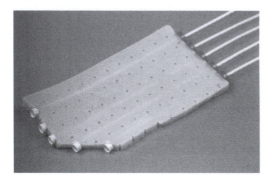

FIGURE 28.1 Pad used as an applicator (5 mm thick, Fit type) in the
Rotterdam Erasmus MC–Daniel den Hoed Cancer Centre. (Courtesy
of J.J. Nuyttens.)

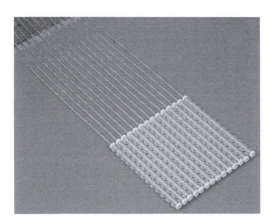

FIGURE 28.2 Commercially available applicator material by
Nucletron, Veenendaal, the Netherlands. (Courtesy of Nucletron.)

28.2.3 Application Technique

A collaboration between the surgeon and the radiation oncologist is essential. The surgeon explains to the radiation oncologist what kind of resection he or she will be performing. He or
she shows the tumor bed and also the extension of the macroscopic tumor if applicable and/or the region(s) with microscopic
positive margins. In that way, they will both define the target
area and, if possible, demarcate the margins with radio-opaque
fiducial markers or hemoclips to identify the target volume on
radiographs for future external beam radiotherapy (EBRT)
planning. He or she also will show the organs at risk (e.g., small
or large bowel, stomach, heart, ureter, kidney, bladder, etc.) close
to the target area. If possible, they retract or shield these organs
at risk with lead. Some centers do not place the markers but use
standard geometries instead in the treatment planning. The
size and shape of the pads are adjusted to the target surface. In
Figure 28.3, the size of the flexible applicator is checked, and if
necessary, the applicator can be cut to fit the tumor bed.

After positioning, the applicator is pressed against the target
area by filling the operation field with gauze pads. This is done
to avoid the bolus effect from blood and/or surgical fluid during IORT and to push critical organs away from the applicator.
Figure 28.4 shows an applicator in place. Most centers load the
whole pad and take two orthogonal radiographs to document
the treatment. The treatment planning is performed using the
standard geometries present in the treatment planning system.
After the treatment, the applicator is removed from the treatment site, and the surgeon closes the incision. The dose at the
clips, if placed, can be calculated using the reconstructed template geometry and the actual treatment times. These calculations can be used as a quality check of the treatment. At some
centers, tissue expanders are temporarily placed over the tumor
bed to displace uninvolved normal tissues from the tumor bed
during postoperative external beam irradiation.

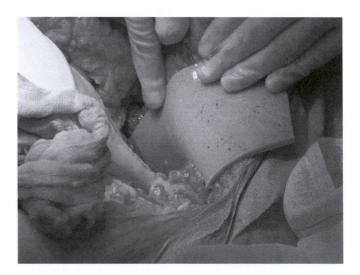

FIGURE 28.3 Size of the flexible applicator is checked. (Courtesy of
J.J. Nuyttens.)

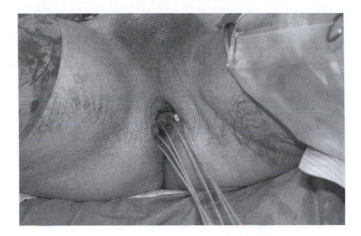

FIGURE 28.4 FIT-type applicator is in place. (Courtesy of J.J. Nuyttens.)

28.3 Dose, Fractionation, and Prescription

There is a wide variation in the boost doses used for HDR-IORT of rectal cancers in the literature. With the HAM applicator, usually, a single dose of 12 Gy is given at 0.5 cm from the applicator surface after negative resection margins and a single dose of 15 Gy for positive margins (Shibata et al. 2000; Harrison et al. 1995). However, higher single doses up to 18 Gy have been given to irradiate resected tumors (Alektiar et al. 2000a). Nag et al. (1998) used different types of applicators and used a single dose of 10–15 Gy for microscopic residual disease or close margins and 17.5–20 Gy for patients with gross residual disease. They prescribed to 0.5 cm from the surface of the applicator. Huber et al. (1996) prescribed a single dose of 15 Gy to the surface of the applicator. Nuyttens et al. (2004) prescribed a single dose of 10 Gy to 1 cm from the surface of the applicator. Dose distributions are shown in Figures 28.5 and 28.6.

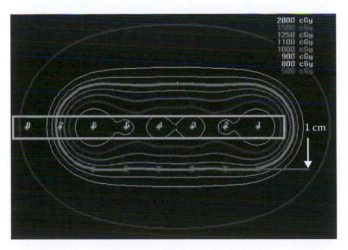

FIGURE 28.5 **(See color insert.)** Dose distribution in the axial plane. Catheters 7 and 8 are not loaded. The blue rectangle represents the applicator. A single dose of 10 Gy is prescribed to 1 cm from the surface of the applicator. (Courtesy of J.J. Nuyttens.)

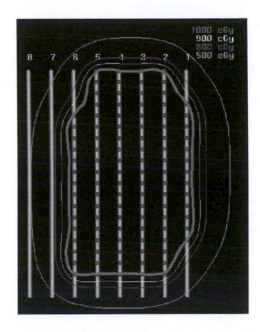

FIGURE 28.6 **(See color insert.)** Dose distribution in the sagittal plane. (Courtesy of J.J. Nuyttens.)

28.4 Clinical Results

Four centers reported their results with HDR-IORT in rectal cancer.

28.4.1 Memorial Sloan-Kettering Experience

From the Memorial Sloan-Kettering Hospital, Harrison and Alektiar reported their results. Harrison et al. (1998) treated 68 patients between November 1992 and December 1996 with primary unresectable or recurrent rectal cancer with HDR-IORT and reported the results of 66 patients. There were 22 patients with primary unresectable disease and 46 patients who presented with recurrent disease. In general, the patients with primary unresectable disease received preoperative chemotherapy with 5-fluorouracil (5-FU) and leucovorin and external beam irradiation to 4500–5040 cGy, followed by surgical resection and HDR-IORT (1000–2000 cGy). In general, the patients with recurrent disease were treated with surgical resection and HDR-IORT (1000–2000 cGy) alone. The median follow-up was 17.5 months (1–48 months). In primary cases, the actuarial 2-year local control was 81%. For patients with negative margins, the local control was 92% versus 38% for those with positive margins ($p < 0.002$). The 2-year actuarial disease-free survival was 69%, 77% for patients with negative margins versus 38% for patients with positive margins ($p < 0.03$). For patients with recurrent disease, the 2-year actuarial local control rate was 63%. For patients with negative margins, it was 82%, while it was 19% for those with positive margins ($p < 0.02$). The disease-free survival was 47% (71% for negative margins and 0% for positive margins) ($p < 0.04$).

Alektiar et al. (2000a) reported the results of 74 patients with locally recurrent rectal cancer who were treated with surgery and HDR-IORT. Additional EBRT was given to 29 patients, and 33 patients received 5-fluorouracil–based chemotherapy. Twenty-one of 74 had positive microscopic margins. The dose of HDR-IORT ranged from 10 to 18 Gy. With a median follow-up of 22 months, the 5-year local control, distant metastasis disease-free, disease-free, and overall survival rates were 39%, 39%, 23%, and 23%, respectively. The only predictor of improved local control was a negative margin of resection with a 5-year local control rate of 43%, compared to 26% in those with positive margin ($p < 0.02$). For overall survival, a negative microscopic margin ($p < 0.04$) and the use of IORT with EBRT ($p < 0.04$) were significant predictors of improved survival.

28.4.2 Rotterdam Erasmus MC Experience

Nuyttens et al. (2004) from the Erasmus MC-Daniel den Hoed Cancer Center reported the results of 37 patients with locally advanced or recurrent rectal cancer. Eighteen patients had locally advanced rectum cancer and 19 had locally recurrent rectal cancer. All patients received preoperative radiotherapy, most of them 50 Gy in 2 Gy/fraction. The median follow-up of the surviving patients was 3 years. Overall, 12 patients (32%) had local recurrence, 5 (14%) of which were in the HDR-IORT field. The 3-year local failure rate for primary tumors and recurrent tumors was 19% and 52%, respectively ($p \le 0.0042$). The 3-year local failure rate was 37% for negative margins and 26% for positive margins ($p < 0.51$). Out-of-field failures were seen earlier than in-field failures (median time 16 versus 31 months, $p < 0.077$). Eight of the 12 local failures were located in the posterior pelvis. The median distance of the out-of-field recurrence to the area treated with the applicator was 2 cm (range 1–5 cm). Four recurrences were found growing in the sacrum or sacral foramina. During surgery, the tumor bed was clipped, and the dose to the clips was calculated: a high mean dose at the clip (17.3 Gy) was found. The overall survival was significantly different for primary versus recurrent tumors, stage, and grade.

28.4.3 Ohio State University Experience

Martinez-Monge et al. from the Ohio State University reported on 80 patients with locally recurrent or para-aortic local relapses of colorectal cancer (Martinez-Monge et al. 1999; Nag et al. 1998). Twenty-three patients were treated with HDR-IORT; the others were treated with IOERT or ^{125}I brachytherapy. Thirteen of the 27 previously unirradiated patients received postoperative radiotherapy. Some previously irradiated patients received additional postoperative external beam to 20–30 Gy. After a median follow-up of 47 months (range 19–75 months), the 5-year local control rate for the whole group was 26%, with a median of 12 months. The 5-year local control rate for para-aortic relapses was 65% versus 11% for pelvic relapses. There was a trend toward better local control for patients with no prior radiation therapy, smaller treatment volume, absence of metastatic disease, and HDR-IORT.

28.4.4 Technische Universität, München, Experience

Huber et al. (1996) from the Technische Universität, München, treated 19 patients staged as T3 tumors (group 1) and 19 patients staged as T4 tumors (group 2). Patients in group 1 received postoperative radiochemotherapy, whereas patients in group 2 received preoperative radiochemotherapy. The mean follow-up was 25.5 months. Local recurrence developed in 16% of the patients in group 1 and in 10% in group 2. All recurrences were out of the IORT field.

28.5 Complications

It is difficult to separate clearly treatment-related complications from disease-related complications in patients with recurrent rectal cancers. Therefore, all authors report the complications that are grouped together with no attempt to separate between surgery- and radiation-related complications.

The most common type of toxicity reported by Alektiar et al. (2000a) was wound complication (24%), followed by ureter (23%) and bladder complications (20%). Peripheral nerve damage, which represents the dose-limiting toxicity for IORT, occurred in 16% (12/74) of patients with no grade 4 toxicity. There was no correlation between the IORT tumor dose and peripheral nerve damage. With an IORT dose of 12 Gy or less, the rate was 20% (5/25) compared to 14% (7/49) with IORT dose 15 Gy or more ($p = 0.5$).

Harrison et al. (1998) reported in 38% of the patients grade 3 or higher complications. Wound complications and infections were the most frequent (22% of the patients). Bladder and urinary complications were seen in 6%. Nerve complications were seen in two patients: one patient had a foot drop, and the other had poor sexual functioning.

Nuyttens et al. (2004) also reported many complications: a delay in wound healing in 46%, abscesses in 16%, leakage at the anastomosis site in 5%, and fistulas in 8%. Plexopathy was found in 14% of the patients. Only three patients had late complications: one had chronic diarrhea (RTOG grade 1), another had chronic pain in the pelvis (RTOG grade 2), and the last had radiating pain to the lower extremities (RTOG grade 2). Sacral necrosis was not found.

Martinez-Monge et al. (1999) reported severe complications (RTOG grade 4–5) in 19% of the patients. There were 14 cases of fistula formation (11 enteric, 3 urinary) in 12 patients. Ten were postoperative, and four occurred later during the follow-up period. Eight patients died as a consequence of complications (six cases of fistulae, one case of intraoperative myocardial infarction, and one case of intraoperative death after correction of ischemic ileostomy).

Huber et al. (1996) reported a higher complication rate in preoperative patients (84%) than in postoperative patients (53%).

Shibata et al. (2000) reported on functional outcome and quality of life in 18 patients with rectal cancer after combined modality therapy, intraoperative radiation therapy, and sphincter preservation. Fifty-six percent of the patients reported unfavorable (poor or fair) function of the sphincter. A quality-of-life satisfaction score based on social, professional, and recreational restrictions demonstrated 56% of patients to be dissatisfied with their bowel function.

28.6 Conclusions

The numerous potential advantages of HDR-IORT make it a useful addition to the radiation therapy armamentarium. The target volume can be accurately defined visually and irradiated directly, minimizing the risk of a geographical miss; radiosensitive normal tissues can usually be retracted away from the volume to be irradiated; tissues that cannot be retracted can often be shielded to reduce normal tissue toxicities, unless they are part of the target volume; radiation can be given during surgery, hence eliminating delay in treatment; radiation can be delivered in addition to a moderate dose of EBRT, thus allowing a greater total radiation dose to the tumor; finally, the procedure is brief (1 to 2 h).

However, HDR-IORT does have its disadvantages. First is the radiobiology of a large single dose with potential late toxicity to normal tissues; however, this disadvantage can be minimized if the HDR-IORT can be given as a boost to the immediate tumor bed to supplement moderate doses of EBRT to a greater volume. HDR-IORT requires a shielded OR or transportation of an anesthetized patient, hence limiting the widespread use of the technique.

While HDR-IORT is useful, it is to be noted that a comprehensive intraoperative program should also additionally have intraoperative electron beam radiation therapy (IOERT) and perioperative brachytherapy facilities available to optimally treat all sites. Interstitial brachytherapy is preferable for the treatment of gross residual tumor; IORT (IOERT for accessible sites and HDR-IORT for poorly accessible sites) is added to intraoperatively irradiate the surrounding margins after gross resection, and fractionated EBRT could be used in moderate doses postoperatively to irradiate the entire area of potential microscopic disease. Hence, depending on the volume and location of the tumor, and the available expertise and equipment, IOERT or HDR-IORT and/or perioperative brachytherapy could be used along with EBRT and surgery for the optimal management of malignancies. The best results of HDR-IORT are obtained when used as a moderate dose-conformal boost to the tumor bed after resection in conjunction with a supplementary moderate dose of EBRT.

References

Alektiar, K.M., Hu, K., Anderson, L. et al. 2000b. High-dose-rate intraoperative radiation therapy (HDR-IORT) for retroperitoneal sarcomas. *Int J Radiat Oncol Biol Phys* 47:157–63.

Alektiar, K.M., Zelefsky, M.J., Paty, P.B. et al. 2000a. High-dose-rate intraoperative brachytherapy for recurrent colorectal cancer. *Int J Radiat Oncol Biol Phys* 48:219–26.

Harrison, L.B., Enker, W.E., and Anderson, L.L. 1995. High-dose-rate intraoperative radiation therapy for colorectal cancer. *Oncology (Williston Park)*. 9:737–41.

Harrison, L.B., Minsky, B.D., Enker, W.E. et al. 1998. High dose rate intraoperative radiation therapy (HDR-IORT) as part of the management for locally advanced primary and recurrent rectal cancer. *Int J Radiat Oncol Biol Phys* 42:325–30.

Huber, F.T., Stepan, R., Zimmermann, F. et al. 1996. Locally advanced rectal cancer: Resection and intraoperative radiotherapy using the flab method combined with preoperative or postoperative radiochemotherapy. *Dis Colon Rectum* 39:774–9.

Martinez-Monge, R., Nag, S., and Martin, E.W. 1999. Three different intraoperative radiation modalities (electron beam, high-dose-rate brachytherapy, and iodine-125 brachytherapy) in the adjuvant treatment of patients with recurrent colorectal adenocarcinoma. *Cancer* 86:236–47.

Nag, S., Martinez-Monge, R., Mills, J. et al. 1998. Intraoperative high dose rate brachytherapy in recurrent or metastatic colorectal carcinoma. *Ann Surg Oncol* 5:16–22.

Nuyttens, J.J., Kolkman-Deurloo, I.K., Vermaas, M. et al. 2004. High-dose-rate intraoperative radiotherapy for close or positive margins in patients with locally advanced or recurrent rectal cancer. *Int J Radiat Oncol Biol Phys* 58:106–12.

Shibata, D., Guillem, J.G., Lanouette, N. et al. 2000. Functional and quality-of-life outcomes in patients with rectal cancer after combined modality therapy, intraoperative radiation therapy, and sphincter preservation. *Dis Colon Rectum* 43:752–8.

Strassmann, G., Walter, S., Kolotas, C. et al. 2000. Reconstruction and navigation system for intraoperative brachytherapy using the flab technique for colorectal tumor bed irradiation. *Int J Radiat Oncol Biol Phys* 47:1323–9.

VII

Radiation Protection and Quality Management in Brachytherapy

Radiation Protection in Brachytherapy

Panagiotis Papagiannis
University of Athens Medical School

Ning J. Yue
UMDNJ-Robert Wood Johnson Medical School

29.1 Introduction: Brachytherapy and the International Recommendations for a System of Radiological Protection

The task of summarizing the key elements of radiation protection in brachytherapy might appear intimidating due to the variety of radionuclides, dose rates, and techniques in clinical use. Still, brachytherapy does not represent but one of the plethora of sources (a term used for any physical entity or procedure that results in a potentially quantifiable radiation dose to a person; ICRP 2007) for which the international recommendations for a system of radiological protection apply.

These recommendations stem from work performed by a number of independent, international committees, mainly the International Committee on Radiological Protection (ICRP), and two members of the United Nations family of entities, the Scientific Committee on the Effects of Atomic Radiation (UNSCEAR) and the International Atomic Energy Agency (IAEA). This is a dynamic process based on the continuous review of sources of ionizing radiation and its effects on human health, that is, global/regional radiation exposures, evidence of radiation-induced health effects in exposed groups including atomic bomb survivors, and advances in the understanding of biological mechanisms involved in the occurrence of radiation-induced health effects (UNSCEAR 2008; NAS/NRC 2006; ICRP 2005a). Updated, general recommendations for a system of radiological protection are prepared by the ICRP (2007). These are further disseminated internationally through the publication of basic safety standards in the IAEA safety series and European Atomic Energy Community (Euratom) directives, which are based heavily on ICRP recommendations.* Ultimately, these recommendations form the basis of national regulations, and they are further particularized with respect to sources by national and international scientific and professional bodies.

In ICRP terminology, a brachytherapy program is a *source* leading to *planned exposure situations* that involve the deliberate introduction and operation of the program and are characterized by the ability to predict the magnitude and extent of the exposures and plan the radiological protection in advance. In planned exposures, all categories of exposure can occur, which include occupational exposure, public exposure, and medical exposure of patients or individuals that consent to undertake their comfort and care. It must be emphasized that planned exposure situations include both *normal exposures* that are reasonably expected to occur [i.e., medical exposures of patients, medical exposures of comforters or public exposures in permanent low-dose-rate (LDR) implants, occupational exposures in applications involving source handling] and *potential exposures* that may result from deviation from normal operating procedures (i.e., accidents). Although the focus of this chapter is on

---

* At the time of writing, following ICRP Publication 103 in 2007 that replaced the previous 1990 recommendations (ICRP 1991), the replacement of the 1996 Basic Safety Standards by the IAEA (1996) was published as General Safety Requirements Part 3 (Interim), and the European Commission had just adopted a proposal for a new directive replacing, inter alia, directive 96/29/Euratom of May 13, 1996.

brachytherapy planned exposures, during their operation, *emergency exposure situations* might occur due to the operation itself or malevolent action (i.e., breach of source security). Emergency exposure situations cannot be predicted; they may require urgent protective actions, and they could even result in *existing exposure situations* in the form of prolonged exposure situations following emergencies.

In terms of biological dose response to radiation, deterministic effects are characterized by dose thresholds so that there is absence of risk for tissue reactions at doses lower than about 100 mSv (updated information on dose thresholds in the form of dose resulting in about 1% incidence is provided in ICRP 2007). Although the introduction of a practical dose threshold has been proposed for stochastic effects, it is deemed that current data support that at low doses (below about 100 mSv), the incidence of cancer and heritable effects rises proportionally to the increase of radiation dose over the background. [ICRP (2007) reports detriment-adjusted nominal risk coefficients of 5.5×10^{-2} and 4.1×10^{-2} Sv^{-1} for the whole population and adult workers, respectively, for cancer, and 0.2×10^{-2} and 0.1×10^{-2} Sv^{-1} for the whole population and for adult workers, respectively, for heritable disease up to the second generation.] The adoption of this linear no threshold (LNT) model implies that a finite risk is associated with any exposure to radiation, however small, and protection must include considerations of what level of risk is deemed acceptable. Hence, any system of radiological protection is based on the following three fundamental principles: justification, optimization, and application of dose limits.

29.1.1 Justification

Any decision that alters the radiation exposure situation should do more good than harm. (ICRP 2007)

This principle is source-related and applies to all exposure situations. With regard to occupational and public exposures, planned exposures such as those from a brachytherapy program should not be introduced if a net benefit to individuals or to society is not warranted, and any decision to reduce further exposures from emergency and existing exposure situations should be justified in the sense that more good than harm will come from it. The justification process in the above situations is a broad process wherein radiation detriment serves as one of the many necessary inputs. For medical exposures, brachytherapy planned exposure situations are justified at a first level since the medical use of radiation is widely accepted as doing more good than harm. At a second level, the potential of new brachytherapy techniques to improve treatment should be justified by competent national and international professional bodies and authorities. At the third level, referral criteria and patient groups must be established to facilitate the justification of medical exposure to any particular individual by physicians, who should be well aware of the risks and benefits of particular brachytherapy procedures as well as potential treatment alternatives.

29.1.2 Optimization of Protection

The likelihood of incurring exposure, the number of people exposed, and the magnitude of their individual doses should all be kept as low as reasonably achievable, taking into account economic and societal factors. (ICRP 2007)

This is also a source-related and universal principle, that is, it applies to all exposure situations that have been justified. It is the cornerstone of any system for radiological protection, and its application involves a process of prospective and iterative character with the aim of preventing or reducing future exposures. This process includes the review of an exposure situation, the restriction of doses likely to be delivered to a nominal individual, the listing of options available for protection, the selection of the best option(s) under the circumstances using quantitative methods and cost-effect analysis, the implementation of the optimization option decided on, and its dynamic evaluation. Dose restriction is applied in the form of *dose constraints* for planned exposure situations (except for medical exposure of patients) and *reference levels* for emergency and existing exposure situations that represent upper bounds of dose predicted in optimizing protection from a particular source and must be defined in the planning stage of optimization. It must be kept in mind that optimization is not minimization of dose but rather a balancing of radiation detriment and resources available for protection. Dose constraints and reference levels do not represent a limit between "safe" and "dangerous" conditions, and they are by no means prescriptive limits or target values. Their excess should trigger an investigation, and optimization of protection should aim at establishing acceptable dose levels below them.

For medical exposures of patients, optimization assumes the form of a set of measures to ensure that dose is in accordance with the medical purpose. For diagnosis, *reference dose levels* are set that are not individual patient dose constraints but levels of dose indicating that patient dose is neither too high nor too low for a particular procedure. (Patient dose management information for imaging techniques used in conjunction with brachytherapy, such as fluoroscopically guided interventional procedures, computed tomography, and digital radiology, are provided in the literature; see ICRP 2000a,b, 2004.) Reference dose levels do not apply to radiation therapy where optimization involves the delivery of the prescribed dose to the target while also planning the protection of healthy tissues (ICRP 1985). This is related to quality assurance (QA) procedures and accident prevention that are dealt with in Chapters 30–32 of this book (see also Venselaar and Pérez-Calatayud 2004; ICRP 2005b; IAEA 2008).

29.1.3 Application of Dose Limits

The total dose to any individual from regulated sources in planned exposure situations other than medical exposure of patients should not exceed the appropriate limits specified by the Commission. (ICRP 2007)

In contrast to the previous two principles, this one is individual-related and applies to planned exposure situations only (except for medical exposures of patients). ICRP recommendations for dose limitation are summarized in Table 29.1. Regulatory dose limits are set by national competent authorities taking into account these recommendations, and variations may occur [i.e., NCRP (1993) recommends an occupational effective dose limit of 50 mSv per year with the provision that the cumulative effective dose does not exceed the product of the worker's age by 10 mSv per year].

Members of the public involved in patient comforting and caregiving are not subject to the dose limit for public exposure (a source-specific, prospectively defined, reference level should apply). For female workers that declare pregnancy, additional controls must be implemented to ensure protection for the embryo similar to that for the public.

Dose limitation is a means of regulatory control to prevent the occurrence of deterministic effects and confine the risk of stochastic effects to levels currently deemed acceptable. The importance of optimization cannot be overstressed. Source-specific dose constraints and reference levels for planned exposures of workers and members of the public cannot be greater than dose limits and are an aid to ensuring that the latter will not be exceeded due to exposure from different sources. In the special case of potential exposures in planned situations, where the probability and magnitude of exposure cannot be accurately estimated, *risk constraints* are defined with the aim to equate the health risk to that implied by the corresponding dose constraints and reference levels. [In view of detriment-adjusted nominal risk coefficients mentioned above and average occupational and public doses, occupational and public risk constraints of 2×10^{-4} and 10^{-5} per year are recommended in ICRP (2007), respectively.]

Neither the above brief description nor the remainder of this chapter seeks to replace the need for a thorough review of the literature on recommendations for radiological protection and their implementation to brachytherapy.

The introduction of remote afterloading devices has brought a radical change in brachytherapy practice as a whole, as well as in radiation protection in particular. Occupational doses have been drastically reduced (Grigsby et al. 1991) to levels comparable to those in external beam therapy. Still, brachytherapy reserves its special niche due to the storage of sealed sources emitting radiation constantly, permanent implants, and the need for source manipulation in permanent implant applications (i.e., for prostate treatment) or, less frequently, applications involving manual source loading (i.e., eye plaque applicators for ocular melanoma treatment or use of pin and wire sources). For example, the use of sealed radionuclides for brachytherapy and unsealed radionuclides for radiopharmaceutical therapy are collectively identified as radionuclide therapy in the NCRP 155 report, which offers guidance for the management of radionuclide therapy patients (NCRP 2006).

This chapter aims at summarizing selected technical and procedural information to aid the process of optimization of protection in brachytherapy and highlighting some key elements of this process by elaborating on aspects of radiation protection in practice.

Finally, a word of caution is in order. This chapter does not attempt to summarize regulatory or licensing requirements. This would be impractical since international bodies leave it to national protection bodies to formulate regulations appropriate to their individual countries. Data and methods included in this chapter should therefore be considered informative and carefully checked against local regulations prior to adoption.

29.2 Design Considerations of Brachytherapy Facilities

A brachytherapy program normally requires the availability of physical plant facilities that include an operating room, radiograph imaging room, patient treatment room, and/or a hot lab (IAEA 2008). Each of these facilities has its own functionality and plays a unique role in a brachytherapy program, as described in the following. For radiation protection purposes, they may need to be designated controlled or supervised areas (ICRP 2007). Controlled areas are those where specific protection measures are required to control normal exposures and prevent/limit potential exposures. These measures include a clear delineation of the area, appropriate labeling, access restriction, radiological surveillance, and the establishment of written procedures and working instructions. Supervised areas, often surrounding controlled areas, are those where protection measures are not normally needed but occupational exposures are kept under review. Areas may be designated according to the magnitude of expected exposure, that is, in the United Kingdom and several other countries, areas where annual dose could be greater than 6 and 1 mSv are designated controlled and supervised areas, respectively (IAEA 2006). Brachytherapy treatment rooms and source preparation/storage rooms ("hot labs"), as well as imaging rooms, are controlled areas; operating rooms could fall into a designated area category depending on application and source loading procedure; and high-dose-rate (HDR) treatment consoles are supervised areas.

The operating room is used for catheter/applicator placement and sometimes for radioactive source "hot loading." Depending on the type of brachytherapy procedures, the operating room may require sterilization conditions. The availability of an

TABLE 29.1 Dose Limits Recommended by ICRP (2007)

	Public Exposure	Occupational Exposure
Effective dose limit	1 mSv per year	20 mSv per year averaged over 5 years and less than 50 mSv in any single year
Eye lens equivalent dose limit	15 mSv	150 mSv
Skin equivalent dose limit	50 mSv	500 mSv
Hands and feet equivalent dose limit	–	500 mSv

imaging room/equipment is critical to the success of a brachytherapy program. The radiograph imaging room/equipment is typically used to acquire patient 2D/3D images on which the target volumes, critical organs, and/or catheters/applicators can be localized for accurate dosimetry calculations and treatment planning. A CT/MR scanner room can be used for this purpose, although it is important to keep in mind that catheters/applicators compatible with the imaging modality are desirable to avoid image artifacts. A conventional 2D simulator is a reasonable option, although it is difficult or impossible to identify soft tissue type target volumes from the images. In some cases, a C-arm–type fluoroscopy machine is used in the operating room to ensure precise placement of brachytherapy instruments. The images acquired with the C-arm machine can be directly utilized for the dosimetry calculations and treatment planning. However, care must be taken in using the correct image geometric parameters, such as source–axis distance and source–film distance, in geometry reconstruction and dose calculations.

The patient treatment room must be designed and prepared appropriately for a brachytherapy program. The requirements for the treatment room are different between an LDR program and an HDR program. More details in this respect will be specifically given in the following sections. A hot lab should be designed and made available for long-term or temporary storage of radioactive sources. In addition, many physics-related procedures, such as seed assay and preparation, wipe test, etc., should be conducted in the hot lab. The hot lab should be equipped with a well-shielded source storage cabinet, lead bricks, leaded glass, and other accessories. It is desirable to install a radiation area monitor in the hot lab to constantly monitor the radiation level.

Radiation survey meters with different levels of sensitivity should be included in the facility design and budget. These survey meters are important to ensure the integrity and accuracy of radiation safety and documentation. Depending on the type of brachytherapy procedure to be implemented, considerations should be given to survey meters that can be used for particles other than photons as well.

Emergency radiation safety kits, such as a lead container, should also be considered and purchased for a brachytherapy program (IAEA 2008).

Aspects of brachytherapy facility design are reviewed in the literature (IAEA 2008; Glasgow 2005; NCRP 2006; IAEA 2001, 2006, 2008; IPEM 1997). Some important features are given below for LDR and HDR treatment rooms.

29.2.1 LDR Brachytherapy Treatment Room

The LDR brachytherapy procedures can be performed by one of the two ways: (1) placement of afterloaders (e.g., applicators) first in an operating room, and then loading radioactive sources into the afterloaders after the completions of dosimetry calculations and treatment planning; and (2) loading the radioactive sources directly into or around the target volumes with or without the applicators in the operating room. Most temporary brachytherapy procedures follow the first procedural flow to minimize

unnecessary radiation exposures to the members of medical staff. On the other hand, many permanent brachytherapy procedures, such as prostate implants, are performed by directly implanting radioactive sources into the target volumes in the operating room.

As discussed previously, for temporary LDR brachytherapy procedures, the radioactive sources will be most likely loaded into the afterloaders in a patient treatment room. Since the procedures usually last for hours or days, the treatment room should be made as comfortable and convenient as possible. This patient treatment room does not have to be dedicated to brachytherapy procedures. It can be a regular inpatient room when there are no brachytherapy procedures. However, the room should be carefully selected in terms of its location to minimize radiation exposures to other patients, the general public, and medical staff, especially for the brachytherapy procedures that involve high-energy photon-emitting sources. It is ideal to select a room that is located in a corner or a room that is remote to populated areas. In many cases, shielding blocks are required in the patient treatment room to reduce the radiation levels in the surrounding areas or rooms. These shielding blocks are heavy and need to be stored in a close-by room when no brachytherapy procedures are underway. It is desirable to have the type of shielding blocks that can be easily moved from one location to another (e.g., on wheels).

29.2.2 HDR Brachytherapy Treatment Room

Since the HDR brachytherapy procedures normally deliver a large dose to patient target volumes within a relatively short period of time, many safety features and measures have to be taken into consideration in the design of an HDR treatment room.

An HDR treatment room has to be properly shielded in its construction (see the next section). The shielding design should follow the recommendations of international/national organizations that are recognized for providing scientific bases for radiation protection and safety. A radiation exposure survey needs to be performed by qualified physicists after the construction is completed to ensure that the radiation exposure levels do not exceed the limits set by the government agencies.

Safety interlocks, such as door interlock, power interlock, and backup system, and procedure interruption systems should be installed. The procedure interruption systems should be placed at locations that can be easily reached by the medical staff members during a treatment. Almost all commercially available HDR systems are equipped with some interruption systems, and they should be periodically checked and evaluated. Radiation monitoring systems, which are independent of the HDR unit, should be installed inside an HDR treatment room and at the door of the room to monitor the radiation levels. The monitoring systems inside the room should preferably be muted but give out clearly visible signals (e.g., flashing light) when the radiation exposure exceeds a certain level. On the other hand, the monitoring systems installed at the door should preferably give out not only visible signals but also audible signals to further heighten the awareness of potential loss of the radioactive source from the HDR unit and room.

A warning sign should be installed at the room entrance to clearly show that an HDR procedure is in progress. This type of system can be placed right above or beside the room door and should be automatically turned on when an HDR treatment starts.

It is important to include a video monitoring system and an intercom audio system in the design of an HDR treatment room so that during HDR treatments, the patient and treatment procedure can be constantly and visually monitored, and a verbal communication channel is always available between patient and medical staff members.

It is not very uncommon to utilize a conventional external beam treatment room as an HDR room, especially for clinics where HDR procedures are relatively rare and/or budget/space is limited. If a conventional external beam treatment is used as an HDR treatment room, the room shielding is normally adequate, although careful radiation survey is highly recommended, and the room may need to be modified to be equipped with all the above-mentioned safety and interlock features.

29.3 Brachytherapy Facility Shielding

When time and distance are not adequate to ensure acceptable levels of protection, recourse to the third option for radiation protection in practice, shielding, has to be made. This can be simply described as the process where one needs to calculate the thickness of a particular material that must be interposed to a radiation field in order to achieve a reduction of the expected exposure to a shielding design goal exposure at a point of interest (POI).

This immediately raises a question about the physical quantity used for the expected and shielding design goal exposures. The latter will of course be a fraction of the set dose limits expressed in units of effective dose. Protection quantities, however, such as equivalent and effective dose, cannot be directly measured. Therefore, air kerma is the quantity of choice for expressing expected exposure at a POI from photons and demonstrating compliance with the design goal (NCRP 2004a) since it can be directly measured using appropriately calibrated instruments and thus facilitates postconstruction radiation survey. Moreover, air kerma will always be an overestimate of the effective dose since dose to water is about 10% greater than air kerma (the mean mass energy absorption coefficient ratio of tissue to air is 1.099), but the radiation weighting factor for photons is unity (ICRP 2007), and average air kerma to effective dose conversion factors in the 0.05-keV to 1-MeV photon energy range have been found safely lower than unity in various standardized geometries of adult anthropomorphic computational phantom exposure (ICRU 1998). Hence, for the purpose of shielding calculations where a "safe side" approach is traditionally employed, it can be conservatively assumed that 1 R of exposure equals 1 cGy of air kerma (air kerma can be obtained by dividing exposure measurements using appropriately calibrated instruments by 114 in the 0.05-keV to 1-MeV photon energy range) or dose to tissue at the same point, which in turn can be considered equal to 10 mSv of equivalent or effective dose (i.e., 1 R ≈ 1 cGy ≈ 1 cSv = 10 mSv).

Returning to the description of the shielding calculation problem, let us now turn to the quantification of the shielding design goal exposure at a POI. Obviously, this cannot be greater than the regulatory dose limits for occupational and public exposure expressed in millisieverts per year (see Table 29.1). For example, in the United States, NCRP recommends annual effective dose shielding design goals of 5 and 1 mSv for controlled and public areas, respectively (NCRP 2004b, 2005). The corresponding U.K. recommendations are 6 and 0.3 mSv, respectively (IAEA 2006). Based on the assumptions of a 50-week working year and a brachytherapy facility workload that is evenly distributed in time, weekly derivative shielding design goals are traditionally used (i.e., United States: 0.1 mSv/week and 20 µSv/week in controlled and public areas, respectively; United Kingdom: 0.12 mSv/week and 6 µSv/week in controlled and public areas, respectively).

In view of the above, the expected exposure at a POI, commonly referred to as the facility workload, needs to be determined in units of air kerma per week of operation. In order to estimate the workload, one needs the air-kerma strength of the source(s) to be used in units of U (1 U = 1 µGy m^2 h^{-1}), as well as an estimate of the number of patients treated per week and the average treatment time, so that the product will yield the workload in units of grays per week at 1 m from the source(s). The estimate of the number of patients should be based on sound data (such as average national or state figures available) and at minimum meet the viability standards set by the hospital management or national regulations. For example, the NHS suggests that a brachytherapy center patient throughput should, at minimum, be 50 patients per year overall, further elaborated to at least 10 intrauterine insertions, 10 of each of low throughput treatment sites (head-and-neck interstitials, bronchial and esophageal intraluminals, breast and rectal interstitials, and cervical applicator insertions), or 25 permanent prostate interstitial implants, as applicable (NCPR 2011). Treatment time is of course site- and technique-specific, and the order of magnitude of the source(s) S_K is known (i.e., 4×10^4 U for a typical ^{192}Ir HDR source and 0.4–3 U for ^{125}I and ^{103}Pd LDR sources). It should be noted that the workload estimate should also include periodic quality control procedures and measurements. The workload should be appropriately weighted for the distance from the source to the POI using the inverse square law. This distance should be the shortest one (if applicable), and the POI is usually taken at 30 cm behind any physical barrier/wall. Since we are seeking to protect not the area but the most exposed individual occupying it, the workload must also be weighted by an appropriate occupancy factor, which corresponds to the average fraction of working time during which the POI is occupied by the single person who spends the most time there. The occupancy factor is usually determined by facility employees and assumes a value of 1 for controlled areas [for a list of indicative occupancy factors for shielding calculations, refer to NCRP (2005) or IAEA (2006)].

In summary, the reduction factor, R, by which the expected exposure at a POI must be reduced to comply with a shielding design goal exposure, is given by

$$R = \frac{P}{WF/d^2} = \frac{Pd^2}{S_K tNF} \qquad (29.1)$$

where P is the shielding design goal in units of effective or equivalent dose per week (Sv per week); W is the facility workload in units of air kerma per week at 1 m from the source(s) (Gy m² per week where air kerma can be conservatively assumed equal to effective or equivalent dose, i.e., Gy ≈ Sv); F is the occupancy factor at the POI (usually denoted by the capital letter T, not used here to avoid confusion with the transmission factor, $T = R^{-1}$, defined in the following); d is the distance from the source(s) to the POI (m); S_K is the source air-kerma strength (Gy m² h⁻¹, to be multiplied by the number of sources used in procedures not employing remote afterloaders of a single stepping source); t is the time per treatment (h, i.e., the sum of source dwell times in remote afterloading procedures); and N is the number of treatments per week.

For exceedingly low workloads, the assumption of a workload evenly distributed in time made above does not apply. Shielding calculations using Equation 29.1 might therefore prove inadequate due to high instantaneous dose rates. For example, due to the time necessitated for catheter placement and the limited availability of operating rooms, brachytherapy treatment rooms are not used as frequently as those for external beam treatments, and while the average workload might be low, the dose rate will be zero most of the time and high during treatment sessions. In order to preclude such scenarios, regulatory shielding design goals are also set for the instantaneous dose rate (IDR, the dose per hour measured over a time interval sufficient with regard to the dosimeter response time, usually about 1 min for a survey chamber). Attention must be drawn to the fact that these IDR regulatory limits are not derived from annual effective dose shielding design goals, a practice that would lead to excessive shielding requirements. For example, according to U.K. recommendations, the IDR should not exceed 7.5 µSv h⁻¹ in uncontrolled areas, which is higher than the value of 0.15 µSv h⁻¹ derived from the 0.3 mSv per year shielding design goal assuming a 50-week year of 40 working hours per week. Also, according to U.S. regulations, the dose equivalent should not exceed 20 µSv h⁻¹ in any 1 h, which is higher than the value of 0.5 µSv h⁻¹ derived as above from the 1 mSv per year shielding design goal.

Having calculated the reduction factor, R, or equivalently the transmission factor, $T = R^{-1}$, defined as the ratio of air-kerma rate at a POI with and without the interposition of a shielding barrier, one needs to translate this transmission to a shielding thickness for a given material. As shown in Figure 29.1, the use of narrow beam conditions employing material attenuation coefficients [i.e., solving equation $T(x) = I(x)/I_0 = \exp(-\mu x)$ for x] is unacceptable due to the buildup factor. The exact form of the buildup factor is not calculable (if the validity of this statement

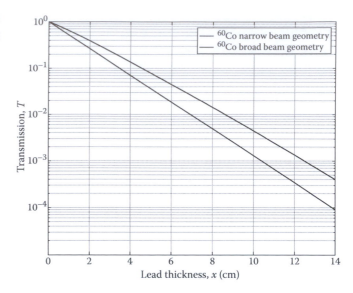

FIGURE 29.1 ⁶⁰Co radiation transmission through a lead shielding barrier in narrow and broad beam geometry conditions. (Courtesy of P. Papagiannis.)

is not immediately taken for granted, please refer to Section 7.2 of Chapter 7).

In order to augment outdated experimental transmission data for selected brachytherapy radionuclide/material combinations in broad beam conditions, the numerical method of Monte Carlo simulation has been employed in a series of publications (Granero et al. 2006; Lymperopoulou et al. 2006; Papagiannis et al. 2008). This approach, however, produces discrete values of transmission versus material thickness in tabular form or graphs, thus necessitating interpolations. Often, recourse is made to the use of transmission curve indices such as the half and tenth value layer (HVL and TVL, respectively) for particular radionuclide–material combinations (NCRP 2006; IAEA 2006). These indices, however, do not remain constant due to spectral variation with increasing shielding barrier thickness (Lymperopoulou et al. 2006; Papagiannis et al. 2008). This is evidently shown, for example, in Figure 29.2.

It has been suggested (IAEA 2006) to use the first HVL or TVL at small R (or large T) and HVL_e or TVL_e at large R (or small T) for the calculation of material shielding thickness to achieve a given R or T value. HVL_e and TVL_e, called "hard" or equilibrium values, correspond to that penetrating region where the radiation directional and spectral distributions are practically independent of thickness, so that a single value of the HVL or TVL is valid. This suggestion, however, has been shown to introduce potentially significant errors (Papagiannis et al. 2008). It is therefore both convenient and accurate to fit an ad hoc analytical representation to transmission data for use in practical dose calculations and shield designs. Within the framework of a BRAPHYQS activity (BRAPHYQS, an acronym for brachytherapy physics quality system, is a working group within GEC-ESTRO), Monte Carlo simulation was used to generate radiation

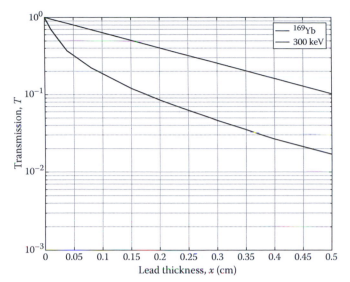

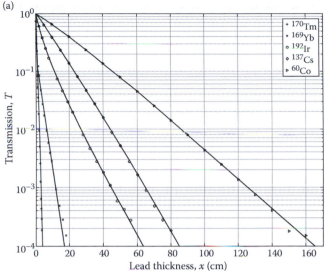

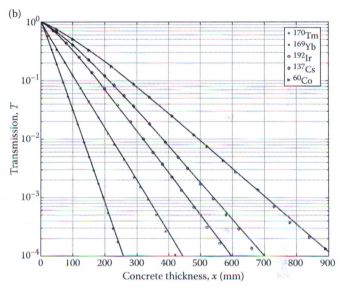

FIGURE 29.2 ^{169}Yb radiation transmission through a lead shielding barrier in broad beam geometry conditions. Note the change of slope in the transmission curve at small thickness due to the combined effect of increased attenuation of the lower photon energy emissions by ^{169}Yb and buildup. The transmission curve for monoenergetic photons of energy close to the maximum energy emitted by ^{169}Yb is also presented for comparison. (Courtesy of P. Papagiannis.)

broad beam transmission data for various radionuclide–material combinations (^{60}Co, ^{137}Cs, ^{198}Au, ^{192}Ir, ^{169}Yb, ^{170}Tm, ^{131}Cs, ^{125}I, and ^{103}Pd photons through concrete, stainless steel, lead, as well as lead glass and baryte concrete). Besides the tabulation of first and equilibrium HVL/TVL values, a three-parameter analytical expression was fitted to results to facilitate accurate and simple radiation shielding calculations (Papagiannis et al. 2008). The Monte Carlo calculated data sets as well as fitting coefficient results are available online at http://www.estro.org/estroactivities/Pages/RadiationProtectionData.aspx. Indicative data are presented in Figure 29.3. This figure also provides a hint on the selection of the shielding material. A high-Z/high-density material such as lead would require a smaller shielding thickness, which could be favorable if space is limited. Depending on the facility workload and layout, however, support issues might ensue that could be dealt with by using a combination of materials (i.e., concrete and lead or stainless-steel sheets) or a maze to reduce door shielding. Cost, time to completion, and other technical aspects are also pertinent parameters; the solution to the shielding problem is neither unique nor generally applicable.

The responsibility for shielding design goes beyond the calculation of required thickness of a given material and includes the supervision of construction (i.e., the materialization of the shielding as planned needs to be verified in general and in particular: ducts for cable, plumbing and ventilation, concrete composition, density that varies with local aggregate, etc.) and postconstruction survey. More information can be found in IPEM (1997), Glasgow (2005), NCRP (2006), and IAEA (2006).

FIGURE 29.3 Radiation transmission through a lead (a) and a concrete (b) shielding barrier in broad beam geometry conditions for selected radionuclides employed in brachytherapy. Presented lines correspond to a fit of the three-parameter model introduced by Archer et al. (1983) to the Monte Carlo simulation calculated data. Data and fitting results are available online at http://www.estro.org/estroactivities/Pages/RadiationProtectionData.aspx for various radionuclide-shielding material combinations. (Courtesy of P. Papagiannis.)

29.4 Radiation Protection Issues Associated with Specific Techniques

Although both LDR and HDR brachytherapy procedures are treatments of patients with radioactive sources, LDR procedures differ from HDR procedures in several aspects:

(1) Dose rates at prescription points are significantly different. These dose rates in LDR brachytherapy procedures

range from 4 to 200 cGy/h, while the dose rates in HDR procedures are higher than 1200 cGy/h.

(2) LDR brachytherapy can be permanent and temporary, while HDR treatments are always temporary.

(3) Temporary LDR brachytherapy may require continuous treatment for hours or days, while HDR treatments rarely last more than 1 h.

(4) Patients treated with LDR brachytherapy may be released with implanted radioactive sources, while HDR patients should never be released with the radioactive source(s).

These differences should be taken into account in radiation protection and should be addressed specifically.

29.4.1 LDR Brachytherapy

In many LDR brachytherapy treatments, radioactive sources need to be ordered specifically for the treatments in terms of source strength and source quantity. Wipe test should be first performed after source delivery to check the source integrity. The sources should also be carefully assayed independently from manufacturers before the start of treatment to ensure the accuracy of source strength and quantity. For radioactive sources of long half-life (e.g., ^{137}Cs) in storage, periodical wipe test and assay should be conducted.

Patients should be surveyed before LDR treatment to establish a radiation baseline. During the treatment procedures, the sources—if not in an afterloader—need to be visually checked so that they are all accounted for. After the procedures, radiation surveys should be conducted in the treatment rooms to prevent source loss. The survey should cover the entire room, including trash bins, equipment, and outfits of medical staff members. Since the source strength or the energy levels of emitted particles from the sources can be relatively low, a highly sensitive radiation detector may be needed for this purpose. A radiation survey should also be conducted for the patient to evaluate whether it is safe to release the patient. If the patient is to stay in a room, such as the treatment room, the recovery room, or the designated inpatient room, for a period of time, a radiation survey should be performed after the patient moves into the room, and radiation signs and safety instructions for visitors, patient, and medical staff should be posted on the room door. During the patient's stay, it should be clearly documented in the patient chart that the patient is undergoing brachytherapy treatment, and labels with radiation signs should be attached to the patient. Radiation emergency equipment should be placed in the room in case of unexpected events. After the completion of treatments, additional radiation surveys should be conducted for the patient, equipment, and room. Related radiation signs should be then removed from the room and the patient. For temporary implants, the removed sources should be counted, logged, and returned to storage in the hot lab. For permanent implants, patients should be surveyed before release to ensure that the implanted radioactive sources do not pose potential radiation risks to the general public. The criteria for the patient release are that based on the

patient radiation survey, the exposure to the public from the implanted radioactive source should not exceed the annual limit to the public. Radiation safety instructions should be provided to the patients, including special instructions to protect pregnant women and children from being exposed to excessive radiation. The patients should be advised to minimize contact with pregnant women and children. They should also be provided contact information so that their radiation-related concerns can be answered after their release. More information can be found in ICRP (2005c) and NCRP (2006).

29.4.2 HDR Brachytherapy

The HDR source is periodically replaced. The frequency of replacement is based on the source half-life. Wipe tests should be performed for both new and old sources during the replacement. Source strength should be independently measured during the source replacement and periodically thereafter. Since most of HDR procedures are delivered with computerized systems, accuracy of source strength decay should be manually checked and confirmed before each treatment. Emergency procedure instructions should be posted in the control area.

Radiation monitors and safety interlocks and interruption buttons should be checked before each treatment. Patients should be surveyed prior to treatment to establish a radiation baseline. Radiation emergency safety equipment should be made ready to use before the start of treatment and through the treatment procedure. Patient and treatment should be closely and continuously monitored via video monitors and audio communication devices during procedures. After the treatment is finished, patient and equipment should be surveyed to ensure that the HDR radioactive source returns to the HDR unit safe. More information can be found in IAEA (2001, 2008).

29.5 Source Storage and Transportation

Sources are delivered to the brachytherapy department in specially designed containers that are clearly marked (isotope, activity, transport index) and accompanied by proper shipping documents, all subject to international regulations for their safe transport (IAEA 2009). Beginning with the acceptance of the sources by authorized personnel, specific measures have to be taken to ensure both optimal radiation protection and source security. For radiation protection, these measures consist of facility design elements, the configuration of specific procedures, and the use of appropriate instrumentation, all based on the three basic rules for radiation protection in practice (exposure time reduction, increase in distance from a source, and use of shielding). Source security is ensured by exercising continuous accountability and control of the sources through appropriate procedures for the safe exchange and movement of radioactive sources within the institution and controls to prevent theft, loss, unauthorized withdrawal or damage of sources, or entrance of unauthorized personnel to controlled areas (Podgorsak 2005).

HDR brachytherapy procedures using remote afterloaders do not require source preparation. The source is safely stored in the afterloader when not used for treatment since most afterloaders are certified as transportable radioactive containers (IAEA 2001). In terms of source security, the afterloader is usually kept in the treatment room, which is a controlled area, and consequently, access is restricted. The room should be locked when not in use, and it is even suggested that the capability to detect unauthorized access or removal of the source should be implemented (IAEA 2006). A set of long forceps or tongs and an emergency source container (sufficiently large to accept the entire applicator assembly containing the source) must be available in the treatment room in the event of source retraction failure during use. A dedicated storage room might still be required if unused sources are temporarily kept in the department to decay to safe levels. This room is also a controlled area subject to all requirements for appropriate labeling, access restriction, radiological surveillance, and the establishment of written procedures and working instructions (ICRP 2007; Podgorsak 2005).

A dedicated source preparation and storage room (commonly called a "hot" lab) is an essential prerequisite to LDR and manual brachytherapy. The hot lab must include an area where all LDR sealed sources can be safely stored in an orderly fashion, space and facilities for receiving and returning sources, source inventory, source calibration and quality control, space and equipment for source preparation for treatments, and adequate space for QA and treatment aids and, if necessary, for storage of short-lived sources or temporary storage of unused long-lived sources (NCRP 2006). A work area should also be available, in or near the hot lab, where records can be prepared and stored without danger of radioactive contamination (NCRP 2006).

Following their acceptance, sources and source containers should be marked using a unique identification scheme, entered in the department's source register (listing the id, location, activity of all sources, as well as dates and results of checks and periodic inventories) and put in a storage safe (Podgorsak 2005). Different safes are used for different source types. These must be fire resistant, equipped with a lock and shielded. Regulations will generally require a maximum exposure rate of less than 1 μSv h^{-1} at 10-cm distance from the container surface, and most commercially available equipment ensures a surface dose rate of about 25 μSv h^{-1} (Venselaar and Pérez-Calatayud 2004). Their functionality and their place in the room should ensure that occupational doses are maintained as low as reasonably achievable. For example, a safe for a particular source type should be close to the area reserved for the manipulation of the sources for preparation or quality control purposes, and a diagram on the source safe showing the exact location of each source within it aids in reducing the time to locate and identify a source (Podgorsak 2005).

Source manipulation, for any purpose, should be done using forceps or tongs and never directly by hand (Podgorsak 2005). The use of lead gloves or other protective garments serves little to reduce exposure to high-energy sources and might unnecessarily prolong the manipulation time (Venselaar and Pérez-Calatayud

2004). However, personnel shielding such as a lead "L-block" or a structure made from interlocking lead blocks must be provided. including a leaded window of sufficient thickness (NCRP 2006) or at least a mirror system. A magnifying viewer might be required in the shielded preparation area since sources for manual brachytherapy must be inspected visually for damage after each use, especially if they are subject to possible damage from heat, abrasion, chemical attack, and mechanical stress for cleaning and sterilization purposes (Podgorsak 2005). In general, sources should never be left on preparation surfaces, and the work surface should be easily cleaned and brightly lit to facilitate finding dropped sources (Podgorsak 2005; IAEA 2008). Especially for ^{192}Ir wires, equipment for cutting and handling should be properly decontaminated and any radioactive waste collected and stored (Podgorsak 2005; IAEA 2008). Wipe tests for source leakage and preparation area contamination need to be periodically performed and results documented (Podgorsak 2005). For HDR units, these wipe tests are obviously performed on the afterloading drive assembly and not the high activity sources.

The hot lab is of course a controlled area (ICRP 2007; Podgorsak 2005). It should be locked at all times, and a latch to automatically lock the door is recommended (NCRP 2006). An area radiation monitor should be visible on entering the room. While manipulating the sources, radiation protection measures and their effectiveness should be periodically reevaluated, and periodic area surveys should be performed around the room (Podgorsak 2005; IAEA 2008). The hot lab and treatment, treatment planning, operating, and patient rooms have to be located as close as possible to each other to reduce distances over which patients and sources have to be transported, particularly avoiding use of elevators (IAEA 2008).

Source transport should be made using a mobile, shielded container reducing exposure to less than 2 mSv h^{-1} in contact and 1 mSv h^{-1} at 1 m (Venselaar and Pérez-Calatayud 2004) or as required by national regulations. The transport container should be clearly marked for its radioactive content, never left unattended, and surveyed both before and after brachytherapy procedures. The movement of sources should be entered to a source movement log with a signed record of the date of removal from the safe, patient name, and the return of the source (Podgorsak 2005). Besides access restriction, source registry, and periodic inventories, additional measures for source control might be in order according to the specific applications. For example, HDR brachytherapy patients should be surveyed to ensure source return to its safe; all linen, dressing, clothing, equipment, and refuse collectors should not be removed from the brachytherapy patient room until they are checked for stray sources and results are documented; and a filter should be used wherever there is a risk for loss of a source through a drain (i.e., applicator cleaning, LDR seed dislodging from the prostate).

29.6 Organizational Issues

Among other aspects of setting up a brachytherapy program, a radiation protection program must be established to ensure compliance with regulations of radiation safety and protection enacted by government agencies (IAEA 2001, 2008). The radiation protection program should be designed and developed to ensure doses to the medical staff and general public are as low as is reasonably achievable (see Section 29.1). A qualified radiation safety officer, who is responsible for implementing the radiation protection program, should be identified and officially appointed. The radiation safety officer should be given sufficient administrative authority, in writing, to supervise the program. A radiation safety committee should also be established if the institution plans to start a brachytherapy program that involves multiple types of procedures and multiple types of radioactive sources. For example, as required by the U.S. Nuclear Regulatory Commission (U.S. NRC; http://www.nrc.gov), a radiation safety committee "must include an authorized user of each type of use permitted by the license, the Radiation Safety Officer, a representative of the nursing service, and a representative of management who is neither an authorized user nor a Radiation Safety Officer. The Committee may include other members the licensee considers appropriate."

The radiation protection program should clearly document and specify the medical staff members or the groups of medical staff members who are permitted to be involved in brachytherapy procedures and their respective responsibilities in the procedures. Adequate training should be required, provided, and documented. A typical medical team for a brachytherapy program consists of radiation oncologists, medical physicists, dosimetrists, radiation therapists, nurses, and radiation safety officers. It is critical to establish the rules that only the qualified radiation oncologists, qualified medical physicists, and qualified radiation safety officers be permitted in conducting brachytherapy procedures. In the United States, qualified radiation oncologists, qualified medical physicists, and qualified radiation safety officers are normally those who have been certified by the recognized medical specialty boards and who have received training in brachytherapy procedures and radiation safety, although some may become qualified based on their experiences and trainings.

All of the team members should receive adequate training in the brachytherapy procedures to be implemented before being authorized to perform the procedures. The training should cover not only the operations of brachytherapy systems but also radiation safety and protection, including the principles, regulations, safety features of brachytherapy delivery machines and treatment room, radiation surveys, radiation monitoring, records, reports, and emergency procedures and measures. The training should be provided to new staff members and periodically to the current staff members.

The radiation protection program should establish policies for radiation surveys to protect medical staff members, patients, and the general public and to be compliant with government regulations. The radiation surveys should be conducted after receiving and before shipping out radioactive sources, before and after patient treatments, and periodically on brachytherapy units and radioactive sources in stock. The survey results should be documented for future reference.

The radiation protection program should also establish policies for radiation monitoring. Individual radiation monitoring devices should be provided to the medical staff members who are involved in brachytherapy procedures and to the individuals who may potentially receive radiation doses higher than the set limits.

Solid security systems should be put in place for brachytherapy units and hot labs where radioactive materials are stored. Appropriate radiation warning signs should be placed to prevent accidental entries.

Emergency procedures should be established for each individual brachytherapy procedure. The emergency procedures should be distributed and taught to each of the brachytherapy team members before the start of the program. The procedures should also be posted to areas where the brachytherapy treatments are conducted. Periodical training should be provided to the team members.

The radiation protection program should establish policies for radiation-related accidents or errors. The policies should be compliant with respective government regulations and provide guidelines for procedures after accidents, including patient care and instructions, reporting channels and steps, and corrective actions.

A comprehensive QA program should be set up and strictly enforced for brachytherapy programs. As clearly stated in the "Code of Practice for Brachytherapy Physics" (Nath et al. 1997), "the goal of the brachytherapy quality assurance program is to maximize the likelihood that each individual treatment is administered consistently, that it accurately realizes the radiation oncologist's clinical intent, and that it is executed with regard to safety of the patient and others who may be exposed to radiation during the course of treatment." A well-designed and -executed QA program not only safeguards the qualities of patient care but also protects patients, medical staff, and the general public from unnecessary radiation exposures and minimizes radiation accidents. A typical brachytherapy QA program should address processes of radioactive source handlings and assays, applicator preparation and insertion, treatment design and planning, treatment delivery, and posttreatment evaluation and patient instructions. On the other hand, since different brachytherapy procedures usually require different types of devices and consist of different components, a QA program should be specifically designed for a brachytherapy procedure based on its features and work flow, characteristics of equipment and radioactive sources, and specific concerns on radiation safety and protection.

Records of the radiation protection program should be systematically maintained and be compliant with the government regulations. The records should include the provisions and policies of the radiation protection program, written directives, radiation surveys, radioactive source logging information, records of individual radiation monitoring results and occupational radiation doses, records of brachytherapy procedures, records of brachytherapy equipment, records of staff credentials and training certificates, and records of dose to individual members of the general public.

Radiation-related accidents and incidents, including medical events and incidents, including medical events and loss and theft of radioactive sources, should be reported to the institutional authority and government agencies that are authorized to regulate the radioactive sources and brachytherapy programs. The reporting process should be in compliance with the government regulations and should be clearly documented in the provisions of the radiation protection program. Instructions and training should be provided to staff members about the reporting process. A report on other radiation-related records and results should also be conducted following the institutional rules and government regulations.

References

Archer, B.R., Thornby, J.I., Bushong, S.C. 1983. Diagnostic X-ray shielding design based on an empirical model of photon attenuation. *Health Phys* 44:507–17.

Glasgow, G.P. 2005. Brachytherapy facility design. In: *Brachytherapy Physics*, 2nd ed. Proc. of the Joint AAPM/ABS Summer School. Medical Physics Publishing, Madison, WI, pp. 127–151.

Granero, D., Pérez-Calatayud, J., Ballester, F., Bos, A.J., and Venselaar, J. 2006. Broad-beam transmission data for new brachytherapy sources, Tm-170 and Yb-169. *Radiat Prot Dosimetry* 118:11–5.

Grigsby, P. W., Perez, C. A., Eichling, J. et al. 1991. Reduction in radiation exposure to nursing personnel with the use of remote afterloading brachytherapy devices. *Int J Radiat Oncol Biol Phys* 20:627–9.

International Atomic Energy Agency (IAEA). 1996. International basic safety standards for protection against ionizing radiation and for the safety of radiation sources. IAEA Safety Series No. 115. International Atomic Energy Agency, Vienna. Available online at http://www-pub.iaea.org/mtcd/publications/pdf/ss-115-web/pub996_web-1a.pdf.

International Atomic Energy Agency (IAEA). 2001. Implementation of microsource high dose rate (mHDR) brachytherapy in developing countries. IAEA-TECDOC-1257. International Atomic Energy Agency, Vienna. Available online at http://www-pub.iaea.org/MTCD/publications/PDF/te_1257_prn.pdf.

International Atomic Energy Agency (IAEA). 2006. Radiation protection in the design of radiotherapy facilities. IAEA Safety Series No. 47. International Atomic Energy Agency, Vienna. Available online at http://www-pub.iaea.org/MTCD/publications/PDF/Pub1223_web.pdf.

International Atomic Energy Agency (IAEA). 2008. Setting up a radiotherapy programme: clinical, medical Physics, radiation protection and safety aspects. International Atomic Energy Agency, Vienna. Available online at http://cancer.iaea.org/documents/Ref5-TecDoc_1040_Design_RT_proj.pdf.

International Atomic Energy Agency (IAEA). 2009. Regulations for the safe transport of radioactive material. IAEA safety requirements. Safety standards No. TS-R-1. International

Atomic Energy Agency, Vienna. Available online at http://www-pub.iaea.org/MTCD/publications/PDF/Pub1384_web.pdf.

International Commission on Radiological Protection (ICRP). 1985. *Protection of the Patient in Radiation Therapy*. ICRP Publication 44. *Ann. ICRP* 15(2). Elsevier Science, New York.

International Commission on Radiological Protection (ICRP). 1991. *1990 Recommendations of the International Commission on Radiological Protection*. ICRP Publication 60. *Ann. ICRP* 21:1–3. Elsevier Science, New York.

International Commission on Radiological Protection (ICRP). 2000a. *Avoidance of Radiation Injuries from Medical Interventional Procedures*. ICRP Publication 85. *Ann. ICRP* 30(2). Elsevier Science, New York.

International Commission on Radiological Protection (ICRP). 2000b. *Managing Patient Dose in Computed Tomography*. ICRP Publication 87. *Ann. ICRP* 30(4). Elsevier Science, New York.

International Commission on Radiological Protection (ICRP). 2004. *Managing Patient Dose in Digital Radiology*. ICRP Publication 93. *Ann. ICRP* 34(1). Elsevier Science, New York.

International Commission on Radiological Protection (ICRP). 2005a. *Low Dose Extrapolation of Radiation-Related Cancer Risk*. ICRP Publication 99. *Ann ICRP* 35(4). Elsevier Science, New York.

International Commission on Radiological Protection (ICRP). 2005b. *Prevention of High-Dose-Rate Brachytherapy Accidents*. ICRP Publication 97. *Ann. ICRP* 35(2). Elsevier Science, New York.

International Commission on Radiological Protection (ICRP). 2005c. *Radiation Safety Aspects of Brachytherapy for Prostate Cancer Using Permanently Implanted Sources*. ICRP Publication 98. *Ann. ICRP* 35(3). Elsevier Science, New York.

International Commission on Radiological Protection (ICRP). 2007. *The 2007 Recommendations of the International Commission on Radiological Protection*. ICRP publication 103. *Ann ICRP* 37:1–332. Elsevier Science, New York.

International Commission on Radiation Units and Measurements (ICRU). 1998. *Conversion Coefficients for Use in Radiological Protection Against External Radiation*. ICRU Report 57. Oxford University Press, Cary, NC.

Institute of Physics and Engineering in Medicine (IPEM). 1997. *The Design of Radiotherapy Treatment Room Facilities*. IPEM Report No. 75. Institute of Physics and Engineering in Medicine, York.

Lymperopoulou, G., Papagiannis, P., Sakelliou, L., Georgiou, E. Hourdakis, C. J., and Baltas, D. 2006. Comparison of radiation shielding requirements for HDR brachytherapy using ^{169}Yb and ^{192}Ir sources. *Med Phys* 33:2541–7.

National Academy of Sciences/National Research Council (NAS/NRC). 2006. *Health Effects from Exposure to Low Levels of Ionizing Radiation: BEIR VII Phase 2*. Board on Radiation

Effects Research. Research Council of the National Academies, Washington.

Nath, R., Anderson, L.L., Meli, J.A., Olch, A.J., Stitt, J.A., and Williamson, J.F. 1997. Code of practice for brachytherapy physics: Report of the AAPM Radiation Therapy Committee Task Group No. 56. *Med Phys* 24:1557–98.

National Cancer Peer Review Programme (NCPR). 2011. Manual for Cancer Services: Radiotherapy Measures Version 4.0. National Cancer Action Team, Part of the National Cancer Peer Review Programme, National Health Service. London. Available online at http://www.cquins.nhs.uk.

National Council on Radiation Protection and Measurements (NCRP). 1993. *Limitation of Exposure to Ionizing Radiation*. NCRP Report No. 116. National Council on Radiation Protection and Measurements, Bethesda, MD.

National Council on Radiation Protection and Measurements (NCRP). 2004a. *Structural Shielding Design for Medical X-Ray Imaging Facilities*. NCRP Report No. 147. National Council on Radiation Protection and Measurements, Bethesda, MD.

National Council on Radiation Protection and Measurements (NCRP). 2004b. *Recent Applications of the NCRP Public Dose Limit Recommendation for Ionizing Radiation*. NCRP Statement No. 10. National Council on Radiation Protection and Measurements, Bethesda, MD.

National Council on Radiation Protection and Measurements (NCRP). 2005. *Structural Shielding Design and Evaluation for Megavoltage X- and Gamma-Ray Radiotherapy Facilities*. NCRP Report No. 151. National Council on Radiation Protection and Measurements, Bethesda, MD.

National Council on Radiation Protection and Measurements (NCRP). 2006. *Management of Radionuclide Therapy Patients*. NCRP Report No. 155. National Council on Radiation Protection and Measurements, Bethesda, MD.

Papagiannis, P. Baltas, D., Granero, D. et al. 2008. Radiation transmission data for radionuclides and materials relevant to brachytherapy facility shielding. *Med Phys* 35:4898–906.

Podgorsak, E.B. (ed.). 2005. *Radiation Oncology Physics: A Handbook for Teachers and Students*. International Atomic Energy Agency, Vienna. Available online at http://www-naweb.iaea.org/nahu/dmrp/syllabus.shtm.

United Nations Scientific Committee on the Effects of Atomic Radiation (UNSCEAR). 2008. *Report to the General Assembly with Scientific Annexes. Sources and Effects of Ionizing Radiation*. United Nations, New York. Available online at http://www.unscear.org/unscear/en/publications.html.

Venselaar, J.L.M. and Pérez-Calatayud, J. (eds.). 2004. *A Practical Guide to Quality Control of Brachytherapy Equipment*. Brussels. ©2004 by ESTRO. Available online at http://www.estro-education.org/publications/Documents/booklet8_Physics.pdf.

30

New Paradigm for Quality Management in Brachytherapy

Susan Richardson
Washington University

Geethpriya Palaniswaamy
Scott & White Healthcare System

Sasa Mutic
Washington University

Eric E. Klein
Washington University

30.1 Introduction

The International Committee on Radiation Protection (ICRP) Publication 97, "Prevention of High-Dose-Rate Brachytherapy Accidents," states that as of 2004, there were over 500 reported high-dose-rate (HDR) brachytherapy incidents, including one fatality (ICRP 2005). It concludes that human error is the primary cause of these accidents. Prevention and mitigation of errors are a result of a good quality management (QM) program. The purpose of the report from the American Association of Physicists in Medicine's (AAPM's) Task Group 100, "Application of Risk Analysis Methods to Radiation Therapy Quality Management," is to lend guidance to individuals involved in establishing a QM program for radiation therapy treatments (AAPM 2009). It provides the methodology of Failure Modes and Effects Analysis (FMEA) to proactively delineate the incidents that, if occurred undetected, would be the most probable to happen and the most detrimental to the patient. Historically, one difficulty for radiation therapy administrators has been to produce an appropriate balance of resources with maximal quality and safety to patients. This method allows concentration of quality assurance efforts on those events with the greatest need for quality control. This chapter seeks to explore historically known brachytherapy incidents and provide three case studies of FMEA applied to modern-day brachytherapy treatments including HDR brachytherapy, low-dose-rate (LDR) prostate implants, and Xoft electronic brachytherapy.

30.2 Background

Incident and error reporting in the radiation oncology setting is becoming a more embraced and necessary practice in the twenty-first century. Several incident reporting databases are available worldwide for the capturing of events, such as the ROSIS network from the European Society for Radiotherapy and Oncology (ESTRO 2008), Patient Safety Net from the Agency for Healthcare Research and Quality (AHRQ 2010), or in-house developed systems (Mutic et al. 2008). The reporting of medical events or near misses allows the department or hospital to perform a root cause analysis (RCA) to determine what, if any, steps should be added to the QM program to reduce the likelihood of the incident from occurring again. The problem with RCA alone is that it is a reactionary process subsequent to an actual event. An FMEA approach encourages evaluation of a process from start to finish and investigates a priori where there is need for improvement in either quality control practices or procedures. This is particularly important to the practice of brachytherapy, although FMEA is challenging for any procedure.

QM is broadly defined as "all activities of the overall management function that determine the quality policy, objectives and responsibilities, and implement them by means such as quality planning, quality control, quality assurance, and quality improvement…" (ISO 1994). Quality assurance is the act of providing evidence to establish that a certain quality is being performed. Quality control is the tools and techniques used to meet the requirements of a quality assurance process. Quality improvement is a formal approach to the analysis of performance and systematic efforts to improve it.

Brachytherapy QM is particularly important in radiation oncology because of several factors. In the case of HDR brachytherapy, patients are typically implanted with an applicator or source delivery device such as a tandem and ovoids. The patients are often medicated and uncomfortable. Therefore, there is a need to treat patients in a timely manner. Whereas an external beam radiotherapy patient may have a week or two between simulation and their first treatment, a brachytherapy patient has the applicators inserted, images taken, their treatment plan calculated, and the treatment delivered all in a few hours. The quality control tasks that must happen occur on a very short time scale, and often, the brachytherapy team is pressured to proceed quickly. Additionally, for HDR treatments, the dose rates are in the order of 2–5 Gy/min, with total treatment times usually under 10 min. This leaves very little margin for error during the actual treatment itself to notice a problem or to stop the treatment prior to delivery of a large dose. Finally, patients often receive only a few fractions of HDR brachytherapy (1–10) compared to external beam patients, who often receive 30 or more separate fractions. This means that the consequences of an error on a single brachytherapy fraction could have large implications on the rest of the patient's treatment, with less potential for errors to "average out" over the course of the rest of the fractions.

For LDR permanent implants such as prostate implants, a patient will receive only a single fraction. Any mistakes that occur and are not caught during the quality assurance procedures will likely reach the patient and will be uncorrectable in many cases. For example, if the wrong seed strength is ordered and this error is undetected, the patient will be implanted with anywhere from 50 to 100 seeds of the wrong activity. Correcting this mistake after it has occurred would probably require the patient to undergo extensive surgery. Ineffective quality assurance practices in the ultrasound unit or in patient positioning in the operating room can lead to seeds being implanted in the incorrect site, such as the perineum or the bladder. Selecting the wrong template from a list of template choices could result in the entire implant being shifted from the desired location.

Additionally, image-based brachytherapy creates new opportunities for errors not usually addressed in traditional QA (Cormack 2008). Most clinics are at least partially using CT or MR simulation for HDR brachytherapy treatments, and the use of ultrasound in prostate implants is well established. However, the focus of a physicist is typically on more tangible products of a brachytherapy treatment, such as source assay or applicator length measurements. Quality assurance procedures must evolve with the technology to be effective.

30.3 Brachytherapy Errors and Accidents

Currently, in 16 states of the United States, the Nuclear Regulatory Commission (NRC) governs the control and medical use of radioactive materials and determines the definition of a medical event. The states that are not under the control of the NRC are called *agreement* states. In agreement states, the state itself, rather than the NRC, defines medical errors and controls radioactive material. In the state of Texas, for example, all radioactive material and licensing is controlled by the Texas Department of Health and Services. The agreement state must

have regulations that are deemed equivalent or more stringent than those of the NRC.

The Code of Federal Regulations 10.35.3045 defines a medical event in the United States as the following for sealed source brachytherapy (NRC 2010):

- Wrong patient treated.
- Wrong isotope used.
- The total dosage delivered differs from the prescribed dosage by 20% or more or falls outside the prescribed dosage range.
- The fractionated dose delivered differs from the prescribed dose, for a single fraction, by 50% or more.
- Treatment with a leaking source.

Additionally, a dose equivalent to any part of the body other than the treatment site that exceeds by 0.5 Sv and 50% or more of the dose expected from the administration defined in the written directive is considered a "wrong site" medical event. A patient falling into any of these categories requires the submission of a report to the NRC (or to the state, for agreement states). Subsequently, each report is reviewed, and a summary is also posted on the NRC or the state's Web site. Hence, brachytherapy errors and accidents are usually well publicized in the United States due to the publication of these summaries by the NRC. Usually, the reporting institutions will be required to do an RCA and determine what QM procedures or policies need to be implemented to prevent the incident from happening again. Occasionally, fines will be levied against an institution, and there is potential for an institution's radioactive material license to be revoked.

Other countries have different rules regarding medical events or, as they are called in England, radiation incidents (RIs). For example, prior to 2000, in England, Wales, and Scotland, there were no legal requirements to report radiotherapy errors other than those caused by equipment failure (The Stationery Office UK 1988). However, these regulations changed to an RI being defined as "where the employer knows or has reason to believe that an incident has or may have occurred in which a person, while undergoing a medical exposure was, otherwise than as a result of a malfunction or defect in equipment, exposed to ionizing radiation to an extent much greater than intended, he shall make an immediate preliminary investigation of the incident and, unless that investigation shows beyond a reasonable doubt that no such overexposure has occurred, he shall forthwith notify the appropriate authority and make or arrange for a detailed investigation of the circumstances of the exposure and an assessment of the dose received" (The Stationery Office 2000). This definition includes both external beam radiotherapy and brachytherapy events.

Learning from one's past mistakes is a classic way to improve a system. However, learning from others' mistakes is more efficient. As Stuart Emslie said in his report on International Perspectives on Patient Safety, "If it's not being measured, it's not being improved" (Emslie 2005). The public posting of medical events is helpful to other radiation therapy departments so an understanding of the types of errors that occur can be learned from and prevented. The International Atomic Energy Agency (IAEA) has training material that was developed with the World

Health Organization titled "Prevention of Accidental Exposures in Radiotherapy" (IAEA 2011). This module contains dozens of both external beam and brachytherapy events that have occurred and includes an RCA of each event. Many reported events are pure human blunders, which are difficult to prevent, but many are also software or hardware faults, which can be improved directly by the equipment manufacturer. For multifaceted errors, the need for a comprehensive QM program is evident.

The most infamous HDR brachytherapy event happened at Indiana Regional Cancer Center in 1992 (NUREG-0090 1992). An elderly patient was being treated for anal carcinoma with five catheters. The source strength of the ^{192}Ir source was 160 GBq (4.3 Ci). She was scheduled for three fractions of 6 Gy per fraction. During one fraction, the first four catheters were treated without incident. While treating the fifth catheter, an error was indicated on the console. After some attempts to clear the error, the treatment was cancelled. The console of the afterloader indicated that conditions were "safe." The transfer guide tubes were disconnected from the patient. It was noted that the radiation alarms in the vault were going off, but they were ignored. A handheld survey meter was available but was not used to clear the room. The patient was sent back to her nursing facility. Unfortunately, the staff did not know the source had broken off from the source train and was lodged inside the patient in the fifth catheter.

Four days later, the patient's catheters fell out and were collected in a red biomedical trash bin at the nursing home. The patient died the following day. The catheters were stored in the bin for 5 days before a waste handler picked up the trash and transferred it to a facility in Carnegie, PA. From there, it was transferred to another facility in Warren, OH. In Warren, radiation detectors found emissions from the trash and ordered it returned back to Carnegie. Five days later, the source was finally found and traced back to the nursing home facility. The nursing home notified the hospital and the physicist, who had been unaware that the source had been missing for 16 days (no further patient treatments or QA were performed on the unit during this time).

Upon investigation by the manufacturer, Omnitron, it was determined that moisture had corroded the Nitinol wire and weakened the weld of the source to the wire. A similar event happened only a few weeks later at a separate facility, but the physicist was aware of the previous event and responded appropriately. Because of this mishap, the manufacturer changed the design of the shipping cask and the design of the source wire to reduce the risk of separation of the source from the wire.

An NRC consultant determined that the patient had received 16,000 Gy and that her death was likely caused by or contributed to by radiation overexposure. Additionally, 94 members of the hospital, nursing facility, visitors, and the trash collection company were all exposed to radiation. The NRC responded by implementing changes to the regulations, including that the licensee shall make a radiation survey of the patient prior to his or her release. Additionally, a physician and a physicist were both required henceforth to be present for all HDR cases.

While this event is the most severe and most discussed error in brachytherapy history, it is not a solitary incident. The NRC regularly presents "abnormal occurrence" reports to the United States Congress. Section 208 of the Energy Reorganization Act of 1974 (Public Law 93-438) defines an "abnormal occurrence" (AO) as "an unscheduled incident or event that the NRC determines to be significant from the standpoint of public health or safety." These AO reports are all available on the Internet from the NRC reading room at http://www.NRC.gov. Table 30.1 summarizes the number of abnormal occurrences reported to the United States Congress between 2005 and 2009.

The table is divided into human events and medical events. Human events typically are those in which radiation reached an unintended person. For the majority, these are populated by ^{131}I ablation patients who either were or became pregnant shortly after receiving their radiotherapy. The uptake by the thyroid of the fetus is considered an unintended exposure. A medical event involves the misadministration of radiation to the patient, such as to the wrong site or for the wrong dose. It should be noted that each occurrence does not necessarily represent a single patient. In 2008, one AO event involved the misadministration of radiation to the wrong site for 35 LDR prostate patients at a Veteran's Affairs (VA) medical center. Also notable is that the number of AOs reported has not decreased over time; that is, based on this measure, we are not getting better at preventing errors from happening. Complexities introduced by multimodality imaging, 3D multiplanar treatment plans, and new treatment devices present new challenges for control processes.

In addition to the catastrophic and abnormal occurrences reported to the United States Congress, there are hundreds of other smaller errors that have been reported to the NRC. According to a 1992 report from the NRC, the rate of misadministration is 4.2 per 100,000 procedures (NUREG-1272 1992). The last pattern-of-care study was conducted in 1995 and estimated that approximately 29,000 brachytherapy procedures are conducted each year (Nag et al. 1995). In 1996, it was reported that there were also 50,000 ^{131}I thyroid ablations in the United States (Gottfried and Penn 1996). In the NRC database, from December 2009 to June 2010, there were 35 separate reported medical events and accidental exposures. A summary of the events is shown in Table 30.2.

In this small sample, the most common type of error was wrong dose; however, many of these incidents could also qualify as wrong site. The most frequent type of radiation error was with LDR prostate brachytherapy. Postplan dosimetry of the implants often shows that many seeds were not implanted into the prostate or that the total dose to the prostate was inadequate ($D_{90} < 90\%$).

TABLE 30.1 Abnormal Occurrences Reported to the U.S. Congress from 2005 to 2009

Year	Total	Human Event	Medical Event
2009	9	2	7
2008	10	2	8
2007	11	1	10
2006	7	2	5
2005	8	1	7

TABLE 30.2 Summary of Reported Events by Radiation Type (Left) and Error Type (Right)

Type of Radiation	No. of Incidents	Type of Error	No. of Incidents
HDR	10	Wrong Site	9
LDR	15	Wrong Dose	17
Gamma Knife	4	Lost Source	4
Radiopharmaceutical	6	Other	5

The data for this time period may be artificially high; after the events with 35 patients from the VA medical center, many institutions performed retrospective analysis of their postplans and submitted previously unreported errors. The most common problem associated with HDR brachytherapy was incorrect applicator length. For example, on a follow-up appointment with her radiation oncologist, a patient treated with a multi-lumen balloon applicator complained of "redness" on her skin. The plan was reviewed by the physician and the physicist, and it was found that the programmed length of the applicator and transfer guide tube combination was 10 cm shorter than the actual length. Consequently, during treatment, the source was positioned very close to the skin surface and not inside the balloon applicator where it was planned to be. Also noteworthy is that this was the first time this particular applicator was used at that facility and that a representative from the applicator manufacturer was present at the treatment. Special caution should be used when implementing new treatment devices, and no comfort should be gained by representatives of the treatment device manufacturer being present. In a report by Thomadsen et al. (2003), the most common failure mode found in an HDR process tree analysis was entering an incorrect treatment length.

The number of reports (35) in only a 6-month period of time is quite large. Removing the radiopharmaceutical and gamma knife events, there were still 25 incidents. That same report by Thomadsen et al. (2003) reported 90 HDR and LDR events in a 21-year period, averaging to 4–5 events per year. The number reported here far exceeds that; as the nature of error reporting becomes more culturally acceptable and technology becomes more complex, it can only be surmised that this number will increase further in years to come.

30.4 Overview of TG-100 Risk Analysis Methodology

The goal of the AAPM's Task Group 100 was to apply modern risk-based analysis techniques to the radiotherapy process and demonstrate the techniques in a way that would identify efficient and effective methods to improve the safety and quality of patient treatment. Specifically, the Task Group looked at the intensity-modulated radiotherapy (IMRT) process and presented process maps, FMEAs, fault trees, and a QM program. In essence, it is to be used as a guide for individuals or hospitals to develop their own risk analyses. A brief summary of the methodology will be given.

30.4.1 Process Mapping

A process map or tree is a visual illustration or a diagram to bring forth a clearer understanding of a process or series of parallel processes. Both the physical and temporal relationships between the steps should be demonstrated. For radiotherapy applications, it should include all processes involving various entities such as schedulers, therapists, dosimetrists, physicists, nurses, and physicians that are involved with the patient from start to finish.

30.4.2 Failure Mode and Effects Analysis

FMEA is a procedure for analysis of potential failure modes within a system for classification by the severity and likelihood of the failures (Stamatis 1995). Each step in the process map should be assessed for three specific parameters:

- Occurrence (O): describes the probability of a particular cause for the occurrence if a failure mode occurs
- Severity (S): describes the severity of the effect on the process outcome assuming it is not detected
- Detectability (D): describes the probability that the failure mode will not be detected

All three of these components can be combined to obtain a single quantitative measure of the risk of a particular event, called the risk priority number (RPN):

$$RPN = O*S*D.$$

Once the RPN has been completed for each step in the process tree, it is easy to determine the areas of greatest concern. High RPN failure modes should be given the highest priority for corrective action. This means that a failure mode with the highest severity may not be given the highest RPN because it is easily detectable, extremely unlikely, or both. Often, lower severity failure modes are deemed more critical because they occur more often and have a higher chance of going undetected. In this work, RPNs were determined by a single individual with expertise in the given area. The benefit of having only a single individual determine RPNs is that it is a very self-consistent ranking; however, there could be omissions. Another way to determine RPNs is by a consensus of a working group. Then RPNs can be averaged or determined by some other means.

For this work, we will use the same scales that TG-100 used for the O, S, and D indices:

- O ranges from 1 (failure unlikely, <1 in 10^4) to 10 (failure highly likely, more than 5% of the time).
- S ranges from 1 (no danger, minimal disturbance of clinical routine) to 10 (catastrophic).
- D ranges from 1 (very detectable, 0.01% or fewer of the events go undetected throughout treatment) to 10 (very hard to detect, > 20% of the failures persist through the treatment course).

30.4.3 Fault Tree Analysis

Fault trees are companions to the process tree. Each failure mode is identified with the corresponding process that could allow the failure. The tree is written out with nodes connected by a set of logic gates (typically AND/OR gates) that describe each possible event involved in the node. Fault tree analysis has been successful in a wide range of industries, including nuclear power (Nomura 1992) and vehicle design (Ivanovic et al. 1994).

30.4.4 QM Program

Each fault tree can be evaluated from a QM standpoint. Nodes that are connected by AND gates provide protection; OR gates are opportunities for error propagation. Each fault mode should have a QM measure between it and each node. The measures can include policies, check lists, procedures, quality assurance, quality control, hardware interlocks, software checks, etc. Interventional solutions for items with RPN scores in the top 10% or over a certain value, like 100, should be considered. Feasible and effective improvements should be identified.

30.5 Case Studies

Several case studies will be given to elucidate the FMEA process. One example, HDR brachytherapy, will be given in its entirety:

process mapping, fault tree development, *OSD* determination, results, and the quality improvement process. The others will show only the *OSD* tables and contain a brief description of the results.

30.5.1 HDR Brachytherapy

30.5.1.1 Process Map

The process map or tree developed pertains to the authors' particular institution. The brachytherapy suite at Washington University/Barnes Jewish Hospital contains two separate vaults with two separate Varian iX HDR afterloaders. It has a dedicated nursing staff that performs the premedication and anesthesia for the patients. Most nursing issues have been omitted from the process tree—only those that directly affect the physical delivery of the radiation have been included. Patients in the clinic receive either 3D CT imaging (breast, interstitial, etc.) or MR (cervix, inoperable endometrial). The suite also has dedicated dosimetrists and brachytherapists who perform the treatment planning and deliver patient treatments, respectively. The physicist is primarily utilized for quality control and plan review as opposed to performing many of the routine patient treatment functions that a smaller clinic's physicist might perform. Nonetheless, many of the items noted here are transferable to large and small clinics. The process map may be seen in Figure 30.1.

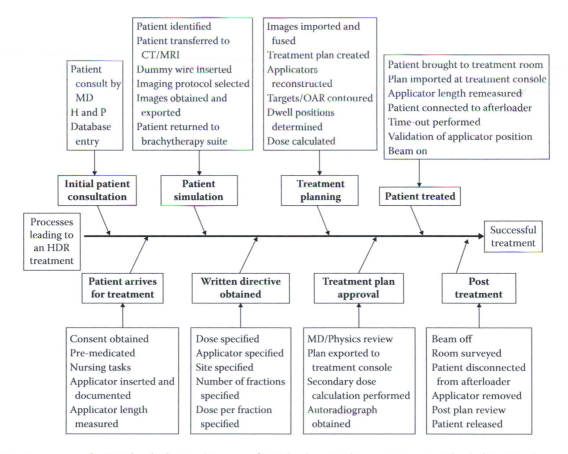

FIGURE 30.1 Process map for HDR brachytherapy. (Courtesy of S. Richardson, Washington University School of Medicine.)

The process map should be read from left to right along the central arrow, starting at "processes leading to an HDR treatment" and ending at "successful treatment." The bold boxes directly connected to the process tree line by an arrow are the main procedures that take place, such as treatment planning. The various components of the process are shown in a separate box connected by a vertical arrow. The process map is neither all-inclusive nor exhaustive, but it contains the main components that influence the patient from start to finish.

30.5.1.2 Failure Mode and Effects Analysis

Next, for each step described in the process map, at least one failure mode will be identified. For each failure mode, the potential cause or mechanism of failure will be described and the RPN determined. For succinctness, only a few processes will be shown in their entirety here. The first process described is the applicator insertion and documentation and is shown in Table 30.3. Three different failure modes were identified: wrong applicator inserted, applicator inserted incorrectly, and applicator documented incorrectly. "Wrong site" refers to a completely unintended area of the patient receiving dose. "Wrong location of dose" means that the correct dose distribution is somewhat offset to the target volume, possibly resulting in an underdose to the target or unplanned irradiations to areas near the target volume. "Wrong dose" means that the wrong dose distribution was calculated or delivered. Each potential failure mode will be described and the thought processes behind the S, O, and D values shown below.

(1) Wrong applicator inserted:
- The failure mode could be that the catheter was not the length desired, for example, 100 cm instead of 150 cm, or the wrong size of ovoids for a patient receiving tandem and ovoids could be inserted (e.g., 1.6-cm diameter instead of 2.0-cm diameter).
- The potential effects of the failure were determined to be either wrong site or wrong dose. For wrong site, a catheter that was incorrectly assigned a length of 150 might result in the patient's nose or throat being irradiated instead of a bronchial lesion. For wrong dose, the wrong ovoid size may result in the dwell times being too long or short for a given diameter, resulting in either an underdose or overdose to the cervix region.

- Both events would have a serious effect on the patient, and a value of 8 was given for severity. For wrong site, this was due to the chance of delivering an entire fractional dose to the wrong region. At our institution, the difference in dwell times between mini ovoids (1.6-cm diameter) and standard ovoids (2.0-cm diameter) would be 35%; a serious overdose.
- The potential causes were attributed to human blunder or error and inadequate training in recognition of the various types and sizes of applicators.
- Both had low probability of occurrence, and thus, a value of 2 was selected. Our institution performs hundreds of tandem and ovoid treatments every year, and multiple people check the size of the ovoids prior to insertion. Catheter lengths are also measured and marked prior to sterilization; therefore, the chances of these mistakes occurring at our clinic were deemed to be low.
- Every patient receives some type of imaging, and on these images, the ovoid diameter can be measured. A catheter is also measured multiple times before treatment. Therefore, it is highly likely that this failure mode would be detected, so a detection value of 3 was given.
- The resulting RPN for each potential effect of wrong applicator inserted was 48. The significance of this value will be discussed later.

(2) Applicator inserted incorrectly
- This failure mode could occur if a tandem perforated the uterus, or an endobronchial catheter ended up in a bronchial path other than the desired one.
- The two potential effects were determined to be wrong site and discomfort to the patient. Wrong site would occur in the both examples, the catheter in the wrong bronchus and the perforation. Discomfort to patient (a mild descriptor) is an additional effect for the example of a perforation. While many potential effects of a failure are medical events, this is not always the case—any ill effect to the patient or the procedure can be a potential effect.
- The discomfort to the patient itself is a minimal severity, and a value of 2 was given. While the two

TABLE 30.3 FMEA for Applicator Insertion and Documentation

Process	Potential Failure Mode	Potential Effect(s) of Failure	Sev	Potential Cause(s)/Mechanism(s) of Failure	Occur	Detect	RPN
Applicator inserted and documented	Wrong applicator inserted	Wrong site	8	Blunder, inadequate training	2	3	48
		Wrong dose	8	Blunder, inadequate training	2	3	48
	Applicator inserted incorrectly	Discomfort to patient	2	Blunder, inadequate training, patient movement	7	3	42
		Wrong site	8	Blunder, inadequate training, patient movement	7	3	168
	Applicator documented incorrectly	Wrong dose	8	Blunder, miscommunication	3	3	72

potential examples have different ranges of seriousness (perforation versus wrong bronchus), an average value must be selected. The wrong site could have serious consequences and was determined to be an 8. If the tandem perforates the uterus, it is directly adjacent to the small bowel and could cause the patient serious problems.

- The potential causes were attributed to human blunder or error, inadequate training, and patient movement.
- Both effects had a moderately high probability of occurrence (7) due to the fact that we cannot totally control patient movement. For example, many endobronchial patients have difficulty breathing and can cough strongly enough to dislodge a catheter from the desired bronchus.
- Since every patient receives some type of imaging, it is likely that this failure mode would be detected, so a detection value of 3 was given.
- The resulting RPN was 42 for discomfort to the patient and 168 for wrong dose.

(3) Applicator documented incorrectly
- This failure mode refers to the incorrect transcription of the applicator that is inserted into the patient for treatment planning purposes. At our institution, this information is documented on the written directive. While similar to the "wrong applicator inserted" failure mode, this will yield treatment planning errors as opposed to patients actually having the wrong device inserted.
- The potential effect was determined to be wrong dose, as described for the first failure mode (wrong applicator inserted).
- The effect would have a serious consequence for the patient, and a value of 8 was given for severity.
- The potential causes were attributed to human blunder often caused by haste.
- This effect had a low probability of occurrence, and thus, a value of 3 was selected. It was chosen as a 3 to show that it had a slightly higher chance of happening than the previous failure mode of physically implanting the wrong applicator. For

TABLE 30.4 FMEA for HDR Failure Modes with Significant RPN Scores

Process	Potential Failure Mode	Potential Effect(s) of Failure	Sev	Potential Cause(s)/ Mechanism(s) of Failure	Occur	Detect	RPN
Applicator inserted and documented	Applicator inserted incorrectly	Wrong site	8	Blunder, inadequate training, patient movement	7	3	168
Images imported and fused	Right patient, wrong data set imported	Wrong site	8	Blunder, not following procedure	3	6	144
	Fusion incorrect	Wrong location of dose	5	Blunder, lack of training	5	5	125
Applicators reconstructed	Wrong digitization of channel	Wrong location of dose	5	Blunder, lack of training, not following procedure	6	6	180
	Applicator length not entered correctly	Wrong location of dose	5	Blunder, lack of training, not following procedure	5	5	125
		Wrong site	8	Blunder, lack of training, not following procedure	5	4	160
Dose calculation	Wrong formalism in TPS	Wrong dose	6	Lack of training	3	7	126
	Source strength incorrect	Wrong dose	9	Source exchange error, not following procedure, blunder	5	3	135
	Wrong dose points used	Wrong dose	8	Lack of training, blunder	4	4	128
Patient connected to afterloader	Wrong transfer-guide-tube selected	Wrong site	8	Blunder, lack of training	4	4	128
	Wrong connection between catheter and TGT	Wrong location of dose	6	Blunder	5	4	120
Applicator length measured	Length measured incorrectly	Wrong site	8	Blunder, lack of training	6	4	192
		Wrong location of dose	6	Blunder, lack of training	6	4	144
	Wrong length documented	Wrong site	8	Blunder, lack of training	6	4	192
Treatment plan created	Wrong prescription entered	Wrong dose	8	Blunder	4	4	128
	Wrong afterloader selected	Wrong dose	9	Blunder	4	4	144
Dummy wire inserted	Dummy wire not fully inserted	Wrong location of dose	6	Blunder, lack of training	4	7	168

example, the radiation therapist may forget to document the applicator at the time of insertion and then recalls the information incorrectly when they are filling out the written directive.

- It is likely that this failure mode would be detected on imaging, so a detection value of 3 was given.
- The resulting RPN was 72.

For the process of "Applicator Insertion and Documentation," there were several different failure modes addressed; however, one has a much higher RPN than the others. Therefore, additionally quality control/assurance steps should be noted for this process (to be discussed later).

The total FMEA process for HDR contained 42 separate processes and 82 different failure modes. The 17 failure modes that resulted in the highest RPN values (>100, chosen arbitrarily) are shown in Table 30.4.

There are several items of note from Table 30.4. One point that was mentioned previously is that some of the most severe errors imaginable are not listed here. For example, if the source broke off in the patient, the severity if not detected would be a 10. However, since the detectability of such an event is so likely ($D = 2$) with

current standard operating procedures and the probability of an occurrence is so low ($0 = 1$), the resulting RPN ($10*20*1 = 20$) does not make the list. Alternatively, one error that shows up frequently with very high RPNs is the measuring and entering of the physical length of the applicator. This is because it is both a severe error and an error that could easily occur. One quality assurance procedure that could reduce the RPN value would be to have two separate individuals each independently measure and document the applicator lengths. Then the RPN value could be recalculated. This is particularly important if the applicator is new to the clinic or if the catheter lengths are patient-specific, for example, trimmed catheters for a sarcoma case. Another method would be to implement a software change. With Varian HDR systems, the default applicator length is 150 cm, the length of their longest catheter. Ideally, the applicator length would default to 0 cm, and therefore, a patient could not be treated without entering in an applicator length.

30.5.1.3 Fault Tree Analysis

The fault tree given in Figure 30.2 illustrates one of the most basic failure effects of treatment: wrong dose. Reading from left to right, the fault tree bifurcates into separate possible causes,

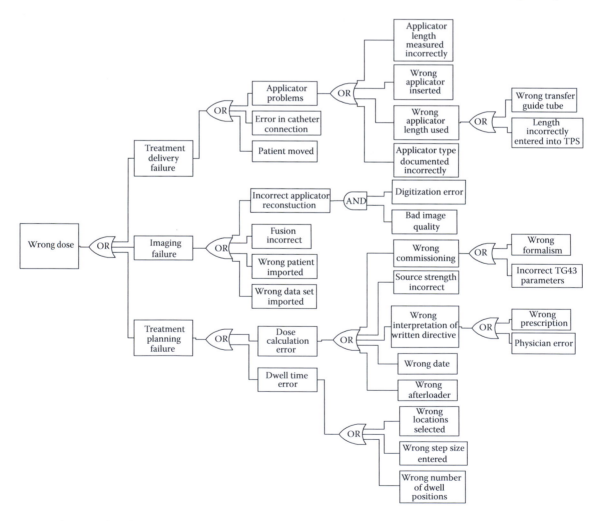

FIGURE 30.2   Fault tree analysis for wrong dose. (Courtesy of S. Richardson, Washington University School of Medicine.)

dividing again until the most fundamental process causing the error is reached. From this fault tree, one can see where quality assurance processes or quality control checks need to be implemented, if not already existing. For example, "Applicator Problems" (highlighted) has four separate processes stemming from it, including "Applicator Length Measured Incorrectly." At our institution, we have two individuals independently measure the applicator length for cases where nonstandard applicators are used (e.g., interstitial, where the catheters are physically cut, or partial breast treatments, where the balloon lengths vary by 1 to 2 mm). Having redundancy will reduce the chance the error will propagate and ultimately cause a wrong dose scenario.

30.5.1.4 Quality Management

According to AAPM's TG-59 "High Dose Rate Brachytherapy Treatment Delivery" (Kubo et al. 1998), the four main tools for a successful HDR QM program are:

(1) Use of written documentation
(2) Development of formal procedures
(3) Exploitation of redundancy
(4) Exploitation of quality improvement techniques

Each tool can be used for a variety of quality assurance and process improvement tasks. In one study of NRC misadministrations, there were eight direct causes of error: lack of policies and procedures, lack of radiation safety oversight, lack of training and experience, lack of supervision, errors in judgment, errors in interpretation, lack of communication, and hardware errors (Ostrom et al. 1994). As shown in the FMEA list, most errors in brachytherapy are caused by human mistakes or blunders rather than mechanical error. Corrective actions cannot be for the person at fault to "try harder" or "pay attention." For high-risk procedures, the easiest way to prevent the failure is to "design" out the failure. If this is not possible, QM is necessary. A QM program must focus on the interception of the failure prior to it occurring and place quality control procedures at those points that are deemed critical. A variety of tools are available to improve human performance, including checklists, redundant checks, and a culture of patient safety.

Another tool is to have an impartial evaluation of on the whole process. An inspection by a colleague at an unrelated institution can provide a fresh view of the processes occurring in the clinic. A wide variety of additional possible HDR QA procedures can be found in Thomadsen's book "Achieving Quality in Brachytherapy" (Thomadsen 2000).

The most common classifications for the possible causes for the failure shown in the HDR fault tree analysis from Figure 30.2 are shown below in Table 30.5.

As stated previously, human error or blunder is the category with the greatest frequency. Inadequate training has the second highest occurrence, and the lack of or deviation from the standard procedures is third. These three categories comprise three quarters of the events.

30.5.2 Prostate Brachytherapy

Prostate brachytherapy made the front page of the *New York Times* in 2009. Unfortunately, it was not about exceedingly high cure rates or new treatment methodologies. The title of the article was *At V.A. Hospital, a Rogue Cancer Unit* (Bogdanich 2009). In this article, Bogdanich described the "botching" of 92 separate prostate brachytherapy cases over a period of 6 years at a VA hospital. A total of 116 brachytherapy cases had been treated yielding a failure rate of almost 80%. According to the article, 57 cases had too little dose to the prostate, and 35 had overdoses to other parts of the body. Many cases are currently under review by the NRC and may be retracted and consequently *not* be medical events. The NRC is also currently reviewing their definition of a medical event for prostate brachytherapy because of difficulties in defining "wrong dose" to an ambiguous target. For the events in this article, it is unclear exactly what went wrong. Some information points to faulty or malfunctioning equipment, while other data indicates say that some of the physicians involved did not have the required training or knowledge to perform good prostate implants. One thing that was certain is that post-implant dosimetry was not done in a timely manner in most cases.

The American College of Radiology practice guidelines state that "post-implant dosimetry is mandatory for each patient" (ACR 2010). While the optimal timing of post-implant dosimetry is debated (1 day, 1 week, 1 month post-implant), it is the standard that it *should* be done. The logic for calculating post-implant treatment plans is that it provides a measure of the quality of the implant. By reviewing the postplan, an institution may use figures of merit and/or dose distributions to describe the quality of their implants. It is used as a check of the overall system, including CT and ultrasound imaging, patient positioning reproducibility, as well as seed placement accuracy. Constraints that are often reported include the V_{100} (defined as the percentage of the target volume delineated on the post-implant CT receiving 100% of the prescribed dose), D_{90} (dose received by 90% of the target volume), and organ-at-risk (OAR) parameters such as R_{100} (volume of rectum receiving 100% of the prescription dose). While post-implant dosimetry provides a lot of information regarding the quality of the implant, they do not provide complete QA of the entire brachytherapy process; there are still many items that require separate verification. These include source receipt and assay, source modeling within the treatment planning system, and dose calculation accuracy.

TABLE 30.5 Categories of Errors in HDR Fault Tree Analysis

Category	Occasions
Human failures (blunders, lack of attention)	51
Lack of standardized procedures or failure to follow existing procedures	24
Inadequate training	29
Inadequate communication	5
Hardware/software/equipment failure	14
Inadequate commissioning	3
Patient-related factors	10

At our institution, a patient receives a volume study, followed by the development of a treatment plan. The sources are ordered, assayed, and then sterilized. Typically, we use preloaded stranded needles and order an additional 10% of the total seeds to be loose. The implant itself is monitored with intraoperative software tools that allow real-time dose calculation during the procedure. Once the patient is positioned in the operating room, another volume study and ultrasound data set are generated and fused with the contours from the preplan. During the implant, the dose is calculated in real time, and any small positioning errors in either the patient setup or the needle/source position are accounted for. If necessary, the extra seeds can be implanted to account for any cold spots in the treatment volume. The physician will often use fluoroscopy during implantation to verify needle placement. Once the implant is complete, the patient receives a CT and/or an MR that day. A postplan is completed as soon as possible.

An FMEA of our intraoperative prostate brachytherapy program is described in Table 30.6. For succinctness, only the implant process itself will be described.

In this process, three potential failure modes resulted in RPN values over 100: needles placed in the wrong location in the template, wrong retraction plane used, and needle deflection. The first item is strongly physician-dependent and relies on their attentiveness and training. In our procedure, the location of the needle is called out to the physician (e.g., C-5). The physician places the tip of the needle into the template and calls back out "C-5." Then the needle is pushed into the patient. No other individual currently checks the placement of the needle within the template. This is why the likelihood of detection is less than others noted above. During the implant as well as in other processes in the prostate FMEA, wrong retraction plane was a potential failure mode that resulted in a high RPN. Needles can be loaded several different ways—some clinicians prefer to have all of their needles inserted to the base; thus, the first few centimeters of a needle may not contain any other seeds. Others may prefer to only have to push the needle in as deep as required to place the most superior seed. This must be specified in the treatment plan

and during the source ordering. The vendor must appropriately interpret and build the needles according to the loading specified. The physician must also know during the procedure exactly what retraction plane is being used to ensure correct positioning of the needle. Due to this complexity and the multiple levels of communication required, it had a high RPN. Needle deflection is also a common issue due to patient's anatomy causing the needles not to go straight into the desired location. In our experience, it happens quite often but can also be detected fairly easily with the use of ultrasound and fluoroscopy. If the needle deflects too much, it is retracted and reinserted.

30.5.3 Xoft Brachytherapy

Electronic brachytherapy using Xoft is a relatively new technology and was implemented in our clinic in 2009. The Xoft Axxent brachytherapy system (iCad, Inc., Sunnyvale, CA) consists of a miniature water-cooled x-ray tube that produces 50-kVp photons. It has been used primarily as an alternative to ^{192}Ir brachytherapy with applications in endometrial cancer and breast (Dickler et al. 2008, 2009; Hiatt et al. 2008). It has also been used as a superficial x-ray unit for surface lesions using a skin applicator system (Rong 2010). Performing an FMEA for a new process or system can be beneficial in that it can help establish necessary equipment, personnel, and time requirements for appropriate quality assurance tests to be done. This differs slightly from performing FMEA for an existing process in which one is seeking to improve the process rather than establishing baseline QA. In light of that reasoning, an FMEA for Xoft has been included here.

Following the acceptance testing and commissioning of Xoft and establishment of guidelines and procedures for different treatment and quality assurance procedures, an FMEA was performed on several processes as a retrospective study. A partial list of failure events analyzed for treatment delivery process has been shown in Table 30.7. The FMEA was performed assuming that there was no quality assurance procedures established. This demonstrates almost no chance for detection of errors, making

TABLE 30.6 FMEA for Intraoperative Prostate Brachytherapy

Process	Potential Failure Mode	Potential Effect(s) of Failure	Sev	Potential Cause(s)/Mechanisms of Failure	Occur	Detect	RPN
Implant	Patient positioned incorrectly	Slightly wrong dose	5	Lack of training	5	2	50
	Apex and/or base set incorrectly	Wrong dose, wrong site	8	Lack of training, blunder	3	3	72
	Intraop slice spacing wrong	Wrong dose, wrong site	9	Lack of training, blunder	3	3	81
	Intraop slices dropped	Wrong dose, wrong site	7	Software or hardware malfunction	1	3	21
	Intraop images fused incorrectly to preplan images	Wrong dose, wrong site	7	Lack of training, blunder	3	3	63
	Needles placed in wrong location in template	Wrong location of dose, wrong site	8	Blunder	4	5	160
	Wrong retraction plane used	Wrong dose, wrong site	9	Lack of training, blunder	4	4	144
	Needle deflection	Slightly wrong dose, wrong site	6	Patient anatomy, lack of training	6	3	108
	Seed migration	Slightly wrong dose, wrong site	5	Patient anatomy, lack of training	5	2	50
	Equipment or needle malfunction	Slightly wrong dose, wrong site	7	Lack of training, hardware error	3	3	63

TABLE 30.7 FMEA on Xoft for Treatment Delivery Processes with No Quality Assurance Procedures Included

Process	Potential Failure Mode	Potential Effect(s) of Failure	Sev	Potential Cause(s)/Mechanism(s) of Failure	Occur	Detect	RPN
Treatment delivery	System timer error	Wrong dose	8	Hardware error	1	10	80
	Calibration error			Hardware error, blunder, lack of training	5	10	400
	Incorrect dwell file			Blunder, software error	3	10	240
	Barcode error	Wrong site	10	Blunder, hardware error	2	10	200
	Incorrect applicator length			Blunder, lack of training	3	5	150
	Source positioning error			Hardware error	2	10	200
	Applicator not in the patient			Blunder, lack of training	4	6	240
	Source transit error	Treatment interrupt	5	Treatment planning error	2	10	100
			5	Incorrect dwell position/applicator length	3	10	150

the RPNs much higher than those discussed in the previous sections. The RPN values were based on the experience of a medical physicist with expertise in brachytherapy.

The different potential effects of failure listed in the table are analyzed as follows:

(1) Wrong dose
- The potential causes for delivering a wrong dose to the target could be due to system timer error, calibration error, or incorrect dwell file. The probability of an error occurring in the system timer is very minimal, although it cannot be ignored; hence, a minimal value of 1 is chosen. As it might result in a very high or very low dose to the target, leading to tumor recurrence or treatment-related complications to the patient, a higher severity value of 8 is given. The probability of an error occurring in the calibration procedure varies depending on the training provided as it is a new procedure to be followed in the clinic. Hence, there is a higher chance of error happening, at least until the therapists/physicists are familiar with the procedure. Hence, a moderate value of 5 was chosen for this error. Familiarity with the training and the procedure should reduce this value in future. The probability of an incorrect dwell file pertaining to the same patient being transferred is not negligible, although less than the probability of a calibration error happening. Hence, a value of 3 was chosen during the analysis.
- When any of these errors occur, the probability of them being detected is very low since no quality control processes are in place, and therefore, the largest value of 10 has been assigned for all these errors. Comparing the RPN scores, it was found that the calibration error had the highest RPN score of 400.

(2) Wrong site
- Since wrong site can result in an unacceptable amount of radiation to a normal structure,

resulting in normal tissue complications, the largest severity value of 10 is chosen.
- Potential causes: there might be several potential causes for this kind of failure to occur.
  a. The Xoft system requires a barcode scan which communicates to the Xoft system what type of applicator is being used during the patient treatment. An incorrect barcode could be printed, or the wrong barcode could be scanned. Although a possibility, it is fairly unlikely, and an occurrence value of 2 was assigned here. When such an error occurs, the probability of detection becomes highly unlikely, and a higher value of 10 is assigned.
  b. If the applicator length is incorrect (which could be due to an error that occurred while applicator acceptance or at the time of patient treatment), the source dwells at the wrong place resulting in a wrong site of treatment. The probability of such an error occurring is assigned a value of 3, and detection, a value of 5, as these values are verified against standard values obtained during commissioning. The source is measured only once, during source acceptance; therefore, it is unlikely this error would be caught once the source is approved for clinical use.
  c. An error in source positioning might occur if the mechanical system malfunctions. A small value of 2 was assigned due to very few malfunctions experienced during commissioning. However, when such an error occurs, detection is highly unlikely until the next scheduled quality assurance procedure is performed. This results in a high value of 10 for detection.
  d. Applicator not in patient: This kind of error has a higher frequency of occurrence (4) but an even higher probability of being detected (6).

(3) Treatment interrupt

- This failure can occur if there is a system error due to dwell position and applicator length not being in correspondence with each other. For example, when the dwell position entered is longer than the actual length of the applicator, a source transit error occurs that results in a treatment interrupt as the system fails to position the source at the requested dwell position. Such a failure results in repeating the entire process of measurement, calculation, and planning; increased treatment time; and patient discomfort. The probability of occurrence is reasonably low (3) but highly unlikely to be detected before treatment begins (10). A treatment planning error could also result in a delayed treatment as the software has a system of internal checks, which must clear prior to treatment.

This FMEA accentuates the need for reducing the probability that an event will go undetected. Quality assurance procedures and double checks must be built in to the program to reduce this number to acceptable values and result in much lower RPN values. As experience is gained with the Xoft system, the RPN calculations should be updated. Routine FMEA on a regular basis (such as annually) will allow for new experiences to influence the values of the RPN system and for new QA processes that have been implemented to increase detection.

30.6 Summary

This chapter sought to explain the methodology for implementing FMEA for brachytherapy applications. One example was covered in depth: HDR brachytherapy. Xoft and prostate brachytherapy were briefly described to provide guidance for individuals performing their own FMEA. Conventional quality control practices are time-consuming and often do not provide quality assurance on those processes that are most important to a patient. FMEA seeks to redistribute physicist, physician, therapist, dosimetrist, and managerial workloads onto those processes that will directly benefit the patient the most. Hopefully, this will lead to fewer radiation accidents and medical events in the future. Several facilities that experienced serious accidents are mentioned by name in this chapter. This information is part of the public domain and was used as such to illustrate the subject material at hand. There are no hospitals or clinics with a perfect patient safety record, and we only hope by distributing this information that those involved in patient care and safety will become more educated individuals in regard to quality assurance.

References

AAPM. 2009. Method for evaluating QA needs in radiation therapy. American Association of Physicists in Medicine. Task Group Report 100 (unpublished). http://www.aapm.org/org/structure/report_stats_notes.asp?committee_code = TG100.

ACR. 2010, ACR-ASTRO Practice Guideline for Transperineal Permanent Brachytherapy of Prostate Cancer. American College of Radiology. [Online] http://www.acr.org/SecondaryMainMenuCategories/quality_safety/guidelines/ro.aspx.

AHRQ. 2010. Patient Safety Network (PSN). Agency for Healthcare Research and Quality. [Online available] http://psnet.ahrq.gov/Accessed: October 14, 2010.

Bogdanich, W. 2009. At VA Hospital, a Rogue Cancer Unit. *The New York Times* 20 June 2009. Newspaper on-line. [Online] http://www.nytimes.com. Accessed: November 15, 2010.

Cormack, R.A. 2008. Quality assurance issues for computed tomography-, ultrasound-, and magnetic resonance imaging-guided brachytherapy. *Int J Radiat Oncol Biol Phys* 70(Suppl.):136–41.

Dickler, A. Kirk, M.C., Coon, A. et al. 2008. A dosimetric comparison of Xoft Axxent electronic brachytherapy and iridium-192 high-dose-rate brachytherapy in the treatment of endometrial cancer. *Brachytherapy* 7:351–4.

Dickler, A., Ivanov, O., and Francescatti, D. 2009. Intraoperative radiation therapy in the treatment of early-stage breast cancer utilizing Xoft Axxent electronic brachytherapy. *World J Surg Oncol* Mar 2;7:24.

Emslie, S. 2005. International Perspectives on Patient Safety: Report to the National Audit Office, England.

ESTRO. 2008. Radiation Oncology Safety Information System (ROSIS). European Society for Radiotherapy and Oncology. [Online] Available at http://www.clin.radfys.lu.se/ROSIS_data_intro.asp. Accessed: October 14, 2010.

Gottfried, K.D. and Penn, G. 1996. Radiation in Medicine: A Need for Regulatory Reform. Committee for Review and Evaluation of the Medical Use Program of the Nuclear Regulatory Commission, National Academies of Science, Washington, DC. http://books.nap.edu/openbook.php?record_id=5154&page=R1 Accessed: November 1, 2010.

Hiatt, J.J.R., Cardarelli, G., Hepel, J., Wazer, D., and Sternick, E. 2008. A commissioning procedure for breast intracavitary electronic brachytherapy systems. *J Appl Clin Med Phys* 9:2775.

IAEA. 2011. Prevention of Accidental Exposure in Radiotherapy. International Atomic Energy Agency. Vienna. [Online] www.iaea.org/newscenter/focus/radiationprotection/exposure.pdf.

ICRP. 2005. Prevention of high-dose-rate brachytherapy accidents. International Committee on Radiation Protection. Report 97. *Annals ICRP* 35:1–52.

ISO (International Organization for Standardization). *The ISO 9000 Handbook*. ISO 8402. New York: McGraw-Hill, 1994. Quoted by Peach R.W., 1992.

Ivanovic, G., Popovic, P., Stojovic, M., and Djokic, M. 1994. Fault-tree analysis applied to vehicle design. *Int Vehicle Des* 16:416–24.

Kubo, D.H., Glasgow, G.P., Pethel, T.D., Thomadsen, B.R., and Williamson, J.F. 1998. High dose rate brachytherapy treatment delivery: Report of the AAPM Radiation Therapy Committee Task Group No. 59. *Med Phys* 25:375–403.

Mutic, S., Parikh, P., Klein, E., Drzymala, R., and Michalski, M. 2008. An error reporting and tracking database tool for process improvement in radiation oncology. *Med Phys* 35:2984.

Nag, S., Owen, J., Farnan, N. et al. 1995. Survey of brachytherapy practice in the United States: A report of the clinical research committee of the American Endocurietherapy Society. *Int J Radiat Oncol Biol Phys* 31:103–7.

Nomura, Y. 1992. Fault tree analysis of loss of cooling to a Halw storage tank. *J Nucl Sci Technol* 29:813–23.

NRC. 2010. Nuclear Regulatory Commission 10 CFR. [Online] http://www.nrc.gov/reading-rm/doc-collections/cfr/part035/part035-3045.html. Accessed: October 15, 2010.

NUREG-0090. 1992. Report to Congress on Abnormal Occurrences. NUREG-0090 92-18. Loss of Iridium-192 Source and Medical Therapy Misadministration at Indiana Regional Cancer Center in Indiana, Pennsylvania. U.S. Nuclear Regulatory Commission. Vol 15, No. 4. Washington D.C.

NUREG-1272. 1992. Analysis and Evaluation of Operational Data. Nuclear Regulator Commission. Washington D.C.

Ostrom, L.T., Leahy, T.J., and Novack, S.D. 1994. Summary of 1991 and 1992 misadministration events investigations. EG&G-2707, NUREG/CR-6088.

Rong, Y.Y. 2010. SU-GG-T-43: Physics commissioning in Xoft Axxent® electronic brachytherapy (eBx) for the primary treatment of non-melanoma skin cancer. *Med Phys* (Lancaster). 37:3193.

Stamatis, D.H. 1995. *Failure Mode and Effect Analysis: From Theory to Execution*. ASQ Quality Press, Milwaukee, WI.

The Stationery Office. 1988. Ionising Radiation (Protection of Persons Undergoing Medical Examination or Treatment) Regulations. London, England (SI 1988/788). [Online] http://www.legislation.gov.uk/uksi/1988/778/contents/made.

The Stationery Office. 2000. The Ionizing Radiation (Medical Exposure) Regulations. London: SI 2000/1059. http://www.opsi.gov.uk/si/si2000/20001059.htm. Accessed: October 10, 2010.

Thomadsen, B.R. 2000. *Achieving Quality in Brachytherapy*. Institute of Physics Publishing. Medical Science Series. Bristol and Philadelphia.

Thomadsen, B.R., Lin, S.W., Laemmrich, P. et al. 2003. Analysis of treatment delivery errors in brachytherapy using formal risk analysis techniques. *Int J Radiat Oncol Biol Phys* 57:1492–508.

31

Quality Audits in Brachytherapy

Carlos E. de Almeida
Rio de Janeiro State University

David S. Followill
M.D. Anderson Cancer Center

Claudio H. Sibata
21st Century Oncology

31.1 Introduction

Brachytherapy refers to the delivery of radiation therapy such that the radioactive sealed sources are placed close to or directly within the volume of interest to be treated. Sealed sources can be inserted within body cavities (e.g., the uterus, vagina, bronchus, esophagus, and rectum), placed on the surface of superficial tumors near the skin, or placed directly into a tissue by interstitial techniques as used in the head-and-neck region, prostate, and breast. The radioactive sources can be placed near the target volume either by means of a manual insertion or through the use of an afterloader. The objective of brachytherapy is to accurately locate the sources in the target so as to accurately and safely deliver the prescribed dose to the target volume in a conformal manner while avoiding unnecessary dosage to surrounding healthy tissue (Hoskin and Bownes 2006).

In order to ensure the optimal treatment of patients, much effort is required in the commissioning phase of new brachytherapy equipment [including radioactive sources, treatment delivery devices, dosimetry equipment, and treatment planning systems (TPSs)] and maintenance during their clinical lifetime. In order to ensure accuracy and consistency, an institution must develop a proper quality assurance (QA) program for brachytherapy sources and equipment.

In 2000, the International Atomic Energy Agency (IAEA) published its Report No. 17, "Lessons Learned from Accidental Exposures in Radiotherapy" (IAEA 2000). Although brachytherapy is applied only in about 5% of all radiotherapy cases, 32 of the 92 accidents reported in this booklet were related to the use of brachytherapy sources. Errors in the specification of the source activity, dose calculation, or the quantities and units resulted in doses that were up to twice the prescribed dose. In many instances, human error played a key role in the accidents.

The synopsis of incidents given in IAEA Safety Report No. 17 (IAEA 2000) and the ICRP Report No. 112 (ICRP 2009) demonstrates clearly the need for a well-designed program of QA for brachytherapy. The main topics to be considered in a QA program involve the consistency of the administration of each individual treatment; the realization of the clinical prescription by the radiation oncologist; safe execution of the treatment with regard to the patient and to others who may be involved with, or exposed to, the sources during treatment; and education and training in the techniques used in brachytherapy. All of these topics must be included in the QA program, including openness to report any errors discovered so that they can be corrected without fear of reprimand (Venselaar and Pérez-Calatayud 2004).

Independent quality audits are an essential component of a QA program, and there is a strong tendency for them to become an operational requirement by the licensing and regulatory agencies. The quality audit is a way of independently verifying the suitability and usefulness of a QA program. This audit should be performed by qualified medical physicists (IAEA 2002).

One such entity that conducts independent audits is the Radiological Physics Center (RPC), located at the University of Texas M. D. Anderson Cancer Center, which has been funded by the National Cancer Institute (NCI) continuously since 1968 to provide quality auditing of dosimetry practices at institutions participating in NCI cooperative clinical trials (RPC 2011). The RPC QA program consists of various audit tools, but the most comprehensive audit is the on-site dosimetry review visit. A visit to an institution is conducted by an RPC medical physicist to ensure that the dosimetric practices of the institution are according to accepted protocols and to resolve any discrepancies that might exist between the institution and the audit. Similar services are provided by the ESTRO Quality Assurance Network (EQUAL), which was funded by the European Society for Radiotherapy and Oncology (ESTRO) (Roué et al. 2006), but now is privately owned; the Radiotherapy Quality Control Program (Programa de Qualidade em Radioterapia [PQRT]) (de Almeida 1998) and the Laboratório de Ciências Radiológicas (LCR) (Ochoa et al. 2007), both in Brazil; and recently, the IAEA, through its promotion of quality audits worldwide as part of its Quality Assurance Team for Radiation Oncology (QUATRO) program (IAEA 2007).

Other organizations providing various forms of quality audits are the American College of Radiology (ACR) (Hulick and Ascoli 2005; Ellerbroek et al. 2006) and the American College of Radiation Oncology (ACRO) (Dobelbower et al. 2001; Cotter and Dobelbower 2005). These two programs are much more involved with the complete radiotherapy treatment process and consist of site visits to the radiation oncology department where procedures and documentation are reviewed, but no independent dosimetry measurements are made. They typically rely on the RPC for any quality audit dosimetry measurements. Peer review for radiation oncologists by the ACR and ACRO has also been recognized as an important practice in radiation oncology (Johnstone et al. 1999; Halvorsen et al. 2005). This philosophy is now being recognized by the medical physics professional associations as an important course of action to be adopted by the medical physics community

as part of the QA program (AAPM 1998). A successful QA program for both brachytherapy and external beam radiotherapy must involve the radiation oncologist, medical physicist, dosimetrist, and therapist as a team. The consequences of errors by any of these personnel in a radiation oncology clinical environment, as shown in Figure 31.1, may result in the patient not receiving a safe or appropriate radiotherapy treatment.

Therefore, the type of mistakes made by each staff member is different, as well as the magnitude of the impact to the patient, for example, as specified below:

- If a physician makes a mistake, it usually affects one patient.
- If a dosimetrist makes a mistake, it affects one patient or one location of the tumor.
- If a technologist makes a mistake, it normally affects one fraction of the treatment.
- However, if a physicist makes a mistake, it can be appalling since it may affect all patients in the clinic during the time the mistake is in place.

As a result of several accidents recently involving the use of high technology in radiation oncology, independent quality audits are viewed as a key preventive action. A sound QA program must be in place where these mistakes can be revealed. A quality audit will hopefully address these potential avoidable errors, and it shall include in the peer review process a team of radiation oncologists and medical physicists in order to review the entire radiation oncology process.

31.2 Scope of Audit Process

31.2.1 Guiding Principles

The audit protocol is in general based on a set of guiding principles, as described:

- *Commitment:* verifies the coherence between the established vision, mission, and policy statements of the

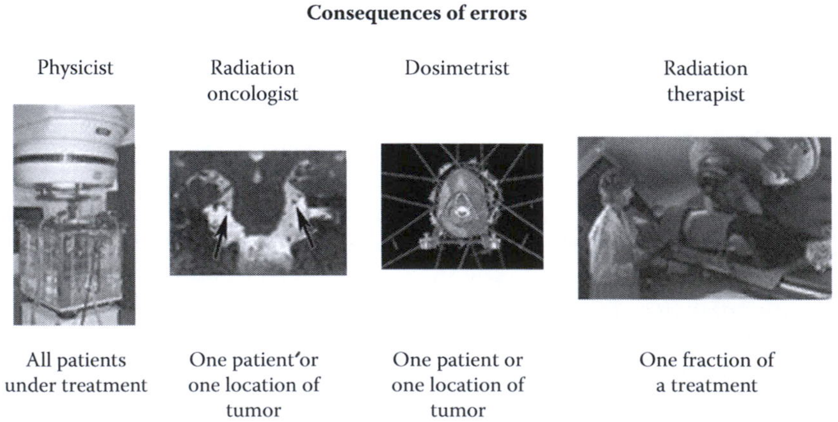

Consequences of errors

Physicist	Radiation oncologist	Dosimetrist	Radiation therapist
All patients under treatment	One patient or one location of tumor	One patient or one location of tumor	One fraction of a treatment

FIGURE 31.1 Consequences of errors made by members of the team in a radiation oncology clinical environment. (Courtesy of C.H. Sibata. Modified from Sibata, C.H. and Gossman, M.S. Role of quality audits: View from North America. In: *Quality and Safety in Radiotherapy (Imaging in Medical Diagnosis and Therapy)*. T. Pawlicki, P. Dunscombe, A.J. Mundt, and P. Scaillet (eds.). Taylor & Francis, Boca Raton, FL, pp. 161–65, 2010.)

institution with their relationship to the day-to-day operations

- *Plan assessment and implementation:* verifies the actual results of operations compared to the planned ones
- *System support:* reviews resources and activities needed to support QA implementation so as to achieve the desired objectives
- *Monitoring:* procedures and actions implemented to assess the compliance and effectiveness of the operations
- *Achievement of objectives:* periodic and timely assessments made in order to determine whether objectives are being achieved

31.2.2 Objectivity, Independence, and Competence

To ensure the impartiality of the audit process and its findings or conclusions, members of the audit team must be independent of the activities they audit. Each member of an audit team must be independent of any governmental organization, regulatory agency, licensing authority, or entity that is being subjected to the audit. There should not be any potential conflicts of interest.

31.2.3 Professional Care

In the execution of the audit, the auditor must use the care, attentiveness, skill, and judgment expected of any other auditor in similar circumstances. Auditors should comply with all applicable legislation and policy in the conduct of the audit.

The relationship between the auditor/auditee must be one of confidentiality and discretion. The auditor/audit team will not disclose information or documents obtained during the audit to any third party without the expressed approval of the auditee.

31.2.4 Audit Data and Findings

The auditor will collect, analyze, interpret, and document appropriate information and findings to be used in an examination and evaluation process to determine whether performance criteria are met.

The audit data should be of such a quality and quantity that competent auditors working independently of each other would reach similar findings from evaluating the same parameters or processes.

Sufficient objective data, as outlined in the audit protocol, must be collected to assess whether the audit criteria and procedures have been met. The procedures in the audit protocol are intended to serve as a guide to the auditor and are not necessarily intended to be all-encompassing. The quality of the audit findings relates to their relevance and reliability.

31.2.5 Audit Conclusions

The auditing process is designed to provide the institution and auditor with the desired level of confidence in the reliability of the audit findings and any audit conclusions regarding the correspondence between audit findings, criteria, and procedures.

All current procedures/processes have to be reviewed at the time of the audit. The audit findings represent the state of the current processes or conditions at the facility at the time of the audit. If condition changes and new procedures/processes are introduced, another audit is required.

31.2.6 Roles and Responsibilities

The *lead auditor* of the audit team leads the audit process and is responsible for ensuring the efficient and effective conduct, completion, and confidentiality of the audit process. The *audit team members* are generally responsible for conducting the audit following the directions of and supporting the lead auditor, with each member typically in charge of a specific area such as treatment planning, clinical application, dose calculations, etc.

The institution shall take all necessary steps to guarantee the success of the quality audit, such as informing employees about the objectives and scope of the audit; providing the facilities needed for the audit team; appointing responsible and competent staff who may accompany members of the audit team, acting as guides to their activities and to ensure that auditors are aware of health, safety, and other appropriate requirements; providing access to the applicable personnel and relevant data material as requested for the audit; attending meetings as considered necessary by the auditor; and providing any necessary input to action plan and status report development, through the auditee, particularly for any recommendations that may be directed.

The purpose of the brachytherapy independent audit is to:

- Assess the brachytherapy dosimetry (including source strengths, source positioning, and dose calculations) at the institution
- Assess whether actual results are comparable with planned results and whether those actual results are accurately tabulated in the reporting process
- Assess the institution's QA program in place and to what extent the medical and physics activities comply with the program
- Assess the effectiveness in meeting the objectives set out in the QA plan, as measured in relation to the criteria established for the audit
- Assess the effectiveness of any action plans implemented to remedy shortcomings revealed by a previous audit
- Provide feedback to the institution as to any corrective actions that might be required as a result of the quality audit

The audit considers the compliance aspect of planning through a comprehensive review of the management plan, a review of actual operations, and the monitoring and reporting requirements thereon. For purposes of this audit, the effectiveness of the management activities is examined based on planned versus actual results as verified during field audit sampling.

31.2.7 Audit Guideline

The general procedures to be used for brachytherapy quality auditing visits must include:

- Preparation for the quality audit visit
- Verification of licensing: current authorized users, medical physicist, current radiation protection information
- Dosimetry audits including source strength measurements
- Verification of imaging and applicator reconstruction
- Verification of treatment planning dose calculations
- Assessment of source localization
- Examining and resolving discrepancies linked to the brachytherapy processes
- Reviewing the institution's QA program, including the clinical and physical aspects

The audit components listed above include physical measurements and personnel interviews that cover operation and organization of the brachytherapy process and radiation safety precautions. Physical measurements should include verification of the source strength, verification of dose calculation procedures, reconstruction of implant geometry, verification of the source placement if using afterloaders, and completion of brachytherapy benchmark cases.

The auditor should be prepared to perform a quality audit that includes the brachytherapy procedures associated with all sources that the institution uses.

Typical brachytherapy techniques to be evaluated include manual loading, manual afterloading, and remotely controlled afterloading. The specific quality audit procedures are determined by the techniques and/or equipment available at and clinically used by the institution. When there is more than one afterloader unit in operation at the institution, the expert will test each unit during the visit.

31.2.8 Preparing for Audit

31.2.8.1 Information Gathering

Due to the fact that the physical forms of the various brachytherapy sources may be significantly different from each other, the auditor must be prepared to take measurements for the various brachytherapy sources at the institution. The auditor should ascertain from the institution what brachytherapy sources are used at the institution prior to conducting the audit visit. Pre-audit preparations may include obtaining catheters or well-type chamber inserts as appropriate for the specific brachytherapy sources in use at the institution.

Additional information regarding the TPS used, applicators used, and imaging and applicator reconstruction procedures should be gathered prior to making the audit visit. Gathering this information will assist the auditor in understanding the brachytherapy processes and procedures prior to the visit, making the audit more comprehensive.

During the quality audit visit, information should be collected in the course of interviews, examination of documents,

and observation of activities and conditions. Indications of non-conformity to the audit criteria should be recorded. They should be verified by acquiring supporting information where possible, such as observations, records, and results of existing measurements. An additional component of the independent audit is to assess to what extent management planning activities comply with the QA plan and also to assess the effectiveness of medical and physical management activities in meeting the objectives set out in the QA plan. Therefore, it is expected that a mutually cooperative spirit will prevail in the gathering and exchange of information between the auditor and the institution's staff.

While the institution's staff needs to ensure that they meet their obligations to provide or make information available, audit teams will make an effort to gather and distribute information, and conduct their work, in a manner that minimizes demands on the auditee's time.

Based on the information previously gathered, the audit team must be equipped with suitable instrumentation: (1) calibrated well-type chamber (calibration for each brachytherapy source type), calibrated electrometer, barometer, and thermometer; (2) inserts for the well-type chamber suitable for insertion of the afterloader catheters; (3) catheters to connect to the different types of afterloaders; (4) a caliper and tape measure for measuring distances as anticipated by the auditor (typically for less than 10 cm and for approximately 1 m); (5) relevant documents including chamber calibration certificates and guidelines such as the ESTRO Booklet 8 (Venselaar and Pérez-Calatayud 2004); and (6) analysis software and a series of data sheets to be used as reference data during the site visit.

31.2.8.2 Institutional Organization

It is important to clearly determine the organizational structure of the institution and department to be audited. A key part of any independent audit is the institution's leadership's willingness to allow the audit to proceed with their full support. An understanding of the institution's administrative structure will enhance and facilitate the communication of any audit findings at the completion of the audit visit.

Information needed should include the department identification and complete address; contact person; personnel along with their credentials and professional accreditations; reporting structure within the department and institution; and responsibilities of the institution's staff in regard to the brachytherapy treatment process.

31.2.8.3 Clinical Procedures

The auditor will identify the special brachytherapy procedures currently used at the institution, such as the computed tomography (CT) and magnetic resonance imaging (MRI) acquisition protocols, imaging reconstruction techniques, and data transfer to the TPS. The auditor should question the institution as to their dose validation methods and documentation of the results (IAEA 2004).

The auditor should verify that the institution's policy and procedures include all clinical procedures currently employed by the institution, simulation procedures, treatment techniques, and QA procedures. The QA procedures should include the frequency of each test, training requirements, continued education, evaluation of each patient's treatment record, and incident reports. A reasonable number of clinical patient treatment records should be reviewed by the auditor for patients under treatment and completed treatment to verify that the records are properly maintained, procedures agree with stated institutional policies and procedures, and QA procedures are performed in a timely manner.

31.2.8.4 Physical Procedures

The complex set of physical data taken from tables, publications, guidelines, measurements, or protocols must be well documented. The auditor will verify the documentation and QA procedures for any changes in the physical data used to treat patients. This verification should be very comprehensive since any change in the physical data may have a significant impact on the delivered dose to many patients. This verification may include aspects of patient and staff safety, machine acceptance and commissioning, source calibration, patient data (protocols for imaging acquisition and data entry to the TPS), dose calculations to the patient, dose validation, patient treatment record check, and the performance of the QA program.

31.2.8.5 Safety Aspects

The patient and staff radiation protection are important aspects to be considered during an audit with the aim of providing the appropriate treatment to the patient without endangering the staff. Especially with brachytherapy, limiting the exposure of an institution's staff without compromising the patient's treatment is necessary (AAPM 1998).

The audit review should verify, but not be limited to, the following items:

- The radiation safety officer credentials and license issued by the licensing authority, date of expiration, and the specification for what radioactive materials are allowed to be used
- The compliance of the present user's equipment list with the one in the operational license
- The QA program related to the treatment and simulation machines
- Personal monitoring records
- Periodic radiation surveys of the rooms where the brachytherapy sources are located
- The physical conditions as well as calibration certificates of the physics equipment (ionization chambers, barometers, electrometers, film scanners, in vivo dosimetry equipment, thermometers, survey meters, and any other special equipment used)
- Emergency procedures and drills, that is, if the source has to be retracted manually (actions taken related to unusual events and their registration)
- The source storage and disposal

- Posting of warning signs
- Maintenance of the source inventory and the results of the leakage wipe tests
- Electrical interlocks at the entrance to the equipment room and source exposure indicator lights on the afterloader
- Viewing and intercom systems
- Emergency response equipment
- Radiation monitors to indicate source position
- Timer accuracy, clock (date and time) in unit's computer, and decayed source activity in unit's computer

31.2.8.6 Source Strength

The auditor will verify, under the observation of the institution's staff, the source calibration (in terms of the reference air-kerma rate) for at least one source of each isotope used at the institution for brachytherapy procedures. This may include a source from a remote afterloading unit or a sample of individual sources (wires or seeds for manual afterloading). The auditor's well-type chamber must be calibrated to have a reference air-kerma calibration factor for the specific sources whose source strengths are verified (Goetsch et al. 1991; Di Prinzio and de Almeida 2009; IAEA 1999).

In order to place the brachytherapy source in the center of the most sensitive spot, the "sweet spot," of the well chamber, inserts to the well-type chamber must be available to place the source(s) centrally in the chamber. Depending on the type of well chamber, a correction for ambient temperature and air pressure must be applied. The measured source strength will be compared against the institution's clinical value. Agreement between the measured source strength and that used clinically by the institution should be within ±5%. Any discrepancy and its consequences have to be resolved.

The auditor will review the institution's calibration procedure, carefully taking note of the step-by-step procedure and parameters used by the institution. A review of the institution's well chamber calibration certificates should also be conducted. The expert must receive a copy of the source strength certificate from the vendor of the source(s). Especially on older documents (with long-lived sources), special attention must be paid to units of source strength specification on the certificate and the way these units are used in the clinic.

31.2.8.7 Treatment Planning Dose Calculation

The institution's physicist is asked to prepare some simple source dose distributions as calculated by the TPS. The source configurations are typically a single-source and then a multiple-source calculation. The dose distributions in multiple planes for each source configuration are determined and printed for comparison against standard data that an auditor should have access to. The dose distributions should be calculated using the same technique as used for brachytherapy patients and should include isodose lines at various distances from the source in all three major planes. Taking into account the actual source strength, the auditor can compare the results of the single-source calculations with data from an along-and-away table typical for the source.

A two-source configuration determines the basic ability of the TPS to calculate dose distributions in a multiple-source situation. In this configuration, the conversion of the dose prescription into a treatment time can easily be evaluated.

Any discrepancies noted by the auditor should be discussed with the institution's physicist and resolved during the audit visit.

31.2.8.8 QA Procedures

Since the regulations in each country may differ, the institution's medical physicist must establish a QA program that will comply with those specific regulations. Recommendations made by the AAPM (1998) and ESTRO (Venselaar and Pérez-Calatayud 2004) are intended to provide the medical physicist with the appropriate guidance to guarantee that brachytherapy procedures are carried out in safety and with due consideration to the licensing authorities regulations.

The auditing process must review applicator QA procedures; typical periodic spot-check; full source calibration (Rivard et al. 2004); the integrity of the transfer tube/applicator system; treatment planning QA (benchmarked clinical cases) and staff training; the treatment QA program; use of port films for each treatment session; geometrical reconstruction verifications; in vivo measurements if any; the final physics checks done for all patients under treatment; and if the treatments are performed in the presence of the physicist.

31.2.8.9 End-to-End Test

The auditor should preferably verify the total brachytherapy procedure while at the institution. This is called an *end-to-end test* to the therapy practices. Phantoms are, however, often not suited for the full procedure. For example, the verification of the reconstruction step of an implant is normally accomplished with an accurately designed geometrical type of phantom such as the 3D matrix phantom ("Baltas" phantom) with predetermined marker locations within the phantom (Roué et al. 2006). The institution's physicist is asked to reconstruct the phantom as if it were a patient using whatever imaging modality they would normally use in the clinic for brachytherapy treatments. The images are then transferred to the TPS either electronically or using a digitizer. The institution's physicist will, using the TPS, register the markers in the phantom images, noting their coordinates. These coordinates will then be compared to the actual known location of the markers in the phantom. It is important that this work be performed by the institution's physicist so that the standard institutional procedure is followed as if it were a real brachytherapy patient. Deviations in the location of the two sets of coordinates are normally indicated in the form of the mean deviation. This technique will usually identify magnification errors, improper imaging, or improper use of the reconstruction aspect of the treatment planning software (Roué et al. 2006).

The "Baltas" phantom cannot be used for verification of the dose delivery as it does not contain any channels for the applicator and measurement device. For this purpose, a specifically designed ideal or an anthropomorphic phantom can be used. Such a phantom is then "treated" as a patient by using the normal imaging steps, transfer of this information into the TPS, reconstruction of the dwell positions and points of interest, the creation of a plan, and the transfer of data to the afterloader. In the treatment delivery to the phantom, the dose is measured. An example of such an anthropomorphic phantom was described by de Almeida et al. (2002).

The combination of information on the reconstruction steps and the dose-delivery check forms the end-to-end test of a given brachytherapy procedure. Any deviations noted should be discussed and resolved between the auditor and the institution's physicist. Any clinical consequence of a deviation will need to be followed up by the institution.

31.2.9 External Evaluations

The complexity of new technology has produced new avenues for mistakes, software flaws, defective programming, and outdated safety procedures, including QA techniques and clinical protocols. Even very experienced and meticulous physics staff frequently have a learning curve to keep up-to-date with the demands of new technology due to the possibility that some of the routine QA needs are postponed or ignored.

The auditing process as an outside independent medical physics group review service must double-check randomly selected patient charts; provide a second opinion on decisions made on procedures; and work with the local staff to provide an independent professional review. It should also evaluate and confirm the calibration and source data integrity; remeasure and/or double check beam models; evaluate on-site QA tools; review/establish QA processes and documentation records; build problem-solving and early-detection systems; provide written custom clinical guidelines/protocols; verify and validate benchmark tests done; and verify the consistency of the incident reporting system. Some experiences with external evaluations are cited below.

The RPC has long provided the NCI and institutions participating in NCI-funded clinical trials assurance that institutions prescribe and deliver radiation doses that are comparable and consistent. One of its many audit programs includes the retrospective recalculation of doses delivered to patients entered onto clinical trials. The RPC is the only QA group to perform this type of recalculation for patients treated with brachytherapy within the NCI clinical trial arena. The brachytherapy recalculations include the use of high-dose-rate (HDR) and low-dose-rate (LDR) sources for tandem and ovoid applicators, tandem and ring applicators, and vaginal cylinders for gynecological treatments as well as multicatheter prostate implants for genitourinary treatments.

The recalculation is performed to each dose specification point for each insertion to remove any differences in dose computation to a given point caused by using different point definitions or calculation algorithms by the many participating institutions. An in-house–developed program developed at the RPC known as RadComp is used to perform the recalculations. The RadComp program utilizes a numerical solution to the Sievert integral to perform the dose calculation. More recently, two Varian Eclipse Brachyvision planning systems (Varian Associates, Palo Alto, CA) have been added to the RPC's recalculation capability. These Eclipse systems were benchmarked against the RadComp program to determine consistency of the brachytherapy dose calculation algorithms. The RPC recalculates doses to point A, point B, the rectum, and the bladder for the gynecological treatments. The definitions for point A and point B were those described by the Manchester System (ICRU 1985; see also Chapter 24). The RPC uses a 15% criterion established in conjunction with the NCI study groups for agreement between the brachytherapy calculated dose value and the value reported by the institution. The 15% criterion corresponds to a 2-mm distance difference in point A due to the dose gradient.

The RPC, in a recent study, reviewed 182 HDR brachytherapy (tandem and ovoid/tandem and ring) implants and compared the institution's bladder and rectum point locations and doses to those determined by the RPC, strictly adhering to protocol specifications (ICRU 38) and using its independent dose calculation algorithm. The RPC also analyzed its own uncertainty in defining these two points. The dose agreement criterion of ±15% was used as mentioned above (Lowenstein et al. 2011). The RPC disagreed with the reported bladder and rectal doses in 25% and 45% of the 182 implants, respectively. The RPC's own uncertainty in defining the bladder and rectal points was 1 ± 0.1 (s.d.) mm, which in a worst-case scenario might account for 7% of the dose disagreement. The majority of the dose disagreements were due to the institution's incorrect localization of the bladder and rectal points, by greater than 5 and 4 mm, respectively, away from the ICRU 38 (1985) defined location. There were no differences noted, whether the applicator used was a tandem and ovoid or tandem and ring. The RPC has found through its quality audits that most errors result from institutions incorrectly defining the bladder and rectum dose calculation points per ICRU 38 (1985). Additional education, timely clinical and dosimetric reviews of implant data, and communication with institutions are needed to reduce the number of these discrepancies.

Another experienced group is the LCR located at the Rio de Janeiro University. The LCR uses a phantom to measure two important dosimetric parameters needed to assure the desired level of accuracy in the dose delivered to the patient in brachytherapy (Ochoa et al. 2007): first, the source strength at 10 cm from the source in air, and second, the absorbed dose at 2 cm in water. The latter point was made coincident with the historical point A used in gynecological brachytherapy. The phantom has three cylindrical polyethylene capsules (20-mm inner length, 3-mm inner diameter, and 1-mm wall thickness; the outer length including the plug is 28 mm) containing about 160 mg of thermoluminescent dosimeter (TLD) lithium fluoride. In order to avoid the unacceptable dose gradient values across the TLD mass, the cylindrical polyethylene capsules were made curved, with circumferences with radii of 10 and 2 cm. The correction factors required to take into account the specificities related to the geometry and the phantom materials have been assessed using the PENELOPE Monte Carlo code and experimental methods (Ochoa et al. 2007). This dedicated phantom,

used as part of an independent QA program, allows simultaneous irradiation of three TLD capsules, requiring only one source stop (dwell position). The phantom is mailed to the radiotherapy institutions, and the results demonstrate its effectiveness in verifying the source air-kerma rate and the correctness of treatment planning dose calculation in a water phantom. The comparisons made between the phantom measurements, the well-type ionization chamber, and source specifications stated by the hospital (most of the times provided by the source manufacturer) agreed within 3%. The phantom has been used by request in several radiotherapy institutions in Brazil, together with a reference well-type ionization chamber (Standard Imaging model HDR Plus) and the irradiation instructions. The irradiation times were calculated from formulas provided with those instructions. Two sets of irradiations are performed at each institution, representing a total of six capsules in air and six capsules in water. From each capsule, five thermoluminescent readings were obtained. Therefore, each final result (an average of 18 selected readings) represented a statistical analysis of the 30 readings. At the local institution, the irradiation is performed by the local physicist, who also performed the measurements with the well-type ionization chamber in order to compare the results with those of TLD measurements.

31.2.10 Discussion

In recent years, advanced technology has been continuously introduced into the field of radiotherapy, including new innovations in brachytherapy that have revolutionized its practices. As this technology becomes more widely available, and based on the lessons learned from incidents and accidents, the need for a strong continuous education program becomes paramount (see Chapter 32). The auditing process must review the contents of the continuing education program that should include radiation safety procedures and emergency training; the frequency of the courses and whether the whole staff attends; the records that document the training; the names of the individuals who conducted the training; the names of the individuals who received the training; the dates and duration of the training; and the list of the topics covered.

It is common to review a department and to determine that no medical event has taken place in the department for many years, which is a good indication that the QA program is working well. It is troublesome, however, to find out that in an institution, no treatment variance has ever occurred during that time. A robust QA program should reveal mistakes from time to time if it is implemented rigorously. If no mistakes were ever detected, then it can be an indicator that the QA program was not adequate or that the program was not adequately implemented. If no mistake has ever been found in a brachytherapy program, the auditor should take this into account and determine the full extent of the institution's QA program and whether it is comprehensive enough to adequately ensure that patients received accurate treatments and no one was put at risk.

Another issue frequently found is that the institution uses a very simple QC test (such as a single-source configuration) for low-frequency testing of the brachytherapy TPS or/and the institution checks manually the printed plan data for each clinical case and signs this plan. However, individual brachytherapy plans often deviate significantly from routine, and the combination of these two QC steps may prove to be insufficient to detect all errors. A proper QA program should consist of a dedicated set of steps, tailored to the clinical routines of the institution.

31.3 Recommendations

Training of auditors is essential in order to have a meaningful audit of a brachytherapy program. Determining the importance of the key components of the brachytherapy program for review during the audit is crucial when auditing an institution's brachytherapy program. Time is limited, and spending time on regulatory issues may not be the best use of the available time. Looking for holes in the QA program that may allow medical and appalling events is more important than spending time on regulatory issues. With the advent of record-and-verify systems and the push for electronic charting, QA programs may be spending a lot of time on worthless issues when more useful tests could be done. Electronic charting also makes chart checks much more complicated since most record-and-verify systems cannot handle the information for each patient adequately, for example, frequently, the treatment plan is scanned in the record-and-verify system, and the review of the plan is much more difficult.

One of the many reporting anecdotes on the accreditation process is the nonuniformity of the reviewing process. The accreditation bodies and the quality auditors have to have the same degree of accounting as the clinics being audited to maintain the quality of the accreditation or audit. This means that some other organization such as ISO, IAEA, or a joint effort from professional bodies should oversee these entities and that a rigorous "quality label" or standards be applied to the accreditation and auditing programs. This chapter's appendix is a modified form of the ACRO Physics Survey Form adapted for brachytherapy survey only (C.H. Sibata, 2011, personal communication), which provides the reader with an example of a list of contents of an audit procedure.

Minimum standards for practicing of brachytherapy should be as follows: the use of afterloading techniques; adequate dosimetric equipment; minimum number of cases to maintain clinical experience; and adequate QA program encompassing both equipment and procedures.

External checks of brachytherapy sources, similar to the remote audit checks performed by the RPC and IAEA for linac beam outputs, should be made available to clinical departments. Although a dosimetric measurement in a quality audit is a costly process to be conducted during a normal quality audit, it will provide a second check on the activities of the brachytherapy sources being used clinically.

31.4 Future Directions to Be Considered in Auditing Process

Brachytherapy is set to continue its development, embracing, in particular, the advances in imaging and computer dosimetry

capabilities. As a result, the scope of the auditing procedures must include all new developments in clinical brachytherapy use. The following aspects should be considered in that respect.

Imaging. Recent recommendations have confirmed the role of a sectional imaging-based approach, preferably with MRI, instead of the traditional (orthogonal) radiograph approach. However, although MRI offers improved accuracy of the target volume and critical structures, the brachytherapy applicators are difficult to accurately reconstruct. Many centers in Europe tend to reconstruct the applicator on CT scans or films and then register the applicator onto the MRI that is used for organ and target delineation.

Functional biological imaging may also play a role, in particular, in magnetic resonance spectroscopy imaging (MRSI) and positron-emission tomography. If a raised (choline + creatine)/citrate ratio > 1 on MRSI is considered suggestive of an intraprostatic tumor, two target volumes may then be used: a biological target volume and a physical target volume defined on conventional sectional imaging.

Dosimetry. The TG-43 (Rivard et al. 2004) dose formalism is now regarded as the standard dosimetry formalism for brachytherapy dose calculations. Current planning systems assume that the medium around the source is water and compute a dose to water. The next generation of 3D dose calculation methods in brachytherapy needs to be able to deal with tissue inhomogeneities, finite human body dimensions, applicator material, and shielding materials if used. Monte Carlo techniques would be the gold standard for 3D dose calculations, but it still needs to be simplified and faster.

Bioeffect dose modeling is of considerable interest, and up to this moment, it has been used only for research purposes.

31.5 Conclusions

Even though quality management program brachytherapy annual reviews have always been adopted by the U.S. Nuclear Regulatory Commission regulations, quality external audits are not a common practice in brachytherapy or even teletherapy. ACR and ACRO in the United States have been doing practice accreditations in which a quality audit is done both for teletherapy and brachytherapy procedures but not involving any dosimetric measurements due to cost issues involving practice accreditations. ESTRO/Equal in Europe and the RPC have included in their quality audits these dosimetric measurements.

In this chapter, the background material in this topic has been discussed as well as current practices and future recommendations.

Brachytherapy quality audits have the potential to improve the quality of care for patients by raising the bar of the QA programs in radiotherapy departments that use them. Furthermore, by promoting the dissemination of information gathered by a reporting system for errors and near-miss events, which is normally an item that is reviewed in an audit, it may improve the safety practice of clinical brachytherapy (de Almeida et al. 1999).

References

AAPM. 1998. Report of AAPM TG 53, Quality assurance for clinical radiation treatment planning. American Association of Physicists in Medicine. *Med. Phys* 25:1773–829.

Cotter, G.W. and Dobelbower, Jr. R.R. 2005. Radiation oncology practice accreditation: The American College of Radiation Oncology, Practice Accreditation Program, guidelines and standards. *Crit Rev Oncol Hematol* 55:93–102.

de Almeida, C.E. 1998. *The Brazilian National Quality Program in Radiation Oncology*. Technical Document. National Institute of Cancer, Rio de Janeiro, Brazil.

de Almeida, C. E., Pereira, A.J., Marechal, M.H. et al. 1999. Intercomparison of calibration procedures for ^{192}Ir HDR sources in Brazil. *Phys Med Biol* 44:N31–8.

de Almeida, C.E., Rodriguez, M., Vianello, E., Ferreira, I.H., and Sibata, C. 2002. An anthropomorphic phantom for quality assurance and training in gynaecological brachytherapy. *Radiother Oncol* 63:75–81.

Di Prinzio, R., and de Almeida, C.E. 2009. Air kerma standard for calibration of well type chambers in Brazil using ^{192}Ir HDR sources and its traceability. *Med Phys* 36:953–60.

Dobelbower, R.R., Cotter, G., Schilling, P.J. et al. 2001. Radiation oncology practice accreditation. *Rays* 26:191–8.

Ellerbroek, N.A., Brenner, M., Hulick, P. et al. 2006. Practice accreditation for radiation oncology: Quality is reality. *J Am Coll Radiol* 3:787–92.

Goetsch, S.J., Attix, F.H., Pearson, D.W., and Thomadsen, B.R. 1991. Calibration of ^{192}Ir high-dose-rate afterloading systems. *Med Phys* 18:462–7.

Halvorsen, P.H., Das, I.J. Fraser, M. et al. 2005. AAPM Task Group 103 report on peer review in clinical radiation oncology physics. *J Appl Clin Med Phys* 6:50–64.

Hoskin, P.J. and Bownes, P. 2006. Innovative technologies in radiation therapy. *Brachyther Semin Radiat Oncol* 16:209–17.

Hulick, P.R. and Ascoli, F.A. 2005. Quality assurance in radiation oncology. *J Am Coll Radiol* 7:613–6.

IAEA. 1999. Calibration of brachytherapy sources: Guidelines of standardized procedures for the calibration of brachytherapy sources at secondary standard dosimetry laboratories (SSDL) and hospital. TECDOC 1079. International Atomic Energy Agency, Vienna.

IAEA. 2000. Lessons Learned from Accidental, Exposures in Radiotherapy. International Atomic Energy Agency. Safety Report Series No. 17, Vienna.

IAEA. 2002. Standardized Quality Audit Procedures for On-site Dosimetry Visits to Radiotherapy Hospitals. International Atomic Energy Agency. SSDL Newsletter No. 46, Vienna.

IAEA. 2004. Commissioning and Quality Assurance of Computerized Treatment Planning Systems for Radiation Treatment of Cancer, Technical Reports Series No. 430. International Atomic Energy Agency, Vienna.

IAEA. 2007. Comprehensive Audits of Radiotherapy Practices: A Tool for Quality Improvement, Quality Assurance Team for Radiation Oncology (QUATRO). International Atomic Energy Agency, Vienna.

ICRP. 2009. Preventing Accidental Exposures from New External Beam Radiation Therapy Technologies. International Commission on Radiological Protection. ICRP Publication 112. *Ann. ICRP* 39(4).

ICRU. 1985. *Dose and Volume Specification for Reporting and Recording Intracavitary Therapy in Gynecology*. International Commission on Radiation Units and Measurements. Report 38, ICRU Publications. Bethesda, MD.

Johnstone, P.A., Rohde, D.C. May, B.C. et al. 1999. Peer review and performance improvement in a radiation oncology clinic. *Qual Manag Health Care.* 8:22–8.

Lowenstein, J., Roll, J., Harris, I., Hall, F., Hollan, A., and Followill, D. 2011. Cervix brachytherapy dosimetry: Inconsistencies in defining bladder and rectal points. *Med Phys* 38:3572.

Ochoa, R., Gomes. F., Ferreira. I.H., Gutt. F., and de Almeida, C.E. 2007. Design of a phantom for the quality control of high dose rate 192-Ir source in brachytherapy. *Radiother Oncol* 82:222–8.

Rivard, M.J., Coursey, B.M., DeWerd, L.A. et al. 2004. Update of AAPM Task Group No. 43 Report: A revised AAPM protocol for brachytherapy dose calculations. *Med Phys* 31:633–74.

Roué, A, Ferreira, I., Van Dam, J., Svensson, H., and Venselaar, J. 2006. The EQUAL-ESTRO audit on geometric reconstruction techniques in brachytherapy. *Radiother Oncol* 78:78–83.

RPC. 2011. Radiological Physics Center History. Available from: http://rpc.mdanderson.org/RPC/home.htm.

Sibata, C.H. and Gossman, M.S. 2010. Role of quality audits: View from North America. In: *Quality and Safety in Radiotherapy (Imaging in Medical Diagnosis and Therapy)*. T. Pawlicki, P. Dunscombe, A.J. Mundt, and P. Scaillet (eds.). Taylor & Francis, Boca Raton, FL, pp. 161–65.

Venselaar, J.L.M. and Pérez-Calatayud, J. (eds.). 2004. *A Practical Guide to Quality Control of Brachytherapy Equipment*, 1st ed., ESTRO, Brussels. Available at http://www.estro-education.org/publications/Documents/booklet8_Physics.pdf.

Appendix

Indicative Auditing Outline

Facility name being surveyed: _____ Reaccreditation: _____

(Hospital Affiliation: _____)

Location: _____ State_____ Date_____

	Board Certified			Board Certified
Physicists: _____	☐	Dosimetrists:	_____	☐
Physicians: _____	☐	Administrator:	_____	☐
Technologists: _____	☐	Technologists:	_____	☐
Technologists: _____	☐	Other:	_____	☐

Facility Info

Number of patients treated per day _____   per year on average: _____

Does facility participate in Cooperative Clinical Trials, e.g. RTOG   ☐ yes   ☐ no

Comments: _____

Room Surveys

Radiation Room surveys adequate and complete   ☐ yes   ☐ no

Room _____   Source _____

Physics Equipment List

Item	Make	Model	Serial #	Year calibrated	Comments
Electrometer	_____	_____	_____	_____	_____
Well Ion Chamber	_____	_____	_____	_____	_____
Film scanner	_____	_____	_____	_____	_____
Survey Meter	_____	_____	_____	_____	_____
Water Phantom	_____	_____	_____	_____	_____

Review available physics equipment. Determine adequacy and shortcomings. Include water phantoms, film scanners, morning QA devices, ionization chambers, electrometers, etc.

Comments: _____

Brachytherapy

Review general radiation safety policies. Are they adequate?   ☐ yes   ☐ no

RSO available/listed?   ☐ yes   ☐ no

Name _____   Phone (no. sign) _____

Is the facility located in an   NRC ☐   or   Agreement state? ☐

State license current   ☐ yes   ☐ no   Exp. Date: _____   License #_____

Postings

Required Postings Adequate   ☐ yes   ☐ no
☐ Notice card   ☐ Section 206 equiv   ☐ RSO notice (may be listed on Notice to Employee)
☐ State required postings, "Notice to Employees"
Proper location of notices adequate   ☐ yes   ☐ no
i.e., easily and readily observable
Comments: _____

LDR Brachytherapy

Number of Brachytherapy Charts Reviewed:

Treatment Planning

Manufacturer _____   Model _____   Version _____

Commissioning: _____
TPS QA _____
Medical physicist listed on license: _____   for _____
Current to actual site physicist   ☐ yes   ☐ no   ☐ n/a
Authorized Users List updated: ☐ yes   ☐ no
Procedure types performed: ☐ Sr-90 (eye)
                     LDR using: _____
Procedure Manual for LDR Brachytherapy including: ☐ yes   ☐ no
Sources used and their physical data   ☐ yes   ☐ no
Location of sources   ☐ yes   ☐ no
All forms used in LDR Brachytherapy   ☐ yes   ☐ no
QMP for each procedure   ☐ Gyn Implant   ☐ yes   ☐ no
                        ☐ Interstitial Implant   ☐ yes   ☐ no
                        ☐ Other _____   ☐ yes   ☐ no
Verify patient by at least two means   ☐ yes   ☐ no
Method used: _____
Second Checks done: _____

Prostate Implant

Number of Brachytherapy Charts Reviewed: _____

Treatment Planning

Manufacturer _____   Model _____   Version _____

Commissioning: _____
TPS QA: _____
Procedure types performed: ☐ Prostate (☐ I-125, ☐ Pd-103)
Are seeds assayed as per TG-43 recommendations, i.e. 10%   ☐ yes   ☐ no
Are Cs-137 sources, etc. assayed in a well type chamber prior to first use?   ☐ yes   ☐ no
Prostate Seed Post Implant Planning done?   ☐ yes   ☐ no   Using CT?   ☐ yes   ☐ no
How: _____   How soon after implant: _____
Comments: _____
QMP for procedure   ☐ Prostate Implant   ☐ yes   ☐ no

Miscellaneous Procedures

Any procedures for I-131, Sr-89, Sm-153, etc. or other radiopharmaceuticals? _____
Exposure monitoring records reviewed, signed, current   ☐ yes   ☐ no
Annual exposure totals distributed to those monitored under req. rules,
e.g. Form 5 report   ☐ yes   ☐ no

Regulatory Miscellaneous

Quarterly source storage location surveys completed	☐ yes	☐ no	☐ n/a
Quarterly inventory of all radiation sources complete	☐ yes	☐ no	☐ n/a
Semi-annual wipe tests completed satisfactorily	☐ yes	☐ no	☐ n/a
Package surveys/wipe tests completed according to	☐ yes	☐ no	☐ n/a
License app procedures followed and/or DOT rules (Title 49 CFR)			
Are shippers trained according to DOT rules?	☐ yes	☐ no	☐ n/a

Equipment/Storage Security Issues

Radiation Materials Locked and secured properly	☐ yes	☐ no	☐ n/a
Observe hotlab			
Storage locations postings appropriate	☐ yes	☐ no	☐ n/a
Machine keys removed when not in use	☐ yes	☐ no	☐ n/a

HDR Brachytherapy   Not applicable ☐

Number of Brachytherapy Charts Reviewed: _____

Treatment Planning

Manufacturer _____   Model _____   Version _____

Commissioning: _____

TPS QA: _____

Procedure Manual for HDR Brachytherapy

Sources used and their physical data	☐ yes	☐ no
Location of source	☐ yes	☐ no
All forms used in HDR Brachytherapy	☐ yes	☐ no
Source exchange tests completed properly and prior to patient treatment	☐ yes	☐ no ☐ n/a

Method used to calibrate source: ☒ well chamber   ☐ Goetsch method   ☐ Other

Calibration reports observed by this reviewer	☐ yes	☐ no
Observe quarterly checks	☐ yes	☐ no
Observe treatment day checks	☐ yes	☐ no
Observe proper emergency procedures	☐ yes	☐ no
Review emergency procedure	☐ yes	☐ no
Review QMP Program for HDR	☐ yes	☐ no
Verify a second check is performed prior to treatment	☐ yes	☐ no
Verify patient by at least two means	☐ yes	☐ no

Method used: _____

Second Checks done: _____

Summary of Brachytherapy

Review of QMP

Date: _____

Findings: _____

   What was done to correct the findings?

Last State/NRC Inspection

Date: _____

Findings: _____

   What was done to correct the findings? _____

Policies & Staffing

Review credentials of staff and continuing education, i.e. RTTs used, CMD's used

Board certification credentials on site?	☐ yes	☐ no
Technologist licensure current and on file	☐ yes	☐ no
All licenses appropriate to duties (RTT, RT)	☐ yes	☐ no
Annual training requirements, commitments fulfilled	☐ yes	☐ no
Competency in equipment operation complete	☐ yes	☐ no
Full time physics coverage	☐ yes	☐ no

If no, describe physics services: _____

Describe staffing: _____

Weekly chart rounds performed with physics involvement	☐ yes	☐ no

When? _____

Physician peer review performed	☐ yes	☐ no

When? _____

Are incident reports logged/QI meetings/incidents evaluated	☐ yes	☐ no

If yes, describe: _____

Does staff know what to do if outside established tolerances?	☐ yes	☐ no
Are these tolerances written in the policy and procedure manual?	☐ yes	☐ no

If yes, describe them: _____

Training Records

Initial Training for New Technologies:
Continued Training

Policies and Procedure Manual

Policy and procedure manual reviewed (annually) by physicist of record, on date: _____

Complies with annual review requirement:	☐ yes	☐ no

Minimally the Policy and Procedure Manual should have the following sections:

After title, a statement about the what changes were made and by whom and that the changes were discussed in the annual

inservice to the clinical personnel	☐ yes	☐ no
List of all procedures done clinically	☐ yes	☐ no
Simulation procedures	☐ yes	☐ no
Planning Procedures	☐ yes	☐ no
Treatment Procedures	☐ yes	☐ no
Weekly checks by physics	☐ yes	☐ no
QA procedures	☐ yes	☐ no
Commissioning of all equipment including TPS	☐ yes	☐ no
Daily, Monthly and Annual QA on all equipment including special procedures	☐ yes	☐ no
Equipment calibration frequency	☐ yes	☐ no
ID of patient by two means	☐ yes	☐ no
Annual Inservice	☐ yes	☐ no
Changes in the Procedure Manual	☐ yes	☐ no
Radiation Safety review	☐ yes	☐ no
Emergency procedures	☐ yes	☐ no
Competency annual evaluation	☐ yes	☐ no
Charts Rounds	☐ yes	☐ no
Peer Review Policy	☐ yes	☐ no
Incidents reports (may or not be misadministrations)	☐ yes	☐ no

Miscellaneous

Is there a preventative maintenance/repair program for equipment   □ yes   □ no
Machine fault log book utilized for recording machine fault history   □ yes   □ no
Assessment of the effectiveness of the QA Program
Recommendations for any needed modifications or improvements
Evidence previous audit's findings were promptly reviewed
Documentation of the action taken
If the recommendations not acted upon then the reasons why are given

Expert Training in Brachytherapy

Alex Rijnders
Europe Hospitals

Zoubir Ouhib
Lynn Cancer Institute

György Kovács
University of Lübeck

Peter Grimm
Prostate Cancer Center of Seattle

32.1 Introduction

In recent years, brachytherapy has undergone important changes due to different factors:

- New isotopes have been introduced, expanding the scope of possible procedures.
- Developments allowing changes in source type, source geometry, source strengths, dwell positions, and dwell times have allowed wider utilization of automatic afterloading systems.
- Image guided brachytherapy (IGBT) has become a common technique in most of the applications, while utilizing computed tomography (CT), magnetic resonance (MR), and/or ultrasound (US) imaging.
- Introduction of new hardware (devices, applicators, and treatment delivery systems) and software in the clinical setting.

The aim for a high-precision and a more accurate dose delivery and the use of advanced technology have increased the complexity of the treatments. Modern brachytherapy procedures require the close collaboration of several individuals: radiation oncologists and other medical specialists, physicists, medical dosimetrists, radiotherapy technologists (RTTs), and nurses. All these experts should have received adequate training in using the equipment and applying the procedures in order to prevent errors and to ensure an optimal treatment.

In Chapter 30, some examples were given on serious incidents that have occurred in the past during brachytherapy procedures. In many of these incidents, such as in the 1992 high-dose-rate (HDR) brachytherapy incident at Indiana Regional Cancer Center or the low-dose-rate (LDR) prostate brachytherapy incident at the VA Medical Center in Philadelphia, inadequate training was shown to be one of the root causes of these incidents.

This has also been stressed, for example, in Safety Report 17 of the IAEA (2000a) and in several ICRP publications (ICRP 97, 2005; ICRP 112, 2009).

In the example of a failure modes and effects analysis (FMEA) discussed by the authors of Chapter 30 on HDR brachytherapy, inadequate training or lack of training was shown to be the second most frequent potential cause of failure, after human error or blunder. In this chapter, we will limit the discussion to the specific expert training in brachytherapy for clinicians and physicists. In general, this expert training is preceded by a general training in radiotherapy.

32.2 Training Responsibility

It is the responsibility of several groups to make sure that adequate training has been met before brachytherapy treatment procedures are performed:

1. *Brachytherapists:* All radiation oncologists interested in starting practicing brachytherapy have the responsibility to seek out and complete the appropriate training prior to performing such a treatment. This training should be done regardless of their previous education/training, unless their documented experience explicitly confirms that they have had up-to-date and appropriate training for the specific modalities being provided. Experienced brachytherapists have the responsibility to maintain their expertise by frequent attendance on courses and seminars.

2. *Medical institutions and private radiation oncology groups:* Institutions providing brachytherapy services have the responsibility to ensure that their physicians are initially qualified and receive updated training in the techniques and equipment being used. No patient should be treated with a new modality (new device, new software, etc.) that

has not been tested and approved by experts in the brachytherapy community. In addition, they are also responsible for initial necessary and ongoing required training to all staff involved in patient care (current and future employees). These institutions should provide periodic support to their staff for education and training in brachytherapy. Or, as stated in the summary as a responsibility for *administrators* in the IAEA TecDoc 1257, "Specialised training is required for the radiation oncologist, medical physicist, and technician before HDR can be introduced. Training for the oncologist and medical physicist is an ongoing process as new techniques or sites of treatment are introduced" (IAEA 2001).

3. *Manufacturers:* Manufacturers are responsible for appropriate testing of any new device as well as the appropriate training in the use of those devices. Full documented training of all users should be required prior to its use. To ensure implementation of this training, the training should be part of the cost of the device/software and should not be a removable line item in their quotes.

32.3 Organization of Expert Training Programs

Every country has implemented a training program for the radiation oncologist and medical physics expert in radiotherapy. Usually, brachytherapy is a part of this general training scheme; however, differences in approach may exist between countries and even within a country between different training institutions.

32.3.1 In Europe

Some countries may have set up a specialized brachytherapy educational/training program. In Germany, for instance, the German Society of Radiation Oncology (DEGRO) started in 2009 a special training program in "Interventional Radiotherapy (Brachytherapy)," while keeping general brachytherapy knowledge as part of the curriculum for being board certified. After completing the requirements of this special training program (including performing a certain number of applications in various anatomic sites) and passing an oral examination, the candidate becomes a "Certified Brachytherapy Expert."

In general, however, the national training programs offer only a basic training in brachytherapy, and additionally, a more specialized training in modern brachytherapy techniques is highly recommended.

In Europe, in the frame of the European Society for Radiotherapy and Oncology (ESTRO), the International School of Radiotherapy and Oncology (ESRO) develops and coordinates all teaching and education activities. The school is governed by ESTRO's Education and Training Committee (ETC). ESRO aims to improve, professionalize, and standardize knowledge and practice in radiation oncology and associated professions. The most prominent ESTRO educational activities are the teaching

courses, which are designed to meet the needs of the radiotherapy team including physicists, physicians, and RTTs. Since the first ESTRO teaching course in 1985, ESTRO has welcomed several thousand participants to the courses and has had the benefit of a large number of international experts as tutors. Many informal contacts and networks have been created that enhance the practice of radiation oncology and underline ESTRO as an international organization.

Different forms of teaching and learning are applied in the educational activities of the ESTRO School, such as up-front lectures, case discussions, literature discussion, interactive group work, tutorials, hands-on sessions, visits to RT departments, and instructions by brachytherapy vendors. All educational activities organized by the School are continuously monitored through the course evaluation forms completed by the participants and the course examinations.

The main knowledge reference structures for the educational activities of the ESTRO School are the ESTRO *Core Curricula* (Eudaldo et al. 2004). It is the responsibility of the School to make sure that the ESTRO educational program covers these curricula comprehensively with appropriate educational activities. The Core Curricula are updated on a regular basis. Revisions of the curricula are a joint responsibility of ESTRO, the European Union of Medical Specialists (UEMS), and the National Societies for Radiation Oncology in Europe.

In 2010, ESTRO updated the core curriculum for radiation oncologists—the third edition (ESTRO 2010). Brachytherapy is an integrated part of this core curriculum, and brachytherapy items are incorporated in several areas and learning outcomes. Every radiation oncologist should get solid training in brachytherapy, allowing him or her to start and build up a practice in brachytherapy.

32.3.2 In the United States

In the United States, physicians performing brachytherapy are required to have certification in Radiology by the American Board of Radiology and in Radiation Oncology by the American Board of Radiology or Osteopathic Board of Radiology. In addition, the radiation oncologist must have formal training in brachytherapy. Currently, many of these trainees receive their training in residency programs; however, training outside the residency setting for US-guided, CT-guided, or MRI-guided brachytherapy is acceptable. Residents in programs with a few cases are encouraged to participate in fellowship programs offered by the American Brachytherapy Society (ABS). Those receiving training outside a residency also are required to participate in hands-on workshops and be proctored for their first five cases.

32.4 Example of Brachytherapy Medical Physics Training Program

The European Federation of Organisations for Medical Physics (EFOMP) and ESTRO issued jointly in 2004 guidelines for the education and training of medical physicists within radiation oncology (Eudaldo et al. 2004). Recently, these guidelines have been updated by the two organizations to accommodate the contemporary requirements for the knowledge or competency needs in this rapidly evolving field of medicine (ESTRO 2011a). The core curriculum discusses general competencies with respect to professionalism and professional attitude and specific fundamental knowledge and skills that a radiotherapy physicist should be able to demonstrate in different areas of physics and oncology.

For the field of brachytherapy, core curriculum items cover equipment (sources, applicators, afterloading systems, imaging systems, and source calibration equipment), source specification quantities and measurement methods, treatment techniques and methods, dose specification and reference volumes, treatment planning and dose calculation, quality assurance, and radiation protection.

With respect to brachytherapy physics, the competences as shown in Table 32.1 have been defined.

TABLE 32.1 Competences as Defined by ESTRO (2011a) for Medical Physics Training Program

Ability to report on the results of source calibration, QA controls, clinical dosimetry, etc., in written and oral presentations

Demonstrate an understanding of the basic operation of the afterloading systems commercially available and of the locally available systems

Ability to assess the advantages and limitations of the locally available afterloading systems and brachytherapy sources

Ability to apply calibration protocols for the brachytherapy sources used locally and to determine the uncertainties of the measurement

Ability to assess the functional characteristics of the source calibration equipment and to perform quality control of this equipment

Ability to participate in the overall clinical process of brachytherapy from operating theater through simulator localization, treatment planning, and treatment delivery

Ability to discuss the use of the different closed/sealed brachytherapy sources

Demonstrate an understanding of the dosimetry systems for intracavitary brachytherapy and interstitial brachytherapy (Manchester, Paris, image-based dosimetry, GEC–ESTRO recommendations)

Ability to assist in the preparation of brachytherapy sources for clinical use

Demonstrate an understanding of the basic principles of imaging systems for brachytherapy

Demonstrate an understanding of the TG-43 dose calculation algorithm and modern model-based algorithms

Demonstrate understanding of the use and limitations of optimization techniques in brachytherapy treatment planning

Ability to perform independent verifications of the calculated treatment times of intracavity insertions and interstitial implants using manual methods

Ability to set up a quality control program of the brachytherapy sources, applicators, and equipment, including the TPS

Ability to handle basic radiation safety procedures, such as leakage tests on the sources, disposal of sources, prevention, and actions in case of source loss

Ability to respond to an emergency situation (HDR) in the event of a system malfunction

Ability to discuss national and international regulations for the use and transport of radioactive materials

In the United States, the competency requirements are fairly similar to the European ones as described above. In addition, there are several task group reports by the American Association of Physicists in Medicine (AAPM) that have addressed some of the responsibilities of the medical physicist in brachytherapy: AAPM TG-43 (Nath et al. 1995; Rivard et al. 2004, 2007), TG-56 (Nath et al. 1997), TG-59 (Kubo et al. 1998), and the AAPM Monograph #31 (AAPM 2005). The AAPM has also regularly organized the Summer School for Brachytherapy (see, e.g., the AAPM Monograph #31), where all physics and clinical aspects of brachytherapy were presented. This was strictly a didactic program. Fellowship programs for medical physicists have been introduced in the last few years and are used to provide individuals with more hands-on clinical experience in radiation therapy, including brachytherapy.

32.5 Advanced Training Facilities and Courses

As mentioned earlier, brachytherapy procedures and the underlying physics have become more and more complex, and additional specific training is highly recommended. Such advanced training is offered by several national and international scientific and professional organizations such as ESTRO, EFOMP, ABS, and AAPM. Also, the International Atomic Energy Agency (IAEA) is organizing regional training courses on brachytherapy, targeting specifically practitioners from the developing countries (ESTRO 2011b; EFOMP 2011; ABS 2011; AAPM 2011; IAEA 2000b). All companies that are offering equipment and software applications will provide training for their customers. These training sessions can be organized locally in the hospital, in classroom sessions at the company, or over the Internet as a Webinar.

Advanced training courses that are available at present include the following.

- *EFOMP* is organizing yearly training courses for physicists covering all aspects of medical physics. In these sessions, 1 week is devoted to brachytherapy. While intended for trainees, this course aims to cover all aspects of brachytherapy physics: calibration of sources, QA of equipment and procedures, dosimetry systems and dose calculation algorithms, optimization models, different clinical applications, and notions on radiation protection.
- *GEC–ESTRO* drives several courses aiming at education in brachytherapy. In general, these courses aim at attracting a mixture of radiation oncologists, physicists, as well as technologists, and often, it is promoted that the entire brachytherapy team should jointly attend the course. The courses are held annually for 3–4 days and generally move through Europe from year to year. Occasionally, the courses are duplicated to be held in other parts of the world (ESTRO 2011b):
  (1) *Modern brachytherapy techniques.* This course aims to cover the basics and general principles of

brachytherapy. Historical notes on evolution of brachytherapy, sources, afterloading systems, imaging for brachytherapy, dosimetry, and radiobiology of different dose rates as well as radioprotection and organization of a dedicated brachytherapy unit will be discussed. Different aspects of intraluminal, interstitial, and endocavital brachytherapy will be covered. Interactive discussions on gynecological (cervix and endometrium), head and neck (oral cavity, oropharynx), urology [prostate LDR and HDR/pulsed-dose-rate (PDR)], skin, soft tissue sarcomas, pediatric malignancies, and brachytherapy of benign diseases are performed. The course is highly advised for trainees in radiotherapy (physicians and physicists). It is intended as an essential basis for the advanced courses of different brachytherapy techniques. In many countries, participation in this course is accepted as a part of the national training program.

(2) *Brachytherapy for prostate cancer.* This very first interdisciplinary course was originally introduced and is still driven as a joint course of ESTRO and the European Association of Urology (EAU) starting about a decade ago. Target groups are those who may be involved in the team required to set up prostate brachytherapy, that is, urologists, radiation oncologists, physicists, radiologists who are sometimes responsible for the transrectal US imaging, as well as nurse specialists. It covers an overview on epidemiology and treatment options for localized prostate cancer and gives an adequate introduction to prostate brachytherapy. Patient selection for both LDR and HDR/PDR brachytherapy is discussed. Staffing as well as radiation protection issues are also covered. Special steps of working, like volume study, treatment planning, and different implant techniques, are presented in the frame of interactive discussions on teaching application videos. Critical comparison of different techniques and management of PSA failure are interdisciplinarily discussed.

(3) *3D image-based brachytherapy in gynecological malignancies.* The target group is gynecological oncologists/radiotherapists as well as physicists and RTTs involved in modern 3D image-based gynecologic brachytherapy. Basic knowledge of gynecologic oncology and brachytherapy is required. The aim of the course is to provide a comprehensive overview on the whole field of gynecological brachytherapy involving cervix, corpus, vulva, and vaginal malignancies. An introductory guide is given to introduce 2D/3D image-based volume concepts of gross tumor volume (GTV), clinical target volume (CTV), and planning target volume (PTV) in this anatomic site. Normal and pathologic anatomy as well as related imaging of the female pelvis and brachytherapy indications, technology/performance, as well as combination with external beam

radiation are discussed. Special attention is given to advanced treatment planning methods, recording, and reporting of the treatment. Therapeutic outcome on different levels of evidence is discussed.

(4) *Interdisciplinary teaching course on head-and-neck brachytherapy.* This "hands-on" course is not a full ESTRO teaching course but is endorsed by ESTRO as well as by the European Federation of Oto-Rhino-Laryngological Societies (EU.F.O.S.) and the Working Groups of Oncology/Brachytherapy of the German ORL (DGHNO) as well as Radiotherapy (DEGRO) Societies. Targeted are experienced groups of head-and-neck oncological surgeons and radiotherapists/physicists with basic knowledge in interstitial brachytherapy. Imaging, indication, and different implantation techniques are presented and interdisciplinarily discussed in a very interactive way. Video transmission and moderated discussion of different live implantations as well as practical implantation exercises on pig heads are involved. The advantage of interdisciplinary cooperation in treatment performance, and functional and cosmetic outcome is discussed. A knowledge update in the high evidence level of treatment possibilities (radiochemotherapy) complementary to brachytherapy is given.

The ABS (formerly AES, American Endocurietherapy Society) offered in the past training modules for body sites for radiation oncologists/physicists. These courses were too short (2 to 3 days) and could not provide adequate training and generally required additional support and training. In the last few years, the ABS has implemented several successful schools for prostate, GYN, and breast brachytherapy. These sessions are didactic and hands-on courses and do provide a good base for brachytherapy knowledge. The ABS has also managed to create some residency programs for radiation oncologists at one or two institutions where individuals can learn about different brachytherapy applications. The duration is usually a week of training, and it is being organized through a sponsorship program. In addition, some individual oncology centers organize a teaching program that is open for participation by practitioners from elsewhere. A good example is the one in Seattle (Prostate Cancer Center of Seattle), which has an excellent prostate brachytherapy program (2 days) where radiation oncologists, physicists, and nurses can learn about the procedure (didactic and observing cases). In addition, proctoring and physics planning support is available through a treatment planning program (ProQura). The ABS has also recently added a quality management school for brachytherapy. During recent ABS annual conferences, several practical sessions were very successfully organized to teach, for example, 3D contouring for gynecology and prostate cases in order to learn the most recent recommendations.

32.6 Conclusion

Brachytherapy is evolving in multiple treatment sites, and therefore, physicians and physicists interested in managing these patients require specialized knowledge and training. Completion of expert training is a necessary requisite for proficiency, quality assurance, and safety. Physicists and physicians should be required by their institutions to familiarize themselves with all aspects of care and document their experience before embarking on independent management of these patients. Programs in brachytherapy should offer education in all aspects of management. Expert training courses and fellowships, which are available through professional organizations in Europe and the United States, should be encouraged. Practical information can be found in the links to the Internet pages of the international organizations, listed in the References section of this chapter.

References

AAPM. 2011. Meetings and courses: http://www.aapm.org/meetings/default.asp.

AAPM. 2005. Monographs No. 31. *Brachytherapy Physics*, 2nd ed. B. Thomadsen, M. Rivard, and W. Butler (eds.). Medical Physics Publishing, Madison, WI.

ABS. 2011. Meetings and courses: http://www.americanbrachytherapy.org/meetings/index.cfm.

Eudaldo, T., Huizenga, H., Lamm, I.L. et al. 2004. Guidelines for education and training of medical physicists in radiotherapy. Recommendations from an ESTRO/EFOMP working group. *Radiother Oncol* 70:125–35.

ESTRO. 2010. Core Curriculum: http://www.estro-education.org/europeantraining/Pages/EuropeanCurricula.aspx.

ESTRO. 2011a. Core Curriculum: http://www.estro-education.org/europeantraining/Pages/EuropeanCurricula.aspx.

ESTRO. 2011b. Courses: http://www.estro-education.org/courses/Pages/default.aspx.

EFOMP. 2011. Meetings and courses: http://www.efomp.org/conferences-meetings-courses.html.

IAEA. 2000a. Lessons learned from accidental exposures in radiotherapy. International Atomic Energy Agency. Safety Reports Series No. 17. IAEA, Vienna.

IAEA. 2000b. Clinical Training of Medical Physicists Specializing in Radiation Oncology. International Atomic Energy Agency. Training Course Series 37. IAEA, Vienna. http://www-pub.iaea.org/MTCD/publications/PDF/TCS-37_web.pdf.

IAEA. 2001. Implementation of microsource high dose rate (mHDR) brachytherapy in developing countries. International Atomic Energy Agency. IAEA-TECDOC-1257. IAEA, Vienna.

ICRP. 2005. Prevention of high-dose-rate brachytherapy accidents. International Commission on Radiological Protection. ICRP Publication 97. In: *Ann ICRP* 35 (2).

ICRP. 2009. Preventing accidental exposures from new external beam radiation therapy technologies. International Commission on Radiological Protection, ICRP Publication 112. In: *Ann ICRP* 39 (4).

Kubo, H.D., Glasgow, G.P., Pethel, T.D. et al. 1998. High dose-rate brachytherapy treatment delivery: Report of the AAPM Radiation Therapy Committee Task Group No. 59. *Med Phys* 25:375–403.

Nath, R., Anderson, L.L., Luxton, G. et al. 1995. Dosimetry of interstitial brachytherapy sources: Recommendations of the AAPM Radiation Therapy Committee Task Group No. 43. American Association of Physicists in Medicine. *Med Phys* 22:209–34.

Nath, R., Anderson, L.L., Meli, J.A. et al. 1997. Code of practice for brachytherapy physics: AAPM Radiation Therapy Committee Task Group 56. *Med Phys* 24:1557–98.

Rivard, M.J., Coursey, B.M., DeWerd, L.A. et al. 2004. Update of AAPM Task Group No. 43 Report: A revised AAPM protocol for brachytherapy dose calculations. *Med Phys* 31:633–74.

Rivard, M.J., Butler, W.M., DeWerd, L.A. et al. 2007. Supplement to the 2004 update of the AAPM Task Group No. 43 Report. *Med Phys* 34:2187–205.

Index